Texts and Monographs in Physics

Springer
Berlin
Heidelberg
New York
Barcelona
Budapest
Hong Kong
London
Milan
Paris
Santa Clara
Singapore
Tokyo

Texts and Monographs in Physics

Series Editors: R. Balian W. Beiglböck H. Grosse E. H. Lieb
N. Reshetikhin H. Spohn W. Thirring

Edgard Elbaz

Quantum

The Quantum Theory of Particles, Fields, and Cosmology

With 23 Figures

 Springer

Professor Dr. Edgard Elbaz

Institut de Physique Nucléaire de Lyon IN2P3-CNRS
Université Claude Bernard Lyon – 1
43, bd. du 11 Novembre 1918
F-69622 Villeurbanne Cedex, France

Title of the original French edition:
Quantique, published by Ellipses © Édition Marketing S. A. 1995

ISSN 0172-5998
ISBN-13: 978-3-642-64327-9   e-ISBN-13:978-3-642-60266-5
DOI:10.1007/978-3-642-60266-5

Library of Congress Cataloging-in-Publication Data
Elbaz, Edgard, 1937– [Quantique. English] Quantum: the quantum theory of particles, fields, and cosmology / E. Elbaz. p. cm. – (Texts and monographs in physics) Includes bibliographical references and index. ISBN-13: 978-3-642-64327-9 (Berlin: alk. paper) 1. Quantum theory. I. Title. II. Series. QC174.12.E4413 1998 530.12–dc21 97-33659

Typesetting: Data conversion by Jörg Steffenhagen, Königsfeld
Cover design: *design & production* GmbH, Heidelberg
SPIN: 10552588   55/3144-5 4 3 2 1 0

To my wife
my children and grand-children
Ephraïm, Levana, Shoshana,
Eliott, Yakov,
Rachel.

Preface

I consider teaching to have been one of my greatest privileges in life. I have always attemped to make a matter, sometimes difficult, sometimes in constant evolution, simpler and easier to understand. There comes a time then, when one feels the necessity to write and publish a book. For this reason, I have already published (in french) several lecture books on Classical Mechanics, Quantum Mechanics, Matrix Optics, Electroweak Interaction, General Relativity and Gravitation, Cosmology. Teaching quantum theory has been a particular delight as its constant evolution and enlargement, embraces all domains of physics. Even if the general layout remains relatively unchanged, the evolution of our understanding of the physical world imposes the introduction of new approaches. It therefore seemed that a physics textbook, even one with a graduate readership in mind, had to introduce the Dirac Electron Theory and some rudimentary material on quantum field theory. This holds for also for explanations of the spontaneous symmetry breaking of Higgs scalar fields that gives mass to the bosons involved in short range interactions. We have deliberately cut down on material on some "classical" topics of quantum theory to make space for less known methods. Examples of the latter include the Feynman path integral "3rd quantization" and the interpretation of quantum mechanics in terms of phase focusing coherence.

After a review of the most important results of research in analytical mechanics, it is shown that the description of a system of material particles is possible either by using a "discrete" formalism, such as Poisson equations, or a "continuous " formalism, such as Hamilton-Jacobi wave equations. Next, drawing the parallel between mechanics and optics serves as the pretext to introduce an equation for the propagation of mechanical waves and, with the help of a further hypothesis, the Schrödinger equation for the wave-mechanical description of a classical system of particles with definite mass. The next chapter develops wave-mechanical theory from the stand point of the correspondence principle. It then goes on to briefly show the relationship between the Heisenberg uncertainty relation and the different interpretations of wave mechanics and wave function, interpretations stemming from the Schrödinger equation: the pilot wave theory of Louis de Broglie and the quantum potential theory of D. Böhm, the theory of universal Brownian motion of E. Nelson, the probability theory of Max Born and Niels Bohr of the

Copenhagen School. Mention is also made of the EPR (Einstein-Podolsky-Rosen) paradox as well as the Bell inequalities which where tested by Alain Aspect and Philippe Grangier. Their experiments have definitely established that quantum theory is a non-separable theory, that is without any hidden variables.

Chapter 3 is a little more technical, describing classical stationary one-dimensional quantum systems, the potential well, the potential barrier and the semi-classical approach (W.K.B.). The following three chapters return to the Dirac formalism of nonrelativistic quantum mechanics and the two representations it has generated: wave mechanics by projection on the configuration space and matrix mechanics by projection on the vector column space.

After the study of quantum dynamics and the Schrödinger or Heisenberg pictures of the evolution of a nonrelativistic quantum mechanical system, we return to the interpretation of quantum mechanics in terms of the coherence of the quantum states underlying all forms of observer-independent interpretation. For the Copenhagen School, measurement determines the history of a quantum system with which a probability can be associated. When, as in quantum cosmology, it is practically impossible to separate the observer from the object being observed, the meaning of the probability postulate must be stretched if the theory is to retain its internal coherence. In such a situation, the Copenhagen School interpretation becomes an approximation of a more general framework characteristic of measurement and the decoherence resulting from the observer's measuring apparatus.

Chapter 7 develops the so-called 3rd quantization based on the Feynman path integrals. This method has the advantage of being adapted to all forms of interaction, including gravitation. It also emphasizes the importance of the phase function in wave function. As with the Huyghens' principle, the probability amplitude $\psi(x', t')$ is given by the superposition in the whole space of the probability amplitude at an earlier instant. This underscores the importance of the propagator developed with the Feynman conjecture from classical data of the system. This approach to quantum physics sheds new light on the role of action in quantum physics and its classical limit, the Hamiltonian action.

Next, we examine the properties of symmetry and invariance in quantum physics generally: translation, rotation, parity, time reversal. Much space has been alloted to non-local gauge transformations and to their associated symmetry groups, the Abelian groups, just as in the gauge transformation of the electromagnetic field gauge or the non-Abelian for Yang-Mills fields. These transformations have, in the recent years, assumed great importance in the description of electromagnetic and electroweak interactions in a unified model. This point is also featured in the discussion of the standard theory of electroweak interactions in Chapter 19.

Five chapters have been devoted to the angular momentum theory in the standard representation, the theorems of addition of two or more angular momenta, rotation matrices and spherical harmonics and, finally, to irreducible tensor operators. The innovation is in the relatively early introduction (Chapter 10) of the Clebsch-Gordan graphical representation and "$3jm$" coefficients. This facilitates the visualization of calculations and the understanding of the techniques of Racah algebra.

In the chapters on angular momentum (Chapters 9 through 13), the graphical representation has been used to supplement analytical calculation. It seemed to us an effective method for handling irreducible tensor operators and the Wigner-Eckart theorem. It should also facilitate access to some particularly difficult calculations, such as those for angular distributions and particle polarization. We have also sought to extend this graphical approach to classical vector calculus. This should simplify statements involving vector products and vector inner products (these points are developed in the appendix of Chapter 13).

Another innovation in the chapters on angular momentum is the introduction of the electroweak isospin of leptons and quarks, and of the magnetic moment of atoms, particles and quarks, as applications of the standard representation and the Wigner-Eckart theorem.

We have brought together in Chapter 14 the various methods of approximation in nonrelativistic quantum physics: stationary perturbations of degenerate and non-degenerate states, the variational method, and time-dependent variation. The Feynman diagram method is introduced to illustrate perturbative methods.

Feynman diagrams facilitate the understanding of the processes by which relativistic or nonrelativistic particles are scattered.

The problems associated with scattering are studied in Chapter 15, using the transition operator and its relationship with the transition amplitude and the differential cross section of elastic scattering. The Lippmann-Schwinger statement on the transition amplitude with respect to the interaction potential inevitably brings us back to Feynman. Next, we seek to extend the scope of the problem to the calculation of angular distribution and polarization in direct nuclei interaction processes and it should be possible to expand the results using G.S.A. graphical techniques without great difficulties.

In Chapter 16, "Second Quantization", we decided to treat together one-dimensional and n-dimensional harmonic oscillators and the second quantization formalism for boson and fermion systems. Again, it seemed to us that the formalism of creation and annihilation operators, when applied to the harmonic oscillator, constituted an ideal introduction to the notion of second quantization. We have also included a review of the present state of knowledge on the structure of matter with fermions that are emitters and receivers of the bosons which mediate in fundamental interactions.

Classical and quantum field theory has unfortunately been too often neglected in physics curricula. We have therefore devoted the next two chapters to this subject.

In the chapter on boson fields, we have included a section on Higgs scalar field and the process of spontaneous symmetry breaking which confers mass to exchange bosons. Chapter 18 describes fermion fields. After a brief excursion into the Dirac equation of a free particle, we return to the problem of gauge invariance and, using the Noether theorem, try to show that a law of conservation of quadricurrent can be inferred from overall gauge invariance. Next, we examine the interaction of a charged fermion with the electromagnetic field before going on to redefine the conditions for local gauge invariance. It is thus possible to introduce a little more naturally the Lagrangian electrodynamic quantum density. After describing the property of charge conjugation, we apply the Dirac equation to a detailed study of the hydrogen atom, taking into consideration both the relativistic effects and the electron spin.

The next chapter is entilted "Quantum X-dynamics". It describes in very brief terms quantum electrodynamic methods and the extensions quantum chromodynamics has brought to these theories. The standard model (G.S.W.) of electroweak interaction with the process of spontaneous symmetry breaking of Higgs fields which confers mass to the bosons carrying the weak interaction (the intermediate bosons $W^{\pm}$ and Z^0) receives a more detailed presentation. This chapter will no doubt appear a little more difficult than the preceding ones. However, it should be remembered that the unification of electromagnetic and electroweak interactions constituted such a stride in the advancement of knowledge that one cannot do justice to the topics discussed here without treating the method used and the spectacular results obtained through the determination of the mass of intermediate bosons.

Finally, the last chapter is dedicated to the problems of quantum cosmology. This is an introduction to a particularly difficult but thought-provoking subject which conditions the very foundations of the fundamental notions of space and time has been added at the suggestion of some colleagues. As at the time of writing, there are no quantum physics textbooks that can be considered accessible to cosmologists or cosmology textbooks that are accessible to quantum physicists. Considering that we have had published a textbook in cosmology which closes with a chapter on quantum cosmology, it seemed natural that the present one too to close with a similar chapter.

We have therefore given a brief review of general relativity and the standard models of Cosmology (the big-bang model) before going on to describe the Hamiltonian formulation of Einstein's equations which, after quantization, lead to the fundamental equation of quantum cosmology: the Wheeler-DeWitt equation. The use of this equation in well-chosen minisuperspaces often leads to the (W.K.B.) expression of the Universal wave function. This brief introduction is not intended as a substitute for specialized texts on the question but rather to give an idea of the issues raised and, indeed, to incite curiosity to find out more and perhaps also follow it up with a higher degree and research in an almost virgin area.

This text is intended to be self-contained in itself without necessarily laying claims to being exhaustive, to be clear without seeking to over-simplify. Concerning the title: Quantum Mechanics? or Quantum Physics? or Quantum Theory? We finally settled for Quantum which contains all of them. To facilitate the retention of the ideas introduced, each chapter closes with a table reproducing the main points where they are deemed to constitute an aid in the understanding of the text. We can only hope then that this book will be of real assistance to the target public: undergraduate students (Chapters 1 through 6 or, at best, through 8), masters students (Chapters 6 through 18) and first-year doctoral students (the whole of the text). Because this textbook derives in part from an M.Sc. Physics course I run at Université Claude Bernard - Lyon 1, it has greatly benefitted from numerous observations by the students I have had over the years.

I would like to thank Mr Z. Hernaus whole-heartedly for handling the illustrations. My profound gratitude also goes to my colleagues in Lyon, and especially to P. Desgrolard, for the patience and care with which he went through the original manuscript and for the numerous suggestions that have been instrumental in clarifying the ideas expressed in the text. I am grateful too to Mrs S. Florès for keying in the original text in English, to M. Chartoire for his technical assistance with the computer and to Uzoma Chukwu and Swanny Prakash for translating the text from French.

Lyon, August 1997   *E. Elbaz*

Contents

Chapter 1
Particles and Waves

1. Generalized Coordinates

Classical mechanics, whether relativistic or nonrelativistic, is built around the notion of the particle. It is essentially corpuscular mechanics to which has been added the intuitive notion of point particle, an infinitely small area in space where energy is concentrated in the form of mass, a point endowed with intrinsic properties such as electric charge and spin.

The position of a particle is therefore defined as a vector function $\vec{r}\,(t)$ in three-dimensional Euclidean space $\mathbb{R}^3$ and here we will deal with nonrelativistic classical mechanics, that is the mechanics of particles with speeds very low compared to the speed of light.

Nonrelativistic mechanics therefore specifies the position of a point particle with the function $\vec{r}\,(t)$ and its velocity with $\dot{\vec{r}}\,(t)$.

To specify the position of a system of N point particles, we will need to know N vector positions that are dependent on $3N$ coordinates. In the absence of any special constraints, we can say that the system has $3N$ degrees of freedom q_i ($i = 1, \ldots, 3N$), termed generalized coordinates, and $\dot{q}_i = dq_i/dt$ generalized velocities. We then note that the position of N point particles is defined by a vector in a $3N$-dimensional space, its configuration space:

$$\vec{r} = \vec{r}\,(q_1, \ldots, q_{3N}) \ \in \ \mathbb{R}^{3N} \qquad \text{$3N$-dimensional configuration space} \quad (1.1)$$

and by differentiating the function $\vec{r}$ with respect to each of the variables q_i on which it is dependent,

$$\dot{\vec{r}} = \frac{d\,\vec{r}}{d\,t} = \sum_i \frac{\partial\,\vec{r}}{\partial\,q_i}\,\dot{q}_i\,, \qquad (1.2)$$

we obtain following differentiation with respect to $\dot{q}_i$ the result

$$\frac{\partial\,\dot{\vec{r}}}{\partial\,\dot{q}_i} = \frac{\partial\,\vec{r}}{\partial\,q_i}\,. \qquad (1.3)$$

2. Energy Conservation

The fundamental law governing Newtonian dynamics links the acceleration $\ddot{\vec{r}}(t)$ of a point particle in a three-dimensional space to the force producing the movement, with this force being defined a priori as a vector quantity $\vec{F}$:

$$\vec{F} = m\,\ddot{\vec{r}}(t) = \frac{d\,\vec{p}(t)}{d\,t} = \dot{\vec{p}} \qquad \text{with} \qquad \vec{p} = m\,\dot{\vec{r}}(t) = m\,\vec{v}(t). \qquad (1.4)$$

Defined with reference to an "absolute" system (the Copernican system), the law holds for any point in a rectilinear uniform translation (Galilean system) since if $\vec{r} = \vec{r}' + \vec{v}\,t$, then the second derivatives are equal to $\ddot{\vec{r}} = \ddot{\vec{r}}'$.

It should perhaps be added that this fundamental law of Newtonian dynamics takes for granted the existence, on the one hand, of a universal and absolute time and, on the other, of a time-independent mass m, two assumptions that are not valid in relativistic mechanics.

If the point particle is moving along part of the curve $M_1 M_2$ of a trajectory (C), the work done by the force $\vec{F}$ in the course of the movement of the point particle from M_1 to M_2 is given as the curvilinear integral

$$\begin{aligned}
W &= \int_{\widehat{M_1 M_2}} \vec{F} \cdot d\,\vec{s} = \int_{M_1}^{M_2} \frac{d\,m\,\vec{v}}{d\,t} \cdot d\,\vec{s} = \int_{t_1}^{t_2} \frac{d\,m\,\vec{v}}{d\,t} \cdot \frac{d\,\vec{s}}{d\,t}\,d\,t \\
&= \int_{v_1}^{v_2} d\,m\,\vec{v} \cdot \vec{v} = \int_{v_1}^{v_2} m\,\vec{v} \cdot d\,\vec{v} = \int_{v_1}^{v_2} d\left(\frac{1}{2}\,m\,v^2\right) \\
&= \frac{1}{2}\,m\,v_2^2 - \frac{1}{2}\,m\,v_1^2 = T_2 - T_1, \qquad\qquad\qquad (1.5)
\end{aligned}$$

where v_1 and v_2 are the velocities at points M_1 and M_2 and T_1 and T_2 the kinetic energies of the particle at points M_1 and M_2.

If the force $\vec{F}$ derives from a potential $\vec{F} = -\vec{\nabla}V$, then we can write

$$W = \int_{M_1}^{M_2} \vec{F} \cdot d\,\vec{s} = -\int_{M_1}^{M_2} \vec{\nabla}V \cdot d\,\vec{s} = V_1 - V_2, \qquad (1.6)$$

which, when equated to the previous value, yields

$$W = T_2 - T_1 = V_1 - V_2$$

and by grouping together functions with the same subscript

$$T_1 + V_1 = T_2 + V_2 = \text{constant } E. \qquad (1.7)$$

The constant E is termed total energy. Let us stress here that E is an observable (a measurable quantity) whereas $\vec{F}$ is not. The most that can be done is to measure the effect of $\vec{F}$ on the mass m for a given acceleration.

3. Principle of Virtual Work

The virtual displacement $\delta \vec{r}_i$ of a system is an infinitesimal displacement during which the forces applied to the system do not undergo any change. In contrast, for a true displacement, these forces may undergo changes.

If the system of N particles is in equilibrium, $\vec{F}_i = 0 \ \forall \ i$ and hence $\vec{F}_i \cdot \delta \vec{r}_i = 0 \ \forall \ i$, or better still

$$\sum_i \vec{F}_i \cdot \delta \vec{r}_i = 0. \tag{1.8}$$

Let us separate the force $\vec{F}_i$ acting on a particle i into an applied force F_i^a that is external to the system and a force internal to the system $\vec{f}_i$:

$$\sum_i \vec{F}_i^{\,a} \cdot \delta \vec{r}_i + \sum_i \vec{f}_i \cdot \delta \vec{r}_i = 0. \tag{1.9}$$

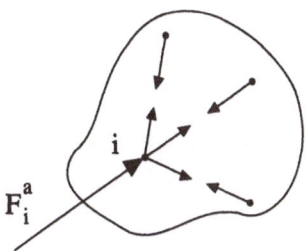

The system is undeformable, if the work of the internal forces is equal to zero (rigid bodies for instance). We thus obtain the principle of virtual work

$$\sum_i \vec{F}_i^{\,a} \cdot \delta \vec{r}_i = 0. \tag{1.10}$$

In any form of virtual displacement, the total work of the forces applied is equal to zero for an undeformable system in equilibrium. If we consider that $\vec{F}_i = \dot{\vec{p}}_i$ and, therefore, that $\left(\vec{F}_i - \dot{\vec{p}}_i \right) = 0$, we obtain, by replacing $\vec{F}_i$ with $(\vec{F}_i^{\,a} + \vec{f}_i - \dot{\vec{p}}_i)$ in (1.8) ,

$$\sum_i \left(\vec{F}_i^{\,a} + \vec{f}_i - \dot{\vec{p}}_i \right) \cdot \delta \vec{r}_i = 0$$

and by eliminating the null work of the internal forces of an undeformable system

$$\sum_i \left(\vec{F}_i{}^a - \dot{\vec{p}}_i \right) \cdot \delta \vec{r}_i = 0 \quad , \tag{1.11}$$

This is d'Alembert's principle defining the motion of a set of N particles in terms of the virtual work of the forces that are external to the system.

4. Lagrange Equations

Let us assume that $\vec{r}_i$ is a function of n independent generalized coordinates q_α and of the time coordinate t:

$$\vec{r}_i = \vec{r}_i \ (q_1, ..., q_n, t) \ . \tag{1.12}$$

The velocity $\vec{v}_i$ is calculated by differentiating this function with $(n + 1)$ variables with respect to time:

$$\vec{v}_i = \frac{d \vec{r}_i}{d t} = \frac{\partial \vec{r}_i}{\partial q_\alpha} \dot{q}_\alpha + \frac{\partial \vec{r}_i}{\partial t} \tag{1.13}$$

(the summation over repeated indices α is understood). The variation of the function $\vec{r}_i$ of a variation δq_α of each of the variables q_α can also be determined:

$$\delta \vec{r}_i = \frac{\partial \vec{r}_i}{\partial q_\alpha} \delta q_\alpha \ . \tag{1.14}$$

We infer from this another form of the principle of virtual work (1.10):

$$\sum_i \vec{F}_i{}^a \cdot \delta \vec{r}_i = \sum_{ij} \vec{F}_i{}^a \cdot \frac{\partial \vec{r}_i}{\partial q_j} \delta q_j = \sum_j Q_j \delta q_j \tag{1.15}$$

with Q_j being the generalized force acting on the particle j:

$$Q_j = \sum_i \vec{F}_i{}^a \cdot \frac{\partial \vec{r}_i}{\partial q_j} \ . \tag{1.16}$$

Now, let us use relation (1.14) to evaluate the second term of d'Alembert's principle in (1.11):

$$\sum_i \dot{\vec{p}}_i \cdot \delta \vec{r}_i = \sum_i m_i \ddot{\vec{r}}_i \cdot \delta \vec{r}_i = \sum_{ij} m_i \ddot{\vec{r}}_i \cdot \frac{\partial \vec{r}_i}{\partial q_j} \delta q_j \ , \tag{1.17}$$

but the sum over i can be rearranged by writing:

$$\sum_i m_i \ddot{\vec{r}}_i \cdot \frac{\partial \vec{r}_i}{\partial q_j} = \sum_i \frac{d}{dt}\left(m_i \ddot{\vec{r}}_i \cdot \frac{\partial \vec{r}_i}{\partial q_j}\right) - m_i \ddot{\vec{r}}_i \cdot \frac{d}{dt}\frac{\partial \vec{r}_i}{\partial q_j}$$

$$= \sum_i \frac{d}{dt}\left(m_i \ddot{\vec{r}}_i \cdot \frac{\partial \vec{r}_i}{\partial q_j}\right) - m_i \ddot{\vec{r}}_i \cdot \frac{d}{dq_j}\frac{\partial \vec{r}_i}{\partial t}. \quad (1.18)$$

By changing the order of differentiation in the last term, we notice that

$$\frac{d}{dq_j}\frac{\partial \vec{r}_i}{\partial t} = \frac{\partial \vec{v}_i}{\partial q_j}.$$

Because $\partial \dot{r}_i/\partial \dot{q}_j = \partial r_i/\partial q_j$ following Eq. (1.3), Eq. (1.18) can equally take the following form:

$$\sum_i m_i \ddot{\vec{r}}_i \cdot \frac{\partial \vec{r}_i}{\partial q_j} = \sum_i \frac{d}{dt}\left(m_i \vec{v}_i \cdot \frac{\partial \vec{v}_i}{\partial \dot{q}_j}\right) - m_i \vec{v}_i \cdot \frac{\partial \vec{v}_i}{\partial q_j}. \quad (1.19)$$

By inserting (1.15) and (1.19) into the d'Alembert's principle, we obtain the following relation:

$$0 = \sum_i \vec{F}_i^{\,a} \cdot \delta \vec{r}_i - \vec{p}_i \cdot \delta \vec{r}_i$$

$$= \sum_j \left[Q_j - \sum_i \frac{d}{dt}\left(m_i \vec{v}_i \cdot \frac{\partial \vec{v}_i}{\partial \dot{q}_j}\right) - m_i \vec{v}_i \cdot \frac{\partial \vec{v}_i}{\partial q_j}\right]\delta q_j. \quad (1.20)$$

Given that the total kinetic energy is $T = \frac{1}{2}\sum_i m_i v_i^2$, we obtain by rearranging the second and the third terms

$$\left[\sum_j \frac{d}{dt}\frac{\partial}{\partial \dot{q}_j}\left(\sum_i \frac{1}{2}m_i v_i^2\right) - \frac{\partial}{\partial q_j}\left(\sum_i \frac{1}{2}m_i v_i^2\right) - Q_j\right]\delta q_j = 0$$

$$\sum_j \left(\frac{d}{dt}\frac{\partial T}{\partial \dot{q}_j} - \frac{\partial T}{\partial q_j} - Q_j\right)\delta q_j = 0.$$

(1.21)

Because q_j are independent variables, each of the terms of the sum is equal to zero and relation (1.21) leads to the equations of motion

$$\boxed{\frac{d}{dt}\frac{\partial T}{\partial \dot{q}_j} - \frac{\partial T}{\partial q_j} = Q_j} \quad . \qquad (1.22)$$

For forces derived from the time-independent potential V (termed conservative forces)

$$\vec{F}_i = -\vec{\nabla}_i V.$$ (1.23)

We obtain a simple statement of the generalized force (1.16):

$$Q_j = \sum_i \vec{F}_i \cdot \frac{\partial \vec{r}_i}{\partial q_j} = -\sum_i \vec{\nabla}_i V \cdot \frac{\partial \vec{r}_i}{\partial q_j} = -\frac{\partial V}{\partial q_j}.$$ (1.24)

A more general time- or velocity-dependent potential can also be introduced and the generalized forces defined using the relation:

$$Q_j = \frac{d}{dt}\frac{\partial V}{\partial \dot{q}_j} - \frac{\partial V}{\partial q_j}.$$ (1.25)

This leads to the Lagrange equations

$$\frac{d}{dt}\frac{\partial T}{\partial \dot{q}_j} - \frac{\partial T}{\partial q_j} = \frac{d}{dt}\frac{\partial V}{\partial \dot{q}_j} - \frac{\partial V}{\partial q_j},$$

or, by grouping together similar terms

$$\frac{d}{dt}\frac{\partial (T-V)}{\partial \dot{q}_j} - \frac{\partial (T-V)}{\partial q_j} = 0.$$ (1.26)

By defining the Lagrangian function as $L = T - V$, we obtain the n second-order Euler–Lagrange equations of motion:

$$\boxed{\frac{d}{dt}\frac{\partial L}{\partial \dot{q}_j} - \frac{\partial L}{\partial q_j} = 0}.$$ (1.27)

If we define the generalized canonical momentum to the coordinate q_j to be

$$p_j = \frac{\partial L}{\partial \dot{q}_j},$$ (1.28)

we obtain another formulation of Lagrange Eqs. (1.27):

$$\frac{d}{dt} p_j = \frac{\partial L}{\partial q_j}.$$ (1.29)

If we set $F_j = \partial L / \partial q_j$, then we obtain once more the fundamental law of Newtonian dynamics

$$\frac{d p_j}{dt} = F_j.$$ (1.30)

5. Hamilton's Principle of Least Action

It is also possible to derive Lagrange equations from the principle of least action. The assumption here is that every mechanical system can be characterized by a function

$$L\ (q_\alpha,\ \dot{q}_\alpha,\ t) \equiv L\ (q,\ \dot{q},\ t) \qquad (1.31)$$

termed Lagrangian function and the Hamiltonian action will be defined as

$$S = \int_{t_1}^{t_2} L\ (q_\alpha,\ \dot{q}_\alpha,\ t)\ d\,t. \qquad (1.32)$$

5.1 Principle of Least Action

The principle of least action, or Hamilton's principle stipulates that every motion in the system tends toward an extremum for the variable in the action. In other words, the variation of the action is equal to zero for every motion:

$$\delta\,S = \delta \int_{t_1}^{t_2} L\ (q,\ \dot{q},\ t)\ d\,t = 0. \qquad (1.33)$$

This can also be formulated as

$$\delta\,S = \int_{t_1}^{t_2} \delta\,L\ (q,\ \dot{q},\ t)\ d\,t = 0.$$

The variation of the Lagrangian function can be expressed using the variation of the variables q, $\dot{q}$ and t on which it is dependent:

$$\delta\,L = \frac{\partial L}{\partial q}\,\delta\,q + \frac{\partial L}{\partial \dot{q}}\,\delta\,\dot{q} + \frac{\partial L}{\partial t}\,\delta\,t.$$

However, the last term is equal to zero because of the absence of variation in time (time being universal in classical nonrelativistic mechanics). Only the positions of the particles in the system can vary:

$$\delta\,L = \frac{\partial L}{\partial q}\,\delta\,q + \frac{\partial L}{\partial \dot{q}}\,\delta\,\dot{q} = \frac{\partial L}{\partial q}\,\delta\,q + \frac{\partial L}{\partial \dot{q}}\,\frac{d}{d\,t}\,\delta\,q,$$

and hence with a change of order in the last differential:

$$\delta\,S = \int \frac{\partial L}{\partial q}\,\delta\,q\,d\,t + \int \frac{\partial L}{\partial \dot{q}}\,\frac{d}{d\,t}\,\delta\,q\,d\,t.$$

By integrating the last integral by parts we obtain

$$\int_{t_1}^{t_2} \frac{\partial L}{\partial \dot{q}}\,d\,\delta\,q = \left[\frac{\partial L}{\partial \dot{q}}\,\delta\,q \right]_{t_1}^{t_2} - \int_{t_1}^{t_2} \frac{d}{d\,t}\,\frac{\partial L}{\partial \dot{q}}\,\delta\,q\,d\,t.$$

The variation at the limits of the integral is equal to zero, $\delta(q_1) = \delta(q_2) = 0$, if q_1, the departure point at time t_1, and q_2, the arrival point at time t_2, are taken to be fixed. This, in the final analysis, leaves us with:

$$\delta S = \int_{t_1}^{t_2} \left(\frac{\partial L}{\partial q} - \frac{d}{dt} \frac{\partial L}{\partial \dot{q}} \right) \delta q \, dt = 0. \tag{1.34}$$

Because the integral must be equal to zero for every variation δq of the generalized coordinates, the integrand itself must be equal to zero, thus yielding again the Euler–Lagrange equations

$$\frac{d}{dt} \frac{\partial L}{\partial \dot{q}_\alpha} - \frac{\partial L}{\partial q_\alpha} = 0. \tag{1.35}$$

For ease of notation, no mention has been made of the variable α in the demonstration above. However, for every degree of freedom α, there is an associated Lagrange equation of the above type describing the evolution of the functions for the particular degree of freedom.

5.2 Determination of the Lagrangian Function

If the mechanical system is composed of the two parts, A and B, with the Lagrangian functions L_A and L_B, and if the two parts are sufficiently far apart for there to be no possibility of interaction between them, then the Lagrangian function for the whole is the sum of the Lagrangian functions of each of its parts $L = L_A + L_B$. Now, let us first determine this function for an isolated point particle.

a) Free Point Particle

Let us place ourselves in a Galilean coordinate frame in which space is homogeneous and isotropic (i.e., identical properties of space in all directions) and time uniform (i.e., time is unique). The Lagrangian function can therefore contain neither $\vec{r}$ nor t. It can only contain $\dot{\vec{r}} = \vec{v}$, but because space is isotropic, it cannot be dependent on the direction of $\vec{v}$ but only on its magnitude or, rather, on $\dot{q}^2$. We can therefore proceed to write in the simplest case

$$L(q, \dot{q}, t) = L(\dot{q}^2) = \alpha \dot{q}^2 + \beta. \tag{1.36}$$

The use of the Lagrange Eq. (1.35) coupled with the fact that $\partial L / \partial q = 0$ leads to the equation

$$\frac{d}{dt} \frac{\partial L}{\partial \dot{q}} = 0$$

resulting in

$$\frac{\partial L}{\partial \dot{q}} = \text{constant}$$

and the form (1.36) makes it possible to evaluate the partial differential

$$\frac{\partial L}{\partial \dot{q}} = 2\, \alpha\, \dot{q} = \text{constant}, \tag{1.37}$$

showing that $\dot{q} = \text{constant}$.

Every motion of a free particle in a Galilean frame is at a constant velocity. This is the law of inertia.

The Lagrangian function for a free point particle is equal to its kinetic energy and the corresponding Lagrange equation necessarily leads to a rectilinear uniform motion with velocity v:

$$L = \frac{1}{2}\, m\, v^2$$

$$\frac{\partial L}{\partial \dot{q}} = \frac{\partial L}{\partial v} = m\, v \quad \text{and} \quad \frac{\partial L}{\partial q} = 0 \tag{1.38}$$

and the Lagrange equations become:

$$\frac{d}{dt}\frac{\partial L}{\partial \dot{q}} - \frac{\partial L}{\partial q} = \frac{d}{dt}\, m\, v = 0 \quad \text{or} \quad m\, v = \text{constant}.$$

The Lagrangian function of free point particles is therefore the sum of their kinetic energy or of the total kinetic energy of the system:

$$L = T = \sum_i \frac{1}{2}\, m_i\, v_i^2. \tag{1.39}$$

b) System of Interacting Point Particles

If we subtract from the Lagrangian function of the free point particles a function of the coordinates of the point particles termed potential energy V (see Sect. 4), we obtain the definition (1.26) of the Lagrangian:

$$L = T - V\ (r_1,\ r_2\ \ldots). \tag{1.40}$$

It should be noted that it is enough to change the position of one of the points to modify the function V. A change in the position of one of the points has an immediate effect on the others. Thus, in Newtonian classical mechanics, the propagation of interaction is instantaneous.

6. Conservation Laws

6.1 Conservation of Energy

Because time is uniform in all Galilean systems, the Lagrangian of a closed system will show no explicit dependence in t:

$$L = L\,(q_1 ... q_\alpha,\ \dot{q}_1 ... \dot{q}_\alpha ...) \equiv L\,(q,\ \dot{q})\,. \qquad (1.41)$$

The total time derivative of the function L will take the form

$$\frac{d\,L}{d\,t} = \sum_\alpha \frac{\partial\,L}{\partial\,q_\alpha}\,\dot{q}_\alpha + \frac{\partial\,L}{\partial\,\dot{q}_\alpha}\,\ddot{q}_\alpha\,.$$

If we isolate $\partial\,L\,/\,\partial\,q_\alpha$ from Lagrange Eq. (1.35), we are left with

$$\frac{\partial\,L}{\partial\,q_\alpha} = \frac{d}{d\,t}\frac{\partial\,L}{\partial\,\dot{q}_\alpha}\,,$$

which we insert into the foregoing total derivative:

$$\frac{d\,L}{d\,t} = \sum_\alpha \frac{d}{d\,t}\frac{\partial\,L}{\partial\,\dot{q}_\alpha}\,\dot{q}_\alpha + \frac{\partial\,L}{\partial\,\dot{q}_\alpha}\,\ddot{q}_\alpha = \sum_\alpha \frac{d}{d\,t}\left(\frac{\partial\,L}{\partial\,\dot{q}_\alpha}\,\dot{q}_\alpha\right)\,.$$

By bringing together the two parts of the equation, we obtain

$$\frac{d}{d\,t}\left(\sum_\alpha \frac{\partial\,L}{\partial\,\dot{q}_\alpha}\,\dot{q}_\alpha - L\right) = 0\,, \qquad (1.42)$$

showing that the term in brackets is a constant:

$$\sum_\alpha \frac{\partial\,L}{\partial\,\dot{q}_\alpha}\,\dot{q}_\alpha - L = \text{ constant } = E\,. \qquad (1.43)$$

This constant of motion is called energy. Mechanical systems in which energy is conserved are termed conservative systems.

We have also seen that $L = T - V = T\,(\dot{q}^{\,2}) - V(q)$, where T is a function of the square of the velocities. We can therefore evaluate $\partial\,L\,/\,\partial\,\dot{q}$ and write

$$\sum_\alpha \dot{q}_\alpha\,\frac{\partial\,L}{\partial\,\dot{q}_\alpha} = \sum_\alpha \dot{q}_\alpha\,\frac{\partial\,T}{\partial\,\dot{q}_\alpha} = 2T\,, \qquad (1.44)$$

because, by using the usual form of the kinetic energy T, we obtain

$$T = \frac{1}{2}\,m\,\dot{q}^{\,2}\,, \quad \frac{\partial\,T}{\partial\,\dot{q}} = m\,\dot{q} \quad \text{and} \quad \dot{q}\,\frac{\partial\,T}{\partial\,\dot{q}} = m\,\dot{q}^{\,2} = 2T\,.$$

Energy can therefore be developed as the sum of the kinetic energy and the potential energy since definition (1.43) and (1.40) lead to

$$E = \sum_\alpha \frac{\partial\,L}{\partial\,\dot{q}_\alpha}\,\dot{q}_\alpha - L = 2T - L = 2T - (T - V) = T + V\,. \qquad (1.45)$$

6.2 Conservation of Momentum

Because of the homogeneity of space in Galilean coordinate systems, the mechanical properties of a system are invariant for a translation $\delta\,q_\alpha = \varepsilon$ in space, that is if q_α is replaced with $q_\alpha + \varepsilon$. Thus

$$\delta L = \sum_\alpha \frac{\partial L}{\partial q_\alpha}\,\delta\,q_\alpha = \varepsilon \sum_\alpha \frac{\partial L}{\partial q_\alpha} = 0 \quad \text{and} \quad \sum_\alpha \frac{\partial L}{\partial q_\alpha} = 0. \quad (1.46)$$

But the Lagrange Eqs. (1.35) yield the partial differential L with respect to the variable q_α and Eq. (1.46) reduces to

$$0 = \sum_\alpha \frac{\partial L}{\partial q_\alpha} = \sum_\alpha \frac{d}{dt}\frac{\partial L}{\partial \dot{q}_\alpha} = \frac{d}{dt} \sum_\alpha \frac{\partial L}{\partial \dot{q}_\alpha}. \quad (1.47)$$

Let us introduce the canonical conjugate momentum with definition (1.28):

$$p_\alpha = \frac{\partial L}{\partial \dot{q}_\alpha} \quad (1.48)$$

and the conjugate momentum or total momentum P,

$$P = \sum_\alpha \frac{\partial L}{\partial \dot{q}_\alpha} = \sum_\alpha p_\alpha. \quad (1.49)$$

Equation (1.47) will then express the conservation of total momentum:

$$\frac{d}{dt} \sum_\alpha p_\alpha = \frac{d}{dt} P = 0 \quad \text{and} \quad P = \text{constant}. \quad (1.50)$$

Because of the homogeneity of space, the total momentum is conserved in a Galilean coordinate system.

6.3 Conservation of Angular Momentum

Because of the isotropy of the Galilean space, the mechanical properties of a system are invariant for a complete rotation of the space. Let us define such a rotation using the infinitesimal rotation vector $\delta\,\vec{\omega}$ such that the position $\vec{r}_i$ changes by $\delta\,\vec{r}_i$ and the velocity $\vec{v}_i$ by $\delta\,\vec{v}_i$. These can be expressed as

$$\delta\,\vec{r}_i = \delta\,\vec{\omega} \wedge \vec{r}_i,$$
$$\delta\,\vec{v}_i = \delta\,\vec{\omega} \wedge \vec{v}_i. \quad (1.51)$$

The invariance under rotation of the Lagrangian is hence written in vector form as

$$\delta L = \sum_i \frac{\partial L}{\partial \vec{r}_i} \cdot \delta\,\vec{r}_i + \frac{\partial L}{\partial \vec{v}_i} \cdot \delta\,\vec{v}_i = 0 \quad (1.52)$$

or, if we include the vector rotation in (1.51),

$$\sum_i \frac{\partial L}{\partial \vec{r}_i} \cdot \delta \vec{\omega} \wedge \vec{r}_i + \frac{\partial L}{\partial \vec{v}_i} \cdot \delta \vec{\omega} \wedge \vec{v}_i = 0. \tag{1.53}$$

However, written in vector form, Lagrange Eq. (1.35) will yield

$$\frac{\partial L}{\partial \vec{r}_i} = \frac{d}{dt}\frac{\partial L}{\partial \vec{v}_i} = \frac{d}{dt}\vec{p}_i$$

and combining it with (1.53):

$$\sum_i \frac{d}{dt}\vec{p}_i \cdot \delta \vec{\omega} \wedge \vec{r}_i + \vec{p}_i \cdot \delta \vec{\omega} \wedge \vec{v}_i = 0.$$

If we isolate the rotation vector, which is independent of the index i, we obtain

$$0 = \delta \vec{\omega} \cdot \left[\sum_i \dot{\vec{p}}_i \wedge \vec{r}_i + \vec{p}_i \wedge \dot{\vec{r}}_i \right] = \delta \vec{\omega} \cdot \left[\frac{d}{dt} \sum_i \vec{p}_i \wedge \vec{r}_i \right].$$

The total angular momentum is the vector defined by

$$\vec{\ell} = \sum_i \vec{\ell}_i = \sum_i \vec{r}_i \wedge \vec{p}_i. \tag{1.54}$$

Combining this with the preceding equation, we thus obtain

$$\delta \vec{\omega} \cdot \frac{d}{dt} \sum_i \vec{r}_i \wedge \vec{p}_i = \delta \vec{\omega} \cdot \frac{d}{dt}\vec{\ell} = 0. \tag{1.55}$$

Because this equation remains valid irrespective of the vector rotation $\delta \vec{\omega}$, we have

$$\frac{d\vec{\ell}}{dt} = 0 \quad \text{and} \quad \vec{\ell} = \text{constant}. \tag{1.56}$$

The total kinetic energy of a Galilean coordinate system is always conserved because space is isotropic.

7. The Center of Gravity Theorem

The center of gravity (or center of mass) of a mechanical system plays a very important role in the description of the system's dynamics. It permits classification of the properties of systems with a large number of degrees of freedom.

The position of the center of mass $\vec{R}$ of a set of k point particles with mass m_k is given by the relation

$$\vec{R} = \sum_k m_k \vec{r}_k / M \quad \text{where} \quad M = \sum_k m_k. \qquad (1.57)$$

By differentiating the product $M\vec{R}$ with respect to time, we obtain

$$M\dot{\vec{R}} = M\vec{V} = \sum_k m_k \dot{\vec{r}}_k = \sum_k m_k \vec{v}_k = \sum_k \vec{p}_k. \qquad (1.58)$$

The momentum of the center of gravity is (by definition) equal to the geometrical sum of the momenta of all the particles comprising the system.

The law of Newtonian dynamics which introduces the internal forces $\sum_j \vec{F}_{ij}$ and the external forces $\vec{F}_i{}^e$ is written for a particle i

$$m_i \frac{d^2 \vec{r}_i}{d t^2} = \vec{F}_i{}^e + \sum_j \vec{F}_{ij}. \qquad (1.59)$$

If we introduce the principle of action and reaction to show that the action of i on j is equal and opposite in sign to that of j on i, we obtain the relations

$$\vec{F}_{ij} = -\vec{F}_{ij} \quad \text{and} \quad \sum_{ij} \vec{F}_{ij} = 0. \qquad (1.60)$$

By summing over i and applying the principle of action and reaction, the relation becomes

$$\sum_i m_i \frac{d^2 \vec{r}_i}{d t^2} = \sum_i \vec{F}_i{}^e = \vec{F}^e.$$

But the first term is simply the first derivative of $M\dot{\vec{R}}$, the total momentum assigned to the center of mass and defined in (1.58). The foregoing relation hence takes the form

$$M \frac{d^2 \vec{R}}{d t^2} = \vec{F}^e. \qquad (1.61)$$

The motion of the center of mass of a system is the motion of a point particle assigned the mass of the entire system to which is applied the external force $\vec{F}^e$.

The notion of point particle, which was originally an abstract and theoretical entity (a dimensionless point with mass), has now acquired a precise meaning in classical mechanics. We begin with a set of particles each of finite dimension with mass m_k and define the center of mass as point particle assigned the total mass of the particles. The center of mass of a system is in many regards like a true particle. If the position $\vec{r}_i'$ of the ith particle is defined with respect to the center of mass:

$$\vec{r}_i = \vec{r}'_i + \vec{R} \tag{1.62}$$

then, following differentiation with respect to time,

$$\dot{\vec{r}}_i = \dot{\vec{r}}'_i + \dot{\vec{R}} \quad \text{or} \quad \vec{v}_i = \vec{v}'_i + \vec{V}, \tag{1.63}$$

we obtain the total kinetic energy

$$T = \sum_i \frac{1}{2} m_i v_i^2 = \sum_i \frac{1}{2} m_i v_i'^2 + \sum_i \frac{1}{2} m_i V^2 + \vec{V} \cdot \sum_i m_i \vec{v}'_i. \tag{1.64}$$

When the position of the center of mass is considered as the origin of the relative coordinates $\vec{r}'_i$, we infer with (1.58) that

$$\sum_i m_i \vec{v}'_i = 0.$$

The statement of the kinetic energy therefore reduces to

$$T = T' + T_{CM}. \tag{1.65}$$

The kinetic energy of a system is therefore the sum of the kinetic energy of the center of mass and the kinetic energy relative to the center of mass:

$$T = \sum_i \frac{1}{2} m_i v_i^2 \quad T' = \sum_i \frac{1}{2} m_i v_i'^2 \quad \text{and} \quad T_{CM} = \frac{1}{2} M V^2. \tag{1.66}$$

Let us consider the example of a system of two particles with mass m_1 and m_2, since we will be using this example in our discussions on quantum mechanics.

The center of mass is defined by the vector

$$\vec{R} = \frac{m_1 \vec{r}_1 + m_2 \vec{r}_2}{M}. \tag{1.67}$$

Now let us introduce as second coordinate the vector $\vec{r}$, which is the difference between the position vectors of the two particles,

$$\vec{r} = \vec{r}_1 - \vec{r}_2. \tag{1.68}$$

The velocity of the center of mass is obtained from definition (1.57) as

$$\dot{\vec{R}} = \frac{m_1}{M} \dot{\vec{r}}_1 + \frac{m_2}{M} \dot{\vec{r}}_2. \tag{1.69}$$

If we term $\vec{r}'_1$ and $\vec{r}'_2$ the center-of-mass coordinates of the particles m_1 and m_2, we obtain

$$\begin{aligned} \vec{r}_1 &= \vec{r}'_1 + \vec{R}, \\ \vec{r}_2 &= \vec{r}'_2 + \vec{R}, \end{aligned} \tag{1.70}$$

and by subtraction:

$$\vec{r}_2 - \vec{r}_1 = \vec{r} = \vec{r}\,'_2 - \vec{r}\,'_1 . \tag{1.71}$$

Thus, expressing r'_1 and r'_2 in terms of $\vec{r}$,

$$\vec{r}\,'_1 = \vec{r}_1 - \vec{R} = -\frac{m_2}{M}\,\vec{r},$$
$$\vec{r}\,'_2 = \vec{r}_2 - \vec{R} = \frac{m_1}{M}\,\vec{r}. \tag{1.72}$$

By the application of theorem (1.65), the kinetic energy becomes

$$T = \frac{1}{2}\,(m_1 + m_2)\,\dot{R}^2 + \frac{1}{2}\,m_1\,\dot{r}\,'^2_1 + \frac{1}{2}\,m_2\,\dot{r}\,'^2_2 . \tag{1.73}$$

The relative kinetic energy is easily found by differentiating (1.72):

$$T = \frac{1}{2}\,M\,\dot{R}^2 + \frac{1}{2}\,\mu\,\dot{r}^2 , \tag{1.74}$$

with the reduced mass μ given by

$$\frac{1}{\mu} = \frac{1}{m_1} + \frac{1}{m_2} \quad \text{or} \quad \mu = \frac{m_1\,m_2}{M} . \tag{1.75}$$

Example. It is interesting to evaluate the Lagrangian of the system of two interacting particles used in our example when the interaction potential is dependent only on the relative distance r between the particles:

$$L = T - V(r) = \frac{1}{2}\,M\,\dot{R}^2 + \frac{1}{2}\,\mu\,\dot{r}^2 - V(r) = L\,(R,\,\dot{R}) + L(r,\,\dot{r}). \tag{1.76}$$

This Lagrangian can be split into the Lagrangian of a "free point particle" with mass $M = m_1 + m_2$ (the center of mass of the system) and an interacting system with the reduced mass μ.

The Lagrange equations of the center of mass will be of the form

$$0 = \frac{d}{dt}\frac{\partial L}{\partial \vec{R}} = \frac{d}{dt}\,M\,\dot{\vec{R}}. \tag{1.77}$$

This yields the equation of motion

$$M\,\dot{\vec{R}} = M\,\dot{\vec{R}}_0 = \text{const,}$$

and by expanding the coordinate $\vec{R}(t)$ of the center of mass,

$$\vec{R} = \dot{\vec{R}}_0\,t + \vec{R}_0 . \tag{1.78}$$

The motion of the center of mass is that of a free particle with mass M. In the absence of external forces, it is a rectilinear uniform motion.

The relative motion is obtained using the Lagrange equations in the variable $\vec{r}$ and in spherical coordinates $\vec{r} = (r\sin\theta\cos\varphi,\ r\sin\theta\sin\varphi,\ r\cos\theta)$; it is of the form:

$$L\,(r,\ \dot{r}) = \frac{1}{2}\,\left(\dot{r}^{\,2} + r^2\,\dot{\theta}^{\,2} + r^2\,\sin^2\theta\,\dot{\varphi}^{\,2}\right) - V\,(r) \qquad (1.79)$$

or, written with Newton's equation of relative motion

$$\mu\,\frac{d^2\,\vec{r}}{dt^2} = -\vec{\nabla}\,V\,. \qquad (1.80)$$

The total momentum P can be expressed as the sum of the momenta of each constituent:

$$\vec{P} = \vec{p}_1 + \vec{p}_2 = m_1\,\dot{\vec{r}}_1 + m_2\,\dot{\vec{r}}_2 = M\,\dot{\vec{R}}\,. \qquad (1.81)$$

By introducing the relative momentum

$$\vec{p} = \frac{m_2}{M}\,\vec{p}_1 - \frac{m_1}{M}\,\vec{p}_2\,, \qquad (1.82)$$

the total kinetic energy can be written as

$$T = \frac{p_1^2}{2m_1} + \frac{p_2^2}{2m_2} = \frac{p^2}{2\mu} + \frac{P^2}{2M}\,, \qquad (1.83)$$

leading to a total kinetic energy of the form:

$$E = T + V$$
$$= \frac{p_1^2}{2m_1} + \frac{p_2^2}{2m_2} + V(r) = \frac{P^2}{2M} + \frac{p^2}{2\mu} + V(r)\,. \qquad (1.84)$$

This statement will be called upon again and again in our formulation of the quantum mechanics of the motion of two interacting particles.

8. Symmetry and Invariance

The law of classical Newtonian mechanics can be written with (1.30) as

$$\dot{\vec{p}} = \vec{F} \qquad \text{or} \qquad \dot{\vec{p}}_k = \vec{F}_k \qquad \text{with} \qquad \vec{p}_k = m_k\,\dot{\vec{x}}_k \qquad (1.85)$$

for the kth particle of the system. If we introduce time-independent forces deriving from the potential V:

$$\vec{F}_k = -\vec{\nabla}_k\,V\,(\vec{x}_1, \ldots \vec{x}_k, \ldots \vec{x}_N)\,, \qquad (1.86)$$

Newton's law for the dynamics of the kth point particle will then assume the form:

$$\dot{\vec{p}}_k = -\vec{\nabla}_k\,V\,(\vec{x}_1, \ldots \vec{x}_k, \ldots \vec{x}_N)\,. \qquad (1.87)$$

8.1 Translation Invariant Potential

If a translation of $\lambda \, \vec{a}$ is made for the entire system, the interaction potential will remain unchanged. This is expressed mathematically by the relation.

$$V\left(\vec{x}_1 + \lambda \, \vec{a}, \ldots \vec{x}_k + \lambda \, \vec{a}, \ldots\right) = V\left(\vec{x}_1, \ldots \vec{x}_k, \ldots\right) . \qquad (1.88)$$

A Taylor expansion around the positions $\vec{x}_1 \ldots \vec{x}_k \ldots$ of the potential will otherwise give:

$$V\left(\vec{x}_1 + \lambda \, \vec{a}, \ldots \vec{x}_k + \lambda \, \vec{a}, \ldots\right) = V(x_1, \ldots x_k) + \lambda \, a \, \frac{\partial V}{\partial \vec{x}_1} + \lambda \, a \, \frac{\partial V}{\partial \vec{x}_2} + \ldots$$

$$= V\left(\vec{x}_1, \ldots \vec{x}_k\right) + \lambda \, \vec{a} \cdot \sum_k \vec{\nabla}_k \, V + \ldots .$$
$$(1.89)$$

The invariance by translation (1.88) therefore imposes the condition

$$\vec{a} \cdot \sum_k \vec{\nabla}_k \, V = 0 . \qquad (1.90)$$

By summing over all the particles k, Newton's law (1.59) becomes

$$\sum_k \dot{\vec{p}}_k = \dot{\vec{P}} = \sum_k -\vec{\nabla}_k \, V \qquad (1.91)$$

and, combined with (1.90), the scalar product of the translation with the vector $\vec{a}$ gives

$$\vec{a} \cdot \dot{\vec{P}} = -\vec{a} \cdot \sum_k \vec{\nabla}_k \, V = 0 . \qquad (1.92)$$

Because the relation remains valid irrespective of $\vec{a}$, we have

$$\dot{\vec{P}} = 0 \qquad \text{and} \qquad \vec{P} = \text{constant} = \vec{P}_0 . \qquad (1.93)$$

The invariance under translation of the interaction potential corresponding to the homogeneity property of space brings about conservation of total momentum.

By setting
$$\vec{P} = M \, \dot{\vec{X}} \qquad \text{with} \qquad M = \sum_k m_k \qquad (1.94)$$

we derive the velocity of the center of mass from (1.93):

$$\dot{\vec{X}} = \frac{\vec{P}_0}{M} = \text{constant} , \qquad (1.95)$$

and by integrating each term:

$$\vec{X} = \frac{\vec{P}_0}{M} \, t + \vec{X}_0 . \qquad (1.96)$$

When the interaction potential is invariant under translation, the total momentum is conserved and the center of mass, which effectively carries this total momentum, will have a rectilinear uniform motion.

8.2 Rotationally Invariant Potential

If $\vec{n}$ is the unit vector of the rotation, i.e.,

$$\vec{\omega} = \omega \, \vec{n} \, , \tag{1.97}$$

then the variation by rotation of the position of a point $\vec{x}$ will be

$$\delta \, \vec{x} = \vec{\omega} \wedge \vec{x} \, . \tag{1.98}$$

The invariance of the potential under rotation implies

$$\begin{aligned} V \, (\vec{x}_1, \dots \vec{x}_k, \dots) &= V \, (\vec{x}_1 + \delta \, \vec{x}_1, \dots \vec{x}_k + \delta \, \vec{x}_k, \dots) \\ &= V \, (\vec{x}_1 + \vec{\omega} \wedge \vec{x}_1, \dots \vec{x}_k + \vec{\omega} \wedge \vec{x}_k, \dots) \, . \end{aligned} \tag{1.99}$$

The condition for invariance of the potential under rotation is therefore obtained with a Taylor expansion around the positions $\vec{x}_1 \dots \vec{x}_k$:

$$\vec{\omega} \cdot \sum_k \vec{x}_k \wedge \vec{\nabla}_k V = 0 \, . \tag{1.100}$$

Because this relation remains valid irrespective of the vector rotation ω, it can be inferred that

$$\sum_k \vec{x}_k \wedge \vec{\nabla}_k V = 0 \tag{1.101}$$

expresses the condition of invariance by rotation.

By vector multiplication of Newton's relation (1.87) by $\vec{x}_k$, we obtain

$$\vec{x}_k \wedge \vec{p}_k = -\vec{x}_k \wedge \vec{\nabla}_k V \, , \tag{1.102}$$

and by summing over k, the second term becomes zero whereas the first gives

$$\sum_k \vec{x}_k \wedge \vec{p}_k = \frac{d}{dt} \left(\sum_k \vec{x}_k \wedge \vec{p}_k \right) = 0 \, . \tag{1.103}$$

The angular momentum of the kth particle about the center of mass,

$$\vec{\ell}_k = \vec{x}_k \wedge \vec{p}_k \, , \tag{1.104}$$

leads to the total angular momentum:

$$\vec{\ell} = \sum_k \vec{\ell}_k = \sum_k \vec{x}_k \wedge \vec{p}_k \, . \tag{1.105}$$

Relation (1.103) represents the conservation of total angular momentum:

$$\frac{d}{dt} \vec{\ell} = 0 \qquad \text{implies} \qquad \vec{\ell} = \vec{\ell}_0 \, . \tag{1.106}$$

The invariance of the potential under rotation corresponds to the property of isotropy of space and therefore leads to the conservation of total angular momentum.

8.3 Time Invariant Potential

If the potential is time independent, $\partial V / \partial t = 0$, then

$$\frac{dV}{dt} = \sum_k \frac{\partial V}{\partial x^k} \frac{d x^k}{dt} = \sum_k \vec{\nabla}_k V \, \dot{x}_k \, . \qquad (1.107)$$

Multiplying Newton's relation (1.87) by the velocity, and following summation over k, we obtain

$$\sum_k \dot{x}_k \, \vec{p}_k = -\sum_k \dot{x}_k \vec{\nabla}_k \, V = -\frac{dV}{dt} \, . \qquad (1.108)$$

The first term of (1.108) is simply the differential of the total kinetic energy. Thus we can write

$$\frac{d}{dt} \sum_k \frac{1}{2} \, m_k \, v_k^2 = \frac{dT}{dt} = -\frac{dV}{dt} \qquad \text{hence} \qquad \frac{d}{dt} \, (T + V) = 0 \, . \quad (1.109)$$

The time invariance of the interaction potential leads to the conservation of total energy $T + V = E$.

9. Hamilton's Equations

9.1 Conservative Systems

Consider, for ease of exposition, the example of a system subjected to a velocity-independent interaction (conservative forces) and in which the Lagrangian function is not explicitly time dependent:

$$L = L \, (q_\alpha, \, \dot{q}_\alpha) \, .$$

We have adopted the Einstein's convention as a way of simplifying the notation implying the summation over repeated indices.

The total differential of L is thus written

$$\frac{dL}{dt} = \frac{\partial L}{\partial q_\alpha} \, \dot{q}_\alpha + \frac{\partial L}{\partial \dot{q}_\alpha} \, \ddot{q}_\alpha \qquad (1.110)$$

and by replacing $\partial L / \partial q_\alpha$ with $(d/dt)\, (\partial L/\partial \dot{q}_\alpha)$ taken from the Lagrange equation

$$\frac{dL}{dt} = \frac{d}{dt}\frac{\partial L}{\partial \dot{q}_\alpha}\dot{q}_\alpha + \frac{\partial L}{\partial \dot{q}_\alpha}\ddot{q}_\alpha = \frac{d}{dt}\left(\dot{q}_\alpha \frac{\partial L}{\partial \dot{q}_\alpha}\right)$$

and bringing together the differentials, we have

$$\frac{d}{dt}\left(L - \dot{q}_\alpha \frac{\partial L}{\partial \dot{q}_\alpha}\right) = 0\,. \tag{1.111}$$

If we introduce the Hamiltonian function with the relation

$$H = \dot{q}_\alpha \frac{\partial L}{\partial \dot{q}_\alpha} - L = \dot{q}_\alpha\, p_\alpha - L\,(q_\alpha,\,\dot{q}_\alpha)\,, \tag{1.112}$$

it will have a zero total differential following (1.111):

$$\frac{dH}{dt} = 0 \qquad \text{and} \qquad H = \text{constant}\,. \tag{1.113}$$

In a conservative system, the Hamiltonian is constant.

It is also worth noting that the constant is simply the total energy, as was pointed out in a preceding section.

$$H = T + V = E \qquad \text{in a conservative system}\,. \tag{1.114}$$

Without attempting to offer any demonstration, we should also like to note that in the relativistic case and by introducing the restmass m and the relation $\beta = v/c$, we obtain, to within an additive constant,

$$L = m\, c^2\, \sqrt{1 - \beta^2} - V\,,$$

$$H = \frac{m\, c^2}{\sqrt{1 - \beta^2}} + V\,. \tag{1.115}$$

9.2 The General Case

The principle of least action can be written in a very general form:

$$\delta \int_{t_1}^{t_2} L\,(q,\,\dot{q},\,t)\,dt = \delta \int_{t_1}^{t_2} [\dot{q}_\alpha\, p_\alpha - H\,(p_\alpha,\,q_\alpha)]\;dt = 0\,. \tag{1.116}$$

Using the Hamiltonian as departure point and taking the total differential, we have

$$H\,(p,\,q,\,t) = \sum_\alpha p_\alpha\, \dot{q}_\alpha - L\,(q_\alpha,\,\dot{q}_\alpha,\,t)\,.$$

The only variables in the first term are p_α, q_α and t, and hence

$$d\,H = \frac{\partial H}{\partial p_\alpha}\,d\,p_\alpha + \frac{\partial H}{\partial q_\alpha}\,d\,q_\alpha + \frac{\partial H}{\partial t}\,d\,t. \qquad (1.117)$$

The second term introduces the variables p_α, $\dot q_\alpha$, q_α and t and we obtain the total differential

$$d\,H = \sum_\alpha d\,p_\alpha\,\dot q_\alpha + p_\alpha\,d\,\dot q_\alpha - \frac{\partial L}{\partial q_\alpha}\,d\,q_\alpha - \frac{\partial L}{\partial \dot q_\alpha}\,d\,\dot q_\alpha - \frac{\partial L}{\partial t}\,d\,t.$$

After simplification of the relation, we can use the definition of the generalized momenta $p_\alpha = \partial L/\partial \dot q_\alpha$ and the Lagrange equations $\dot p_\alpha = \partial L/\partial p_\alpha$ to obtain

$$d\,H = \sum_\alpha \dot q_\alpha\,d\,p_\alpha - \dot p_\alpha\,d\,q_\alpha - \frac{\partial L}{\partial t}\,d\,t. \qquad (1.118)$$

If we compare with Eq. (1.117) for the total differential of H,

$$d\,H = \sum_\alpha \frac{\partial H}{\partial p_\alpha}\,d\,p_\alpha + \frac{\partial H}{\partial q_\alpha}\,d\,q_\alpha + \frac{\partial H}{\partial t}\,d\,t,$$

we obtain Hamilton's equations:

$$\dot q_\alpha = \frac{\partial H}{\partial p_\alpha},$$
$$\dot p_\alpha = -\frac{\partial H}{\partial q_\alpha}, \qquad (1.119)$$
$$\frac{\partial L}{\partial t} = -\frac{\partial H}{\partial t}.$$

If L is not explicitly time dependent,

$$0 = \frac{\partial L}{\partial t} = -\frac{\partial H}{\partial t}. \qquad (1.120)$$

In deriving the Hamilton equation, we obtain

$$\frac{\partial \dot q_\alpha}{\partial q_\alpha} = \frac{\partial^2 H}{\partial p_\alpha \partial q_\alpha} \quad \text{and} \quad \frac{\partial \dot p_\alpha}{\partial p_\alpha} = -\frac{\partial^2 H}{\partial p_\alpha \partial q_\alpha}.$$

The conjugate variables q_α and p_α thus verify the condition:

$$\frac{\partial \dot q_\alpha}{\partial q_\alpha} + \frac{\partial \dot p_\alpha}{\partial p_\alpha} = 0. \qquad (1.121)$$

A time-independent potential implies a Lagrangian and a Hamiltonian that are not explicitly time dependent and, hence, a conservative system. Let us combine the Hamilton Eqs. (1.119) with the total time differential of H:

$$\frac{d H}{d t} = \frac{\partial H}{\partial q_\alpha} \dot{q}_\alpha + \frac{\partial H}{\partial p_\alpha} \dot{p}_\alpha + \frac{\partial H}{\partial t}$$

$$= \frac{\partial H}{\partial q_\alpha} \frac{\partial H}{\partial p_\alpha} - \frac{\partial H}{\partial p_\alpha} \frac{\partial H}{\partial q_\alpha} + \frac{\partial H}{\partial t} = \frac{\partial H}{\partial t} . \qquad (1.122)$$

We therefore obtain generally

$$\frac{d H}{d t} = \frac{\partial H}{\partial t} \qquad (1.123)$$

and if $L = L(q, \dot{q})$ for an explicitly time-independent potential:

$$\frac{\partial L}{\partial t} = -\frac{\partial H}{\partial t} = 0 \qquad \text{then} \qquad \frac{d H}{d t} = 0$$

implies $H = E$. The system is therefore conservative (see Sect. 9.1 below).

10. Maupertuis' Least-Action Principle

Maupertuis' action A is introduced via the relation

$$A = \int_{t_1}^{t_2} \sum_i p_i \dot{q}_i \, d t \qquad (1.124)$$

and Maupertuis' least-action principle stipulates that in a conservative system the variation of A is equal to zero for every motion in the system:

$$\delta A = \delta \int_{t_1}^{t_2} \sum_i p_i \dot{q}_i \, d t = 0 . \qquad (1.125)$$

The variation δ implies a true succession of displacement in the interval $d t$ and may therefore include a variation in time, even at the extremes where the variation of q_i is equal to zero.

If the Hamiltonian is not explicitly time dependent, we can write

$$A = \int_{t_1}^{t_2} (L + H) \, d t = \int_{t_1}^{t_2} L \, d t + H (t_2 - t_1) . \qquad (1.126)$$

In nonrelativistic mechanics, it is easily noticed that

$$\sum_i p_i \dot{q}_i = L + H = T - V + T + V = 2T$$

and Maupertuis' action is the integral of the kinetic energy:

$$A = \int_{t_1}^{t_2} 2T\, d\, t \,.$$

Maupertuis' least-action principle can therefore also be written

$$\delta \int T\, d\, t = 0 \,. \tag{1.127}$$

If, in addition, no external forces are applied to the system (a rigid body with no net applied forces, for example) then the kinetic energy T as well as the Hamiltonian H are conserved and the least-action principle takes the form

$$\delta\, (t_2 - t_1) = 0 \,. \tag{1.128}$$

Thus, among all possible paths between two points, the system will follow the path that will minimize time in order to conserve energy.

Writing the kinetic energy T as

$$T = \frac{1}{2}\, m\, v^2 = \frac{1}{2}\, m\, v\, \frac{d\, s}{d\, t} = \frac{1}{2}\, \sqrt{2m\, T}\, \frac{d\, s}{d\, t} \tag{1.129}$$

and because the kinetic energy can also be written

$$T = \frac{p^2}{2m} = \frac{m^2\, v^2}{2m} \qquad \text{gives} \qquad m\, v = \sqrt{2m\, T} \tag{1.130}$$

Maupertuis' least-action principle takes the form

$$\delta \int T\, d\, t = \delta \int \frac{1}{2}\, \sqrt{2m\, T}\, \frac{d\, s}{d\, t}\, d\, t = \delta \int \frac{1}{2}\, \sqrt{2m\, T}\, d\, s = 0$$

$$\text{or} \quad \delta \int \sqrt{2m\, T}\, d\, s = 0 \,. \tag{1.131}$$

Stated in this form, Maupertuis' principle strongly reminds us of Fermat's principle of geometrical optics as will be shown in Sect. 14 below.

11. Canonical Transformations

11.1 Generating Function

Consider two functions Q_i and P_i of the variables p, q, t appearing in the Hamiltonian formalism:

$$Q_i = Q_i\, (q,\, p,\, t)$$
$$P_i = P_i\, (q,\, p,\, t) \,. \tag{1.132}$$

If there exists a function $K\,(Q, P, t)$ such that the equations of momentum in the new system of coordinates can still be written:

$$\dot{Q}_i = \frac{\partial K}{\partial P_i}$$
$$\dot{P}_i = -\frac{\partial K}{\partial Q_i} \qquad (1.133)$$

then the transformation (1.133) is said to be canonical and K assumes the role of the Hamiltonian, that is the Hamiltonian least-action principle can simultaneously be written for the variables $(q,\ p,\ t)$:

$$\delta \int_{t_1}^{t_2} \left[\sum_i p_i\,\dot{q}_i - H\,(q,\ p,\ t) \right]\, d\,t = 0\,, \qquad (1.134)$$

and for the variables $(Q,\ P,\ t)$:

$$\delta \int_{t_1}^{t_2} \left[\sum_i P_i\,\dot{Q}_i - K\,(Q,\ P,\ t) \right]\, d\,t = 0\,. \qquad (1.135)$$

This does not mean that the integrands are equal but rather that they differ at most by a total time derivative of an arbitrary function F, termed transformation generating function. This function can therefore be described using two independent variables chosen from p, q, P, Q and we thus obtain four different combinations depending on the choice of pair of variables.

Case No. 1: $F = F_1\,(q,\ Q,\ t)$

$$\sum_i p_i\,\dot{q}_i - H = \sum_i P_i\,\dot{Q}_i - K + \frac{d\,F_1}{d\,t}. \qquad (1.136)$$

If we expand the total differential of the generating function $F_1(q,\ Q,\ t)$ as

$$\frac{d\,F_1}{d\,t} = \sum_i \frac{\partial F_1}{\partial\,q_i}\,\dot{q}_i + \sum_i \frac{\partial F_1}{\partial\,Q_i}\,\dot{Q}_i + \frac{\partial F_1}{\partial\,t}\,,$$

we obtain, by comparing the preceding equation term-by-term,

$$p_i = \frac{\partial F_1}{\partial\,q_i}\,,$$
$$P_i = -\frac{\partial F_1}{\partial\,Q_i}\,, \qquad (1.137)$$
$$K = H + \frac{\partial F_1}{\partial\,t}\,.$$

Case No. 2: $F = F_2\,(q,\,P,\,t)$

The foregoing case permits us to write

$$P_i = -\frac{\partial F_1}{\partial Q_i}.$$

This gives, after integration,

$$F_1 = -\sum_i P_i\,Q_i + F_2 \qquad \text{and} \qquad F_2 = F_1 + \sum_i P_i\,Q_i. \qquad (1.138)$$

Elsewhere (1.136), we saw that it is possible to write

$$\sum_i p_i\,\dot{q}_i - H = \sum_i P_i\,\dot{Q}_i - K + \frac{d\,F_1}{d\,t}.$$

If we use (1.138) to express the total differential as

$$\frac{d\,F_1}{d\,t} = \sum_i -\dot{P}_i\,Q_i - P_i\,\dot{Q}_i + \frac{d\,F_2}{d\,t},$$

then we are led to the following:

$$\sum_i p_i\dot{q}_i - H = \sum_i P_i\,\dot{Q}_i - K - \dot{P}_i\,Q_i - P_i\,\dot{Q}_i + \frac{d\,F_2}{d\,t}$$

$$= -\sum_i \dot{P}_i\,Q_i - K + \frac{d\,F_2}{d\,t}.$$

We develop the total differential of $F_2\,(q,\,P,\,t)$ and thus

$$\sum_i p_i\,\dot{q}_i - H = -\sum_i \dot{P}_i\,Q_i - K + \frac{\partial F_2}{\partial q_i}\,\dot{q}_i + \frac{\partial F_2}{\partial P_i}\,\dot{P}_i + \frac{\partial F_2}{\partial t}$$

we obtain, by comparing terms

$$p_i = \frac{\partial F_2}{\partial q_i},$$

$$Q_i = \frac{\partial F_2}{\partial P_i}, \qquad (1.139)$$

$$K = H + \frac{\partial F_2}{\partial t}.$$

The two other combinations of variables in the generating function will not be used here. They lead to statements for the missing variables in terms of the partial differentials of the generating functions. In all the possible cases, we come up with the same relation between K, H and the generating function:

$$K = H + \frac{\partial F}{\partial t}. \qquad (1.140)$$

11.2 Lagrange Brackets

The dynamical state of a system is defined as a point in a Cartesian space with $2n$ coordinates $q_1...q_n$, $p_1...p_n$, termed phase space. Poincaré has shown that the integral

$$J_1 = \int\int_S \sum_i dq_i\, dp_i \tag{1.141}$$

is invariant for every canonical transformation. The surface S indicates that J_1 should be evaluated over an arbitrary two-dimensional surface in the phase space:

$$q_i = q_i\,(u,\,v) \qquad \text{and} \qquad p_i = p_i\,(u,\,v)\,. \tag{1.142}$$

The differential element $dp_i\, dp_i$ combines with the Jacobian $\partial(q_i,\,p_i)/\partial(u,\,v)$ to yield $du\, dv$, the determinant obtained from the partial differentials

$$dq_i\, dp_i = \frac{\partial\,(q_i,\,p_i)}{\partial\,(u,\,v)}\, du\, dv\,. \tag{1.143}$$

The Jacobian involved in the transformation of the variables $(p,\,q)$ into $(u,\,v)$ is

$$\frac{\partial\,(q_i,\,p_i)}{\partial\,(u,\,v)} = \det \begin{vmatrix} \frac{\partial\,q_i}{\partial\,u} & \frac{\partial\,q_i}{\partial\,v} \\ \frac{\partial\,p_i}{\partial\,u} & \frac{\partial\,p_i}{\partial\,v} \end{vmatrix} = \frac{\partial\,q_i}{\partial\,u}\frac{\partial\,p_i}{\partial\,v} - \frac{\partial\,q_i}{\partial\,v}\frac{\partial\,p_i}{\partial\,u}\,. \tag{1.144}$$

To say that J_1 is invariant for every canonical transformation is to say that

$$\int\int_S \sum_i dq_i\, dp_i = \int\int_S \sum_k dQ_k\, dP_k\,, \tag{1.145}$$

and hence

$$\int\int_S \sum_i \frac{\partial\,(q_i,\,p_i)}{\partial\,(u,\,v)}\, du\, dv = \int\int_S \sum_k \frac{\partial\,(Q_k,\,P_k)}{\partial\,(u,\,v)}\, du\, dv\,.$$

This leads to the following relationship between Jacobians:

$$\sum_i \frac{\partial\,(p_i,\,q_i)}{\partial\,(u,\,v)} = \sum_k \frac{\partial\,(Q_k,\,P_k)}{\partial\,(u,\,v)}\,, \tag{1.146}$$

which is easily expanded into

$$\sum_i \frac{\partial\,q_i}{\partial\,u}\frac{\partial\,p_i}{\partial\,v} - \frac{\partial\,p_i}{\partial\,u}\frac{\partial\,q_i}{\partial\,v} = \sum_i \frac{\partial\,Q_i}{\partial\,u}\frac{\partial\,P_i}{\partial\,v} - \frac{\partial\,P_i}{\partial\,u}\frac{\partial\,Q_i}{\partial\,v}\,.$$

Each of these expressions may we written as a Lagrange bracket:

$$\{u,v\}_{qp} = \sum_i \frac{\partial\,q_i}{\partial\,u}\frac{\partial\,p_i}{\partial\,v} - \frac{\partial\,p_i}{\partial\,u}\frac{\partial\,q_i}{\partial\,v} \tag{1.147}$$

and the invariance under canonical transformation of J_1 leads to

$$\{u, v\}_{qp} = \{u, v\}_{QP}.$$ (1.148)

Lagrange brackets are invariant for every canonical transformation.

They satisfy the special relations:

$$
\begin{aligned}
\{q_i,\ q_j\} &= 0 \\
\{p_i,\ p_j\} &= 0 \\
\{q_i,\ p_j\} &= \delta_{ij}.
\end{aligned}
$$ (1.149)

The Kronecker symbol δ_{ij} is given by

$$
\delta_{ij} =
\begin{cases}
1 & \text{if}\quad i = j \\
0 & \text{if}\quad i \neq j.
\end{cases}
$$ (1.150)

12. Poisson's Equations

12.1 Poisson Brackets

Consider a point of a two-dimensional surface in phase space with the parameters u and v:

$$
\begin{aligned}
u &= u\,(q_i,\ p_i), \\
v &= v\,(q_i,\ p_i).
\end{aligned}
$$ (1.151)

The Poisson bracket is defined by the relation

$$[u,\ v] = \sum_k \frac{\partial u}{\partial q_k}\frac{\partial v}{\partial p_k} - \frac{\partial u}{\partial p_k}\frac{\partial v}{\partial q_k}.$$ (1.152)

Poisson brackets satisfy analogous relations to those satisfied by Lagrange brackets. The following relations are thus easily verified:

$$
\begin{aligned}
[p_i,\ p_j] &= 0, \\
[q_i,\ q_j] &= 0, \\
[q_i,\ p_j] &= \delta_{ij}.
\end{aligned}
$$ (1.153)

We next introduce the fundamental relation between Lagrange and Poisson brackets:

$$\sum_\ell \{u_\ell,\ u_i\}\,[u_\ell,\ u_j] = \delta_{ij},$$ (1.154)

in which each function u is a function of the $2n$ coordinates $q_1 \cdots q_n, p_1 \cdots p_n$.

Now, let us evaluate the Poisson bracket for the two arbitary functions F and G:

$$[F,\ G] = \sum_j \frac{\partial F}{\partial q_j} \frac{\partial G}{\partial p_j} - \frac{\partial F}{\partial p_j} \frac{\partial G}{\partial q_j}. \tag{1.155}$$

If we consider that q_j and p_j are functions of the canonically transformed variables P_k and Q_k, we obtain

$$[F,\ G] = \sum_{ij} \left[\frac{\partial F}{\partial q_j} \left(\frac{\partial G}{\partial Q_k} \frac{\partial Q_k}{\partial p_j} + \frac{\partial G}{\partial P_k} \frac{\partial P_k}{\partial p_j} \right) \right. $$
$$\left. - \frac{\partial F}{\partial p_j} \left(\frac{\partial G}{\partial Q_k} \frac{\partial Q_k}{\partial q_j} + \frac{\partial G}{\partial P_k} \frac{\partial P_k}{\partial q_j} \right) \right]. \tag{1.156}$$

By rearranging the terms, it emerges that

$$[F, G]_{q,p} = \sum_k \frac{\partial G}{\partial Q_k} [F,\ Q_k] + \frac{\partial G}{\partial P_k} [F,\ P_k]. \tag{1.157}$$

If in particular, we choose Q_k as the function F, we obtain the Poisson bracket $[Q_k,\ F]$:

$$\left. \begin{array}{c} F = Q_k \\ G = F \end{array} \right\} \implies [Q_k, F]_{q,p} = \sum_j \frac{\partial G}{\partial Q_j} [Q_k,\ Q_j] + \frac{\partial F}{\partial P_j} [Q_k,\ P_j].$$

Considering relations (1.153), we obtain

$$[F,\ Q_k] = -\frac{\partial F}{\partial P_k}. \tag{1.158}$$

In the same way, by changing the function F into P_k and G into F, we obtain

$$\left. \begin{array}{c} F = P_k \\ G = F \end{array} \right\} \implies [F,\ P_k] = \frac{\partial F}{\partial Q_k}. \tag{1.159}$$

If the results are carried over into Poisson brackets (1.157), we obtain

$$[F,\ G]_{q,p} = \sum_k -\frac{\partial G}{\partial Q_k} \frac{\partial F}{\partial P_k} + \frac{\partial G}{\partial P_k} \frac{\partial F}{\partial Q_k} = [F,\ G]_{Q,P}$$

$$\boxed{[F,\ G]_{q,\,p} = [F,\ G]_{Q,\,P}}. \tag{1.160}$$

Poisson brackets are therefore invariant with respect to any canonical transformation.

12.2 Dynamical Law

The following relations are easily verified using the definition (1.152) of Poisson brackets:

$$[u, v] = -[v, u],$$
$$[u, u] = 0,$$
$$[u, c] = 0,$$
$$[u + v, w] = [u, w] + [v, w],$$
$$[u, v\,w] = [u, v]\,w + v\,[u, w].$$

(1.161)

The equations of motion assume a very simple form if we use Poisson brackets. If we use the arbitrary function F, the Hamiltonian function H and the transformed variables p and q as variables in Eqs. (1.158) and (1.159), we obtain

$$[f, p_k] = \frac{\partial f}{\partial q_k} \quad \text{and} \quad [p_k, f] = -\frac{\partial f}{\partial q_k},$$

$$[f, q_k] = -\frac{\partial f}{\partial p_k} \quad \text{and} \quad [q_k, f] = \frac{\partial f}{\partial p_k},$$

and for $f = H$ and with Hamiltonian Eqs. (1.67), we obtain

$$[q_i, H] = \frac{\partial H}{\partial p_i} = \dot{q}_i,$$
$$[p_i, H] = -\frac{\partial H}{\partial q_i} = \dot{p}_i.$$

(1.162)

This gives a simple and remarkable form to the equations of the motion of a point particle in classical mechanics:

$$\frac{d\,q_i}{d\,t} = [q_i, H] \quad \text{and} \quad \frac{d\,p_i}{d\,t} = [p_i, H].$$

(1.163)

More generally speaking, it is easy to obtain the evolution equation of any observable u described by a function of the variables q and p in phase space by simply writing its total differential as

$$\frac{d\,u}{d\,t} = \sum_i \frac{\partial u}{\partial q_i}\,\dot{q}_i + \frac{\partial u}{\partial p_i}\,\dot{q}_i + \frac{\partial u}{\partial t},$$

(1.164)

or by using the Hamilton equations

$$\frac{d\,u}{d\,t} = \sum_i \frac{\partial u}{\partial q_i}\frac{\partial H}{\partial p_i} - \frac{\partial u}{\partial p_i}\frac{\partial H}{\partial q_i} + \frac{\partial u}{\partial t}.$$

(1.165)

We obtain Poisson's dynamical law:

$$\boxed{\frac{d\,u}{d\,t} = [u,\ H] + \frac{\partial\,u}{\partial\,t}}\ . \tag{1.166}$$

It should be noted that if u is the Hamiltonian itself, then we once more obtain

$$\frac{d\,H}{d\,t} = \frac{\partial\,H}{\partial\,t}. \tag{1.167}$$

An observable u that is explicitly time independent (that is, such that $\partial u/\partial t = 0$) will be termed a constant of the motion if its Poisson bracket with the Hamiltonian is equal to zero.

12.3 Example of Application

Let us describe the motion of a particle with mass m and electric charge e in the electric field $\vec{E} = -\vec{\nabla}\,\phi$ and the magnetic field $\vec{B} = \vec{\nabla} \wedge \vec{A}$ using the Hamiltonian

$$H = \frac{1}{2m}\ (\vec{p} - e\,\vec{A})^2 + e\,\phi. \tag{1.168}$$

Combined with (1.160) and definition (1.152), the evolution of the position variables will be

$$\frac{d\,x_i}{dt} = [x_i,\ H] = \frac{1}{m}\ (p_i - e\,A_i). \tag{1.169}$$

It is only natural then to introduce the kinematic momentum

$$\vec{\pi} = \vec{p} - e\,\vec{A} \tag{1.170}$$

whence the foregoing dynamical law becomes

$$m\ \frac{d\,\vec{x}}{dt} = m\ \vec{v} = \vec{\pi} = \vec{p} - e\,\vec{A}, \tag{1.171}$$

while the Hamiltonian (1.168) takes the now familiar form

$$H = \frac{\pi^2}{2m} + e\,\phi. \tag{1.172}$$

The Poisson bracket for kinematic momenta is evaluated using the definition of $\vec{\pi}$ and of the magnetic field $\vec{B}$:

$$[\pi_i,\ \pi_j] = -e\ (\partial_j\,A_i - \partial_i\,A_j) = e\ \varepsilon_{ijk}\,B_k\,, \tag{1.173}$$

where ε_{ijk} is the antisymmetrical Levi–Civita symbol. We can then go ahead to infer the second dynamical law:

$$\frac{d\,\pi_i}{dt} = [\pi_i,\ H] = -\frac{e}{m}\,(p_j - e\,A_j)\,(\partial_j\,A_i - \partial_i\,A_j) - e\,\partial_i\,\phi, \qquad (1.174)$$

which is none other than an expression for the of Lorentz force:

$$\frac{d\,\vec{\pi}}{dt} = \vec{F} = e\,(\vec{E} + \frac{1}{m}\,\vec{\pi} \wedge \vec{B}) = e\,(\vec{E} + \vec{v} \wedge \vec{B}). \qquad (1.175)$$

Before we leave corpuscular mechanics to look at the situation in wave mechanics, it might be helpful to summarize the different formulations of particle mechanics for a force $\vec{F}$ deriving from a potential V.

$$\vec{F} = -\vec{\nabla}\,V \qquad F_i = -\frac{\partial\,V}{\partial\,q_i} \qquad i = 1, 2, 3$$

The particle with mass m possesses a kinetic energy

$$T_i = \frac{1}{2}\,m\,\dot{q}_i^2\,.$$

Newtonian formulation: $T = \sum_i \frac{1}{2}\,m\,\dot{q}_i^2$

$$p_i = \frac{\partial\,T}{\partial\,\dot{q}_i} \qquad \frac{d\,p_i}{d\,t} = -\frac{\partial\,V}{\partial\,q_i}\,, \qquad \text{or} \quad \vec{F} = m\,\vec{\gamma}.$$

Lagrangian formulation: $L = T - V$

$$p_i = \frac{\partial\,L}{\partial\dot{q}_i} \qquad \frac{d\,p_i}{d\,t} = \frac{\partial\,L}{\partial\,q_i} \quad \Longrightarrow \quad \frac{d\,p_i}{d\,t} = -\frac{\partial\,V}{\partial\,q_i}.$$

Hamiltonian formulation: $H = T + V$ and $T = \sum_i p_i^2/2m$

$$\frac{d\,p_i}{d\,t} = -\partial\,H/\partial\,q_i \quad \Longrightarrow \quad \frac{d\,p_i}{d\,t} = -\frac{\partial\,V}{\partial\,q_i}$$

$$\frac{d\,q_i}{d\,t} = \frac{\partial\,H}{\partial\,p_i} \quad \Longrightarrow \quad \frac{d\,q_i}{d\,t} = \frac{p_i}{m}.$$

Poisson formulation:

$$[u, \, v] = \sum_k \frac{\partial u}{\partial q_k} \frac{\partial v}{\partial p_k} - \frac{\partial u}{\partial p_k} \frac{\partial v}{\partial q_k}$$

$$[q_i, \, H] = \frac{\partial H}{\partial p_i} \qquad [p_i, \, H] = -\frac{\partial H}{\partial q_i}$$

$$\frac{dF}{dt} = [F, \, H] + \frac{\partial F}{\partial t} \implies \dot{p}_i = [p_i, \, H] = -\frac{\partial H}{\partial q_i} = -\frac{\partial V}{\partial q_i}$$

$$\dot{q}_i = [q_i, \, H] = \frac{\partial H}{\partial p_i}.$$

13. Hamilton–Jacobi Equations

To ensure that P_i and Q_i, the canonically transformed variables of p_i and q_i, are constants over time, we choose a transformed Hamiltonian K, which is identically equal to zero, $K \equiv 0$ such that relations (1.166) will always yield:

$$\frac{\partial K}{\partial P_i} = \dot{Q}_i = 0 \qquad \text{that is} \qquad Q_i = \text{constant},$$

$$-\frac{\partial K}{\partial Q_i} = \dot{P}_i = 0 \qquad \text{that is} \qquad P_i = \text{constant}. \tag{1.176}$$

We saw in (1.140) that the functions K and H are linked by the relation

$$K = H + \frac{\partial F}{\partial t}, \tag{1.177}$$

such that with $K \equiv 0$ we are left with an equation with the partial differentials

$$H(q, \, p, \, t) + \frac{\partial F}{\partial t} = 0. \tag{1.178}$$

Let us use the second determination of the generating function

$$F = F_2(q, \, P, \, t) = S(q, \, P, \, t). \tag{1.179}$$

$S(q, \, P, \, t)$ is termed Hamilton's principal function. Equations (1.139) give the missing variables:

$$p_i = \frac{\partial S}{\partial q_i} \qquad \text{and} \qquad Q_i = \frac{\partial S}{\partial P_i}, \tag{1.180}$$

whereas relation (1.178) becomes what has been termed the Hamilton–Jacobi equation:

$$H\left(q_i,\ \frac{\partial S}{\partial q_i},\ t\right) + \frac{\partial S}{\partial t} = 0 \quad . \tag{1.181}$$

This is a partial differential equation with $(n+1)$ variables. A complete solution to the equation would call into play $(n+1)$ independent integration constants $\alpha_1...\alpha_{n+1}$.

If S is a solution, then $S+\alpha$ is also a solution of (1.181). One of the $(n+1)$ constants should, therefore, be simply additive and a complete solution to (1.181) would take the form

$$S = S\,(q_1...q_n,\ \alpha_1...\alpha_n,\ t)\,, \tag{1.182}$$

in which none of the constants is additive. We can, therefore, freely choose n constants α_i which are, for example, equal to the new momenta P_i (which are constants according to (1.176)):

$$P_i = \alpha_i\,. \tag{1.183}$$

Equation (1.180) therefore shows that $Q_i = \partial\,S\,/\,\partial\,P_i = \text{constant}$ and, when combined with (1.183),

$$Q_i = \frac{\partial\,S\,(q_i,\ \alpha_i,\ t)}{\partial\,\alpha_i} = \beta_i\,. \tag{1.184}$$

The constants α_i and β_i can be determined using the initial conditions, and equation (1.184) makes it possible to come up with values for q_i as a function of α_i, β_i and time t:

$$q_i = q_i\,(\alpha_i,\ \beta_i,\ t)\,. \tag{1.185}$$

We have thus been able to obtain a solution q with respect to the initial conditions and to time and, thereby, an equation of motion of the system.

S is dependent only on q_i and time since P_i are constants. The total differential may, therefore, be written and the partial differentials of S expressed using (1.180) and (1.181). In this manner, we obtain

$$\frac{d\,S}{\partial\,t} = \sum_i \frac{\partial\,S}{\partial\,q_i}\,\dot{q}_i + \frac{\partial\,S}{\partial\,t} = \sum_i P_i\,\dot{q}_i - H = L\,, \tag{1.186}$$

and, following integration, we obtain*

$$S = \int L \, dt + S_0 \, . \tag{1.187}$$

Hamilton's principal function is, to within a constant, the same as the Hamiltonian action.

An especially interesting case of the Hamilton–Jacobi equation is that in which the Hamiltonian is not explicitly time dependent:

$$\frac{\partial S}{\partial t} + H \left(q_i, \frac{\partial S}{\partial q_i} \right) = 0 \, . \tag{1.188}$$

If we separate space variables from time variables by setting

$$S \left(q_i, \, \alpha_i, \, t \right) = W \left(q_i, \, \alpha_i \right) - \beta_1 \, t \, , \tag{1.189}$$

we obtain with Eq. (1.178)

$$H \left(q_i, \frac{\partial W}{\partial q_i} \right) = \beta_1 \, , \tag{1.190}$$

which is no longer time dependent. One of the integration constants in (1.182), the constant β_1, is therefore the total energy E and, if the system is conservative, we obtain

$$S \left(q, \, P, \, t \right) = W \left(q, \, P \right) - E \, t \, , \quad \text{and}$$
$$H \left(q_i, \frac{\partial W}{\partial q_i} \right) = E \, . \tag{1.191}$$

14. Propagation Equation of Geometric Optics

So far, the assumption that matter is concentrated at certain points in space has underlain our description of particle mechanics and our objective has been essentially to follow the movement of these point particles in time. Wave mechanics rejects the concept of localization in favor of that of continuous systems. Each point in a continuous medium is therefore part and parcel of the motion of the whole and the complete motion can only be described by detailing the position coordinates of all the points constituting the system at the same time. The wave propagation velocity is therefore assumed to be constant in time and characteristic of the medium in which the waves propagate.

---

* As pointed out by Thirring, this feature characterizes integrable systems. Even though the equations always have local solutions, global solutions exist only in special cases.

In optics, when the medium in which the electromagnetic waves propagate is the vacuum, the velocity of propagation (a universal constant) is represented by c.

We can, therefore, proceed to use Maxwell's equations of E.M. (electromagnetic) fields to write the equation for the propagation of this field.

Maxwell's equations in the absence of charges and current are written in the S.I. system, as follows:

$$\vec{\nabla} \wedge \vec{E} + \frac{\partial \vec{B}}{\partial t} = 0,$$
$$\vec{\nabla} \cdot \vec{B} = 0,$$
$$\vec{\nabla} \wedge \vec{H} - \frac{\partial \vec{D}}{\partial t} = 0,$$
$$\vec{\nabla} \cdot \vec{D} = 0. \tag{1.192}$$

The permittivities ε and μ serve to link the electric and magnetic fields, $\vec{E}$ and $\vec{B}$ respectively, to the vectors $\vec{D}$ and $\vec{H}$ with the aid of the relations:

$$\vec{D} = \varepsilon \vec{E} \quad \text{and} \quad \vec{B} = \mu \vec{H}. \tag{1.193}$$

Now let us differentiate with respect to time, the third Maxwell equation:

$$\frac{\partial}{\partial t} \vec{\nabla} \wedge \vec{H} - \frac{\partial}{\partial t} \frac{\partial \vec{D}}{\partial t} = 0,$$

which introduces the time-differentials of the electric and magnetic fields:

$$\vec{\nabla} \wedge \frac{\partial \vec{H}}{\partial t} - \varepsilon \frac{\partial^2 \vec{E}}{\partial t^2} = 0 \tag{1.194}$$

but the first Maxwell equation yields $\partial \vec{H} / \partial t$:

$$\frac{\partial \vec{H}}{\partial t} = \frac{1}{\mu} \frac{\partial \vec{B}}{\partial t} \quad \text{and} \quad \frac{\partial \vec{B}}{\partial t} = -\vec{\nabla} \wedge \vec{E}.$$

Inserting this into Eq. (1.194) we obtain

$$-\vec{\nabla} \wedge (\vec{\nabla} \wedge \vec{E}) - \mu \varepsilon \frac{\partial^2 \vec{E}}{\partial t^2} = 0,$$

but $\vec{\nabla} \wedge (\vec{\nabla} \wedge \vec{E}) = \vec{\nabla}(\vec{\nabla} \cdot \vec{E}) - \vec{\nabla} \cdot \vec{\nabla} \vec{E}$ and because $\vec{\nabla} \cdot \vec{E} = \frac{1}{\varepsilon} \vec{\nabla} \cdot \vec{D} = 0$ we are left with the equation for the propagation of the electric field:

$$\nabla^2 \vec{E} - \varepsilon \mu \frac{\partial^2 \vec{E}}{\partial t^2} = 0. \tag{1.195}$$

We further introduce the refractive index of the medium by setting

$$\varepsilon\,\mu = n^2/c^2\,, \tag{1.196}$$

where n is the refractive index and c the propagation velocity in vacuum.
    In other words, if v is the propagation velocity in the medium considered

$$v = \frac{c}{n} = \frac{1}{\sqrt{\varepsilon\,\mu}}\,. \tag{1.197}$$

The equation for the propagation of the electric field is then written

$$\nabla^2\,\vec{E} - \frac{n^2}{c^2}\,\frac{\partial^2\,\vec{E}}{\partial\,t^2} = \nabla^2\,\vec{E} - \frac{1}{v^2}\,\frac{\partial^2\,\vec{E}}{\partial\,t^2} = 0\,. \tag{1.198}$$

An identical equation is obtained for the propagation of the magnetic field $\vec{H}$.
To avoid confusion with the energy E, we introduce the field ϕ. The equation
of the propagation of the E.M. field will therefore take the following form:

$$\nabla^2\,\phi - \frac{n^2}{c^2}\,\frac{\partial^2\,\phi}{\partial\,t^2} = 0 \tag{1.199}$$

where $\phi = \vec{E}_i$ or $\vec{H}_i$ with $i = 1, 2, 3$.
If the refractive index n is constant in the medium under consideration, then
we may want to look for a solution in the following form

$$\phi = \phi_0\,e^{i\,(\vec{k}\cdot\vec{r} - \omega\,t)}\,. \tag{1.200}$$

The field wave ϕ satisfies the propagation Eq. (1.199) if

$$k = \frac{n}{c}\,\omega = \frac{2\,\pi\,n}{\lambda} = k_0\,n \tag{1.201}$$

with a relation between the angular velocity ω, the frequency ν and the
wavelength λ:

$$\omega = 2\pi\,\nu = \frac{2\pi}{T} = \frac{2\pi\,c}{\lambda}\,. \tag{1.202}$$

Consider, for the sake of simplicity, the wave vector $\vec{k}$ in the Oz direction. If
we use relation (1.201), we obtain:

$$\phi = \phi_0\,e^{i\,k_0\,(n\,z - c\,t)}\,, \tag{1.203}$$

which is the solution of the propagation equation.
    The reader's attention is drawn here to the presence in the above solution
of the optical path $n\,z$.
    Now, let us imagine that the index n varies slowly in space and look for
a solution similar to the preceding one, that is of the form

$$\phi = e^{A(r) + i\,k_0\,[L(r) - c\,t]} \tag{1.204}$$

in which $A(r)$ represents the amplitude (real) and L the optical path traveled (also real). We obtain by inserting (1.204) into (1.199):

$$i\,k_0\,\left[\nabla^2\,L + 2\vec{\nabla}\,A.\vec{\nabla}\,L\right]\,\phi + \left[\nabla^2\,A + (\vec{\nabla}\,A)^2 - k_0^2\,(\vec{\nabla}\,L)^2 + n^2\,k_0^2\right]\,\phi = 0\,. \tag{1.205}$$

By separately eliminating the real and imaginary parts of the equation, we come up with two coupled equations, each with terms in L and in A:

$$\nabla^2\,A + (\vec{\nabla}\,A)^2 + k_0^2\,\left[n^2 - (\vec{\nabla}\,L)^2\right] = 0, \tag{1.206}$$

$$\nabla^2\,L + 2\,\vec{\nabla}\,L \cdot \vec{\nabla}\,A = 0\,. \tag{1.207}$$

Let us now use the approximation of geometric optics, that is we assume that n does not vary much and, in particular, that n does not change significantly over distances that are of the order of the wavelength.

Geometric optics assumes that the wavelength λ is small in comparison to any change in the medium.

The dominant term in Eq. (1.206) becomes the term in k_0^2 and we obtain the relation known as the eikonal equation of geometrical optics:

$$(\vec{\nabla}\,L)^2 = n^2 \quad \text{so that} \quad |\,\vec{\nabla}\,L\,| = n \quad \text{and} \quad L = \int n\,ds\,. \tag{1.208}$$

Fermat's principle therefore states that the variation in the optical path L is an extremum, that is:

$$\delta \int n\,ds = 0\,, \tag{1.209}$$

and Malus' theorem tells us that the rays of geometric optics are nothing but lines that are orthogonal to wave surfaces (equiphase surfaces).

15. Mechanics and Waves

Let us return to definition (1.191) of the Hamilton–Jacobi function when H is not explicitly time dependent:

$$S\,(q,\,P,\,t) = W\,(q,\,P) - E\,t\,. \tag{1.210}$$

The surfaces $S = $ constant may be considered as wave fronts propagating in the configuration space:

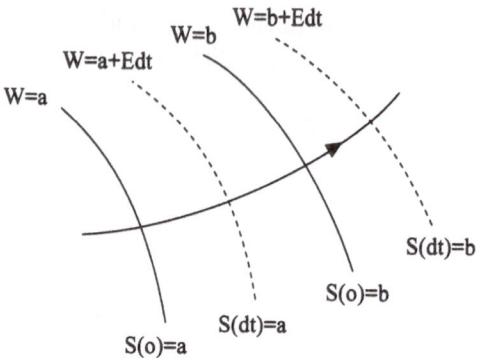

Now, let us compute the propagation velocity for these wave surfaces by paying particular attention to the fact that for $S = $ constant $dS = 0$, which leads to the statement

$$dS = \vec{\nabla}S \cdot d\vec{x} + \frac{\partial S}{\partial t}\, dt = 0\,. \qquad (1.211)$$

We introduce the unit vector $\vec{n}$ normal to the surface S with $\vec{n} = \vec{\nabla}S/|\,\vec{\nabla}S\,|$ and the element with length ds on this normal $ds = \vec{n} \cdot d\vec{x}$, thus transforming the foregoing relation as follows:

$$|\,\vec{\nabla}S\,|\, \frac{ds}{dt} + \frac{\partial S}{\partial t} = 0\,. \qquad (1.212)$$

With (1.210), the propagation velocity of "mechanical waves" will hence be

$$u = \frac{ds}{dt} = -\frac{\partial S}{\partial t}/|\,\vec{\nabla}S\,| = E/|\,\vec{\nabla}W\,|\,. \qquad (1.213)$$

If we consider a system consisting of only one particle and take the cartesian position coordinates as the generalized coordinates, the Hamilton–Jacobi equation will then yield the value of the gradient W:

$$H = \frac{p^2}{2m} + V = E \qquad \text{and} \qquad p_i = \frac{\partial W}{\partial q_i} = (\vec{\nabla}\,W)_i\,. \qquad (1.214)$$

This leads to the equation of conservation of energy

$$\frac{1}{2m}\left[\left(\frac{\partial W}{\partial x}\right)^2 + \left(\frac{\partial W}{\partial y}\right)^2 + \left(\frac{\partial W}{\partial z}\right)^2\right] + V = E\,, \qquad (1.215)$$

which by introducing the kinetic energy T can also be written as follows:

$$(\vec{\nabla}\,W)^2 = 2m\,(E - V) = 2m\,T\,. \qquad (1.216)$$

The propagation velocity of mechanical waves will then be

$$u = \frac{E}{\sqrt{2m\,T}}. \tag{1.217}$$

Notice that $p_i = \partial W / \partial q_i$ and, hence, $\vec{p} = \vec{\nabla}\,W$ will give the following form to the propagation velocity of wave surfaces:

$$u = \frac{E}{p} = \frac{E}{m\,v} = \frac{E}{\sqrt{2m\,T}}. \tag{1.218}$$

Attention is drawn to the close resemblance between Eq. (1.216) and Eq. (1.208) for the optical path in geometric optics:

$$(\vec{\nabla}\,W)^2 = 2m\,T\,,$$
$$(\vec{\nabla}\,L)^2 = n^2\,. \tag{1.219}$$

This resemblance is also found in Fermat's variational principle

$$\delta \int n\,ds = 0 \qquad \text{Fermat} \tag{1.220}$$

as well as in Maupertuis' variational principle given in (1.131), with the medium's refractive index with respect to particles with mass m and kinetic energy T being $n = \sqrt{2m\,T} = m\,v$

$$\delta \int \sqrt{2m\,T}\,ds = 0 \qquad \text{Maupertuis}\,. \tag{1.221}$$

Attention is further drawn to the fact that (1.218) leads to a constant product of the velocities u (of the surface S) and v (of the particle), which is equal to E/m, and to c^2 if $E = mc^2$.

16. Wave Mechanics

Let us pursue the equivalence between Fermat's principle, on the one hand, and Maupertuis' principle, on the other, by assuming that W is proportional to L and, therefore, that S is proportional to $k_0\,(L - ct)$

$$k_0\,(L - ct) = \frac{2\pi}{\lambda_0}\,(L - ct) = \frac{2\pi}{\lambda_0}\,L - 2\pi\,\nu\,t\,,$$
$$S = W - E\,t\,. \tag{1.222}$$

This presupposes term-by-term proportionality between W and L on the one hand, and between E and ν, on the other, and leads to the Einstein relation

$$\boxed{E = h\,\nu}\ .\tag{1.223}$$

But for the time being, h is simply a constant with nothing to do with Planck's constant. We also know that the wavelength λ is linked to the vibration frequency through the propagation velocity u:

$$\lambda = \frac{u}{\nu} = \frac{E}{p}\frac{1}{\nu} = \frac{h\nu}{p}\frac{1}{\nu}\quad\text{and thus}\quad\lambda = \frac{h}{p}.\tag{1.224}$$

Let us recapitulate in term of waves the description of classical mechanics with respect to the Hamilton–Jacobi formalism.

It has been shown that instead of following the trajectories of the particles constituting the system, we could follow the motion of a "mechanical wave", the surface described by the Hamilton–Jacobi equation $S\ (q,\ P,\ t)$. This surface propagates at the speed $u = E\,/\,p$ and at the frequency $\nu = E/h$ or a wavelength of $\lambda = h\,/\,p$. Corresponding to the optical wave

$$\phi_{\text{opt}} = e^{A(r)}\ e^{i\,k_0(L-ct)}\tag{1.225}$$

we may therefore introduce a mechanical wave:

$$\phi_{\text{mech}} = e^{A(r)}\ \exp\left[\frac{i}{\hbar}(W - Et)\right] = e^{A(r)}\ \exp\left[\frac{i}{\hbar}S\right].\tag{1.226}$$

Does such a wave actually exist or is it just an imaginary wave corresponding to a mathematical model? This is an open question which will come up again and again in our discussions of the interpretation of wave mechanics and of Feynman's formulation of quantum mechanics.

The formal analogy between ϕ_{opt} and ϕ_{mech} can be used to determine the time-independent propagation equation (that is, "propagation in space" of the wave ϕ_{opt}) and then to introduce the scalar function ψ corresponding to wave mechanics and satisfying the same form of time-independent equation.

16.1 The Schrödinger Equation

What then becomes of the equation for the propagation of geometric optics, given the hypothesis $E = h\,\nu$?

Let us return to the propagation Eq. (1.198):

$$\nabla^2\,\phi - \frac{1}{v^2}\,\frac{\partial^2\,\phi}{\partial\,t^2} = 0,$$

with the form (1.200) for the solution:

$$\phi = e^{A(r)} \, e^{ik_0 \, (L(r) - ct)} .$$

The differential with respect to time is written

$$\frac{\partial^2 \phi}{\partial t^2} = (-ik_0 \, c)^2 \, \phi = -k_0^2 \, c^2 \, \phi$$

with the expression for the wave number

$$k = k_0 \, n = k_0 \, \frac{c}{v} = \frac{2\pi}{\lambda} \qquad \text{or} \qquad k_0 = \frac{2\pi \, v}{c \, \lambda} , \qquad (1.227)$$

thus yielding the second differential with respect to time

$$\frac{\partial^2 \phi}{\partial t^2} = -\frac{4\pi^2 \, v^2}{\lambda^2} \, \phi . \qquad (1.228)$$

This leads to the time-independent differential equation:

$$\nabla^2 \, \phi + \frac{4\pi^2}{\lambda^2} \, \phi = 0 \qquad (1.229)$$

for the field ϕ, the optical wave amplitude. There is, therefore, a quantity (termed the wave function) which should satisfy an equation analogous to (1.229) and we will proceed to set $\phi_{\text{mech}} = \psi$. This gives:

$$\nabla^2 \, \psi + \frac{4\pi^2}{\lambda^2} \, \psi = 0 , \qquad (1.230)$$

or, if we use the value of the wavelength (1.224),

$$\lambda = \frac{h}{p} = \frac{h}{\sqrt{2m \, (E - V)}} , \qquad (1.231)$$

we obtain the wave equation for ψ,

$$\nabla^2 \, \psi + \frac{8\pi^2 \, m}{h^2} \, (E - V) \, \psi = 0 . \qquad (1.232)$$

We set $\hbar = h/2\pi$ and write the foregoing equation as follows:

$$\boxed{-\frac{\hbar^2}{2m} \, \nabla^2 \, \psi + V \, \psi = E \, \psi} . \qquad (1.233)$$

This is the Schrödinger equation which is the basis of wave mechanics, where the constant h is Planck's constant.

In the context presented above, the wave function ψ describes the propagation (in space and not in time) of the Hamilton–Jacobi surface. This would suggest that classical mechanics is to wave mechanics what geometric optics is to physical optics, that is an approximation corresponding to the smallness of $\lambda \ (= h / p)$ relative to the possible changes in the medium.

To regain the conditions of classical mechanics would then mean to cause h to tend to zero in the solutions of the Schrödinger equation.

At this level wave mechanics requires an unnatural hypothesis that will be justified in quantum dynamics: the evolution in time of the function ψ is obtained by replacing the energy E of the second term of the Schrödinger equation by the time dependent differential operator $i\hbar\partial/\partial t$ acting on the wave function ψ:

$$i\hbar\frac{\partial}{\partial t}\psi = \left(-\frac{h^2}{2m}\nabla^2 + V\right)\psi.$$

Properly applied to form (1.226) of the wave function $\psi = \phi_{mech}$, we once more obtain the Schrödinger Eq. (1.233).

To make sure that we obtain the Jacobi equation of classical mechanics at the limit $h \rightarrow 0$, we should seek solutions of the type $\exp[(i/\hbar)\,S]$ by setting:

$$\psi\,(x,\,y,\,z,\,t) = e^{\frac{i}{\hbar}\,S(x,\,y,\,z,\,t)}. \tag{1.234}$$

The differentials with respect to x and t are:

$$\frac{\partial\psi}{\partial x} = \frac{i}{\hbar}\,e^{\frac{i}{\hbar}\,S(x,\,y,\,z,\,t)}\,\frac{\partial S}{\partial x},$$

$$\frac{\partial^2\psi}{\partial x^2} = \left[\frac{i}{\hbar}\frac{\partial^2 S}{\partial x^2} - \frac{1}{\hbar^2}\left(\frac{\partial S}{\partial x}\right)^2\right]e^{\frac{i}{\hbar}\,S(x,\,y,\,z,\,t)}, \tag{1.235}$$

$$\frac{\partial\psi}{\partial t} = \frac{i}{\hbar}\frac{\partial S}{\partial t}\,\exp\left(\frac{i}{\hbar}\,S(x\,y\,z\,t)\right).$$

By inserting (1.235) into (1.233), we obtain:

$$\begin{aligned}-\frac{\partial S}{\partial t} &= -\frac{\hbar^2}{2m}\left[\frac{i}{\hbar}\nabla^2 S - \frac{1}{\hbar^2}(\vec{\nabla}S)^2\right] + V \\ &= \frac{1}{2m}(\vec{\nabla}S)^2 + V - \frac{i\hbar}{2m}\nabla^2 S.\end{aligned} \tag{1.236}$$

If $\hbar \rightarrow 0$, we quite naturally regain the Jacobi equation of classical mechanics:

$$-\frac{\partial S}{\partial t} = \frac{1}{2m}\left(\vec{\nabla}S\right)^2 + V. \tag{1.237}$$

Is classical mechanics then a limiting case of wave mechanics? The question is a fundamental one and we intend to take it up on several occasions.

16.2 The Correspondence Principle

Let us try to find a better formalism for this result by establishing a correspondence principle between classical mechanics and wave mechanics based
on the Schrödinger equation. The Hamilton–Jacobi equation

$$\frac{\partial S}{\partial t} + H \left(q_i, \frac{\partial S}{\partial q_i}, t \right) = 0 \qquad (1.238)$$

of a point particle in Cartesian coordinates becomes:

$$\frac{\partial S}{\partial t} + \frac{1}{2m} \left(p_x^2 + p_y^2 + p_z^2 \right) + V(x, y, z, t) = 0 \qquad (1.239)$$

using the correspondence in classical mechanics:

$$p_x = \frac{\partial S}{\partial x} . \qquad (1.240)$$

The Schrödinger equation is, for its part, written as follows:

$$i\hbar \frac{\partial \psi}{\partial t} = -\frac{\hbar^2}{2m} \nabla^2 \psi + V \psi . \qquad (1.241)$$

If we bear in mind the possibility of writing

$$(p_x)^2 = \left(-i\hbar \frac{\partial}{\partial x} \right)^2 = -\hbar^2 \frac{\partial^2}{\partial x^2} , \qquad (1.242)$$

we see that we can go from the Hamilton–Jacobi equation to the Schrödinger
equation with the correspondence:

$$\boxed{\begin{array}{l} E \;\rightarrow\; i\hbar \dfrac{\partial}{\partial t} \\[2mm] p_i \;\rightarrow\; -i\hbar \dfrac{\partial}{\partial q_i} \end{array}} . \qquad (1.243)$$

This correspondence is often set up as a principle, the correspondence principle, between classical mechanics and wave mechanics, by setting:

$$\vec{p} = -i\hbar \vec{\nabla} \qquad \text{in Cartesian coordinates} . \qquad (1.244)$$

There is another possible correspondence which consists in replacing the Poisson bracket by the commutator divided by $i\hbar$:

$$[A, H] \;\rightarrow\; \frac{1}{i\hbar} [A, H] . \qquad (1.245)$$

The advantage is that it leads directly to the equation of evolution in time since

$$\dot{A} = \frac{dA}{dt} = [A, H] + \frac{\partial A}{\partial t} \qquad \text{in classical mechanics}, \qquad (1.246)$$

combined with the above correspondence principle, becomes

$$\dot{A} = \frac{dA}{dt} = \frac{1}{i\hbar} [A, H] + \frac{\partial A}{\partial t} \qquad \text{in quantum mechanics}. \qquad (1.247)$$

As a matter of fact, these two formulations of the correspondence principle are rigorously equivalent. We have already shown (1.162) that for the Poisson brackets

$$[p_i, H] = -\frac{\partial H}{\partial q_i}. \qquad (1.248)$$

Let us calculate the commutator $[p_i, H]$ by applying it to a wave function and by using the correspondence principle $p_i \rightarrow -i\hbar \, \partial/\partial q_i$. The commutator of two operators, A and B, is the mathematical object $[A, B] = AB - BA$. By applying $[p_i, H]$ to ψ, we then obtain

$$[p_i, H] \, \psi = p_i \, H \, \psi - H \, p_i \, \psi = -i\hbar \frac{\partial}{\partial q_i} (H \, \psi) - H \left(-i\hbar \frac{\partial}{\partial q_i} \psi \right)$$

$$= -i\hbar \frac{\partial H}{\partial q_i} \psi - i\hbar H \frac{\partial \psi}{\partial q_i} + i\hbar H \frac{\partial \psi}{\partial q_i}. \qquad (1.249)$$

After simplification, we obtain a relation that is valid irrespective of the wave function ψ:

$$[p_i, H] \, \psi = -i\hbar \frac{\partial H}{\partial q_i} \psi \qquad (1.250)$$

leading to the following relation between operators:

$$[p_i, H] = -i\hbar \frac{\partial H}{\partial q_i}. \qquad (1.251)$$

By comparing with relation (1.248), we further obtain

$$-\frac{\partial H}{\partial q_i} = \frac{1}{i\hbar} [p_i, H] \qquad \text{in wave mechanics}$$

$$= [p_i, H] \qquad \text{in classical mechanics} \qquad (1.252)$$

thus establishing the correspondence:

$$[A, H] \rightarrow \frac{1}{i\hbar} [A, H]. \qquad (1.253)$$

The generalization of this result to all the conjugate operators constitutes the correspondence principle of wave mechanics

$$\boxed{[A,\ B] \ \rightarrow \ \frac{1}{i\ \hbar}\ [A,\ B]}\ . \tag{1.254}$$

In classical mechanics, A and B are two canonically conjugate variables whereas, in wave mechanics, they are two conjugate operators acting in the Hilbert space of the wave functions. The Poisson brackets (1.153) thus lead to the commutation relations between position operator X_i and momentum operator p_j:

$$[X_i,\ X_j] = 0 \qquad [p_i,\ p_j] = 0 \quad \text{and} \quad [X_i,\ p_j] = i\ \hbar\ \delta_{ij} \tag{1.255}$$

Attention is drawn to a delicate point arising from the use of the time and energy variables, t and E respectively: One would expect that the correspondence principle (1.243) should also take the form of a commutation relation $[t,\ E] = i\ \hbar$ as one is wont to suspect by replacing E by $i\ \hbar\ \partial\ /\ \partial\ t$ and by causing the commutator to act on ψ. Yet t and E are not operators associated with observable phenomena since E is, in reality, the eigenvalue of the Hamiltonian H whereas t is a parameter linked to the evolution of the wave functions. We will return to this point in due course.

16.3 Example of Application

Let us return to the earlier example of a particle with electric charge e and mass m, in an electromagnetic field, the example treated within the framework of classical mechanics in Sect. 12. We can associate the operator with the kinematic momentum $\vec{\pi}$ thus:

$$\vec{\pi} = -i\ \hbar\ \vec{\nabla} - e\ \vec{A}, \tag{1.256}$$

whereas the Poisson bracket $[\pi_i,\ \pi_j]$ with the value $e\ \varepsilon_{ijk}\ B_k$ in classical mechanics will be replaced by the operator $i\hbar e\varepsilon_{ijk}\ B_k$ in wave mechanics, since the commutator of the associated operators $\vec{\pi}$ is:

$$[\pi_i,\ \pi_j] = i\ \hbar\ e\ \varepsilon_{ijk}\ B_k\ , \tag{1.257}$$

or, written as a vectoroperator:

$$\vec{\pi} \wedge \vec{\pi} = i\ \hbar\ e\ \vec{B}\ . \tag{1.258}$$

The associated wave function will then be obtained by solving the Schrödinger equation written from the classical Hamiltonian:

$$H = \frac{\pi^2}{2m} + e\ \Phi\ . \tag{1.259}$$

This form of the Hamiltonian will also be used in determining gauge invariance in wave mechanics (see Chap. 7).

17. Analytical Mechanics – Schematic Summary

Lagrangian

$$L\left(q_\alpha,\ \dot{q}_\alpha\right) = T\left(\dot{q}_\alpha^2\right) - V\left(q_\alpha\right)$$

Hamiltonian action

$$S = \int_{t_1}^{t_2} L\left(q_\alpha,\ \dot{q}_\alpha\right) dt$$

Euler–Lagrange equations

$$\delta S = 0 \implies \frac{d}{dt}\frac{\partial L}{\partial \dot{q}_\alpha} - \frac{\partial L}{\partial q_\alpha} = 0$$

Hamiltonian function

$$H = \dot{q}_\alpha\, p_\alpha - L \ \text{ with }\ p_\alpha = \frac{\partial L}{\partial \dot{q}_\alpha}$$

$$\frac{dH}{dt} = \frac{\partial H}{\partial t}$$

Hamilton's equations

$$\dot{q}_\alpha = \frac{\partial H}{\partial p_\alpha} \ \text{ and }\ \dot{p}_\alpha = -\frac{\partial H}{\partial q_\alpha}$$

Poisson brackets

$$[u,\ v] = \sum_i \frac{\partial u}{\partial q_i}\frac{\partial v}{\partial q_i} - \frac{\partial u}{\partial p_i}\frac{\partial v}{\partial p_i}$$

$$[p_i,\ p_j] = 0 \quad [q_i,\ q_j] = 0 \quad [q_i,\ p_j] = \delta_{ij}$$

Poisson's equations

$$\frac{du}{dt} = [u,\ H] + \frac{\partial u}{\partial t}$$

Hamilton Jacobi formalism

$$p_i = \frac{\partial S}{\partial q_i} \quad \text{and} \quad H\left(q_i, \frac{\partial S}{\partial q_i}, t\right) + \frac{\partial S}{\partial t} = 0$$

$$\text{If} \quad \frac{\partial H}{\partial t} = 0 \quad \frac{\partial S}{\partial t} + H\left(q_i, \frac{\partial S}{\partial q_i}\right) = 0$$

$$S\left(q_i, \alpha_i, t\right) = W\left(q_i, \alpha_i\right) - E\,t$$

$$H\left(q_i, \frac{\partial W}{\partial q_i}\right) = E$$

Wave mechanics

$$dW = |\vec{\nabla} W|\, ds = E\, dt \implies u = \frac{ds}{dt} = \frac{E}{|\vec{\nabla} W|}$$

$$\frac{1}{2m}\left[\left(\frac{\partial W}{\partial x}\right)^2 + \left(\frac{\partial W}{\partial y}\right)^2 + \left(\frac{\partial W}{\partial z}\right)^2\right] + V(x\,y\,z) = E$$

$$u = E/\sqrt{2m\,T} = E/p$$

Chapter 2
Wave Mechanics

The Hamilton–Jacobi description of a system of particles with mass m_i and kinetic energy $T_i = P_i^2/2m_i$ permits the interpretation of the real trajectories of such particles as lines orthogonal to the surfaces $S\,(q,\,\vec{\nabla}\,W,\,t)$. By taking these surfaces to analogous to optical wave surfaces, it is possible to introduce an "equation of the mechanical wave" or wave mechanics describing the evolution of a wave function $\psi\,(\vec{r},\,t)$ with frequency given via $E = h\,\nu$ and with wavelength $\lambda = h/p$. This wave function, which plays the role of the optical wave $\vec{E}$ or $\vec{H}$, obeys a fundamental equation of propagation, the Schrödinger equation, written from the Hamiltonian formulation of a classical system using a correspondence principle between the classical variables momentum, $\vec{p}$, and energy, E, and operators acting on the wave function sought.

Let us then postulate the existence of the wave functions $\psi\,(\vec{r},\,t)$ obeying the partial differential equation written with the correspondence principle in Cartesian coordinates

$$H\,(\vec{r},\,-i\,\hbar\,\vec{\nabla},\,t)\,\psi\,(\vec{r},\,t) = i\,\hbar\,\frac{\partial}{\partial t}\,\psi\,(\vec{r},\,t)\,, \qquad (2.1)$$

where H is the operator associated with the classical Hamiltonian,

$$H = T + V = \frac{p^2}{2m} + V \qquad (2.2)$$

for a particle with mass m in a potential $V\,(r)$. This boils down to postulating the existence of the Schrödinger equation for determining the wave function $\psi\,(\vec{r},\,t)$, a complex function of space and time:

$$\left[-\frac{\hbar^2}{2m}\,\nabla^2 + V\,(\vec{r}) \right]\,\psi\,(\vec{r},\,t) = i\,\hbar\,\frac{\partial}{\partial t}\,\psi\,(\vec{r},\,t)\,. \qquad (2.3)$$

A priori, this would appear to be simply an extension of the models for describing mechanical systems. However, in reality, it constitutes a revolution in the concepts of mechanics, also in physics, and even in science and scientific thought as a whole. The justification of the Schrödinger equation has led to the questioning of the concept of measurement and the idea of interaction between the objects observed and the observer. It also takes to task the notion of physical space in which the objects modeling physical systems are defined, as well as the notion of real space in which the measurements of such systems are made. Finally, it has led to different possible interpretations of the wave functions associated with particles. Wave mechanics, and more generally speaking, quantum mechanics, of which the former constitutes a special description, have so far not been faulted in their description of microscopic physical systems, from molecules to systems of the scale of elementary particles. Yet the theoretical foundations have continued to inspire numerous theoretical works and different interpretations. The probabilistic interpretation is today the most generally accepted interpretation amongst users of this new mechanics, as we hope to show in this chapter.

1. Conservative Systems

In a conservative system, V and hence H, are not explicitly time dependent. The Hamiltonian H is therefore a constant, equal to the total energy. A solution to the Schrödinger equation can be obtained by separating space and time variables:

$$\psi\left(\vec{r},\, t\right) = \psi\left(\vec{r}\right) A\left(t\right).$$

It is thus possible to take the partial differential with respect to time:

$$i\,\hbar\, \frac{\partial\,\psi\left(\vec{r},\, t\right)}{\partial\, t} = i\,\hbar\, \frac{\partial\, A}{\partial\, t}\, \psi\left(\vec{r}\right). \tag{2.4}$$

Inserting this into Eq. (2.1), we obtain:

$$H\,\psi\left(\vec{r}\right) A\left(t\right) = i\,\hbar\, \frac{\partial\, A}{\partial\, t}\, \psi\left(\vec{r}\right),$$

$$\frac{H\,\psi\left(\vec{r}\right)}{\psi\left(\vec{r}\right)} = \frac{i\,\hbar}{A\left(t\right)}\, \frac{\partial\, A}{\partial\, t} = E. \tag{2.5}$$

The first term is not time dependent and the second is not dependent on the coordinates. We can replace the partial differential of A with respect to time by the total differential to obtain a first-order differential equation which easily determines $A\left(t\right)$:

$$A\left(t\right) = \exp\left(\frac{-i}{\hbar}\, E\, t\right)$$

and the general form of the wave function of a conservative system will take the following form:

$$\psi\left(\vec{r},\, t\right) = \psi\left(\vec{r}\right)\, \exp\left(\frac{-i}{\hbar}\, E\, t\right). \tag{2.6}$$

The stationary state $\psi\,(\vec{r})$ of the energy E will be described by the time independent Schrödinger equation:

$$\boxed{\left[-\frac{\hbar^2}{2m}\,\nabla^2 + V\,(\vec{r})\right]\psi\,(\vec{r}) = E\,\psi\,(\vec{r})}\,. \tag{2.7}$$

It is often convenient to write this equation in a system of coordinates other than Cartesian. We obtain the Laplacean in curvilinear coordinates from the length element:

$$d\,s^2 = \sum_i h_i^2\,d\,x_i^2 \quad \text{with} \quad x_1 = x,\ x_2 = y \text{ and } x_3 = z\,, \tag{2.8}$$

and by using the general expression for the Laplacean in curvilinear coordinates:

$$\nabla^2 = \frac{1}{h_1\,h_2\,h_3}\left[\frac{\partial}{\partial x_1}\left(\frac{h_3\,h_2}{h_1}\,\frac{\partial}{\partial x_1}\right) + \frac{\partial}{\partial x_2}\left(\frac{h_1\,h_3}{h_2}\,\frac{\partial}{\partial x_2}\right)\right.$$
$$\left. + \frac{\partial}{\partial x_3}\left(\frac{h_2\,h_1}{h_3}\,\frac{\partial}{\partial x_3}\right)\right] \tag{2.9}$$

Expressed in spherical coordinates, for example, the Laplacean operator becomes,

$$\Delta = \nabla^2 = \frac{1}{r^2}\frac{\partial}{\partial r}\,r^2\,\frac{\partial}{\partial r} + \frac{1}{r^2\,\sin\theta}\frac{\partial}{\partial \theta}\,\sin\theta\,\frac{\partial}{\partial \theta} + \frac{1}{r^2\,\sin^2\theta}\frac{\partial^2}{\partial \varphi^2}\,. \tag{2.10}$$

2. Interpretations of the Wave Function

2.1 Quantum Potential

Let us choose, a priori, the wave function $\psi\,(\vec{r},\,t)$ to have a real amplitude function $a\,(\vec{r},\,t)$ and a real phase function $\varphi\,(\vec{r},\,t)$, that is, we set

$$\psi\,(\vec{r},\,t) = a\,(\vec{r},\,t)\,\exp\left(\frac{i}{\hbar}\,\varphi\,(\vec{r},\,t)\right)\,. \tag{2.11}$$

Now let us insert this into the time dependent Schrödinger equation. The partial differential with respect to time will be written

$$i\hbar\,\frac{\partial\,\psi}{\partial\,t} = \left(i\hbar\,\frac{\partial\,a}{\partial\,t} - a\,\frac{\partial\,\varphi}{\partial\,t}\right)\,\exp\left(\frac{i}{\hbar}\,\varphi\,(\vec{r},\,t)\right).$$

The differential with respect to the position variables will, for its part, be

$$\vec{\nabla}\,\psi = \vec{\nabla}\,a\,\exp\left(\frac{i}{\hbar}\,\varphi\right) + \frac{i}{\hbar}\,\vec{\nabla}\,\varphi\,a\,\exp\left(\frac{i}{\hbar}\,\varphi\right),$$

$$\nabla^2\,\psi = \left[\nabla^2\,a + \frac{2\,i}{\hbar}\,\vec{\nabla}\,a\cdot\vec{\nabla}\,\varphi + \frac{i\,a}{\hbar}\,\nabla^2\,\varphi - \frac{1}{\hbar^2}\,(\nabla\,\varphi)^2\,a\right]\,\exp\left(\frac{i}{\hbar}\,\varphi\right).$$

By inserting these terms into the Schrödinger Eq. (2.3), we obtain the equation linking the functions $a\,(\vec{r},\,t)$ and $\varphi\,(\vec{r},\,t)$:

$$\left(i\hbar\,\frac{\partial\,a}{\partial\,t} - a\,\frac{\partial\,\varphi}{\partial\,t}\right)$$
$$= -\frac{\hbar^2}{2m}\left[\nabla^2 a + \frac{2\,i}{\hbar}\,\vec{\nabla}a\cdot\vec{\nabla}\varphi + \frac{i\,a}{\hbar}\,\nabla^2\,\varphi - \frac{a}{\hbar^2}\,(\nabla\varphi)^2\right] + V a. \quad (2.12)$$

If we equate the real and the imaginary parts separately, we obtain two coupled equations:

$$-\frac{\partial\,\varphi}{\partial\,t} = -\frac{1}{2m}\left[\hbar^2\,\frac{\nabla^2\,a}{a} - (\nabla\,\varphi)^2\right] + V, \qquad (2.13)$$

$$\frac{\partial\,a}{\partial\,t} = -\frac{1}{m}\,\vec{\nabla}\,a\cdot\vec{\nabla}\,\varphi - \frac{1}{2m}\,a\,\nabla^2\,\varphi. \qquad (2.14)$$

Let us introduce the quantum potential

$$Q = -\frac{\hbar^2}{2m}\,\frac{\nabla^2\,a}{a} \qquad (2.15)$$

following Böhm's terminology. This transforms the first equation as follows:

$$\frac{\partial\,\varphi}{\partial\,t} + \left[\frac{1}{2m}\,\left(\vec{\nabla}\,\varphi\right)^2 + V\right] + Q = 0. \qquad (2.16)$$

When the quantum potential is deemed to be negligible (for example, in classical mechanics when $\hbar \to 0$), we obtain the classical Hamilton–Jacobi equation

$$H\,\left(= \frac{(\vec{\nabla}\,S)^2}{2m} + V\right) + \frac{\partial\,S}{\partial\,t} = 0. \qquad (2.17)$$

This is the situation described in the last chapter in which we were simply interested in the phase function $e^{(i/\hbar)\,S}$ of the wave function. The quantum potential Q has been calculated in Young's two-slit experiment (Phillipidis et al., 1979) showing that it might be responsible for the interferential aspect observed on the screen. It is most interesting to note here that the quantum potential combines all the elements of the experiment: mass, velocity of particles, and the width of, and the spacing between, slits. In fact, in this particular case, space does not just constitute an inert framework in the experiment. It is structured in a manner as to impose certain constraints on the experiment itself.

2.2 Probability Current

The amplitude function $a(\vec{r}, t)$ plays an important role in the wave function $\psi(\vec{r}, t)$ and leads to an entirely different interpretation of the latter.

Let us multiply the two terms of the second equation, that is (2.14) by $2a$ and, bearing in mind that $2a\,\partial a/\partial t = \partial a^2/\partial t$, write

$$\frac{\partial a^2}{\partial t} = -\frac{1}{m}\left[2\,a\,\vec{\nabla}\,a\cdot\vec{\nabla}\,\varphi + a^2\,\nabla^2\,\varphi\right]. \tag{2.18}$$

The term in brackets represents a divergence given that

$$\vec{\nabla}\cdot(a^2\,\vec{\nabla}\,\varphi) = a^2\,\nabla^2\,\varphi + \vec{\nabla}\,a^2\cdot\vec{\nabla}\,\varphi$$
$$= a^2\,\nabla^2\,\varphi + 2a\,\vec{\nabla}\,a\cdot\vec{\nabla}\,\varphi, \tag{2.19}$$

and the second coupled Eq. (2.14) reduces to

$$\frac{\partial a^2}{\partial t} = -\frac{1}{m}\,\vec{\nabla}\cdot(a^2\,\vec{\nabla}\,\varphi). \tag{2.20}$$

Recollect that $\vec{p} = \vec{\nabla}\,S$ in the Hamilton–Jacobi canonical transformation. It then follows that

$$\vec{v} = \frac{\vec{p}}{m} = \frac{1}{m}\,\vec{\nabla}\,S. \tag{2.21}$$

Given that following the coupled equations, the function φ plays the role of the Jacobi function S, it is only natural to write

$$\frac{\vec{\nabla}\,\varphi}{m} = \vec{v} \tag{2.22}$$

and (2.20) will then take the form of a hydrodynamical continuity equation expressing the conservation of the flux of a fluid with density $\varrho = a^2$ and velocity $\vec{v}$:

$$\frac{\partial \varrho}{\partial t} + \vec{\nabla}\cdot(\varrho\,\vec{v}) = 0. \tag{2.23}$$

We can therefore attribute to the square of the amplitude function, or more generally speaking to the square modulus, (which eliminates the phase function) of wave function (2.1) the interpretation of a density of a fluid whose flux is conserved over time. We will then set

$$\varrho\,(\vec{r},\,t) = \psi^*\,(\vec{r},\,t)\,\psi\,(\vec{r},\,t) = |\,\psi\,(\vec{r},\,t)\,|^2\,. \tag{2.24}$$

The Copenhagen interpretation of N. Bohr and M. Born attributes to $\varrho\,(\vec{r},\,t)$ the meaning of probability density of the presence of a particle with mass m, or of a set of particles at point $\vec{r}$ and instant t,

$$\varrho\,(\vec{r},\,t) = |\,\psi\,(\vec{r},\,t)\,|^2 = P_r\,. \tag{2.25}$$

If we accept this interpretation, with the probability of the presence of particles in the entire space at the instant t being certain (where, of course, such particles exist), we are led to write

$$\int \varrho\,(\vec{r},\,t)\,d\vec{r} \equiv \int |\,\psi\,(\vec{r},\,t)\,|^2\,d\vec{r} = 1\,. \tag{2.26}$$

This therefore imposes a special form on the wave function. The function $\psi\,(\vec{r}\ t)$, solution to the Schrödinger equation (2.11), must be a square-integrable function normalized to unity.

In electromagnetism, the propagation of the wave vector $\vec{E}$ or $\vec{B}$ is such that the density of the electric (or magnetic) energy is proportional to $|\,E\,|^2$ or $|\,B\,|^2$. In wave mechanics, the probability for the presence of a particle with mass m (or of a set of particles with mass m) is given by the wave function $\Psi\,(\vec{r},\,t)$. It is only natural then for the square modulus to be linked to the probability density for the presence of such particles.

Now let us use the Schrödinger equation (2.3) and the conjugate equation to evaluate the probability current from the wave function $\psi\,(\vec{r},\,t)$, that is the term $\varrho\,\vec{v}$ in (2.26). By computing the difference between $\psi^*\,\partial\,\psi/\partial\,t$ and $\psi\,\partial\,\psi^*/\partial\,t$, we obtain

$$\left(i\,\hbar\,\psi^*\,\frac{\partial\,\psi}{\partial\,t}\right) - \left(-i\,\hbar\,\psi\,\frac{\partial\,\psi^*}{\partial\,t}\right) = \frac{\hbar^2}{2m}\,(\psi^*\,\nabla^2\,\psi - \psi\,\nabla^2\,\psi^*)\,. \tag{2.27}$$

The first term is equal to $i\,\hbar\,\frac{\partial}{\partial\,t}\,(\psi^*\,\psi) = i\,\hbar\,\frac{\partial}{\partial\,t}\,\varrho$, leaving us with

$$\frac{\partial\,\varrho}{\partial\,t} + \frac{\hbar}{2m\,i}\,(\psi^*\,\nabla^2\,\psi - \psi\,\nabla^2\,\psi^*) = 0\,. \tag{2.28}$$

We introduce the divergence of the current $\vec{J} = \varrho\,\vec{v}$ by setting

$$\vec{J} = \frac{\hbar}{2m\,i}\,(\psi^*\,\vec{\nabla}\,\psi - \psi\,\vec{\nabla}\,\psi^*) = \mathrm{Re}\,\{\frac{\hbar}{m\,i}\,\psi^*\,\vec{\nabla}\,\psi\}\,. \tag{2.29}$$

The hydrodynamical continuity equation of the probability density current is therefore

$$\boxed{\frac{\partial \varrho}{\partial t} + \vec{\nabla} \cdot \vec{J} = 0 \qquad \text{with} \qquad \vec{J} = \text{Re} \left\{ \frac{\hbar}{m\, i} \, \psi^* \, \vec{\nabla} \, \psi \right\}} \quad . \tag{2.30}$$

In the stationary case in particular:

$$\frac{\partial \varrho}{\partial t} = 0 \qquad \text{and} \qquad \vec{\nabla} \cdot \vec{J} = 0 \,. \tag{2.31}$$

To determine $\vec{J}$, it is often necessary to use the gradient operator expressed in spherical coordinates:

$$\vec{\nabla} = \vec{e}_r \, \frac{\partial}{\partial r} + \vec{e}_\theta \, \frac{1}{r} \, \frac{\partial}{\partial \theta} + \vec{e}_\varphi \, \frac{1}{r \sin \theta} \, \frac{\partial}{\partial \varphi} \,. \tag{2.32}$$

For a spherical wave of the type

$$f\,(\theta)\, \frac{e^{i\,k\,r}}{r} \,, \tag{2.33}$$

the probability current becomes, for example, with definitions (2.29) and (2.32),

$$\vec{J} = \frac{\hbar\, k}{m} \, \frac{|\, f\,(\theta)\, |^2}{r^2} \,, \tag{2.34}$$

whereas for a plane wave:

$$e^{i\,\vec{k}\cdot\vec{r}} = e^{i\,k\,r\,\cos\,\theta} \,, \tag{2.35}$$

the probability current will have two components:

$$\vec{J} = \vec{e}_r \, m\, k\, \cos\, \theta + \vec{e}_\theta \left(-\frac{k\, r\, \sin\, \theta}{m} \right) \,. \tag{2.36}$$

2.3 The Phase Function

An important point emerged in the definition of the probability density for particles: it is the square modulus of the wave function $\psi\,(\vec{r},\, t)$, and not the wave function itself, that leads to the results of the observations, to the measurable quantities. Should we therefore conclude that if the wave function is written as $\psi\,(\vec{r},\, t) = a\,(\vec{r},\, t)\, \exp[\,(i/\hbar)\, S\,(\vec{r},\, t)]$ the phase function will cease to have a role since $|\,\Psi\,(\vec{r},\, t)\,|^2 = a^2\,(\vec{r},\, t)$ in the square modulus? Far from it! It is exactly this phase function that is responsible for the wave mechanical character of the observations.

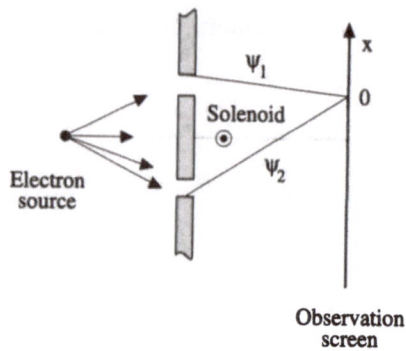

Young's double-slit experiment with electrons

Consider a system involving the superposition of two states. This is the case, for example, for Young's slit experiment involving particles such as electrons described in the diagram above.

The wave function $\psi_i\,(x,\,t)$ describes the probability amplitude for particles passing through slit i $(= 1$ or $2)$.

These wave functions can be written with a phase function:

$$\psi_i\,(x,\,t) = a_i\,(x,\,t)\,\exp\left(\frac{i}{\hbar}\,S_i\right)\quad\text{and}\quad \varrho_i\,(x,\,t) = \mid a_i\,(x,\,t)\mid^2.\quad (2.37)$$

The observation at position x on the screen does not involve the superposition of the densities ϱ_1 and ϱ_2 but rather the superposition of the wave functions:

$$\psi\,(x,\,t) = \psi_1\,(x,\,t) + \psi_2\,(x,\,t).\quad (2.38)$$

By evaluating the square modulus, this will yield

$$\varrho\,(x,\,t) = \varrho_1\,(x,\,t) + \varrho_2\,(x,\,t) + 2\,\mathrm{Re}\,\{\psi_1\,(x,\,t)\,\psi_2^*\,(x,\,t)\}\quad (2.39)$$

The last term is responsible for the interferential character of the observation. The situation is identical in optics where the optical state is described by the electric field $\vec{E}$ (or the magnetic $\vec{H}$), which is the solution of the propagation equation $\Box\,\vec{E} = 0$ of the form $\vec{E} = e^{A\,(\vec{r})}\,e^{i\,k_0\,(L-ct)}$ whereas the energy is proportional to the vector's square modulus. The phase shift of the waves $\vec{E}_1$ and $\vec{E}_2$ arriving at the point of observation, and corresponding to the differences in the optical path will lead to interference phenomenal if we evaluate

$$\mid E\mid^2 = \mid \vec{E}_1 + \vec{E}_2\mid^2 = \mid E_1\mid^2 + \mid E_2\mid^2 + 2\,\mathrm{Re}\{\vec{E}_1\,\vec{E}_2^*\}.\quad (2.40)$$

It has been possible to directly verify the importance of the phase function $\exp\left[\left(\frac{i}{\hbar}\right)\,S\right]$ in the wave function $\psi\,(x,\,t)$ by applying a magnetic field produced by the flow of electric current in a solenoid (see preceding diagram). If there is no magnetic field, we will notice an interference figure that is typical of wave mechanical phenomena. If the magnetic field is applied, there will be a further phase shift

$$\frac{1}{\hbar} \left(S_2 - S_1 \right) = -\frac{e}{\hbar\, c}\, \phi = -\frac{e}{\hbar\, c} \int \vec{H} \cdot \vec{ds}, \qquad (2.41)$$

where ϕ is the flux of the magnetic field in the solenoid. There is a global shift of the interference fringes which are easily suppressed by adjusting the electric current such that the magnetic flux brings about a phase shift $(e/\hbar\, c)\, \phi = 2\pi\, n$, leading to a phase function equal to unity.

This is the Aharonov–Bohm effect to which we shall return in our discussion of gauge invariance.

3. Mean Values of Observables

Given the probability $P\,(\vec{r})$ of finding a particle or a set of particles at the point with coordinates $\vec{r}$, the mean value of a function $F\,(\vec{r})$ will be

$$\langle F\,(\vec{r}) \rangle = \int F\,(\vec{r})\, P\,(\vec{r})\, d\,\vec{r}. \qquad (2.42)$$

By replacing $P\,(\vec{r})$ by $\psi^*\,(\vec{r})\, \psi\,(\vec{r})$, we obtain

$$\langle F\,(\vec{r}) \rangle = \int \psi^*\,(\vec{r})\, F\,(\vec{r})\, \psi\,(\vec{r})\, d\,\vec{r}. \qquad (2.43)$$

The mean value of the position of a particle for example will be

$$\langle \vec{r} \rangle = \int \psi^*\, \vec{r}\, \psi\,(\vec{r})\, d\,\vec{r} \qquad (2.44)$$

and the mean value of its momentum

$$\langle \vec{p} \rangle = \int \psi^*\,(\vec{r})\, (-i\hbar\, \vec{\nabla})\, \psi\,(\vec{r})\, d\,\vec{r} = -i\,\hbar \int \psi^*\,(\vec{r})\, \vec{\nabla}\, \psi\,(\vec{r})\, d\,\vec{r}. \qquad (2.45)$$

Attention is drawn here to a fact that may not yet be obvious and which we intend to demonstrate in the course of the transition from quantum to wave mechanics: The mean value of $F\,(\vec{r})$ is obtained by setting the function between $\psi^*\,(\vec{r})$ and $\psi\,(r)$ (and not the reverse, nor by writing $F\,(r)\, \psi^*\, \psi$) in the integral over r.

4. Heisenberg's Uncertainty Relation

Let us consider the one-dimensional case and compute the positive definite integral dependent on a real parameter λ from $-\infty$ to $+\infty$, that is:

$$I\left(\lambda\right) = \int \left| x\,\psi + \lambda\,\hbar\,\frac{d\,\psi}{d\,x} \right|^2 d\,x \geq 0. \tag{2.46}$$

By taking the square modulus $\left(\psi^*\,x + \lambda\,\hbar\,\frac{d\,\psi^*}{d\,x}\right)\left(x\,\psi + \lambda\,\hbar\,\frac{d\,\psi}{d\,x}\right)$, we obtain the following integral:

$$I\left(\lambda\right) = \int \psi^*\,x^2\,\psi\,d\,x + \lambda\,\hbar\,\int \left(\frac{d\,\psi^*}{d\,x}\,x\,\psi + \psi^*\,x\,\frac{d\,\psi}{d\,x}\right) d\,x$$

$$+ \lambda^2\,\hbar^2 \int \left|\frac{d\,\psi}{d\,x}\right|^2 d\,x. \tag{2.47}$$

The first integral is the expectation value of x^2. Considering the normalization relation, we integrate by parts the second and the third integrals to obtain the following result:

$$-\hbar^2 \int \psi^*\,\frac{d^2}{d\,x^2}\,\psi\,d\,x = \int \psi^*\left(-i\,\hbar\,\frac{d}{d\,x}\right)^2 \psi\,d\,x$$

$$= \int \psi^*\,p^2\,\psi\,d\,x = \langle p^2 \rangle. \tag{2.48}$$

The positive definite integral $I\left(\lambda\right)$ will then be written as follows:

$$I\left(\lambda\right) = \langle x^2 \rangle - \hbar\,\lambda + \langle p^2 \rangle\,\lambda^2 \geq 0. \tag{2.49}$$

For this inequality to be fulfilled irrespective of λ, it is necessary and sufficient that the discriminant of the second order equation in λ be negative or zero:

$$\hbar^2 - 4\,\langle x^2 \rangle\,\langle p^2 \rangle \leq 0. \tag{2.50}$$

This inequality can also take the following form:

$$\langle x^2 \rangle\,\langle p^2 \rangle \geq \frac{\hbar^2}{4}. \tag{2.51}$$

If we introduce the mean square deviations of x and p by setting

$$(\Delta x)^2 = \langle x^2 \rangle - \langle x \rangle^2 = \langle (x - \langle x \rangle)^2 \rangle, \tag{2.52}$$

$$(\Delta p)^2 = \langle p^2 \rangle - \langle p \rangle^2 = \langle (p - \langle p \rangle)^2 \rangle. \tag{2.53}$$

By replacing x by $x - \langle x \rangle$ and p with $p - \langle p \rangle$ in (2.51), we obtain the inequality

$$(\Delta x)^2\,(\Delta p_x)^2 \geq \frac{\hbar^2}{4}. \tag{2.54}$$

This leads to Heisenberg's uncertainty relation

$$\boxed{(\Delta x)\,(\Delta p_x) \geq \frac{\hbar}{2}}\;. \qquad (2.55)$$

The variables x and p_x are canonically conjugate and Heisenberg's uncertainty relation is valid for every pair of canonically conjugate variables. We will return to this point in Chap. 3.

5. Interpretations of Wave Mechanics

There are two major schools of thought as far as the interpretation of the wave function of wave mechanics is concerned: The mechanical interpretation of L. de Broglie or the stochastic one of E. Nelson and the probabilistic interpretation of the Copenhagen School (N. Bohr and M. Born).

5.1 De Broglie's Pilot Wave and the Quantum Potential

As we have already seen, it is possible to associate to the motion of a particle or set of particles a description in terms of waves. The waves. $\psi\,(\vec{r},\,t)$ is then a pilot wave with an amplitude $a\,(\vec{r},\,t)$ and a phase $\exp\,[(i/\hbar)\,S\,(\vec{r},\,t)]$. We should therefore grant the existence of a quantum potential $Q = -(\hbar^2/2m)\,(\nabla^2\,a/a)$ which has no effect from a classical point of view because the term $\nabla^2\,a/a$ oscillates rapidly and if the detector is of very large dimensions compared to the wavelength $\lambda = h/p$ of the pilot wave, the function $S\,(\vec{r},t)$ will obey a classical Hamilton–Jacobi equation. Conversely, in this interpretation, it is the quantum potential that determines the passage of particles through one or the other of the slits in Young's slit experiment involving particles; and leads to the wave character of mechanics (Holland, 1993).

In this approach, a wave packet with the extension Δx in configuration space will lead to a wave packet with the extension Δp_x in phase space since the wave function of phase space is the Fourier transform of the wave function of configuration space:

$$\phi\,(\vec{p}) = (2\pi\,\hbar)^{-3/2} \int \exp\left(-\frac{i}{\hbar}\,\vec{p}\cdot\vec{r}\right)\,\psi\,(\vec{r})\,d\,\vec{r}. \qquad (2.56)$$

The widths of the wave packets Δx and Δp_x are thus related by

$$\Delta x\,\Delta p_x \geq \frac{\hbar}{2}\,. \qquad (2.57)$$

Heisenberg's uncertainty relation expresses the fact that the phase space is the Fourier transform of the configuration space and that an extremely narrow wave packet in configuration space leads to an extremely spread-out packet in phase space.

To better appreciate this fact, it might be helpful to recall that a wave packet can be constructed by adding a wave with a lower frequency and another with a higher frequency to the wave with frequency ν. If there is a sufficiently large number of frequencies, we obtain a true impulse, a wave packet confined in a small area in space. The duration of the pulse is a function of the width $\Delta \nu$ of the frequency range through the inequality:

$$\Delta \nu \, \Delta t \geq \frac{1}{4\pi}.$$

The more we seek to reduce the size of the wave packet (Δt), the larger the frequency band must be. If we introduce Einstein's relation (1.223) between the frequency and the transported energy $E = h \nu$, we obtain the following inequality:

$$\frac{\Delta E}{h} \, \Delta t \geq \frac{1}{4\pi},$$

which is another way of writing Heisenberg's uncertainty relation

$$\Delta E \, \Delta t \geq \frac{\hbar}{2},$$

and which shows that E and t should behave like conjugate variables, but we will return to this point in the next two chapters.

In de Broglie's interpretation of wave mechanics, $\mid \psi \mid^2$ is interpreted as the probability density of a particle being at a certain position, in contrast to the Copenhagen School interpretation which considers $\mid \psi \mid^2$ as the probability that a particle will be found at position x during an adequate measurement. It should also be noted that the quantum potential Q is dependent solely upon the form of ψ and not on its amplitude. It depends on the quantum system considered as an indivisible whole. Let us consider the example of two particles with mass m:

$$i \hbar \frac{\partial \psi}{\partial t} = -\frac{\hbar^2}{2m} \left(\Delta_1^2 + \Delta_2^2 + V \right) \psi. \tag{2.58}$$

By taking $\psi = a \, \exp \left(\frac{i}{\hbar} S \right)$ and $\varrho = a^2 = \mid \psi \mid^2$, and calculating as in the previous case (see (2.13), (2.14) and (2.15)), we obtain

$$\frac{\partial \varrho}{\partial t} + \vec{\nabla}_1 \cdot \left(\frac{\varrho}{m} \vec{\nabla}_1 S \right) + \vec{\nabla}_2 \cdot \left(\frac{\varrho}{m} \vec{\nabla}_1 S \right) = 0, \tag{2.59}$$

$$\frac{\partial S}{\partial t} + \frac{1}{2m} \left(\vec{\nabla}_1 S \right)^2 + \frac{1}{2m} \left(\vec{\nabla}_2 S \right)^2 + V + Q = 0, \tag{2.60}$$

with the quantum potential:

$$Q = -\frac{\hbar^2}{2m} \left(\frac{\nabla_1^2\, a}{a} + \frac{\nabla_2^2\, a}{a} \right).$$
(2.61)

This potential depends on particles 1 and 2. In an n-particle problem, the quantum potential depends on all the particles and on the overall amplitude $a\,(\vec{r},\,t)$ of the wave function of the whole. This quantum potential can therefore describe an interaction of the non-local type.

This pilot wave and quantum potential type interpretation of Bohm, however, poses problems of coherence (the non-local quantum interaction introduces, for example, an instantaneous modification in the response, and this is in contradiction with the special relativity theory in which information is propagated at most at the speed of light). The problem of coherence is also present with respect to the physical interpretation (the non-spreading in time of the wave packets constituting the particles). This is why the probabilistic interpretation is today the most widely accepted interpretation amongst physicists using wave mechanics to interpret microscopic phenomena at molecular levels on smaller scales.

5.2 Nelson's Stochastic Mechanics – Universal Brownian Motion

Nelson (1966) was able to demonstrate that the Schrödinger equation of wave mechanics could be obtained with a purely classical approach by using the hypothesis that each particle with mass m is subjected to a Brownian motion described by a Markov–Wiener process with a scattering coefficient $\eta = \hbar/2m$ independent of the medium in which the particles are assumed to be in motion. This means that the particles have classical but random trajectories and that the wave function of wave mechanics is simply the probability amplitude and described by stochastic mechanics of random processes in which classical observables are the expectation values of the corresponding classical objects. If we introduce a probability density $\varrho\,(\vec{x},t)$, a real and differentiable function, such that at any given time

$$\int \varrho\,(\vec{x},t)\, d^3\,x = 1\,,$$
(2.62)

the expectation value of a classical observable A is defined by the relation

$$\langle A \rangle \;=\; \int A\,\varrho\,(\vec{x},t)\, d^3\,x\,.$$
(2.63)

In studying Brownian motion as elaborated by Einstein, the position $\vec{x}$ of a particle with mass m is a random variable, and the function $\vec{x}(t)$ a non-differentiable function. Nelson defines the forward $d + /dt$ and backwards mean differentials $d - /dt$ by setting:

$$\frac{d\pm}{dt}\, \vec{y}\,(t) = \lim_{\Delta t \to 0\pm} \left\langle \frac{\vec{y}\,(t + \Delta t) - \vec{y}\,(t)}{\Delta t} \right\rangle \tag{2.64}$$

and the application to the function $\vec{x}(t)$ defines the mean forward $\vec{b}_+\,(\vec{x}(t), t)$, and backwards $\vec{b}_-\,(\vec{x}(t), t)$ velocities. It goes without saying that if the function $\vec{x}(t)$ is differentiable then $\vec{b}_+ = \vec{b}_-$. In any form of stochastic process, and Brownian motion happens to be such a process the mean forward and backward velocities are different from each other.

Let $\vec{v}(t)$ the current mean velocity, be the half-sum of $\vec{b}_+$ and $\vec{b}_-$ and $\vec{u}(t)$, the osmotic velocity, the half-difference. Hence, we set:

$$\vec{v}(t) = \frac{1}{2}\,(\vec{b}_+ + \vec{b}_-), \qquad \vec{u}(t) = \frac{1}{2}\,(\vec{b}_+ - \vec{b}_-). \tag{2.65}$$

It might be convenient to define the complex differentiation operator (Nottale, 1993)

$$D = \frac{1}{2}\,\left[(d_+ + d_-) - i\,(d_+ - d_-)\right] = d_v - i\,d_u \tag{2.66}$$

and by applying this to the function $\vec{x}(t)$, we obtain the complex mean velocity $\vec{w}(t)$:

$$\vec{w}(t) = \frac{D}{dt}\,\vec{x}(t) = \frac{d_v}{dt}\,\vec{x}(t) - i\,\frac{d_u}{dt}\,\vec{x}(t) = \vec{v} - i\,\vec{u}. \tag{2.67}$$

The complex second differential introduces the complex mean acceleration:

$$\begin{aligned}
\vec{\Gamma}(t) &= \frac{D\,\vec{w}(t)}{dt} = \frac{d_v - i\,d_u}{dt}\,(\vec{v} - i\,\vec{u}) \\
&= \frac{d_v\,\vec{v} - d_u\,\vec{u}}{dt} - i\,\frac{d_v\,\vec{u} + d_u\,\vec{v}}{dt} \\
&= \frac{1}{2}\,\left(\frac{d_+\,d_- + d_-\,d_+}{dt^2} - i\,\frac{d_+^2 + d_-^2}{dt^2}\right)\,\vec{x}(t).
\end{aligned} \tag{2.68}$$

If an external force $\vec{F}$ is applied to a particle with mass m, Newton's law $\vec{F} = m\,\vec{\gamma}(t)$ will only be relevant to the real part of the acceleration:

$$\mathrm{Re}\{\vec{\Gamma}(t)\} = \vec{\gamma}(t) = \frac{1}{2}\,\frac{d_+\,d_- + d_-\,d_+}{dt^2}\,\vec{x}(t) = \frac{1}{2}\,\frac{d_+\,\vec{b}_- + d_-\,\vec{b}_+}{dt}. \tag{2.69}$$

The generalization of Newton's law is termed the Newton–Nelson law and is written

$$\vec{F} = m\,\vec{\Gamma} = m\,\frac{D\,\vec{w}(t)}{dt}. \tag{2.70}$$

By noting that the imaginary part of $\vec{\Gamma}$ is zero and describes the stochastic aspect of $\vec{x}(t)$ linked to the non-differentiability of the function, we notice that the real part of $\vec{\Gamma}$, that is $\vec{\gamma}(t)$, describes the classical mean acceleration that occurs within the framework of Newton's law.

The differential of the function $\vec{x}(t)$ can be described with the mean forward and backward velocities and a random function $\vec{\xi}(t)$ by setting

$$d\,\vec{x}(t) = \vec{b}_{\pm}\,(\vec{x}(t), t)\,dt + d\,\vec{\xi}_{\pm}(t)\,. \tag{2.71}$$

The signs of $\vec{b}$ and $\vec{\xi}$ are identical, $(+)$ for the forward velocities and $(-)$ for the backward velocities. In a Wiener process, the functions $d\,\vec{\xi}(t)$ are of the Gaussian type and have a null expectation value, are mutually independent and have a mean square proportional to the element dt:

$$\langle d\,\xi_{\pm\,i}\,d\,\xi_{\pm\,j}\rangle = \pm\,2\eta\,\delta_{ij}\,dt\,. \tag{2.72}$$

Just as in (2.71), the signs are identical. For particles with mass m in suspension in a fluid at temperature T, the Markov–Wiener scattering coefficient η will be proportional to T and inversely proportional to the mass m. It is written $\eta = k_B\,T/m\,\beta$, where k_B is the Boltzman constant and β^{-1} the relaxation time of the scattering process.

The Fokker–Planck hydrodynamical continuity equations for a forward scattering process (sign $+$) or a backward scattering process (sign $-$) are written by Nelson as follows:

$$\partial_t\,\varrho = -\vec{\nabla}\cdot(\vec{b}_{\pm}\,\varrho)\pm\eta\,\Delta\,\varrho\,. \tag{2.73}$$

The half-sum of the forward and backward scattering equations eliminates the stochastic scattering term $\eta\,\Delta\,\varrho$ and restores the Fokker–Planck hydrodynamical continuity for a current mean velocity $\vec{v}$ of the particles in Brownian motion:

$$\partial_t\,\varrho = -\vec{\nabla}\cdot(\vec{v}\,\varrho)\,. \tag{2.74}$$

The term-to-term half-difference of the forward and backward scattering equations introduces a constraint equation on the osmotic velocity $\vec{u}$:

$$0 = -\vec{\nabla}\cdot(\vec{u}\,\varrho)+\eta\,\Delta\,\varrho\,. \tag{2.75}$$

By noting that the probability density is a real function that is entirely continuous and differentiable, we can group the foregoing equations by introducing the complex velocity $\vec{w}$:

$$\begin{aligned}\partial_t\,\varrho &= -\vec{\nabla}\cdot(\vec{w}\,\varrho)+i\,\eta\,\Delta\,\varrho\\ &= -\vec{\nabla}\cdot((\vec{w}+i\,\eta\,\vec{\nabla})\,\varrho)\,.\end{aligned} \tag{2.76}$$

By returning to the zero imaginary part, that is to Eq. (2.75), we obtain

$$0 = \vec{\nabla}\cdot(-\vec{u}\,\varrho+\eta\,\vec{\nabla}\,\varrho)\,. \tag{2.77}$$

This leads to the trivial solution

$$\vec{u} = \eta \, \frac{\vec{\nabla} \varrho}{\varrho} = \eta \, \vec{\nabla} \, \ell_n \, \varrho = \frac{2\eta}{\hbar} \, \vec{\nabla} \, R \qquad (2.78)$$

by setting $R = (\hbar/2) \, \ell_n \, \varrho$ with $\hbar = $ constant, so far without reference to Planck's constant. This boils down to defining the probability density by the exponential function

$$\varrho = \exp \left(\frac{2R}{\hbar} \right). \qquad (2.79)$$

To determine the evolution of the osmotic velocity with time, we simply use the Fokker–Planck Eq. (2.74) and introduce the form (2.78) of the osmotic velocity:

$$\partial_t \, \varrho = - \varrho \left(\vec{\nabla} \cdot \vec{v} + \frac{1}{\eta} \, \vec{v} \cdot \vec{u} \right), \qquad (2.80)$$

and then go on to evaluate $\partial_t \, (\vec{u} \, \varrho)$ with (2.78). We infer from this the driving equation of the velocity $\vec{u}$:

$$\partial_t \, \vec{u} = - \vec{\nabla} \, (\vec{v} \cdot \vec{u}) - \eta \, \vec{\nabla} \, (\vec{\nabla} \cdot \vec{v}). \qquad (2.81)$$

By using the real function R with (2.78), we obtain, to an arbitrary phase function $\alpha(t)$, the equivalent form:

$$\partial_t \, R = - \vec{v} \cdot \vec{\nabla} \, R - \frac{\hbar}{2} \, \vec{\nabla} \cdot \vec{v} \qquad (2.82)$$

which we choose equal to zero for the sake of simplicity.

A stationary state corresponding to a time-independent probability density $\varrho(\vec{x})$, defines a time-independent function R, obtained from (2.82) with a zero current mean expectation value. Conversely, a zero current mean expectation value defines a time independent probability density function, that is, leads to a stationary state.

Let us, by analogy, define with (2.78) a non-zero current mean velocity, the gradient of a real function $S(\vec{x}(t), t)$, by setting:

$$\vec{v} = \frac{2\eta}{\hbar} \, \vec{\nabla} \, S. \qquad (2.83)$$

This will transform the driving Eq. (2.81) of the velocity $\vec{u}$ into

$$\partial_t \, R = - \frac{2\eta}{\hbar} \, \vec{\nabla} \, R \cdot \vec{\nabla} \, S - \eta \, \Delta \, S. \qquad (2.84)$$

To determine the driving equation of the current mean velocity $\vec{v}$, we carry out a second-order Taylor expansion of an arbitrary function $f(\vec{x}, t)$:

$$df = \left(dt \, \partial_t + \vec{dx} \cdot \vec{\nabla} + \frac{1}{2} \sum_{ij} d \, x_i \, d \, x_j \, \frac{\partial^2}{\partial \, x_i \, \partial \, x_j} \right) f. \qquad (2.85)$$

If we use the expectation value of this equation, we obtain the forward d_+, or backwards d_- differentials, bearing in mind that the second-order term introduces the Markov–Wiener scattering coefficient (2.71) whereas $d\vec{x}/dt = \vec{v}$ is replaced by $\vec{b}_+$ or by $\vec{b}_-$. This will then yield

$$d_\pm \, f/dt = \left(\partial_t + \vec{b}_\pm \cdot \vec{\nabla} \pm \eta \, \Delta\right) f \, . \tag{2.86}$$

It is convenient to introduce the complex time-derivative operator (2.66) grouping together Eq. (2.86) into the single equation

$$D/dt = \left(\partial_t + \vec{w} \cdot \vec{\nabla} - i \, \eta \, \Delta\right) . \tag{2.87}$$

By setting $f = \vec{b}_+$ and $f = b_-$ in (2.86) the half-difference will give a null acceleration of the stochastic scattering process, that is the driving Eq. (2.81) of the osmotic velocity, whereas the half-sum will lead to the mean acceleration $\vec{\gamma}(t)$, the real part of $\vec{\Gamma}$. It is however simpler to directly write the complex acceleration from (2.87):

$$\vec{\Gamma} = D \, \vec{w}/dt = \left(\partial_t + \vec{w} \cdot \vec{\nabla} - i \, \eta \, \Delta\right) \vec{w} \, . \tag{2.88}$$

By expanding the real part, we further obtain:

$$\vec{\gamma}(t) = \partial_t \, \vec{v} + (\vec{v} \cdot \vec{\nabla}) \, \vec{v} - (\vec{u} \cdot \vec{\nabla}) \, \vec{u} - \eta \, \Delta \, \vec{u} \, . \tag{2.89}$$

This is the driving equation of the current mean velocity $\vec{v}$. It can also take a different form with the identity

$$\vec{\nabla} \, a^2 = \vec{\nabla} \, (\vec{a} \cdot \vec{a}) = 2 \, (\vec{a} \cdot \vec{\nabla}) \, \vec{a} \, . \tag{2.90}$$

By isolating the partial derivative of $\vec{v}$ in (2.89), we obtain

$$\partial_t \, \vec{v} = \vec{\gamma}(t) + \frac{1}{2} \, \vec{\nabla} \, (u^2 - v^2) + \eta \, \Delta \, \vec{u} \, . \tag{2.91}$$

If an external force $\vec{F}$ deriving from the potential V is acting on the particle:

$$\vec{F} = m \, \vec{\gamma} = -\vec{\nabla} \, V \, , \tag{2.92}$$

the formulas (2.78) and (2.83) of the velocities $\vec{u}$ and $\vec{v}$ will lead to the driving equation of the real function S:

$$\partial_t \, S = -\frac{\hbar}{2\eta \, m} \, V + \frac{\eta}{\hbar} \, \left[(\vec{\nabla} \, R)^2 - (\vec{\nabla} \, S)^2\right] + \eta \, \Delta \, R \, . \tag{2.93}$$

Let us note that the driving Eq. (2.88) of the velocities $\vec{u}$ and $\vec{v}$ are simply expressions of the Newton–Nelson equations for a Markov–Wiener stochastic scattering process, if we introduce a truly real interaction potential V:

$$-\vec{\nabla}\, V = m\, D\, \vec{w}/dt = D\, m\, \vec{w}/dt = D\, \vec{P}/dt\,. \tag{2.94}$$

The complex conjugate momentum will therefore be of the form:

$$\vec{P} = m\, \vec{w} = m\, (\vec{v} - i\, \vec{u}) = -\frac{2i\, m\, \eta}{\hbar}\, \vec{\nabla}\, (R + i\, S)\,. \tag{2.95}$$

Let us introduce a complex function Ψ whose amplitude is the square root of the probability density by setting

$$\Psi = \exp\left(\frac{1}{\hbar}\, (R + i\, S)\right) \quad \text{and} \quad |\,\Psi\,|^2 = \exp\left(\frac{2R}{\hbar}\right) = \varrho\,. \tag{2.96}$$

For the sake of simplicity, Ψ is termed the probability amplitude of the Brownian motion. The gradient of the function Ψ will be expressed as follows:

$$\frac{\vec{\nabla}\, \Psi}{\Psi} = \vec{\nabla}\, \ell_n\, \Psi = \frac{1}{\hbar}\, \vec{\nabla}\, (R + i\, S)\,. \tag{2.97}$$

By inserting this result into the complex conjugate momentum (2.95), we obtain the following relation

$$\vec{P}\, \Psi = -2i\, m\, \eta\, \vec{\nabla}\, \Psi\,. \tag{2.98}$$

In Brownian motion, we cause the operator $-2i\, m\, \eta\, \vec{\nabla}$ to correspond to the operator $\vec{P}$ acting on the probability amplitude Ψ. This is the correspondence principle. Now, let us a priori define the vector $\vec{J}$ from the complex function Ψ by setting

$$\vec{J} = \frac{\eta}{i\, \hbar}\, \left(\Psi^*\, \vec{\nabla}\, \Psi - \Psi\, \vec{\nabla}\, \Psi^*\right)\,. \tag{2.99}$$

Using definitions (2.96) of Ψ and (2.83) of $\vec{v}$, we obtain

$$\vec{J} = \frac{2\eta}{\hbar}\, \varrho\, \vec{\nabla}\, S = \varrho\, \vec{v}\,. \tag{2.100}$$

The vector $\vec{J}$ is therefore the probability density current verifying the Fokker–Planck hydrodynamical continuity Eq. (2.74) for the classical scattering of a fluid with density ϱ and velocity $\vec{v}$:

$$\partial_t\, \varrho + \vec{\nabla} \cdot \vec{J} = 0\,. \tag{2.101}$$

The time derivative of the probability amplitude Ψ is expanded with (2.96) by using the driving Eqs. (2.93) and (2.84), yielding

$$i\,\hbar\,\frac{\partial_t\,\Psi}{\Psi} = i\hbar\,\partial_t\,\ell_n\,\Psi = i\,\partial_t\,(R+i\,S)$$

$$= -\eta\,\left[\Delta\,(R+i\,S) + \frac{2i}{\hbar}\,\vec{\nabla}\,R\cdot\vec{\nabla}\,S + \frac{1}{\hbar}\,\left[(\vec{\nabla}\,R)^2 - (\vec{\nabla}\,S)^2\right]\right]$$

$$+\,\frac{\hbar}{2\eta\,m}\,V\,. \tag{2.102}$$

We next use the identity

$$\frac{\Delta\,a}{a} \equiv \Delta\,\ell_n\,a + (\vec{\nabla}\,\ell_n\,a)^2 \tag{2.103}$$

to evaluate $\Delta\,\Psi\,/\,\Psi$ from definition (2.97):

$$\frac{\Delta\,\Psi}{\Psi} = \Delta\,\ell_n\,\Psi + (\vec{\nabla}\,\ell_n\,\Psi)^2$$

$$= \frac{1}{\hbar}\,\Delta\,(R+i\,S) + \frac{1}{\hbar^2}\,\left(\vec{\nabla}\,(R+i\,S)\right)^2$$

$$= \frac{1}{\hbar}\,\left\{\Delta\,(R+i\,S) + \frac{1}{\hbar}\,\left[(\vec{\nabla}\,R)^2 - (\vec{\nabla}\,S)^2 + 2i\,\vec{\nabla}\,R\cdot\vec{\nabla}\,S\right]\right\}\,. \tag{2.104}$$

The comparison of Eqs. (2.104) and (2.102) leads to the following equation:

$$i\,\frac{\partial_t\,\Psi}{\Psi} = -\eta\,\frac{\Delta\,\Psi}{\Psi} + \frac{1}{2m\,\eta}\,V. \tag{2.105}$$

The complex function $\Psi = \sqrt{\varrho}\,\exp\,[(i/\hbar)\,S]$, which may be termed the probability, amplitude, since $|\,\Psi\,|^2 = \varrho$, therefore describes a Brownian motion with Markov–Wiener scattering coefficient η of a particle with mass m subjected to a potential V. The fundamental Eq. (2.105) can also be written

$$i\,\partial_t\,\Psi = \left(-\eta\,\Delta + \frac{1}{2m\,\eta}\,V\right)\,\Psi\,. \tag{2.106}$$

We notice that the constant $\hbar$ introduced to express the velocities $\vec{u}$ and $\vec{v}$ in the gradient form of the real functions R and S, no longer appears in the fundamental Eq. (2.106) of Brownian motion. Just like Nelson, we could have also refrained from introducing it in the expression (2.96) for the function Ψ. The objective is to obtain the same form $\exp\,[(i/\hbar)\,S]$ of the phase function as in wave mechanics, because we need not insist on the fundamental Eq. (2.106) of the Brownian motion being formally analogous to the Schrödinger equation of wave mechanics.

For a medium-independent Markov–Wiener scattering coefficient in which the particle with mass m is in suspension, the scattering coefficient η should be inversely proportional to the mass. We can therefore follow Nelson and set

$$\eta = \frac{\hbar}{2m},$$

(2.107)

which yields the equation of universal Brownian motion

$$i\hbar\, \partial_t\, \Psi = \left(-\frac{\hbar^2}{2m}\, \Delta + V\right)\, \Psi$$

with $\quad |\, \Psi\, |^2 = \varrho \quad$ and $\quad \Psi = \sqrt{\varrho}\, \exp\left(\frac{i}{\hbar}\, S\right).$

(2.108)

If the constant $\hbar$ introduced in (2.107) is Planck's constant, the equation of universal Brownian motion will be the Schrödinger equation. The current mean velocity then becomes $\vec{v} = (1/m)\, \vec{\nabla}\, S$ and relation (2.98) the correspondence principle of wave mechanics.

Should we therefore conclude that wave mechanics is simply the description of a universal Brownian motion to which the particles are subjected? Does this Brownian motion reflect the fact that the elementary particles (fermions) are no more than emitter-receivers of particles (bosons) carrying fundamental interactions (see Chaps. 17, 18, and 19)?

The function Ψ would become the complex probability amplitude of this universal Brownian motion, hence of a wave mechanical nature, a non-local function (i.e., conditioned by the presence of all the particles) or local (i.e., linked solely to the description of the random motion on a microscopic scale less than De Broglie's wavelength $\lambda = \hbar/mc$)? Nelson's stochastic mechanics therefore leads to a mechanical and statistical interpretation of wave mechanics, in which the particles retain their corpuscular character whereas the probability of observing any property whatsoever, linked to the probability density through expectation values (2.63) of observables, is wave mechanical following Eq. (2.108) of the universal Brownian motion. This interpretation separates the corpuscular aspects from the wave mechanical aspects that exist side by side in De Broglie's interpretation. Another significant result stemming from this approach is that the path traveled by a particle, i.e., $\vec{x}(t)$ is a non-differentiable function. This is a property we will return to in the course of our discussion of Feynman path integrals in Chap. 7.

5.3 The Probability Theory of the Copenhagen School

The interpretation of Heisenberg, Born and Bohr (the Copenhagen School) considers $|\, \psi\, |^2$ as the probability of the presence of a particle or a set of particles during a measurement. The Heisenberg uncertainty principle therefore reflects the fact that we cannot simultaneously measure two conjugate variables such as position and momentum with the same high degree of precision. In the probabilistic interpretation, a statistical interpretation can be given to the uncertainty relations. They are expressions of a relation between statistical dispersions of the measurements of two conjugate observables around their expectation values.

5.4 Statistical Dispersion

Consider the successive impacts of shells fined from the same canon (Fig. 1) with identical pitch of the canon. The velocity $\vec{v}_0$ of each projectile, the resistance of air for each shot and the exact pitch of the canon cannot be defined with extreme precision. This leads to a dispersion ellipse for the impact of the shells.

The dispersion ellipse for the shell impacts expresses the relation between the deterministic theory constituted by classical mechanics and the statistical character of the results of a set of measurements prepared in an identical manner.

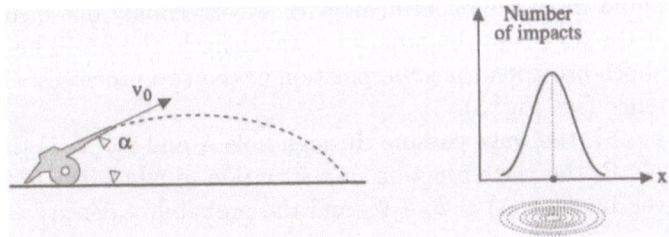

Figure 1

Let us imagine the same experiment with a machine gun, firing at a wall in which are two holes. The bullets are received by a stop screen. If one of the holes is closed, the number of impacts passing through a hole will be described by a bell curve. If both holes are open, the number of impacts will correspond to the sum of the impacts for each of the holes. It is the statistical expression of a set of completely deterministic measures. Each bullet passes through a hole following a classical trajectory.

If we term $\varrho_1 (x, t)$ the Gaussian probability density that the machine gun bullets will pass through hole number 1 and $\varrho_2 (x, t)$ that of hole number 2, the probability density will read:

$$\varrho (x, t) = \varrho_1 (x, t) + \varrho_2 (x, t) \qquad (2.109)$$

This is shown in Fig. 2. The different probability densities are simply added together.

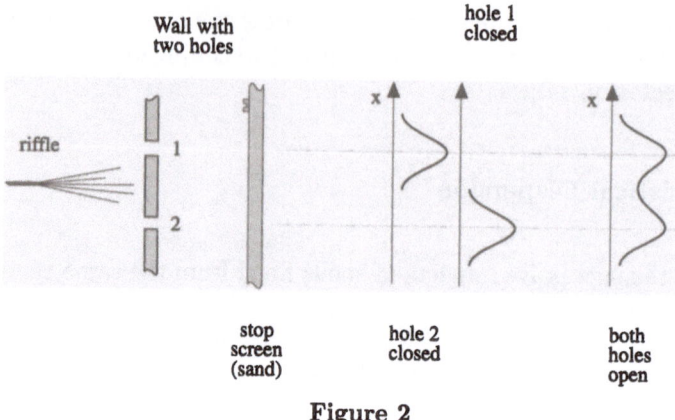

Figure 2

5.5 Interference of Wave Functions

Let us imagine the same experiment with waves. If only one of the holes is open and if the hole is of the order of a wavelength, there will be diffraction whereas if both are open the superposition of the two processes will produce an interference (see Fig. 3).

If $\Psi_1 (x, t)$ is the wave passing through hole A and $\Psi_2 (x, t)$ that passing through hole B, the wave function in a situation in which the two holes are open is given by $\Psi (x, t) = \Psi_1 + \Psi_2$ and the probability density will be

$$| \Psi |^2 = | \Psi_1 |^2 + | \Psi_2 |^2 + 2 \, \mathrm{Re} \, \{\Psi_1 \, \Psi_2^*\} . \tag{2.110}$$

The experiment involving an electron beam and a crystal displays a typical wave-mechanical aspect. Are we dealing with a real "pilot" wave, linked to the

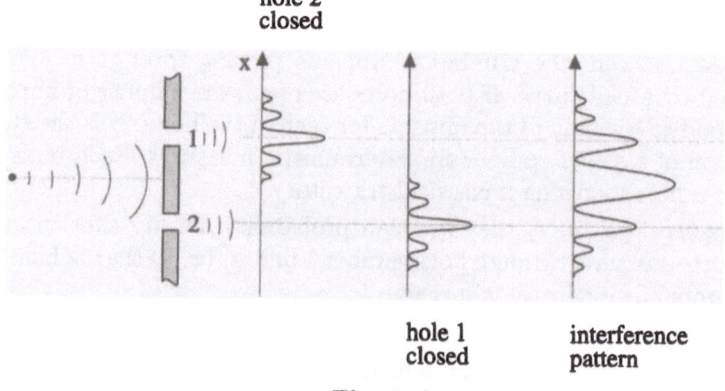

Figure 3

propagation of electrons? Or, is it the revelation of the statistical character of the experiment due to the fact that the probability of an electron impact at a given point is a function of the wave amplitude? Now, let us imagine the electron source to be sufficiently weak as to send electrons one at a time through the slits. What will happen if one of the slits is closed? Are we going to obtain a diffraction pattern of increasing darkness or electron impacts progressively reconstituting the diffraction pattern? And by opening the two slits, are we going to have an interference pattern that is progressively sharper with increase in the number of electron projectiles or electron impacts progressively reconstituting an interference pattern? In the first case, this would mean that there is a pilot wave associated with each particle and in the second that the trajectories of each electron are deterministic, if not determined, and that the probability of an electron projectile passing through one or the other of the slits and of impact at a given point is a mathematical wave-mechanical function.

The result of the experiment is as follows. With an extremely weak particle beam, we observe impacts localized on the photographic plate of the detector and if the experiment is left to run long enough, these impacts will be ordered following an interference pattern typical of the wave-mechanical phenomenon (Fig. 4).

Does our poor understanding of the initial conditions cause a statistical dispersion of the results? In other words, are there hidden variables that make wave mechanics nondeterministic? Or is there an intrinsic non-determination of the trajectories of particles and is the wave function simply a mathematical intermediate whose square modulus permits the computation of the probability that a set of particles will be in a state revealed by the observation? This is a crucial issue that necessitated several years of discussion before being resolved by the crucial experiment conducted by Alain Aspect at the Institut d'Optique in Paris.

5.6 Hidden Variables

The scenario reflecting the existence of a deterministic theory with hidden variables assumes that quantum mechanics is an incomplete and probabilistic theory. The random element exists because of our lack of knowledge of certain hidden variables. This is Einstein's interpretation. Bohr's interpretation is completely different and assumes that chance is intrinsic and that it is even the essence of the phenomena and, therefore, that quantum mechanics is a complete theory in which chance intervenes intrinsically.

An example here would be the difference in the possible interpretations of the same result with or without hidden variables in the following experiment: Take two cards, one red and the other black, and put them in two different sealed envelopes. Shuffle the envelopes and send one of them to a point A and the other to a point B such that the observers at point A and at point B

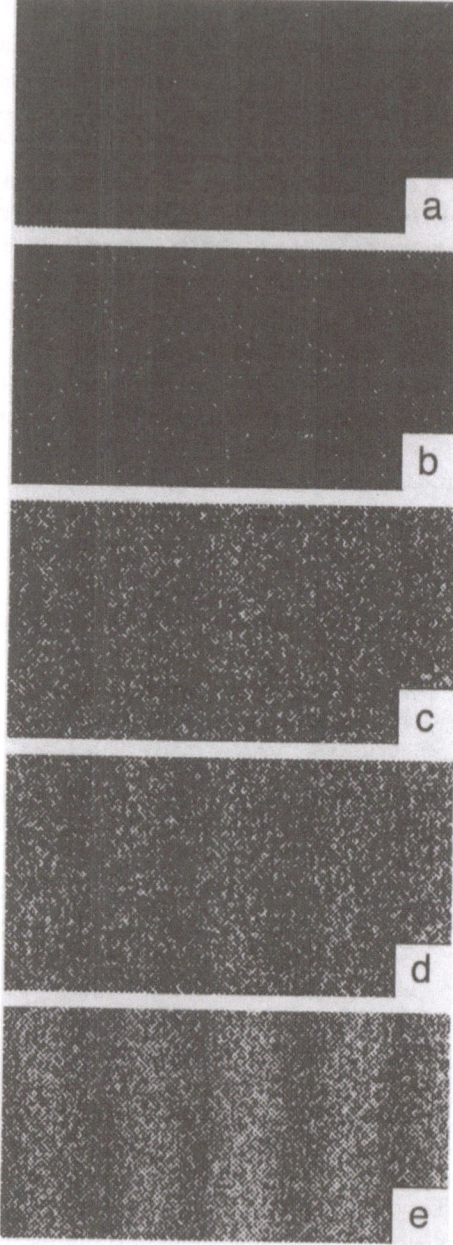

Figure 4. Young's slit experiment with single electrons impinging one after another. The different figures represent impacts corresponding to an increasing number of incident electrons: 10, then 100, 3000, 20,000 and 70,000. The impacts are ordered to finally build up the electron interference pattern (Tonomura et al., 1989)

receive their envelopes at the same time. If the observer at A has a red card, he can go ahead to state that the observer at B has a black card and vice versa. This experiment is simple and quite clear but it can be interpreted in entirely different ways depending on whether or not one admits the existence of a hidden variable, in this case color.

Interpretation No. 1 (Einstein's)
The colors are real and determined as soon as the envelopes are sealed but they are hidden and the determination of one necessarily leads to that of the other. Initially, we were dealing with a system with two components but each component was determined by a color variable which was hidden during the movement of the envelopes. This interpretation is founded on a realist, deterministic and separable (or local) theory of a hidden variable: color.

Interpretation No. 2 (Bohr's)
When the envelope were being shuffled, we lose the information we had concerning the color of the cards and the probability that an envelope will contain a red or black card is equal to 1/2. The envelope "is therefore in a brown state", the superposition of a red state and a black state. If an envelope is opened, the color is determined by measurement and, with it, the color of the other card is fixed. This theory is positivist, probabilistic and non-separable and links observable colors to one another. The first interpretation may appear to be intuitively more representative of reality, and the second to be a model for the easy understanding of the result. Yet it required Aspect's experiments on Bell inequalities (see the inequalities defined in Bell et al., 1964) conducted in the 1980's for the issue to be resolved.

5.7 The Bell Inequalities

Consider again the card experiment but, this time, instead of putting a card in each envelope, let us put two randomly chosen cards. Let α_1 and α_2 be the colors of the cards in envelope A and β_1 and β_2 those of the cards in envelope B. If a card is red, we will assign it the value $+1$ and if it is black -1.

What is the correlation between the colors of the cards in envelopes A and B, considering that the colors are determined at the time the cards are introduced into the envelopes? To solve this problem, we will need to determine a correlation coefficient or Bell parameter defined by the relation:

$$S = \alpha_1\,\beta_1 + \alpha_1\,\beta_2 + \alpha_2\,\beta_1 - \alpha_2\,\beta_2 . \tag{2.111}$$

Setting up a table for the possible values of $\alpha_1 \, \alpha_2 \, \beta_1 \, \beta_2$, we obtain:

α_1	+1								−1							
α_2	+1				−1				+1				−1			
β_1	+1		−1		+1		−1		+1		−1		+1		−1	
β_2	+	−	+	−	+	−	+	−	+	−	+	−	+	−	+	−
$\alpha_1 \beta_1$	+	+	−	−	+	+	−	−	−	−	+	+	−	−	+	+
$\alpha_1 \beta_2$	+	−	+	−	+	−	+	−	−	+	−	+	−	+	−	+
$\alpha_2 \beta_1$	+	+	−	−	−	−	+	+	+	+	−	−	−	−	+	+
$\alpha_2 \beta_2$	+	−	+	−	−	+	−	+	+	−	+	−	−	+	−	+
S	2	2	−2	−2	2	−2	2	−2	−2	2	−2	2	−2	−2	2	2

The Bell correlation function is equal to ± 2.

If α and β can take values between -1 and 1 and no longer just $+1$ or -1, then S will lie between -2 and $+2$.

$$\text{If} \quad \left. \begin{array}{c} -1 \leq \alpha \leq +1 \\ -1 \leq \beta \leq +1 \end{array} \right\} \quad -2 \leq S \leq 2 \qquad (2.112)$$

A probabilistic theory with hidden variables satisfies Bell inequalities: $-2 \leq S \leq 2$

5.8 The EPR Paradox

In 1935, Einstein, Podolsky, and Rosen imagined an experiment leading to a paradox known as the EPR paradox and which may be described as follows.

Let us consider a source S that emits particles with spin 0 which disintegrate into two particles with spin 1/2 dispersing in opposite directions (Fig. 5):

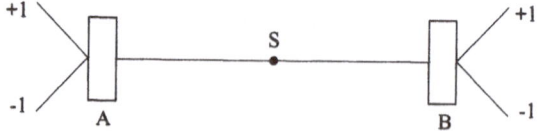

Figure 5

In A, the spin will be $+\frac{1}{2}$ or $-\frac{1}{2}$ (we will retain the notation $+1$ or -1 in order to have the same notation as with the red and black cards) and in B the spin will be similary $\pm\frac{1}{2}$. Now, if we measure the spin in A, it should be possible to determine the spin in B, since the sum of the spins of A and B should be equal to 0.

For Einstein, the spin of each particle is a hidden variable existing at the moment of disintegration, and measurement merely reveals the value of the spin. If A and B are separated, then A will cease to have any influence on B. For Bohr, on the other hand, particles from S are identical, a mixture of states $\pm\frac{1}{2}$. They constitute a non-separable state until they reach detectors A and B. At the time of the measurement, A and B are both determined. The quantum state constitutes an indivisible whole.

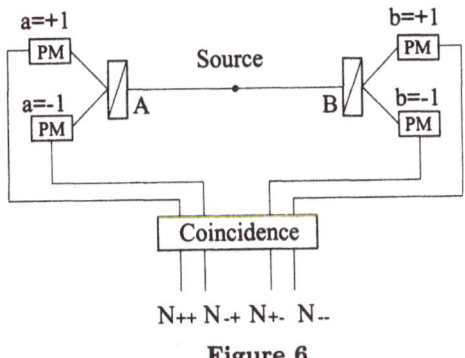

Figure 6

Aspect's experiment (Fig. 6) uses two different geometries in which detectors of A and B constitute an angle Θ or $3\,\Theta$. Because the correlation coefficient is proportional to the cosine of the double angle of their relative direction (this result will be demonstrated in Chapter 5), we obtain:

$$E\,(a_1,\,b_1) = \cos 2\Theta \quad E\,(a_2,\,b_1) = \cos 2\Theta, \qquad (2.113)$$
$$E\,(a_1,\,b_2) = \cos 2\Theta \quad E\,(a_2,\,b_2) = \cos 6\Theta.$$

The Bell parameter is defined by the relation:

$$S = E\,(a_1,\,b_1) + E\,(a_2,\,b_1) + E\,(a_1,\,b_2) - E\,(a_2,\,b_2). \qquad (2.114)$$

This gives, with the preceding results (this relation will be demonstrated in Chap. 5, Sect. 2.2):

$$S = 3\cos 2\Theta - \cos 6\Theta. \qquad (2.115)$$

Bell's inequality $-2 \le S \le +2$
would be satisfied if hidden variables existed

5.9 Aspect's Experiment

In the early 1980s, a group of researchers at the Institut d'Optique d'Orsay (Alain Aspect, Philippe Grangier, Gérard Roger and Jean Dalibard) carried out a study of calcium-emitted polarized photons. A pulse of calcium atoms excited by two laser beams with frequencies ν' and ν'' led to a level $J = 0$, which de-excites in two phases, going through an intermediate level $J = 1$ which, itself, is de-excited by emitting two photons with wavelengths $\lambda_1 = 4227$ Å and $\lambda_2 = 5513$ Å (see Fig. 7).

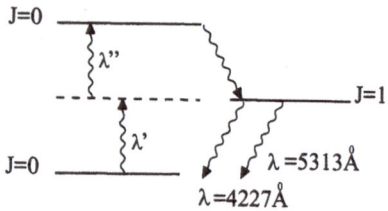

Figure 7

The polarization of photons A and B can be visualized by the diagram in Fig. 8

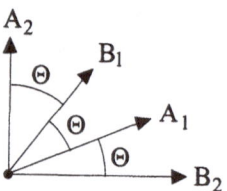

Figure 8

The total number of photons collected will be

$$N = N_{++} + N_{--} + N_{-+} + N_{+-} . \tag{2.116}$$

The polarization probabilities will, for their part, be

$$P_{++} = \frac{N_{++}}{N} \quad P_{--} = \frac{N_{--}}{N} \quad P_{-+} = \frac{N_{-+}}{N} \quad P_{+-} = \frac{N_{+-}}{N} \tag{2.117}$$

and the correlation coefficient $E\,(a,\,b)$ will be written as follows:

$$E\,(a,\,b) = \frac{N_{++} - N_{-+} - N_{+-} + N_{--}}{N} = P_{++} - P_{-+} - P_{+-} + P_{--}\,, \tag{2.118}$$

leading to the Bell parameter

$$S = E\,(a_1,\,b_1) + E\,(a_2,\,b_1) + E\,(a_1,\,b_2) - E\,(a_2,\,b_2). \tag{2.119}$$

The experiment clearly demonstrated that the Bell inequality is violated, i.e., that the theory of hidden variables is not verified:

$$S = 3\cos 2\,\Theta - \cos 6\,\Theta \qquad (2.120)$$

yields for $\Theta = 22.5°$, $S = 2\sqrt{2}$, and $-2\sqrt{2}$ for $\Theta = 67.5°$.

There is no doubt about the violation of Bell's inequality, as is clearly illustrated in Fig. 9

$$\text{At} \quad 22.5° \begin{cases} S_{\text{measured}} = 2.70 \pm 0.015 \\ S_{\text{theoretical}} = 2.82 . \end{cases}$$

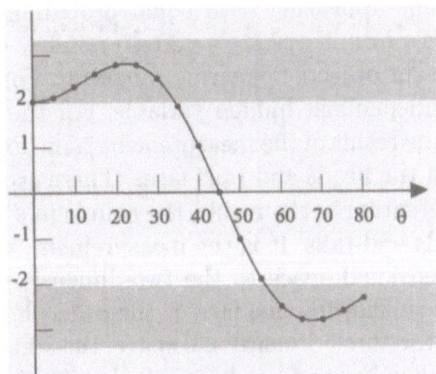

Figure 9

To ascertain that the detector had no influence on the result of the measurement, a second experiment introduced a 10-nanosecond periodical commutation of the detectors but the result remained practically the same, showing that Bell's inequalities were violated. The theories of hidden variables are not suitable and quantum mechanics is the appropriate theory for describing these phenomena as we can see from the points positioned with high precision on the curve, outside the range of values allowed by Bell's inequalities.

A comparison of Einstein's interpretation of quantum mechanics with that of Bohr can be made as follows:

Einstein	Bohr
hidden variables	no hidden variables
realist	positivist
deterministic	probabilistic
separable (localizable)	non-separable (non-localizable)

Aspect's experiment therefore shows that it is Bohr's interpretation that is correct.

If a quantum system that is properly defined by its initial conditions evolves without any external perturbation, all the points of the system will remain correlated even if they are separated by a space-like interval.

The non-separability postulate of quantum mechanics emerges then as being as important as the postulates of special relativity.

The non-separability postulate is perhaps more easily understood with the following simple example. Consider an observer in a two-dimensional space, examining the appearance of figures on coins (heads or tails) of a player in a three-dimensional space (time is the 4th dimension indispensable to the player in the three-dimensional space, and the 3rd dimension for the observer in the two-dimensional space). The observer in the two-dimensional space will see heads or tails appearing with equal probability if the observation lasts long enough. For him, there are a certain number of heads objects and another number of tails objects appearing randomly. For this observer, there is a measurement-independent hidden variable. For the player in the three-dimensional space, the result of the measurement is no doubt the same but the tossed coin has both the heads and tails faces. There are no hidden variables and, before the observation of the result, the coin is in a state that is a linear combination of heads and tails. It is the measurement, that is the transition from the three-dimensional space to the two-dimensional space that forces the coin to come down showing one face or the other, hence determining the state of the coin. In the three-dimensional space, the state of the coin is heads and tails whereas it has become heads or tails in the two-dimensional space.

The quantum state is neither of the two; it is both and it is nature, measurement, that reveals the wave or corpuscular state. This is the Copenhagen interpretation. There is yet another interpretation based on the notion of coherence and event branches that we will introduce in Chapter 6 after having studied quantum dynamical equations.

Finally, Nelson's stochastic interpretation in terms of the universal Brownian motion separates the corpuscular aspect related to the very existence of particles, from the wave mechanical aspect related to the probability amplitude function, that is to the randomness of the trajectories and non-differentiability of the position function of the particles. If we grant that this function is conditioned by the entire system, then non locality, an essential aspect of quantum theory, would have been verified. However, one of the major difficulties of current theories is due to quantum gravitation, given the need to reconcile a theory that is fundamentally local, the general relativity theory, with one that is not local, quantum theory. Is there a theory that simultaneously takes into account these two aspects?

6. Wave Mechanics – Schematic Summary

Einstein–de Broglie

$$E = h\,\nu = \hbar\,\omega \text{ and } \lambda = \frac{h}{p}$$

Schrödinger

$$-\frac{\hbar^2}{2m}\,\nabla^2\,\psi + V\,\psi = i\,\hbar\,\frac{\partial\,\psi}{\partial\,t}$$

Wave function

$$\psi\,(\vec{r},\,t) = a\,(\vec{r},\,t)\,\exp\left[\frac{i}{\hbar}\,\varphi\,(\vec{r},\,t)\right]$$

$$-\frac{\partial\,\varphi}{\partial\,t} = -\frac{1}{2m}\left[\hbar^2\,\frac{\nabla^2\,a}{a} - (\nabla\,\varphi)^2\right] + V$$

$$\frac{\partial\,a^2}{\partial\,t} = -\frac{1}{m}\,\vec{\nabla}\cdot(a^2\,\vec{\nabla}\,\varphi)$$

$$\vec{\nabla}\,\varphi = m\,\vec{v} \qquad \varrho = |\,\psi\,|^2$$

$$\vec{J} = \text{Re}\left\{\frac{\hbar}{i\,m}\,\psi^*\,\nabla\,\psi\right\}$$

$$\partial_t\,\varrho + \vec{\nabla}\cdot(\varrho\,\vec{v}) = \partial_t\,\varrho + \vec{\nabla}\cdot\vec{J} = 0$$

Interpretation of ψ

$$\int |\,\psi\,(\vec{r},\,t)\,|^2\,d\,\vec{r} = 1$$

de Broglie → pilot wave + quantum potential
Nelson → universal Brownian motion
Born → Pr (presence of one particle in $\vec{r}$ at t)
Statistical → Pr (set of particles in $\vec{r}$ at t)

Young's slit experiment, Bell's inequality and EPR paradox

Quantum theory is a non-local
theory without hidden variables,
and non-separable

Chapter 3
One-Dimensional Systems

1. Motion of the Center of Mass

If two bodies of mass m_1 and m_2 interact with a force that is solely dependent on their separation $|\vec{r}| = r = |\vec{r}_1 - \vec{r}_2|$, it is possible in classical mechanics to isolate the motion of the center of mass, $\vec{R} = (m_1 \vec{r}_1 + m_2 \vec{r}_2)/m$ from the relative motion.

The correspondence principle applied to the Hamiltonian of the two bodies will read

$$H = \frac{p_1^2}{2m_1} + \frac{p_2^2}{2m_2} + V\left(|\vec{r}_1 - \vec{r}_2|\right) \tag{3.1}$$

and will lead to the Schrödinger equation

$$\left(-\frac{\hbar^2}{2m_1}\nabla_1^2 - \frac{\hbar^2}{2m_2}\nabla_2^2 + V\left(|\vec{r}_1 - \vec{r}_2|\right)\right) \psi\left(\vec{r}_1, \vec{r}_2, t\right)$$

$$= i\hbar \frac{\partial}{\partial t} \psi\left(\vec{r}_1, \vec{r}_2, t\right) \tag{3.2}$$

and to the normalized wave function $\psi\left(\vec{r}_1, \vec{r}_2, t\right)$:

$$\int |\psi\left(\vec{r}_1, \vec{r}_2, t\right)|^2 \, d\vec{r}_1 \, d\vec{r}_2 = 1. \tag{3.3}$$

It is, in fact, difficult in the general case to separate the variables $\vec{r}_1$ and $\vec{r}_2$ in the potential and it is more expedient to isolate the motion of the center of mass by setting

$$\vec{P} = \vec{p}_1 + \vec{p}_2 \quad \text{and} \quad \vec{p} = \frac{m_2}{m}\vec{p}_1 - \frac{m_1}{m}\vec{p}_2, \tag{3.4}$$

which leads to the Hamiltonian

$$H = \frac{P^2}{2m} + \frac{p^2}{2\mu} + V(r), \tag{3.5}$$

with the total mass m and the reduced mass μ of the particles

$$m = m_1 + m_2 \quad \text{and} \quad \frac{1}{\mu} = \frac{1}{m_1} + \frac{1}{m_2}. \tag{3.6}$$

This will transform the Schrödinger Eq. (3.2) into:

$$\left(-\frac{\hbar^2}{2m}\nabla_R^2 - \frac{\hbar^2}{2\mu}\nabla_r^2 + V(r)\right)\psi(\vec{R}, \vec{r}, t) = i\hbar\frac{\partial}{\partial t}\psi(\vec{R}, \vec{r}, t). \quad (3.7)$$

Let us look for a solution that factorizes into functions of the variables $\vec{R}$ and $\vec{r}$ by setting

$$\psi(\vec{r}, \vec{R}, t) = f(\vec{R}, t)\,\psi(\vec{r}, t).$$

Equation (3.7) will lead to two coupled partial differential equations:

$$-\frac{\hbar^2}{2R}\nabla_R^2 f(\vec{R}, t) = i\hbar\frac{\partial}{\partial t}f(\vec{R}, t),$$

$$\left(-\frac{\hbar^2}{2\mu}\nabla_r^2 + V(r)\right)\psi(\vec{r}, t) = i\hbar\frac{\partial}{\partial t}\psi(\vec{r}, t). \quad (3.8)$$

Because the potential is not explicitly time dependent, the time dependence of the solutions is easily obtained in the usual form:

$$f(\vec{R}, t) = f(\vec{R})\,e^{-\frac{i}{\hbar}E_R t},$$

$$\psi(\vec{r}, t) = \psi(\vec{r})\,e^{-\frac{i}{\hbar}E_r t}, \quad (3.9)$$

leading to the general solution:

$$\psi(\vec{R}, \vec{r}, t) = \psi(\vec{r})\,f(\vec{R})\,e^{-\frac{i}{\hbar}E t} \quad\text{with}\quad E = E_r + E_R. \quad (3.10)$$

The total kinetic energy E is, like in classical mechanics, the kinetic energy of the center of mass E_R plus the relative kinetic energy E_r.

The motion of the center of mass is the motion of a free particle with mass $m = m_1 + m_2$ obeying the differential equation:

$$-\frac{\hbar^2}{2m}\nabla_R^2 f(\vec{R}) = E_R f(\vec{R}). \quad (3.11)$$

It has as solution a plane wave $\exp[(i/\hbar)\vec{P}\cdot\vec{R}]$ with $E_R = P^2/2m$.

The relative motion will be described by the stationary wave function $\psi(\vec{r})$ satisfying the differential equation:

$$\left(-\frac{\hbar^2}{2\mu}\nabla_r^2 + V(r)\right)\psi(\vec{r}) = E_r\,\psi(\vec{r}). \quad (3.12)$$

Let us begin by resolving this equation for different one-dimensional potential wells and postpone the study of three-dimensional cases until after the introduction of angular momentum and spherical harmonics.

Let us, however, first point out an important fact: consider a system of n identical particles, with mass m. By defining step by step the motion of the center of mass of two particles to which is added a third, then at the center of mass thus obtained a fourth particle, etc., we end up with the motion of the center of mass for the set of n particles with total mass $M = n\, m$. If we apply the same procedure to the relative motion, we obtain a reduced mass $1/\mu = (1/m) + 1/((n-1)m)$ that is $\mu = m - m/n$. The larger the system, the easier it is to approximate μ by m. Let us now examine the effect of this result on Eqs. (3.11) and (3.12). For an isolated system (a system that is not in interaction with other arbitrarily delimited systems), the motion of the center of mass is always that of a free particle with mass $M = nm$, whereas the relative motion of a particle with mass m is always that of a two-body particle with mass μ in a two-body interaction with the center of mass of the entire system. All the particles of the system can therefore be considered as individual particles in a potential V resulting from the sum of possible interactions. The relative wave function is, therefore, never negligible compared to that of the center of mass. In classical mechanics, we can ignore the relative motion of the particles constituting the system; in quantum mechanics, on the other hand, the relative wave function is always present, in the entire space and the observer has to choose between measuring the relative observables or ignoring them but the system's wave function is global. The difference between classical and quantum mechanics does not lie in the size of the system but rather depends on whether we choose to isolate a part of the system or to consider it in interaction with its medium.

2. General Properties of the Wave Function

Consider the case of a particle with mass m moving in a one-dimensional space. Let x be the variable for the space coordinates and let us assume that the force $F(x)$ acting on the system derives from a potential $V(x)$, that is,

$$F(x) = -\frac{\partial}{\partial x} V(x).$$
(3.13)

In classical mechanics, the total energy of a particle in interaction will be written

$$E = \frac{p^2}{2m} + V(x),$$
(3.14)

and the state of such a particle described by its trajectory

$$x(t) \in \mathbb{R}^1.$$
(3.15)

In wave mechanics, on the other hand, the state of a particle at instant t is described by a wave function $\psi(x, t)$ satisfying the normalization condition:

$$\int_{-\infty}^{+\infty} |\psi(x, t)|^2\, dx = 1.$$
(3.16)

As a function of t, $\psi(x, t)$ is assumed to be continuous and infinitely differentiable. In addition, $\psi(x, t)$ should have a continuous first and second derivative. $\psi\,(x,\,t)$ may then be considered for any given instant t as an element of the Euclidean space $C^{(1)}\,(\mathbb{R})$ of the square-integrable complex functions:

$$\int_{-\infty}^{+\infty} |\, f\,(x)\,|^2\; dx \; < \; +\infty. \tag{3.17}$$

The inner product (scalar product) in this space will take the form

$$\langle\, f,g\rangle = \int_{-\infty}^{+\infty} f^*\,(x)\; g\,(x)\; dx \tag{3.18}$$

and we recognize in $\langle\psi,\,\psi\rangle = \|\,\psi\,\|^2$ the square of the norm of the wave function, the normalization condition deriving from the probability postulate

$$\|\,\psi\,\|^2 = \int_{-\infty}^{+\infty} |\,\psi\,(x,\,t)\,|^2\; dx = 1\,. \tag{3.19}$$

In wave mechanics, $\psi\,(x,\,t)$ satisfies the Schrödinger equation

$$i\,\hbar\,\frac{\partial\,\psi\,(x,\,t)}{\partial\,t} = \left[-\frac{\hbar^2}{2m}\,\frac{\partial^2}{\partial\,x^2} + V\,(x)\right]\,\psi\,(x,\,t)\,, \tag{3.20}$$

and thus preserves the normalization condition over time:

$$\frac{d}{d\,t}\int_{-\infty}^{+\infty} |\,\psi\,(x,\,t)\,|^2\; dx = \int \left(\frac{\partial\,\psi^*}{\partial\,t}\,\psi + \psi^*\,\frac{\partial\,\psi}{\partial\,t}\right)\,dx\,.$$

If we extract the partial derivatives of the Schrödinger equation (3.20), the integral will reduce to

$$\frac{\hbar}{2i\,m}\int \left(\frac{\partial^2\,\psi^*}{\partial\,x^2}\,\psi - \psi^*\,\frac{\partial^2\,\psi}{\partial\,x^2}\right)\,dx\,.$$

And if we integrate by parts over an interval $2a$, which we then cause to tend to infinity, we will obtain

$$-\frac{\hbar}{2i\,m}\,\lim_{a\to\infty}\left[\frac{\partial\,\psi^*}{\partial\,x}\,\psi - \psi^*\,\frac{\partial\,\psi}{\partial\,x}\right]_{x=-a}^{x=+a}$$
$$-\int_{-\infty}^{+\infty}\left(\frac{\partial\,\psi^*}{\partial\,x}\,\frac{\partial\,\psi}{\partial\,x} - \frac{\partial\,\psi^*}{\partial\,x}\,\frac{\partial\,\psi}{\partial\,x}\right)\,dx = 0\,. \tag{3.21}$$

This shows that the normalization (3.19) determined at instant t is conserved over time.

3. Stationary Schrödinger Equation

The stationary solutions of the Schrödinger equation are obtained by separating space and time variables. This leads to solutions of the type

$$\psi\left(x,\, t\right) = \psi\left(x\right) e^{-\frac{i}{\hbar} E t}, \tag{3.22}$$

where $\psi\left(x\right)$ obeys the time-independent equation:

$$-\frac{\hbar^2}{2m} \frac{d^2\,\psi\left(x\right)}{dx^2} + V\left(x\right)\psi\left(x\right) = E\,\psi\left(x\right). \tag{3.23}$$

We introduce a reduced potential $U\left(x\right)$ and a reduced energy ε:

$$U\left(x\right) = \frac{2m}{\hbar^2} V\left(x\right),$$

$$\varepsilon = \frac{2m}{\hbar^2} E. \tag{3.24}$$

The Schrödinger equation (3.23) is a second-order differential equation and can therefore be written as follows:

$$\boxed{\frac{d^2\,\psi}{d\,x^2} + \left(\varepsilon - U\right)\psi\left(x\right) = 0} \,. \tag{3.25}$$

We are interested in the finite and continuous solutions of a Sturm–Liouville type equation that are differentiable from $-\infty$ to $+\infty$.

3.1 Potential Jump

Let us consider a particle with mass m and reduced energy ε arriving at a potential barrier with height $U_2 - U_1$, that is whose reduced potential is defined by

$$U\left(x\right) = \begin{cases} U_2 & \text{for} \quad x < 0 \quad \text{zone II} \\ U_1 & \text{for} \quad x > 0 \quad \text{zone I} \end{cases}$$

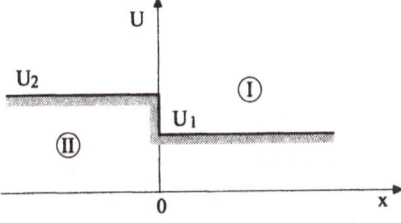

Figure 1

Let us assume that $U_2 > U_1$ (Fig. 1).

Case No. 1: $U_1 < \varepsilon < U_2$
Zone I:

$$\frac{d^2 \psi}{d x^2} + (\varepsilon - U_1)\, \psi = 0 \quad \text{and} \quad \varepsilon - U_1 = k_1^2 > 0. \tag{3.26}$$

The solution of the Schrödinger equation can, in this case, be written as follows:

$$\psi_1 = A_1 \sin (k_1\, x + \varphi). \tag{3.27}$$

Zone II:

$$\frac{d^2 \psi}{d x^2} + (\varepsilon - U_2)\, \psi = 0 \quad \text{and} \quad \varepsilon - U_2 = -\chi_2^2 < 0 \tag{3.28}$$

and the solution ψ will take the form

$$\psi_2 = A_2\, e^{\chi_2\, x}. \tag{3.29}$$

Instead of writing the continuity of the function ψ and its derivative at the point $x = 0$, we will write the continuity of the function and its logarithmic derivative $(1/\psi)\,(d\,\psi/d\,x)$:

- continuity of the function at $x = 0$:

$$A_1 \sin \varphi = A_2 \quad \text{and} \quad \frac{A_2}{A_1} = \sin \varphi \tag{3.30}$$

- continuity of the logarithmic derivative at $x = 0$:

$$k_1 \cot \varphi = \chi_2 \quad \text{and} \quad \varphi = \tan^{-1} \frac{k_1}{\chi_2} \tag{3.31}$$

where $\tan^{-1}$ is the determination of the arc in the interval $\left(-\frac{\pi}{2}, \frac{\pi}{2}\right)$. By extracting φ of (3.31) and carrying over in (3.30), we obtain:

$$\frac{A_2}{A_1} = \sin \varphi = \frac{k_1}{\sqrt{k_1^2 + \chi_2^2}} = \sqrt{\frac{\varepsilon - U_1}{U_2 - U_1}} \tag{3.32}$$

Case No. 2: $\varepsilon > U_2$

The solution is of the oscillatory type in the entire space. To each value of energy ε then will correspond two linearly independent eigenfunctions: the energy is said to be degenerate and its degeneracy to be of second order.

Let us choose a solution of the type $e^{-i\,k_2\,x}$ in zone II describing a traveling wave attached to a particle moving initially from $-\infty$ to the origin whereas in zone I we may have an incident wave of the type $e^{-i\,k_1\,x}$ (of unit amplitude) and a reflected wave with amplitude R. The solutions will then take the form

$$\psi = e^{-i\,k_1\,x} + R\,e^{i\,k_1\,x} \qquad x > 0,$$
$$\psi = S\,e^{-i\,k_2\,x} \qquad x < 0. \tag{3.33}$$

We determine the derivative for $x > 0$:

$$\psi' = -i\,k_1\left(e^{-i\,k_1\,x} - R\,e^{i\,k_1\,x}\right) \tag{3.34}$$

giving for $x = 0$ a logarithmic derivative:

$$\left(\frac{\psi'}{\psi}\right)_{x=0} = -i\,k_1\,\frac{1-R}{1+R}. \tag{3.35}$$

Similarly, the derivative of the wave function for $x < 0$ is

$$\psi' = -i\,k_2\,S\,e^{i\,k_2\,x} \tag{3.36}$$

leading to the logarithmic derivative at the origin:

$$\left(\frac{\psi'}{\psi}\right)_{x=0} = -i\,k_2 \tag{3.37}$$

The continuity of the logarithmic derivative at $x = 0$ will give the amplitude of the reflected wave:

$$R = \frac{k_1 - k_2}{k_1 + k_2} \tag{3.38}$$

whereas the continuity of the function at the point $x = 0$ defines the amplitude of the transmitted wave

$$S = 1 + R = \frac{2\,k_1}{k_1 + k_2}. \tag{3.39}$$

The probability of finding the particle in the reflected wave is equal to $\mid R \mid^2$ and that of finding it in the transmitted wave will be equal to $1 - \mid R \mid^2$, which gives:

$$1 - \mid R \mid^2 = \frac{k_2}{k_1}\mid S \mid^2 \tag{3.40}$$

Now, let us introduce the transmission coefficient τ of the barrier:

$$\tau = \frac{k_2}{k_1}\mid S \mid^2 = \frac{4k_1\,k_2}{(k_1 + k_2)^2} \tag{3.41}$$

and the reflection coefficient $\mathcal{R}$ of the barrier, the square modulus of the amplitude:

$$\mathcal{R} = \mid R \mid^2 = \frac{(k_1 - k_2)^2}{(k_1 + k_2)^2}, \tag{3.42}$$

which naturally gives:

$$\mathcal{R} + \tau = 1. \tag{3.43}$$

We notice here the strong analogy with the propagation of light in a non-absorbing medium with the index varying in a discontinuous manner at the origin.

3.2 Tunneling

Let us now consider the problem of the transmission of a particle across a square barrier:

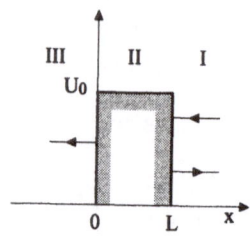

$$U(x) = \begin{cases} 0 & \text{for} \quad x > L \quad \text{zone I} \\ U_0 & \text{for} \quad 0 < x < L \quad \text{zone II} \\ 0 & \text{for} \quad x < 0 \quad \text{zone III} \end{cases}$$

The solution of the Schrödinger equation will lead to the following wave functions

Zone I:

$$\psi = e^{-i\sqrt{\varepsilon}\,x} + R\,e^{i\sqrt{\varepsilon}\,x} \tag{3.44}$$

Zone III:

$$\psi = S\,e^{-i\sqrt{\varepsilon}\,x} \tag{3.45}$$

Zone II:

$$\text{if} \quad \varepsilon < U_0 \quad \psi = A\,e^{\chi\,x} + B\,e^{-\chi\,x} \quad \text{with} \quad \chi = \sqrt{U_0 - \varepsilon} \tag{3.46}$$

$$\text{if} \quad \varepsilon > U_0 \quad \psi = C\,e^{i\,k\,x} + D\,e^{-i\,k\,x} \quad \text{with} \quad k = \sqrt{\varepsilon - U_0}. \tag{3.47}$$

The transmission coefficient can be calculated as $\tau = (k_3/k_1)\mid S\mid^2 = \mid S\mid^2$ since the wave vector $\sqrt{\varepsilon}$ is present zones in I and III. The continuity of the derivatives and the functions at the points $x = 0$ and $x = L$ will yield the following result:

$$\tau = \begin{cases} \dfrac{4\varepsilon\,(\varepsilon - U_0)}{4\varepsilon\,(\varepsilon - U_0) + U_0^2\,\sin^2 KL} & \text{if} \quad \varepsilon > U_0 \\[4mm] \dfrac{4\varepsilon\,(U_0 - \varepsilon)}{4\varepsilon\,(U_0 - \varepsilon) + U_0^2\,\sinh^2 \chi L} & \text{if} \quad \varepsilon < U_0 \end{cases} \tag{3.48}$$

The most interesting result concerns the case $\varepsilon < U_0$, that is $\varepsilon/U_0 < 1$.

Whereas a particle in classical mechanics cannot go through the barrier, we notice in wave mechanics that even in this case the transmission coefficient is not zero. If ε increases from 0 to U_0, the transmission coefficient increases from 0 to $\left(1 + (U_0\,L^2/4)\right)^{-1}$. This is the phenomenon tunneling.

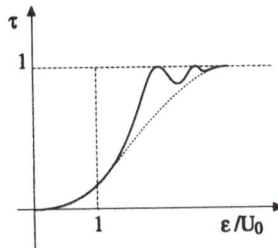

The lower the height and thickness of the barrier, the greater is the tunnel effect.

If $\varepsilon/U_0 > 1$ complete transmission is only possible for certain values of energy (as may be noticed in the above figure, showing τ as a function of $f(\varepsilon/U_0)$), those for which $k\,L = n\,\pi$.

3.3 The Square-Well Potential – Resonances

Let us now consider a square potential well with finite height before going on to consider the case of an infinitely deep well:

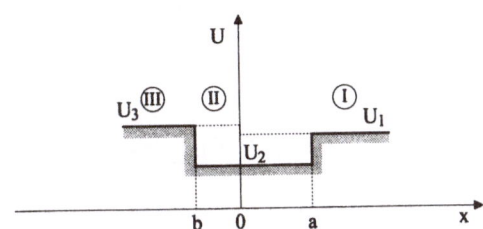

$$U(x) = \begin{cases} U_1 & \text{for} & x > a \\ U_2 & \text{for} & b < x < a \\ U_3 & \text{for} & x < b \end{cases} \qquad \text{with } U_2 < U_1 < U_3 .$$

Case No. 1: $U_2 < \varepsilon < U_1$
The solution is of the following form:

$$\psi(x) = \begin{cases} A_1\, e^{-\chi_1\, x} & \text{zone I} \\ A_2\, \sin(k_2\, x + \varphi) & \text{zone II} \\ A_3\, e^{\chi_3\, x} & \text{zone III} \end{cases} \qquad (3.49)$$

The continuity of the logarithmic derivative at points $x = a$ and $x = b$ will lead to the following two equations:

$$k_2\, \cot(k_2\, a + \varphi) = -\chi_1 , \qquad (3.50)$$
$$k_2\, \cot(k_2\, b + \varphi) = \chi_3 . \qquad (3.51)$$

In other words, equation (3.50) defines the angle φ with:

$$\varphi = -k_2\, a - \tan^{-1} \frac{k_2}{X_1} + n\, \pi \qquad n \text{ integer } \geq 0 \qquad (3.52)$$

and (3.51) similarly defines this same angle φ:

$$\varphi = -k_2\, b + \tan^{-1} \frac{k_2}{X_3}\,. \qquad (3.53)$$

If we equate the values of φ, we obtain:

$$n\,\pi - k_2\,(a-b) = \tan^{-1} \frac{k_2}{X_1} + \tan^{-1} \frac{k_2}{X_3}\,. \qquad (3.54)$$

Then making the following changes of variable:

$$L = b - a \qquad K = \sqrt{U_1 - U_2} \qquad \text{and} \qquad \xi = \frac{k_2}{K}\,, \qquad (3.55)$$

$$\cos\,\gamma = \sqrt{\frac{U_1 - U_2}{U_3 - U_4}} \qquad \text{with} \qquad 0 < \gamma < \frac{\pi}{2}\,, \qquad (3.56)$$

equation (3.54) reduces to

$$n\,\pi - \xi\,K\,L = \sin^{-1}\,\xi + \sin^{-1}\,(\xi\,\cos\,\gamma)\,. \qquad (3.57)$$

This equation can be solved graphically by looking for the intersection of the three straight lines D_n:

$$n - \frac{K\,L}{\pi}\,\xi = y(\xi)\,, \qquad (3.58)$$

with the curve C defined by the equation

$$f(\xi) = \frac{1}{\pi}\,\left(\sin^{-1}\,\xi + \sin^{-1}\,(\xi\,\cos\,\gamma)\right)\,. \qquad (3.59)$$

We obtain a discrete set of solutions for ξ, hence, for the energy ε.

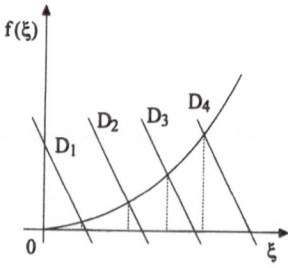

It is enough for the integer n to be sufficiently small for $y(\xi)$ and $f(\xi)$ to intersect:

$$K\,L \geq (n-1)\,\pi + \gamma. \tag{3.60}$$

Case No. 2: $U_1 < \varepsilon < U_3$

The general solution will take the form

$$\psi = \begin{cases} e^{-i\,k_1\,x} + e^{i(k_1\,x + 2\varphi_1)} & \text{zone I} \\[2ex] 2A\,e^{i\,\varphi_1}\,\sin\left(k_2\,x + \varphi_2\right) & \text{zone II} \\[2ex] 2B\,e^{i\,\varphi_1}\,e^{\chi_3\,x} & \text{zone III} \end{cases} \tag{3.61}$$

The continuity of the logarithmic derivatives defines the phase φ_1 through the relation

$$\varphi_1 = -k_1\,a - \frac{\pi}{2} + \tan^{-1}\left(\frac{k_1}{k_2}\,\tan\left(k_2\,L + \tan^{-1}\frac{k_2}{\chi_1}\right)\right), \tag{3.62}$$

whereas the phase φ_2 will be determined by the equation:

$$\varphi_2 = -k_2\,b + \tan^{-1}\frac{k_2}{\chi_3}. \tag{3.63}$$

Let us assume that $U_3 - \varepsilon \gg \varepsilon - U_2$, that is,

$$k_2 \ll \chi_3,$$
$$k_1 \ll \chi_3.$$

It is as if the barrier in zone III were infinitely high such that $B = 0$.

Let us, for the sake of simplicity, take on the other hand $a = 0$ and $b = -L$ and then set

$$\eta = \frac{k_1}{K} = \sqrt{\xi^2 - 1} \qquad \xi = \sqrt{\eta^2 + 1}. \tag{3.64}$$

By neglecting the term in $\frac{k_2}{\chi_3}$, equation (3.62) becomes:

$$\varphi_1 = \tan^{-1}\left(\frac{\eta}{\xi}\,\tan \xi\,K\,L\right) - \frac{\pi}{2} \tag{3.65}$$

The continuity of the function makes it possible, on the other hand, to determine the amplitude:

$$A^2 = \frac{\eta^2}{\eta^2 + \cos^2 \xi\,K\,L}. \tag{3.66}$$

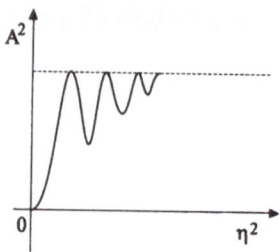

The curve yielding $A^2 = f(\eta^2)$ shows a series of sharp peaks with width $4\eta/(K\,L)$ which are separated from one another by $2\pi/(K\,L)$.

In this case, we say that a resonance phenomenon has been obtained. In some restricted energy domains (with width of $4\eta/K\,L$), the intensity of the wave in the internal zone is equal to 1. These resonance energies are obtained for $\varphi_2 = \left(n + \frac{1}{2}\right)\,\pi$. Outside these zones, the intensity of the wave, that is, the square modulus of the wave amplitude is very small.

3.4 Infinitely Deep Well

Consider the case of a particle with mass m in an infinitely deep well:

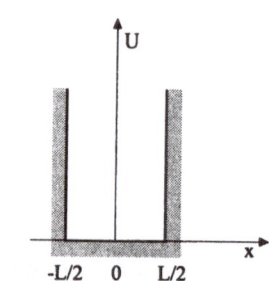

$$U = \begin{cases} 0 & \text{for} \quad -\dfrac{L}{2} < x < \dfrac{L}{2} \\[2mm] \infty & \text{for} \quad x < -\dfrac{L}{2} \\[2mm] \infty & \text{for} \quad x > \dfrac{L}{2}\,. \end{cases}$$

We are interested in the solution ψ that goes to zero at the edges of the well and which, within the well, satisfies the Schrödinger equation:

$$\frac{d^2\,\psi}{d\,x^2} + \varepsilon\,\psi = 0\,. \tag{3.67}$$

If we set $k^2 = \varepsilon$, then the general solution of (3.53) will take the form

$$\psi = A \sin k\,x + B \cos k\,x. \tag{3.68}$$

The continuity conditions near the well will then lead us to write:

$$\psi\left(\frac{L}{2}\right) = \psi\left(-\frac{L}{2}\right) = 0, \tag{3.69}$$

that is, by expanding the wave functions

$$A \sin \frac{k\,L}{2} + B \cos \frac{k\,L}{2} = -A \sin \frac{k\,L}{2} + B \cos \frac{k\,L}{2} = 0. \tag{3.70}$$

Case No. 1:

$$A = 0 \qquad \text{and} \qquad B \cos \frac{k\,L}{2} = 0$$

$$B = 0 \qquad \text{is} \qquad \text{a trivial solution}$$

$$B \neq 0 \quad \text{will yield} \quad \cos \frac{k\,L}{2} = 0$$

and then we extract the possible values

$$\frac{k\,L}{2} = \frac{\pi}{2} + n\,\pi \qquad \text{and} \qquad k = (2n+1)\,\frac{\pi}{L} \tag{3.71}$$

which can be written

$$k = \frac{N\,\pi}{L} \quad \text{with } N \text{ odd}. \tag{3.72}$$

The solutions are therefore of the type

$$\Psi_N = \cos \frac{N\,\pi}{L}\,x \qquad \text{with } N \text{ odd}. \tag{3.73}$$

Case No. 2:

$$B = 0 \qquad \text{and} \qquad A \sin \frac{k\,L}{2} = 0$$

$$A = 0 \qquad \text{is a trivial solution} \tag{3.74}$$

$$A \neq 0 \qquad \text{will yield} \qquad \sin \frac{k\,L}{2} = 0$$

imposing the values

$$\frac{k\,L}{2} = N\,\pi \qquad \text{with } N \text{ even}. \tag{3.75}$$

The solutions are therefore of the type

$$\psi_N = \sin \frac{N\,\pi}{L}\,x \qquad \text{with } N \text{ even}. \tag{3.76}$$

Hence, the reduced energy can be expressed in every one of these cases as

$$k_N^2 = \varepsilon_N = \frac{N^2\,\pi^2}{L^2} \qquad N = 1, 2, ..., \infty. \tag{3.77}$$

The energy E is quantized. It can only take discrete values:

$$E_N = \frac{\hbar^2}{2m} \varepsilon_N = \frac{\hbar^2}{2m} \frac{N^2 \pi^2}{L^2}. \tag{3.78}$$

If we consider a particle with mass m in a box of dimensions L_x by L_y by L_z, the quantization along each axis will yield a (3.78) type energy, giving overall

$$E = \frac{\pi^2 \hbar^2}{2m} \left(\frac{n_x^2}{L_x^2} + \frac{n_y^2}{L_y^2} + \frac{n_z^2}{L_z^2} \right), \tag{3.79}$$

whereas the wave function is a stationary three-dimensional wave. For the even values of the quantum numbers n_x, n_y, and n_z, it will be of the form

$$\psi = A \, \sin \left(\frac{\pi n_x}{L_x} x \right) \, \sin \left(\frac{\pi n_y}{L_y} y \right) \, \sin \left(\frac{\pi n_z}{L_z} z \right). \tag{3.80}$$

The number of nodes of the eigenfunctions (zeros of the eigenfunctions excluding the extremities) increases with the eigenvalue of the energy. It increases by one unit when we move from one eigenvalue to the next higher one. The fundamental has no nodes whereas the $(n-1)$th excited state ψ_n has $(n-1)$ nodes. There is a complete analogy here with the stationary states of vibrating strings.

3.5 The Simple Harmonic Oscillator

In classical mechanics, a particle with mass m subjected to a restoring force $-K^2 x$ is termed a harmonic oscillator. Its oscillatory motion is defined by the equation

$$x = x_0 \, \sin \left(\omega \, t + \varphi_0 \right) \quad \text{with} \quad \omega^2 = \frac{K^2}{m}. \tag{3.81}$$

Its potential energy $V = \int K^2 x \, d x$ can be evaluated as

$$V = \frac{K^2 x^2}{2} = \frac{1}{2} m \, \omega^2 \, x_0^2 \, \sin^2 \left(\omega \, t + \varphi_0 \right). \tag{3.82}$$

The same is true of its kinetic energy T,

$$T = \frac{1}{2} m \, v^2 = \frac{1}{2} \, m \, \omega^2 \, x_0^2 \, \cos^2 \left(\omega \, t + \varphi_0 \right) \tag{3.83}$$

and, hence, of its total energy (obviously constant over time):

$$T + V = E = \frac{1}{2} \, m \, \omega^2 \, x_0^2 \, \left[\sin^2 \left(\omega_0 \, t + \varphi_0 \right) + \cos^2 \left(\omega \, t + \varphi_0 \right) \right]$$

$$= \frac{1}{2} \, m \, \omega^2 \, x_0^2 = \frac{1}{2} \, K^2 \, x_0^2 \,. \tag{3.84}$$

Let us now determine the solution of the same problem in wave mechanics by solving the Schrödinger equation

$$\frac{d^2 \, \psi}{d \, x^2} + \frac{2m}{\hbar^2} \left(E - \frac{K^2 \, x^2}{2} \right) \psi = 0 \,. \tag{3.85}$$

By writing $\psi(x)$ as

$$\psi(x) = e^{-\alpha \, x^2} F(x) \qquad \alpha > 0 \tag{3.86}$$

the Schrödinger equation becomes

$$\frac{d^2 \, F}{d \, x^2} - 4\alpha \, x \, \frac{d \, F}{d \, x} + \left[\frac{2m \, E}{\hbar^2} - 2\alpha + x^2 \left(4\alpha^2 - \frac{m \, K^2}{\hbar^2} \right) \right] F = 0 \,. \tag{3.87}$$

The parameter α can be chosen so as to eliminate the term in x^2:

$$\alpha = \frac{(m \, K^2)^{1/2}}{2\hbar} \,. \tag{3.88}$$

In addition, let us set:

$$\frac{2m \, E}{\hbar^2} - 2\alpha = b \,, \tag{3.89}$$

which leaves us with the second-order differential equation

$$\frac{d^2 \, F}{d \, x^2} - 4\alpha \, x \, \frac{d \, F}{d \, x} + b \, F = 0 \,. \tag{3.90}$$

Let us look for a solution in the form of a series:

$$F(x) = \sum_p a_p \, x^p \,. \tag{3.91}$$

The coefficient of the term in x^p in (3.90) will give

$$(p + 1)(p + 2) \, a_{p+2} - 4\alpha \, p \, a_p + b \, a_p = 0 \,, \tag{3.92}$$

leading to the recurrence relation

$$a_{p+2} = \frac{-b + 4\alpha \, p}{(p + 1)(p + 2)} a_p \,. \tag{3.93}$$

The general solution is of the form

$$F = a_0 \left[1 + \frac{a_2}{a_o} x^2 + \frac{a_4}{a_o} x^4 + \ldots \right] + a_1 \left[x + \frac{a_3}{a_1} x^3 + \ldots \right]. \tag{3.94}$$

This, however, is not appropriate because it is not square-integrable. To avoid this problem, two possible cases are distinguished:

$$
\begin{array}{llll}
\text{i)} & a_1 = 0 & a_0 \neq 0 & b = 4\alpha\,p \quad p \text{ even}, \\
\text{ii)} & a_0 = 0 & a_1 \neq 0 & b = 4\alpha\,p \quad p \text{ odd}.
\end{array} \tag{3.95}
$$

The solution for $F(x)$ then reduces to a p-degree polynomial and $\psi(x)$, which is the product of the polynomial with $e^{-\alpha x^2}$, will tend to zero for very large x and become square-integrable.

Using conditions (3.95), the differential Eq. (3.90) reduces to

$$\frac{d^2 F}{d x^2} - 4\alpha\,x\,\frac{d F}{d x} + 4\alpha\,p\,F = 0. \tag{3.96}$$

If we change the variable:

$$(2\alpha)^{1/2}\,x = U, \tag{3.97}$$

the differential Eq. (3.96) becomes

$$\frac{d^2 F}{d u^2} - 2u\,\frac{d F}{d u} + 2p\,F = 0. \tag{3.98}$$

This is a differential equation whose solutions $H_p(u)$ are the pth order Hermite polynomials

By normalizing the function, we end up with the following solution:

$$\psi(x) = \left[2^p\,p!\pi^{1/2} \right]^{-1/2} (2\alpha)^{1/4}\,H_p\left(x\,\sqrt{2\alpha} \right) e^{-\alpha x^2}. \tag{3.99}$$

The energy levels can be calculated using the parameter b given in (3.89) and relation (3.95):

$$b = 4\alpha\,p = \frac{2m\,E}{\hbar^2} - 2\alpha \quad \text{and} \quad \frac{2m\,E}{\hbar^2} = 4\alpha \left(p + \frac{1}{2} \right).$$

The energy levels of the harmonic oscillator are quantized as follows:

$$\boxed{E_n = \left(n + \frac{1}{2} \right) \left(\frac{K^2}{m} \right)^{1/2} \hbar = (n + 1/2)h\,\nu \quad n = 0, 1, 2, \ldots} \tag{3.100}$$

In wave mechanics, energy cannot be zero because $E_{\text{min}} = \frac{1}{2}h\,\nu$, contrary to what holds in classical mechanics where a particle at the origin with zero velocity, has zero energy.

3.6 Periodic Potential Well

A one-dimensional crystal of identical atoms is schematically represented by a periodic potential well of the form

$$V(x) = V(x + a),\tag{3.101}$$

where a represents the distance between two atoms of the crystal.

The translation operator $T_a = e^{ika}$ transforms the wave function at the point x into the wave function at the point $x + a$:

$$T_a\,\psi(x) = e^{ika}\,\psi(x) = \psi(x + a);\tag{3.102}$$

k is a real number and $0 \le ka \le 2\pi$.

We introduce the Bloch wave

$$u_k(x) = e^{-ikx}\,\psi(x).\tag{3.103}$$

It is easy then to write the translation invariance of the Bloch wave:

$$u_k(x + a) = e^{-ik(x+a)}\,\psi(x + a) = e^{-ik(x+a)}\,e^{ika}\,\psi(x)$$
$$= e^{-ikx}\,\psi(x) = u_k(x).\tag{3.104}$$

The Bloch wave is periodic, with the same periodicity as the one-dimensional crystal.

Let us describe the one-dimensional crystal with a Kronig-Penney potential:

$$V(x) = V_0 \sum_{n=-\infty}^{+\infty} \delta(x - na)\qquad V_0 > 0.\tag{3.105}$$

It is a potential with a second-order singularity, implying the non-continuity of the derivative of the wave function $\psi(x)$.

The wave function in the region between two atoms $(V(x) = 0)$ is written from the Schrödinger equation as follows:

$$\psi(x) = A\,e^{iqx} + B\,e^{-iqx}\qquad\text{with}\qquad q^2 = \frac{2m\,E}{\hbar^2},\tag{3.106}$$

whereas the Bloch function becomes

$$u_k(x) = A\,e^{i(q-k)x} + B\,e^{-i(q+k)x}.\tag{3.107}$$

The periodicity of the function $u_k(x)$ implies that for $x = 0$

$$u_k(0) = u_k(a)\tag{3.108}$$

and, when combined with (3.107), we obtain the relation

$$A + B = A\,e^{i(q-k)a} + B\,e^{-i(q+k)a}.\tag{3.109}$$

To match the derivatives $\psi'(x)$ let us use the Schrödinger equation as follows:

$$\frac{\hbar^2}{2m} \frac{d^2}{dx^2} \psi + E\,\psi = V_o \sum_n \delta\,(x - na)\,\psi\,(x) \tag{3.110}$$

and integrate each of the terms between $-\varepsilon$ and $+\varepsilon$, bearing in mind that because the wave function $\psi\,(x)$ is continuous,

$$\int_{-\varepsilon}^{+\varepsilon} \psi\,(x)\,dx \to 0 \quad \text{if} \quad \varepsilon \to 0. \tag{3.111}$$

The integration of the second term of the Schrödinger Eq. (3.110) will then yield

$$V_0 \sum_n \int_{-\varepsilon}^{+\varepsilon} \delta\,(x - na)\,\psi\,(x) \to \Psi\,(0) \quad \text{if} \quad \varepsilon \to 0. \tag{3.112}$$

We then obtain with the first term:

$$\frac{\hbar^2}{2m} \left[\left(\frac{d\psi}{dx}\right)_\varepsilon - \left(\frac{d\psi}{dx}\right)_{-\varepsilon} \right] = V_0\,\psi\,(0). \tag{3.113}$$

This clearly shows that with a singular potential of the Kronig–Penney type, the derivative of the wave function ceases to be continuous.

Equation (3.106) leads to the derivative:

$$\frac{d\psi}{dx} = i\,q\,\left(A\,e^{i\,q\,x} - B\,e^{-i\,q\,x}\right), \tag{3.114}$$

which gives, as ε tends to zero,

$$\left(\frac{d\psi}{dx}\right)_\varepsilon \to i\,q\,(A - B) \quad \text{if} \quad \varepsilon \to 0. \tag{3.115}$$

By using relation (3.102) to translate the wave function a distance a, we obtain the translation of the derivative:

$$e^{i\,k\,a}\,\psi'\,(-\varepsilon) = \psi'\,(-\varepsilon + a), \tag{3.116}$$

that is to say, the derivative at $-\varepsilon$ can be written

$$\left(\frac{d\psi}{dx}\right)_{-\varepsilon} = e^{-i\,k\,a}\,\left(\frac{d\psi}{dx}\right)_{a-\varepsilon}, \tag{3.117}$$

and by allowing ε to tend to zero, we obtain

$$\left(\frac{d\psi}{dx}\right)_{-\varepsilon} = i\,q\,e^{-i\,k\,a}\,\left(A\,e^{i\,q\,a} - B\,e^{-i\,q\,a}\right). \tag{3.118}$$

We thus obtain the two continuity equations:

$$A + B = A\, e^{i\,(q-k)\,a} + B\, e^{-i\,(q+k)\,a}$$

$$V_0\,(A+B) = \frac{\hbar^2}{2m}\, i\,q\,\left(A - B - A\, e^{i\,(q-k)\,a} + B\, e^{-i\,(q+k)\,a}\right), \tag{3.119}$$

or, by changing variables and setting,

$$\gamma = \frac{\hbar^2}{2m}\,\frac{i\,q}{V_o}\qquad \alpha = e^{i\,(qk)\,a}\qquad \text{and}\qquad \beta = e^{-i\,(q+k)\,a}, \tag{3.120}$$

a system of two equations and two unknowns, A and B,

$$A\,(1-\alpha) + B\,(1-\beta) = 0,$$
$$A\,(1-\gamma+\gamma\,\alpha) + B(1+\gamma-\gamma\,\beta) = 0. \tag{3.121}$$

There exists a solution if, and only if, the determinant is zero:

$$\det\begin{vmatrix} 1-\alpha & 1-\beta \\ 1-\gamma+\gamma\alpha & 1+\gamma-\gamma\beta \end{vmatrix} = 0. \tag{3.122}$$

This imposes the condition:

$$2\gamma\,(1-\alpha)\,(1-\beta) + \beta - \alpha = 0, \tag{3.123}$$

an equation we can write using (3.120) as

$$2\gamma\,\left(1 - e^{i\,q\,a}\, e^{-i\,k\,a}\right)\left(1 - e^{-i\,q\,a}\, e^{-i\,k\,a}\right) + e^{-i\,k\,a}\,\left(e^{-i\,q\,a} - e^{i\,q\,a}\right) = 0.$$

And with a few transformations, condition (3.123) can be written:

$$\cos ka = \cos qa + \frac{m\,V_0\,a}{\hbar^2}\,\frac{\sin qa}{q\,a}. \tag{3.124}$$

Because the function $\cos ka$ can only be defined between -1 and $+1$, it imposes the same limits on the right-hand side of the above equation:

$$-1 \leq \cos qa + \frac{m\,V_o\,a}{\hbar^2}\,\frac{\sin qa}{q\,a} \leq 1. \tag{3.125}$$

Let us depict this result on a diagram, with the periodicity qa of the crystal on the x-axis and, on the y-axis, the $\cos ka$ expressed as a function of qa using Eq. (3.124).

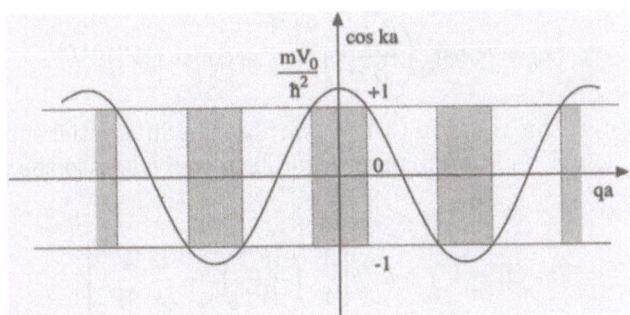

The possible values of the energy are therefore continuous and within allowed bands whereas there are a certain number of forbidden bands (the discontinuous bands in the figure). There are an infinite number of energy solutions for every value of $\cos ka$:

$$E = \frac{\hbar^2}{2m} q^2 = f\,(k) \qquad -\frac{\pi}{a} \leq k \leq +\frac{\pi}{a}. \qquad (3.126)$$

This example provides the foundation for the Kronig–Penny band theory in solid state physics.

4. Semi-classical (W.K.B.) Approximation

In the last chapter, with the aid of form (2.11), we showed the importance of the phase function of the wave function by using a particular form of the latter, that is

$$\Psi\,(\vec{r},\,t) = a(\vec{r},\,t)\,\exp\left[\frac{i}{\hbar}\,S\,(\vec{r},\,t)\right]. \qquad (3.127)$$

In a one-dimensional problem, the wave function of a conservative system may be written in a similar manner by using the classical Hamiltonian (1.191):

$$\Psi\,(x,\,t) = a\,(x)\,\exp\left[\frac{i}{\hbar}\,S\,(x,\,t)\right] = a\,(x)\,\exp\left[\frac{i}{\hbar}\,(W\,(x) - E\,t)\right]. \qquad (3.128)$$

Using the stationary form (3.23) of the Schrödinger equation as the departure point, we obtain the following coupled equations:

$$\left(\frac{\partial W}{\partial x}\right)^2 + 2m\,(E - V\,(x)) = \hbar^2\,\frac{1}{a}\,\frac{\partial^2 a}{\partial x^2}, \qquad (3.129)$$

$$2\,\frac{\partial W}{\partial x}\,\frac{\partial a}{\partial x} + a\,\frac{\partial^2 W}{\partial x^2} = 0. \qquad (3.130)$$

The functions $W\,(x)$ and $a\,(x)$ are functions of the variable x, which enables us to obtain a solution for the continuity Eq. (3.130):

$$a\,(x) = \text{const}\,\left(\frac{\partial W}{\partial x}\right)^{-1/2} = \text{const}\,(W')^{-1/2}. \qquad (3.131)$$

By inserting over this result in (3.129), we obtain a differential equation defining the function $W\,(x)$ in a manner rigorously equivalent to the Schrödinger equation:

$$W'^2 = 2m\,(E - V) + \hbar^2\,\left[\frac{3}{4}\,\left(\frac{W''}{W'}\right)^2 - \frac{1}{2}\,\frac{W''}{W}\right]. \qquad (3.132)$$

The separation of Eqs. (3.129) and (3.130), which is possible in the one-dimensional case, is impossible in the general case because the continuity condition (3.130) becomes

$$\vec{\nabla} \cdot (a^2 \, \vec{\nabla} \, W) = 0 \, . \tag{3.133}$$

This does not permit the straightforward extraction of the function W from the equation generalizing form (3.129):

$$(\vec{\nabla} \, W)^2 + 2m \, (E - V) = \hbar^2 \, \frac{\nabla^2 \, a}{a} \, . \tag{3.134}$$

The semi-classical approximation developed by G. Wentzel, A. Kramers, and L. Brillouin, (W.K.B. method), consists in expanding W in increasing powers of $\hbar^2$ by setting

$$W \, (x) = W_0 + \hbar^2 \, W_1 + \ldots \tag{3.135}$$

and retaining only the zero-order terms. This means neglecting the terms in $\hbar^2$ in (3.132), leaving us the semi-classical result

$$\left(\frac{\partial \, W}{\partial \, x} \right)^2 = 2m \, (E - V \, (x)) \, . \tag{3.136}$$

Case No. 1: $E > V \, (x)$
The semi-classical solution is easily obtained as

$$W \, (x) = \pm \int^x \, dx' \, \sqrt{2m \, (E - V \, (x'))} \, , \tag{3.137}$$

leading to a stationary W.K.B. solution:

$$\Psi \, (x) \simeq \frac{\text{const}}{[E - V \, (x)]^{1/4}} \, \exp \left[\pm \frac{i}{\hbar} \int^x \, dx' \, \sqrt{2m \, (E - V \, (x'))} \right] \, . \tag{3.138}$$

Case No. 2: $E < V \, (x)$
This zone, prohibited from the standpoint of classical mechanics, has the following semi-classical W.K.B. solution:

$$\Psi \, (x) \simeq \frac{\text{const}}{[V \, (x) - E]^{1/4}} \, \exp \left[\pm \frac{1}{\hbar} \int^x \, dx' \, \sqrt{2m \, (V \, (x') - E)} \right] \, , \tag{3.139}$$

provided that $\hbar \, \sqrt{2m \, (V - E)}$ is small compared to the characteristic dimensions of the system, that is, compared to variations in the potential $V \, (x)$.

The W.K.B. solutions expanded in (3.138) and (3.139) have general validity except at the turning point x_0 defined by

$$V \, (x_0) = E \, . \tag{3.140}$$

The procedure for linking semi-classical wave functions is quite elaborate. It consists in linearizing the potential $V(x)$ around the turning point x_0 with a Taylor expansion, and then solving the Schrödinger equation for $E = 0$ in x_0, that is:

$$-\frac{\hbar^2}{2m} \frac{d^2 \Psi}{d x^2} + (x - x_0) \left(\frac{d V}{d x}\right)_{x_0} = 0 \qquad (3.141)$$

We follow up by linking this solution to the foregoing W.K.B. solutions by adequately adjusting the integration constants.

The advantage in using the W.K.B. method lies exclusively in yielding of the behavior of solutions far away from the turning points. It is a method widely used with simple systems as well as in quantum cosmology.

Chapter 4
Dirac Formalism

All physical theories may be considered as comprising three parts:

1. the formalism
2. the dynamical laws
3. the correspondence rules.

The formalism is the set of symbols and inference rules for formulating hypotheses and propositions which, upon application, yield other symbols and inference rules, and so on. The primary concepts underlying the formalism are those of physical states and observables. In a classical system, the state of the system is the set of all the trajectories followed by all the material points constituting the system. The formalism used may be that of analytical mechanics based on Hamilton's variational principle and leading to the different formulations described in Chap. 1. In the general relativity theory of gravitation, on the other hand, the state of the system is constituted by the space-time geometry determined from the system's physical content. The formalism employed is that of tensor analysis and Einstein's law is used to link the space-time curvature to the physical content, matter or radiation. The notion of state is more abstract in quantum mechanics and rather refers to a mathematical space vector with certain well-defined properties, that is, a Hilbert space. The observables here will be the symbols associated with the experimental procedures for measuring such observables and will be linked to special mathematical entities, hermitian operators, acting in the Hilbert space. This formalism, introduced by the English physicist P.A.M. Dirac, has been termed the Dirac formalism. It is also called the first quantization formalism because of the existence of two other formalisms which take different points of view but which, nevertheless, lead to the same results. The great advantage of the Dirac formalism is that it makes very explicit the relationship between the abstract Hilbert space and the true physical space in which we live.

By projecting the quantum states of the Hilbert space onto the states of the true, or configuration, space and by simply introducing the correspondence principle, we obtain a wave function, the wave mechanical function, obeying the Schrödinger equation. The second quantization formalism assumes that physical states, vectors of the Hilbert space in the first formalism, have all been taken into account and that it is only a question of indicating whether one of the states is empty or occupied. Using a vacuum quantum state $| 0\rangle$, which must be clearly distinguished from the vacuum as absence of matter with which we are familiar, it is possible to fill certain quantum states using creation operators acting on this vacuum. Conversely, it is possible to destroy some occupied states and empty them of their content by using other operators. This formalism, essentially based on an operator algebra, permits in particular the description of physical systems with many quantum states, e.g. atoms and nuclei, and, in particular, to easily take into account a foundational principle in particle physics, namely, the Pauli exclusion principle. This second formalism of quantum mechanics will be discussed in Chap. 16. The third quantization formalism is gaining in importance, especially in quantum field theory. It was first formulated by Feynman and is based on a conjecture. The wave function at a point x' and time t' is dependent on the knowledge of the wave functions at an earlier instant t at every point in space as well as on a function propagating the wave function from a particular point at instant t to another at instant $t' > t$. This result, for example, proceeds directly from the Dirac formalism. Feynman's conjecture enables us to calculate the propagator between two instants and two infinitely close points from the classical phase shift function $\exp{(i\,S/\hbar)}$ with a normalization adjusted to the wave function of a free particle. This permits one to directly obtain the phase shift function which is of prime importance as already shown in Chap. 2. We notice, however, that these three formalisms lead to the same result and are all based on Dirac-style quantization.

Dynamical laws are explicitly time-dependent mathematical relations (differential equations, integral equations, or analyticity conditions) that must satisfy the basic elements of the formalism. The dynamical laws of classical mechanics are Newton's, Hamilton's, and Poisson's equations. In general relativity, the dynamical law is contained in Einstein's equation, expressed in a four-dimensional space-time and, hence, including time evolution. In non-relativistic wave mechanics, the dynamical law is the time-dependent Schrödinger equation introduced artificially in Chap. 1 to justify the classical limit of the solution. The Dirac formalism permits us to obtain the dynamical law either by allowing the evolution of quantum states and leaving the operators for transforming these states invariant (this is Schrödinger's point of view), or by fixing the quantum states and then causing the operators associated with the dynamical states to evolve (the latter is the completely different point of view adopted by Heisenberg). Heisenberg's point of view is naturally that adopted by the second quantization whereas, using Schrödinger's picture as departure point, the third quantization includes the evolution of the wave function in the propagator itself. No matter which point of view is adopted, we end up with the expected values of the operators, i.e., measurable quantities. This problem is due to the special role assigned to time in quantum mechanics.

In non-relativistic quantum mechanics, time is absolute and universal, and is used to measure the evolution of our observations during our lifetime as observers. Special relativity theory introduces a local time, specific to each observer and thus eliminates the notions of absolute time and instantaneous interactions. In this way, it transforms time into one of the four coordinates defining the position of particles in space and in time. General relativity theory does not discard the space-time concepts of special relativity but it also needs a reference time for describing the evolution of the Universe. It therefore introduces a cosmic time associated with the co-moving coordinates, that is a parameter for defining the "Universal scale factor" for any observer whatsoever. In non-relativistic quantum mechanics, time is introduced as an observable analogous to position although it is in actual fact not one. The position of a particle is, in fact, the eigenvalue of the hermitian operator associated with the position obeying the rules defined by quantum mechanics, and the conjugate variable is the momentum, such that there is an uncertainty relation between these variables. Time is not an observable because it is not an eigenvalue of a hermitian operator. There is, however, an uncertainty relation between the conjugate variables energy and time. This peculiarity of the role of time in non-relativistic quantum dynamics is a great problem in the quantum physics of systems under observation because measurement is always made with a classical apparatus for which the notion of time or, more properly speaking, of time intervals is empirical although conceptually easy to apprehend. The problem becomes more serious when we turn to quantum cosmology which deals with the wave function of the Universe just after its creation by the Big Bang. The Universis wave function deriving from quantum cosmology is time independent whereas the observable Universe is quite naturally time dependent. The interpretation usually proposed is that the very existence of an observer reduces the total wave function of the Universe to that part describing the world observed by him. This is the position of the Copenhagen School, a position whose foundations have been shaken by the transition from classical to quantum physics. We will return to this point in connection with quantum dynamics and the density matrix and in quantum cosmology.

Finally, correspondence rules assign an empirical meaning to some objects or symbols used in the formalism. Correspondence rules in classical mechanics draw on a direct intuition that offers no option of interpretation. In general relativity theory, the effect of gravitation on all physical laws is introduced via the principle of co-variance. These laws are, in fact, correspondence rules for dealing with gravitation by replacing Minkowski space-time with a Riemann space-time and ordinary derivatives with co-variant derivatives. In quantum mechanics, correspondence rules make it possible to write the operators associated with observables in the configuration space and, hence, to go from the Hilbert physical space to the true space. These rules define correspondence between the Poisson brackets of the functions of the observables of classical mechanics and the operator commutators associated with these observables in quantum mechanics, or directly between the variables associated with these observables of classical mechanics and the corresponding operators in quantum mechanics. The correspondence principle of quantum mechanics can lead to different interpretations of the mathematical objects it induces. This accounts for both the difficulty and the conceptual richness of this new physics.

1. Postulates of Quantum Mechanics

The dynamical law and the correspondence principle of quantum mechanics will be treated in the next chapter. In the meantime, let us attempt a description of the Dirac formalism of quantum mechanics. It rests on four postulates.

Postulate No. 1. The state of a given physical system at any given instant t_0 is defined by the data of a ket $| \psi (t_0) \rangle$, termed quantum state and belonging to a linear complex vector space with a decomposable (separable) and complete scalar product, the Hilbert space $\mathcal{H}$.

Postulate No. 2. Any measurable physical property $\mathcal{A}$ is described by a Hermitian operator A, termed observable and acting in $\mathcal{H}$.

It is thus possible to associate with the position of a particle the position operator x_j ($j = 1, 2, 3 = x, y, z$), and with its momentum, the operator p_j.

The space of the states $\mathcal{H}$ is totally characterized by the knowledge of all the algebraic relations between the members of the complete set of linear operators acting in $\mathcal{H}$. This set is such that none of its members can commute with all the others but, rather, only with multiples of the identity operator.

Commutation relations $[x_j,\ x_k]$, $[p_j,\ p_k]$ and $[x_j,\ p_k]$, for example, totally characterize the space of the states of a particle without internal variables (spin, electric charge, etc.).

Postulate No. 3. The measurement of a physical property $\mathcal{A}$ can only yield one of the eigenvalues of the corresponding observable A.

If $| n \rangle$ is the non-degenerate eigenstate of a Hermitian operator A associated with the observable $\mathcal{A}$, the observable eigenvalue a_n will be determined by the characteristic equation:

$$A \mid n \rangle = a_n \mid n \rangle .$$

Postulate No. 4. If we measure the physical property $\mathcal{A}$ of a system in a normalized quantum state $\mid \psi \rangle$, the probability $P\left(a_n\right)$ of obtaining the eigenvalue a_n of the observable A associated with $\mathcal{A}$ for a non-degenerate eigenstate $\mid n \rangle$ of A will be equal to:

$$P_r\left(a_n\right) = \mid \langle\, n \mid \psi \rangle \mid^2 .$$

2. Dirac Formalism

2.1 Ket Vectors

To every dynamic state is associated a contra-vector or ket vector (or, simply, ket) written $\mid\, \rangle$. This is Dirac's notation.

The labels (continuous or discrete) characterizing the dynamical states are inserted in the ket.

Kets combine to form a vector space ξ, written

$$\mid u_1 \rangle \in \xi \qquad \text{and} \qquad \mid u_2 \rangle \in \xi$$

Every linear combination of these kets also belongs to the vector space ξ,

$$\lambda_1 \mid u_1 \rangle + \lambda_2 \mid u_2 \rangle \in \xi \tag{4.1}$$

We can thus define a ket dependent on a continuous index in some domain $(x_1,\ x_2)$ by writing the integral

$$\mid w \rangle = \int_{x_2}^{x_1} \lambda\left(x\right) \mid x \rangle d\, x . \tag{4.2}$$

It is further postulated that the space ξ is a Hermitian positive (see Sect. 2.3) and separable or, rather, decomposable Hilbert space, that is a complex space with the following characteristics:

- vectorial   if $a \in \xi$ $b \in \xi$ $a + b \in \xi$ $0 + x = x$ $1x = x$
- unitary   it is possible to define an inner product in ξ
- normalized   the scalar product of an element with itself is positive
- complete   any normal Cauchy series converges in ξ
- separable   there is a denumerable orthonormal base in ξ

Care should be taken not to give separability the same meaning in physics as in mathematics. Separability in physics has to do with the independence of an object with respect to its components. The motion of a system is said to be separable into its variables if the equation corresponding to each of these variables (e.g. the motion of the center of mass and relative motion) can be made uncoupled. In mathematics, on the other hand, a space is considered separable if it can be decomposed into direct sums of the subspaces $\xi = \xi_1 \oplus \xi_2 \oplus ... \xi_N$.

Thus, to every pure quantum state will be associated a corresponding ket $|u\rangle$ belonging to the Hilbert space, with a denumerable (finite or infinite) set of linearly independent base vectors spanning such a state.

2.2 Bra Covectors

To every vector space ξ can be associated a dual covector space ξ^* such that the product of any element of the vector space with any element of its dual space is an scalar quantity.

Covectors (or bras) are written as $\langle \chi |$, and are vectors of the dual space ξ^*. To every vector $|u\rangle$ of ξ can be associated $\langle \chi |$ such that $\langle \chi | u \rangle$ is a scalar product. Thus, we will say that a covector is zero:

$$\langle \theta | = 0 \quad \text{if} \quad \forall | u \rangle \quad \langle \theta | u \rangle = 0 \tag{4.3}$$

and that two covectors are equal:

$$\langle \theta_1 | = \langle \theta_2 | \quad \text{if} \quad \forall | u \rangle \quad \langle \theta_1 | u \rangle = \langle \theta_2 | u \rangle . \tag{4.4}$$

The space ξ and its dual, ξ^*, are of the same dimension, that is, they have the same number of linearly independent bra vectors (or covectors).

We show that there is a one-to-one correspondence between the vectors of ξ and the vectors of ξ^*. Such vectors are said to be mutually conjugate and $\langle u |$, the conjugate of $| u \rangle$, will be written:

$$| u \rangle \in \xi \quad \text{is the conjugate of} \quad \langle u | \in \xi^* .$$

The correspondence between the ket vectors and the bra covectors is antilinear, that is, if

$$| u \rangle = \lambda_1 | u_1 \rangle + \lambda_2 | u_2 \rangle \quad \text{in} \quad \xi , \tag{4.5}$$
$$\langle u | = \lambda_1^* \langle u_1 | + \lambda_2^* \langle u_2 | \quad \text{in} \quad \xi^* , \tag{4.6}$$

where λ_1^* and λ_2^* are complex conjugates of λ_1 and λ_2.

2.3 Inner Product

An inner, or scalar, product on a complex vector space ξ (that is, one based on the complex numbers $\mathbb{C}$) is a definite positive Hermitian form.

By Hermitian form, we mean a mapping of the set $\xi \otimes \xi$ into the set $\mathbb{C}$ of complex numbers which, to every vector pair (xy), associates a complex number written $\langle x \mid y \rangle$ such that the following conditions are satisfied:

$$\langle x \mid \alpha \, y + \beta \, z \rangle = \alpha \langle x \mid y \rangle + \beta \langle x \mid z \rangle \begin{cases} \forall \, y, z \in \xi \\ \forall \, \alpha, \beta \in \mathbb{C} \end{cases}$$

$$\langle x \mid y \rangle^* = \langle y \mid x \rangle . \tag{4.7}$$

A Hermitian form is said to be positive if the norm $\langle x \mid x \rangle \geq 0 \;\; \forall \, x \in \xi$. On the other hand, it will be said to be positive definite if $\langle x \mid x \rangle = 0$ implies $\mid x \rangle = 0$.

We obtain from properties (4.6) and (4.7) the relation

$$\langle \alpha \, y + \beta \, z \mid x \rangle = \alpha^* \langle y \mid x \rangle + \beta^* \langle z \mid x \rangle . \tag{4.8}$$

Two vectors $\mid x \rangle$ and $\mid y \rangle$ are said to be orthogonal if $\langle x \mid y \rangle = 0$. More generally speaking, two sub-sets, F and G, contained in ξ are orthogonal if $x \perp y$ for any $x \in F$ and $y \in G$. We therefore note that $F \perp G$ implies:

$$\langle x \mid y \rangle = 0 \;\; \forall \mid x \rangle \in F \quad \text{and} \quad \mid y \rangle \in G . \tag{4.9}$$

It should be noted that two orthogonal subspaces cannot have non-zero vectors in common because their norm as well as the vector itself would be zero.

The orthogonal subspace $\xi_1^\perp$ is termed the complementary subspace of ξ_1.

It can be shown (see (4.60) below) that any vector of ξ can be described in one and only one way, as the sum of a vector $\mid u_1^\perp \rangle \in \xi_1^\perp$ and a vector $\mid u_1 \rangle \in \xi_1$ orthogonal to $\mid u_1 \rangle$:

$$\mid u \rangle = \mid u_1 \rangle + \mid u_1^\perp \rangle ,$$
$$\langle u_1 \mid u_1^\perp \rangle = 0 \quad \text{with} \quad \xi_1 \perp \xi_1^\perp . \tag{4.10}$$

2.4 Schwartz' Inequality

On a complex vector space a positive Hermitian form satisfies Schwartz' inequality

$$\mid \langle x \mid y \rangle \mid^2 \leq \langle x \mid x \rangle \langle y \mid y \rangle \qquad \forall \mid x \rangle \text{ and } \mid y \rangle . \tag{4.11}$$

The definition of a positive Hermitian form (Sect. (2.3). above) can, in fact, be used to set

$$\langle \lambda \, x + y \mid \lambda \, x + y \rangle \geq 0 \qquad \mid x \rangle \quad \text{and} \quad \mid y \rangle \in \xi \quad \text{and} \quad \lambda \in \mathbb{C} . \tag{4.12}$$

Using (4.5) and (4.6), this will be written as

$$\lambda \, \lambda^* \, \langle x \mid x \rangle + \lambda^* \, \langle x \mid y \rangle + \lambda \langle x \mid y \rangle^* + \langle y \mid y \rangle \geq 0 . \tag{4.13}$$

We introduce the following form of the parameter λ:

$$\lambda = t \, \frac{\langle x \mid y \rangle}{\mid \langle x \mid y \rangle \mid} \qquad \text{where} \qquad t \in \mathbb{R} \tag{4.14}$$

to transform Eq. (4.13) into

$$t^2 \, \langle x \mid x \rangle + 2t \, \langle x \mid y \rangle + \langle y \mid y \rangle \geq 0 \qquad \forall \, t \in \mathbb{R}. \tag{4.15}$$

This inequality is satisfied if the reduced discriminant of the second-order equation in t is negative, that is, if

$$\mid \langle x \mid y \rangle \mid^2 - \langle x \mid x \rangle \, \langle y \mid y \rangle \leq 0. \tag{4.16}$$

This leads to Schwartz' inequality:

$$\boxed{\mid \langle x \mid y \rangle \mid^2 \leq \langle x \mid x \rangle \, \langle y \mid y \rangle} \, . \tag{4.17}$$

The equality can only hold if the vectors $\mid x \rangle$ and $\mid y \rangle$ are proportional:

$$\mid x \rangle = \lambda \mid y \rangle. \tag{4.18}$$

2.5 Operators

If to every $\mid u \rangle \in \xi$ can be associated $\mid v \rangle \in \xi$, then we say $\mid v \rangle$ is obtained by the action of the operator A on $\mid u \rangle$:

$$\mid v \rangle = A \mid u \rangle. \tag{4.19}$$

If the vectors of ξ are normalized, then the operator A can be formally written as:

$$A = \mid v \rangle \langle u \mid \tag{4.20}$$

as can be seen with definition (4.19).

Because the space ξ is a linear vector space, A is linear and, therefore, has the following properties:

i) Nullity

$$A = 0 \qquad \text{if} \qquad \forall \mid u \rangle \qquad \mid v \rangle = A \mid u \rangle = 0$$

This can also take the following form:

$$\langle u \mid A \mid u \rangle = 0 \qquad \forall \mid u \rangle \in \xi \tag{4.21}$$

ii) Equality

$$A = B \qquad \text{if} \qquad \langle u \mid A \mid u \rangle = \langle u \mid B \mid u \rangle \qquad \forall \mid u \rangle \in \xi \tag{4.22}$$

iii) Sum

$$\text{commutative} \qquad A + B = B + A \tag{4.23}$$

$$\text{associative} \qquad A + B + C = A + (B + C) \tag{4.24}$$

iv) Product

$$\text{associative} \qquad A\,B\,C\,D = A\,(B\,C)\,D \tag{4.25}$$

distributive with
respect to the sum
$$A\,(B + C) = AB + AC \tag{4.26}$$

$$\text{non-commutative} \qquad AB \neq BA \quad \text{and} \quad [A, B] = AB - BA \neq 0 \tag{4.27}$$

Let us note some useful relations involving the commutator of operators A and B:

$$[\lambda A, B] = \lambda [A, B]$$
$$[A, B] = -[B, A]$$
$$[A, BC] = [A, B]C + B[A, C]$$
$$[A^n, B] = \sum_{s=0}^{n-1} A^s\,[A, B]A^{n-s-1}$$

v) Inverse

If $|\,v\rangle = A\,|\,u\rangle$ and $|\,u\rangle = B\,|\,v\rangle\ \forall\,|\,u\rangle$ and $|\,v\rangle \in \xi$ operators A and B are mutually inverse: $AB = BA = \mathbb{1}$. We will therefore write $B = A^{-1}$ and then observe that

$$|\,v\rangle = A\,|\,u\rangle \qquad \text{and} \qquad |\,u\rangle = A^{-1}\,|\,v\rangle. \tag{4.29}$$

It should be noted that the inverse of a product is equal to the product of the inverses:

$$(PQ)^{-1} = Q^{-1}\,P^{-1} \tag{4.30}$$

given that $(PQ)(PQ)^{-1} = \mathbb{1}$ and by left-multiplying by $Q^{-1}P^{-1}$, we obtain with (4.26), relation (4.30).

2.6 Example of Application

Show that two operators A and B which do not commute with each other but commute with their commutator $[A, B]$ verify the Baker–Campbell relation.

$$\exp(A)\,\exp(B) = \exp\left(A + B + \frac{1}{2}\,[A, B]\right) \tag{4.28}$$

Hint: Start with the function $f(x) = \exp(xA)\,\exp(xB)$ and determine the first derivative df/dx in the form

$$\frac{df}{dx} = (A+B)\ \exp(xA)\ \exp(xB) + [\exp(xA), B]\ \exp(xB)$$

and show that $[\exp(xA), B] = x\ [A, B]\ \exp(xA)$ leads to the first-order differential equation

$$\frac{df}{dx} = (A + B + x\ [A, B])\ f(x)$$

giving after integration the Baker–Campbell relation.

2.7 Adjoint Operators

Assuming that the operator A acts in the space ξ, what is its action in the dual space ξ^*?

Consider the bra $\langle \chi \mid$ of the dual space and the ket $A \mid u\rangle$ of the space ξ. This defines the inner product $\langle \chi \mid (A \mid u)\rangle$. We can operate in the same manner for every vector $\mid u\rangle$, that is, in the final analysis, define some bra $\langle \eta \mid$ which, multiplied by $\mid u\rangle$, leads us once more to the inner product $\langle \chi \mid (A \mid u)\rangle$:

$$\langle \eta \mid u\rangle = \langle \chi \mid (A \mid u)\rangle. \tag{4.31}$$

We can therefore associate a bra $\langle \eta \mid$ to every bra $\langle \chi \mid$ through the action of an operator B by noting:

$$\langle \eta \mid\ =\ \langle \chi \mid B \quad \text{and} \quad \langle \chi \mid\ \in\ \xi^*. \tag{4.32}$$

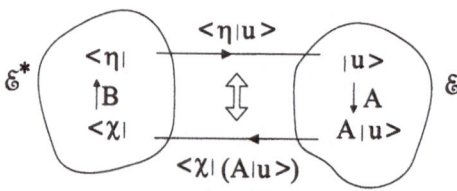

Thus, by knowing the action of A in ξ, we also know the action of B in ξ^*. If we set

$$\langle \eta \mid\ =\ \langle \chi \mid B, \tag{4.33}$$

$\mid \eta\rangle$ will be antilinearly dependent on $\langle \chi \mid$, hence, linearly dependent on $\mid \chi\rangle$. Operator B^+, the transpose of B, will be defined by the relation

$$\mid \eta\rangle = B^+ \mid \chi\rangle. \tag{4.34}$$

It goes without saying that if $B = 0$ $B^+ = 0$ and vice versa.

Now, let us use (4.7) in Sect. 2.3 above to write

$$\langle t \mid v \rangle = \langle v \mid t \rangle^* . \tag{4.35}$$

Using relations (4.33) and (4.34):

$$| v \rangle = A^+ | u \rangle \qquad \text{and} \qquad \langle v | = \langle u | A , \tag{4.36}$$

the first term of (4.35) reduces to

$$\langle t \mid v \rangle = \langle t \mid A^+ \mid u \rangle ,$$

and the second to

$$\langle v \mid t \rangle^* = \langle v \mid A \mid t \rangle^* .$$

This gives the defining relation for the operator A^+ acting in the space ξ^* if A is acting in the dual space ξ:

$$\boxed{\langle t \mid A^+ \mid u \rangle = \langle u \mid A \mid t \rangle^* \quad \forall \mid u \rangle \quad \text{and} \quad | t \rangle \in \xi} . \tag{4.37}$$

The operators A and A^+ are said to be mutual transposes or adjoints and the following fundamental properties should be noted

$$\left(A^+ \right)^+ = A , \tag{4.38}$$

$$\left(c\, A \right)^+ = c^* \, A^+ , \tag{4.39}$$

$$\left(A + B \right)^+ = A^+ + B^+ , \tag{4.40}$$

$$\left(AB \right)^+ = B^+ \, A^+ , \tag{4.41}$$

$$\left(\langle v \mid \, \right)^+ = | v \rangle , \tag{4.42}$$

$$\left(\mid u \rangle \right)^+ = \langle u \mid .$$

Note, in particular with (4.41), that the transposition of an operator product reverses the order of the operators.

Thus, to transpose any expression whatsoever, the order of notation of the different symbols needs to be reversed and then replaced as follows:

i) numbers with their complex conjugates
ii) kets with bras and vice versa
iii) operators with their transposes

Thus, for example, the adjoint (or transpose) of the following expression will be written:

$$(A\, b \mid c) \, \langle w \mid k \rangle^+ = k^* \mid w \rangle \, \langle c \mid b^* \, A^+$$

An operator is said to be Hermitian if it is equal to its adjoint:

$$H = H^+ . \tag{4.43}$$

An operator is said to be unitary if its adjoint is equal to its inverse:

$$U^+ = U^{-1}. \tag{4.44}$$

2.8 Eigenvalues and Eigenvectors of Hermitian Operators

a) Definitions

The complex number a is said to be the eigenvalue of the operator A corresponding to the eigenvector (or eigenket) $| u \rangle$ if

$$A | u \rangle = a | u \rangle. \tag{4.45}$$

Eq. (4.45) is termed the eigenvalue equation of the operator A. It is easily noticed that if $| u \rangle$ is the eigenvector of A, then any ket $\lambda | u \rangle$ is also an eigenvector of A, since

$$A \left(\lambda | u \rangle \right| = \lambda A | u \rangle = \lambda a | u \rangle = a \left(\lambda | u \rangle \right). \tag{4.46}$$

b) Degeneracy

If there are several linearly independent eigenvectors corresponding to the same eigenvalue a of an operator A, then any combination of these eigenvectors is also an eigenvector of A corresponding to the same eigenvalue a. This is easily demonstrated, for if by hypothesis

$$A | u_i \rangle = a | u_i \rangle \quad \forall \, i, \tag{4.47}$$

and we set up the linear combination

$$| u \rangle = \sum_i \lambda_i | u_i \rangle, \tag{4.48}$$

the following result is easily obtained:

$$A | u \rangle = \sum_i \lambda_i A | u_i \rangle = \sum_i \lambda_i a | u_i \rangle$$

$$= a \sum_i \lambda_i | u_i \rangle = a | u \rangle$$

which yields the following eigenvalue equation:

$$A | u \rangle = a | u \rangle. \tag{4.49}$$

Thus, the set of eigenvectors of A constitutes an N-dimensional vector space ξ_a (considering that there are N linearly independent vectors $| u_i \rangle$ in ξ_a).

If $N = 1$ the eigenvalue is not degenerate. There is only one eigenvector corresponding to the eigenvalue a.

If $N \neq 1$ the eigenvalue is degenerate and N is the order of degeneracy of the eigenvalue a. This then means that there are N eigenvectors corresponding to the eigenvalue a.

c) Reality

If the operator A is Hermitian, then its eigenvalue a is necessarily real. This can be demonstrated by using definition (4.49) as departure point and by multiplying the two terms of the bra $\langle u \mid$, leading to the eigenvalue:

$$a = \frac{\langle u \mid A \mid u \rangle}{\langle u \mid u \rangle}. \tag{4.50}$$

Operator A is Hermitian, $A^+ = A$, and by transposing (4.49),

$$\langle u \mid A = \langle u \mid a^*,$$

we obtain, by multiplying the eigenvalue a^* by the ket $\mid u \rangle$:

$$a^* = \frac{\langle u \mid A \mid u \rangle}{\langle u \mid u \rangle}. \tag{4.51}$$

If we compare (4.51) and (4.50), we naturally obtain the condition

$$a = a^*, \tag{4.52}$$

demonstrating that the eigenvalue of a Hermitian operator is a real.

2.9 Orthogonality

To any two different eigenvalues of a Hermitian operator correspond two orthogonal eigenvectors. In effect:

$$\begin{aligned} A \mid u \rangle &= a \mid u \rangle \quad \text{and} \quad \langle v \mid A \mid u \rangle = a \langle v \mid u \rangle \\ A \mid v \rangle &= b \mid v \rangle \quad \text{and} \quad \langle v \mid A = \langle v \mid b \end{aligned} \tag{4.53}$$

and, following multiplication by $\mid u \rangle$ of the last of these equations,

$$\langle v \mid A \mid u \rangle = b \langle v \mid u \rangle. \tag{4.54}$$

The term-by-term difference between (4.53) and (4.54) thus leaves us with:

$$(a - b) \langle v \mid u \rangle = 0. \tag{4.55}$$

If $a \neq b$, $\langle v \mid u \rangle = 0$ and thus $\mid v \rangle$ and $\mid u \rangle$ are orthogonal. This proves the theorem.

2.10 Projectors

a) Definition

Consider an operator A with an eigenvector $\mid a \rangle$ corresponding to the non-degenerate eigenvalue a and the operator $P_a = \mid a \rangle \langle a \mid$ termed the projector onto the subspace ξ_a of the eigenvalue.

Let us assume that the vector $\mid a \rangle$ is normalized, that is,

$$\langle a \mid a \rangle = 1. \tag{4.56}$$

The projector has two characteristic properties:

i) P_a is Hermitian $P_a = P_a^+$

$$P_a = \mid a \rangle \langle a \mid \quad \text{and} \quad P_a^+ = (\mid a \rangle \langle a \mid)^+ = \mid a \rangle \langle a \mid. \tag{4.57}$$

ii) $P_a^2 = P_a$

$$P_a^2 = \mid a \rangle \langle a \mid a \rangle \langle a \mid = \mid a \rangle \langle a \mid = P_a. \tag{4.58}$$

Generally speaking:

The only necessary and sufficient condition for a Hermitian operator P to be a projector is that it satisfies the condition $P^2 = P$.

b) Eigenvalues

What are the possible eigenvalues p of a projector P?

$$P \mid p \rangle = p \mid p \rangle.$$

If the operator P is applied again, we obtain

$$P^2 \mid p \rangle = p^2 \mid p \rangle,$$

and by calculating the term-by-term difference of the foregoing equations:

$$(P^2 - P) \mid p \rangle = (p^2 - p) \mid p \rangle = p(p - 1) \mid p \rangle.$$

However, $P^2 - P = 0$ and $\mid p \rangle \neq 0$ by definition, implying

$$p(p - 1) = 0, \tag{4.59}$$

an equation with 0 and 1 as roots.

The only possible eigenvalues of a projector are, therefore, the real numbers 0 and 1.

c) Projection Space

It is easily noted that any vector whatsoever in the Hilbert space can be decomposed as

$$| u \rangle = P | u \rangle + (1 - P) | u \rangle, \qquad (4.60)$$

and the vectors $P | u \rangle$ and $(1 - P) | u \rangle$ are orthogonal since

$$\langle u | P (1 - P) | u \rangle = \langle u | P^2 - P | u \rangle = 0. \qquad (4.61)$$

i) The ket $P | u \rangle$ is the eigenvector of P corresponding to the eigenvalue 1, since we can write

$$P (P | u \rangle) = P^2 | u \rangle = P | u \rangle = 1(P | u \rangle). \qquad (4.62)$$

ii) The ket $(1 - P) | u \rangle$ is the eigenvector of P corresponding to the eigenvalue 0, since

$$P ((1 - P) | u \rangle) = (P - P^2) | u \rangle = 0 = 0(1 - P) | u \rangle. \qquad (4.63)$$

Now, if we return to the projector P_a on the subspace ξ_a, we notice that $P_a = | a \rangle \langle a |$ is the projector onto the (one-dimensional) space ξ_a whereas $1 - P_a = 1 - | a \rangle \langle a |$ is the projector onto the space $\xi_a^\perp$ orthogonal to ξ_a. The projection of $| u \rangle$ onto ξ_a will hence be noted:

$$| u_a \rangle = P_a | u \rangle = | a \rangle \langle a | u \rangle. \qquad (4.64)$$

Since $\langle a | u \rangle$ represents a number $c = \langle a | u \rangle$, we arrive again at the fact that $| u_a \rangle = c | a \rangle$, i.e., that the projection of any vector $| u \rangle$ whatsoever onto the one-dimensional subspace ξ_a implied by the base vector $| a \rangle$ is necessarily proportional to this vector.

d) Projector onto an N-Dimensional Space

Consider a set of N orthonormal vectors $| e_1 \rangle | e_2 \rangle \ldots | e_N \rangle$ spanning an N-dimensional space ξ_N:

$$\langle e_i | e_j \rangle = \delta_{ij} \qquad \text{orthonormality}. \qquad (4.65)$$

The projector of the space ξ_N will be

$$P_N = \sum_{i=1}^{N} | e_i \rangle \langle e_i |. \qquad (4.66)$$

It is easily demonstrated that relations (4.57) and (4.58) make P_N a projector since $P_N = P_N^+$ and $P_N^2 = P_N$.

e) Three-Dimensional Space

The projector onto a three-dimensional space is written

$$P_3 = \sum_{i=1}^{3} | e_i \rangle \langle e_i |. \tag{4.67}$$

The unit vectors $\vec{e}_1$, $\vec{e}_2$, and $\vec{e}_3$ along the coordinate axes are easily identified with the basis vector $| e_1 \rangle$, $| e_2 \rangle$, and $| e_3 \rangle$ by setting

$$| e_i \rangle = \vec{e}_i \quad (i = 1, 2, 3 = x, y, z). \tag{4.68}$$

Orthonormality relations (4.65) therefore show that the axes Ox_i are orthogonal and that the basis vectors are unit vectors:

$$\langle e_i | e_j \rangle = \vec{e}_i \cdot \vec{e}_j = \delta_{ij} \quad i, j = 1, 2, 3. \tag{4.69}$$

The projector P_3 onto the three-dimensional space will determine the components of any vector whatsoever on each of the coordinates axes. If the vector $\vec{V}$ is decomposed into its components, we have

$$\vec{V} = \sum_i \vec{e}_i X_i = [\vec{e}_1 \ \vec{e}_2 \ \vec{e}_3] \begin{bmatrix} X_1 \\ X_2 \\ X_3 \end{bmatrix}. \tag{4.70}$$

It is clear that the orthonormality of the base vectors $\vec{e}_i$ will yield an expression for the components of the form

$$X_i = \vec{V} \cdot \vec{e}_i. \tag{4.71}$$

Now, let us use the Dirac notation for the Hilbert space vector $| V \rangle$ and set

$$\vec{V} = P_3 | V \rangle = \sum_i | e_i \rangle \langle e_i | V \rangle = \sum_i \vec{e}_i X_i. \tag{4.72}$$

By comparing expressions (4.70) and (4.72), we arrive once more at the X_i components of the vector:

$$X_i = \vec{e}_i \cdot \vec{V} = \langle e_i | V \rangle. \tag{4.73}$$

We further notice that the base vector line constitutes the base S:

$$S = [\vec{e}_1 \ \vec{e}_2 \ \vec{e}_3] = [\ | e_1 \rangle \ | e_2 \rangle \ | e_3 \rangle \ ], \tag{4.74}$$

whereas the elements of the projection column are the components along the coordinates axes:

$$X = \begin{bmatrix} X_1 \\ X_2 \\ X_3 \end{bmatrix} = \begin{bmatrix} \vec{e}_1 \cdot \vec{V} \\ \vec{e}_2 \cdot \vec{V} \\ \vec{e}_3 \cdot \vec{V} \end{bmatrix} = \begin{bmatrix} \langle e_1 | V \rangle \\ \langle e_2 | V \rangle \\ \langle e_3 | V \rangle \end{bmatrix}. \tag{4.75}$$

The projection of the vector $\vec{V}$ onto the base S will yield the column X, that is,

$$V = S\,X. \qquad (4.76)$$

The vector $|\,V\rangle$ is said to be projected in the space ξ_3, that is the ket $|\,V\rangle$ has been represented in the base $\{e_i\}$. The above results can naturally be generalized to an N-dimensional space.

3. Observables

During an observation, the observer measures the value of a dynamic variable at a given instant. The ensuing result is necessarily a real number. It will therefore be assumed that the result of a measurement is none other than the eigenvalue (real value) of a Hermitian operator representing the dynamic variable in question.

If the dynamical system is associated with an eigenstate $|\,a\rangle$ of a dynamic variable A (represented by a Hermitian operator) corresponding to the eigenvalue a, then the measurement of A will necessarily result in the number a and vice versa.

In other words, if $A\,|\,a\rangle = a\,|\,a\rangle$ is a pure state (that is, a non-degenerate state), the eigenvalue a will be obtained in the form

$$\langle A \rangle = a = \frac{\langle a\,|\,A\,|\,a\rangle}{\langle a\,|\,a\rangle}. \qquad (4.77)$$

This represents the measurement of the dynamic variable A. This is the mean value of the Hermitian operator A with only one associated eigenvector $|\,a\rangle$.

If there are several eigenvectors $|\,n\rangle$ corresponding to different eigenvalues a_n and any state whatsoever can be expressed with respect to the states $|\,n\rangle$, then the vector series $|\,n\rangle$ constitutes a complete basis if all projection subspaces taken together cover the Hilbert space of the states:

$$A = A^+ \quad \text{and} \quad A\,|\,n\rangle = a_n\,|\,n\rangle. \qquad (4.78)$$

The orthonormal and linearly independent vectors $|\,n\rangle$ will be mathematically said to form a complete orthonormal basis if they obey the following two fundamental relations:

$$
\begin{array}{ll}
\text{orthonormality} & \langle n\,|\,n'\rangle = \delta_{nn'} \\[2mm]
\text{closure} & \sum_n |\,n\rangle\,\langle n\,| = \mathbb{1}
\end{array}
\qquad (4.79)
$$

The identity operator has been noted $\mathbb{1}$. This is because, when applied to $|\,u\rangle$, it returns us to $|\,u\rangle$.

The completeness relation (4.79) shows that any vector $\mid u \rangle$ can be written as a linear combination of the base vectors $\mid n \rangle$ since

$$\mid u \rangle = \mathbb{1} \mid u \rangle = \sum_n \mid n \rangle \langle n \mid u \rangle = \sum_n c_n \mid n \rangle, \qquad (4.80)$$

with $c_n = \langle n \mid u \rangle$, the projection of $\mid u \rangle$ onto $\mid n \rangle$.

With the fourth postulate of quantum mechanics, the probability of obtaining the eigenvalue a_n will be

$$p_n = \mid \langle n \mid u \rangle \mid^2 = \mid c_n \mid^2,$$

and the expectation value of an operator A will b

$$\langle A \rangle = \sum_n p_n \langle n \mid A \mid n \rangle = \sum_n \mid \langle n \mid u \rangle \mid^2 \langle n \mid A \mid n \rangle. \qquad (4.81)$$

This can be written by expanding the square modulus and using the eigenvalue Eq. (4.78) and the completeness relation (4.79):

$$\langle A \rangle = \langle u \mid A \mid u \rangle. \qquad (4.82)$$

The expectation value of the observable A is independent of the basis used. It depends solely on the current quantum state $\mid u \rangle$ of the observable.

By observable, we mean any physical quantity that can be represented by a Hermitian operator whose projection subspaces yield a sum that covers the entire Hilbert state space. The set of eigenvectors of this Hermitian operator therefore constitutes a complete system.

If the operator A, which is Hermitian ($A = A^+$), has the series $a_1 ... a_n$ of discrete eigenvalues (this reasoning is also tenable if these eigenvalues are continuous), $\xi_1 ... \xi_n$ will be termed subspaces relative to these eigenvalues and $P_1 ... P_n$ the projectors onto the subspaces.

If the eigenvalues a_n are not degenerate, each subspace ξ_n will be a one-dimensional subspace and the projector P_n of the form $\mid n \rangle \langle n \mid$. If the eigenvalues a_n are r-fold degenerate,

$$A \mid n\,r \rangle = a_n \mid n\,r \rangle, \qquad (4.83)$$

the subspaces ξ_n will be r-dimensional and the projectors P_n of the form

$$P_n = \sum_r \mid n\,r \rangle \langle n\,r \mid. \qquad (4.84)$$

The dynamic variable represented by the operator A is an observable if the following conditions are fulfilled by its eigenvectors:

orthonormality	$\langle n\,r \mid n'\,r' \rangle = \delta_{nn'}\,\delta_{rr'}$	
closure	$\sum_{nr} \mid n\,r \rangle \langle n\,r \mid = \mathbb{1}$	(4.85)

The operator A can be expressed in terms of its eigenvectors $\mid n\ r\rangle$ and the corresponding eigenvalues a_n since definition (4.83):

$$A \mid n\ r\rangle = a_n \mid n\ r\rangle$$

multiplied by the conjugate bra gives

$$A \mid n\ r\rangle\ \langle n\ r \mid = a_n \mid n\ r\rangle\ \langle n\ r \mid,$$

and summing over the subscripts n and r yields

$$A \sum_{nr} \mid n\ r\rangle\ \langle n\ r \mid = \sum_{nr} a_n \mid n\ r\rangle\ \langle n\ r \mid.$$

The completeness relation (4.85) thus defines the expression for the operator:

$$A = \sum_{nr} \mid n\ r\rangle\ a_n\ \langle n\ r \mid. \tag{4.86}$$

This form readily leads to the expectation value (4.81):

$$\langle A \rangle = \langle u \mid A \mid u \rangle = \sum_{nr} \langle u \mid n\ r\rangle\ a_n\ \langle n\ r \mid u \rangle = \sum_{n} p_n\ a_n \tag{4.87}$$

with the probability p_n taking the value

$$p_n = \sum_{r} \mid \langle n\ r \mid u \rangle \mid^2. \tag{4.88}$$

Hermitian operators with a continuous spectrum of eigenvalues play an important role in the projection of the Hilbert space onto continuous space, the configuration space or phase space. We will return to this in greater detail in the next chapter. In the meantime, let us simply add that the orthonormality relations (4.85) replace the Kronecker symbol with a Dirac delta function*

---

* The definition of the Dirac delta function is of the form

$$\delta\ (x - a) = \begin{matrix} 0 & \text{if} & x \neq a \\ \infty & \text{if} & x = a \end{matrix} \quad \text{and} \quad \int_{-\infty}^{+\infty} \delta\ (x - a)\ dx = 1$$

and its principal characteristics

$$\int f(x)\ \delta\ (x - a)\ dx = f(a) \qquad \delta\ (a\ x) = \frac{1}{\mid a \mid}\ \delta(x) \qquad x\ \delta(x) = 0$$

The delta function can be written as the Fourier transform:

$$\delta(x) = \frac{1}{2\pi\ \hbar} \int_{-\infty}^{+\infty} e^{-\frac{i}{\hbar}\ p\ x}\ dp$$

and that the summation over the quantum numbers n is replaced with an integration:

$$X = X^+ \text{ and } X \mid x\rangle = x \mid x\rangle,$$
$$\langle x \mid x'\rangle = \delta(x - x'),$$
$$\int \mid x\rangle \, dx \, \langle x \mid = \mathbb{1}. \tag{4.89}$$

If the variable x defines the position of an N-dimensional space, the Dirac function $\delta^{(N)}(x-x')$ will also be defined in the same space, and the integration will involve an element of this space with volume $d^N x$.

4. Complete Set of Compatible Observables

If two observables, A and B, commute, they are said to be compatible observables

$$[A, B] = 0, \tag{4.90}$$

and if $\mid a\rangle$ is the eigenvector of A corresponding to the eigenvalue a, then $B \mid a\rangle$ is also the eigenvector A corresponding to the same eigenvalue a. It should in fact be possible to write

$$0 = (AB - BA) \mid a\rangle = AB \mid a\rangle - BA \mid a\rangle$$
$$= AB \mid a\rangle - aB \mid a\rangle$$

which corresponds exactly to the eigenvalue equation

$$A(B \mid a\rangle) = a(B \mid a\rangle). \tag{4.91a}$$

Similarly we can write

$$B(A \mid b\rangle) = b(A \mid b\rangle), \tag{4.91b}$$

that is, we can construct a vector $\mid ab\rangle$ which is an eigenvector common to the commuting operators A and B:

$$A \mid ab\rangle = a \mid ab\rangle \quad \text{and} \quad B \mid ab\rangle = b \mid ab\rangle, \tag{4.92}$$

and by applying operators A or B, we obtain

$$AB \mid ab\rangle = ab \mid ab\rangle \quad \text{and} \quad BA \mid ab\rangle = ba \mid ab\rangle. \tag{4.93}$$

Because the eigenvalues are real numbers (for Hermitian operators) in the commutation relation, we infer that

$$[A, B] \mid ab\rangle = 0. \tag{4.94}$$

More generally speaking:

A series A, B, C of observables constitutes a complete set of compatible observables if the observables commute two by two. Such a series has one, and only one, common base system. The determination of the Hilbert space is therefore conditioned by the knowledge of the complete set of compatible observables acting in this space.

5. Heisenberg's Uncertainty Relations

Heisenberg's uncertainty relations derive directly from the correspondence principle, that is, the commutation relations between two canonically conjugate variables A and B, and from Schwartz' inequality inherent in the Hilbert state space in which these operators are defined.

Consider two canonically conjugate operators A and B, i.e., variables satisfying the commutation relation

$$[A, B] = i\,\hbar. \tag{4.95}$$

Let us introduce the standard deviation into the measurement of these observables:

$$\hat{A} = A - \langle A \rangle \quad \text{and} \quad \hat{B} = B - \langle B \rangle \tag{4.96}$$

which yields the same commutation relation between $\hat{A}$ and $\hat{B}$ since $\langle A \rangle$ and $\langle B \rangle$ are expectation values and, hence, scalar quantities commuting with the operators A and B:

$$[\hat{A}, \hat{B}] = i\,\hbar. \tag{4.97}$$

The mean square deviations are the expectation values of the squares of the standard deviations and, hence, will be written as follows:

$$\begin{aligned}
(\Delta A)^2 &= \langle \hat{A}^2 \rangle = \langle A^2 \rangle - \langle A \rangle^2, \\
(\Delta B)^2 &= \langle \hat{B}^2 \rangle = \langle B^2 \rangle - \langle B \rangle^2.
\end{aligned} \tag{4.98}$$

Now, let us apply Schwartz' inequality (4.17) to the vectors $\mid x \rangle = \hat{A} \mid u \rangle$ and $\mid y \rangle = \hat{B} \mid u \rangle$ of the Hilbert space:

$$\mid \langle u \mid \hat{A}\,\hat{B} \mid u \rangle \mid^2 \leq \langle u \mid \hat{A}^2 \mid u \rangle \langle u \mid \hat{B}^2 \mid u \rangle, \tag{4.99}$$

or, by introducing the expectation values with relation (4.82),

$$\mid \langle \hat{A}\,\hat{B} \rangle \mid^2 \leq \langle \hat{A}^2 \rangle \langle \hat{B}^2 \rangle. \tag{4.100}$$

Let us separate the product $\hat{A}\,\hat{B}$ into a Hermitian part $\frac{1}{2}(\hat{A}\,\hat{B} + \hat{B}\,\hat{A})$ and an anti-Hermitian part $\frac{1}{2}(\hat{A}\,\hat{B} - \hat{B}\,\hat{A}) = \frac{1}{2}i\,\hbar$ and then expand its square modulus:

$$\left| \left\langle \frac{\hat{A}\,\hat{B} + \hat{B}\,\hat{A}}{2} + \frac{i\,\hbar}{2} \right\rangle \right|^2 = \left| \frac{\langle \hat{A}\,\hat{B} + \hat{B}\,\hat{A} \rangle}{2} \right|^2 + \frac{\hbar^2}{4}. \tag{4.101}$$

Using (4.98) and (4.101), Schwartz' inequality (4.100) then reduces to

$$\left| \frac{\langle \hat{A}\,\hat{B} + \hat{B}\,\hat{A} \rangle}{2} \right|^2 + \frac{\hbar^2}{4} \leq (\Delta A)^2 (\Delta B)^2, \tag{4.102}$$

and because the sum of the two positive terms is necessarily greater than each of them, we obtain the following inequality:

$$(\Delta A)^2 \ (\Delta B)^2 \ \geq \ \frac{\hbar^2}{4}.$$
(4.103)

This leads to Heisenberg's uncertainty relation:

$$\boxed{\Delta A \ \Delta B \ \geq \ \frac{\hbar}{2}}$$
(4.104)

which is simply the mathematical formulation of Schwartz' inequality, as obeyed by two canonically conjugate operators, A and B.

This is the case, for example, of the position operator x_i and the momentum operator p_j associated with the position and momentum observables. They satisfy the commutation relation $[x_i, \ p_j] = i \ \hbar \ \delta_{ij}$. This leads to Heisenberg' uncertainty relations of the following type:

$$\Delta x \ \Delta p_x \ \geq \ \frac{\hbar}{2}.$$
(4.105)

Now, let us return briefly to a point made in Chap. 1, i.e., that the time and energy variables, t and E respectively, which are not Hermitian operators and, hence not observables, satisfy the commutation relation $[t, E] = i \ \hbar$. The energy E is the eigenvalue of the Hermitian operator H, whereas time is a parameter introduced a priori to describe the evolution of the system. Is this time the proper time of the reference frame associated with the observer or the proper time associated with the reference frame of the particle in motion? Yet, reference is constantly made to the uncertainty relation (see Chap. 2)

$$\Delta E \ \Delta t \ \simeq \ \hbar$$

linking the width of a wave packet with an energy uncertainty ΔE to its mean lifetime Δt. We will return to this point in the discussion on time-dependent perturbations and give the answer to the above question in the next chapter, in connection with the correspondence principle.

In conclusion then, the Dirac formalism of quantum mechanics assumes that a quantum state, an abstract mathematical entity or a basic physical entity, is a vector of a state space, namely, the Hilbert space, whereas a physical measurement is associated with the eigenvalue of a Hermitian operator acting in this space. To obtain these special operators, we use a correspondence principle between the Poisson' brackets of the classical dynamic variables and the commutator of their associated operators. Does the quantum state change with time or is it immutable? Is there just one quantum state, decomposable into quantum sub-systems or is every classical physical system (arbitrarily delimited by the observer) associated with a special quantum state of the Hilbert space? Are the operators acting in the Hilbert space immutable in time or do they evolve with time? The correspondence principle might provide an answer to this question since the Poisson dynamical law allows direct description of the evolution in time of operators associated with classical dynamic variables even if it says nothing about the evolution of quantum states. These questions no doubt raise the issue of the interpretation of quantum mechanics in its different representations as well as that of (classical) measurements of the observables of a quantum system. These issues will be examined in Chaps. 6 and 7, which are devoted to quantum dynamics and path integrals, respectively.

6. Dirac Formalism – Schematic Summary

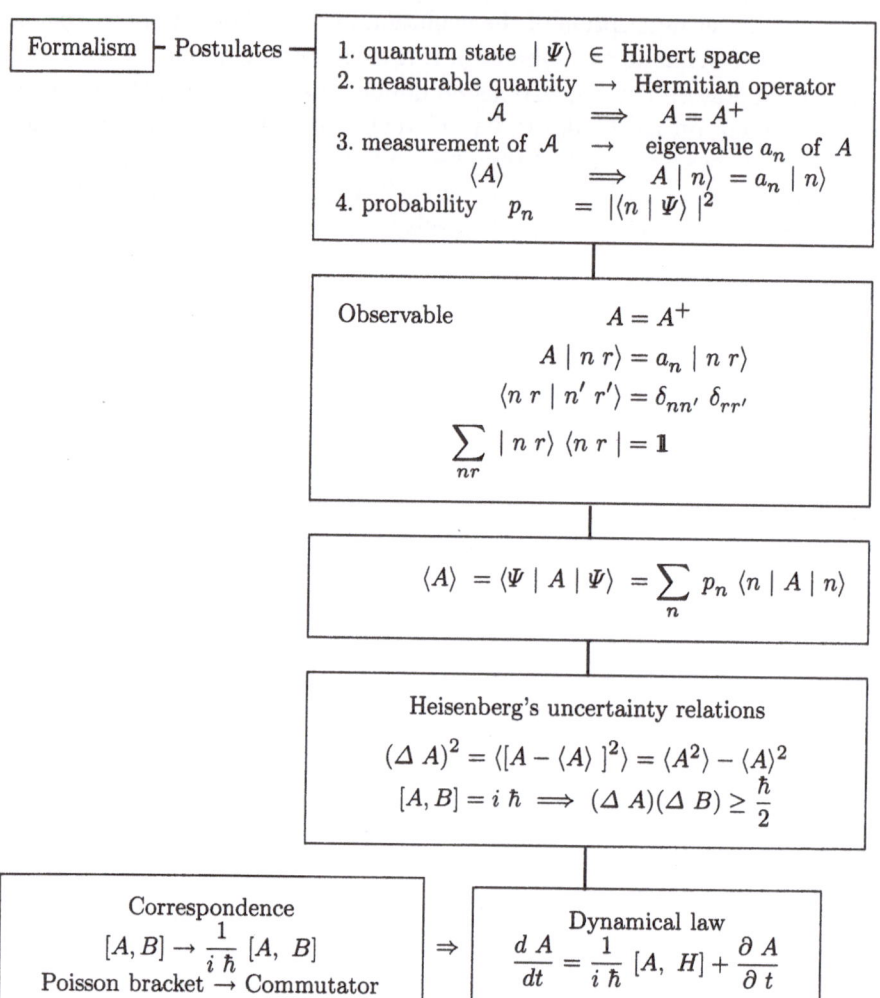

Formalism — Postulates —

1. quantum state $|\Psi\rangle \in$ Hilbert space
2. measurable quantity $\rightarrow$ Hermitian operator
$$\mathcal{A} \qquad \Longrightarrow \qquad A = A^+$$
3. measurement of $\mathcal{A}$ $\rightarrow$ eigenvalue a_n of A
$$\langle A \rangle \qquad \Longrightarrow \qquad A \mid n\rangle = a_n \mid n\rangle$$
4. probability $p_n = |\langle n \mid \Psi \rangle|^2$

Observable $\qquad\qquad A = A^+$
$$A \mid n\,r\rangle = a_n \mid n\,r\rangle$$
$$\langle n\,r \mid n'\,r'\rangle = \delta_{nn'}\,\delta_{rr'}$$
$$\sum_{nr} \mid n\,r\rangle \langle n\,r \mid = \mathbf{1}$$

$$\langle A \rangle = \langle \Psi \mid A \mid \Psi \rangle = \sum_n p_n \langle n \mid A \mid n \rangle$$

Heisenberg's uncertainty relations
$$(\Delta A)^2 = \langle [A - \langle A \rangle]^2 \rangle = \langle A^2 \rangle - \langle A \rangle^2$$
$$[A, B] = i\,\hbar \implies (\Delta A)(\Delta B) \geq \frac{\hbar}{2}$$

Correspondence
$$[A, B] \rightarrow \frac{1}{i\,\hbar} [A, B]$$
Poisson bracket $\rightarrow$ Commutator

$\Rightarrow$

Dynamical law
$$\frac{d A}{dt} = \frac{1}{i\,\hbar} [A, H] + \frac{\partial A}{\partial t}$$

Chapter 5
Representations

The space of quantum states, Hilbert space, is an abstract physical space. The observer and the measuring apparatus are defined in a concrete geometrical space. Hence it is necessary to give a concrete representation of the quantum states and the operators acting on these states in order to obtain unambiguous interpretable measurement results. The possible formulations are termed Representations of Quantum Mechanics. Two of them stem directly from Dirac formalism: Wave mechanics, first introduced by de Broglie and Schrödinger, and matrix mechanics called as such by Heisenberg and Jordan. These two representations lead to completely different interpretations of the actual process of physical measurement and of the transition from quantum to classical mechanics.

1. Wave Mechanics

1.1 {X} Representation (Configuration Space)

To simplify the problem at hand, let us consider the case of a particle with mass m moving in the direction Ox.

The position x of the particle will be the eigenvalue of the Hermitian position operator X (we will use the upper case to distinguish the operator from its eigenvalue). The spectrum of the eigenvalues of X is therefore real and continuous and extends from $-\infty$ to $+\infty$. The eigenvectors $\mid x \rangle$ of X will be orthonormalized and, because the operator X is an observable, they will form a complete system and thus define the $\{X\}$ representation of quantum mechanics:

$$\{X\} \quad \text{representation} \quad \left\{ \begin{array}{cr} X = X^+ & \\ X \mid x \rangle = x \mid x \rangle & (5.1) \\ \langle x' \mid x \rangle = \delta\left(x' - x\right) & (5.2) \\ \int \mid x \rangle \, dx \, \langle x \mid = \mathbb{1}. & (5.3) \end{array} \right.$$

A quantum state $|\,a\rangle$ can be expanded within this representation by projection onto the $\{X\}$ representation, the configuration space:

$$|\,a\rangle \,=\, \int |\,x\rangle \, d\,x \, \langle x\,|\,a\rangle. \qquad (5.4)$$

The projection will be written as follows:

$$\langle x\,|\,a\rangle \,=\, \Psi_a(x) \qquad (5.5)$$

emphasizing the fact that $\langle x\,|\,a\rangle$ does not represent an inner product since $\langle x\,|$ and $|\,a\rangle$ do not belong to dual spaces. It is rather the projection of the quantum state $|\,a\rangle$ onto the one-dimensional configuration space. A quantum state $|\,a\rangle$ will be expressed in the basis $\{X\}$ by an integral:

$$|\,a\rangle \,=\, \int |\,x\rangle \, d\,x \, \Psi_a(x). \qquad (5.6)$$

Using definition (5.5) of the function $\Psi_a(x)$, the Dirac formalism brings us back to the notation of the wave mechanical function space.

Thus, the inner product of two states will be

$$\langle b\,|\,a\rangle \,=\, \int \langle b\,|\,x\rangle \, d\,x \, \langle x\,|\,a\rangle \,=\, \int \Psi_b^*(x) \, \Psi_a(x) \, d\,x. \qquad (5.7)$$

The norm of a function $\Psi_a(x)$ will, in turn, be written

$$\langle a\,|\,a\rangle \,=\, \int \langle a\,|\,x\rangle \, d\,x \, \langle x\,|\,a\rangle \,=\, \int \Psi_a^*(x) \, \Psi_a(x) \, d\,x. \qquad (5.8)$$

By multiplying definition (5.1) by the bra $\langle x'\,|$ and using orthonormality (5.2), we obtain

$$\langle x'\,|\,X\,|\,x\rangle \,=\, \langle x'\,|\,x\,|\,x\rangle \,=\, x \, \langle x'\,|\,x\rangle$$
$$=\, x \, \delta\,(x'-x). \qquad (5.9)$$

This relation holds for every function of the position operator.

Every operator function is diagonal in its eigenrepresentation.

$$\boxed{\langle x'\,|\,f(X)\,|\,x\rangle \,=\, \delta\,(x'-x)\,f(x)}. \qquad (5.10)$$

The result of a measurement i.e., the expectation value of the Hermitian operator A, can therefore be expressed with the normalized functions $\Psi_a(x)$:

$$a \,=\, \langle A\rangle \,=\, \langle a\,|\,A\,|\,a\rangle \,=\, \int \langle a\,|\,x\rangle \, d\,x \, \langle x\,|\,A\,|\,x'\rangle \, d\,x' \, \langle x'\,|\,a\rangle$$

and by using relation (5.10) for the observable A expressed with only operators of the configuration space X:

$$\langle x \mid A \mid x' \rangle = \delta (x' - x) A (x). \tag{5.11}$$

With the characteristic property of a delta function:

$$\int \delta (x' - x) f (x') \, d x' = f (x), \tag{5.12}$$

we obtain the following well-known result of wave mechanics, with the operator $A (x)$ acting on the function $\Psi_a (x)$:

$$a = \langle A \rangle = \int \Psi_a^* (x) A (x) \Psi_a (x) \, d x. \tag{5.13}$$

We draw attention here to the problem (Chap. 2, Sect. 3) involved in writing $\langle A \rangle$ in the function space by assigning the meaning of a probability density to the square modulus of the wave function:

$$\varrho (x) = \mid \Psi_a (x) \mid^2 \quad \text{and} \quad \int \varrho (x) \, d x = 1. \tag{5.14}$$

This problem is resolved since relation (5.13) shows that $\langle A \rangle$ cannot be written as follows:

$$\int \varrho (x) A (x) \, d x \tag{5.15}$$

but must be written as in (5.13).

It is, however, necessary that the function $\Psi_a(x)$ thus introduced be the wave function $\Psi_a(x)$ of wave mechanics, that is, it will have to be shown that the function $\Psi_a(x)$ satisfies the Schrödinger equation.

1.2 Translation Operator

Consider the operator D_ξ translating the coordinate x into $x + \xi$:

$$\boxed{D_\xi \mid x \rangle = \mid x + \xi \rangle}. \tag{5.16}$$

It is easily shown, by applying this operator to the state $\mid x \rangle$, that

$$D_\eta D_\xi = D_\xi D_\eta = D_{\eta + \xi}, \tag{5.17}$$

and, by carrying out n successive translations, that

$$(D_\xi)^n = D_{n \, \xi}. \tag{5.18}$$

The operator D_ξ should conserve the orthonormality and closure relations and, hence, be a unitary operator. This operator can be written with the

Hermitian operator $A^+ (\xi) = A (\xi)$:

$$D_\xi = e^{i\, A\, (\xi)}. \tag{5.19}$$

Relation (5.18) leads us to set

$$n\, A\, (\xi) = A\, (n\, \xi), \tag{5.20}$$

showing in this way that the function $A\, (\xi)$ is a first-order homogenous function.

The Hermitian operator K, generator of the infinitesimal translations satisfying relation (5.18), is defined through

$$D_\xi = e^{-i\,\xi\, K}, \quad \text{that is}, \quad e^{-i\,K\,\xi}\, |\, x \rangle = |\, x + \xi \rangle. \tag{5.21}$$

We will then note that the Hermitian operator K makes D_ξ a unitary operator and

$$D_\xi^+ = D_\xi^{-1} = D_{-\xi}, \tag{5.22}$$

such that by transposing definition (5.16):

$$\langle x\, |\, D_\xi^+ = \langle x + \xi\, |, \tag{5.23}$$

or by replacing D_ξ^+ with $D_{-\xi}$, we obtain

$$\langle x\, |\, D_{-\xi} = \langle x + \xi\, |.$$

We only have to change ξ into $-\xi$ to obtain the action of D_ξ on the bra $\langle x\, |$:

$$\boxed{\langle x\, |\, D_\xi = \langle x - \xi\, |}. \tag{5.24}$$

The action of D_ξ on the quantum state $|\, a \rangle$ will be obtained by projecting this quantum state onto the space $\{X\}$, since D_ξ acts on the variable x simply to translate it by the real quantity ξ:

$$e^{-i\, K\, \xi}\, |\, a \rangle = e^{-i\, K\, \xi} \int |\, x' \rangle\, dx'\, \langle x'\, |\, a \rangle$$

$$= \int |\, x' + \xi \rangle\, dx'\, \langle x'\, |\, a \rangle. \tag{5.25}$$

By changing variables with $x' + \xi = x$, the foregoing equation becomes

$$e^{-i\, K\, \xi}\, |\, a \rangle = \int |\, x \rangle\, dx\, \langle x - \xi\, |\, a \rangle$$

$$= \int |\, x \rangle\, dx\, \psi_a\, (x - \xi). \tag{5.26}$$

Let us make a series expansion of the exponential function as well as the function $\psi_a(x - \xi)$ around the value x:

$$\sum_n \frac{(-i K \xi)^n}{n!} \mid a \rangle = \int \mid x \rangle \, dx \sum_n \frac{(-\xi)^n}{n!} \left(\frac{\partial}{\partial x} \right)^n \Psi_a(x). \qquad (5.27)$$

By equating similar powers of ξ, we obtain the fundamental relation:

$$K^n \mid a \rangle = \int \mid x \rangle \, dx \left(\frac{1}{i} \frac{\partial}{\partial x} \right)^n \Psi_a(x) \,. \qquad (5.28)$$

It then becomes easy to determine the commutator of the Hermitian operators X and K. We first observe that

$$X K \mid a \rangle = X \int \mid x' \rangle \, dx' \frac{1}{i} \frac{\partial}{\partial x'} \Psi_a(x')$$

$$= \int x' \mid x' \rangle \, dx' \frac{1}{i} \frac{\partial}{\partial x'} \Psi_a(x'). \qquad (5.29)$$

On the other hand, the action of $K X$ on the ket $\mid a \rangle$ is calculated using (5.11), thus yielding:

$$K X \mid a \rangle = \int K \mid x \rangle \, dx \, \langle x \mid X \mid x' \rangle \, dx' \, \langle x' \mid a \rangle$$

$$= \int K \mid x \rangle \, dx \, x \, \Psi_a(x). \qquad (5.30)$$

By applying fundamental relation (5.28), we obtain for $a = x$ and $n = 1$:

$$K \mid x \rangle = \int \mid x' \rangle \, dx' \frac{1}{i} \frac{\partial}{\partial x'} \delta \, (x' - x) \qquad (5.31)$$

and by inserting this result into the preceding relation:

$$K X \mid a \rangle = \int \mid x \rangle \, dx \frac{1}{i} \frac{\partial}{\partial x} x \, \Psi_a(x)$$

$$= \int \mid x \rangle \, dx \frac{1}{i} \Psi_a(x) + \int x \mid x \rangle \, dx \frac{1}{i} \frac{\partial}{\partial x} \Psi_a(x). \qquad (5.32)$$

The term-by-term difference between eqs. (5.32) and (5.29) thus gives the commutator $[X, K]$ applied to the quantum state $\mid a \rangle$:

$$[X, K] \mid a \rangle = -\frac{1}{i} \int \mid x' \rangle \, dx' \, \Psi_a(x') = i \int \mid x' \rangle \, dx' \, \langle x' \mid a \rangle$$

$$= i \mid a \rangle \,. \qquad (5.33)$$

Because the relation must be valid irrespective of the quantum state $| a \rangle$, we infer from it the value of the commutator:

$$\boxed{[X, K] = i}$$. $\qquad$ (5.34)

If we compare this value with the commutator resulting from the correspondence principle $[X, p_x] = i\,\hbar$, we notice that the generator of infinitesimal translations is, to within $\hbar$, the particle's momentum:

$$\boxed{p_x = \hbar\, K_x}$$. $\qquad$ (5.35)

We notice that $p_x = -i\,\hbar\,\partial_x$ implies that $K = -i\,\partial_x$ and the translation operator will take the following form in the one-dimensional space:

$$\boxed{D_\xi = e^{-i\,\xi\,K} = e^{-\xi\,\partial_x}}$$. $\qquad$ (5.36)

By applying this result to a wave function $\Psi_a(x) = \langle x \mid a \rangle$, we obtain

$$D_\xi\,\psi_a(x) = e^{-\xi\,\partial_x}\,\Psi_a(x) = \left(1 - \xi\,\frac{\partial}{\partial x} + \frac{\xi^2}{2!}\left(\frac{\partial}{\partial x}\right)^2 + \dots\right)\psi_a(x)$$

$$= \Psi_a(x - \xi) \qquad (5.37)$$

The series expansion of the exponential function in fact yields a Taylor expansion of the function $\Psi_a(x - \xi)$ around the point x. In other words, we can write

$$D_\xi\,\langle x \mid a \rangle = \langle x \mid D_\xi \mid a \rangle = \langle x - \xi \mid a \rangle . \qquad (5.38)$$

Because this relation must hold irrespective of $| a \rangle$, we are left with the result (5.24):

$$\langle x \mid D_\xi = \langle x - \xi \mid . \qquad (5.39)$$

1.3 Time-Independent Schrödinger Equation

Let us use as departure point the eigenvalue equation of the Hamiltonian:

$$H \mid a \rangle = E \mid a \rangle , \qquad (5.40)$$

in which the operator H is the classical Hamiltonian

$$H = \frac{P^2}{2m} + V = \frac{\hbar^2}{2m}\,K^2 + V(x) . \qquad (5.41)$$

Now, let us project (5.40) onto the configuration space:

$$\langle x \mid H \mid a \rangle = E \langle x \mid a \rangle = E\,\Psi_a(x) . \qquad (5.42)$$

By using form (5.41) of the Hamiltonian, this yields:

$$\langle x \mid H \mid a \rangle = \frac{\hbar^2}{2m} \langle x \mid K^2 \mid a \rangle + \langle x \mid V \mid a \rangle . \qquad (5.43)$$

The form (5.40) presupposes that we are in a stationary energy state for a time-independent potential that depends only on the position variable x:

$$\langle x \mid V \mid a \rangle = \int \langle x \mid V(X) \mid x' \rangle \, dx' \, \langle x' \mid a \rangle$$

$$= \int V(x') \, \delta(x' - x) \, dx' \, \Psi_a(x')$$

$$= V(x) \, \Psi_a(x) . \qquad (5.44)$$

Let us further calculate the term in $\langle x \mid K^2 \mid a \rangle$ of (5.43) using the fundamental relation (5.28) for $n = 2$:

$$\langle x \mid K^2 \mid a \rangle = \int \langle x \mid x' \rangle \, dx' \left(\frac{1}{i} \frac{\partial}{\partial x'} \right)^2 \Psi_a(x')$$

$$= \int \delta(x - x') \, dx' \left(-\frac{d^2}{d x'^2} \right) \Psi_a(x')$$

$$= -\frac{d^2}{d x^2} \, \Psi_a(x) . \qquad (5.45)$$

By inserting (5.44) and (5.45) into the projection of the vector $H \mid a \rangle$ onto the configuration space:

$$\langle x \mid H \mid a \rangle = \left(-\frac{\hbar^2}{2m} \frac{d^2}{d x^2} + V(x) \right) \Psi_a(x) , \qquad (5.46)$$

and by equating this with (5.42), we regain the time-independent Schrödinger equation

$$\left[-\frac{\hbar^2}{2m} \frac{d^2}{d x^2} + V(x) \right] \Psi_a(x) = E \, \Psi_a(x) . \qquad (5.47)$$

The Schrödinger equation is the expression of the projection of the quantum state (or quantum field) in the configuration space at a given point and at a particular instant. It corresponds to the projection of the quantum state $\mid a \rangle$ onto the observer space $\mid \vec{r} \rangle$ at the instant t.

1.4 The {P} Representation

The wave functions can also be defined in the {P} representation, that is, in phase space.

The definition of the {P} representation is analogous to that of the {X} representation and, considering that the Hermitian operator $P = \hbar K$, the definition of the $\{P\}$ representation will take the form

$$\{P\} \quad \text{representation} \qquad \left\{ \begin{array}{l} K \mid k\rangle = k \mid k\rangle \\[2mm] \langle k \mid k'\rangle = \delta\,(k - k') \\[2mm] \displaystyle\int \mid k\rangle \; dk \; \langle k \mid \; = \mathbb{1} \,. \end{array} \right. \qquad (5.48)$$

If we place ourselves in a one-dimensional phase space as we should, the projection of the quantum state $\mid a\rangle$ on this space will define the wave function

$$\phi_a(k) = \langle k \mid a\rangle \,. \qquad (5.49)$$

We must now seek the relation between the wave functions $\Psi_a(x)$ of the configuration space and $\phi_a(k)$ of the phase space.

If we set $a = k'$ in the fundamental relation (5.28), we obtain

$$K \mid k'\rangle = k' \mid k'\rangle = \frac{1}{i} \int \mid x'\rangle \; dx' \; \frac{\partial}{\partial x'} \; \langle x' \mid k'\rangle \,. \qquad (5.50)$$

By multiplying the second and third expressions by the bra $\langle x'' \mid$:

$$k' \, \langle x'' \mid k'\rangle = \frac{1}{i} \int \langle x'' \mid x'\rangle \; dx' \; \frac{\partial}{\partial x'} \; \langle x' \mid k'\rangle \,, \qquad (5.51)$$

and because of the orthonormality of the position vectors, we are left with

$$k' \, \langle x' \mid k'\rangle = \frac{1}{i} \frac{d}{dx'} \; \langle x' \mid k'\rangle \,. \qquad (5.52)$$

This first-order differential equation with separate variables has the following solution:

$$\langle x' \mid k'\rangle = g(k') \; e^{i\,k'\,x'} \,. \qquad (5.53)$$

The normalization of the vectors $\mid k\rangle$ determines that the function $g(k')$ can be defined thus:

$$\begin{aligned} \langle k'' \mid k'\rangle &= \delta(k'' - k') = \int \langle k'' \mid x'\rangle \; dx' \; \langle x' \mid k'\rangle \\[2mm] &= \int g^*(k'') \; e^{-i\,k''\,x'} \; dx' \; g(k') \; e^{i\,k'\,x'} \\[2mm] &= g^*(k'') \; g(k') \int e^{i(k'-k'')\,x'} \; dx' \,. \end{aligned} \qquad (5.54)$$

Hence, the relation:

$$\delta(k'' - k') = g^*(k'') \; g(k') \int e^{i\,(k'-k'')\,x'} \; dx' \,. \qquad (5.55)$$

The integral is the unit Fourier transform and, therefore, the Dirac delta function:

$$\int \exp\left[i(k' - k'')\, x'\right]\, dx' = 2\pi\, \delta\,(k' - k'')\,.$$
(5.56)

This determines the function $g(k')$ to be

$$g(k') = \frac{1}{\sqrt{2\pi}}\, e^{i\,\alpha(k')}\,,$$
(5.57)

although, for the sake of simplicity, we choose the zero phase and thus obtain

$$\langle x' \mid k' \rangle = \frac{1}{\sqrt{2\pi}}\, \exp\,(i\,k'\,x')$$

and

$$\langle x' \mid p' \rangle = \frac{1}{\sqrt{2\pi\,\hbar}}\, \exp\,(\frac{i}{\hbar}\,p'\,x')$$
(5.58)

It is easy then to obtain the relation between $\phi_a(k)$ and $\Psi_a(x)$ by simply projecting one space onto the other, for example

$$\phi_a(k) = \langle k \mid a \rangle = \int\, \langle k \mid x \rangle\, dx\, \langle x \mid a \rangle$$

$$\phi_a(k) = \frac{1}{\sqrt{2\pi}}\, \int\, e^{-i\,k\,x}\, \Psi_a(x)\, dx\,.$$
(5.59)

The wave function of the phase space is therefore the Fourier transform of the wave function in the configuration space. It is a result we have already alluded to in the course of discussing the interpretation of Heisenberg's uncertainty relation which assigns to a particle the meaning of a wave packet.

If we term Q the Hermitian operator defining the eigenvalue q as a position variable, then $Qu(q, x) = q\, u(q, x)$ is solved for any arbitrary real value of q if the corresponding eigenfunction is

$$u\,(q, x) = \delta\,(x - q)\,.$$
(5.60)

If $P = -i\,\hbar\,\partial_x$, then the eigenvalue equation

$$P\, v(p,\, x) = p\, v(p,\, x)$$
(5.61)

is solved for any arbitrary real value of p if the eigenfunction is the plane wave

$$v\,(p,\, x) = \exp\,\left(\frac{i}{\hbar}\,p\,x\right)\,.$$
(5.62)

This gives an indefinite position for a fixed momentum or, conversely, a fixed momentum for an indefinite position.

The Schrödinger equation may be easily written in phase space by projecting the energy eigenvalue equation onto k, instead of x:

$$\langle k \mid H \mid a \rangle = E \langle k \mid a \rangle = E\,\phi_a(k)$$

$$= \langle k \mid \frac{\hbar^2}{2m} K^2 \mid a \rangle + \langle k \mid V \mid a \rangle. \tag{5.63}$$

But $\langle k \mid K^2 = \langle k \mid k^2$ and the first term of the preceding equation immediately yields

$$\frac{\hbar^2}{2m} k^2 \langle k \mid a \rangle = \frac{\hbar^2}{2m} k^2\, \phi_a(k). \tag{5.64}$$

Because the potential V is diagonal in the $\{X\}$ representation, the last term in (5.63) will have to be modified as follows:

$$\langle k \mid V \mid a \rangle = \int \langle k \mid x \rangle dx \langle x \mid V \mid x' \rangle dx' \langle x' \mid k' \rangle dk' \langle k' \mid a \rangle$$

$$= \frac{1}{2\pi} \int e^{-i k x}\, V(x)\, \delta\,(x - x')\, dx'\, e^{i k' x'}\, dk'\, \phi_a(k')$$

$$= \frac{1}{2\pi} \int\int e^{i(k'-k) x}\, V(x)\, \phi_a(k')\, dk'. \tag{5.65}$$

If we introduce the Fourier transform of the potential $V(x)$:

$$v\,(k' - k) = \frac{1}{\sqrt{2\pi}} \int V(x)\, e^{i(k'-k) x}\, dx, \tag{5.66}$$

we obtain the projection of the vector $V \mid a \rangle$ onto the phase space:

$$\langle k \mid V \mid a \rangle = \frac{1}{\sqrt{2\pi}} \int v\,(k' - k)\, \phi_a(k')\, dk'. \tag{5.67}$$

In the $\{P\}$ representation, the Schrödinger equation is no longer a differential equation but rather an integral equation:

$$\frac{\hbar^2}{2m} k^2\, \phi_a(k) + \frac{1}{\sqrt{2\pi}} \int v\,(k' - k)\, \phi_a(k')\, dk' = E\,\phi_a(k). \tag{5.68}$$

1.5 The Klein–Gordon Equation

The projection of the quantum state $\mid a \rangle$ onto the "observer's quantum state", determined by the latter's position at instant t in the vector of the observer's state $\mid x,\ t \rangle$ determines the system's wave function, accessible to the observer through its square modulus, that is, the probability density of the presence of the quantum state $\mid a \rangle$ at the point $(x,\ t)$. The time entering the wave function $\Psi_a(x,\ t) = \langle x\ t \mid a \rangle$ is therefore the proper time for the observer-related reference frame.

Let us use the position x^μ (defined in $c = 1$ natural units, that, is with the coordinates $x^\mu = (x^0,\ x^i) = (t,\ x^i)$), to introduce the space-time relating to the observer's eigenvalues of the observer's Hermitian position operator in the four-dimensional Minkowski space-time with metric $\eta_{\mu\nu} = (-+++)$. The configuration space-time will be defined by the usual relations:

$$X^\mu \mid x^\mu\rangle = x^\mu \mid x^\mu\rangle \quad \mu = 0, 1, 2, 3\,,$$

$$\langle x^\mu \mid x^{\mu'}\rangle = \delta^4\left(x^\mu - x^{\mu'}\right)\,, \tag{5.69}$$

$$\int \mid x^\mu\rangle\, d^4 x\, \langle x^\mu \mid\, = \mathbb{1}\,.$$

The observer's proper time is now an observable and the correspondence principle associates it with an energy conjugate variable, the eigenvalue p^0 of the operator P^μ canonically conjugate to X^μ:

$$P^\mu \mid p^\mu\rangle = p^\mu \mid p^\mu\rangle \quad \text{with} \quad p^\mu = (p^0,\ p^i) = (E,\ p^i)\,. \tag{5.70}$$

The correspondence principle next associates the configuration space operator ∂^μ with the energy-momentum operator P^μ:

$$P^\mu = -i\, \partial^\mu = -i\, \eta^{\mu\nu}\, \partial_\nu = \left(i\, \partial_t,\ -i\, \vec{\nabla}\right)\,, \tag{5.71}$$

whereas the conjugate Hermitian will yield

$$P_\mu = -i\, \partial_\mu = \left(-i\, \partial_t, -i\, \vec{\nabla}\right)\,, \tag{5.72}$$

thus generalizing the commutation relations and legitimizing the relation between conjugate variables x^0 and p^0, i.e., t and E:

$$[X^\mu, P_\nu] = -i\, \delta^\mu_\nu\,, \tag{5.73}$$

while noting that $x^0 = t$, $x_0 = -t$, $p^0 = E$ and $p_0 = -E$ whereas $p^i = p_i$.

Because the second postulate of quantum mechanics associates the operator $-i\, \partial^\mu$ with the energy-momentum four-vector P^μ, the eigenvalue equation can, for a spinless particle, be written

$$P^2 \mid a\rangle = -m^2 \mid a\rangle\,, \tag{5.74}$$

where $-m^2$ is a real constant, an eigenvalue of the operator $P^2 = P^\mu P_\mu$. This is an eigenvalue equation corresponding to the following classical relation in relativistic mechanics:

$$0 = P_\mu P^\mu + m^2 = P_0 P^0 + P_i P^i + m^2 = -E^2 + \vec{p}^{\,2} + m^2\,, \tag{5.75}$$

that is, the well-known Einstein equation

$$E^2 = \vec{p}^{\,2} + m^2\,. \tag{5.76}$$

For a spinless particle in an electromagnetic field, this equation is written with the kinematic momentum $\vec{\pi} = \vec{p} - e\, \vec{A}$ and the scalar potential Φ:

$$(E - e\, \Phi)^2 = \pi^2 + m^2\,. \tag{5.77}$$

For nonrelativistic particles, the energy E comprises the kinetic energy and the mass energy since

$$E = m \left(\frac{p^2}{m^2} + 1 \right)^{1/2} \simeq m + \frac{p^2}{2m}. \tag{5.78}$$

Using the correspondence principle, the projection of the eigenvalue equation onto the configuration space x^μ will give

$$\langle x^\mu \mid P^2 \mid a \rangle = \langle x^\mu \mid P_\mu P^\mu \mid a \rangle = (-i\, \partial_\mu)(-i\, \partial^\mu) \langle x^\mu \mid a \rangle. \tag{5.79}$$

And if we use the d'Alembertian expression of the Minkowski space:

$$\partial_\mu\, \partial^\mu = \Box = (-\partial_t^2 + \partial_i^2) = \left(-\frac{\partial^2}{\partial\, t^2} \right) + (\nabla^2)\,, \tag{5.80}$$

the projection onto the configuration space of the quantum state associated with a spinless particle will lead to the Klein–Gordon propagation equation:

$$(\Box - m^2)\, \Psi_a(x^\mu) = 0\,. \tag{5.81}$$

This is the propagation equation of a free scalar field with mass m.

An interaction with an external potential V boils down to adding the term V to the classical definition (5.75):

$$V = P_\mu\, P^\mu + m^2 \tag{5.82}$$

and allowing the associated operators to act on the quantum state $\mid a \rangle$ of the particle:

$$(P_\mu\, P^\mu + m^2)\, \mid a \rangle = V \mid a \rangle\,. \tag{5.83}$$

This leads to the Klein–Gordon equation with interaction:

$$(\Box - m^2)\, \Psi_a(x^\mu) = V(x^\mu)\, \Psi_a(x^\mu)\,, \tag{5.84}$$

which introduces the current

$$J(x^\mu) = V(x^\mu)\, \Psi_a(x^\mu) \tag{5.85}$$

in the second term. In this way, a particle with mass m and zero spin is described exclusively by the scalar field $\Psi_a(x^\mu)$ satisfying the Klein–Gordon propagation equation.

The non-relativistic approximation defines the total energy of a free particle via relation (5.78) and the application of the operators associated with the quantum state $\mid a \rangle$ will lead to the equation

$$i\, \hbar\, \partial_t\, \Psi_a(x) = m\, \Psi_a(x) - \frac{\hbar^2}{2m}\, \nabla^2\, \Psi_a(x)\,. \tag{5.86}$$

Notice that the constant term $m\, \Psi_a(x)$ does not feature in the probability current $\vec{J}$ defined in (2.29).

The wave function $\psi_a(x^\mu)$ emerges then as a scalar field associated with the observation in x^μ of the quantum state $\mid a\rangle$ of a particle (or a set of particles). This scalar field uses the fourth postulate of quantum mechanics to define the probability of existence in x^μ of the quantum state $\mid a\rangle$:

$$P_r\ (\mid a\rangle \text{ at } x^\mu) = \mid \langle x^\mu \mid a\rangle\mid^2 = \mid \Psi_a(x^\mu)\mid^2. \tag{5.87}$$

The wave aspect of the quantum state is therefore linked to the probability of observation of this state and not intrinsically to the particle with mass m.

It should be noted that the choice of correspondence principle depends on the metric $\eta_{\mu\,\nu}$ of the Minkowski space-time used.

If we use form (1.242) of the correspondence principle $(E, \vec{p}) \rightarrow (i\hbar\partial_t, -i\hbar\vec{\nabla})$ to write the correspondence principle in space-time,

$$P^\mu \rightarrow -i\,\hbar\,\partial^\mu \quad \text{and} \quad P_\mu \rightarrow i\,\hbar\,\partial_\mu, \tag{5.88}$$

the metric will have to be $\eta^{\mu\nu} = \eta_{\mu\nu} = (-+++)$ since,

$$\begin{aligned} P^0 &= -i\,\hbar\,\partial^0 = -i\,\hbar\,\partial_0\,\eta^{00} = i\,\hbar\,\partial_0 = +i\,\hbar\,\partial_t \\ P^k &= -i\,\hbar\,\partial^k = -i\,\hbar\,\partial_j\,\eta^{j\,k} = -i\,\hbar\,\partial_k. \end{aligned} \tag{5.89}$$

For this particular metric,

$$P^2 = P_\mu\,P^\mu = -m^2. \tag{5.90}$$

If we prefer the metric $\eta_{\mu\,\nu} = (+---)$ as is the case in field theory (see Chap. 16), the correspondence principle reduces to

$$P^\mu \rightarrow i\,\hbar\,\partial^\mu \quad \text{and} \quad P_\mu \rightarrow i\,\hbar\,\partial_\mu. \tag{5.91}$$

It then becomes easy to expand the components of the momentum four-vector:

$$\begin{aligned} P^0 &= i\,\hbar\,\partial^0 = i\,\hbar\,\partial_0\,\eta^{00} = i\,\hbar\,\partial_t \\ P^k &= i\,\hbar\,\partial^k = i\,\hbar\,\partial_j\,\eta^{j\,k} = -i\,\hbar\,\partial_k. \end{aligned} \tag{5.92}$$

In such a metric $P^2 = m^2$ and the correspondence principle (5.91) will yield

$$(\Box + m^2)\,\Psi_a\,(x^\mu) = 0 \tag{5.93}$$

for the Klein–Gordon propagation equation for a spinless particle with mass m. If we set $\Box = \partial_t^2 - \nabla^2$, eqs. (5.81) and (5.93) become identical.

2. Matrix Mechanics

The term matrix mechanics has often been used to describe the matrix formalism of quantum mechanics introduced by Heisenberg and Jordan.

2.1 Matrix Representations

Now, let us choose a basis $\{N\}$ in which the basis vectors $| \, 0 \rangle \, | \, 1 \rangle \ldots | \, n \rangle$ are dependent on the discrete index n and are eigenvectors of a given Hermitian operator N. The $\{N\}$ representation will be defined by the following relations:

$$\{N\} \text{ representation} \quad \left\{ \begin{array}{l} N \, | \, n \rangle \; = q_n \, | \, n \rangle \, , \\[2mm] \langle n \, | \, m \rangle \; = \delta_{nm} \, , \\[2mm] \displaystyle\sum_n \, | \, n \rangle \, \langle n \, | \; = \mathbf{1} \, . \end{array} \right. \qquad (5.94)$$

We assumed, for the sake of simplicity, that the states were not degenerate and, hence, that the basis vectors depended exclusively on the quantum number $n = 0, 1, 2, \ldots$.

Any ket $| \, u \rangle$ of the (separable) Hilbert space associated with a quantum state can be expressed as a linear combination of the basis vectors of $\{N\}$:

$$| \, u \rangle = \sum_n \, | \, n \rangle \, \langle n \, | \, u \rangle \; = \sum_n \, x^n \, | \, n \rangle \, , \qquad (5.95)$$

where x^n represents the projection of the state $| \, u \rangle$ onto the space $\{N\}$:

$$x^n = \; \langle n \, | \, u \rangle \, . \qquad (5.96)$$

As already noted in relation to the three-dimensional space, the basis S can be defined as the row of the basis vectors $| \, 1 \rangle \ldots | \, n \rangle$:

$$S = [\, | \, 0 \rangle \; | \, 1 \rangle \ldots] \qquad (5.97)$$

and the column X by the following elements:

$$X = \begin{bmatrix} x^0 \\ x^1 \\ \vdots \\ x^n \end{bmatrix} . \qquad (5.98)$$

The vector $| \, u \rangle$ will then be the product SX of the column multiplied by the basis:

$$| \, u \rangle \; = SX = [\, | \, 0 \rangle \ldots | \, n \rangle \,] \begin{bmatrix} x^0 \\ \vdots \\ x^n \end{bmatrix} = \sum_n \, x^n \, | \, n \rangle \, . \qquad (5.99)$$

In addition, it should be noted that the relation $V = SX$ is a general one and states that a column (belonging to the column space) is linked to a vector (belonging to the vector space) via an operator S termed basis which is the row of the basis vectors.

The relation among the three commonly used vector spaces (number space, vector space, and column space) can be represented diagramatically.

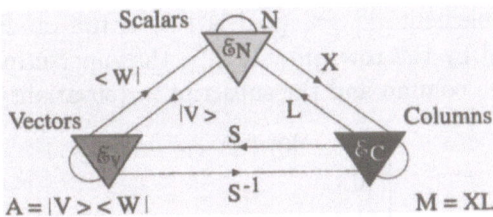

Figure 1

We notice then that the vectors, covectors, columns, and rows can be expressed as operators making the transition from one space to another (Fig. 1). By inserting (5.96) into (5.99), we notice that the vector $| u \rangle$ is represented in the space $\{N\}$ by the column

$$X = \begin{bmatrix} \langle 0 \mid u \rangle \\ \vdots \\ \langle n \mid u \rangle \end{bmatrix}. \tag{5.100}$$

The superscript n in $x^n = \langle n \mid u \rangle$ therefore represents the row index.

We can similarly represent a bra $\langle v \mid$ in base $\{N\}$, bearing in mind that $\langle v \mid n \rangle = \langle n \mid v \rangle^*$. We thus write

$$\langle v \mid = \sum_n \langle v \mid n \rangle \langle n \mid = \sum_n y_n \langle n \mid. \tag{5.101}$$

The bra $\langle v \mid$ is therefore represented in the space $\{N\}$ by the row L:

$$L = [\langle \dot{v} \mid 0 \rangle \ \cdots \ \langle v \mid n \rangle]. \tag{5.102}$$

The index n in y_n therefore represents the column index.

The inner product $\langle v \mid u \rangle$ is easily expressed in the space $\{N\}$ with the closure relation

$$\langle v \mid u \rangle = \sum_n \langle v \mid n \rangle \langle n \mid u \rangle = \sum_n y_n x^n = \sum_n y^{n*} x^n. \tag{5.103}$$

The inner product $\langle v \mid u \rangle$ is, in fact, represented by the product LX in the matrix space:

$$LX = [y_0 \cdots y_n] \begin{bmatrix} x^0 \\ \vdots \\ x^n \end{bmatrix} = \sum_n y_n x^n. \tag{5.104}$$

A linear operator will, in turn, be represented in the space $\{N\}$ by a matrix, since, by using the completeness relation, we obtain

$$A = \sum_m \sum_n | m \rangle \langle m \mid A \mid n \rangle \langle n \mid = \sum_{mn} | m \rangle A_m^n \langle n \mid \tag{5.105}$$

and the matrix element $A^n_m = \langle m \mid A \mid n \rangle$ is the product of the column $A \mid n \rangle$ multiplied by the row $\langle m \mid$. In A^n_m, the superscript n (contravariant index) defines the column and the subscript m (covariant index), the row:

$$
A = \begin{bmatrix}
 & \overset{\displaystyle |0\rangle \;\; |1\rangle \;\cdots \quad\quad |n\rangle \;\cdots}{} & \\
\langle 0 \mid & & \\
\langle 1 \mid & & \\
\vdots & & \vdots \\
\langle m \mid & \cdots\cdots\cdots \quad \langle m \mid A \mid n \rangle & \\
\vdots & & \vdots \\
\vdots & &
\end{bmatrix} . \tag{5.106}
$$

Let us illustrate this with the example of the operator a^+ defined by the relation

$$
a^+ \mid n \rangle = (n+1)^{1/2} \mid n+1 \rangle , \tag{5.107}
$$

with the matrix element

$$
\begin{aligned}
\left(a^+\right)^n_m = \langle m \mid a^+ \mid n \rangle &= (n+1)^{1/2} \langle m \mid n+1 \rangle \\
&= (n+1)^{1/2} \, \delta^{n+1}_m .
\end{aligned} \tag{5.108}
$$

The only non-zero elements are those for which the row index m is equal to the column index $n+1$.

The operator a^+ will then be represented by the matrix

$$
a^+ = \begin{bmatrix}
0 & 0 & 0 & \cdots \\
\sqrt{1} & 0 & 0 & \cdots \\
0 & \sqrt{2} & 0 & \\
\cdot & & \sqrt{3} & 0 \\
\cdot & & & \\
\cdot & & &
\end{bmatrix} . \tag{5.109}
$$

It should be noted that, in its eigenrepresentation, the operator N is represented by a diagonal matrix

$$
\langle m \mid N \mid n \rangle = q_n \, \langle m \mid n \rangle = q_n \, \delta^n_m , \tag{5.110}
$$

and that the same is true of every function of this operator:

$$
\langle m \mid f(N) \mid n \rangle = f\,(q_n)\,\langle m \mid n \rangle = f\,(q_n)\,\delta^n_m . \tag{5.111}
$$

Any operator X commuting with the operator N is also represented by diagonal matrices in $\{N\}$ since

$$
\langle m \mid XN - NX \mid n \rangle = 0 = (q_n - q_m)\,\langle m \mid X \mid n \rangle \tag{5.112}
$$

and for any $q_m \neq q_n$

$$\langle m \mid X \mid n \rangle = 0. \tag{5.113}$$

Let us add that relation (5.94) shows that the eigenvalues of the operator N are diagonal elements of the matrix N representing it in the matrix space. In other words, the search for the eigenvalues of any given operator is equivalent to the diagonalization of the matrix representing the operator.

2.2 Bell's Correlation Function

Mention has already been made of the correlation functions between the polarization of the photons emitted by the calcium atoms in Alain Aspect's experiment to test Bell's inequalities (See Sect. 5.9 of Chap. 2). The computation of these functions is a direct application of the matrix representation of the Dirac formalism.

Consider the polarization states $\mid e_x \rangle$ and $\mid e_y \rangle$ of a photon along the Ox and Oy axes. These states form an orthonormal system:

$$\langle e_i \mid e_j \rangle = \delta_{ij}. \tag{5.114}$$

With the preceding polarization states this gives

$$\langle e_x \mid e_x \rangle = \langle e_y \mid e_y \rangle = 1 \quad \text{and} \quad \langle e_x \mid e_y \rangle = 0. \tag{5.115}$$

The system will be considered complete when the closure relation is obeyed:

$$\sum_i \mid e_i \rangle \langle e_i \mid = \mathbb{1} \quad \text{or} \quad \mid e_x \rangle \langle e_x \mid + \mid e_y \rangle \langle e_y \mid = \mathbb{1}. \tag{5.116}$$

Let us decompose a polarization state $\mid \phi \rangle$ in the plane Oxy in terms of the linearly polarized states $\mid e_x \rangle$ and $\mid e_y \rangle$:

$$\mid \phi \rangle = \sum_i \mid e_i \rangle \langle e_i \mid \phi \rangle$$

gives

$$\mid \phi \rangle = \mid e_x \rangle \langle e_x \mid \phi \rangle + \mid e_y \rangle \langle e_y \mid \phi \rangle. \tag{5.117}$$

The projections of $\mid \phi \rangle$ onto the axes Ox and Oy can be expressed with the angle θ (Fig. 2):

$$\langle e_x \mid \phi \rangle = \cos \theta,$$
$$\langle e_y \mid \phi \rangle = \sin \theta. \tag{5.118}$$

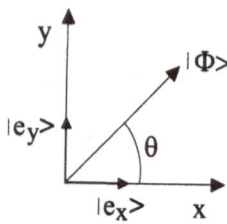

Figure 2

Now let us introduce a polarization state $| \Psi \rangle$, perpendicular to $| \phi \rangle$. By decomposing along the basis vectors $| e_x \rangle | e_y \rangle$, we obtain a similar result as in (5.118):

$$\langle e_x | \Psi \rangle = \cos \left(\theta + \frac{\pi}{2} \right) = - \sin \theta,$$
$$\langle e_y | \Psi \rangle = \sin \left(\theta + \frac{\pi}{2} \right) = \cos \theta. \tag{5.119}$$

The polarization states $| \phi \rangle$ and $| \Psi \rangle$ are therefore expressed with the rectilinear polarizations $| e_x \rangle$ and $| e_y \rangle$ as in (5.117), using projections (5.118) and (5.119), that is,

$$| \phi \rangle = \cos \theta | e_x \rangle + \sin \theta | e_y \rangle,$$
$$| \Psi \rangle = - \sin \theta | e_x \rangle + \cos \theta | e_y \rangle. \tag{5.120}$$

It is easily verified that the states $| \phi \rangle$ and $| \Psi \rangle$ form another complete orthonormal system, since

$$\langle \phi | \phi \rangle = \langle \Psi | \Psi \rangle = 1 \quad \text{and} \quad \langle \phi | \Psi \rangle = 0$$
$$| \phi \rangle \langle \phi | + | \Psi \rangle \langle \Psi | = \mathbb{1}. \tag{5.121}$$

We can, on the other hand, use the above relations to express the polarizations $| e_x \rangle$ and $| e_y \rangle$ with $| \phi \rangle$ and $| \Psi \rangle$:

$$| e_x \rangle = | \phi \rangle \langle \phi | e_x \rangle + | \Psi \rangle \langle \Psi | e_x \rangle,$$
$$| e_y \rangle = | \phi \rangle \langle \phi | e_y \rangle + | \Psi \rangle \langle \Psi | e_y \rangle. \tag{5.122}$$

Projections (5.118) and (5.119) therefore lead us to write

$$| e_x \rangle = \cos \theta | \phi \rangle - \sin \theta | \Psi \rangle,$$
$$| e_y \rangle = \sin \theta | \phi \rangle + \cos \theta | \Psi \rangle. \tag{5.123}$$

Hence, there exists a 2×2 matrix for making the transition from the column

$$\Phi = \begin{bmatrix} | \phi \rangle \\ | \Psi \rangle \end{bmatrix} \quad \text{to the column} \quad X = \begin{bmatrix} | e_x \rangle \\ | e_y \rangle \end{bmatrix}. \tag{5.124}$$

Relations (5.120) then read

$$\Phi = R\,X \qquad \text{with} \qquad R = \begin{bmatrix} \cos\theta & \sin\theta \\ -\sin\theta & \cos\theta \end{bmatrix}, \qquad (5.125)$$

whereas relations (5.123) yield the inverse matrix:

$$X = R^{-1}\,\Phi \qquad \text{with} \qquad R^{-1} = \begin{bmatrix} \cos\theta & -\sin\theta \\ \sin\theta & \cos\theta \end{bmatrix}. \qquad (5.126)$$

The matrix R is, in fact, the rotation matrix of the physical system in the two-dimensional space.

Let us now consider a two-photon state $|\,\Phi\rangle$ polarized in a similar manner:

$$|\,\Phi\rangle \;=\; \frac{1}{\sqrt{2}} \left(|\,e_x^A\,e_x^B\rangle + |\,e_y^A\,e_y^B\rangle \right). \qquad (5.127)$$

This state is normalized to unity:

$$\langle\,\Phi\,|\,\Phi\,\rangle \;=\; 1. \qquad (5.128)$$

By using the polarization states of each of the photons, we can calculate $|\,\Phi\rangle$ as a function of ϕ_A, ϕ_B, Ψ_A, Ψ_B, θ_A, and θ_B. For example,

$$|\,e_y^A\,e_y^B\rangle \;=\; (\sin\theta_A\,|\,\phi_A\rangle + \cos\theta_A\,|\,\Psi_A\rangle)\,(\sin\theta_B\,|\,\phi_B\rangle + \cos\theta_B\,|\,\Psi_B\rangle) \qquad (5.129)$$

By introducing the relative angle $\Theta = \theta_A - \theta_B$ of the directions of the detectors A and B, we obtain the two-photon polarization state

$$
\begin{aligned}
|\,\Phi\rangle \;=\; \frac{1}{\sqrt{2}} \big[&|\,\phi_A\,\phi_B\rangle \cos\Theta - (|\,\Psi_A\,\phi_B\rangle + |\,\phi_A\,\Psi_B\rangle)\, \sin\Theta \\
&+ |\,\Psi_A\,\Psi_B\rangle \cos\Theta \big].
\end{aligned}
\qquad (5.130)
$$

The probabilities of detecting the polarizations are therefore:

$$
\begin{aligned}
P_{++} &= |\,\langle\Phi\,|\,\phi_A\,\phi_B\rangle\,|^2 = \frac{1}{2}\cos^2\Theta, \\[4pt]
P_{--} &= |\,\langle\Phi\,|\,\Psi_A\,\Psi_B\rangle\,|^2 = \frac{1}{2}\cos^2\Theta = P_{++}, \\[4pt]
P_{+-} &= |\,\langle\Phi\,|\,\phi_A\,\Psi_B\rangle\,|^2 = \frac{1}{2}\sin^2\Theta = P_{-+}.
\end{aligned}
\qquad (5.131)
$$

The correlation coefficient $E\,(\vec{a},\,\vec{b})$ is the expectation value of the polarizations in the directions $\vec{a}$ or $\vec{b}$ of the detectors:

$$E\,(\vec{a},\,\vec{b}) = \langle A(\vec{a})\,B(\vec{b})\rangle = \sum P_{ab}\,A(a)\,B(b). \qquad (5.132)$$

If $(+1)$ and (-1) are the upward and downward orientations respectively of the detector A, we easily obtain:

$$E\,(\vec{a},\,\vec{b}) = P_{++}(+1)(+1) + P_{-+}(-1)(+1) + P_{+-}(+1)(-1) + P_{--}(-1)(-1)$$
$$= P_{++} - P_{-+} - P_{+-} + P_{--}$$
$$= 2\,(P_{++} - P_{+-})\,. \tag{5.133}$$

This, with (5.131), will lead to the correlation coefficient

$$E\,(\vec{a},\,\vec{b}) = \cos 2\,\Theta \tag{5.134}.$$

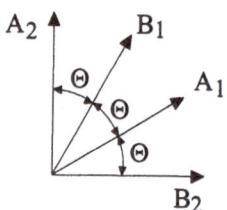

Figure 3

In the experiment by Aspect and Granger, the relative position of detectors A and B gives the correlation coefficients (Fig. 3)

$$\begin{array}{ll} E\,(a_1, b_1) = \cos 2\,\Theta & E\,(a_1, b_2) = \cos 2\,\Theta \\ E\,(a_2, b_1) = \cos 2\,\Theta & E\,(a_2, b_2) = \cos 6\,\Theta \end{array} \tag{5.135}$$

leading to the Bell parameter

$$S = E\,(a_1, b_1) + E\,(a_2, b_1) + E\,(a_1, b_2) - E\,(a_2, b_2)$$
$$= 3\,\cos 2\,\Theta - \cos 6\,\Theta\,.$$

We have already seen that Bell's inequality $-2 \leq S \leq +2$ was decisively violated, obliging us to abandon the hypothesis of hidden variables.

2.3 Density Matrix

The dynamical state of a quantum system will be fully known if we can determine precisely the parameters of one of the complete sets of compatible variables attached to it. It can then be represented by a normalized vector $|\,u\rangle$ and the expectation value of an observable will be

$$\langle A \rangle = \langle u \,|\, A \,|\, u \rangle\,. \tag{5.136}$$

If the information we have is incomplete, we will be content to say that the system has the probability p_1 of being in the state $|\,1\rangle$, p_2 of being in the $|\,2\rangle$, etc., the states being the eigenvectors of a Hermitian operator N constituting a complete orthonormalized basis. The expectation value of the observable A will therefore have the probability p_1 of being evaluated from the state $|\,1\rangle$, the probability p_2 from the state $|\,2\rangle$, etc., and we can then proceed to write

$$\langle A\rangle_m = \langle m\,|\,A\,|\,m\rangle. \tag{5.137}$$

The result of the actual measurement of the observable A is therefore the statistical sum of the expectation values $\langle A\rangle_m$, weighted with their existence probabilities:

$$\langle A\rangle = \sum_m p_m \langle A\rangle_m = \sum_m p_m \langle m\,|\,A\,|\,m\rangle. \tag{5.138}$$

The values of the weighting coefficients p_m satisfy the axioms of the probabilities:

$$p_m \leq 1 \quad \text{and} \quad \sum_m p_m = 1. \tag{5.139}$$

If we assume that the actual quantum state of the system $|\,u\rangle$ is not known, the projection of the state $|\,u\rangle$ onto the complete basis $\{N\}$ will give

$$\langle A\rangle = \langle u\,|\,A\,|\,u\rangle = \sum_{mn} \langle u\,|\,m\rangle \langle m\,|\,A\,|\,n\rangle \langle n\,|\,u\rangle. \tag{5.140}$$

If the operator A is written in its eigenrepresentation, only the diagonal elements of the matrix representing A in basis $\{N\}$ will be called into play and we obtain

$$\langle A\rangle = \sum_m |\langle m\,|\,u\rangle|^2 \langle m\,|\,A\,|\,m\rangle, \tag{5.141}$$

which is consistent with the fourth postulate of quantum mechanics, given that

$$p_m = |\langle m\,|\,u\rangle|^2. \tag{5.142}$$

Let us introduce an operator ϱ of the density of states $|\,m\rangle$:

$$\varrho = \sum_m |\,m\rangle\, p_m\, \langle m\,|. \tag{5.143}$$

If the state is a pure one, capable of confirming that the quantum state is really in the state $|\,m\rangle$, then $p_m = 1$ and $\varrho = \sum_m |\,m\rangle \langle m\,| = \mathbb{1}$, then the density operator will be the identity operator.

Let us introduce another system with a complete orthonormalized basis [states denoted $|n\rangle$ to distinguish them from the states $|\,m\rangle$] defining the mixture of states constituting the unknown quantum state and the eigenrepresentation of the operator A:

$$\begin{cases} \langle n\,|\,n'\rangle = \delta_{nn'} \\ \sum_n |\,n\rangle\langle n\,| = \mathbb{1}. \end{cases} \tag{5.144}$$

Let us evaluate the trace of the density operator in the basis $\{n\}$:

$$\mathrm{Tr}\,\varrho = \sum_n (n \mid \varrho \mid n) = \sum_{n\,m} (n \mid m)\, p_m\, (m \mid n) = \sum_{n\,m} (m \mid n)(n \mid m)\, p_m\,.$$

By summing over n with (5.144), we obtain the following result:

$$\boxed{\mathrm{Tr}\,\varrho = \sum_m p_m = 1}\,. \tag{5.145}$$

We thus obtain the following necessary and sufficient condition: the hermitian operator $\varrho = \sum_m \mid m)\, p_m\, (m \mid$ represents the density of states $\mid m)$ if and only if its trace is equal to 1.

The expectation value of any observable whatsoever can be expressed in the form of a trace with definition (5.138) as departure point and by projecting onto the basis $\{\mid n)\}$:

$$\langle A \rangle = \sum_m p_m\, (m \mid A \mid m)$$

$$= \sum_{mnn'} p_m\, (m \mid n)(n \mid A \mid n')(n' \mid m)\,. \tag{5.146}$$

By commuting the projection matrix elements of basis $\mid n)$ with those of basis $\mid m)$, we obtain:

$$\langle A \rangle = \sum_{mnn'} (n' \mid m)\, p_m\, (m \mid n)(n \mid A \mid n')\,. \tag{5.147}$$

By summing over n and using (5.143) to introduce the density ϱ, we obtain the following important result:

$$\boxed{\langle A \rangle = \sum_{n'} (n' \mid \varrho\, A \mid n') = \mathrm{Tr}\,(\varrho\, A)}\,. \tag{5.148}$$

The projection of the density of states operator $\mid m)$ onto a complete basis $\{\mid n)\}$ leads to the diagonal matrix elements:

$$\varrho_{nn} = (n \mid \varrho \mid n) = \sum_m (n \mid m)\, p_m\, (m \mid n)$$

$$= \sum_m p_m \mid (n \mid m) \mid^2 = \sum_m p_m\, p_n = p_n\,. \tag{5.149}$$

The diagonal elements of the density matrix define the population of quantum states $\mid n)$. They represent the probability that a quantum system will be in the state $\mid n)$.

If the basis $\{ \mid n) \}$ is constructed from the eigenstates of the Hamiltonian $H \mid n) = E_n \mid n)$, then the evolution equation of the density operator, obtained from the Poisson Eq. (1.116) and the correspondence principle (1.142) will take the form

$$i \hbar \frac{d \varrho}{dt} = [H, \varrho].$$

(5.150)

For the diagonal matrix elements this gives

$$i \hbar \frac{d}{dt} \varrho_{nn} = (n \mid [H, \varrho] \mid n) = (E_n - E_n) (n \mid \varrho \mid n) = 0.$$

(5.151)

The populations of the quantum states $\mid n)$ will therefore be constants.

The non-diagonal elements can also be evaluated:

$$\varrho_{nk} = (n \mid \varrho \mid k) = \sum_m (n \mid m) \, p_m \, \langle m \mid k).$$

(5.152)

They reflect the coherences of the quantum states. By using the eigenstates of the Hamiltonian, we obtain

$$i \hbar \frac{d}{dt} \varrho_{nk} = (E_n - E_k) \, \varrho_{nk}.$$

(5.153)

This showes that the coherences oscillate at the Bohr frequency of the system:

$$E_n - E_k = h(\nu_n - \nu_k) = \hbar \, (\omega_n - \omega_k) = \hbar \, \omega_{nk}.$$

(5.154)

The differential Eq. (5.153) will then have the solution

$$\varrho_{nk}(t) = \exp \left(i \, \omega_{kn} \, t \right) \varrho_{nk}(0).$$

(5.155)

Generally speaking, if the states are macroscopically defined, the density matrix will make it possible to represent the state of a statistical mixture of N systems of which n_1 are in state E_1, n_2 in E_2, n_p in E_p, with a $p \times p$-matrix whose diagonal elements are np/N and the off-diagonal ones zero. In quantum mechanics, the off-diagonal terms ϱ_{nk} are non-zero; they represent the correlations between the $\mid n \rangle$ states and the $\mid k \rangle$ states, which cannot be interpreted macroscopically. After a measurement, the N states will be well defined and lead to a statistical mixture described by a diagonal matrix. The problem is therefore that of explaining the disappearance of the off-diagonal elements of the density matrix during measurement. W. Zurek (1981, 1982) has termed this phenomenon coherence but I will continue to use the term decoherence, borrowing from J.B. Hartle's terminology (see end of following chapter). Quantum mechanics calls into play off-diagonal and, hence, coherent elements whereas classical mechanics avoids them: it decoheres quantum states. It should be emphasized that the measurement of a physical quantity presupposes that the corresponding density matrix is diagonal. This means that as long as the basis in which the density ϱ can be diagonalized is not defined, the quantum formalism cannot give the physical quantity measured with the classical apparatus.

Zurek's works have clearly underscored the primordial role played by the medium in the measurement process. The interaction with the medium is responsible for the diagonalization of the density matrix. Diagonal elements are never zero but are simply sufficiently small for their effects to be unobservable, with a decoherence time depending on the type of interaction. Zurek therefore proposes to consider as quantum systems only those systems that are isolated while classical systems should be considered as open systems. This means that reality is always quantum even if, on the surface, it appears to be classical because the off-diagonal elements are too small for their effects to be observable for all practical purposes.

In the Copenhagen interpretation (see Chap. 2), if there is a quantum state $| a \rangle$ of the Universe (matter, radiation, measurement apparatus and observers) and quantum states $| b \rangle$ linked to the observer, such states can never form a complete system because they are already contained in $| a \rangle$. This means that, strictly speaking, we cannot write

$$\sum_b | b \rangle \langle b | = \mathbb{1}$$

$$| a \rangle = \sum_b \langle b | a \rangle | b \rangle . \tag{5.156}$$

The overlap $\langle b | a \rangle$ of the quantum state of the Universe with that of the observer represents what the observer can measure in the state of the Universe (with its square modulus). If the observer were not part of the Universe, then he would be in a position to determine the wave function $| a \rangle$ (which is independent of x and t) by measuring all possible observations and then reconstituting the quantum state $| a \rangle$ with (5.156). But because he is part and parcel of the Universe, such a reconstitution is impossible. This goes to show that there is a principle of incompleteness of our knowledge of the Universe that is in some way in agreement with Gödel's theorem (Gödel, 1931) which states that the consistency of a formal mathematical system cannot be proved within that system; in other words, there exists an unprovable statement within every sufficiently complicated formal system.

We would like to close the discussion in this chapter with two quotations on the relationship between Man and Nature. The first is taken from M. Felden (1992): "Science is made by man for man. It is the scientist who decides and chooses what goes into the system under study; in other words, he fixes the limits and possibilities of formalization, that is the hypotheses with which he works.". The second is taken from N. Bohr and goes even further: "It is not correct to say that the object of Physics is to discover what constitutes Nature. Physics is rather concerned with what we can say about Nature".

3. Wave Mechanics – Schematic Summary

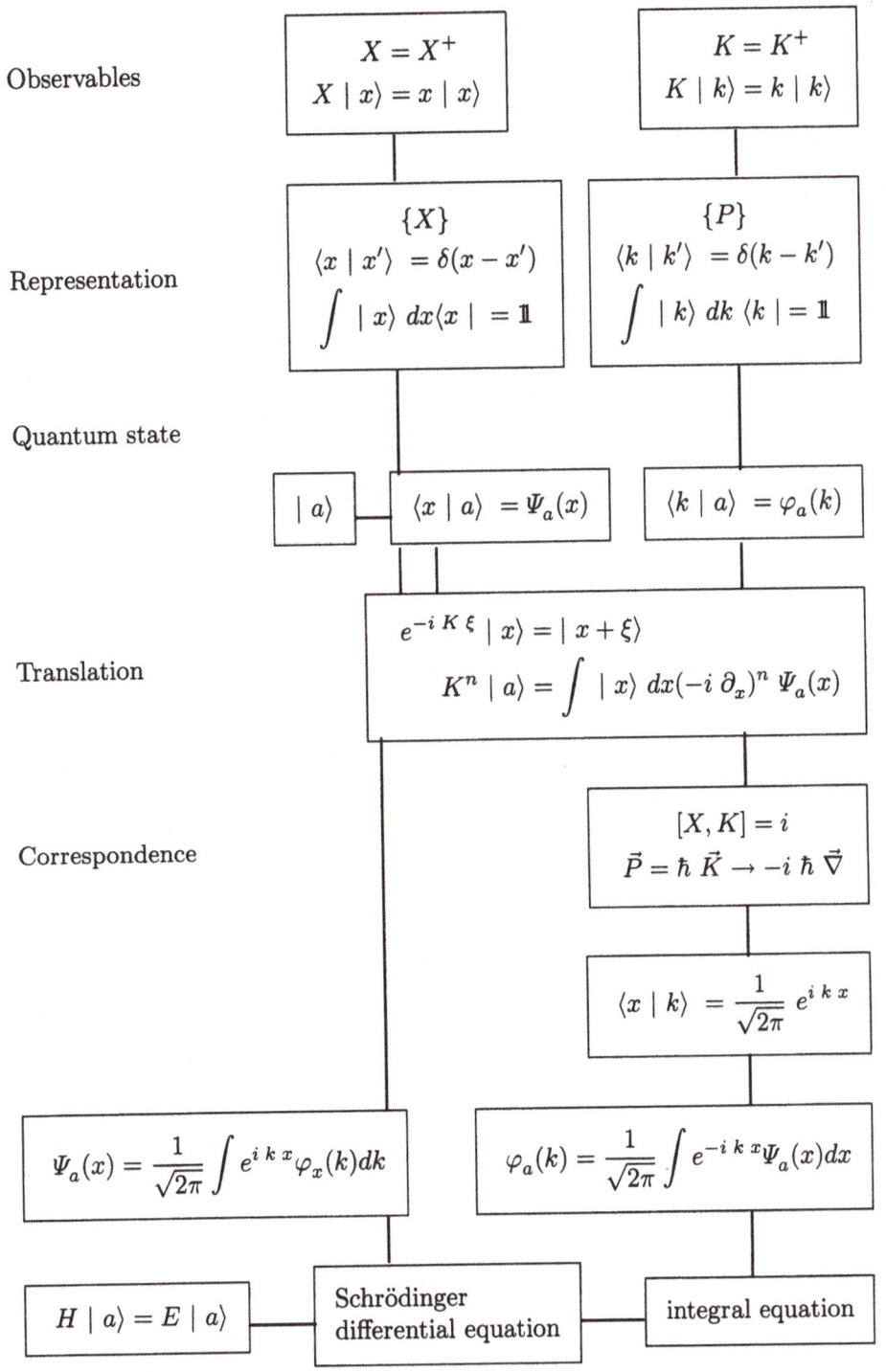

4. Matrix Mechanics – Schematic Summary

Observable

$$N = N^+$$
$$N \mid n) = q_n \mid n)$$

Representation

$$\{N\}$$
$$\langle n \mid n' \rangle = \delta_{nn'}$$
$$\sum_n \mid n \rangle \langle n \mid = \mathbb{1}$$

Quantum state

$$\mid a)$$

Column X element $x^n = \langle n \mid a \rangle$

Matrix $A = \sum_{mn} \mid m \rangle A_m^n \langle n \mid$

$$H \mid a) = E \mid a)$$

$(H - E) X = 0$ matrix diagonalization H

Density of states $\mid m)$

$$\varrho = \sum_n \mid m \rangle p_m \langle m \mid$$
$$p_m \leq 1 \quad \sum_m p_m = 1$$

Observable

$$\langle A \rangle = \langle u \mid A \mid u \rangle = \mathrm{T_r} \, (\varrho A)$$
$$\mathrm{T_r} \, \varrho = 1$$

$$\varrho_{nn} = (n \mid \varrho \mid n) \quad \text{population}$$
$$\varrho_{nk} = (n \mid \varrho \mid k) \quad \text{coherence}$$

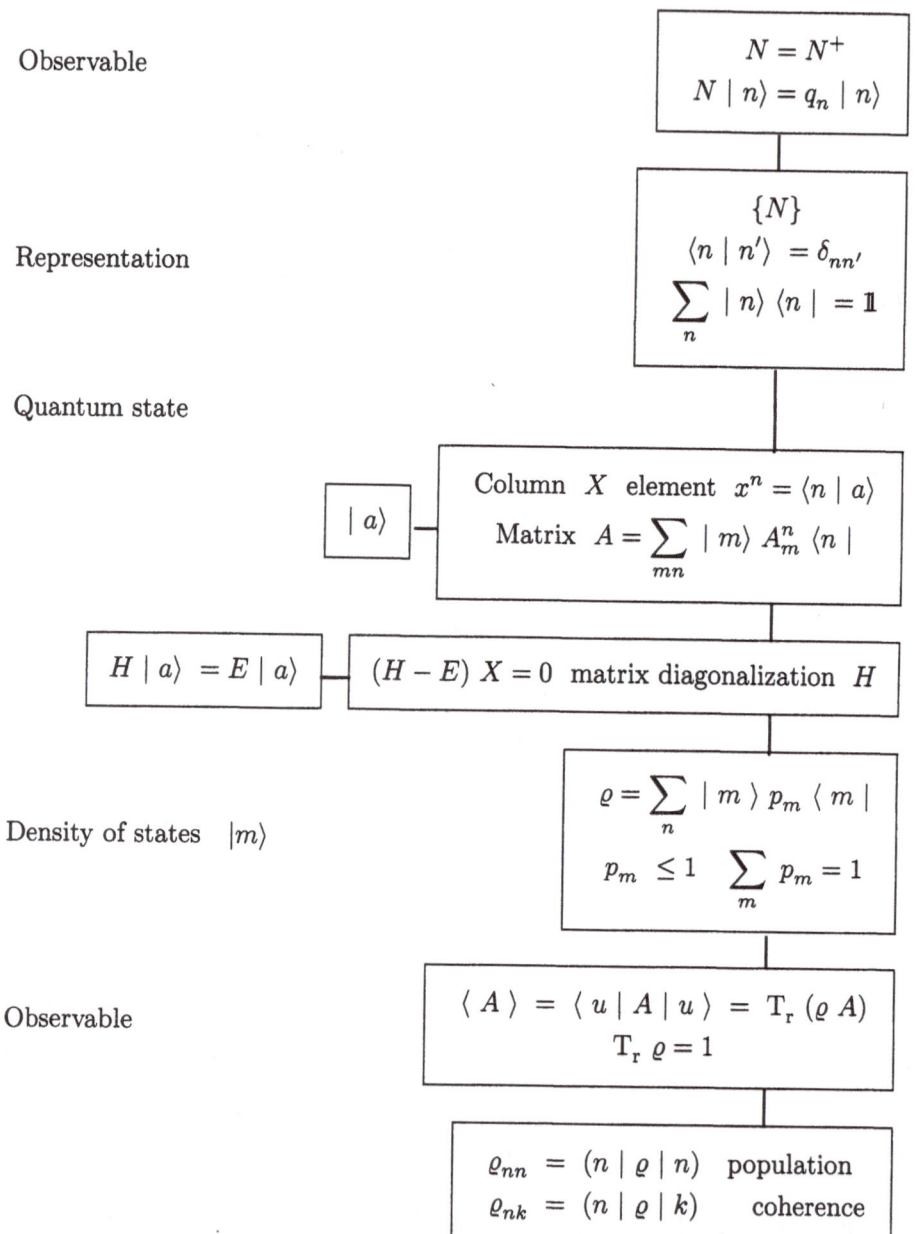

Chapter 6
Quantum Dynamics

As we have already seen in the Dirac formulation of quantum mechanics, dynamics is an essential characteristic of the formalism. It determines the time evolution of quantum states and, hence, of the Hilbert space of states or operators acting in this space. The evolution is considered to be causal, that is, if a quantum state $| a \rangle$ is determined at instant t_0, it will evolve in a predictable manner toward another quantum state at instant t:

$$| a\, (t_0) \rangle \; \rightarrow \; | a(t) \rangle . \tag{6.1}$$

Any linear combination of quantum states at instant t_0 remains the same combination at instant t given that quantum states are linear space vectors:

$$\lambda_1 \, | a(t_0) \rangle \; + \lambda_2 \, | b(t_0) \rangle \; \rightarrow \; \lambda_1 \, | a(t) \rangle \; + \lambda_2 \, | b(t) \rangle . \tag{6.2}$$

This is on the assumption that there is an evolution operator $U(t,\, t_0)$ modifying the quantum states between instants t_0 and t:

$$| a(t) \rangle \; = U(t,\, t_0) \, | a(t_0) \rangle . \tag{6.3}$$

We will therefore examine the properties of this operator before going on to determine its general form.

1. Time-Evolution Operator

1.1 Definition

Because time is not an observable to which can be associated an eigenvalue (in non-relativistic quantum mechanics), it is not strictly correct to write $| a, t \rangle$ even though we will continue to use this form for the sake of convenience. Relation (6.3) then becomes

$$| a, t \rangle \; = U(t,\, t_0) \, | a, t_0 \rangle . \tag{6.4}$$

It should be noted that instants t and t_o are written on the same side in the operator U and in the kets introduced to ease comprehension. For $t = t_0$, we can easily conclude that

$$U(t,t) = U(t_0,t_0) = \mathbb{1}\,. \tag{6.5}$$

We can also introduce the intermediate time, t_1, between t_0 and t and thus obtain, through the application of definition (6.4),

$$\mid a,\ t\rangle\ = U(t,\ t_1)\mid a,\ t_1\rangle \qquad \text{and} \qquad \mid a,\ t_1\rangle\ = U(t_1,\ t_0)\mid a, t_0\rangle\,.$$

Following substitution, this will give the equality:

$$\mid a,\ t\rangle\ = U(t,t_1)\ U(t_1,t_0)\mid a,t_0\rangle\ = U(t,t_0)\mid a,t_0\rangle\,.$$

This results in the relation of inclusion of an intermediate time:

$$U(t,t_0) = U(t,t_1)\ U(t_1,t_0)\,. \tag{6.6}$$

The inversion of the time corresponds to the inversion of the evolution operator itself since, by using relations (6.5) and (6.6), we obtain

$$U(t,t) = \mathbb{1} = U(t,t_0)\ U(t_0,t)\,, \tag{6.7}$$

and by multiplying the two terms by the inverse operator, we have

$$U(t,t_0) = U^{-1}(t_0,t)\,. \tag{6.8}$$

If the vectors $\mid a_n,\ t\rangle$ satisfying at any instant the orthonormalization and completeness relations span the Hilbert space of the states:

$$\langle a_n,t\mid a_{n'},t\rangle\ =\ \langle a_n,t_0\mid a_{n'},t_0\rangle\ =\ \delta_{nn'}\,,$$
$$\sum_n\mid a_n,t\rangle\ \langle a_n,t\mid\ =\ \sum_n\mid a_n,t_0\rangle\ \langle a_n,t_0\mid\ =\ \mathbb{1}\,, \tag{6.9}$$

then the evolution operator is formally written in the form of a discrete sum:

$$U(t,\ t_0) = \sum_n\mid a_n,t\rangle\ \langle a_n,t_0\mid\,, \tag{6.10}$$

whereas for a quantum state related to the position (of the observer, for example)

$$\langle x,\ t\mid x',\ t\rangle\ =\ \langle x,\ t_0\mid x',\ t_0\rangle\ =\ \delta\ (x-x')\,,$$
$$\int\mid x,\ t\rangle\ dx\ \langle x,\ t\mid\ =\ \int\mid x,\ t_0\rangle\ dx\ \langle x,\ t_0\mid\ =\ \mathbb{1}\,. \tag{6.11}$$

This will define the evolution operator in the integral form:

$$U(t,\ t_0) = \int\mid x,\ t\rangle\ dx\ \langle x,\ t_0\mid\,. \tag{6.12}$$

Expressions (6.10) and (6.12), applied to the vectors $\mid a_n,\ t_0\rangle$ or $\mid x',\ t_0\rangle$ respectively, will lead to the vectors $\mid a_n,t\rangle$ or $\mid x',t\rangle$.

1.2 Unitarity

The form (6.10) or (6.12) of the evolution operator shows that we are dealing with a unitary operator:

$$U^+ U = \mathbb{1} \quad \text{and} \quad U^+ = U^{-1}. \tag{6.13}$$

For example, this appears very clearly in form (6.10):

$$U^+(t, t_0) \, U(t, t_0) = \sum_{nn'} | \, a_n, t_0 \rangle \, \langle a_n, t \, | \, a_{n'}, t \rangle \, \langle a_{n'}, t_0 \, |$$

$$= \sum_{n} | \, a_n, t_0 \rangle \, \langle a_n, t_0 \, | \, = \mathbb{1}. \tag{6.14}$$

Furthermore, unitarity property is indispensable if we want to conserve the norm of the Hilbert space vectors:

$$\langle a, t_0 \, | \, U^+(t, t_0) \, U(t, t_0) \, | \, a, t_0 \rangle \; = \; \langle a, t \, | \, a, t \rangle \; = \; \langle a, t_0 \, | \, a, t_0 \rangle \; = 1. \tag{6.15}$$

1.3 The Fundamental Equation

If the time is incremented by infinitesimal amounts ε from a given time t, we obtain:

$$U(t, t) = \mathbb{1},$$
$$U(t + \varepsilon, t) = \mathbb{1} + \varepsilon \, B(t). \tag{6.16}$$

Let us introduce the time t as an intermediate time between $t + \varepsilon$ and t_0:

$$U(t + \varepsilon, t_0) = U(t + \varepsilon, t) \, U(t, t_0) = (\mathbb{1} + \varepsilon \, B(t)) \, U(t, t_0). \tag{6.17}$$

Following expansion and transition to the limit $\varepsilon \to 0$, this will yield:

$$\lim_{\varepsilon \to 0} \frac{U(t + \varepsilon, t_0) - U(t, t_0)}{\varepsilon} = B(t) \, U(t, t_0). \tag{6.18}$$

The first term defines the time derivative of the evolution operator (partial derivative if the function U depends on variables other than t):

$$\frac{\partial \, U(t, t_0)}{\partial \, t} \equiv \partial_t \, U(t, t_0) = B(t) \, U(t, t_0). \tag{6.19}$$

By applying this relation to the ket $| \, a, t \rangle \; = U(t, t_0) \, | \, a, t_0 \rangle$, we easily obtain the evolution of the ket itself:

$$\frac{\partial \, | \, a, t \rangle}{\partial \, t} = B(t) \, | \, a, t \rangle \tag{6.20}$$

and by considering the Hermitian conjugate of each term:

$$\frac{\partial}{\partial \, t} \langle a, t \, | \; = \; \langle a, t \, | \, B^+(t). \tag{6.21}$$

We will have to force the norm of the vector to remain constant (equal to one) during the time evolution, that is,

$$\langle a, t \,|a, t\rangle = 1 \quad \Longrightarrow \quad \frac{\partial}{\partial t} \langle a, t \,|\, a, t\rangle = 0. \tag{6.22}$$

This leads, with the previous relations, to the constraint

$$\langle a, t \,|\, B^+ + B \,|\, a, t\rangle = 0. \tag{6.23}$$

The operator $B(t)$ should therefore be anti-Hermitian (with imaginary eigenvalues). We regain a Hermitian operator denoted H by setting

$$B = \pm \frac{i}{\hbar} H, \tag{6.24}$$

such that the evolution operator obeys the first-order differential equation

$$\begin{aligned}\frac{\partial U(t, t_0)}{\partial t} &= \pm \frac{i}{\hbar} H(t)\, U(t, t_0) \\ U(t_0, t_0) &= U(t, t) = \mathbb{1}.\end{aligned} \tag{6.25}$$

We can also give the integral form of the operator which includes the initial condition:

$$U(t, t_0) = \mathbb{1} \pm \frac{i}{\hbar} \int_{t_0}^{t} H(t')\, U(t', t_0)\, dt'. \tag{6.26}$$

The choice of the sign $(+)$ or $(-)$ is arbitrary, because it is linked to the correspondence principle, so that we will use the choice immediately to remain consistent with the solution adopted:

$$\vec{p} \rightarrow -i\,\hbar\,\vec{\nabla}, \tag{6.27}$$

that is, the sign $(-)$, which gives the fundamental equation

$$\boxed{i\,\hbar\,\partial_t\, U(t, t_0) = H(t)\, U(t, t_0)}. \tag{6.28}$$

For the time being, $H(t)$ is a Hermitian operator, not yet identified with the system's Hamiltonian. The integral form of the fundamental Eq. (6.28) of quantum dynamics is written as in (6.26):

$$\boxed{U(t, t_0) = \mathbb{1} - \frac{i}{\hbar} \int_{t_0}^{t} H(t')\, U(t', t_0)\, dt'}. \tag{6.29}$$

By iterating the solution of this integral equation, we obtain

$$U_1(t, t_0) = \mathbb{1} - \frac{i}{\hbar} \int_{t_0}^{t} H(t_1) \, dt_1,$$

$$U_2(t, t_0) = \mathbb{1} - \frac{i}{\hbar} \int_{t_0}^{t} H(t_1) \, U_1(t_1, t_0) \, dt_1$$

$$= \mathbb{1} - \frac{i}{\hbar} \int_{t_0}^{t} H(t_1) \left\{ \mathbb{1} - \frac{i}{\hbar} \int_{t_0}^{t_1} H(t_2) \, dt_2 \right\} dt_1$$

$$= \mathbb{1} - \frac{i}{\hbar} \int_{t_0}^{t} H(t_1) \, dt_1 + \left(-\frac{i}{\hbar}\right)^2 \int_{t_0}^{t} H(t_1) \int_{t_0}^{t_1} H(t_2) \, dt_1 \, dt_2.$$

$$(6.30)$$

An n-fold iteration will thus lead to the Dyson series as the expression for the evolution operator:

$$U(t, t_0) = \mathbb{1} + \left(-\frac{i}{\hbar}\right) \int_{t_0}^{t} H(t_1) \, dt_1 + \left(-\frac{i}{\hbar}\right)^2 \int_{t_0}^{t} H(t_1) \, dt_1 \int_{t_0}^{t_1} H(t_2) \, dt_2$$

$$+ \ldots$$

$$+ \left(-\frac{i}{\hbar}\right)^n \int_{t_0}^{t} dt_1 \int_{t_0}^{t_1} dt_2 \ldots \int_{t_0}^{t_{n-1}} dt_n \, H(t_1) H(t_2) \ldots H(t_n),$$

$$(6.31)$$

with $t_1 < t_2 < t_3 \ldots < t_{n-1}$.

Two points need special emphasis here. The first is that differential Eq. (6.28) between operators does not permit us to simply write

$$\frac{d \, U(t, t_0)}{U(t, t_0)} = -\frac{i}{\hbar} H(t) \, dt \qquad (6.32)$$

and following integration over time:

$$U(t, t_0) = \exp\left[-\frac{i}{\hbar} \int H(t') \, dt'\right]. \qquad (6.33)$$

Equation (6.28) should, in fact, take the form

$$[d \, U(t, t_0)] \, U^{-1}(t, t_0) = -\frac{i}{\hbar} H(t) \, dt \qquad (6.34)$$

and there is no way of directly integrating the first term of this equation. The second important remark concerns the iteration series (6.31). It looks like the series expansion of the exponential (6.33) except that the coefficient $1/n!$ is missing. This is because $[H(t_1), \, H(t_2)] \neq 0$ and, therefore, we cannot interchange the terms of the series expansion and weight the rank n term with $1/n!$. This results from the fact that in expansion (6.30), the intermediate times are chronologically ordered. The introduction of a chronology (time-ordering) operator, τ, responsible for ordering the intermediate times and, as a result, for multiplying the rank n term by $n!$ enables us then to write the evolution operator, in its series development, in the following correct form:

$$U(t,\ t_0) = \tau\ \exp\left(-\frac{i}{\hbar}\int_{t_0}^{t}H(t')\ dt'\right)\ . \tag{6.35}$$

which is strictly equivalent to (6.31).

When we refer to a conservative system, we will mean a system in which the Hermitian operator H (which will be subsequently identified with the Hamiltonian) is time independent. The integration over time can then be performed and the evolution operator will assume the following simpler form:

$$U(t,\ t_0) = \exp\left(-\frac{i}{\hbar}\ (t-t_0)\ H\right)\ . \tag{6.36}$$

The chronology operator is no longer necessary since there are no more intermediate times to order.

Equation (6.4), in conjunction with (6.28), therefore makes it possible to also write the time evolution of a quantum state thus:

$$\begin{aligned} i\,\hbar\,\partial_t\,U(t,\ t_0) &= H(t)\,U(t,\ t_0)\\ i\,\hbar\,\partial_t\,|\,a,\ t\rangle\ &= H(t)\,|\,a,\ t\rangle\\ \text{with } U(t_0,\ t_0) &= \mathbb{1} \end{aligned} \qquad . \tag{6.37}$$

2. Two Different Descriptions of Quantum Dynamics

Realations (6.37) show that the evolution of the quantum state of a system or, put differently, of the operator $U(t,\ t_0)$, at a given point depends on the knowledge of the Hermitian operator $H(t)$. It is therefore necessary to link this operator to data accessible by observation, that is, to the result of measurements that can be made by any observer whatsoever.

Let us therefore consider a physical quantity that can be measured by an observer. The postulates of quantum mechanics state that a Hermitian operator A acting in the vector space of the quantum states can be associated to such a quantity and that the result of the measurement will yield either the eigenvalue a of the operator if the quantum state is unique (the pure case), or the eigenvalue weighted with the probability of existence of a quantum state p_k if there are several eigenvalues of A.

In a pure case, we will therefore write

$$\langle A \rangle = a = \langle a \mid A \mid a \rangle, \tag{6.38}$$

and in the general case

$$\langle A \rangle = \sum_k p_k \, a_k = \sum_k p_k \langle a_k \mid A \mid a_k \rangle. \tag{6.39}$$

The measurement of the observable can change with time, that is, we will, in fact, obtain $\langle A(t) \rangle$. For ease of comprehension, let us imagine that we are dealing with a pure case so we can use relation (6.38). Which will evolve with time: the quantum state or the operator or both of them simultaneously? These different points of view lead to entirely different descriptions of quantum dynamics.

2.1 Schrödinger's Description

Schrödinger's point of view is that the state of the Hilbert space evolves with time whereas the operators acting in it remain fixed, and are time independent except explicitly. This is expressed as:

$$\frac{dA}{dt} = \frac{\partial A}{\partial t}, \tag{6.40}$$

$$\mid a, \, t \rangle = U(t, \, t_0) \mid a, \, t_0 \rangle, \tag{6.41}$$

$$i \, \hbar \, \partial_t \mid a, \, t \rangle = H(t) \mid a, \, t \rangle. \tag{6.42}$$

This means that the observer is linked to the a state $\mid x \rangle$, which is constant in the configuration space, and

$$\langle x \mid a, t \rangle = \Psi_a(x, t). \tag{6.43}$$

By substituting (6.41) into the mean value (6.38) of the observable A, we obtain

$$\boxed{\frac{d \langle A(t) \rangle}{dt} = \frac{1}{i\hbar} \langle [A, \, H] \rangle + \langle \partial_t \, A \rangle}. \tag{6.44}$$

The measurement of the velocity of a particle will thus yield

$$\frac{d \langle X(t) \rangle}{dt} = \frac{dx(t)}{dt} = \dot{x}(t) = \frac{1}{i\hbar} \langle [X, H] \rangle. \tag{6.45}$$

It should be noted that the mean value of A (its eigenvalue) will obey the evolution equation of a classical dynamic variable (see (1.166)) if we use the correspondence principle (1.243) and if the operator H defining the evolution of the system is the classical Hamiltonian, expressed in the configuration space. If the observable is in fact the energy, i.e., the eigenvalue of the Hamiltonian, then relation (6.44) will give

$$\frac{d \langle H \rangle}{dt} = \langle \frac{\partial H}{\partial t} \rangle . \qquad (6.46)$$

In a conservative system, ($\partial H/\partial t = 0$), the energy is conserved over time. The projection on the configuration space of relation (6.42) will yield

$$i \hbar \, \partial_t \, \Psi_a(x, t) = H(x) \, \Psi_a(x, t) , \qquad (6.47)$$

which corresponds to the fundamental Schrödinger Eq. (2.3) in wave mechanics.

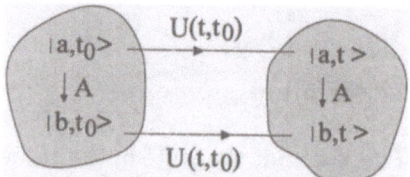

Schrödinger's description

2.2 Heisenberg's Description

We can also assume that the quantum state $|a\rangle$ is unchanging and that it is the operator A that changes with time. This is Heisenberg's point of view which expresses Eq. (6.38) thus:

$$\langle A \rangle = \langle a, t \mid A \mid a, t \rangle = \langle \bar{a} \mid \bar{A}(t) \mid \bar{a} \rangle . \qquad (6.48)$$

Assuming that the quantum state is unchanging, it can be fixed at its value at the initial instant t_0 and relation (6.41) used to express $\bar{A}(t)$ with respect to A (independent of t, except explicitly):

$$|\bar{a}\rangle = |a, t_0\rangle = U(t_0, t) \mid a, t\rangle = U^+(t, t_0) \mid a, t\rangle . \qquad (6.49)$$

By carrying this over into (6.48), we obtain the equality

$$\langle A \rangle = \langle \bar{a} \mid \bar{A}(t) \mid \bar{a} \rangle = \langle \bar{a} \mid U^+(t, t_0) A U(t, t_0) \mid \bar{a} \rangle . \qquad (6.50)$$

The operators in Heisenberg's description are linked to the operators in Schrödinger's description by the relation

$$\bar{A}(t) = U^+(t, t_0) A U(t, t_0) , \qquad (6.51)$$

and by using Eq. (6.40) and the fundamental relation (6.28), we have

$$\boxed{\frac{d\bar{A}(t)}{dt} = -\frac{1}{i\hbar}\left[\bar{A},\,\bar{H}\right] + \overline{\partial_t\,A}}\,. \qquad (6.52)$$

The Hermitian operator H defining the evolution of the system evolution (see relation (6.28)) commutes with itself, thus showing, with definition (6.51), that it does not vary with the description adopted:

$$\bar{H} = H\,,$$
$$\frac{d\bar{H}}{dt} = \frac{\partial\bar{H}}{\partial t} = \frac{\partial H}{\partial t}\,. \qquad (6.53)$$

The Hamiltonian is a time-independent (except explicitly) operator and a conservative system ($\partial H/\partial t = 0$) leads to an energy that is conserved over time.

By taking the mean value of (6.52) between time-independent states $|\bar{a}\rangle$, we quite naturally arrive back at the relation (6.44) concerning the observable mean values.

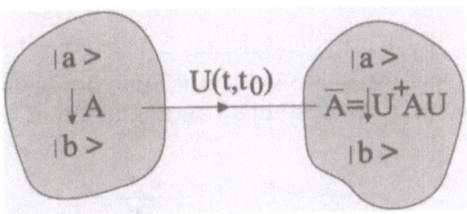

Heisenberg's description

Heisenberg's position operator:

$$\bar{X}(t) = U^+(t,\,t_0)\,X\,U(t,\,t_0) \qquad (6.54)$$

will have time-dependent eigenvalues identical to the eigenvalues of the Schrödinger position operator X. We can, in fact, write

$$\bar{X}(t)\,|\,\bar{x}\rangle = U^+(t,\,t_0)\,X\,U(t,\,t_0)\,|\,x,\,t_0\rangle$$
$$= U^+(t,\,t_0)\,X\,|\,x,\,t\rangle = \bar{x}\,|x,\,t_0\rangle\,. \qquad (6.55)$$

By multiplying by $U(t,t_0)$ from the left, we easily obtain:

$$X\,|\,x,\,t\rangle = \bar{x}\,|\,x,\,t\rangle\,. \qquad (6.56)$$

This is the eigenvalue equation of X and, hence, $x = \bar{x}$, which simply repeats definition (6.48) of $\langle X\rangle$.

The position operator X and momentum operator P are not explicitly time dependent:

$$\frac{\partial X}{\partial t} = 0 \quad \text{and} \quad \frac{\partial P}{\partial t} = 0 . \tag{6.57}$$

They satisfy evolution Eq. (6.52):

$$\frac{d\bar{X}}{dt} = \frac{1}{i\hbar} [\bar{X}, \bar{H}] \quad \text{and} \quad \frac{d\bar{P}}{dt} = \frac{1}{i\hbar} [\bar{P}, \bar{H}] . \tag{6.58}$$

It goes without saying that the central operator of quantum theory is the Hermitian operator H, which is unchanging both in Schrödinger's and in Heisenberg's description. All operators that commute with H are, following (6.52), constants of motion and make it possible to construct a common representational basis with this operator.

3. The Correspondence Principle

3.1 Classical Systems

In Chap. 1 (see (1.166)), we showed that a mechanical system represented by a function A of the position operator q_i and the momentum operator p_i (conjugate momentum of the system's Lagrangian function) obeys the law of classical dynamics:

$$\frac{dA}{dt} = [A, H] + \frac{\partial A}{\partial t} , \tag{6.59}$$

where $[A, H]$ represents the Poisson bracket of the function A and the classical Hamiltonian function:

$$H = \sum_i p_i \dot{q}_i - L (q_i, \dot{q}_i, t) . \tag{6.60}$$

In a conservative system, relation (6.59) will thus yield

$$\frac{\partial L}{\partial t} = 0 \implies \frac{\partial H}{\partial t} = 0 \implies \frac{dH}{dt} = 0 \implies H = \text{const} = E . \tag{6.61}$$

The position variables q_i and the conjugate momenta p_i obey, in particular, the dynamical Eq. (6.59) leading to the Hamiltonian equations of classical mechanics:

$$\frac{dq_i}{dt} = [q_i, H] = \frac{\partial H}{\partial p_i} ,$$

$$\frac{dp_i}{dt} = [p_i, H] = -\frac{\partial H}{\partial q_i} . \tag{6.62}$$

3.2 Quantum Systems

In Heisenberg's description, a quantum system is associated with an unknown time-independent quantum state represented by an unchanging vector $\mid a\rangle$ of the Hilbert space of the states. In this space, linear operators will act on $\mid a\rangle$ to modify the eigenvalue, that is, a measurement is made of the observable associated with the operator A whose $\mid a\rangle$ is an eigenstate. With the exception of H, these operators may change with time. It is H that permits the construction of a time-dependent unitary operator rendering every other operator, with the exception of H itself and the identity $\mathbb{1}$, time dependent.

3.3 Correspondence Principle

By comparing the classical dynamical law (6.59) and the quantum law (6.52), we can establish a correspondence principle leading to the following quantization:

$$
\begin{array}{|ccc|}
\hline
\text{dynamic variable} & & \text{Heisenberg's operator} \\
A(p_i, \ q_i) & \overset{\longrightarrow}{} & \bar{A}(P_i, \ Q_i) \\
\text{Poisson bracket [ ]} & \longrightarrow & \dfrac{1}{i\,\hbar}\,[\ ]\ \text{commutator} \\
\hline
\end{array}
\qquad . \qquad (6.63)
$$

We obtain, in particular, the commutation relations

$$
[X_i, \ X_j] = 0 \qquad [P_i, \ P_j] = 0 \qquad [X_i, \ P_i] = i\,\hbar\,\delta_{ij}\,, \qquad (6.64)
$$

which are valid in Heisenberg's description in keeping with the principle propounded. Using Poisson brackets, relations (6.58) also lead to the commutation relations

$$
[X_i, \ H] = i\,\hbar\,\frac{\partial H}{\partial p_i} \qquad \text{and} \qquad [P_i, \ H] = -i\,\hbar\,\frac{\partial H}{\partial x_i}\,, \qquad (6.65)
$$

where x and p_i are the eigenvalues of the operators X_i and P_i. We have, in addition, shown in the preceding chapters that the correspondence principle (6.63) can be obtained in an equivalent manner in non-relativistic wave mechanics with

$$
\vec{P} \ \longrightarrow \ -i\,\hbar\,\vec{\nabla}\,. \qquad (6.66)
$$

3.4 Ehrenfest's Theorem

By using (6.44) or (6.52) to expand the mean values of the operators X_i or P_i, we obtain with commutation relations (6.65)

$$\frac{d \langle X_i \rangle}{dt} = \frac{d x_i}{dt} = \dot{x}_i = \frac{1}{i\hbar} \langle [X_i,\, H] \rangle = \langle \frac{\partial H}{\partial p_i} \rangle,$$

$$\frac{d \langle P_i \rangle}{dt} = \frac{d p_i}{dt} = \dot{p}_i = \frac{1}{i\hbar} \langle [P_i,\, H] \rangle = \langle -\frac{\partial H}{\partial x_i} \rangle. \tag{6.67}$$

If we suppose that the measurable objects $\dot{x}_i$ and $\dot{p}_i$ are classical quantities obeying Hamiltonian equations of classical dynamics (6.62), we obtain Ehrenfest's theorem, which reads:

$$\frac{\partial H_{\text{classical}}}{\partial p_i} = \left\langle \frac{\partial H}{\partial p_i} \right\rangle_{\text{quantum}},$$

$$\frac{\partial H_{\text{classical}}}{\partial x_i} = \left\langle \frac{\partial H}{\partial x_i} \right\rangle_{\text{quantum}}. \tag{6.68}$$

The measurements made by the classical observer lead to the mean values of the Hermitian operators associated with the corresponding dynamic variables.

Let us emphasize here the fact that Ehrenfest's theorem, which replaces the mean value $\langle -\partial V/\partial x \rangle$ with $-\partial V(\langle x \rangle)/\partial x$, is only valid for particular quantum states: namely, those that can be described by a narrow wave packet. For such states, successive observations of the position over time will show the classical correlations predicted by the equations of motion, provided that these observations are rough enough such that the replacement of the mean value of the potential with the potential of the mean position does not affect the observations themselves. A precise determination of the position, for example, will produce a completely delocalized wave packet at a given future instant and Ehrenfest's substitution will cease to be a good approximation.

4. Intermediate Description

It is sometimes convenient to cause the quantum states and operators to evolve if the Hamiltonian of the quantum system can be separated into a known part H_0 and a perturbation V. This is the basis of the theory of time-dependent perturbations and Feynman diagrams to which we will return in Chaps. 14 and 15.

Let us imagine that the Hamiltonian can be separated thus:

$$H = H_0 + V \tag{6.69}$$

and then associate with the unperturbed Hamiltonian H_0 an evolution operator U_0 obeying the usual conditions

$$U_0(t, t) = U_0(t_0, t_0) = \mathbb{1},$$

$$i \hbar \, \frac{\partial \, U_0(t, t_0)}{\partial t} = H_0(t) \, U_0(t, t_0).$$

(6.70)

We can associate the total Hamiltonian H with another evolution operator $U(t, t_0)$ that will obey analogous relations:

$$U(t, t) = U(t_0, t_0) = \mathbb{1},$$

$$i \hbar \, \frac{\partial \, U(t, t_0)}{\partial t} = H(t) \, U(t, t_0).$$

(6.71)

Let us further determine the evolution operator $U_I(t, t_0)$ associated with the perturbation by setting

$$U(t, t_0) = U_0(t, t_0) \, U_I(t, t_0).$$

(6.72)

Inserting this into relation (6.71), we obtain, with (6.69),

$$i \hbar \, (\partial_t \, U_0) \, U_I + i \hbar \, U_0 \, \partial_t \, U_I = (H_0 + V) \, U_0 \, U_I$$

multiplying from the left by U_0^+ and then using relation (6.70), we obtain

$$i \hbar \, \partial_t \, U_I = U_0^+ \, V \, U_0 \, U_I.$$

(6.73)

We further introduce the operator $H_I(t, t_0)$, the transform of $V(t, t_0)$ in the transformation $U_0(t, t_0)$, that is,

$$H_I(t, t_0) = U_0^+(t, t_0) \, V \, (t, t_0) \, U_0(t, t_0)$$

(6.74)

The evolution operator associated with the perturbation V obeys the evolution Eq. (6.73) which will now take the following form:

$$\boxed{\begin{array}{c} i \hbar \, \partial_t \, U_I(t, t_0) = H_I(t, t_0) \, U_I(t, t_0) \\ U_I(t, t) = U_I(t_0, t_0) = \mathbb{1} \end{array}}$$

(6.75)

The description is said to be intermediate because the operators expressed with respect to time according to form (6.74):

$$A_I(t, t_0) = U_0^+(t, t_0) \, A \, U_0(t, t_0)$$

(6.76)

evolve just like in Heisenberg's picture, since

$$i\hbar \frac{d A_I}{dt} = \left(i\hbar\, \partial_t\, U_0^+\right)\, A\, U_0 + i\hbar\, U_0^+\, \left(\partial_t\, A\right)\, U_0 + i\hbar\, U_0^+\, A\, \partial_t\, U_0$$

will yield, with relations (6.70) and (6.76),

$$\boxed{i\hbar \frac{d A_I}{dt} = [A_I,\ H_I^o] + i\hbar\, \left(\partial_t\, A\right)_I} \,. \tag{6.77}$$

The ket vectors of the intermediate representation evolve, for their part, just like in Schrödinger's picture.

To define the intermediate states, we introduce (6.72):

$$\mid I,t\rangle\ = U_0^+ \mid a,t\rangle\ = U_0^+\, U_0\, U_I \mid a,t_0\rangle\ = U_I \mid a,t_0\rangle \tag{6.78}$$

and determine the time evolution of these vectors:

$$\begin{aligned} i\hbar\, \partial_t \mid I,t\rangle\ &= i\hbar\, \partial_t\, U_I(t,\, t_0) \mid a,t_0\rangle \\ &= H_I(t,\, t_0)\, U_I(t,\, t_0) \mid a,t_0\rangle \,. \end{aligned} \tag{6.79}$$

We then obtain a Schrödinger-type evolution:

$$\boxed{i\hbar\, \partial_t \mid I,t\rangle\ = H_I(t,\, t_0) \mid I,t\rangle} \,. \tag{6.80}$$

Let us assume that the intermediate states constitute a complete discrete orthonormal system at a given instant:

$$\langle I,t \mid I',t\rangle\ = \delta_{II'}$$
$$\sum_I \mid I,t\rangle\ \langle I,t\mid\ = \mathbb{1} \,. \tag{6.81}$$

This amounts to the assumption that the operator $U_I(t,\, t_0)$ is unitary and that the states $\mid a,t\rangle$ also constitute a complete orthonormal system.

The evolution operator of the intermediate representation can therefore be written as in (6.10), that is

$$U_I(t,t_0) = \sum_I \mid I,t\rangle\ \langle I,t_0 \mid, \tag{6.82}$$

and the matrix element of the operator, considered between two quantum states at instant t_0, will yield with (6.78)

$$\langle I',\, t_0 \mid U_I(t,\, t_0) \mid I,\, t_0\rangle\ = \langle I',t_0 \mid I, t\rangle \,. \tag{6.83}$$

The fourth postulate of quantum mechanics defines the transition probability from the state $I'(t_0)$ to the state $I(t)$ as the square modulus of the above matrix element:

$$P_r \left(I'(t_0) \rightarrow I(t) \right) = \left| \langle I', t_0 \mid I, t \rangle \right|^2$$
$$= \left| \langle I', t_0 \mid U_I(t, t_0) \mid I, t_0 \rangle \right|^2 . \qquad (6.84)$$

If we write the states $\mid I, t_0 \rangle$ and $\mid I', t_0 \rangle$ as $\mid \alpha \rangle$ and $\mid \beta \rangle$:

$$\begin{aligned}
\mid I, t_0 \rangle &= \mid a, t_0 \rangle = \mid \alpha(t_0) \rangle , \\
\mid I', t_0 \rangle &= \mid a', t_0 \rangle = \mid \beta(t) \rangle ,
\end{aligned} \qquad (6.85)$$

we obtain the transition probability

$$P_r \left(\alpha \rightarrow \beta \right) = \left| \langle \beta \mid U_I(t, t_0) \mid \alpha \rangle \right|^2 . \qquad (6.86)$$

This is a very important result to which we will return when discussing time dependent perturbations and Feynman diagrams (Chap. 14).

5. Interpretation of Decoherence in Quantum Mechanics

The interpretation of quantum mechanics adopted by the Copenhagen School (see Chap. 2) is based on the separation of a system into a first part governed by classical mechanics (the observer and his measuring apparatus) and a second (the physical system being observed) governed, through essential hypothesis, by quantum mechanics. To date, no experiment or observation has suggested that this procedure and the interpretation of wave function in terms of the probability amplitude are wrong. Everett (1957) was the first to propose that quantum mechanics be considered not as a model but rather as a theoretical expression in mathematical terms for describing the Universe and, therefore, to apply the theory to the Universe considered as a whole, as a closed system. This raises the problem of the observer and his role in choosing the observation and the measurement. Numerous subsequent authors have taken up the problem of "quantum cosmology" and, more generally speaking, the interpretation of the quantum mechanics of an overall system in which the observer and the object being observed are analogous and equivalent parts of the system. We will introduce the main foundational aspects of this study using J.B. Hartle's excellent paper (Hartle, 1993) as a departure point.

First, let us consider again Young's fundamental two-slit experiment involving particles (see Chap. 2) and its interpretation by the Copenhagen School. The wave functions $\Psi_1(y)$ and $\Psi_2(y)$ represent the probability amplitude that a particle detected at point y on the screen has passed through slit 1 or slit 2 respectively. The measurement in y of the number of particles collected is related to the square modulus of the total probability amplitude:

$$| \Psi_1(y) + \Psi_2(y) |^2 \neq | \Psi_1(y) |^2 + | \Psi_2(y) |^2 . \tag{6.87}$$

The interferential aspect of the measurement is correctly given by the missing term. If we could have measured through which slit the electron had passed, then one of the probability amplitudes would have been zero and the interferential aspect destroyed. In the Copenhagen interpretation, the probabilities are not assigned in a general manner but, rather, to the different histories of a sub-system measured by an observer. This would then mean that quantum mechanics is impossible without the presence of an observer. This cannot be the case in quantum cosmology since the Universe exists prior to any measurement by an observer. It is therefore necessary to define a rule for assigning measurement-independent probabilities to the different possible histories of a closed system.

Let P_1 be the projector of the Hilbert space associated with the probability that a particle will pass through slit 1. The eigenvalues of P_1 are 0 or 1, indicating that the particle either does not pass or passes through slit 1. If $| \vec{x}, s \rangle$ is a localized quantum state in $\vec{x}$ with the internal quantum numbers s, (spin, isospin, etc.), then the projector P_1 will take the following form:

$$P_1 = \sum_s \int_1 | \vec{x}, s \rangle \, d^3 x \, \langle \vec{x}, s | . \tag{6.88}$$

The integral extends over a (small) volume surrounding slit 1. The probability of passing through slit 2 will be denoted P_2 and the fact that P_1 and P_2 are mutually exclusive expressed by their complementarity:

$$P_1^2 = P_1 \qquad P_2^2 = P_2 \qquad P_1 P_2 = 0 \qquad P_1 + P_2 = \mathbb{1} . \tag{6.89}$$

The probabilities of the particle reaching different y points on the screen are also linked to the projectors P_y. Using (6.51), Heisenberg's description enables us to define the projector P at instant t or the time-probability $P(t)$ from what we know about the (time-independent) Hamiltonian H of the quantum sub-system consisting of the particle under observation and which is in interaction with the measurement apparatus, i.e., slits 1 and 2 and the observation screen:

$$P(t) = \exp\left(\frac{i}{\hbar} H t\right) P \exp\left(-\frac{i}{\hbar} H t\right) . \tag{6.90}$$

What is the probability that a particle in an initial normalized quantum state $| \Psi(t_0) \rangle$ has passed through slit 1 at instant t_1 (for example) and reached point y at instant t_2?

The evolution from instant t_0 to instant t_1 of the quantum state is given by relation (6.36) applied to the initial state $| \Psi(t_0) \rangle$:

$$| \Psi(t_1) \rangle = \exp\left(-\frac{i}{\hbar} H (t_1 - t_0) \right) | \Psi(t_0) \rangle . \qquad (6.91)$$

The probability that the result of the measurement at instant t_1 will show the particle passing through slit 1 will then be

$$P_r(\text{slit } 1) = | P_1 | \Psi(t_1) \rangle |^2 . \qquad (6.92)$$

After the measurement (assumed not to perturb the particle), the quantum state (still normalized to unity) at instant t_1 will be

$$\frac{P_1 | \Psi(t_1) \rangle}{| P_1 | \Psi(t_1) \rangle |} . \qquad (6.93)$$

The evolution from instant t_1 to the arrival instant t_2 on the screen will give the quantum state

$$| \Psi(t_2) \rangle = \exp\left(-\frac{i}{\hbar} H (t_2 - t_1) \right) \frac{P_1 | \Psi(t_1) \rangle}{| P_1 | \Psi(t_1) \rangle |} . \qquad (6.94)$$

The probability of detecting the particle at y at instant t_2, conditioned in the meantime by the passage through slit 1, will therefore be

$$P_r(y/ \text{ slit } 1) = | P_y | \Psi(t_2) \rangle |^2 . \qquad (6.95)$$

The probability that a particle passes through slit 1 and is detected at point y is the product of (6.92) and (6.95), that is,

$$
\begin{aligned}
P_r(y \text{ and slit } 1) &= P_r(y/ \text{ slit } 1) \times P_r(\text{slit } 1) \\
&= \left| P_y \exp\left(-\frac{i}{\hbar} H (t_2 - t_1) \right) \frac{P_1 | \Psi(t_1) \rangle}{| P_1 | \Psi(t_1) \rangle |} \right|^2 \\
&\quad \times | P_1 | \Psi(t_1) \rangle |^2 .
\end{aligned}
\qquad (6.96)
$$

The denominator is eliminated and $| \Psi(t_1) \rangle$ expressed with (6.91), thus leading to the joint probability

$$P_r(y \text{ and slit } 1) = | P_y U(t_2, t_1) P_1 U(t_1, t_0) | \Psi(t_0) \rangle |^2 . \qquad (6.97)$$

By introducing time-probabilities (6.90), this reduces to

$$\boxed{P_r(y \text{ and slit } 1) = | P_y(t_2) P_1(t_1) | \Psi(t_0) \rangle |^2} \ . \qquad (6.98)$$

Everything in this expression (projector, state vector, Hamiltonian) refers to the Hilbert space of a sub-system, the sub-system of the particle being

measured. In the Copenhagen interpretation, it is the measurement that determines the sub-system history to which can be attributed a probability such
as (6.98). For a closed system, such as the Universe considered in its entirety,
there is no separation between measured and measurable sub-systems so that
the foregoing interpretation will have to be revised.

Everett's theory (Everett, 1957) takes into account all simultaneously
possible histories. It is not because the observer decides to measure only
those particles passing through slit 1 that there will be no particles passing
through slit 2 or elsewhere. The observer and the object being measured are
on the same evolution branch of the Universe, a fact that in no way excludes
the possibility of the existence of other parallel worlds resulting from different
evolutions.

For a closed system, a set of possible histories will have to be envisaged.
For example, in Young's two-slit experiment, we could envisage:

1   the possibility of measuring or not measuring the slit through which the
    particle passes,
2   the possibility that the particle will pass through slit 1 or slit 2,
3   the possibility that the particle will reach one of the positions
    $y_1 \, y_2 \cdots y_9 \cdots y_N$ for each of the preceding eventualities.

These different possible histories are represented by different branches of
the following diagram:

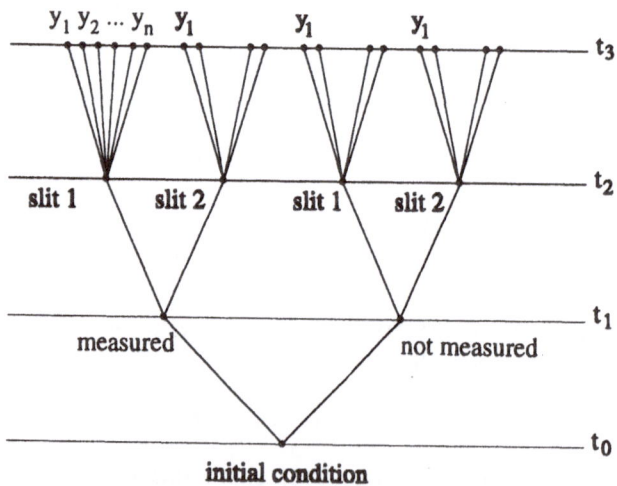

Everett's diagram of the branches of possible histories

Different alternatives at a given instant are represented by orthogonal
projection operators which are time-probabilities in Heisenberg's picture
$P^k_{\alpha_k} \, (t_k)$, where t_k is the time in question, k the set of possible alternatives

at instant t_k (for example, the positions $y_1 \ldots y_N$ on the screen) and α_k a path corresponding to a particular scenario leading, for example, to position y_9 on the screen. The time-probabilities satisfy the following axioms:

$$\sum_{\alpha_k} P^k_{\alpha_k}(t_k) = \mathbb{1},$$

$$P^k_{\alpha_k}(t_k) \, P^k_{\alpha'_k}(t_k) = \delta_{\alpha_k \, \alpha'_k} \, P^k_{\alpha_k}(t_k). \tag{6.99}$$

An individual history in a given whole corresponds to the sequence $(\alpha_1 \ldots \alpha_n) \equiv \alpha$ and there is a chain of projection operators for each particular history:

$$C_\alpha \equiv P^n_{\alpha_n}(t_n) \ldots P^1_{\alpha_1}(t_1). \tag{6.100}$$

This corresponds to the chain of operators related to any branch in the above Everett's diagram for example:

$$P^3_{\text{position } y}(t_3) \qquad P^2_{\text{slit1}}(t_2) \qquad P^1_{\text{measured}}(t_1). \tag{6.101}$$

If the initial state is a pure one represented by $\mid \Psi \rangle$, then the chain of operators (6.100) and the completeness relation (6.99) can be applied to it:

$$\mid \Psi \rangle = \sum_\alpha C_\alpha \mid \Psi \rangle = \sum_{\alpha_n \ldots \alpha_1} P^n_{\alpha_n}(t_n) \ldots P^1_{\alpha_1}(t_1) \mid \Psi \rangle. \tag{6.102}$$

The Hilbert space vector $C_\alpha \mid \Psi \rangle$ describes the branch corresponding to the history α and (6.102) is none other than the mathematical description of the diagram indicating the different possible branches resulting from the initial state $\mid \Psi \rangle$.

There is decoherence for a set of α and α' histories when

$$\langle \Psi \mid C^+_{\alpha'} \, C_\alpha \mid \Psi \rangle \simeq 0 \qquad \forall \, \alpha'_k \neq \alpha_k \tag{6.103}$$

indicating that branches α and α' are almost orthogonal and are mutually exclusive or, to be more precise, that interference between histories α and α' is small enough for us to be able to attribute a probability to each of the branches α and α'.

The probability of emergence in a future measurement of a history α is given by

$$p(\alpha) = \mid C_\alpha \mid \Psi \rangle \mid^2 \tag{6.104}$$

and the decoherence implies the sum rule of the probabilities:

$$\sum_{\alpha_2} p(\alpha_3, \alpha_2, \alpha_1) = p(\alpha_3, \alpha_1). \tag{6.105}$$

This is easily noticed by expanding the first term of the equation:

$$\sum_{\alpha_2} p(\alpha_3, \alpha_2, \alpha_1) = \sum_{\alpha_2} \langle \Psi \mid P^1_{\alpha_1} \, P^2_{\alpha_2} \, P^3_{\alpha_3} \, P^3_{\alpha_3} \, P^2_{\alpha_2} \, P^1_{\alpha_1} \mid \Psi \rangle. \tag{6.106}$$

The decoherence (6.103) permits us to introduce the history α_2' with a negligible error margin and relation (6.99) to extract the sum over α_2:

$$\sum_{\alpha_2} p\,(\alpha_3, \alpha_2, \alpha_1) \simeq \sum_{\alpha_2\,\alpha_2'} \langle \Psi \mid P_{\alpha_1}^1\, P_{\alpha_2'}^2\, P_{\alpha_3}^3\, P_{\alpha_3}^3\, P_{\alpha_2}^2\, P_{\alpha_1}^1 \mid \Psi \rangle$$

$$= \sum_{\alpha_2\,\alpha_2'} \langle \Psi \mid P_{\alpha_1}^1\, \delta_{\alpha_2\,\alpha_2'}\, P_{\alpha_2}^2\, P_{\alpha_3}^3\, P_{\alpha_3}^3\, P_{\alpha_1}^1 \mid \Psi \rangle$$

$$= \langle \Psi \mid P_{\alpha_1}^1\, P_{\alpha_3}^3 \sum_{\alpha_2} P_{\alpha_2}^2\, P_{\alpha_3}^3\, P_{\alpha_1}^1 \mid \Psi \rangle$$

$$= \langle \psi \mid P_{\alpha_1}^1\, P_{\alpha_3}^3\, P_{\alpha_3}^3\, P_{\alpha_1}^1 \mid \Psi \rangle$$

$$= p\,(\alpha_3, \alpha_1). \tag{6.107}$$

Decoherent histories of the Universe can therefore be used in prediction processes in quantum mechanics because they can be assigned probabilities. Thus, decoherence generalizes and replaces the notion of measurement in the Copenhagen interpretation predicting the probabilities of sub-systems under measurement. The measured quantities are linked to decoherent histories. In Young's two-slit experiment, when a particle interacts with an apparatus that determines the slit through which the particle has passed, it is the decoherence of the different configurations of the apparatus that enables us to assign the existence (or passage) probabilities to the particle. The Copenhagen interpretation is an approximation of a more general framework related to measurement situations in which the decoherence of the different configurations of the apparatus can be idealized as if it were complete and instantaneous. Although measurement processes imply decoherence, they are only special cases of decoherent histories. Probabilities can be assigned to the different alternative decoherences of a system whether or not there is an observer to record the values. This brings a different light to bear on Schrödinger's cat paradox which led Everett to propound the theory of parallel worlds. Perhaps we should briefly recall the paradox here. A cat is locked up in an enclosure in which a random system sets off the discharge of a deadly gas. As long as the observer does not look into the enclosure, there is a 50% chance that the cat is dead and a 50% chance that it is alive. The Copenhagen interpretation asserts that it is the observation that determines the wave function of the cat and the probability that it will be alive at the time of the observation. Everett's interpretation is that there is a branch of the Universe in which the cat is living and another in which it is dead and the observer exists in one or the other of these parallel universes at a time. Decoherence presupposes that the two histories have been decohered and that we can assign a probability (50 %) to each of the histories even if we do not observe the inside of the enclosure. This interpretation is more in tune with what seems logical to the human mind and does not give rise to the inevitable speculations associated with "parallel worlds" or with speculations resulting from the presence of man in the Universe as an indispensable element of every measurement process and, hence, of the existence of the world as such.

6. Quantum Dynamics – Schematic Summary

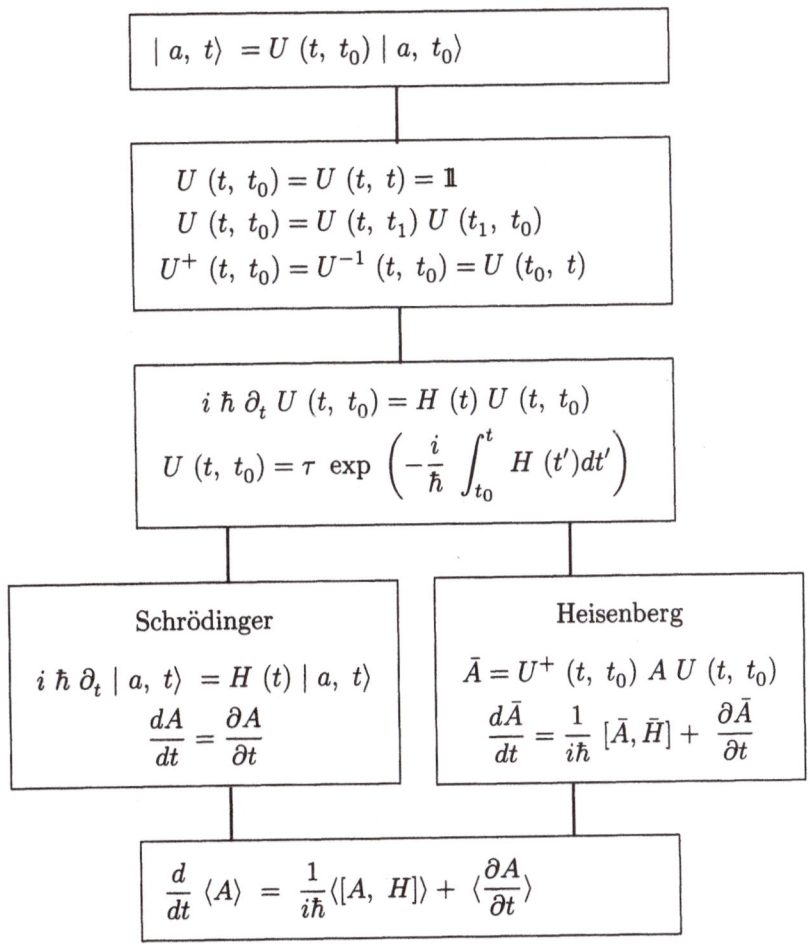

Chapter 7
Path Integrals

The results obtained with quantum dynamics make it possible to give a different description of quantum mechanics and to introduce the more general notion of path integrals. This formulation, introduced by R.P. Feynman in 1965, may be considered as another quantization method. It is today the method of choice in those areas, such as Cosmology, in which other methods are difficult to apply. In the path integral formulation of quantum mechanics a phase is generally attributed to each trajectory. The value of such a phase is determined by classical mechanics. The amplitude for going from point $\vec{X}(t_1)$ to point $\vec{X}(t_2)$ is given by $\exp[(i/\hbar)S]$, where S is the classical action $S = \int L dt$. Because this action is not unambiguously determined, the amplitude may contain a phase factor which may be considered as unobservable and hence arbitrary. This point is much more crucial than one might think. Because the amplitude of the wave function is the sum of all the amplitudes resulting from all possible paths, the result thus obtained is in every respect analogous to the Huygens' principle of wave optics.

The classical action becomes the equivalent of the optical path and $\hbar$ the equivalent of the wavelength. This result is the same as that obtained in Chap. 1 by extending classical mechanics. A quantum process can therefore be described as an optical event, not in ordinary space but rather in configuration space. The phase is then linked to certain properties of integral paths in configuration space.

1. Evolution of the Wave Function

1.1 The Quantum State and the Observer

Configuration space is the space of classical variables permitting the description of the instantaneous state of the system. The wave function is a projection onto configuration space chosen by the observer and hence is linked to the observer. The quantum system will be described in Schrödinger's picture whereas the observer will be in Heisenberg's picture using a fixed ket $| x \rangle$ and observer-linked and time-dependent operators.

quantum system	$i \hbar \, \partial_t \, | a, t \rangle \; = H(t) \, | a, t \rangle$
observer	$\begin{cases} | x \rangle \text{ fixed} \\ \bar{X} = U^+ \, XU \end{cases}$
wave function	$\psi_a(x, t) = \langle x \, | \, a, t \rangle$

$$\text{(7.1)}$$

The wave function of the system is therefore the solution of the time-dependent Schrödinger equation

$$i \hbar \, \partial_t \, \psi_a(x, t) = H(t) \, \psi_a(x, t) \,. \tag{7.2}$$

We may also assume that the quantum state is described in Heisenberg's picture (it is fixed) and the operators evolve with time whereas the observer state is written in Schrödinger's picture and evolves with time.

quantum system	$\begin{cases} | a \rangle \text{ fixed} \\ \bar{A} = U^+ AU \end{cases}$
observer	$- i \hbar \, \partial_t \langle x, t \, | \; = \; \langle x, t \, | \, H(t)$
wave function	$\Psi_a(x, t) = \; \langle x, t \, | \, a \rangle$

$$\text{(7.3)}$$

The evolution of the wave function is therefore described by the Schrödinger equation

$$-i \hbar \, \partial_t \, \Psi_a(x, t) = H(t) \, \Psi_a(x, t) \,. \tag{7.4}$$

We notice that solutions (7.2) and (7.4) are obtained by replacing t with $-t$:

$$\langle x \, | \, a \, (t) \rangle = \langle \, x(-t) \, | \, a \rangle \tag{7.5}$$

or, put differently,

$$\psi_a(x, t) = \Psi_a(x, -t) \,. \tag{7.6}$$

This is a general property which will come under discussion in the next chapter in connection with time-reversal transformation. An analogous result can be obtained with the relativity principle in classical mechanics: the observer is considered stationary and an object in motion is traveling at the speed $\vec{v}$ or the observer travels at the speed $-\vec{v}$ when the object is considered to be stationary.

1.2 Propagator of the Wave Function

Let us place ourselves within the framework of (7.3) of an observer in motion for a stationary quantum state. At the initial instant

$$\Psi_a(x_0, t_0) = \langle x_0, t_0 \mid a \rangle \tag{7.7}$$

and at instant t the wave function becomes

$$\Psi_a(x, t) = \langle x, t \mid a \rangle . \tag{7.8}$$

For the sake of simplicity, we may assume that the quantum state $\mid a \rangle$ is unique and normalized to unity:

$$\begin{aligned} \mid a \rangle \langle a \mid &= \mathbb{1}, \\ \langle a \mid a \rangle &= 1 . \end{aligned} \tag{7.9}$$

Or we may even introduce a series of quantum states without this modifying the problem at hand in any way.

There is an operator termed Green's function which propagates the wave function from (x_0, t_0) to (x, t) and this propagator will be denoted $G(x, t; x_0, t_0)$:

$$\begin{aligned} G(x, t; x_0, t_0) &= \langle x, t \mid a \rangle \langle a \mid x_0, t_0 \rangle \\ &= \Psi_a(x, t)\, \Psi_a^*(x_0, t_0) , \end{aligned} \tag{7.10}$$

or, by using the closure on the quantum system

$$G(x, t; x_0, t_0) = \langle x, t \mid x_0, t_0 \rangle . \tag{7.11}$$

It should be noted that the form obtained in this way looks more like an inner product than an operator but because the vectors are defined at different instants, they are not in dual spaces. Conversely, if $t = t_0$, then we actually obtain dual vectors and their inner product will define the boundary conditions:

$$G(x, t; x_0, t) = G(x, t_0; x_0, t_0) = \delta(x - x_0) . \tag{7.12}$$

Let us determine the wave function at (x, t) by projecting $\Psi_a(x, t)$ onto the observer quantum state $\mid x_0, t_0 \rangle$:

$$\begin{aligned} \Psi_a(x, t) = \langle x, t \mid a \rangle &= \int \langle x, t \mid x_0, t_0 \rangle \, dx_0 \langle x_0, t_0 \mid a \rangle \\ &= \int G(x, t; x_0, t_0)\, \Psi_a(x_0, t_0)\, dx_0 . \end{aligned} \tag{7.13}$$

The wave function at (x, t) is determined by the knowledge of the wave functions at an earlier instant t_0 on every point in space. We see here at work the fundamental property of non-separability in quantum mechanics. This expression of the wave function reminds one of Huygens' principle of optics which states that the optical state at a given point is obtained by the integration of the optical state at an earlier instant (wave front) over the entire space.

In a conservative system, the integration of (7.2) will give the evolution at point x:

$$\Psi_a(x,t) = \exp\left[-\frac{i}{\hbar} H (t - t_0)\right] \Psi_a(x,t_0).\qquad(7.14)$$

This can be expressed completely independently of the observed quantum state $| a \rangle$ representing the observer quantum state in Schrödinger's picture:

$$\langle x,t | = \langle x,t_0 | \exp\left[-\frac{i}{\hbar}(t - t_0) H\right]\qquad(7.15)$$

or in the equivalent but simplified (from the notational point of view) form

$$\langle x,t | = \langle x | \exp\left(-\frac{i}{\hbar} (t - t_0) H (t)\right).\qquad(7.16)$$

The propagator (7.11) of the wave function will then take the form:

$$G(x',t';x,t) = \langle x',t' | x,t \rangle = \langle x' | \exp\left[-\frac{i}{\hbar}(t' - t) H\right] | x \rangle \qquad(7.17)$$

whereas, with (7.13), the wave function becomes

$$\Psi_a(x',t') = \int G(x',t';x,t)\, \Psi_a(x,t)\, dx \qquad(7.18)$$

It is therefore necessary to know the propagator $G(x't'; x,t)$ and an initial state $\Psi_a(x,t)$ (determined by solving the Schrödinger equation or by any other means) to be able to determine the observed wave function corresponding to a quantum state $| a \rangle$ of the physical system. And this is precisely what the path integral method seeks to calculate.

1.3 Examples of Propagators

Consider now the simplest example imaginable, that of the propagation of a particle with mass m. The Hamiltonian $H = p^2/2m$ has the eigenvectors $| p \rangle$ forming a complete orthonormal system. Projected onto the eigenstates, the propagator (7.17) becomes

$$G(x',t';x,t) = \int \langle x' \mid p \rangle \, dp \, \langle p \mid \exp\left(-\frac{i}{\hbar}(t'-t)H\right) \mid p' \rangle dp' \langle p' \mid x \rangle. \quad (7.19)$$

By using the projection of phase space onto configuration space (5.57), that is, the overlap

$$\langle x \mid p \rangle = \frac{1}{\sqrt{2\pi\hbar}} \exp\left(\frac{i}{\hbar} px\right), \quad (7.20)$$

we obtain the integral form of the preceding propagator:

$$G(x',t';x,t) = \frac{1}{2\pi\hbar} \int dp \, \exp\left\{\frac{i}{\hbar}\left[p(x'-x) - (t'-t)\frac{p^2}{2m}\right]\right\}. \quad (7.21)$$

The evaluation of this integral uses the following standard result:

$$\int \frac{dp}{2\pi} \exp\left(-ap^2 - bp\right) = \frac{1}{\sqrt{4\pi a}} \exp\left(\frac{b^2}{4a}\right).$$

We infer from this the integrated form of the propagator of a free particle wave function when setting $\Delta t = \mid t' - t \mid$:

$$
\boxed{
\begin{array}{c}
\text{propagator of a free particle} \\[2mm]
G(x',t+\Delta t;x,t) = \sqrt{\frac{m}{2\pi\,i\,\hbar\,\Delta t}} \; \exp\left[\frac{i\,m(x'-x)^2}{2\hbar\,\Delta t}\right].
\end{array}
} \quad (7.22)
$$

It is also possible to analytically determine the propagator of a harmonic oscillator wave function. The result is naturally not as simple as (7.22) and is only introduced here as a reminder:

$$
\boxed{
\begin{array}{l}
\text{propagator of a harmonic oscillator} \\[2mm]
G(x',t+\Delta t;x,t) = \left[\dfrac{m\,\omega}{2\pi\,i\,\hbar\,\sin\omega\,\Delta t}\right]^{1/2} \\[4mm]
\times \exp\left[\dfrac{i\,m\,\omega}{2\hbar\,\sin\omega\,\Delta t}\left\{(x^2+x'^2)\cos\omega\,\Delta t - 2x\,x'\right\}\right]
\end{array}
} \quad (7.23)
$$

It is interesting to note that for an increase Δx of the position and Δt of the time and by replacing $\sin\omega\Delta t$ with $\omega\Delta t$ and $\cos\omega\Delta t$ with 1 we obtain exactly the same propagator as for a free particle:

$$G(x+\Delta x, t+\Delta t; x, t) = \left(\frac{m}{2\pi\,i\,\hbar\,\Delta t}\right)^{1/2} \exp\left[\frac{i\,m}{2\hbar}\left(\frac{\Delta x}{\Delta t}\right)^2 \Delta t\right]. \quad (7.24)$$

If we replace $\Delta x / \Delta t$ by the velocity $\dot{x}$, we obtain the following expression for the propagator:

$$G(x + \Delta x, t + \Delta t; x, t) = \left(\frac{m}{2\pi i \hbar \Delta t} \right)^{1/2} \exp \left[\frac{i}{\hbar} \frac{1}{2} m \dot{x}^2 \Delta t \right]. \quad (7.25)$$

We may therefore assume that we are dealing with the propagator of a wave function for an infinitesimal movement in space and in time.

2. Path Integrals

2.1 Feynman's Conjecture

The form of the propagator (7.11) shows that it is possible to include as many intermediate states as one wishes:

$$G(x', t'; x, t) = \langle x', t' \mid x, t \rangle = \int \langle x', t' \mid x'', t'' \rangle dx'' \langle x'', t'' \mid x, t \rangle$$

$$= \int G(x', t'; x'' t'') \, G(x'', t''; x, t) \, dx''. \quad (7.26)$$

A finite time lapse $(t' - t)$ can therefore be split into n ε parts and the corresponding propagator written

$$\langle x', t' \mid x, t \rangle = \int \langle x', t' \mid x_{n-1}, t_{n-1} \rangle dx_{n-1} \langle x_{n-1}, t_{n-1} \mid x_{n-2}, t_{n-2} \rangle \cdots$$

$$\cdots dx_1 \langle x_1, t_1 \mid x, t \rangle \quad (7.27)$$

and we only need to know the propagator for the transitions from one point (x_i, t_i) to the next infinitely near point (x_{i+1}, t_{i+1}). By setting $\varepsilon = t_{i+1} - t_i$, we obtain

$$\langle x_{i+1}, t_{i+1} \mid x_i, t_i \rangle = \langle x_{i+1} \mid \exp \left(-\frac{i}{\hbar} \varepsilon H \right) \mid x_i \rangle. \quad (7.28)$$

By projecting onto the phase space, we obtain, as for the free particle,

$$\langle x_{i+1}, t_{i+1} \mid x_i, t_i \rangle = \int \frac{dp_i}{2\pi\hbar} \exp \frac{i\varepsilon}{\hbar} \left[\left(p_i \frac{x_{i+1} - x_i}{\varepsilon} \right) - H \left(p_i, \frac{x_i + x_{i+1}}{2} \right) \right]. \quad (7.29)$$

The path integrals leading to the propagator of a wave function are therefore obtained by the discretization of the quantum system. This means that the continuous paths will be replaced with infinitesimal rectilinear paths. Consider the example of a free particle. The propagator $G(x', t'; x, t)$ can be discretized by introducing n time intervals ε such that at the limit $n \to \infty$ $\varepsilon \to 0$ $n\varepsilon \to t' - t$.

By using form (7.27) of the propagator, we will write

$$G\,(f;i) \equiv G\,(x_f, t_f; x_i, t_i)$$

$$= \int_{\substack{n\to\infty \\ \varepsilon\to 0}} G\,(f; n-1)\,G\,(n-1; n-2)\ldots G\,(1; i)\,dx_{n-1}\ldots dx_1\,,$$

$$(7.30)$$

with the expression of propagator (7.22) for the free particle:

$$G\,(i; i-1) = \left(\frac{m}{2\pi\,i\,\hbar\,\varepsilon}\right)^{1/2}\,\exp\left[\frac{im}{2\hbar\,\varepsilon}\,(x_i - x_{i-1})^2\right].$$

$$(7.31)$$

Now, let us evaluate $G(2; i)$ by integrating over the intermediate variable x_1 alone:

$$G\,(2; i) = \int G\,(2; 1)\,G\,(1; i)\,dx_1$$

$$= \left(\frac{m}{2\pi\,i\,\hbar\,\varepsilon}\right)\,\exp\left[\frac{im}{2\,\hbar\,\varepsilon}[(x_2 - x_1)^2 + (x_1 - x_i)^2]\right]\,dx_1.\quad(7.32)$$

Next, we use the value of the integral over x of a Gaussian function:

$$\int \exp\left[a(x - x_1)^2 + b(x - x_2)^2\right]\,dx = \left(\frac{-\pi}{a+b}\right)^{1/2}\,\exp\left[\frac{ab}{a+b}(x_1 - x_2)^2\right],$$

$$(7.33)$$

which leads to the integrated expression of the propagator $G\,(2; i)$:

$$G\,(2; i) = \left(\frac{m}{2\pi\,i\,\hbar\,(2\varepsilon)}\right)^{1/2}\,\exp\left[\frac{im}{2\hbar\,(2\varepsilon)}\,(x_2 - x_i)^2\right].$$

$$(7.34)$$

This is precisely the propagator $G\,(1; i)$ when x_1 is replaced with x_2 and ε with 2ε. This calculation can be continued by integrating over x_2, and then over x_3... and so on down to x_{n-1} to finally obtain

$$G\,(f; i) = \lim_{\substack{n\to\infty \\ \varepsilon\to 0}} \left(\frac{m}{2\pi\,i\,\hbar\,(n\varepsilon)}\right)^{1/2}\,\exp\left[\frac{im}{2\hbar\,(n\varepsilon)}\,(x_f - x_i)^2\right].\quad(7.35)$$

The passage to the limit is self-evident and consists in replacing $n\varepsilon$ with $t'-t$. By replacing f with (x', t') and i by (x, t), we thus obtain

$$G\,(x', t'; x, t) = \left(\frac{m}{2\pi\,i\,\hbar\,(t'-t)}\right)^{1/2}\,\exp\left(\frac{im}{2\hbar}\frac{(x'-x)^2}{t'-t}\right).\quad(7.36)$$

This is precisely the propagator of a free particle (7.22) although it is for the finite space interval $(x' - x)$ and the finite time interval $(t' - t)$. Similarly, we obtain the propagator $G\,(x', t'; x, t)$ for a particle subjected to an interaction V by substituting (7.29) into (7.27), that is by discretizing the integral and then causing ε to tend toward zero and summing over all the infinitesimal intervals, that is by integrating over a variable t:

$$G(x', t'; x, t) = \int \frac{d\,p_1}{2\pi\,\hbar} \frac{d\,p_2}{2\pi\,\hbar} \cdots \frac{d\,p_{n-1}}{2\pi\,\hbar} \int dx_1 ... dx_{n-1}$$

$$\times \exp \frac{i\,\varepsilon}{\hbar} \left[\sum_i \frac{x_{i+1} - x_i}{\varepsilon} p_i - H \left(p_i, \frac{x_{i+1} + x_i}{2} \right) \right]$$

$$= \lim_{\substack{n \to \infty \\ \varepsilon \to \infty}} \prod_{i=1}^{n-1} \int \frac{d\,p_i}{2\pi\,\hbar} \int \prod_{i=1}^{n} dx_i$$

$$\times \exp \frac{i\,\varepsilon}{\hbar} \left[\sum_{i=1}^{n} \frac{x_{i+1} - x_i}{\varepsilon} p_i - H \left(p_i, \frac{x_{i+1} + x_i}{2} \right) \right] .$$

$$(7.37)$$

By expanding the kinetic energy term $p_i^2/2m$ and the potential term $V[(x_{i+1} + x_i)/2]$ of the Hamiltonian, the integration over dp_i leads to propagator (7.24) of a free particle. The remaining term corresponds to the discretized classical Lagrangian and the replacing the integral over t by a discrete sum introduces the action of the classical Hamiltonian $S(x(t))$.

To simplify the notation we introduce the formal differential element

$$\lim_{\substack{n \to \infty \\ \varepsilon \to 0}} \prod_{i=0}^{n} \frac{d\,x_i}{A_i} = \mathcal{D}x , \qquad (7.38)$$

where A_i is a coefficient yet to be determined. This leads to the propagator of wave function (7.30) written as follows:

$$G(x_f, t_f; x_i t_i) = \int_{x_i, t_i}^{x_f, t_f} \mathcal{D}x\,(t)\,\exp\left\{ \frac{i}{\hbar} \int dt\, L_c\,(x, \dot{x}) \right\} . \qquad (7.39)$$

It is easy to recognize in this integral the Lagrangian of classical mechanics but the preceding Feynman integral concerns operators acting on a wave function at (x_i, t_i) to carry it to (x_f, t_f).

Feynman's conjecture consists in introducing the classical Lagrangian in the integral and in identifying the normalization coefficient A_i with the coefficient of the propagator of a free particle by setting:

$$\mathcal{D}x(t) = \lim_{\substack{n \to \infty \\ \Delta\,t \to 0}} \left(\frac{m}{2\pi\,i\,\hbar\,\Delta\,t} \right)^{\frac{n-1}{2}} dx_1 \ldots dx_n$$

$$G(x_f\,t_f; x_i,\,t_i) = \int_{x_i, t_i}^{x_f, t_f} \mathcal{D}x(t)\,\exp\left[\frac{i}{\hbar}\,S(x(t)) \right]$$

$$(7.40)$$

In classical mechanics, the motion of a particle can be determined from the Hamiltonian action and the principle of least action:

$$\delta\, S = \delta \int_{t_1}^{t_2} L_c(x,\, \dot{x})\, dt = 0 \tag{7.41}$$

or by considering a discretized and small time interval

$$0 = \delta\, S\, (n, n-1) = \delta \int_{t_{n-1}}^{t_n} L_c(\dot{x},\, x)\, dt. \tag{7.42}$$

For a particle with mass m in an interaction potential V,

$$
\begin{aligned}
S(n, n-1) &= \int_{t_{n-1}}^{t_n} \left[\frac{1}{2}\, m\, \dot{x}^2 - V(x) \right] dt \\
&= \Delta t \left\{ \frac{m}{2} \left(\frac{x_n - x_{n-1}}{\Delta t} \right)^2 - V \left(\frac{x_n + x_{n-1}}{2} \right) \right\},
\end{aligned} \tag{7.43}
$$

Feynman's conjecture identifies the classical form with the quantum-mechanical form:

$$\langle x_n, t_n \mid x_{n-1}, t_{n-1} \rangle = \frac{1}{W(\Delta t)}\, \exp \left[\frac{i}{\hbar}\, S(n, n-1) \right] \tag{7.44}$$

and for a free particle $V = 0$ we obtain by setting

$$x_n - x_{n-1} = \Delta x \qquad \text{and} \quad t_n - t_{n-1} = \Delta t$$

the result already expressed in (7.25), that is,

$$G(x + \Delta\, x, t + \Delta t; x, t) = \frac{1}{W(\Delta t)}\, \exp \left[\frac{i}{\hbar}\, \frac{1}{2}\, m \left(\frac{\Delta x}{\Delta t} \right)^2 \Delta t \right]. \tag{7.45}$$

By comparing with the propagator (7.25) for a free particle, it becomes natural to identify $1/W(\Delta t)$ with the value $[m/(2\pi\, i\, \hbar\, \Delta t)]^{1/2}$ and to replace the product of the infinitesimal propagators with the Feynman integral (7.40).

2.2 Example: The Quantum-Mechanical Effect of Gravitation

Because the equivalence principle identifies inertial mass with gravitational mass, the equations of the classical motion of a particle with mass m in a gravitational field are independent of the mass:

$$m\, \ddot{\vec{x}} = -m\, \vec{\nabla}\, \phi \implies \ddot{\vec{x}} = -\vec{\nabla}\, \phi. \tag{7.46}$$

In quantum mechanics, Feynman path integrals lead to a mass-dependent propagation since (7.44) will yield the propagator between instants t_{n-1} and t_n:

$$\langle x_n, t_n \mid x_{n-1}, t_{n-1} \rangle = \sqrt{\frac{m}{2\pi\, i\, \hbar\, \Delta t}}\ \exp\left[\frac{i}{\hbar} \int_{t_{n-1}}^{t_n} \left(\frac{1}{2} m\, \dot{x}^2 - m\, g\, z \right) dt \right]$$

$$= \sqrt{\frac{m}{2\pi\, i\, \hbar\, \Delta t}}\ \exp\left[\frac{i\, m}{\hbar} \int_{t_{n-1}}^{t_n} dt\ \left(\frac{1}{2} \dot{x}^2 - g\, z \right) \right].$$

$$(7.47)$$

This phase shift resulting from gravitational force was first brought to light using a source of thermal neutrons and a device for separating the neutrons along two different paths meeting at a given point D where the neutrons are detected (Fig. 1).

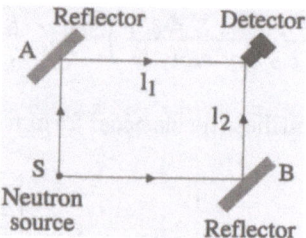

Reflector   Detector

A

l_1

l_2

S   B

Neutron source

Reflector

Figure 1

If points A and B are in the same horizontal plane, the gravitational effect is zero and there is no interference phenomenon indicating the existence of a phase shift δ between the wave function of particles that have traveled different paths. Conversely, if A and B are in the same vertical plane, the quantum mechanical phase shift produced will be

$$\phi_{SAD} - \phi_{SBD} = -\frac{m^2\, g}{\hbar^2}\, \ell_1 \ell_2\, \bar{\lambda}\, \sin\delta = -\frac{m\, g}{\hbar}\, \frac{\ell_1\, \ell_2}{v}\, \sin\delta, \qquad (7.48)$$

where $\bar{\lambda} = \hbar/(m\, v)$ is the de Broglie wavelength of neutrons with mass m and velocity v and ℓ_1, ℓ_2 the length of the horizontal and vertical paths (Fig. 1). The interference between wave functions $\Psi(SAD)$ and $\Psi(SBD)$ will take place at D

$$\Psi(SBD) = \Psi(SAD)\ \exp\left[i(\phi_{SAD} - \phi_{SBD})\right], \qquad (7.49)$$

since the resulting quantum state:

$$\Psi(SAD) + \Psi(SBD) = \Psi(SAD) \left(1 + \exp\left(-\frac{i}{\hbar}\, m\, g\, \frac{\ell_1\, \ell_2}{v}\, \sin\delta \right)\right) \quad (7.50)$$

will, through its square modulus, yield an interference pattern (Fig. 2).

If $\hbar \to 0$, which simply amounts to assuming a classical mechanical stance, the interference patterns become so tightly bunched that the phenomenon ceases to be observable.

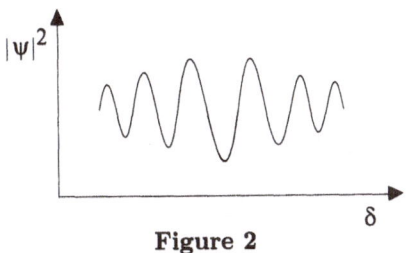

Figure 2

3. Equivalence to the Schrödinger Equation

Feynman's conjecture concerning path integrals may turn out to be another quantization method provided we can show that there is complete equivalence between it and other methods and, in particular, that it is equivalent to wave mechanics.

Let us assume that the wave function at (x, t) is known. Relation (7.18) will then yield, by the evaluation of Feynman's integral, the wave function at x' and at $t')t$:

$$\Psi(x', t') = \int G(x', t'; x, t)\, \Psi(x, t)\, dx \qquad (7.51)$$

and at an instant $t' = t + \varepsilon$, we will write

$$\Psi(x', t + \varepsilon) = \int G(x', t + \varepsilon; x, t)\, \Psi(x, t)\, dx, \qquad (7.52)$$

with propagator (7.43) for the wave function:

$$G(x', t + \varepsilon; x, t) = \left(\frac{m}{2\pi\, i\, \hbar\, \varepsilon}\right)^{1/2} \exp\left[\frac{i}{\hbar}\left(\frac{m}{2\varepsilon}(x' - x)^2 - \varepsilon\, V(\frac{x + x'}{2}, t)\right)\right].$$
$$(7.53)$$

If we set $x = x' - \xi$ and expand the exponential of the potential, we obtain the wave function

$$\Psi(x', t + \varepsilon) = \left(\frac{m}{2\pi\, i\, \hbar\, \varepsilon}\right)^{1/2} \int_{-\infty}^{+\infty} \exp\left(\frac{i\, m\, \xi^2}{2\hbar\, \varepsilon}\right)\left[1 - \frac{i\, \varepsilon}{\hbar}\, V + \dots\right]$$
$$\times \left[\Psi(x', t) - \xi\, \frac{\partial\, \Psi}{\partial\, x'} + \frac{\xi^2}{2}\, \frac{\partial^2\, \Psi}{\partial\, x'^2} + \dots\right]\, d\xi. \qquad (7.54)$$

The expansion to first order in ε of $\Psi(x', t+\varepsilon)$ in fact yields

$$\Psi(x',\ t+\varepsilon) = \Psi(x',\ t) + \varepsilon\frac{\partial\ \Psi(x',\ t)}{\partial\ t}\,. \tag{7.55}$$

The integrals in ξ that make a non-zero contribution to the second term of (7.54) are even in powers of ξ:

$$\int_{-\infty}^{+\infty} \exp\left(\frac{i\ m\ \xi^2}{2\hbar\ \varepsilon}\right)\ d\xi = \left(\frac{2\pi\ i\ \hbar\ \varepsilon}{m}\right)^{1/2}, \tag{7.56}$$

$$\int_{-\infty}^{+\infty} \exp\left(\frac{i\ m\ \xi^2}{2\hbar\ \varepsilon}\right)\ \xi^2\ d\xi = \sqrt{2\pi}\ \left(\frac{i\ \hbar\ \varepsilon}{m}\right)^{3/2}. \tag{7.57}$$

By integrating (7.54) over ξ and then equating to (7.55) we obtain the Schrödinger equation

$$i\ \hbar\ \frac{\partial\ \Psi}{\partial\ t} = -\frac{\hbar^2}{2m}\frac{\partial^2\ \Psi}{\partial\ x^2} + V(x,t)\ \Psi(x,t)\,. \tag{7.58}$$

The above is proof of the equivalence between path integrals and wave mechanics although the Schrödinger equation is only a first-order approximation of the actual result obtained from Eqs. (7.51) and (7.53).

4. Action in Quantum Physics

In classical mechanics, the Hamiltonian action

$$S = \int\ L(x,\ \dot{x},\ t)\ dt \tag{7.59}$$

leads to the Euler–Lagrange equations of motion through the principle of least action.

In quantum mechanics, Young's two-slit experiment shows, for example, that the particles traveling from the source x_0 to the detector x_1 may follow any path whatsoever (not determined by Euler–Lagrange equations) and that the probability amplitude that a particle will go from $(x_0,\ t_0)$ to $(x_1,\ t_1)$ is, for example following definition (7.10), simply

$$A(x_0,\ t_0 \rightarrow x_1,\ t_1) = G(x_1,\ t_1; x_0,\ t_0) = \sum_{\text{paths}} A(x(t))\,, \tag{7.60}$$

where $A(x(t))$ is the probability amplitude corresponding to a particular path obeying the boundary conditions

$$x(t_0) = x_0 \quad \text{and} \quad x(t_1) = x_1\,.$$

In fact, quantum mechanics only gives a probabilistic interpretation to the square modulus of the probability amplitude. We can therefore write:

$$A(x(t)) = N \, \exp \, [i \, \theta(x(t))] \, , \tag{7.61}$$

where N is independent of the path chosen. This amounts to writing

$$G(x_1, \, t_1; x_0, \, t_0) = N \sum_{\text{paths}} \exp \, [i \, \theta(x(t))] \, . \tag{7.62}$$

The classical limit $\hbar \to 0$ should be able to pick out from all the possible paths that which corresponds to the principle of least action.

This boils down to writing the propagator in the form

$$G(x_1, \, t_1; x_0, \, t_0) = N \sum_{\text{paths}} \exp \, \left[\frac{i}{\hbar} \, S(x(t)) \right] \, . \tag{7.63}$$

When $\hbar \to 0$, the phase function $S/\hbar$ becomes too large and the probability amplitude oscillates too fast, wiping out the contributions of most of the integral paths. Only contributions by paths for which S is invariant, even when there is a change in path, will remain:

> When $\hbar \to 0$, only paths corresponding to a stationary action
> contribute to the probability amplitude G, which
> corresponds to the classical limit

.

5. Quantum Paths and Fractals

In Nelson's stochastic interpretation of wave mechanics, the trajectory $\vec{x}(t)$ of a particle with mass m is a non-differentiable continuous function. The path integrals defining the propagator of a wave function from one point in space-time to another are equally non-differentiable continuous functions and this was pointed out by Feynman as early as 1965. Of course Feynman did not use the term fractal, introduced in 1975 by Mandelbrot but we shall be using the approach of Abbot and Wise (1981) to show that quantum paths are two-dimensional fractals.

A fractal curve is a continuous and non-differentiable curve, produced by an infinite series of operations which increase the length by a given factor. The Koch curve, for example (Fig. 3), increases the length of a straight-line segment in each operation by a factor of $\frac{4}{3}$ when Δx becomes $\frac{1}{3} \, \Delta x$ so that the final curve obtained after an infinite series of operations has infinite length.

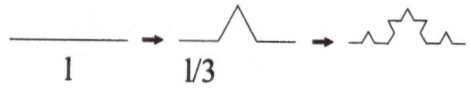

Figure 3

Hausdorff has defined a fractal length L associated with a resolution Δx and a fractal dimension coefficient D ($\neq 1$) by setting:

$$L = \ell \, (\Delta x)^{D-1} . \tag{7.64}$$

The coefficient D is determined by the fact that L must be independent of the resolution Δx, at least in the limit $\Delta x \to 0$.

By a D-dimensional fractal curve, we will be referring to a non-differentiable continuous curve satisfying the Haudsdorff condition for $D \neq 1$.

The discretization of a quantum path into n elements Δt will yield a mean quantum path length

$$\langle \ell \rangle = n \, \langle \Delta \ell \rangle \quad \text{with} \quad t_f - t_i = n \, \Delta t . \tag{7.65}$$

The element with the mean length covered by a quantum particle is determined with Heisenberg's uncertainty relation

$$\langle \Delta \ell \rangle \sim \frac{\hbar}{\langle \Delta p \rangle} = \frac{\hbar}{m} \left\langle \frac{\Delta t}{\Delta x} \right\rangle = \frac{\hbar}{mn} \frac{(t_f - t_i)}{\langle \Delta x \rangle} . \tag{7.66}$$

This will yield the mean fractal length $\langle L \rangle$ with (7.64) and (7.66)

$$\langle L \rangle = \langle \ell \rangle \, \langle \Delta x \rangle^{D-1} = \frac{\hbar}{m} \frac{(t_f - t_i)}{\langle \Delta x \rangle} \, \langle \Delta x \rangle^{D-1} . \tag{7.67}$$

For a Hausdorff length to be independent of Δx, it is necessary and sufficient to have $D = 2$, which shows that the quantum path is a fractal curve (that is continuous but non-differentiable) of fractal dimension 2.

6. Path Integrals – Schematic Summary

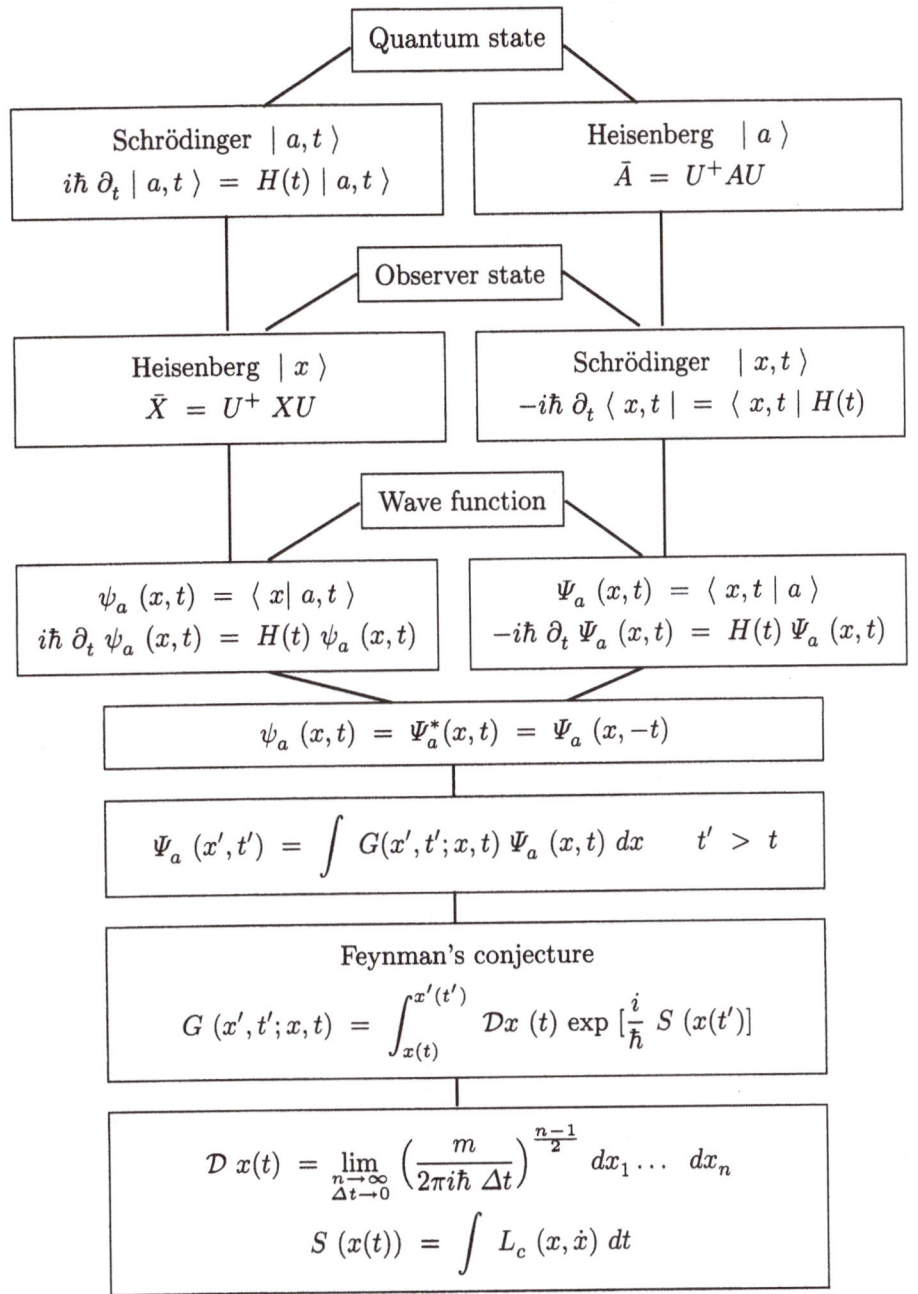

Chapter 8
Symmetries and Invariances

It was shown in Chap. 1 that the conservation laws of classical physics derive from the properties of invariance of the interaction potential with respect to continuous transformations, the translation of space or time coordinates, or the rotation of space coordinates.

Similarly, in quantum physics, the general properties of the wave function describing a particle or a set of particles also derive from the invariance with respect to the continuous or discrete transformations of the Hamiltonian. We will not be describing these translations of the system of coordinates (both space and time) in detail here since that we have already had occasion to reflect upon them in Chaps. 5 and 6. Rotations will be taken up after the introduction of the standard representation base of angular momentum (Chap. 9). We shall be describing two transformations which also play a very important role in quantum physics: the reversal of space coordinates (or parity P) or of time coordinates (reversal of the direction of time T). Charge conjugation (operator C) will be studied in Chap. 17 which is dedicated to quantum electrodynamics. Mention will also be made of the CPT invariance, a fundamental issue in the quantum description of elementary particles.

Electromagnetic field gauge invariance is an essential factor in classical physics. And we shall be looking into how the gauge transformation of the vector potential introduces a phase function in the wave function of a charged particle in interaction with the electromagnetic field. This will lead us to the notion of Abelian and non-Abelian gauge fields and to the generalization of Maxwell's electromagnetic field equations to all gauge fields, that is to the Yang–Mills equations. We shall return to this point in quantum electrodynamics and to its extension to quantum chromodynamics, which is the most advanced of all theories of strong interactions.

1. Unitary Transformations

Consider a transformation $\mathcal{G}$ which transforms the quantum state $| \, \alpha \rangle$ into $| \tilde{\alpha} \rangle$:

$$
\begin{aligned}
| \, \tilde{\alpha} \rangle &= \mathcal{G} \, | \, \alpha \rangle , \\
\langle \tilde{\alpha} \, | &= \langle \alpha \, | \, \mathcal{G}^{+} .
\end{aligned}
\tag{8.1}
$$

If, to conserve the norm, we impose on the transformation:

$$
\langle \tilde{\alpha} \, | \, \tilde{\beta} \rangle \; = \; \langle \alpha \, | \, \beta \rangle ,
\tag{8.2}
$$

this will lead, with relation (8.1), to the following equality:

$$
\langle \alpha \, | \, \mathcal{G}^{+} \, \mathcal{G} \, | \, \beta \rangle \; = \; \langle \alpha \, | \, \beta \rangle \quad \text{and} \quad \mathcal{G}^{+} \, \mathcal{G} = \mathbb{1} .
\tag{8.3}
$$

The operator $\mathcal{G}$ must be a unitary operator. This is the case for the continuous transformations we have already met, such as space or time translations; it is also the case for the rotations we will be looking into a little later.

The transformation of an operator A is obtained from the conservation of the mean value of such an operator, since the mean value is a number resulting from the measurement associated with A. This value is therefore independent of the representation adopted:

$$
\langle A \rangle \; = \; \langle \alpha \, | \, A \, | \, \alpha \rangle \; = \; \langle \tilde{\alpha} \, | \, \tilde{A} \, | \, \tilde{\alpha} \rangle .
\tag{8.4}
$$

This gives, with relation (8.1),

$$
\langle \alpha \, | \, A \, | \, \alpha \rangle \; = \; \langle \alpha \, | \, \mathcal{G}^{+} \, \tilde{A} \, \mathcal{G} \, | \, \alpha \rangle ,
\tag{8.5}
$$

and we can infer from this the transformation of the operator A:

$$
A = \mathcal{G}^{+} \, \tilde{A} \, \mathcal{G} ,
\tag{8.6}
$$

or, by using the unitarity property,

$$
\tilde{A} = \mathcal{G} \, A \, \mathcal{G}^{+} .
\tag{8.7}
$$

In summary then, the unitary transformation U which conserves the norm leads to the following transformations of vectors and operators:

$$
\boxed{
\begin{aligned}
| \, \alpha \rangle \; &\rightarrow \; | \, \tilde{\alpha} \rangle \; = U \, | \, \alpha \rangle \\
A \; &\rightarrow \; \tilde{A} = U \, A \, U^{+} = U \, A \, U^{-1}
\end{aligned}
}
\; .
\tag{8.8}
$$

1.1 The Transformation Generator

Let us assume that the unitary transformation is infinitesimally different from the identity transformation and then set, for the sake of convenience,

$$U = \mathbb{1} - \frac{i\,\varepsilon}{\hbar}\, G\,. \qquad (8.9)$$

The unitarity condition will be written to first order as follows:

$$\mathbb{1} = U^+ U = \left(\mathbb{1} + \frac{i\,\varepsilon}{\hbar}\, G^+\right)\left(\mathbb{1} - \frac{i\,\varepsilon}{\hbar}\, G\right) = \mathbb{1} - \frac{i\,\varepsilon}{\hbar}\,(G - G^+)\,, \quad (8.10)$$

thus forcing the generator G of the symmetry $\mathcal{G}$ described by the unitary operator U into being a Hermitian operator $G = G^+$.

The transformation of operator A will then give to first order

$$\tilde{A} = U\,A\,U^+ = \left(\mathbb{1} - \frac{i\,\varepsilon}{\hbar}\, G\right) A \left(\mathbb{1} + \frac{i\,\varepsilon}{\hbar}\, G\right) = A - \frac{i\,\varepsilon}{\hbar}\,[G, A]\,. \quad (8.11)$$

Any operator that is invariant under symmetry $\mathcal{G}$ should commute with the generator of the corresponding infinitesimal transformation:

$$A = \tilde{A} \implies [G, A] = 0\,. \qquad (8.12)$$

In particular, if the Hamiltonian H is invariant under symmetry $\mathcal{G}$ (translation, rotation), it should commute with the generator of the corresponding infinitesimal transformation:

$$[G, H] = 0 \quad \text{or} \quad [U, H] = 0\,, \qquad (8.13)$$

and Heisenberg equation of motion will imply

$$\frac{d\langle G\rangle}{dt} = 0\,, \qquad (8.14)$$

that is, the Hermitian operator G is a constant of the motion and its eigenvalues are not time dependent.

1.2 Symmetry and Degeneracy

Let $|\,n\rangle$ be the eigenstates of the Hamiltonian corresponding to the eigenvalue E_n:

$$H\,|\,n\rangle = E_n\,|\,n\rangle\,. \qquad (8.15)$$

The operator U commuting with H will transform them (by rotation, translation, or any other continuous transformation capable of conserving the norm) into the form

$$|\,\tilde{n}\rangle = U\,|\,n\rangle\,. \qquad (8.16)$$

These vectors $\mid \tilde{n} \rangle$ are also eigenvectors of H corresponding to the same eigenvalue E_n since, using (8.13), we can write

$$H \mid \tilde{n} \rangle = H U \mid n \rangle = U H \mid n \rangle = U E_n \mid n \rangle = E_n U \mid n \rangle$$
$$= E_n \mid \tilde{n} \rangle \tag{8.17}$$

Because the eigenvectors $\mid n \rangle$ and $\mid \tilde{n} \rangle$ correspond to the same eigenvalue E_n, we infer that the eigenvalue E_n is degenerate.

1.3 Examples of Unitary Continuous Transformations

The exponential of an operator A is defined by its series expansion, that is,

$$\exp(A) = \sum_n \frac{A^n}{n!} = \mathbb{1} + A + \frac{1}{2} A^2 + \dots . \tag{8.18}$$

An operator U acts on the vector $\mid \alpha \rangle$ of a vector space, such as the Hilbert space. The parameter α may represent the time or the position or the angle of observation of a quantum state, and we set by definition

$$\mid \alpha_1 + \alpha_0 \rangle = U(\alpha_1) \mid \alpha_0 \rangle . \tag{8.19}$$

This describes the increase in parameter α_0 by α_1 due to the application of a unitary operator U obeying the following condition:

$$U^+(\alpha_1) = U^{-1}(\alpha_1) = U(\bar{\alpha}_1) . \tag{8.20}$$

By the repeated application of the operator U, it is possible to obtain the relations

$$\mid \alpha_1 + \alpha_2 + \alpha_0 \rangle = U(\alpha_1 + \alpha_2) \mid \alpha_0 \rangle$$
$$= U(\alpha_1) U(\alpha_2) \mid \alpha_0 \rangle$$
$$= U(\alpha_2) U(\alpha_1) \mid \alpha_0 \rangle \tag{8.21}$$

leading to an additional condition satisfied by U:

$$U(\alpha_1 + \alpha_2) = U(\alpha_1) U(\alpha_2) = U(\alpha_2) U(\alpha_1) . \tag{8.22}$$

It goes without saying that this condition is satisfied by an operator $U(\alpha)$ in the form of an exponential:

$$U(\alpha) = C \, \exp(\beta \, \alpha) . \tag{8.23}$$

The unitarity condition (8.20) then reduces to

$$U^{-1}(\alpha) = C^{-1} \, \exp(-\beta \, \alpha) = U^+(\alpha) = C^* \, \exp(\beta^+ \, \alpha) , \tag{8.24}$$

thus imposing the following constraints on the parameters C and β:

$$C^* = C^{-1} \quad \text{and} \quad \beta^+ = -\beta. \qquad (8.25)$$

For the sake of simplicity and to remain consistent with the correspondence principle, we set

$$C = 1 \quad \text{and} \quad \beta = -\frac{i}{\hbar} A \quad \text{with} \quad A^+ = A. \qquad (8.26)$$

The unitary transformation $U(\alpha)$ is therefore of the form

$$\boxed{U(\alpha) = \exp\left(-\frac{i}{\hbar} \alpha A\right) \quad \text{with} \quad A = A^+} \qquad (8.27)$$

This leads to the transformation of vector $\mid \alpha_0\rangle$ by the incrementation of the value α, yielding

$$\mid \alpha + \alpha_0\rangle = \exp\left(-\frac{i}{\hbar} \alpha A\right) \mid \alpha_0\rangle \qquad (8.28)$$

and by differentiating each member with respect to the evolution parameter α:

$$\boxed{\begin{array}{c} \dfrac{\partial \mid \alpha + \alpha_0\rangle}{\partial \alpha} = \dfrac{i}{\hbar} A \mid \alpha + \alpha_0\rangle \\[2mm] \mid \alpha + \alpha_0\rangle = \exp\left(-\dfrac{i}{\hbar} \alpha A\right) \mid \alpha_0\rangle \\[2mm] \text{generator of the transformation:} \\ A = A^+ \end{array}} \qquad (8.29)$$

i)   Generator of Time Translations
If the parameter α is the time,

$$\frac{\partial}{\partial t} \mid t_0 + t\rangle = -\frac{i}{\hbar} H \mid t_0 + t\rangle. \qquad (8.30)$$

The (time-independent) Hamiltonian operator is the generator of time translations in a quantum system.

ii)   Generator of Space Translations

If $\alpha = x$, then the Taylor expansion of a function:

$$\Psi_a(x + \alpha) = \langle x + \alpha \mid a \rangle$$

$$= \Psi_a(x) + \alpha \frac{d}{dx} \Psi_a(x) + \ldots = \left(\langle x \mid + \alpha \frac{d}{dx} \langle x \mid + \ldots \right) \mid a \rangle$$

$$= \langle x \mid \left(\mathbb{1} + \alpha \frac{d}{dx} + \ldots \right) \mid a \rangle \tag{8.31}$$

will transform, with the correspondence principle $p_x = -i \hbar \, (d/dx)$ (with this operator acting on $\langle x \mid$), the generator of space translations as follows:

$$\Psi_a(x + \alpha) = \langle x \mid \exp \left(\frac{i}{\hbar} \alpha \, p_x \right) \mid a \rangle, \tag{8.32}$$

implying the transformation of the coordinates:

$$\langle x + \alpha \mid = \langle x \mid U^+(\alpha) = \langle x \mid \exp \left(\frac{i}{\hbar} \alpha \, p_x \right). \tag{8.33}$$

The operator generating the translations of the coordinates of a ket vector is therefore

$$\boxed{\begin{array}{c} U(\alpha) = \exp \left(-\frac{i}{\hbar} \alpha \, \vec{p} \right) \\ \vec{p} = -i \hbar \, \vec{\nabla} \end{array}} \tag{8.34}$$

The operator $\vec{p}$ is the generator of the translations of the space coordinates of a quantum system.

iii)   Generator of Rotations

Consider a scalar field $\varphi(x, y, z)$ and an infinitesimal rotation $d\alpha$ around the axis Oz. Now, let us use the general form (8.9) to set

$$U(d\alpha) = \mathbb{1} - \frac{i}{\hbar} d\alpha \, J_z. \tag{8.35}$$

The operator J_z is the generator of the infinitesimal rotations of a scalar field around the axis Oz.

The application of U to the scalar field $\varphi(x, y, z)$ will transform the coordinates (x, y, z) into (x', y', z'):

$$\varphi(x', y', z') = U(d\alpha) \, \varphi(x, y, z) \tag{8.36}$$

with the transformed coordinates:

$$\begin{aligned} x' &= x + y \, d\alpha, \\ y' &= -x \, d\alpha + y, \\ z' &= z. \end{aligned} \tag{8.37}$$

Inserting these into the scalar field $\varphi(x', y', z')$ and using a Taylor expansion around (x, y, z), we obtain

$$\begin{aligned} \varphi(x', y', z') &= \varphi(x + y \, d\alpha, -x \, d\alpha + y, z) \\ &= \left[\mathbb{1} + d\alpha \left(y \frac{\partial}{\partial x} - x \frac{\partial}{\partial y} \right) + \dots \right] \varphi(x, y, z). \end{aligned} \tag{8.38}$$

By comparing statement (8.36) with form (8.35) of the operator U, we obtain

$$\frac{i}{\hbar} J_z = x \frac{\partial}{\partial y} - y \frac{\partial}{\partial x} = \frac{i}{\hbar}(x \, p_y - y \, p_x) = \frac{i}{\hbar} L_z. \tag{8.39}$$

The generator of infinitesimal rotations around an axis $O\vec{u}$ for a scalar field is the projection of the orbital momentum $\vec{L} \cdot \vec{u}$ onto the axis in question.

Now let us examine the case of the vector field $\vec{\varphi}(x, y, z)$. Each component φ_x, φ_y or φ_z will undergo the same transformation as the corresponding coordinates, that is, in a form analogous to (8.37),

$$\begin{aligned} \varphi'_x &= \varphi_x(x + yd\alpha, y - xd\alpha, z) + d\alpha \, \varphi_y(x + yd\alpha, y - xd\alpha, z), \\ \varphi'_y &= -d\alpha \, \varphi_x(x + yd\alpha, y - xd\alpha, z) + \varphi_y(x + yd\alpha, y - xd\alpha, z), \\ \varphi'_z &= \varphi_z(x + yd\alpha, y - xd\alpha, z). \end{aligned} \tag{8.40}$$

We further introduce the column of the components of vector fields $\vec{\varphi}$ and $\vec{\varphi}'$:

$$\vec{\varphi} = \begin{pmatrix} \varphi_x \\ \varphi_y \\ \varphi_z \end{pmatrix} \quad \text{and} \quad \vec{\varphi}' = \begin{pmatrix} \varphi'_x \\ \varphi'_y \\ \varphi'_z \end{pmatrix}, \tag{8.41}$$

and transformation law (8.35), that is,

$$\vec{\varphi}' = \left(\mathbb{1} - \frac{i}{\hbar} \, d\alpha \, J_z \right) \vec{\varphi}. \tag{8.42}$$

The Taylor series for the components of $\vec{\varphi}$ and the introduction of the following 3×3 matrices:

$$\begin{aligned} L_z &= -i \, \hbar (x \, \partial_y - y \, \partial_x) \begin{pmatrix} 1 & 0 & 0 \\ 0 & 1 & 0 \\ 0 & 0 & 1 \end{pmatrix} \\ S_z &= -i \, \hbar \begin{pmatrix} 0 & 1 & 0 \\ -1 & 0 & 0 \\ 0 & 0 & 0 \end{pmatrix} \end{aligned} \tag{8.43}$$

will then yield, by comparing with (8.40) and (8.42), the generator of infinitesimal rotations around the Oz axis of a vector field:

$$J_z = L_z + S_z. \tag{8.44}$$

The generator of infinitesimal rotations of a vector field around the $O\vec{u}$ axis is the projection onto this axis of the total angular momentum $\vec{J} \cdot \vec{u}$ with orbital momentum $\vec{L} \cdot \vec{u}$ and spin 1:

$$\vec{J} = \vec{L} + \vec{1}.$$

1.4 The Parity Operator

Consider the unitary operator which changes the sign of a position operator (space inversion):

$$\begin{aligned} |\,\alpha\rangle &\rightarrow |\,\tilde{\alpha}\rangle = P\,|\,\alpha\rangle, \\ \tilde{X} &= P\,X\,P^+ = -X. \end{aligned} \tag{8.45}$$

This can also be written by multiplying from the right by P since $P^+P = \mathbb{1}$:

$$X\,P + P\,X = \{X, P\} = 0. \tag{8.46}$$

The position operator anticommutes with the parity operator.

Let us determine the eigenvalues of the Hermitian unitary operator P:

$$X(P\,|\,x\rangle) = -P\,X\,|\,x\rangle = -P\,x\,|\,x\rangle = (-x)\,(P\,|\,x\rangle). \tag{8.47}$$

The vector $P\,|\,x\rangle$ is an eigenvector of X corresponding to the eigenvalue $(-x)$. Since the eigenvalue equation for X associated with the eigenvector $|-x\rangle$ can also be written as

$$X\,|-x\rangle = (-x)\,|-x\rangle, \tag{8.48}$$

we obtain by comparing (8.47) with (8.48)

$$|\,\tilde{x}\rangle = P\,|\,x\rangle = |-x\rangle \tag{8.49}$$

to within a phase coefficient arbitrarily chosen equal to unity. By applying the parity operator P twice to the vector $|\,x\rangle$, we obtain

$$P^2\,|\,x\rangle = P(P\,|\,x\rangle) = P(|-x\rangle) = |\,x\rangle, \tag{8.50}$$

which simply shows that

$$P^2 = \mathbb{1}. \tag{8.51}$$

The eigenvalues of the unitary and Hermitian operator P are ± 1.

If $|\,\alpha\rangle$ is an eigenstate of P, then we will write

$$|\,\tilde{\alpha}\rangle = P\,|\,\alpha\rangle = \pm\,|\,\alpha\rangle = \varepsilon_\alpha\,|\,\alpha\rangle. \tag{8.52}$$

The parity transformation of a wave function $\Psi_\alpha(x) = \langle x \mid \alpha \rangle$ is easily obtained:

$$\Psi_\alpha(x) \rightarrow \tilde{\Psi}_\alpha(x) = \langle x \mid \tilde{\alpha} \rangle = \langle x \mid P \mid \alpha \rangle. \qquad (8.53)$$

It is written with relation (8.52) as

$$\Psi_\alpha(x) \rightarrow \tilde{\Psi}_\alpha(x) = \pm\, \Psi_\alpha(x). \qquad (8.54)$$

By using the conjugate Hermitian of (8.49), we obtain

$$\tilde{\Psi}_\alpha(x) = \Psi_\alpha(-x). \qquad (8.55)$$

By comparing the above results, we further find

$$\Psi_\alpha(-x) = \pm\, \Psi_\alpha(x) \begin{cases} +\ \text{even parity} = \text{symmetric} \\ -\ \text{odd parity} = \text{antisymmetric}. \end{cases} \qquad (8.56)$$

Expressed in spherical coordinates, the space inversion $\vec{x} \rightarrow -\vec{x}$ corresponds to the variable change $(r,\ \theta, \varphi) \rightarrow (r,\ \pi-\theta,\ \varphi+\pi)$ and the spherical harmonic functions will be parity-transformed according to the following law:

$$Y_{\ell m}(\theta, \varphi) \rightarrow P\, Y_{\ell m}(\theta, \varphi) = Y_{\ell m}(\pi-\theta,\ \varphi+\pi) = (-)^\ell\, Y_{\ell m}(\theta,\ \varphi). \quad (8.57)$$

In such a case we say that the parity of the spherical harmonics is $(-)^\ell$.

If we evaluate the matrix element of the position operator between two parity states ε_α and ε_β, relations (8.45) and (8.52) make it possible to write

$$\langle \beta \mid X \mid \alpha \rangle = \langle \beta \mid P^+ P X P^+ P \mid \alpha \rangle = -\, \langle \tilde{\beta} \mid X \mid \tilde{\alpha} \rangle$$
$$= -\varepsilon_\alpha\, \varepsilon_\beta\, \langle \beta \mid X \mid \alpha \rangle. \qquad (8.58)$$

The matrix element is therefore zero for identical parities ε_α and ε_β.

The projection onto the configuration space will lead to the selection rule

$$\int \Psi_\beta^*(x)\, x\, \Psi_\alpha(x)\, dx = 0 \qquad \text{if} \qquad \varepsilon_\alpha\, \varepsilon_\beta = +1. \qquad (8.59)$$

2. Anti-unitary Transformations

2.1 Definition

Consider the transformation $\mathcal{G}$ conserving only the norm modulus, that is,

$$\begin{aligned} \mid \tilde{\alpha} \rangle &= \mathcal{G} \mid \alpha \rangle \\ \langle \tilde{\beta} \mid \tilde{\alpha} \rangle &= \langle \beta \mid \alpha \rangle^*. \end{aligned} \qquad (8.60)$$

This boils down to assuming that

$$\mid \langle \tilde{\alpha} \mid \tilde{\alpha} \rangle \mid = \mid \langle \alpha \mid \alpha \rangle \mid. \qquad (8.61)$$

We will show that $\mathcal{G}$ is an anti-unitary operator, the product of a unitary operator U and a complex conjugation operator K:

$$K\left(c\mid\alpha\right) = c^{*}\left(K\mid\alpha\right). \tag{8.62}$$

$$\boxed{\begin{array}{c} \mathcal{G} \text{ is anti-unitary if for } \mid u\rangle = a\mid\alpha\rangle + b\mid\beta\rangle \\ \mathcal{G}\mid u\rangle = \mid\tilde{u}\rangle = a^{*}\mid\tilde{\alpha}\rangle + b^{*}\mid\tilde{\beta}\rangle \\ \mathcal{G} = U\,K \end{array}} \tag{8.63}$$

To prove this result it is sufficient to expand the foregoing relations:

$$\begin{aligned} \mathcal{G}\mid u\rangle &= U\,K(a\mid\alpha\rangle + b\mid\beta\rangle) = U\left(a^{*}\,K\mid\alpha\rangle + b^{*}\,K\mid\beta\rangle\right) \\ &= a^{*}\,U\,K\mid\alpha\rangle + b^{*}\,U\,K\mid\beta\rangle = a^{*}\mid\tilde{\alpha}\rangle + b^{*}\mid\tilde{\beta}\rangle. \end{aligned} \tag{8.64}$$

To prove (8.60) we begin by projecting the state $\mid\alpha\rangle$ onto a complete basis $\mid n\rangle$ since we will use (8.62) to cause K to act on $\langle n\mid\alpha\rangle\mid n\rangle$:

$$\begin{aligned} \mid\tilde{\alpha}\rangle &= \mathcal{G}\mid\alpha\rangle = U\,K\mid\alpha\rangle \\ &= U\sum_{n} K\mid n\rangle\,\langle n\mid\alpha\rangle \\ &= \sum_{n}\mid\tilde{n}\rangle\,\langle n\mid\alpha\rangle^{*} = \sum_{n}\mid\tilde{n}\rangle\,\langle\alpha\mid n\rangle. \end{aligned} \tag{8.65}$$

The Hermitian conjugate of this result will yield the transformation of the bra

$$\langle\tilde{\beta}\mid = \sum_{n'}\langle\tilde{n}'\mid\,\langle n'\mid\beta\rangle, \tag{8.66}$$

and by computing the inner product $\langle\tilde{\beta}\mid\tilde{\alpha}\rangle$ we obtain $\langle\tilde{n}'\mid\tilde{n}\rangle = \delta_{nn'}$, which finally leaves us with

$$\langle\tilde{\beta}\mid\tilde{\alpha}\rangle = \sum_{nn'}\delta_{nn'}\,\langle n'\mid\beta\rangle\,\langle\alpha\mid n\rangle = \langle\alpha\mid\beta\rangle = \langle\beta\mid\alpha\rangle^{*}, \tag{8.67}$$

thus proving proposition (8.60).

2.2 The Time-Reversal Operator

The time evolution of a quantum state $\mid\alpha\rangle$ is given by (8.9), with the Hamiltonian acting as generator of the time translation:

$$\mid\tilde{\alpha}\rangle = U(\varepsilon)\mid\alpha\rangle = \left(\mathbb{1} - \frac{i\,\varepsilon}{\hbar}\,H\right)\mid\alpha\rangle. \tag{8.68}$$

If we apply the time-reversal operator T before evolution:

$$| \alpha \rangle \; \rightarrow \; T \,| \alpha \rangle , \qquad (8.69)$$

we obtain for the time ε (after evolution in time)

$$\left(\mathbb{1} - \frac{i \, \varepsilon \, H}{\hbar} \right) T \,| \alpha \rangle . \qquad (8.70)$$

It is the quantum state at instant $(-\varepsilon)$ that will evolve and we can thus proceed to write:

$$\left(\mathbb{1} - \frac{i \, \varepsilon}{\hbar} \, H \right) T \,| \alpha \rangle \; = T \left(\mathbb{1} - \frac{i \, (-\varepsilon)}{\hbar} \, H \right) | \alpha \rangle . \qquad (8.71)$$

This yields the equality between operators:

$$-i \, H \, T = i \, T \, H . \qquad (8.72)$$

If, as has already been alluded to in Sect. 1.2 of this chapter, the operator T is unitary, then the vector $T \,| n \rangle$ should be an eigenvector of the Hamiltonian corresponding to the eigenvalue $-E_n$. This is clearly impossible since the energy spectrum of a free particle will range from zero to infinity. If the operator T is anti-unitary, relation (8.63) will lead us to write (8.72) as follows:

$$T(i \, H \,| \alpha \rangle) = -i(T \, H \,| \alpha \rangle) \quad \forall \,| \alpha \rangle , \qquad (8.73)$$

thus enabling us to eliminate the scalar i and write the equality between $T = UK$ and H operators in (8.72):

$$T \, H = H \, T \qquad (8.74)$$

with the anti-unitary time-reversal operator.

The transformation involving the time-reversal of the operators will then be written as follows:

$$\langle \beta \,| \, A \,| \alpha \rangle \; = \; \langle \tilde{\beta} \,| \, T \, A \, T^{-1} \,| \tilde{\alpha} \rangle . \qquad (8.75)$$

The observable A is either even or odd under the reversal of the direction of time if

$$T \, A \, T^{-1} = \pm \, A \; \begin{cases} + \quad \text{even} \\ - \quad \text{odd} . \end{cases} \qquad (8.76)$$

Inserting this result into (8.75), we obtain for $\alpha = \beta$ the time-reversal transformation of the mean values. If, for example,

$$\langle \alpha \,| \, \vec{p} \,| \alpha \rangle \; = -\langle \tilde{\alpha} \,| \, \vec{p} \,| \tilde{\alpha} \rangle , \qquad (8.77)$$

this implies the time-reversal transformation of the operator $\vec{p}$:

$$T \, \vec{p} \, T^{-1} = -\vec{p} \qquad (8.78)$$

and the eigenvalue equation for $\vec{p}$ reduces, under time-reversal, to

$$\vec{p}\, T \mid p\rangle = -T \,\vec{p}\, T^{-1} \, T \mid p\rangle$$
$$= -p(T \mid \vec{p}\rangle)\,. \tag{8.79}$$

The vector $\mid \vec{p}\rangle$ is an eigenstate of $\vec{p}$ corresponding to the eigenvalue p although under time-reversal $T \mid \vec{p}\rangle$ it is the eigenstate of $\vec{p}$ corresponding to the eigenvalue $(-p)$. If we apply a similar reasoning to other operators of quantum mechanics, we are led to the following transformations:

$$T \, X \, T^{-1} = X \quad \text{and} \quad T \mid x\rangle = \mid x\rangle\,. \tag{8.80}$$

By combining (8.78) with (8.80) we can infer that

$$[X_i, \ P_j] = i\,\hbar\,\delta_{ij}\,,$$
$$T\,[X_i, \ P_i]\,T^{-1} = -i\,\hbar\,\delta_{ij}\,, \tag{8.81}$$

whereas $\vec{L} = \vec{r} \wedge \vec{p}$ will yield, by the reversal of the direction of time,

$$T\,\vec{L}\,T^{-1} = -\vec{L}\,. \tag{8.82}$$

To extend these results to particles with angular momentum $\vec{J}$, we will use the more general transformation

$$T\,\vec{J}\,T^{-1} = -\vec{J}. \tag{8.83}$$

It should be noted here that the Dirac formalism of quantum mechanics implies linear operators whereas the time-reversal transformation leads to antilinear operators. This accounts for the delicate problems raised by this transformation.

The projection of a quantum state onto the configuration space:

$$\mid \alpha\rangle = \int \mid x\rangle \, dx \, \langle x \mid \alpha\rangle\,, \tag{8.84}$$

will yield, under time-reversal,

$$\mid \tilde{\alpha}\rangle = T \mid \alpha\rangle = \int T \mid x\rangle \, dx \, \langle x \mid \alpha\rangle\,. \tag{8.85}$$

We then apply the anti-unitarity condition (8.63) for T to obtain

$$T \mid \alpha\rangle = \int T \mid x\rangle \, dx \, \langle x \mid \alpha\rangle^*\,, \tag{8.86}$$

and, if we use relation (8.80),

$$\mid \tilde{\alpha}\rangle = T \mid \alpha\rangle = \int \mid x\rangle \, dx \, \langle x \mid \alpha\rangle^*\,. \tag{8.87}$$

By projecting onto configuration space, we infer the time-reversal transformation of the wave function:

$$\tilde{\Psi}_\alpha(x) = \langle x \mid \tilde{\alpha} \rangle = \langle x \mid \alpha \rangle^* = \Psi_\alpha^*(x). \tag{8.88}$$

If, in the Schrödinger equation

$$i\,\hbar\,\partial_t\,\Psi = H\,\Psi, \tag{8.89}$$

we change the direction of time, the left-hand side becomes

$$i\,\hbar\,\partial_{-t}\,\Psi = -i\,\hbar\,\partial_t\,\Psi. \tag{8.90}$$

This function is no longer a solution of the Schrödinger equation, whereas by changing t into $-t$ and Ψ into Ψ^* we obtain the Hermitian conjugate of the Schrödinger Eq. (8.89) and, hence, a satisfactory solution:

$$\boxed{T\,\Psi_\alpha(x,t) = \tilde{\Psi}_\alpha(x,t) = \Psi_\alpha^*(x,-t)} \tag{8.91}$$

This amounts to writing

$$\begin{aligned}
\tilde{\psi}_\alpha(x,t) &= \langle x \mid \tilde{\alpha}(t) \rangle = \langle x \mid \alpha(-t) \rangle^* = \psi_\alpha^*(x,-t), \\
\tilde{\Psi}_\alpha(x,t) &= \langle \tilde{x}(t) \mid \alpha \rangle = \langle x(-t) \mid \alpha \rangle^* = \Psi_\alpha^*(x,-t).
\end{aligned} \tag{8.92}$$

In a time-reversal transformation of the wave function, we need to change t to $-t$ and go from the space $\mathcal{E}$ to its dual $\mathcal{E}^*$.
It was shown in the last chapter that

$$\psi_a(x,t) = \Psi_a(x,-t), \tag{8.93}$$

which yields, by using the complex conjugates,

$$\psi_a^*(x,t) = \Psi_a^*(x,-t) = \tilde{\Psi}_a(x,t). \tag{8.94}$$

An invariant Hamiltonian under time-reversal with a system of non-degenerate eigenvectors possesses real eigenfunctions since

$$H(T \mid n \rangle) = T\,H \mid n \rangle = E_n\,(T \mid n \rangle) \tag{8.95}$$

and if E_n is not degenerate, then the eigenstates $\mid n \rangle$ and $T \mid n \rangle$ are identical. By projecting the states $\mid n \rangle$ and $T \mid n \rangle$ onto configuration space, we easily obtain:

$$\Psi_n(x) = \langle x \mid n \rangle = \langle x \mid T \mid n \rangle = \langle x \mid n \rangle^* = \Psi_n^*(x). \tag{8.96}$$

The eigenfunctions of H are therefore real functions.

3. Gauge Transformations

3.1 A Particle in an Electromagnetic Field

It was shown in Chap. 1 that a particle with mass m and electric charge e placed in an electromagnetic field $(\vec{E}, \vec{B}) = (-\vec{\nabla}\phi, \vec{\nabla} \wedge \vec{A})$ exhibits a motion that can be described by the Hamiltonian:

$$H = \frac{\pi^2}{2m} + e\,\phi,\tag{8.97}$$

with the classical kinematic momentum (with $c = 1$)

$$\vec{\pi} = \vec{p} - e\,\vec{A}.\tag{8.98}$$

The application of the correspondence principle will associate the operator

$$\vec{\pi} = -i\,\hbar\,\vec{\nabla} - e\,\vec{A}\tag{8.99}$$

with the kinematic momentum. The eigenvalue equation in Schrödinger's picture,

$$H\mid\alpha,\ t\rangle\ = i\,\hbar\,\partial_t\mid\alpha,\ t\rangle,\tag{8.100}$$

projected onto the configuration space defines the wave function $\psi_\alpha(x,t)$ associated with the motion of a particle in an electromagnetic field:

$$\left[\frac{1}{2m}\left(-i\,\hbar\,\vec{\nabla} - e\,\vec{A}\right)^2 + e\,\phi\right]\psi_\alpha(x,t) = i\,\hbar\,\partial_t\,\psi_\alpha(x,t).\tag{8.101}$$

3.2 Gauge Transformation of the Electromagnetic Field

When we add the gradient of a scalar function $\chi(x)$ to the vector potential $\vec{A}$, we normally say that we are making a gauge transformation of the vector potential:

$$\vec{A}\ \rightarrow\ \tilde{\vec{A}} = \vec{A} + \vec{\nabla}\,\chi(x).\tag{8.102}$$

This transformation does not modify the electromagnetic field, since

$$\begin{aligned}\vec{E}\ &\rightarrow\ \tilde{\vec{E}} = -\vec{\nabla}\,\tilde{\phi} = -\vec{\nabla}\,\phi = \vec{E}\\ \vec{B}\ &\rightarrow\ \tilde{\vec{B}} = \vec{\nabla}\,\wedge\,\tilde{\vec{A}} = (\vec{\nabla}\,\wedge\,(\vec{A} + \vec{\nabla}\,\chi(x))) = \vec{\nabla}\,\wedge\,\vec{A} = \vec{B}.\end{aligned}\tag{8.103}$$

The electromagnetic field is thus gauge-invariant.

How does gauge transformation (8.99) affect the wave function $\psi_\alpha(x,t)$?

3.3 The Gauge-Transformation Operator

Let us determine the operator $\mathcal{G}$ transforming $\mid \alpha, t\rangle$ into $\mid \tilde{\alpha}, t\rangle$ in the gauge transformation (8.102) by setting

$$\mid \tilde{\alpha},\ t\rangle\ = \mathcal{G} \mid \alpha,\ t\rangle\,. \tag{8.104}$$

Because the norm of the quantum state is conserved, the operator $\mathcal{G}$ must be unitary:

$$\langle \alpha,\ t \mid \alpha,\ t\rangle\ = \langle \tilde{\alpha},\ t \mid \tilde{\alpha},\ t\rangle\ = \langle \alpha,\ t \mid \mathcal{G}^{+}\, \mathcal{G} \mid \alpha,\ t\rangle\,. \tag{8.105}$$

This implies the unitarity relation

$$\mathcal{G}^{+}\mathcal{G} = \mathbb{1} \quad \text{and} \quad \mathcal{G}^{+} = \mathcal{G}^{-1}\,. \tag{8.106}$$

The gauge transformation of the kinematic momentum operator is obtained from its mean value which is invariant:

$$\langle \pi\rangle\ = \langle \tilde{\alpha},\ t \mid \tilde{\pi} \mid \tilde{\alpha},\ t\rangle\ = \langle \alpha,\ t \mid \mathcal{G}^{+}\, \tilde{\pi}\, \mathcal{G} \mid \alpha,\ t\rangle\ = \langle \alpha,\ t \mid \pi \mid \alpha,\ t\rangle\,. \tag{8.107}$$

By identifying the operators, we obtain the following relation:

$$\mathcal{G}^{+}\, \tilde{\pi}\, \mathcal{G} = \pi \quad \text{or} \quad \tilde{\pi} = \mathcal{G}\, \pi\, \mathcal{G}^{+}\,, \tag{8.108}$$

and, similarly, the gauge transformation of the Hamiltonian is written

$$\tilde{H} = \mathcal{G}\, H\, \mathcal{G}^{+}\,. \tag{8.109}$$

The position of a particle, mean value of the position operator $\vec{X}$, must be independent of the gauge transformation and depend solely on the electromagnetic field itself:

$$\langle \vec{X}\rangle\ = \langle \tilde{\alpha},\ t \mid \vec{X} \mid \tilde{\alpha},\ t\rangle\ = \langle \alpha,\ t \mid \mathcal{G}^{+}\, \vec{X}\, \mathcal{G} \mid \alpha,\ t\rangle\,. \tag{8.110}$$

This implies gauge invariance:

$$[\vec{X}, \mathcal{G}] = 0\,. \tag{8.111}$$

The operator $\mathcal{G}$ is therefore diagonal in configuration space and should be solely dependent on the variable $\vec{x}$, that is,

$$\langle \vec{x}' \mid \mathcal{G} \mid \vec{x}\rangle\ = \mathcal{G}(\vec{x})\, \delta^{3}(\vec{x}' - \vec{x})\,. \tag{8.112}$$

Now, let us choose the operator $\mathcal{G}$ in its exponential form:

$$\boxed{\mathcal{G}(\vec{x}) = \exp\left[\frac{i\,e}{\hbar}\, \chi(\vec{x})\right]} \tag{8.113}$$

and then determine the transformation of the vector $\vec{p}$ by applying it to any function $\Psi(x)$ whatsoever:

$$
\begin{aligned}
\tilde{\vec{p}}\,\Psi(\tilde{x}) &= \mathcal{G}\,\vec{p}\,\mathcal{G}^+\,\Psi = \mathcal{G}\,(-i\,\hbar\,\vec{\nabla})\,\mathcal{G}^+\,\Psi \\
&= -i\,\hbar\,\mathcal{G}\,\mathcal{G}^+\,\vec{\nabla}\,\Psi - i\,\hbar\,\mathcal{G}\,\Psi\,\vec{\nabla}\,\mathcal{G}^+ \\
&= \vec{p}\,\Psi - i\,\hbar\,\Psi(x)\,\exp\left(\frac{i\,e}{\hbar}\,\chi\right)\left(-\frac{i\,e}{\hbar}\,\vec{\nabla}\,\chi\right)\exp\left(-\frac{i\,e}{\hbar}\,\chi\right) \\
&= \left(\vec{p} - e\,\vec{\nabla}\,\chi(\tilde{x})\right)\Psi(\tilde{x}) \qquad \forall\,\Psi(x)\,.
\end{aligned}
\tag{8.114}
$$

This defines the transformation of the vector $\vec{p}$:

$$
\tilde{\vec{p}} = \vec{p} - e\,\vec{\nabla}\,\chi(\tilde{x})\,.
\tag{8.115}
$$

We next infer the transformation of the kinematic momentum $\vec{\pi}$:

$$
\begin{aligned}
\tilde{\vec{\pi}} &= \mathcal{G}\,\pi\,\mathcal{G}^+ = \mathcal{G}\,\left(\vec{p} - e\,\vec{A}\right)\mathcal{G}^+ \\
&= \mathcal{G}\,\vec{p}\,\mathcal{G}^+ - e\,\vec{A} = \vec{p} - e\,\vec{\nabla}\,\chi(\tilde{x}) - e\,\vec{A} \\
&= \vec{p} - \frac{e}{c}\,(\vec{A} + \vec{\nabla}\,\chi) = \vec{p} - e\,\tilde{\vec{A}}\,.
\end{aligned}
\tag{8.116}
$$

Form (8.113) of the gauge transformation operator is quite compatible with the gauge transformation (8.99) of the potential vector.

The gauge transformation of the wave function is obtained by projecting definition (8.101) onto configuration space:

$$
\tilde{\psi}_\alpha(x,t) = \mathcal{G}(\tilde{x})\,\psi_\alpha(\tilde{x},t) = \exp\left[\frac{i\,e}{\hbar}\,\chi(\tilde{x})\right]\,\psi_\alpha(\tilde{x},t)\,.
\tag{8.117}
$$

The gauge transformation introduces a phase function in the wave function whereas the probability density

$$
\varrho(\tilde{x},t) = |\,\psi_\alpha(\tilde{x},t)\,|^2 = |\,\tilde{\psi}_\alpha(\tilde{x},t)\,|^2
\tag{8.118}
$$

is gauge-invariant.

This again underscores the primordial role of the phase function in wave mechanics as was experimentally demonstrated by Aharonov and Bohm (see Chap. 2).

The gauge transformation $\mathcal{G}(\tilde{x})$ of the electromagnetic field depends on the position $\tilde{x}$ of the charged particle. It is a local gauge transformation. It depends solely on the parameter e (electric charge). The gauge group of the electromagnetic field is the group $U(1)$, a group of one-dimensional unitary transformations.

Gauge invariance has become a central issue in quantum physics in connection with the description of fundamental interactions and has been extended to transformation groups that are larger than $U(1)$, in particular for describing weak interactions (group $SU(2)$) or strong interactions (group $SU(3)$).

3.4 Non-Abelian Gauge Fields

Let $\phi(x)$ be a matter field and a gauge field, the set of vector fields $W_\mu^a(x)$ with $\mu = 0, 1, 2, 3$ and $a = 1, 2, 3 \ldots \dim U$ where U is the continuous unitary group specifying the gauge transformation of the matter field:

$$\phi(x) \;\rightarrow\; \tilde{\phi}(x) = U(x)\,\phi(x)\,, \tag{8.119}$$

$$U(x) = \mathbb{1} - \sum_{a=1}^{\dim U} \frac{i}{\hbar}\, \varepsilon_a(x)\, T^a\,. \tag{8.120}$$

Let $\varepsilon_a(x)$ be the coordinates for the group and T^a the infinitesimal generators of the gauge group U satisfying the following commutation relations:

$$[T_a, T_b] = i\, C_{ab}^c\, T_c\,. \tag{8.121}$$

The structure constants of the group C_{ab}^c are antisymmetrical coefficients:

$$C_{ab}^c = -C_{ba}^c\,, \tag{8.122}$$

satisfying the Jacobi relation

$$C_{MN}^A\, C_{AP}^D + C_{NP}^A\, C_{AM}^D + C_{PM}^A\, C_{AN}^D = 0\,. \tag{8.123}$$

The integrated form of the gauge-transformation operator (8.120) can be written as

$$U(x) = \exp\left(-\frac{i\,g}{\hbar}\, \vec{\theta}(x) \cdot \vec{T} \right)\,, \tag{8.124}$$

where $\vec{\theta}(x)$ are arbitrary functions of x and g is the coupling constant between the gauge field and the matter field.

In the case of the electromagnetic field, the gauge group is $U(1)$ and the group generators the mutually commuting scalar functions. The group is said to be Abelian. The coupling constant of the gauge field $A_\mu(x)$ and the matter field $\psi_\alpha(x, t)$ is the electric charge $-e$. If the generators of the gauge group obey a commutation relation such as (8.121), the group is said to be non-Abelian and the vectors of the gauge field W_μ^a are brought together in one vector field, the Yang–Mills field, by setting

$$W_\mu = W_\mu^a\, T_a = \vec{W}_\mu \cdot \vec{T} \quad \text{with} \quad a = 1, 2 \ldots \dim U\,. \tag{8.125}$$

The derivative operator $\partial_\mu = \partial/\partial\, x^\mu$ remains the same in the gauge transformation. Now, let us introduce a covariant derivative operator D_μ undergoing a transformation just like the matter field:

$$\tilde{D}_\mu\, \tilde{\phi} = U\, D_\mu\, \phi\,. \tag{8.126}$$

By using transformation (8.119) of the matter field, we further obtain

$$\tilde{D}_\mu\, U\, \phi = U\, D_\mu\, \phi \qquad \forall\, \phi(x)\,. \tag{8.127}$$

By multiplying from the left by U^{-1}, we easily obtain the expected result: The operator D_μ is transformed like all other operators, that is,

$$U^{-1} \tilde{D}_\mu U = D_\mu \quad \text{or} \quad \tilde{D}_\mu = U D_\mu U^{-1} \; . \tag{8.128}$$

Let us use the Yang–Mills field W_μ to define the covariant derivative operator:

$$D_\mu = \partial_\mu - \frac{i\,g}{\hbar} W_\mu = \partial_\mu - \frac{i\,g}{\hbar} W_\mu^a T_a = \partial_\mu - \frac{i}{\hbar} \vec{W}_\mu \cdot \vec{T} \tag{8.129}$$

and then determine the gauge transformation of the operator D_μ using (8.128):

$$\tilde{D}_\mu = \partial_\mu - \frac{i\,g}{\hbar} \tilde{W}_\mu = U \left(\partial_\mu - \frac{i\,g}{\hbar} W_\mu \right) U^{-1} \; . \tag{8.130}$$

By applying this to an arbitrary function $\Psi(x)$, we obtain

$$\partial_\mu \Psi - \frac{i\,g}{\hbar} \tilde{W}_\mu \Psi = U \left(\partial_\mu U^{-1} \right) \Psi + U U^{-1} \partial_\mu \Psi - \frac{i\,g}{\hbar} U W_\mu U^{-1} \Psi \; . \tag{8.131}$$

Because this relation is valid for any function $\Psi(x)$, we infer the gauge transformation of the gauge field itself:

$$W_\mu \;\rightarrow\; \tilde{W}_\mu = U W_\mu U^{-1} - \frac{\hbar}{i\,g} U \partial_\mu U^{-1} \; . \tag{8.132}$$

Because in electromagnetism the coupling constant is $g = -e$, the gauge transformation operator is written

$$U(x) = \exp \left[\frac{i\,e}{\hbar} \chi(x) \right] \; . \tag{8.133}$$

It commutes with the vector potential $A_\mu(x)$. The term $U A_\mu U^{-1}$ is therefore equal to A_μ whereas the second term in (8.132) is easily evaluated as follows:

$$U \partial_\mu U^{-1} = \exp \left(\frac{i\,e}{\hbar} \chi \right) \partial_\mu \exp \left(-\frac{i\,e}{\hbar} \chi \right)$$
$$= -\frac{i\,e}{\hbar} \partial_\mu \chi \; , \tag{8.134}$$

which gives, by inserting the expected gauge transformation into (8.128):

$$A_\mu \;\rightarrow\; \tilde{A}_\mu = A_\mu + \partial_\mu \chi(x) \; . \tag{8.135}$$

The covariant derivative operator (8.130) becomes, for its part, with $g = -e$,

$$D_\mu = \partial_\mu + \frac{i\,e}{\hbar} A_\mu \tag{8.136}$$

and by using the correspondence principle (5.91) in a $(+ - --)$ signature,

$$\boxed{i\,\hbar\,D_\mu = \pi_\mu = p_\mu - e\,A_\mu}\ .\qquad (8.137)$$

The covariant derivative operator in the electromagnetic field is the operator associated with the classical kinematic momentum.

The Faraday-like tensor of a non-Abelian gauge field is written

$$
\begin{aligned}
F_{\mu\nu} &= D_\mu\,W_\nu - D_\nu\,W_\mu \\
&= \left(\partial_\mu - \frac{i\,g}{\hbar}\,W_\mu -\right)W_\nu - \left(\partial_\nu - \frac{i\,g}{\hbar}\,W_\nu\right)W_\mu \\
&= \partial_\mu\,W_\nu - \partial_\nu\,W_\mu - \frac{i\,g}{\hbar}\,[W_\mu, W_\nu]\ . \qquad (8.138)
\end{aligned}
$$

In the case of the electromagnetic field (Abelian gauge field) the commutator is zero and we once more find the form of the Faraday-like tensor. Relation (8.138) is termed the Yang–Mills relation of a non-Abelian gauge field. It can also be written (with (8.125) and (8.129)) as follows:

$$F^a_{\mu\nu} = \partial_\mu\,W^a_\nu - \partial_\nu\,W^a_\mu + \frac{g}{\hbar}\,C^a_{bc}\,W^b_\mu\,W^c_\nu\ . \qquad (8.139)$$

By analogy with the electromagnetic field, the Yang–Mills field will be described by a Lagrangian density

$$\mathcal{L} = -\frac{1}{4}\,F^a_{\mu\,\nu}\,F^{\mu\,\nu}_a = -\frac{1}{4}\,\vec{F}_{\mu\,\nu}.\vec{F}^{\mu\,\nu} \qquad a = 1, 2\ldots = \dim U\ . \qquad (8.140)$$

If the gauge group associated with the Yang–Mills field is the group $SU(N)$, which is a sub-group of $U(N)$ of the $N \times N$ unitary matrices with determinant equal to $+1$, it will cause $N^2 - 1$ generators to occur. It is thus that the gauge group $SU(2)$ will involve three generators and the group $SU(3)$ eight generators (or eight gauge bosons). These results will be taken up again in the description of weak interactions (three gauge bosons W^+ W^- and Z^0) and strong interactions (eight gauge bosons, the eight colored gluons G^j_i $i = 1, 2, 3 \neq j$).

The Euler–Lagrange equations leading to the Maxwell equations $\partial_\mu\,F^{\mu\,\nu} = 0$ of the electromagnetic field generalize into Yang–Mills equations

$$\partial_\mu\,F^{\mu\,\nu} - \frac{i\,g}{\hbar}\,[W_\mu, F^{\mu\,\nu}] = 0\ , \qquad (8.141)$$

and we arrive back at the usual form of the Maxwell equations by introducing the Yang–Mills derivative operator

$$\partial_\mu - \frac{i\,g}{\hbar\,c}\,[W_\mu,] = \nabla_\mu\ , \qquad (8.142)$$

which transforms the equations of the gauge field as follows:

$$\boxed{\nabla_\mu F^{\mu\,\nu} = 0}\ .$$

(8.143)

The second group of Maxwell equations,

$$\partial_\tau F_{\mu\,\nu} + \partial_\nu F_{\tau\,\mu} + \partial_\mu F_{\nu\,\tau} = 0$$

(8.144)

for the Yang–Mills field is thus written

$$D_\tau F_{\mu\,\nu} + D_\nu F_{\tau\,\mu} + D_\mu F_{\nu\,\tau} = 0\,,$$

(8.145)

or, by using the fact that,

$$D_\tau F_{\mu\,\nu} = \partial_\tau F_{\mu\,\nu} - \frac{i\,g}{\hbar}\,[W_\tau, F_{\mu\,\nu}]$$

(8.146)

and the Jacobi identity of commutators, we show that the second group of Maxwell equations retains the form of (8.142) for the Yang–Mills field. We will return to gauge fields in due course after the study of quantum electrodynamics and its extension to quantum chromodynamics, or the Glashow–Salam–Weinberg (G.S.W.) standard model of electroweak interactions (see Chap. 19).

Chapter 9
Angular Momentum

1. General Definition

In classical mechanics, a particle with mass m and momentum $\vec{p}$ possesses an angular momentum $\vec{L} = \vec{r} \wedge \vec{p}$, where $\vec{r}$ defines the vector position with respect to a fixed point 0. The transition to quantum mechanics will be made by applying the correspondence principle, that is, by defining an operator $\vec{L}$:

$$\vec{L} = -i \, \hbar \, \vec{r} \wedge \vec{\nabla} = -i \, \hbar \begin{vmatrix} \vec{e}_1 & \vec{e}_2 & \vec{e}_3 \\ x & y & z \\ \frac{\partial}{\partial x} & \frac{\partial}{\partial y} & \frac{\partial}{\partial z} \end{vmatrix} = -i \, \hbar \begin{vmatrix} \vec{e}_1 & \vec{e}_2 & \vec{e}_3 \\ x & y & z \\ \partial_x & \partial_y & \partial_z \end{vmatrix} \tag{9.1}$$

or by expanding the determinant:

$$\vec{L} = - i\hbar \left[\vec{e}_1 \left(y \frac{\partial}{\partial z} - z \frac{\partial}{\partial y} \right) + \vec{e}_2 \left(z \frac{\partial}{\partial x} - x \frac{\partial}{\partial z} \right) \right.$$
$$\left. + \vec{e}_3 \left(x \frac{\partial}{\partial y} - y \frac{\partial}{\partial x} \right) \right] . \tag{9.2}$$

The vector product can also be expressed using the Levi–Civita symbol:

$$\varepsilon_{ijk} = \begin{cases} 1 & \text{if } (i \, j \, k) = (123) = (x \, y \, z) \\ -1 & \text{if } (i \, j \, k) = (213) = (y \, x \, z) \\ 0 & \text{otherwise} \end{cases} \tag{9.3}$$

where $(i \, j \, k)$ designates the circular permutations of subscripts i, j and k. Then we can write

$$\vec{L} = -i \, \hbar \, \varepsilon_{ijk} \, \vec{e}_k \, x_i \, \partial_j . \tag{9.4}$$

The components of this operator in spherical coordinates are obtained by changing variables as follows:

$$x = r \, \sin \, \theta \, \cos \, \varphi ,$$
$$y = r \, \sin \, \theta \, \sin \, \varphi ,$$
$$z = r \, \cos \, \theta .$$

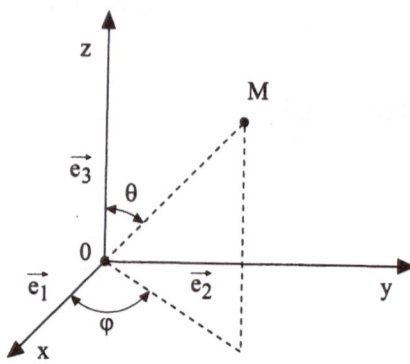

We then obtain the components

$$L_x = i\,\hbar\,\left(\sin\,\varphi\,\frac{\partial}{\partial\,\theta} + cotg\,\theta\,\cos\,\varphi\,\frac{\partial}{\partial\,\varphi}\right)$$

$$L_y = i\,\hbar\,\left(-\cos\,\varphi\,\frac{\partial}{\partial\,\theta} + cotg\,\theta\,\sin\,\varphi\,\frac{\partial}{\partial\,\varphi}\right) \qquad (9.5)$$

$$L_z = -i\,\hbar\,\frac{\partial}{\partial\,\varphi}$$

and the square modulus of the angular momentum reduces to

$$L^2 = L_x^2 + L_y^2 + L_z^2 = -\hbar^2\,\left[\frac{1}{\sin\,\theta}\,\frac{\partial}{\partial\,\theta}\,\left(\sin\,\theta\,\frac{\partial}{\partial\,\theta}\right) + \frac{1}{\sin^2\,\theta}\,\frac{\partial^2}{\partial\,\varphi^2}\right]. \quad (9.6)$$

The position operator x_i of the particle and the momentum operator p_j ($i,\ j = x, y, z$) satisfy the commutation relations

$$[x_i, x_j] = 0 \qquad [p_i, p_j] = 0 \qquad [x_i, p_j] = i\,\hbar\,\delta_{ij}. \qquad (9.7)$$

It is easy then to determine the commutation relations between the components L_x, L_y, L_z of the angular momentum. We could, for example, calculate

$$[L_x, L_y] = (y\,p_z - z\,p_y)\,(z\,p_x - x\,p_z) - (z\,p_x - x\,p_z)\,(y\,p_z - z\,p_y).$$

In view of relations (9.7), we obtain $[L_x, L_y] = i\,\hbar\,L_z$ and, by proceeding in a similar manner:

$$[L_y, L_z] = i\,\hbar\,L_x \qquad [L_z, L_x] = i\,\hbar\,L_y.$$

All three commutation relations may be written in vector notation as follows:

$$\boxed{\vec{L} \wedge \vec{L} = i\,\hbar\,\vec{L}}\,. \qquad (9.8)$$

We draw attention here to the fact that, in classical mechanics, the preceding notation is meaningless since the vector product of two co-linear vectors is zero. In quantum mechanics, on the other hand, the components of the vector operators do not commute with one another. Notation (9.8) is therefore perfectly legitimate in this case and

$$\vec{L} \wedge \vec{L} = \vec{e}_1 \left[L_y, L_z \right] + \vec{e}_2 \left[L_z, L_x \right] + \vec{e}_3 \left[L_x, L_y \right] .$$

It is easily seen here, with the aid of (9.8), that the squared length L^2 of the angular momentum $L^2 = L_x^2 + L_y^2 + L_z^2$ commutes with each of the components of $\vec{L}$:

$$\left[L^2, \vec{L} \right] = 0 . \tag{9.9}$$

Relations (9.8) and (9.9) will be characteristic of an angular momentum in quantum mechanics and any vector operator $\vec{J}$ verifying the commutation relations

$$\vec{J} \wedge \vec{J} = i\,\hbar\,\vec{J} \quad \text{or} \quad [J_k, J_\ell] = i\,\hbar\,\varepsilon_{k\,\ell\,m}\,J_m \tag{9.10}$$

(and hence $\left[J^2, \vec{J} \right] = 0$) will be termed angular momentum. To simplify the notation, we henceforth set $\hbar = 1$ (natural units).

2. Standard Representation

2.1 The Operators J_+ and J_-

Let us introduce the mutually conjugate Hermitian operators J_+ and J_-, and then $J_+^\dagger = J_-$ and $J_-^\dagger = J_+$ with Hermitian operators J_x and J_y:

$$
\begin{aligned}
J_+ &= J_x + i\,J_y \\
J_- &= J_x - i\,J_y
\end{aligned}
\quad \text{or better still} \quad
\begin{aligned}
J_x &= \frac{1}{2}\left(J_+ + J_- \right) \\
J_y &= \frac{1}{2i}\left(J_+ - J_- \right) .
\end{aligned}
\tag{9.11}
$$

It is relatively easy to evaluate the commutation relations between these operators themselves and with J^2 and J_z:

$$
\begin{aligned}
\left[J_z, J_+ \right] &= J_+ \\
\left[J_z, J_- \right] &= -J_- \\
\left[J_+, J_- \right] &= 2J_z \\
\left[J^2, J_+ \right] &= \left[J^2, J_- \right] = \left[J^2, J_z \right] = 0 .
\end{aligned}
\tag{9.12}
$$

By using expression (9.11) for J_x and J_y in terms of J_+ and J_-, the squared length of the angular momentum will be expressed with $J_+\,J_-$ and J_z as follows:

$$J^2 = J_x^2 + J_y^2 + J_z^2 = \frac{1}{2} \left(J_+ \, J_- + J_- \, J_+ \right) + J_z^2 .$$

With commutation relation (9.12), we obtain the following two important relations:

$$J_- \, J_+ = J^2 - J_z \, (J_z + \mathbb{1}) \, ,$$
$$J_+ \, J_- = J^2 - J_z \, (J_z - \mathbb{1}) \, .$$

(9.13)

2.2 Eigenvalues of J^2 and J_z

Now, let us use as basis of the angular momentum operators the eigenbasis common to the commuting operators J^2 and J_z. We will denote the eigenvectors of these operators by $| \, jm \rangle$:

$$J^2 \, | \, jm \rangle \ = a_j \, | \, jm \rangle \qquad \text{and} \qquad J_z \, | \, jm \rangle \ = m \, | \, jm \rangle .$$

(9.14)

The operator J^2 is a Hermitian operator defined positive since J_x, J_y, J_z are Hermitian operators:

$$J^2 = J_x^2 + J_y^2 + J_z^2 .$$

If we evaluate its norm $\langle jm \, | \, J^2 \, | \, jm \rangle \ = a_j$ on the normalized vectors $| \, jm \rangle$:

$$\langle jm \, | \, J_x^2 + J_y^2 + J_z^2 \, | \, jm \rangle = \langle jm \ | \, J_x^+ \, J_x \, | \, jm \rangle \ + \ \langle jm \, | \, J_y^+ \, J_y \, | \, jm \rangle$$
$$+ \ \langle jm \, | \, J_z^+ \, J_z \, | \, jm \rangle ,$$

we obtain the sum of the norms of the vectors:

$$a_j = N \, \left(J_x \, | \, jm \rangle \right) + N \, \left(J_y \, | \, jm \rangle \right) + N \, \left(J_z \, | \, jm \rangle \right) .$$

This shows that $a_j \geq 0$ and if we set $a_j = j(j+1)$, we obtain the following fundamental relation:

$$\boxed{J^2 \, | \, jm \rangle \ = j(j+1) \, | \, jm \rangle \qquad j \geq 0} \ .$$

An eigenstate $| \, jm \rangle$ of J^2 will then be labeled with the real and positive quantum number j, whose corresponding eigenvalue is $j(j+1)$. We will conversely term m the eigenvalue of J_z corresponding to the same eigenvector $| \, jm \rangle$.

The ket $| \, jm \rangle$ will be said to represent an angular momentum state (jm) if the following eigenvalue equations are satisfied:

$$\boxed{\begin{aligned} J^2 \, | \, jm \rangle \ &= j(j+1) \, | \, jm \rangle \qquad j \geq 0 \\ J_z \, | \, jm \rangle \ &= m \, | \, jm \rangle \end{aligned}} \ .$$

(9.15)

2.3 Relationship Between Quantum Numbers m and j

By applying the operators $J_+ J_-$ and $J_- J_+$ defined by Eq. (9.13) to a vector $|\,jm\rangle$, we obtain

$$J_- J_+ |\,jm\rangle = [J^2 - J_z (J_z + \mathbb{1})] |\,jm\rangle = [j(j+1) - m(m+1)] |\,jm\rangle,$$
$$J_+ J_- |\,jm\rangle = [J^2 - J_z (J_z - \mathbb{1})] |\,jm\rangle = [j(j+1) - m(m-1)] |\,jm\rangle. \tag{9.16}$$

Now, let us multiply each of these relations by $\langle jm\,|$. Because the operators J_+ and J_- are Hermitian conjugates of one another and the basis vectors $|\,jm\rangle$ orthonormal, the first relation leads to the norm $J_+ |\,jm\rangle$:

$$\langle jm\,|\,J_- J_+ |\,jm\rangle = N\ (J_+ |\,jm\rangle) = j(j+1) - m(m+1)$$
$$= (j-m)\,(j+m+1) \geq 0, \quad (9.17)$$

whereas the second (9.16) leads to the norm of $J_- |\,jm\rangle$:

$$\langle jm\,|\,J_+ J_- |\,jm\rangle = N\ (J_- |\,jm\rangle) = j(j+1) - m(m-1)$$
$$= (j+m)\,(j-m+1) \geq 0. \quad (9.18)$$

The first inequality shows that $-j-1 \leq m \leq j$ whereas the second imposes $-j \leq m \leq j+1$.

These inequalities are therefore simultaneously obeyed in the domain:

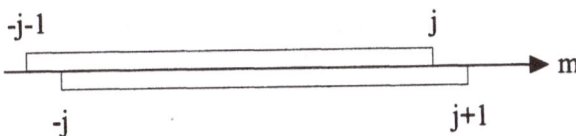

where m is between $-j$ and $+j$. We therefore obtain a first selection rule for the magnetic momenta m:

$$\boxed{-j \leq m \leq j} \,. \tag{9.19}$$

2.4 Selection Rule for j

Let us use the ket $|\,jm\rangle$ as departure point to describe the state with angular momentum $|\,jm\rangle$. It is necessary and sufficient for a vector's norm to be zero for the vector itself to be zero (see Chap. 4). With relation (9.17), we thus obtain:

$$J_+ |\,jm\rangle = 0 \quad \text{if} \quad N\ (J_+ |\,jm\rangle) = (j-m)\,(j+m+1) = 0. \quad (9.20)$$

The value $m = j$, the maximum compatible with (9.19), provides a solution to the problem and

$$J_+ \mid jj\rangle = 0. \tag{9.21}$$

This defines the vector with the highest weight $\mid jj\rangle$.

If m is different from j, then it goes without saying that $J_+ \mid jm\rangle$ is an eigenvector of J^2 and J_z corresponding to the eigenvalue $(m + 1)$ of J_z. In effect, following (9.12), $\left[J^2, J_+\right] = 0$, making it possible for us to write

$$J^2 \left(J_+ \mid jm\rangle\right) = J_+ \left(J^2 \mid jm\rangle\right) = J_+ \left(j(j + 1) \mid jm\rangle\right)$$
$$= j(j + 1) \left(J_+ \mid jm\rangle\right) . \tag{9.22}$$

Let us further use the commutation relation $\left[J_z, J_+\right] = J_+$ to write $J_z\, J_+ = J_+ \left(J_z + 1\right)$. We obtain

$$J_z \left(J_+ \mid jm\rangle\right) = J_+ \left(J_z + 1\right) \mid jm\rangle = J_+ \left(m + 1\right) \mid jm\rangle$$
$$= (m + 1) \left(J_+ \mid jm\rangle\right) . \tag{9.23}$$

The ket $J_+ \mid jm\rangle$ therefore describes a state with angular momentum $(j, m + 1)$. The following result can therefore be stated:

- if $m = j$ $J_+ \mid jj\rangle = 0$
- if $m \neq j$ relation (9.19) imposes $m \leq j$ and $J_+ \mid jm\rangle$ is an eigenstate with angular momentum $(j, m + 1)$
- if $m + 1 = j$ $J_+ \left(J_+ \mid jm\rangle\right) = J_+^2 \mid jm\rangle = 0$, otherwise $m + 1 \leq j$, $J_+^2 \mid jm\rangle$ is an eigenstate with angular momentum $(j, m + 2)$.

This reasoning can be pursued down to the pth place and the $J_+^p \mid jm\rangle$ will be an eigenvector of J^2 corresponding to the eigenvalue $j(j + 1)$ and of J_z corresponding to the eigenvalue $(m + p)$.

- If $m + p = j$, $J_+^{p+1} \mid jm\rangle = 0$.

In other words, $p = j - m$ is a positive integer (it denotes the number of times the operator J_+ is applied) and the p vectors $J_+ \mid jm\rangle$, $J_+^2 \mid jm\rangle$, ... $J_+^p \mid jm\rangle$ eigenvector of J^2 corresponding to the eigenvalue $j(j + 1)$ and of J_z corresponding to the values $m + 1, m + 2, \ldots, m + p = j$.

The same procedure will yield $J_- \mid j - j\rangle = 0$ and $J_z \left(J_- \mid jm\rangle\right) = (m - 1) \left(J_- \mid jm\rangle\right)$. The repeated action of the operator J_- on a vector $\mid jm\rangle$ will yield the eigenvectors of J^2 for the eigenvalue $j(j + 1)$ and of J_z for the eigenvalues

$$m - 1, m - 2, \ldots, m - q = -j \qquad (q \geq 0 \text{ integer}).$$

We can therefore note that p and q, the number of times operators J_+ or J_- are applied, are integers:

$$p = j - m \quad \text{integer} ,$$
$$q = j + m \quad \text{integer} ,$$

By considering the member-by-member sum, we obtain $p + q = \text{integer} = 2j$. Hence, j is a positive real integer or half-integer and this result can be formulated as a theorem.

- The only possible eigenvalues of J^2 are of the form $j(j + 1)$ where j is a positive integer or half-integer: $j = 0, 1/2, 1, \ldots$
- The only possible values of m are the integers or half-integers $m = 0, \pm 1/2, \pm 1, \ldots$
- If $j(j + 1)$ and m are the eigenvalues of J^2 and J_z corresponding to a state of the angular momentum (j, m), we necessarily have: $m = -j$, $-j + 1 \ldots + j$ and there exist $2j + 1$ possible values of m.

2.5 Construction of Subspaces

Let us, for each value of j corresponding to the eigenvalue $j(j + 1)$ of J^2, construct a subspace ξ_j spanned by $(2j + 1)$ vectors $| jm \rangle$. This subspace will therefore be a $(2j + 1)$-dimensional subspace.

The vectors $| jm \rangle$, eigenvectors of J^2 corresponding to the eigenvalue $j(j + 1)$ and of J_z corresponding to the eigenvalue m, are orthonormal, that is,

$$\langle jm' | jm \rangle = \delta_{mm'} . \tag{9.24}$$

The action of the operator J_+ on $| jm \rangle$ will yield a vector belonging to the subspace ξ_j since it is an eigenvector of J^2 for the same eigenvalue $j(j + 1)$, whereas the eigenvalue of J_z has increased by one unit,

$$J^2 \left(J_+ | jm \rangle \right) = j(j + 1) \left(J_+ | jm \rangle \right)$$
$$J_z \left(J_+ | jm \rangle \right) = (m + 1) \left(J_+ | jm \rangle \right) ,$$

but we also note that $J_z | jm + 1 \rangle = (m + 1) | jm + 1 \rangle$. We can therefore conclude that $J_+ | jm \rangle$ is co-linear to $| jm + 1 \rangle$.

Now, let us determine the proportionality coefficient x_m between these vectors:

$$J_+ | jm \rangle = x_m | jm + 1 \rangle \tag{9.25}$$

by calculating the norm of the vector

$$N \left(J_+ | jm \rangle \right) = \langle jm + 1 | x_m^* x_m | jm + 1 \rangle$$
$$= | x_m |^2 \langle jm + 1 | jm + 1 \rangle = | x_m |^2 . \tag{9.26}$$

Equation (9.17) will, in addition, yield the value of this norm, thus defining the coefficient x_m:

$$| x_m |^2 = (j-m)(j+m+1) = j(j+1) - m(m+1). \qquad (9.27)$$

The phase factor of the $| jj \rangle$ vector has been chosen positive and real, so we can write:

$$J_+ | jm \rangle = [(j-m)(j+m+1)]^{1/2} | jm+1 \rangle = x_m | jm+1 \rangle \qquad (9.28)$$

with x_m positive:

$$x_m = [(j-m)(j+m+1)]^{1/2} = [j(j+1) - m(m+1)]^{1/2}. \qquad (9.29)$$

We can similarly set

$$J_- | jm \rangle = x'_m | jm-1 \rangle, \qquad (9.30)$$

which, multiplied by the bra $\langle jm-1 |$, yields

$$x'_m = \langle jm-1 | J_- | jm \rangle. \qquad (9.31)$$

Hermitian conjugation of the ket enables us to write the conjugate bra as follows:

$$\langle jm-1 | J_- = [J_+ | jm-1 \rangle]^+ = [x_{m-1} | jm \rangle]^+ = \langle jm | x_{m-1}$$

because the coefficient x_{m-1} is a real number.

By inserting this result into (9.31), we obtain

$$x'_m = \langle jm | x_{m-1} | jm \rangle = x_{m-1}.$$

We can, in the final analysis then, write the following relations:

$$\begin{aligned} J_+ | jj \rangle &= 0 \\ J_- | j-j \rangle &= 0 \\ J_+ | jm \rangle &= x_m | jm+1 \rangle \\ J_- | jm \rangle &= x_{m-1} | jm-1 \rangle \end{aligned} \qquad (9.32)$$

with the coefficient x_m being given by

$$x_m = [j(j+1) - m(m+1)]^{1/2} = [(j-m)(j+m+1)]^{1/2}.$$

Working from any vector $| jm \rangle$ whatsoever of ξ_j and by the action of the operators J_+ and J_-, it is possible, through the application of relations (9.32), to construct a set of $(2j+1)$ orthonormal vectors $| j\,j \rangle\,| j\,j-1 \rangle \,\cdots\, | j\,-j \rangle$ that are the base vectors of ξ_j. These vectors satisfy eigenvalue Eq. (9.15):

$$\begin{aligned} J^2 | jm \rangle &= j(j+1) | jm \rangle \\ J_z | jm \rangle &= m | jm \rangle \end{aligned}$$

and can be inferred from one another with (9.32). It should be borne in mind that the operators J^2, J_z, J_+, J_- transform vectors of ξ_j into other vectors of the same subspace. Hence, they leave the subspace ξ_j invariant.

2.6 Standard Representations $\{J^2, J_z\}$

Among all the possible representations of states of the angular momentum, those in which J^2 and J_z are diagonal operators are preferable because the manipulation of the angular momentum is especially simple. In these representations, the additional variables corresponding to other characteristics of the quantum state, such as energy and electric charge, are represented by operators which commute with J_+ and J_- (or J_x and J_y) and J^2 J_z. They are called standard representations $\{J^2, J_z\}$. These are representations which, because their base vectors correspond to the same value of the quantum number j, can be grouped into one or several series of $(2j+1)$ vectors that can be inferred from one another by applying relations (9.32).

To each series corresponds a subspace ξ_j and the Hilbert space is formed by the sum of all the ξ_j subspaces of this type.

Among the eigenvectors of J^2 and J_z corresponding to the eigenvalue $j(j+1)$ of J^2, let us choose those for which $m = j$. They constitute a ξ_j subspace of the Hilbert space spanned by a complete set of orthonormal vectors $|\tau\, jj\rangle$. The quantum number τ is characteristic of operators that commute with the angular momentum and form with it a complete set of commuting operators. The orthonormalization relation of these vectors reads:

$$\langle \tau\, jj \mid \tau'\, jj \rangle = \delta_{\tau\tau'}. \qquad (9.33)$$

Let us associate to each of the $|\tau\, jj\rangle$ vectors the $2j$ vectors obtained by the repeated action of the operator J_-. In this way, we constitute a $(2j+1)$-dimensional subspace $\xi_{j\tau}$ spanned by the following basis vectors:

$$|\tau\, jj\rangle \; |\tau\, jj-1\rangle \; \ldots \; |\tau\, j-j\rangle .$$

These vectors satisfy (9.32)-type relations in which the quantum number τ is simply a spectator:

$$\begin{aligned}
J^2 \mid \tau\, j\, \mu\rangle &= j(j+1) \mid \tau\, j\, \mu\rangle , \\
J_z \mid \tau\, \mu\rangle &= \mu \mid \tau\, j\, \mu\rangle , \\
J_+ \mid \tau\, j\, \mu\rangle &= x_\mu \mid \tau\, j\, \mu+1\rangle , \\
J_- \mid \tau\, j\, \mu\rangle &= x_{\mu-1} \mid \tau\, j\, \mu-1\rangle .
\end{aligned} \qquad (9.34)$$

The subspaces $\xi_{j\,\tau}$ and $\xi_{j\,\tau'}$ are orthogonal for $\tau' \neq \tau$ by virtue of (9.33) and, when brought together, they form the subspace ξ_j of the eigenvalue $j(j+1)$ of J^2.

By repeating this operation for each of the j possible values such that the eigenvalue of J^2 is $j(j+1)$, we obtain a standard basis for the whole of the

Hilbert space:

$$\sum_{\tau j \mu} | \tau j \mu \rangle \langle \tau j \mu | = \mathbb{1} \, .$$

In such a representation, the matrices representing the different components of the angular momentum operator take a simple form since

$$\langle \tau j \mu | J_z | \tau' j' \mu' \rangle = \delta_{\tau \tau'} \, \delta_{j j'} \, \delta_{\mu \mu'} \, \mu \, ,$$
$$\langle \tau j \mu | J^2 | \tau' j' \mu' \rangle = \delta_{\tau \tau'} \, \delta_{j j'} \, \delta_{\mu \mu'} \, j(j+1) \, ,$$
$$\langle \tau j \mu | J_+ | \tau' j' \mu' \rangle = \delta_{\tau \tau'} \, \delta_{j j'} \, \delta_{\mu \, \mu'+1} \, x_{\mu'} \, ,$$
$$\langle \tau j \mu | J_- | \tau' j' \mu' \rangle = \delta_{\tau \tau'} \, \delta_{j j'} \, \delta_{\mu \, \mu'-1} \, x_{\mu'-1} \, .$$

(9.35)

Next, we calculate the matrix elements of the operators J_x and J_y using the following relations

$$J_x = \frac{1}{2} \left(J_+ + J_- \right) ,$$
$$J_y = \frac{1}{2i} \left(J_+ - J_- \right) .$$

(9.36)

2.7 Example of a Standard Representation

Let us apply the preceding relations to the case of a particle with angular momentum $\vec{J} = \frac{1}{2}$. The matrices representing the angular momentum operators will be obtained by applying relations (9.35):

$$J^2 = \frac{3}{4} \begin{pmatrix} 1 & 0 \\ 0 & 1 \end{pmatrix} \quad J_z = \frac{1}{2} \begin{pmatrix} 1 & 0 \\ 0 & -1 \end{pmatrix} \quad J_+ = \begin{pmatrix} 0 & 1 \\ 0 & 0 \end{pmatrix} \quad J_- = \begin{pmatrix} 0 & 0 \\ 1 & 0 \end{pmatrix}$$

leading to the values

$$J_x = \frac{1}{2} \begin{pmatrix} 0 & 1 \\ 1 & 0 \end{pmatrix} \quad J_y = \frac{1}{2} \begin{pmatrix} 0 & -i \\ i & 0 \end{pmatrix} .$$

(9.37)

We will return to this example in the course of the discussion of Pauli matrices.

3. Examples of Angular Momenta

3.1 Orbital Angular Momentum

Let us return to the special case of a particle with mass m and classical angular momentum $\vec{L}$. We normally associate with this observable an orbital angular momentum operator $\vec{L}$ satisfying commutation relations (9.8) and the orthonormal kets $|\ell\, m\rangle$ of the standard basis satisfy the following eigenvalue equations:

$$L^2 \,|\,\ell\, m\rangle = \ell\,(\ell+1)\,|\,\ell\, m\rangle\,,$$
$$L_z \,|\,\ell\, m\rangle = m\,|\,\ell\, m\rangle\,,$$

(9.38)

with the usual selection rules

$$\ell \geq 0 \qquad \text{and} \qquad -\ell \leq m \leq \ell\,.$$

(9.39)

Operators L^2 and L_z can be defined with (9.5) in a system of spherical coordinates. By projecting the angular momentum state $(\ell,\, m)$ onto this system of coordinates, we obtain functions of the angles θ and φ:

$$Y_{\ell\, m}\,(\theta,\, \varphi) = \langle\theta,\, \varphi\,|\,\ell\, m\rangle$$

(9.40)

termed spherical harmonics which we will discuss in detail in Chap. 12.

For the time being, let us note that because the operator L_z is, following (9.5), expressed as $-i\,(\partial/\partial\varphi)$, the first-order differential equation can be written

$$-i\,\frac{\partial}{\partial\varphi}\,Y_{\ell\, m}\,(\theta,\, \varphi) = m\,Y_{\ell\, m}\,(\theta,\, \varphi)$$

(9.41)

with the following general solution:

$$Y_{\ell\, m}\,(\theta,\, \varphi) = Z_{\ell\, m}\,(\theta)\,e^{i\,m\,\varphi}\,.$$

(9.42)

The function $Y_{\ell\, m}\,(\theta,\, \varphi)$ is uniform if it is invariant under a 2π-rotation of the angle φ:

$$Y_{\ell\, m}\,(\theta,\, \varphi) = Y_{\ell\, m}\,(\theta,\, \varphi+2\pi)$$
$$Z_{\ell\, m}\,(\theta)\,e^{i\,m\,\varphi} = Z_{\ell\, m}\,(\theta)\,e^{i\,m\,(\varphi+2\pi)}$$

(9.43)

This results in $e^{2\,i\,m\,\pi} = 1$, forcing m to be an integer. Since, in addition, $(\ell+m)$ and $(\ell-m)$ are necessarily integers (see above), ℓ will be an integer.

An orbital angular momentum can therefore only take the positive integer values $\ell = 0, 1, 2, \ldots$.

3.2 Spin

Stern and Gerlach were the first to experimentally provide evidence for the existence for an intrinsic angular momentum, or spin. The principle consists in using an electromagnet in whose air gap the applied force F_z is proportional to the magnetic moment μ of the incident particles:

$$F_z = k\,\mu\,.$$

(9.44)

On a photographic plate serving as detector, it should thus be possible to observe $(2j+1)$ smears corresponding to all the possible values $(-j \leq \mu \leq +j)$ of μ.

Stern and Gerlach injected into the air gap of their electromagnet, silver atoms known to have a zero orbital angular momentum. The photographic plate showed two symmetrical smears with respect to the incident direction, thus providing evidence for the existence of an intrinsic angular momentum of the electrons of the atom. Spin $\vec{S}$ is used here to refer to this intrinsic angular momentum for which

$$2S + 1 = 2, \qquad \text{that is} \qquad \vec{S} = \frac{\vec{1}}{2}.$$

The eigenvalue of the operator S^2 will be $\frac{1}{2}\left(\frac{1}{2} + 1\right) = \frac{3}{4}$ whereas S_z will have the eigenvalues $\pm\frac{1}{2}$. The relations defining the state $|\frac{1}{2}\,\sigma\rangle$ will therefore be written

$$S^2 \mid \frac{1}{2}\,\sigma\rangle = \frac{3}{4}\mid \frac{1}{2}\,\sigma\rangle$$
$$S_z \mid \frac{1}{2}\,\sigma\rangle = \sigma \mid \frac{1}{2}\,\sigma\rangle \quad \text{with} \quad \sigma = \pm\frac{1}{2}. \tag{9.45}$$

$\vec{S}$ is an angular momentum obeying the general commutation relations

$$\vec{S} \wedge \vec{S} = i\,\hbar\,\vec{S}.$$

If we apply the operator S_+ twice to the ket $\mid \frac{1}{2}\,\sigma\rangle$, we obtain:

$$S_+ \mid \frac{1}{2}\,\sigma\rangle = x_\sigma \mid \frac{1}{2}\,\sigma + 1\rangle,$$
$$S_+^2 \mid \frac{1}{2}\,\sigma\rangle = S_+ \left(S_+ \mid \frac{1}{2}\,\sigma\rangle \right) = S_+\, x_\sigma \mid \frac{1}{2}\,\sigma + 1\rangle$$
$$= x_{\sigma+1}\, x_\sigma \mid \frac{1}{2}\,\sigma + 2\rangle.$$

Since the magnetic momentum σ can only take the values $\pm\frac{1}{2}$, we obtain $\sigma + 2 = \frac{5}{2}$ or $\frac{3}{2}$, both forbidden values. Because $S_+^2 \mid \frac{1}{2}\,\sigma\rangle = 0\ \forall\ \sigma$, we infer that

$$S_+^2 = 0. \tag{9.46}$$

Using the same method, it can also be shown that

$$S_-^2 = 0. \tag{9.47}$$

Let us develop these relations by expanding S_+ and S_-:

$$S_+^2 = \left(S_x + i\,S_y\right)^2 = S_x^2 - S_y^2 + i\left(S_x\,S_y + S_y\,S_x\right) = 0,$$
$$S_-^2 = \left(S_x - i\,S_y\right)^2 = S_x^2 - S_y^2 - i\left(S_x\,S_y + S_y\,S_x\right) = 0. \tag{9.48}$$

The term-by-term sum of these equations shows that $S_x^2 = S_y^2$ and because the eigenvalue of $S^2 = \frac{3}{4}$ and that of $S_z^2 = \frac{1}{4}$, we can infer the value of the operators:

$$S_x^2 = S_y^2 = S_z^2 = \frac{1}{4}\, \mathbb{1} \tag{9.49}$$

where $\mathbb{1}$ represents the identity matrix $\mathbb{1} = \begin{pmatrix} 1 & 0 \\ 0 & 1 \end{pmatrix}$. Using (9.49) and the term-by-term difference of Eq. (9.48), we obtain the relation

$$S_x\, S_y + S_y\, S_x = 0\,.$$

This is an anticommutation relation. Because the spin operators also satisfy the commutation relations $\vec{S} \wedge \vec{S} = i\, \vec{S}$, we obtain, for example,

$$
\begin{aligned}
S_x\, S_y + S_y\, S_x &= 0\,, \\
S_x\, S_y - S_y\, S_x &= i\, S_z\,,
\end{aligned}
\tag{9.50}
$$

yielding, after the term-by-term summation,

$$S_x\, S_y = \frac{i}{2}\, S_z\,. \tag{9.51}$$

3.3 Pauli Matrices

We usually set $\vec{S} = \frac{1}{2}\, \vec{\sigma}$ and the matrices $\vec{\sigma}$ representing the spin operators in the standard representation (see 9.37) are termed Pauli matrices:

$$\sigma_x = \begin{pmatrix} 0 & 1 \\ 1 & 0 \end{pmatrix} \qquad \sigma_y = \begin{pmatrix} 0 & -i \\ i & 0 \end{pmatrix} \qquad \sigma_z = \begin{pmatrix} 1 & 0 \\ 0 & -1 \end{pmatrix}. \tag{9.52}$$

It is easily seen here that the matrices σ_x, σ_y, σ_z, and the identity matrix $\mathbb{1} = \begin{pmatrix} 1 & 0 \\ 0 & 1 \end{pmatrix}$ form a complete basis, that is, any 2×2 matrix can be expressed as a linear combination of the former.

Consider, for example, a given 2×2 matrix

$$M = \begin{pmatrix} a & b \\ c & d \end{pmatrix}.$$

Now, let us determine the elements of this matrix by writing the linear combinations of the identity and Pauli matrices:

$$
\begin{aligned}
M &= \lambda_0\, \mathbb{1} + \lambda_x\, \sigma_x + \lambda_y\, \sigma_y + \lambda_z\, \sigma_z \\
&= \lambda_0 \begin{pmatrix} 1 & 0 \\ 0 & 1 \end{pmatrix} + \lambda_x \begin{pmatrix} 0 & 1 \\ 1 & 0 \end{pmatrix} + \lambda_y \begin{pmatrix} 0 & -i \\ i & 0 \end{pmatrix} + \lambda_z \begin{pmatrix} 1 & 0 \\ 0 & -1 \end{pmatrix} \\
&= \begin{pmatrix} \lambda_0 + \lambda_z & \lambda_x - i\, \lambda_y \\ \lambda_x + i\, \lambda_y & \lambda_0 - \lambda_z \end{pmatrix} \equiv \begin{pmatrix} a & b \\ c & d \end{pmatrix},
\end{aligned}
\tag{9.53}
$$

which yields the following matrix elements:

$$a = \lambda_0 + \lambda_z,$$
$$b = \lambda_x - i\,\lambda_y,$$
$$c = \lambda_x + i\,\lambda_y,$$
$$d = \lambda_0 - \lambda_z.$$

from which we can infer the λ_i parameters of the linear combination:

$$\lambda_0 = \frac{a+d}{2} \qquad \lambda_x = \frac{b+c}{2}$$
$$\lambda_y = \frac{c-b}{2i} \qquad \lambda_z = \frac{a-d}{2}. \tag{9.54}$$

This shows that there is a one-to-one relation between the elements of the matrix M and the parameters λ_i of the linear combination of the Pauli matrices and the identity matrix.

Working from definition (9.52), it is easy to show that the Pauli matrices satisfy the following relations by setting $i = (x,\ y,\ z)$:

$$\sigma_i^2 = \mathbb{1}$$
$$\sigma_i\,\sigma_j = -\sigma_j\,\sigma_i = i\,\sigma_k \qquad i,\ j,\ k = x,\ y,\ z$$
$$\sigma_x\,\sigma_y\,\sigma_z = i\,\mathbb{1} \tag{9.55}$$
$$\mathrm{Tr}\,\sigma_i = 0$$
$$\det\,\sigma_i = -1.$$

The simultaneously fulfilled commutation and anticommutation relations (9.50) can be written as follows:

$$[\sigma_i,\ \sigma_j] = 2i\,\varepsilon_{ijk}\,\sigma_k \quad \text{and} \quad \{\sigma_i,\sigma_j\} = 2\delta_{ij}\,\mathbb{1}. \tag{9.56}$$

This yields after a term-by-term summation,

$$\sigma_i\,\sigma_j = \delta_{ij}\,\mathbb{1} + i\,\varepsilon_{ijk}\,\sigma_k. \tag{9.57}$$

By introducing the Pauli matrices

$$\sigma_+ = \frac{1}{2}\,(\sigma_x + i\sigma_y) = \begin{pmatrix} 0 & 1 \\ 0 & 0 \end{pmatrix} \quad \text{and} \quad \sigma_- = \frac{1}{2}\,(\sigma_x - i\sigma_y) = \begin{pmatrix} 0 & 0 \\ 1 & 0 \end{pmatrix},$$

we can easily demonstrate the important relations

$$\{\sigma_+,\ \sigma_-\} = \mathbb{1} \quad (\sigma_+)^2 = 0 \quad (\sigma_-)^2 = 0. \tag{9.58}$$

Finally, two vector operators, $\vec{A}$ and $\vec{B}$, commuting with $\vec{\sigma}$ but not necessarily with one another and which are not acting in the spin space satisfy the relation

$$(\vec{\sigma} \cdot \vec{A}) \, (\vec{\sigma} \cdot \vec{B}) = \vec{A} \cdot \vec{B} + i \, \vec{\sigma} \cdot (\vec{A} \wedge \vec{B}), \qquad (9.59)$$

as is easily seen by using relation (9.56).

This relation has two important consequences for the angular momentum $\vec{p}$ and the kinematic momentum $\vec{\pi} = \vec{p} - e \, \vec{A}$. In the first case we obtain

$$(\vec{\sigma} \cdot \vec{p})^2 = p^2, \qquad (9.60)$$

whereas the kinematic momentum and definition (1.256) yield

$$(\vec{\sigma} \cdot \vec{\pi})^2 = \pi^2 + i \, \vec{\sigma} \cdot (\vec{\pi} \wedge \vec{\pi}) = \pi^2 - e \, \vec{\sigma} \cdot \vec{B}. \qquad (9.61)$$

3.4 Isotopic Spin

In nuclear physics, neutrons and protons are often considered as one and the same entity, simply by introducing isotopic spin (or isobaric spin, or isospin).

The nucleon is hence defined as a particle with isospin $I = \frac{1}{2}$. The proton would then be the isospin state $| \frac{1}{2} \, \frac{1}{2} \rangle$ and the neutron the isospin state $| \frac{1}{2} - \frac{1}{2} \rangle$.

Let us denote the states of the neutron and the proton by $| n \rangle$ and $| p \rangle$ respectively and then the operators for making the transition from one state to the other:

$$\begin{aligned} \tau_+ \, | n \rangle &= | p \rangle \\ \tau_- \, | p \rangle &= | n \rangle. \end{aligned} \qquad (9.62)$$

Insisting that only these two states exist yields the conditions

$$\tau_+ \, | p \rangle = 0 \quad \text{and} \quad \tau_- \, | n \rangle = 0. \qquad (9.63)$$

Now let us look for mutually conjugate operators, that is, ones satisfying the additional condition

$$\tau_- = (\tau_+)^+. \qquad (9.64)$$

The simplest representation of the operators τ_+ and τ_- is obtained with the 2×2 matrices

$$| p \rangle = \begin{pmatrix} 1 \\ 0 \end{pmatrix} \quad | n \rangle = \begin{pmatrix} 0 \\ 1 \end{pmatrix} \quad \tau_+ = \begin{pmatrix} 0 & 1 \\ 0 & 0 \end{pmatrix} \quad \tau_- = \begin{pmatrix} 0 & 0 \\ 1 & 0 \end{pmatrix} \qquad (9.65)$$

satisfying relations (9.62), (9.63), and (9.64).

Using (9.65), we can easily calculate the commutator of τ_+ and τ_- which we note as τ_3. We thus find

$$[\tau_+, \tau_-] = \tau_3 = \begin{pmatrix} 1 & 0 \\ 0 & -1 \end{pmatrix}. \qquad (9.66)$$

By causing τ_3 to act on $\mid p\rangle$ and $\mid n\rangle$ defined in (9.65), we obtain

$$\tau_3 \mid p\rangle \;=\; \mid p\rangle \qquad \text{and} \qquad \tau_3 \mid n\rangle \;=\; -\mid n\rangle. \qquad (9.67)$$

In other words, $\mid p\rangle$ and $\mid n\rangle$ are none other than the eigenstates of τ_3, with the corresponding eigenvalues being $+1$ and -1.

If we introduce the linear combinations

$$\tau_1 = (\tau_- + \tau_+) = \begin{pmatrix} 0 & 1 \\ 1 & 0 \end{pmatrix},$$

$$\tau_2 = i\,(\tau_- - \tau_+) = i \begin{pmatrix} 0 & -1 \\ 1 & 0 \end{pmatrix}, \qquad (9.68)$$

$$\tau_3 = \begin{pmatrix} 1 & 0 \\ 0 & -1 \end{pmatrix},$$

we notice that the matrices τ are Pauli matrices $\tau_1 = \tau_x \;\; \tau_2 = \tau_y \;\; \tau_3 = \tau_z$ and because spin S is defined by $\vec{S} = \frac{1}{2}\,\vec{\sigma}$, we introduce an isospin $\vec{I}$ such that

$$\vec{I} = \frac{1}{2}\,\vec{\tau} \quad \text{with} \quad \vec{I} \wedge \vec{I} = i\,\vec{I} \quad \text{or better still} \quad [I_k,\; I_\ell] = i\,\varepsilon_{k\ell m}\,I_m. \quad (9.69)$$

The isospin states $\pm \frac{1}{2}$ therefore represent the two states of the nucleon:

$$\begin{aligned} \text{proton state} \quad &\mid \tfrac{1}{2}\,\tfrac{1}{2}\rangle \quad I = \tfrac{1}{2} \quad \text{and} \quad I_3 = \tfrac{1}{2} \\[4pt] \text{neutron state} \quad &\mid \tfrac{1}{2}\,-\tfrac{1}{2}\rangle \quad I = \tfrac{1}{2} \quad \text{and} \quad I_3 = -\tfrac{1}{2}. \end{aligned} \qquad (9.70)$$

If must be recalled that the electric charges number Q and the isospin I_3 are linked by the Gell-Mann and Nishijima relation:

$$Q = \frac{1}{2} + I_3. \qquad (9.71)$$

The isotopic invariance of the nuclear interaction can be formulated as the invariance with respect to rotations in the isospace. The scattering amplitude S, for example, is an isoscalar:

$$\langle I'\, I_3' \mid S \mid I\, I_3\rangle \;=\; S_I\,\delta_{I\,I'}\,\delta_{I_3\,I_3'},$$

where S_I is independent of I_3, that is, independent of the type of nucleon. This mathematically formulates the fact that the nuclear interaction acts in the same manner on the proton as on the neutron.

3.5 Electroweak Isospin of Leptons and Quarks

The quark model of particle physics assumes the existence of a common underlying sub-structure of the neutron and the proton termed "quark". The quark u is considered to be a quantum state with electric charge $\frac{2}{3} e$ and the quark d a quantum state with electric charge $-\frac{1}{3} e$. Quarks are fermions with spin $\vec{J} = \frac{1}{2}$ that are sensitive to the four basic interactions, gravitation, strong nuclear interaction, weak nuclear interaction, and electromagnetic interaction. The proton is composed of the quarks $(u\ u\ d)$ and the neutron of the quarks $(d\ d\ u)$. The other elementary constituents of matter, or leptons (electrons and neutrinos), are also fermions with spin $\vec{J} = \frac{1}{2}$ that are not sensitive to strong nuclear interactions.

Leptons and quarks constitute isospin doublets whose two constituents correspond to states of electric charge, hence to different isospin states:

$$
\begin{pmatrix} e^- \\ \nu_e \end{pmatrix} \quad \vec{J} = \frac{1}{2} \quad
\begin{matrix} Q = -1 \\ Q = 0 \end{matrix} \quad
\vec{I} = \frac{1}{2} \begin{cases} I_3 = -\frac{1}{2} \\ I_3 = +\frac{1}{2} \end{cases}
$$
$$
\begin{pmatrix} u \\ d \end{pmatrix} \quad \vec{J} = \frac{1}{2} \quad
\begin{matrix} Q = \frac{2}{3} \\ Q = -\frac{1}{3} \end{matrix} \quad
\vec{I} = \frac{1}{2} \begin{cases} I_3 = \frac{1}{2} \\ I_3 = -\frac{1}{2} \end{cases} .
$$

$$(9.72)$$

We introduce a quantum number L termed lepton number, characteristic of leptons, and then set $L = +1$ for e^- and ν_e and another quantum number B, termed baryon number, characteristic of baryons and then again set $B = \frac{1}{3}$ for u and d (or better still $B = 1$ for $n = (u\ d\ d)$ and for $p = (u\ u\ d)$). The Gell-Mann–Nishijima formula will then take the form

$$ Q = I_3 + \frac{B - L}{2} . \tag{9.73} $$

Isospin is conserved in electromagnetic and weak interactions, which explains why it is termed electroweak isospin, or isospin of leptons and quarks. We can introduce a neutral charge (Elbaz et al., 1981) or weak charge N conferring symmetry properties to the two members in each of the preceding doublets. We will say, for example, that the proton has an electric charge of $Q = 1$, but a weak charge of $N = 0$ whereas the neutron has the electric charge $Q = 0$ and a weak charge of $N = 1$. It is clear then that the projection of the electroweak isospin is linked to the electric and weak charges through

$$ I_3 = \frac{1}{2} \left(Q - N \right), \tag{9.74} $$

and by inserting this value into the Gell-Mann–Nishijima formula, we obtain

$$ B - L = Q + N = Y . \tag{9.75} $$

By hypercharge Y we mean the sum of the electric and weak charges or the total charge of a particle. The Gell-Mann–Nishijima formula will then take the following form:

$$Q = I_3 + \frac{Y}{2}. \tag{9.76}$$

Let us emphasize here that this total charge Y is for the nuclei none other than the number of nucleons $A = Z + N$ or atomic number.

Doublet		J	I	Q	N	Y	B	L	I_3
leptons	e^-	$\frac{1}{2}$	$\frac{1}{2}$	-1	0	-1		1	$-\frac{1}{2}$
	ν_e			0	-1				$\frac{1}{2}$
quarks	d	$\frac{1}{2}$	$\frac{1}{2}$	$-\frac{1}{3}$	$\frac{2}{3}$	$\frac{1}{3}$	$\frac{1}{3}$		$-\frac{1}{2}$
	u			$\frac{2}{3}$	$-\frac{1}{3}$				$\frac{1}{2}$
nucleons	n	$\frac{1}{2}$	$\frac{1}{2}$	0	1	1	1		$-\frac{1}{2}$
	p			1	0				$\frac{1}{2}$

By internal quantum numbers we normally mean the quantum numbers Q, N, Y, B, L and I_3 defining the intrinsic properties of particles. Antiparticles will have the same mass as particles but internal quantum numbers with opposite signs $-Q$, $-N$, $-Y$, $-B$, $-L$ and $-I_3$. Internal quantum numbers are simply additive from the algebraic point of view.

Mesons comprising a quark q (with baryon number $1/3$) and an antiquark $\bar{q}$ (with baryon number $-1/3$) will then be particles with baryon number $B = +\frac{1}{3} - \frac{1}{3} = 0$.

Baryons comprising three quarks q (each with the baryon number $1/3$) will, for their part, have the baryon number $B = +1$.

4. Magnetic Moments

An atom with angular momentum $(j\ \mu)$ in a magnetic field will undergo an interaction decoupling the energy levels that are dependent on its quantum number μ, correctly termed magnetic quantum number. We are therefore going to study the effects of a magnetic field on an atom by introducing the magnetic moment of such an atom with the intrinsic magnetic moment of its electronic components.

4.1 Magnetic Moment of an Atom

An atom with Z electrons will be described by a Hamiltonian H_0, which is the sum of the kinetic energies $p_i^2/2m$ of the Z electrons and the Coulomb potential energy of the electrons with the nucleus (assumed infinitely heavy) and the relative Coulomb interaction between the electrons:

$$H_0 = \sum_{i=1}^{Z} \left(\frac{p_i^2}{2m_e} - \frac{Z\,e^2}{r_i} \right) + \sum_{i\langle j} \frac{e^2}{r_{ij}}.$$ (9.77)

If the atom is in a magnetic field $\vec{B}$, an interaction between the electrons and the field will produce a shift of the energy levels of this atom proportional to a moment termed the magnetic moment $\vec{\mu}$. The Hamiltonian H_0 will then be replaced by a Hamiltonian H obtained by changing p_i into $\pi_i = p_i - (e/c)\,A_i$ ($\vec{A}$ is the vector potential of the electromagnetic field). For a constant magnetic field, we can set

$$\vec{A} = \frac{1}{2}\,(\vec{B} \wedge \vec{r}) = \frac{1}{2}\,\varepsilon_{\ell m j}\,B_\ell\,r_m\,\vec{e}_j.$$ (9.78)

In effect, the preceding form will yield the curl

$$\left(\vec{\nabla} \wedge \vec{A}\right)_k = \varepsilon_{ijk}\,\partial_i\,A_j = \varepsilon_{ijk}\,\partial_i\left(\frac{1}{2}\varepsilon_{\ell m j}\,B_\ell\,r_m\right).$$ (9.79)

By using the crossing rule on the Levi–Civita symbols:

$$\varepsilon_{ijk}\,\varepsilon_{ij'k'} = \delta_{jj'}\,\delta_{kk'} - \delta_{jk'}\,\delta_{kj'}$$ (9.80)

we obtain the following result:

$$\left(\vec{\nabla} \wedge \vec{A}\right)_k = -\frac{1}{2}\,\partial_\ell\,B_\ell\,r_k + \frac{1}{2}\,\partial_m\,B_k\,r_m.$$ (9.81)

Because B is constant and $\partial_\ell\,r_k = \delta_{\ell k}$ whereas $\partial_m\,r_m = 3$, we obtain the desired result:

$$\left(\vec{\nabla} \wedge \vec{A}\right)_k = -\frac{1}{2}\,B_k + \frac{3}{2}\,B_k = B_k.$$ (9.82)

For a constant magnetic field, the preceding form of the vector potential is therefore compatible with the usual form of the magnetic field

$$\vec{B} = \vec{\nabla} \wedge \vec{A}.$$ (9.83)

Let us evaluate the square of the kinematic momentum:

$$\pi^2 = \left(\vec{p} - \frac{e}{c}\,\vec{A}\right)^2 = p^2 - \frac{e}{c}\left(\vec{A}\cdot\vec{p} + \vec{p}\cdot\vec{A}\right) + \frac{e^2}{c^2}\,A^2$$

$$= p^2 - \frac{e}{2c}\left[\left(\vec{B}\wedge\vec{r}\right)\cdot\vec{p} + \vec{p}\cdot\left(\vec{B}\wedge\vec{r}\right)\right] + \frac{e^2}{4c^2}\left(\vec{B}\wedge\vec{r}\right)^2 \qquad (9.84)$$

and introduce $\vec{r}\wedge\vec{p} = \vec{\ell}$, the orbital angular momentum of the electron, and further use the crossing rule (9.80) to evaluate the last term:

$$\left(\vec{B}\wedge\vec{r}\right)^2 = \left(\vec{B}\wedge\vec{r}\right)\cdot\left(\vec{B}\wedge\vec{r}\right) = B^2\,r^2 - \left(\vec{B}\cdot\vec{r}\right)^2 = B^2\,r^2\,(1 - \cos^2\theta)$$

$$= B^2\,r^2\,\sin^2\theta = B^2\,r_\perp^2 \qquad (9.85)$$

by setting $r_\perp = r\,\sin\theta$ the projection of the vector r on the plane perpendicular to the magnetic field. We are thus left with

$$\pi^2 = \left(\vec{p} - \frac{e}{c}\,\vec{A}\right)^2 = p^2 - \frac{e}{c}\,\vec{B}\cdot\vec{\ell} + \frac{e^2}{4c^2}\,B^2\,r_\perp^2\,. \qquad (9.86)$$

The Hamiltonian H_0 in which the momentum $\vec{p}$ has been replaced with kinematic momentum $\vec{\pi}$ then becomes

$$H = \sum_i \frac{1}{2m_e}\left(\vec{p}_i - \frac{e}{c}\,\vec{A}\right)^2 - \frac{Z\,e^2}{r_i} + \sum_{i<j}\frac{e^2}{r_{ij}}\,. \qquad (9.87)$$

Combined with the foregoing results, this will be written as follows:

$$H = H_0 - \frac{e}{2m_e\,c}\,\vec{B}\cdot\vec{L} + \frac{e^2}{8m_e\,c^2}\,B^2\sum_i r_{i\perp}^2\,. \qquad (9.88)$$

To a first approximation, the third term in $e^2/m_e\,c^2$ will be negligible compared to the term in $e/m\,c$, thus leaving us with

$$H = H_0 - \frac{e}{2m_e\,c}\,\vec{B}\cdot\vec{L}\,, \qquad (9.89)$$

where $\vec{L} = \sum_i \vec{\ell}_i = \sum_i \vec{r}_i \wedge \vec{p}_i$ is the total orbital angular momentum of the electrons. It is as if each electron induced a magnetic moment

$$\boxed{\vec{\mu}_i = \frac{e}{2m_e\,c}\,\vec{\ell}_i}\,. \qquad (9.90)$$

Because the total magnetic moment of the atom is the sum of the Z individual magnetic moments:

$$\vec{\mu} = \sum_i \vec{\mu}_i = \sum_i \frac{e}{2m_e\,c}\,\vec{\ell}_i = \frac{e}{2m_e\,c}\sum_i \vec{\ell}_i = \frac{e}{2m_e\,c}\,\vec{L} \qquad (9.91)$$

the interaction between the magnetic field $\vec{B}$ and the electrons of the atom is therefore represented by the term $-\vec{\mu} \cdot \vec{B}$:

$$H = H_0 - \vec{\mu} \cdot \vec{B} \quad \text{with} \quad \vec{\mu} = \frac{e}{2m_e\, c}\, \vec{L} \quad . \tag{9.92}$$

When an atom described by a Hamiltonian H_0 with eigenstates $| \, n \, L \, M \rangle$ and eigenenergy $E_{n\,L}^0$:

$$H_0 \, | \, n \, L \, M \rangle \; = E_n^0 \, | \, n \, L \, M \rangle \tag{9.93}$$

is placed in a magnetic field $\vec{B}$ oriented along the quantization axis Oz, we obtain a set $| \, n \, L \, M \rangle$ of eigenvectors of H such that

$$\begin{aligned}
H \, | \, n \, L \, M \rangle \; &= \left(H_0 - \frac{e}{2m_e\, c} \, L_z \, B_z \right) \, | \, n \, L \, M \rangle \\
&= \left(E_n^0 - \frac{e}{2m_e\, c} \, M \, \hbar \, B_z \right) \, | \, n \, L \, M \rangle \, . \tag{9.94}
\end{aligned}$$

The energy levels of the atom with Z electrons in the magnetic field B_z have therefore been shifted, since

$$E_{n\,L\,M} = E_n^0 - \mu_B^e \, M \, B_z \tag{9.95}$$

We set $\mu_B^e = e\, \hbar/2m_e\, c$, the electronic Bohr magneton. Its value, calculated from the fundamental constants, is $\mu_B^e = 5.78838263 \cdot 10^{-11}$ MeV T^{-1}.

> Under the influence of the magnetic field, each energy level
> E_n^0 will split into $(2L + 1)$ energy levels
> corresponding to the different possible values of M

4.2 Magnetic Moment of Charged Leptons

When applied to silver atoms with orbital momentum $L = 0$, the Stern–Gerlach experiment provides experimental evidence for the existence of an intrinsic angular momentum. To account for this fact, Uhlenbeck and Goudsmit introduced in 1925 the hypothesis of the electron spin $\vec{s} = (1/2)\, \vec{\sigma}$.

To describe the motion of a particle with mass m and spin $\vec{\sigma}$ in the potential V, we can use relation (9.60) and replace $p^2/2m$ with $(\vec{\sigma} \cdot \vec{p})^2/2m$. Relation (9.61) shows that $(\vec{\sigma} \cdot \vec{\pi})^2$ is different from π^2 if this particle is placed in an electromagnetic field. The Hamiltonian of a particle with mass

m and electric charge (e) should hence be written as follows:

$$H = \frac{1}{2m} \left(\vec{\sigma} \cdot \vec{\pi}\right)^2 - \frac{e\,\hbar}{2m}\,\vec{\sigma} \cdot \vec{B} + e\,\phi. \tag{9.96}$$

The Pauli Hamiltonian replaces the Hamiltonian (1.259) in the Schrödinger equation and contains the interaction term of the magnetic field $\vec{B}$ of the electron with spin $\vec{\sigma}$. The magnetic moment of the resulting electron is therefore expressed as

$$\vec{\mu} = \frac{e\,\hbar}{2mc}\,\vec{\sigma}. \tag{9.97}$$

Let us emphasize that the Schrödinger–Pauli eigenvalue equation, that is, eigenvalue equation $H\Psi = i\,\hbar\,\partial_t\,\Psi$ in which H is expressed with (9.96), leads us to define Ψ as a column vector of a two-dimensional space:

$$\Psi\left(\vec{x},\ t\right) = \begin{pmatrix} \Psi_1\left(\vec{x},\ t\right) \\ \Psi_2\left(\vec{x},\ t\right) \end{pmatrix} \tag{9.98}$$

where $\Psi_1\left(\vec{x},\ t\right)$ is the probability amplitude for observing projected spin $\frac{1}{2}$ along Oz for a particle at position $\vec{x}$ and instant t, whereas $\Psi_2\left(\vec{x},\ t\right)$ is the probability amplitude of a particle with projected spin $-\frac{1}{2}$ at $\left(\vec{x},\ t\right)$. The normalization of the wave function will then be written as follows:

$$\int \left(|\,\Psi_1\left(\vec{x},\ t\right)|^2 + |\,\Psi_2\left(\vec{x},\ t\right)|^2\right)\,d^3\,x = 1. \tag{9.99}$$

The electron is therefore a particle with spin $\frac{1}{2}$, mass* 0.511 MeV/c^2, and magnetic moment μ_e.

$$\mu_e = \frac{e\,\hbar}{2m_e\,c} \qquad \text{with} \qquad \begin{aligned} m_e &= 0.510\ 999\ 06 \text{ MeV}/c^2 \\ &= 9.109\ 389\ 7 \cdot 10^{-31} \text{ kg} \end{aligned} \tag{9.100}$$

In addition, Dirac theory shows that any point fermion with mass m and spin $\frac{1}{2}$, and with electric charge of Qe will have an intrinsic magnetic moment

$$\mu = \frac{Q\,e\,\hbar}{2m\,c} \qquad \text{Dirac magnetic moment}.$$

The electron has a magnetic moment that is practically equal to the Dirac moment:

$$\mu_e = (1.001\ 159\ 652\ 193 \pm 0.000\ 000\ 000\ 010)\,\mu_B^e. \tag{9.101}$$

---

* The equivalence between MeV/c^2 and kg was, as of 1988: 1 eV/c^2 = 1.782 662 70 $\cdot$ 10^{-36} kg. The speed of light is fixed at $c = 299\ 792\ 458$ m s^{-1} following the 1983 definition of the meter adopted by the general conference on weights and measures: the distance covered by light in vacuum in 1/299 792 458 s. These figures are taken from the Particle Properties Data Booklet, CERN 1992.

We are therefore dealing with an almost point fermion. Experience has shown that the muon (a second-generation electron) with mass $m_\mu = 105.658389$ MeV/c^2 is also a point particle since:

$$\mu_\mu = (1.001\ 165\ 923 \pm 0.000\ 000\ 008)\ \mu_B^e . \tag{9.102}$$

The measurement of the magnetic moment of τ (third-generation electron) with mass $m_\tau = 1\ 777.1$ MeV/c^2 is still not possible but the chances are that, whenever it is, it will yield a value that is very close to that of the Dirac momentum $\mu_B^\tau = e\ \hbar/2m_\tau\ c$.

4.3 Magnetic Moment of Nucleons

4.3.1 Experimental Results

Experience has shown that nucleons, i.e., protons and neutrons, which are particles with spin 1/2, also have an angular momentum. We thus have, expressed in $e\ \hbar/(2m_p\ c) = \mu_B^N$ nuclear magneton units

$$\mu_p = (2.792\ 847\ 39 \pm 0.000\ 000\ 063)\ \mu_B^N$$
$$\mu_n = (-1.913\ 042\ 7 \pm\ 0.0\ 000\ 000\ 5)\ \mu_B^N .$$

This clearly shows that the nucleon cannot be considered as a point particle. And even more interesting is the fact that the neutron, a particle with zero electric charge, has a high magnetic moment. This naturally lends support to the idea of the existence of a proton and neutron sub-structure. It was working from this idea that Gell-Mann and Zweig in 1974 proposed the existence of the quark u and d as the underlying sub-structure of the proton and the neutron. Results obtained by other researchers have confirmed this prediction which is now generally accepted.

Quarks are fermions with spin 1/2 and electric charge $\frac{2}{3}e$ or $-\frac{1}{3}e$ that are confined inside the nucleus. They are not observable in their free state and may take three different forms (or three colors: r for red, y for yellow or g for green). The only observable objects, such as mesons ($q_c\ \bar{q}_{\bar{c}}'$) or baryons ($q_r\ q_y'\ q_g''$), are colorless (white).

The proton comprising quarks ($d\ u\ u$) has the mass

$$m_p = 1.672\ 623\ 10^{-27}\ \text{kg} = 938.272\ 31\ \text{MeV}/c^2 . \tag{9.103}$$

The neutron comprising quarks ($d\ d\ u$) has the mass

$$m_n = 939.565\ 63\ \text{MeV}/c^2 .$$

4.3.2 Magnetic Moment of Quarks

If we concede that the mass of the quarks u and d is roughly a third of the mass of the nucleon, that is $m_u \simeq m_d \simeq 336$ MeV $\simeq \frac{1}{3} m_p \simeq \frac{1}{3} m_n$, then we can associate a Dirac momentum to these two fermions considered as point-like:

$$\mu_u = \frac{2}{3} \frac{e\,\hbar}{2m_u\,c} \simeq 2\,\frac{e\,\hbar}{2m_p\,c} = 2\mu_B^N\,,$$

$$\mu_d = -\frac{1}{3} \frac{e\,\hbar}{2m_d\,c} \simeq -\frac{e\,\hbar}{2m_p\,c} = -1\,\mu_B^N\,.$$

(9.104)

4.3.3 Magnetic Moment of Nucleons

Now, let us work from the magnetic moments of quarks to evaluate the magnetic moment of the proton and the neutron. We will be using for the purpose a result that will subsequently be demonstrated (see Chap. 13) with the aid of the Wigner–Eckart theorem.

A nucleus comprising three fermions with spin $\vec{s}_1$, $\vec{s}_2$ and $\vec{s}_3$ in a relative orbital momentum $L = 0$ may be considered as comprising a difermion with spin $\vec{S}_{12}$ and a third fermion with spin $\vec{s}_3$:

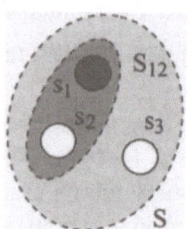

$$\vec{S} = \vec{s}_1 + \vec{s}_2 + \vec{s}_3 = \vec{S}_{12} + \vec{s}_3\,.$$

If we consider μ_1, μ_2 and μ_3 as the magnetic moments of the three fermions constituting the nucleus with spin $\vec{S}$, we obtain two possible values of the magnetic moment of the nucleus depending on the value assumed by the difermion spin:

$$i)\ \ \text{if}\ \ S_{12} = 0 \qquad \mu = \mu_3$$

$$ii)\ \ \text{if}\ \ S_{12} = 1 \qquad \mu = \frac{2}{3}\,(\mu_1 + \mu_2) - \frac{1}{3}\,\mu_3\,.$$

(9.105)

4.3.4 Magnetic Moment of the Proton $p = (u\ u\ d)$

By using the magnetic moments of the quarks u and d, that is,

$$\mu_u = \frac{e\,\hbar}{m_p\,c} = 2\mu_B^N \qquad \text{and} \qquad \mu_d = \frac{1}{2}\,\frac{e\,\hbar}{m_p\,c} = -\mu_B^N\,,$$

(9.106)

it is possible to calculate the magnetic moment of the proton. If the difermion is $(u\ d)$ with a zero spin $S_{12} = 0$, then

$$\mu_p = \mu_u = \frac{e\ \hbar}{m_p\ c} = 2\ \mu_B^N \tag{9.107}$$

and if the spin is $S_{12} = 1$, then

$$\mu_p = \frac{1}{3}\ \mu_u + \frac{2}{3}\ \mu_d = 0\ . \tag{9.108}$$

These two solutions are clearly unacceptable. If the difermion is $(u\ u)$ with spin $S = 0$,

$$\mu_p = \mu_d = -\frac{1}{2}\ \frac{e\ \hbar}{m_p\ c} = -\mu_B^N\ , \tag{9.109}$$

and with spin $S_{12} = 1$

$$\mu_p = 3\ \frac{e\ \hbar}{2m_p\ c} = 3\ \mu_B^N\ . \tag{9.110}$$

Only the last solution corresponds approximately to the experimental value:

$$\mu_p = (2.792\ 847\ 386\ \pm\ 0.000\ 000\ 063)\ \mu_B^N\ . \tag{9.111}$$

We can therefore conclude (Elbaz and Meyer, 1981) that the proton preferentially comprises a diquark $(u\ u)$ with spin 1 and a quark d with spin $1/2$.

4.3.5 Magnetic Moment of the Neutron $n = (d\ d\ u)$

A similar analysis to the above (Elbaz and Meyer, 1981) will lead to a neutron comprising a diquark $(d\ d)$ with spin 1 and a quark u with spin $1/2$. The resulting magnetic moment is therefore

$$\mu_n = \frac{4}{3}\ \mu_d - \frac{1}{3}\ \mu_u = -2\ \frac{e\ \hbar}{2m_p\ c} = -2\ \mu_B^N\ , \tag{9.112}$$

a value that is close to the experimental result

$$\mu_n = (-1.913\ 042\ 75\ \pm\ 0.000\ 000\ 45)\ \mu_B^N\ . \tag{9.113}$$

4.3.6 Quark-Quark and Quark-Antiquark Interactions

André Martin has shown that the interaction between two quarks q could be written as $V_{q\ q} = \lambda\ r$ whereas the interaction between a quark and an antiquark is equal to

$$V_{q\ \bar{q}} = 2\ V_{q\ q} = 2\lambda\ r\ .$$

This shows that a diquark behaves in the same way as an antiquark.

B. Silvestre-Brac (1987) has calculated the probability of the formation of a diquark in a baryon comprising three quarks and their results show that formation of a diquark is favored (see Appendix 2 of Chap. 10).

Chapter 10
Addition Theorems

Quarks are fermions with spin $\frac{\vec{1}}{2}$ and nucleons comprising three quarks should have a spin resulting from the addition of the spins of its constituents. Similarly, atoms made up of an electronic cortège of fermions with spin $\frac{1}{2}$ (electrons) and a nucleus with Z protons and N neutrons, both fermions with spin $\frac{1}{2}$, will have a total angular momentum resulting from the spins and orbital momenta of their constituents. It is therefore necessary to devise a simple method for adding angular momenta (this has been termed Racah algebra). We shall also be presenting together with the analytical method, the foundation of the Graphical Spin Algebra (GSA) used to considerably facilitate this type of computation.

1. Coupling of Two Angular Momenta

1.1 Classical Mechanics: A Few Reminders

In classical mechanics, the angular momentum:

$$\vec{J} = \vec{r} \wedge \vec{p} \tag{10.1}$$

is a pseudo-vector perpendicular to the plane containing $\vec{r}$ and $\vec{p}$, which varies in a continuous manner just like the vectors $\vec{r}$ and $\vec{p}$.

A two-part classical system comprising the angular momenta $\vec{J}_1$ and $\vec{J}_2$ will have a total angular momentum of $\vec{J} = \vec{J}_1 + \vec{J}_2$ and we can proceed to note by projection on the Oz axis that

$$J_z = J_{1z} + J_{2z}. \tag{10.2}$$

The projection of the total angular momentum is the sum of the projections of its components.

The squared length of the total angular momentum will then be

$$J^2 = \left(\vec{J}_1 + \vec{J}_2\right)^2 = J_1^2 + J_2^2 + 2\vec{J}_1 \cdot \vec{J}_2 = J_1^2 + J_2^2 + 2J_1 J_2 \cos\theta. \tag{10.3}$$

If $\theta = 0$, that is if $\vec{J}_1$ and $\vec{J}_2$ are co-linear and point in the same direction $J^2 = (J_1 + J_2)^2$ whereas $J^2 = (J_1 - J_2)^2$ if $\vec{J}_1$ and $\vec{J}_2$ are co-linear and point in opposite directions.

We thus obtain two selection rules:

$$J_z = J_{1z} + J_{2z} ,$$
$$| J_1 - J_2 | \le J \le J_1 + J_2 . \tag{10.4}$$

It goes without saying that the lengths of the vectors $\vec{J}_1$ and $\vec{J}_2$ are arbitrary and are linked to the positions $\vec{r}_1$ and $\vec{r}_2$ as well as to the momenta $\vec{p}_1$ and $\vec{p}_2$.

We will show that selection rules that are formally identical to the above appear hand-in-hand in quantum mechanics with quantization, that is, the quantum numbers m and j satisfy selection rules similar to the above.

1.2 Coupled and Decoupled Bases

It is common in quantum mechanics to find systems comprising two or several parts, each with its own angular momentum and with the relative motion of such parts implying in addition an orbital angular momentum. This is the case, for example, for the deuteron which comprises a proton and a neutron with intrinsic angular momentum $\vec{s}_1 = \vec{1}/2$ and $\vec{s}_2 = \vec{1}/2$ and orbital momentum $\vec{\ell}$. The total angular momentum will therefore be:

$$\vec{J} = \frac{\vec{1}}{2} + \frac{\vec{1}}{2} + \vec{\ell} \tag{10.5}$$

What are the possible values of the angular momentum J and how do we write the state with angular momentum $| j \, m \rangle$ from these constituents? These are two fundamental problems which we shall try to resolve by first taking the simpler of the two cases: the addition of two angular momenta $\vec{J}_1$ and $\vec{J}_2$:

$$\vec{J} = \vec{J}_1 + \vec{J}_2 . \tag{10.6}$$

If the interaction between the two parts is such that it leaves the individual components $J_1^2 \; J_{1z} \; J_2^2 \; J_{2z}$ constants of the motion, a complete set of commuting operators could be built from

$$H , \; J_1^2 , \; J_{1z} , \; J_2^2 , \; J_{2z} , \tag{10.7}$$

and the corresponding eigenfunctions

$$| \alpha \, j_1 \, m_1 \, j_2 \, m_2 \rangle = \sum_{\beta \, \gamma} | \beta \, j_1 \, m_1 \rangle \, | \gamma \, j_2 \, m_2 \rangle , \tag{10.8}$$

will be tensor products of the states $(j_1\ m_1)$, eigenstates of J_1^2 and J_{1z}:

$$J_1^2 \mid j_1\ m_1\ j_2\ m_2\rangle = j_1\ (j_1 + 1) \mid j_1\ m_1\ j_2\ m_2\rangle,$$
$$J_{1z} \mid j_1\ m_1\ j_2\ m_2\rangle = m_1 \mid j_1\ m_1\ j_2\ m_2\rangle,$$
(10.9)

and eigenstates of J_2^2 and J_{2z}:

$$J_2^2 \mid j_1\ m_1\ j_2\ m_2\rangle = j_2\ (j_2 + 1) \mid j_1\ m_1\ j_2\ m_2\rangle,$$
$$J_{2z} \mid j_1\ m_1\ j_2\ m_2\rangle = m_2 \mid j_1\ m_1\ j_2\ m_2\rangle.$$
(10.10)

The ket $\mid j_1\ m_1\ j_2\ m_2\rangle$ is the tensor product of the kets $\mid j_1\ m_1\rangle \mid j_2\ m_2\rangle$. It is a ket defined in the space $\xi = \xi_{j_1} \otimes \xi_{j_2}$. The basis spanned by the vectors $\mid j_1\ m_1\ j_2\ m_2\rangle$ with dimension $(2j_1 + 1)\,(2j_2 + 1)$ is the decoupled basis.

We can choose another complete set with the same number (five) of commuting observables but involving the total angular momentum J^2 and its component J_z along the quantization axis. These five new commuting observables are H, J^2, J_z, J_1^2, J_2^2, with

$$J^2 = \left(\vec{J}_1 + \vec{J}_2\right)^2,$$
$$J_z = J_{1z} + J_{2z}.$$
(10.11)

The eigenfunctions these operators will satisfy the equations

$$J_1^2 \mid (j_1\ j_2)\ jm\rangle = j_1\ (j_1 + 1) \mid (j_1\ j_2)\ jm\rangle,$$
$$J_2^2 \mid (j_1\ j_2)\ jm\rangle = j_2\ (j_2 + 1) \mid (j_1\ j_2)\ jm\rangle,$$
$$J^2 \mid (j_1\ j_2)\ jm\rangle = j\ (j + 1) \mid (j_1\ j_2)\ jm\rangle,$$
$$J_z \mid (j_1\ j_2)\ jm\rangle = m \mid (j_1\ j_2)\ jm\rangle.$$
(10.12)

The basis determined by the vectors $\mid (j_1\ j_2)\ jm\rangle$ is the coupled basis.

2. Clebsch–Gordan Coefficients

2.1 Definition

There exists a unitary transformation linking the two representations above. It is in fact possible to write the coupled states in the decoupled basis

$$\mid (j_1\ j_2)\ jm\rangle = \sum_{m_1\ m_2} \mid j_1\ m_1\ j_2\ m_2\rangle \langle j_1\ m_1\ j_2\ m_2 \mid (j_1\ j_2)\ jm\rangle \quad (10.13)$$

or the decoupled states in the coupled basis:

$$\mid j_1\ m_1\ j_2\ m_2\rangle = \sum_{jm} \mid (j_1\ j_2)\ jm\rangle \langle jm\ (j_1\ j_2) \mid j_1\ m_1\ j_2\ m_2\rangle. \quad (10.14)$$

This defines the vector addition coefficients:

$$\langle j_1 \, m_1 \, j_2 \, m_2 \mid (j_1 \, j_2) \, jm \rangle \qquad \text{and} \qquad \langle (j_1 \, j_2) \, jm \mid j_1 \, m_1 \, j_2 \, m_2 \rangle$$

These coefficients have been termed Clebsch–Gordan coefficients. In order to simplify the notation, we will write $\langle j_1 \, m_1 \, j_2 \, m_2 \mid jm \rangle$ without specifying again that j results from the addition of j_1 and j_2.

2.2 Selection Rules

2.2.1 Addition of Magnetic Momenta

Let us apply the Clebsch–Gordan series (10.13) to the eigenvalue equation for J_z:

$$J_z \mid j \, m \rangle = m \mid j \, m \rangle .$$

By introducing the decoupled basis, this yields, for the first term

$$\sum_{m_1 \, m_2} J_z \mid j_1 \, m_1 \, j_2 \, m_2 \rangle \, \langle j_1 \, m_1 \, j_2 \, m_2 \mid j \, m \rangle$$

$$= \sum_{m_1 \, m_2} m \mid j_1 \, m_1 \, j_2 \, m_2 \rangle \, \langle j_1 \, m_1 \, j_2 \, m_2 \mid j \, m \rangle ,$$

whereas relation (10.4) will yield J_z from the components J_{1z} and J_{2z}:

$$J_z = J_{1z} + J_{2z} \qquad \text{and} \qquad \begin{aligned} J_{1z} \mid j_1 \, m_1 \rangle &= m_1 \mid j_1 \, m_1 \rangle \\ J_{2z} \mid j_2 \, m_2 \rangle &= m_2 \mid j_2 \, m_2 \rangle . \end{aligned}$$

We therefore obtain, by the application of J_z to the second term of (10.13),

$$\sum_{m_1 \, m_2} (m_1 + m_2) \mid j_1 \, m_1 \, j_2 \, m_2 \rangle \, \langle j_1 \, m_1 \, j_2 \, m_2 \mid j \, m \rangle$$

$$= \sum_{m_1 \, m_2} m \mid j_1 \, m_1 \, j_2 \, m_2 \rangle \, \langle j_1 \, m_1 \, j_2 \, m_2 \mid j \, m \rangle , \qquad (10.15)$$

yielding the first selection rule:

$$\boxed{m_1 + m_2 = m} . \qquad (10.16)$$

This corresponds to the quantization of the first selection rule (10.2) for the addition of angular momentum in classical mechanics.

2.2.2 Second Rule: The Fundamental Addition Theorem

If the angular momenta j_1 and j_2 are given, the only possible values of j are those that satisfy the triangle rule:

$$| j_1 - j_2 | \leq j \leq j_1 + j_2 . \tag{10.17}$$

In fact, j is an angular momentum and the selection rule on m gives $-j \leq m \leq j$ with j as the maximum value of m:

$$m_{\text{max}} = j$$

defining the maximum values of m_1 and m_2:

$$m_{1\ \text{max}} = j_1 ,$$
$$m_{2\ \text{max}} = j_2 .$$

If m_1 and m_2 are set to their maximum values, we obtain the maximum maximorum of j with the first selection rule, that is,

$$j_{\text{max}} = j_1 + j_2 . \tag{10.18}$$

To every allowed value of j we can attribute a subspace ξ_j of the $(2j + 1)$-dimensional Hilbert space. By bringing together all these subspaces, we obtain the space ξ itself, that is,

$$\xi_{j\ \text{max}} \oplus \xi_{j\ \text{max}-1} \oplus \cdots \oplus \xi_{j\ \text{min}} = \xi_{j_1} \otimes \xi_{j_2} . \tag{10.19}$$

The dimensions of the subspaces ξ_j expressed with j_1, j_2 or j will therefore satisfy the relation:

$$\sum_{j_{\text{min}}}^{j_{\text{max}}} (2j + 1) = (2j_1 + 1)\,(2j_2 + 1) . \tag{10.20}$$

The sum $\sum_{\alpha}^{\beta} j$ represents the sum of an arithmetic progression with α as the first term and $\beta = \alpha + n - 1$ as the last term and with common difference 1, whose value is therefore

$$\sum_{\alpha}^{\beta} j = \frac{1}{2}\, n\, (\alpha + \beta) = \frac{1}{2}\, (\beta - \alpha + 1)\,(\alpha + \beta)$$

$$= \frac{1}{2}\, \{\beta\,(\beta + 1) - \alpha\,(\alpha - 1)\} . \tag{10.21}$$

The sum $\sum_{\alpha}^{\beta} 1$ is equal to $\beta - (\alpha - 1)$.

By returning to Eq. (10.20) and (10.18), we obtain with (10.21)

$$\sum_{j_{min}}^{j_1+j_2} (2j+1) = 2 \sum_{j_{min}}^{j_1+j_2} j + \sum_{j_{min}}^{j_1+j_2} 1$$

$$= \{(j_1+j_2)(j_1+j_2+1) - j_{min}(j_{min}-1)\} + \{(j_1+j_2) - (j_{min}-1)\}$$

$$= (j_1+j_2)(j_1+j_2+2) - j_{min}^2 + 1.$$

We only have to equate this result to the second term of (10.20):

$$(j_1+j_2)(j_1+j_2+2) - j_{min}^2 + 1 = (2j_1+1)(2j_2+1).$$

This will yield, after simplification,

$$j_{min}^2 = (j_1 - j_2)^2 \tag{10.22}$$

Since j is a number that is always either positive or zero, we can write $j_{min} = |j_1 - j_2|$, thus proving with (10.18) theorem (10.17), that is, the second selection rule:

$$\boxed{|j_1 - j_2| \le j \le j_1 + j_2} \tag{10.23}$$

where j_1, j_2 and j are integers, positive half-integers, or zero.

2.2.3 Example: Deuteron States

What are the possible states of a deuteron comprising a proton and a neutron, each with spin $\frac{\vec{1}}{2}$?

Using the second selection rule, the vector addition of the spins will yield:

$$|\tfrac{1}{2} - \tfrac{1}{2}| \le S \le \tfrac{1}{2} + \tfrac{1}{2}, \qquad \text{that is,} \quad S = 0 \quad \text{or} \quad 1. \tag{10.24}$$

If $S = 0$ we have a singlet state $|00\rangle$.
If $S = 1$ we will have a triplet state $|11\rangle \, |10\rangle \, |1-1\rangle$.

It is possible to expand the coupled states $|00\rangle$ or $|1m\rangle$ with the spin states of the proton and the neutron. The first Clebsch–Gordan series

$$|j\,\mu\rangle = \sum_{m_1\,m_2} |j_1\,m_1\,j_2\,m_2\rangle \langle j_1\,m_1\,j_2\,m_2\,|\,j\,\mu\rangle$$

in fact yields for the singlet state

$$|00\rangle = \sum_{m_1\,m_2} |\tfrac{1}{2}\,m_1\,\tfrac{1}{2}\,m_2\rangle \langle \tfrac{1}{2}\,m_1\,\tfrac{1}{2}\,m_2\,|\,00\rangle$$

and, because $m = 0 = m_1 + m_2$,

$$| 00 \rangle = \sum_{m_1} | \frac{1}{2} m_1 \frac{1}{2} - m_1 \rangle \langle \frac{1}{2} m_1 \frac{1}{2} - m_1 | 00 \rangle . \tag{10.25}$$

The spin projection assumes the values $+\frac{1}{2}$ or $-\frac{1}{2}$, giving for (10.25)

$$\begin{aligned} | 00 \rangle &= | \frac{1}{2} \frac{1}{2} \frac{1}{2} - \frac{1}{2} \rangle \langle \frac{1}{2} \frac{1}{2} \frac{1}{2} - \frac{1}{2} | 00 \rangle \\ &+ | \frac{1}{2} - \frac{1}{2} \frac{1}{2} \frac{1}{2} \rangle \langle \frac{1}{2} - \frac{1}{2} \frac{1}{2} \frac{1}{2} | 00 \rangle . \end{aligned} \tag{10.26}$$

The Clebsch–Gordan coefficients will have the value $\pm 1/\sqrt{2}$ (see table in Sect. 2.3.5), which leaves us with:

$$| 00 \rangle = \frac{1}{\sqrt{2}} \left(| \frac{1}{2} \frac{1}{2} \rangle | \frac{1}{2} - \frac{1}{2} \rangle - | \frac{1}{2} - \frac{1}{2} \rangle | \frac{1}{2} \frac{1}{2} \rangle \right) . \tag{10.27}$$

If the state $| j_1 m_1 \rangle$ represents the proton and $| j_2 m_2 \rangle$ the neutron, then the singlet state is the antisymmetric state

$$| 00 \rangle = \frac{1}{\sqrt{2}} \left(| p \uparrow \rangle | n \downarrow \rangle - | p \downarrow \rangle | n \uparrow \rangle \right) . \tag{10.28}$$

The components of the triplet state are calculated in an analogous manner:

$$\begin{aligned} | 11 \rangle &= (| p \uparrow \rangle | n \uparrow \rangle), \\ | 10 \rangle &= \frac{1}{\sqrt{2}} (| p \uparrow \rangle | n \downarrow \rangle + | p \downarrow \rangle | n \uparrow \rangle), \\ | 1 - 1 \rangle &= (| p \downarrow \rangle | n \downarrow \rangle) . \end{aligned} \tag{10.29}$$

Notice that the deuteron is an isospin singlet state $I = 0$, $I_3 = 0$, constructed using the proton with isospin state $| \frac{1}{2} \frac{1}{2} \rangle$, and the neutron with isospin state $| \frac{1}{2} - \frac{1}{2} \rangle$:

$$| I I_3 \rangle = \sum_{\tau_1 \tau_2} | \frac{1}{2} \tau_1 \frac{1}{2} \tau_2 \rangle \langle \frac{1}{2} \tau_1 \frac{1}{2} \tau_2 | I I_3 \rangle . \tag{10.30}$$

With $\tau_1 = \pm \frac{1}{2}$ and $\tau_2 = \pm \frac{1}{2}$, we obtain

$$| 00 \rangle = | d \rangle = \frac{1}{\sqrt{2}} \left[| \frac{1}{2} \frac{1}{2} \rangle | \frac{1}{2} - \frac{1}{2} \rangle - | \frac{1}{2} - \frac{1}{2} \rangle | \frac{1}{2} \frac{1}{2} \rangle \right] ,$$

that is, by introducing the nature of the nucleons:

$$| d \rangle = \frac{1}{\sqrt{2}} \left[| p \rangle | n \rangle - | n \rangle | p \rangle \right] . \tag{10.31}$$

Care should be taken not to confuse spin states (10.28) or (10.29) with isospin states (10.31) describing the nature of the deuteron independently of its spin.

2.2.4 The Spectroscopist's Notation

Spectroscopists tend to specify the different possible quantum states by their orbital momentum $L = 0, 1, 2, 3$, respectively denoted states $S, P, D, F, G, H, \ldots$ The multiplicity of spin $(2S + 1)$ will be written as a superscript and the value J of the total angular momentum $^{2S+1}L_J$ as a subscript. Thus, the state $^3D_1$ will correspond to $L = 2$, $2S + 1 = 3$, $J = 1$, a triplet spin state.

2.3 Determination of Clebsch–Gordan Coefficients

2.3.1 Constraints

According to the last two subsections, the Clebsch–Gordan (CG) coefficient $\langle j_1\, m_1\, j_2\, m_2 \mid jm \rangle$ exists if the following selection rules are satisfied:

$$\text{i)} \quad m = m_1 + m_2 \,,$$
$$\text{ii)} \quad |j_1 - j_2| \leq j \leq j_1 + j_2 \,.$$

The next step will consist in making a complete calculation of these coefficients. Let us emphasize that definition (10.31):

$$| jm \rangle = \sum_{m_1\, m_2} \langle j_1\, m_1\, j_2\, m_2 \mid jm \rangle \mid j_1\, m_1 \rangle \mid j_2\, m_2 \rangle$$

defines the vectors $\mid jm \rangle$ to within an arbitrary phase. In order to remove this arbitrariness, we will grant a priori that the CG coefficient

$$\langle j_1\, j_1\, j_2\, j - j_1 \mid j\, j \rangle \geq 0 \tag{10.32}$$

is a positive real number.

Let us also remark that if $j = j_1 + j_2$ and $m = j$:

$$| j_1 + j_2\, j_1 + j_2 \rangle = \sum_{m_1\, m_2} \langle j_1\, m_1\, j_2\, m_2 \mid j_1 + j_2\, j_1 + j_2 \rangle \mid j_1\, m_1 \rangle \mid j_2\, m_2 \rangle$$
$$\tag{10.33}$$

with m and j having their maximum value, $m_1 + m_2 = j_1 + j_2$, and hence $m_1 = j_1$ and $m_2 = j_2$. There is therefore only one possible CG coefficient:

$$| j_1 + j_2\, j_1 + j_2 \rangle = \langle j_1 j_1\, j_2\, j_2 \mid j_1 + j_2\, j_1 + j_2 \rangle \mid j_1 j_1 \rangle \mid j_2 j_2 \rangle \tag{10.34}$$

corresponding to the decomposition of the vector with the highest weight, which imposes the following special value:

$$\langle j_1 j_1\, j_2 j_2 \mid j_1 + j_2\, j_1 + j_2 \rangle = 1 \,. \tag{10.35}$$

There exists in effect only one vector with the maximum weight.

These relations make it possible to calculate the CG coefficients using recursion relations with real coefficients, thus showing that all the CG coefficients are real numbers:

$$\langle j_1 \ m_1 \ j_2 \ m_2 \mid jm \rangle \ = \ \langle jm \mid j_1 \ m_1 \ j_2 \ m_2 \rangle . \qquad (10.36)$$

In the final analysis the following three constraints are imposed on the Clebsch-Gordan vector addition coefficients.

i)   CG are all real numbers: $\langle j_1 m_1 \ j_2 m_2 \mid jm \rangle = \langle jm \mid j_1 m_1 \ j_2 m_2 \rangle$

ii) $\langle j_1 j_1 \ j_2 \ j - j_1 \mid jj \rangle \geq 0$ phase choice

iii) $\langle j_1 j_1 \ j_2 j_2 \mid j_1 + j_2 \ j_1 + j_2 \rangle = 1$ maximum weight vector

2.3.2 Recursion Relations

Let us apply the operators $J_+ = J_x + i \ J_y$ and $J_- = J_x - i \ J_y$ to the CG definition relation (10.13), considering that j is an angular momentum by virtue of which it obeys all the rules described in Chap. 9:

$$
\begin{aligned}
J_+ \mid jm \rangle \ &= x_m \mid jm + 1 \rangle \\
&= \sum_{m_1 \ m_2} \langle j_1 \ m_1 \ j_2 \ m_2 \mid jm \rangle \ J_+ \mid j_1 \ m_1 \ j_2 \ m_2 \rangle . \quad (10.37)
\end{aligned}
$$

Because $J_+ = J_{1+} + J_{2+}$ and with the knowledge of the action of J_{1+} and J_{2+} on the kets $\mid j_1 \ m_1 \rangle$ and $\mid j_2 \ m_2 \rangle$, the foregoing equation then becomes:

$$
\begin{aligned}
x_m \mid jm + 1 \rangle \ = \ &\sum_{m_1 \ m_2} \langle j_1 \ m_1 \ j_2 \ m_2 \mid jm \rangle \\
&\{ x_{m_1} \mid j_1 \ m_1 + 1 j_2 \ m_2 \rangle \ + x_{m_2} \mid j_1 \ m_1 \ j_2 \ m_2 + 1 \rangle \} .
\end{aligned}
$$

Now, let us multiply each member by the conjugate bra $\langle j_1 \ m_1' \ j_2 \ m_2' \mid$ in order to introduce the CG:

$$
\begin{aligned}
x_m \ \langle j_1 \ m_1' \ j_2 \ m_2' \mid jm + 1 \rangle \ = \ &\sum_{m_1 \ m_2} \langle j_1 \ m_1 \ j_2 \ m_2 \mid jm \rangle \\
&\times \{ x_{m_1} \langle j_1 \ m_1' \ j_2 \ m_2' | j_1 \ m_1 + 1 \ j_2 \ m_2 \rangle \\
&+ x_{m_2} \langle j_1 \ m_1' \ j_2 \ m_2' | j_1 \ m_1 \ j_2 \ m_2 + 1 \rangle \} .
\end{aligned}
$$

The second term contains two inner products; the first will be different from zero if the following relations are verified:

$$
\begin{aligned}
m_1' &= m_1 + 1 \\
m_2' &= m_2 \ ;
\end{aligned}
$$

the second will be zero if $m_1' = m_1$ and $m_2' = m_2 + 1$. We therefore obtain the recursion relation:

$$x_m \langle j_1 \, m_1' \, j_2 \, m_2' \mid jm+1 \rangle = \langle j_1 \, m_1' - 1 \, j_2 \, m_2' \mid jm \rangle \, x_{m_1'-1}$$
$$+ \langle j_1 \, m_1' \, j_2 \, m_2' - 1 \mid jm \rangle \, x_{m_2'-1} . \quad (10.38)$$

An analogous relation can be proved by applying the operator J_-, yielding the more general recursion relation:

$$\{(j \pm m) \, (j \mp m + 1)\}^{1/2} \, \langle j_1 \, m_1 \, j_2 \, m_2 \mid jm \mp 1 \rangle$$
$$= \{(j_1 \mp m_1) \, (j_1 \pm m_1 + 1)\}^{1/2} \, \langle j_1 \, m_1 \pm 1 \, j_2 \, m_2 \mid jm \rangle$$
$$+ \{(j_2 \mp m_2) \, (j_2 \pm m_2 + 1)\}^{1/2} \, \langle j_1 \, m_1 \, j_2 \, m_2 \pm 1 \mid jm \rangle . \quad (10.39)$$

2.3.3 Example of CG Calculation

Consider a particle with orbital angular momentum $\vec{\ell}$ and spin $\vec{s} = \frac{\vec{1}}{2}$. Now, let us write the CG coefficients for transforming a coupled base $\vec{j} = \vec{\ell} + \vec{s}$ into the decoupled base.

Let $\mid j \, \mu \rangle$ be a ket of the coupled base, $\mid \ell \, m \,\rangle$ and $\mid \frac{1}{2} \, \sigma \rangle$ those of the decoupled base.

The application of relations (10.13) and (10.14) will yield the following coupled states:

$$\mid j \, \mu \rangle = \sum_{m \, \sigma} \langle \ell \, m \, \frac{1}{2} \, \sigma \mid j \, \mu \rangle \mid \ell \, m \, \frac{1}{2} \, \sigma \rangle \quad (10.40)$$

or the decoupled states with the second CG series for the value $\sigma = + \frac{1}{2}$:

$$\mid \ell \, m \, \frac{1}{2} \frac{1}{2} \rangle = \sum_{j \, \mu} \langle \ell \, m \, \frac{1}{2} \frac{1}{2} \mid j \, \mu \rangle \mid j \, \mu \rangle . \quad (10.41)$$

Now, let us cause the operator $J_+ = \ell_+ + s_+$ to act on the second Clebsch-Gordan series (10.41), bearing in mind that the projection of the spin can only take the values $+\frac{1}{2}$ or $-\frac{1}{2}$:

$$J_+ \mid j \, \mu \rangle = x_\mu \mid j \, \mu + 1 \rangle$$
$$\ell_+ \mid \ell \, m \rangle = x_m \mid \ell \, m + 1 \rangle$$
$$s_+ \mid \frac{1}{2} \frac{1}{2} \rangle = 0 . \quad (10.42)$$

By multiplying the two terms by the bra $\langle \ell + \frac{1}{2} \, m + \frac{3}{2} \, |$, we obtain:

$$x_m \, \langle \ell + \frac{1}{2} \, m + \frac{3}{2} \, | \, \ell \, m+1 \, \frac{1}{2}\frac{1}{2} \rangle = \sum_{j\,\mu} \langle \ell \, m \, \frac{1}{2}\frac{1}{2} \, | \, j \, \mu \rangle$$

$$\times \, \langle \ell + \frac{1}{2} \, m + \frac{3}{2} \, | \, j \, \mu+1 \rangle \, x_\mu. \quad (10.43)$$

The inner product will transform the second member as follows:

$$\sum_{j\,\mu} \langle \ell \, m \, \frac{1}{2}\frac{1}{2} \, | \, j \, \mu \rangle \, x_\mu \, \delta_{\ell+\frac{1}{2},\,j} \, \delta_{m+\frac{3}{2},\,\mu+1} \quad (10.44)$$

that is by carrying over in (10.43) and expanding x_m and x_μ:

$$[(\ell - m) \, (\ell + m + 1)]^{1/2} \, \langle \ell + \frac{1}{2} \, m + \frac{3}{2} \, | \, \ell \, m+1 \, \frac{1}{2}\frac{1}{2} \rangle$$

$$= \langle \ell \, m \, \frac{1}{2}\frac{1}{2} \, | \, \ell + \frac{1}{2} \, m + \frac{1}{2} \rangle \, [(\ell - m) \, (\ell + m + 2)]^{1/2}. \quad (10.45)$$

We therefore obtain the recursion relation

$$\langle \ell + \frac{1}{2} \, m + \frac{1}{2} \, | \, \ell \, m \, \frac{1}{2}\frac{1}{2} \rangle = \sqrt{\frac{\ell + m + 1}{\ell + m + 2}} \, \langle \ell + \frac{1}{2} \, m + \frac{3}{2} \, | \, \ell \, m+1 \, \frac{1}{2}\frac{1}{2} \rangle$$

$$\langle \ell + \frac{1}{2} \, m + \frac{3}{2} \, | \, \ell \, m+1 \, \frac{1}{2}\frac{1}{2} \rangle = \sqrt{\frac{\ell + m + 2}{\ell + m + 3}} \, \langle \ell + \frac{1}{2} \, m + \frac{5}{2} \, | \, \ell \, m+2 \, \frac{1}{2}\frac{1}{2} \rangle$$

$- -$

$$\langle \ell + \frac{1}{2} \, \ell - \frac{1}{2} \, | \, \ell\ell - 1 \, \frac{1}{2}\frac{1}{2} \rangle = \sqrt{\frac{2\ell}{2\ell + 1}} \, \langle \ell + \frac{1}{2} \, \ell + \frac{1}{2} \, | \, \ell\ell \, \frac{1}{2}\frac{1}{2} \rangle$$

$$\langle \ell + \frac{1}{2} \, m + \frac{1}{2} \, | \, \ell \, m \, \frac{1}{2}\frac{1}{2} \rangle = \sqrt{\frac{\ell + m + 1}{2\ell + 1}} \, \langle \ell + \frac{1}{2} \, \ell + \frac{1}{2} \, | \, \ell\ell \, \frac{1}{2}\frac{1}{2} \rangle. (10.46)$$

The cross product of these equations gives the result (10.46) and because, following convention (10.35), the CG coefficient $\langle \ell + \frac{1}{2} \, \ell + \frac{1}{2} \, | \, \ell\ell \, \frac{1}{2}\frac{1}{2} \rangle$ is equal to 1, we finally obtain

$$\langle \ell + \frac{1}{2} \, m + \frac{1}{2} \, | \, \ell \, m \, \frac{1}{2}\frac{1}{2} \rangle = \sqrt{\frac{\ell + m + 1}{2\ell + 1}}. \quad (10.47)$$

In a similar manner we can calculate $\langle \ell + \frac{1}{2} \, m - \frac{1}{2} \, | \, \ell \, m \, \frac{1}{2} - \frac{1}{2} \rangle$ by the action of J_- on $| \, \ell \, m \, \frac{1}{2} - \frac{1}{2} \rangle$.

2.3.4 Analytical Expression of Clebsch–Gordon Coefficients

It is sometimes interesting to use an analytical expression for the CG coefficients rather than their tabulated values. This is especially true where the calculation is done on a computer. In such a case, the following expression may be used:

$$
\begin{aligned}
\langle j_1 \ m_1 \ j_2 \ m_2 \mid j_3 \ m_3 \rangle \ &= \Delta \ (j_1 \ j_2 \ j_3) \ [(j_1 + m_1)! \ (j_1 - m_1)! \\
&\times (j_2 + m_2)! \ (j_2 - m_2)! \ (j_3 + m_3)! \ (j_3 - m_3)! \ ]^{1/2} \\
&\times (2j_3 + 1) \sum_t (-)^t \ [t! \ (j_1 + j_2 - t)! \ (j_1 - m_1 - t)! \\
&\times (j_2 + m_2 - t)! \ (j - j_2 + m_1 - t)! \ (j - j_1 - m_2 + t)!]^{1/2}
\end{aligned} \tag{10.48}
$$

with the parameter Δ given by

$$
\Delta \ (abc) = \left[\frac{(a+b-c) \ ! \ (a-b+c) \ ! \ (-a+b+c) \ !}{(a+b+c+1) \ !} \right]^{1/2} .
$$

However, because of the existence of tables of Clebsch–Gordan coefficients and computer routines, this type of formula is only rarely used.

The following table gives the values of the most frequently used Clebsch–Gordan coefficients.

2.3.5 Table of Clebsch–Gordan Coefficients

The reader is reminded that the sign $\sqrt{}$ is implicit in each coefficient. For example, $-\frac{8}{15}$ implies $-\sqrt{\frac{8}{15}}$.

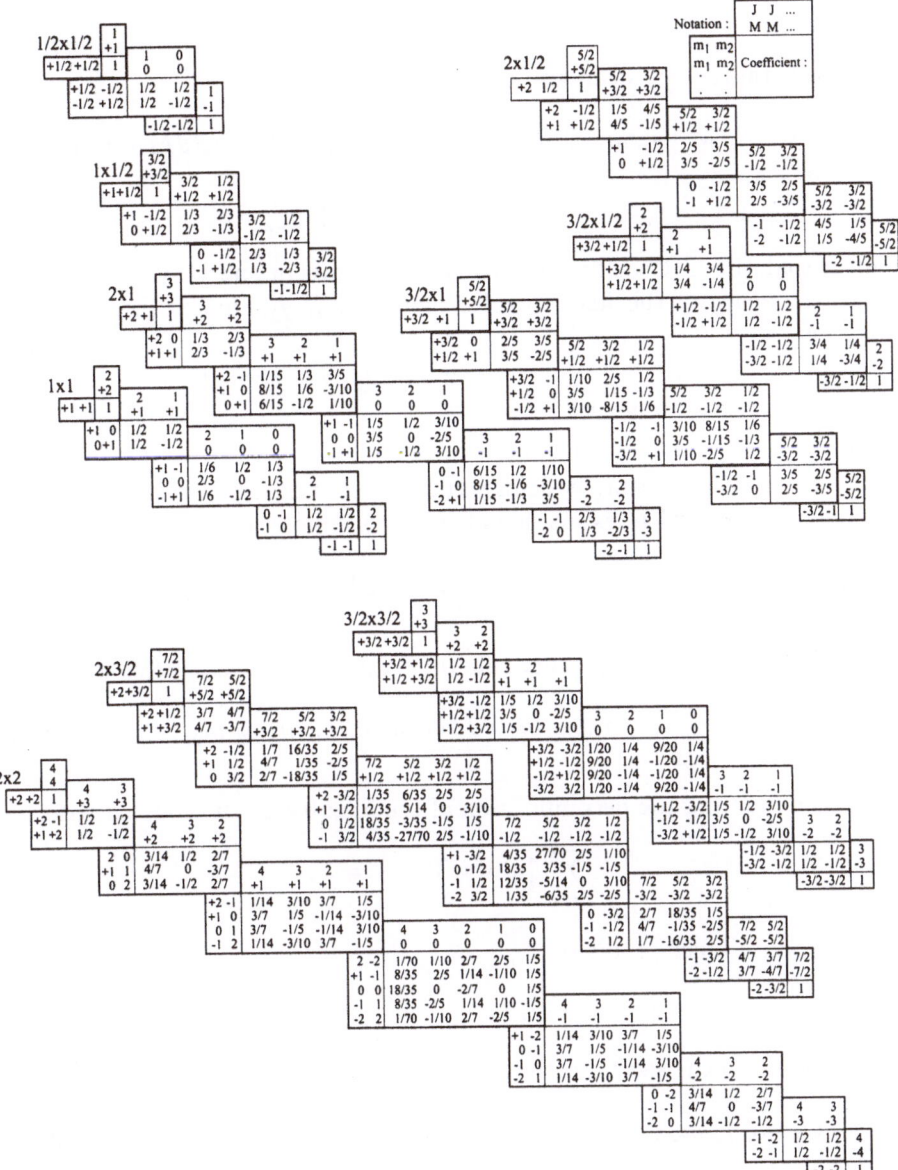

2.4 Orthogonality Relations

Clebsch–Gordan coefficients are coefficients of a unitary transformation and therefore satisfy orthogonality relations:

$$\sum_{m_1 m_2} \langle j_1 m_1 j_2 m_2 \mid JM \rangle \langle j_1 m_1 j_2 m_2 \mid J'M' \rangle = \delta_{JJ'} \delta_{MM'} \{j_1 j_2 J\}.$$

(10.49)

In effect, this relation can also take the form

$$\sum_{m_1 m_2} \langle JM \mid j_1 m_1 j_2 m_2 \rangle \langle j_1 m_1 j_2 m_2 \mid J'M' \rangle = \langle JM \mid J'M' \rangle$$

$$= \delta_{JJ'} \delta_{MM'} \{j_1 j_2 J\} \qquad (10.50)$$

given that the CG coefficients are all real numbers and that the projector onto the space of the fixed angular momenta $j_1 j_2$, that is the decoupled base, is

$$\mathbb{1} = \sum_{m_1 m_2} \mid j_1 m_1 j_2 m_2 \rangle \langle j_1 m_1 j_2 m_2 \mid.$$

The symbol $\{j_1 j_2 j_3\}$ introduced in (10.49) and (10.50) is the triangular delta or "$3j$" coefficient. It defines the existence of the $3jm$ coefficients by giving a mathematical expression of the second selection rule:

$$\{j_1 j_2 J\} = \begin{cases} 1 & \text{if} \quad \mid j_1 - j_2 \mid \leq J \leq j_1 + j_2 \\ 0 & \text{otherwise}. \end{cases} \qquad (10.51)$$

There is a second orthogonality rule that can be expressed as follows:

$$\sum_{JM} \langle j_1 m_1 j_2 m_2 \mid JM \rangle \langle JM \mid j_1 m_1' j_2 m_2' \rangle \qquad (10.52)$$

and by extracting the decomposition of the identity on the coupled base

$$\sum_{JM} \mid JM \rangle \langle JM \mid = \mathbb{1}.$$

We introduce inner products

$$\langle j_1 m_1 \mid j_1 m_1' \rangle \langle j_2 m_2 \mid j_2 m_2' \rangle = \delta_{m_1 m_1'} \delta_{m_2 m_2'}$$

thus leading to the second orthogonality relation:

$$\sum_{JM} \langle j_1 m_1 j_2 m_2 \mid JM \rangle \langle j_1 m_1' j_2 m_2' \mid JM \rangle = \delta_{m_1 m_1'} \delta_{m_2 m_2'}. \qquad (10.53)$$

Orthogonality rules (10.51) and (10.53) play an important role in Racah algebra as well as in the graphical representation of the summation rules for angular momentum.

3. 3jm Coefficients

3.1 Definition

In the Clebsch–Gordan coefficient $\langle j_1\, m_1\, j_2\, m_2 \mid JM \rangle$, the angular momenta $j_1\, j_2$ and the angular momentum J do not play the same role. The first two are derived from a covariant vector and the last from a contravariant vector. This is easily noticed with (10.48) in the following symmetry rules:

$$\langle j_1\, m_1\, j_2\, m_2 \mid JM \rangle = (-)^{j_1+j_2-J} \langle j_2\, m_2\, j_1\, m_1 \mid JM \rangle \quad (10.54)$$

$$= (-)^{j_1-m_1} \sqrt{\frac{2J+1}{2j_2+1}} \langle j_1\, m_1\, J-M \mid j_2 - m_2 \rangle. \quad (10.55)$$

To get round this problem, E.P. Wigner has defined symmetrized vector addition coefficients in which the three angular momenta play the same role and describe the vector addition $\vec{j}_1 + \vec{j}_2 + \vec{j}_3 = 0$. These coefficients are usually described as Wigner's $3j$ coefficients. In place of this notation, which will be reserved for another coefficient, we will be using the term $3jm$ coefficient. Wigner's $3jm$ coefficients are linked to Clebsch–Gordan coefficients by the following definition:

$$\langle j_1\, m_1\, j_2\, m_2 \mid j_3\, m_3 \rangle = \sqrt{2j_3+1}\,(-)^{j_1-j_2+m_3} \begin{pmatrix} j_1 & j_2 & j_3 \\ m_1 & m_2 & -m_3 \end{pmatrix}. \quad (10.56)$$

3.2 Symmetry

The $3jm$ coefficient $\begin{pmatrix} j_1 & j_2 & j_3 \\ m_1 & m_2 & m_3 \end{pmatrix}$ has the following symmetry properties:

i) $m_1 + m_2 + m_3 = 0$,
ii)   it is invariant under circular permutation of the columns,
iii)   it is multiplied by $(-)^{j_1+j_2+j_3}$ if two columns are permuted
iv)   it is multiplied by $(-)^{j_1+j_2+j_3}$ if the signs of the three magnetic momenta are simultaneously exchanged.

Note that the triangle rule will have to be satisfied for the $3j\,m$ to exist, which implies that $j_1 + j_2 + j_3$ is always an integer. This is also true of $(j_i \pm m_i)$. The three simultaneously obeyed inequalities can be written

$$\begin{aligned} \mid j_1 - j_2 \mid \; &\leq \; j_3 \; \leq \; j_1 + j_2, \\ \mid j_1 - j_3 \mid \; &\leq \; j_2 \; \leq \; j_1 + j_3, \\ \mid j_2 - j_3 \mid \; &\leq \; j_1 \; \leq \; j_2 + j_3. \end{aligned} \quad (10.57)$$

3.3 Covariant Momentum

To simplify the notation, we set $[a] = \sqrt{2a+1}$ and use Wigner's covariant notation. By covariant we will be referring to an angular momentum JM appearing in the $3j$ m coefficient with a $(-)^{J-M}$ phase and a negative sign for the magnetic momentum. A covariant moment will therefore be written with the magnetic momentum in the upper position and the angular momentum in the lower position:

$$(-)^{J-M} \begin{pmatrix} J & \cdot & \cdot \\ -M & \cdot & \cdot \end{pmatrix} = \begin{pmatrix} M & \cdot & \cdot \\ J & \cdot & \cdot \end{pmatrix} \tag{10.58}$$

For example, in the following $3jm$ coefficients, the angular momentum j_3 will be covariant:

$$(-)^{j_3-m_3} \begin{pmatrix} j_1 & j_2 & j_3 \\ m_1 & m_2 & -m_3 \end{pmatrix} = \begin{pmatrix} j_1 & j_2 & m_3 \\ m_1 & m_2 & j_3 \end{pmatrix}. \tag{10.59}$$

Relation (10.56) between the Clebsch–Gordan and the Wigner coefficients can also be written in its covariant notation:

$$\langle j_1 \, m_1 \, j_2 \, m_2 \mid j_3 \, m_3 \rangle = (-)^{j_1-j_2-j_3} \, [j_3] \begin{pmatrix} m_1 & m_2 & j_3 \\ j_1 & j_2 & m_3 \end{pmatrix}. \tag{10.60}$$

The angular momenta j_1 and j_2 in the bra of the Clebsch–Gordan are covariant in the $3jm$ coefficients. The same is true if we use the conjugate form of the Clebsch–Gordan coefficient:

$$\langle j_3 \, m_3 \mid j_1 \, m_1 \, j_2 \, m_2 \rangle = (-)^{j_1-j_2+j_3} \, [j_3] \begin{pmatrix} j_1 & j_2 & m_3 \\ m_1 & m_2 & j_3 \end{pmatrix}. \tag{10.61}$$

3.4 Orthogonality Relations

Using the $3jm$ coefficients, the first orthogonality relation reduces to:

$$\sum_{m_1 \, m_2} \begin{pmatrix} j_1 & j_2 & j_3' \\ m_1 & m_2 & m_3' \end{pmatrix} \begin{pmatrix} j_1 & j_2 & j_3 \\ m_1 & m_2 & m_3 \end{pmatrix} = [j_3^{-2}] \, \delta_{j_3 \, j_3'} \, \delta_{m_3 \, m_3'} \, \{j_1 \, j_2 \, j_3\},$$
$$\tag{10.62}$$

whereas the second orthogonality rule will be written

$$\sum_{j_3 \, m_3} [j_3^2] \begin{pmatrix} j_1 & j_2 & j_3 \\ m_1' & m_2' & m_3 \end{pmatrix} \begin{pmatrix} j_1 & j_2 & j_3 \\ m_1 & m_2 & m_3 \end{pmatrix} = \delta_{m_1' \, m_1} \, \delta_{m_2' \, m_2}. \tag{10.63}$$

Using the covariant notation the orthogonality rules will take the following simple form:

$$\begin{pmatrix} j_1 & j_2 & j_3 \\ m_1 & m_2 & m_3 \end{pmatrix} \begin{pmatrix} m_1 & m_2 & m_3' \\ j_1 & j_2 & j_3' \end{pmatrix} = [j_3^{-2}] \, \delta_{j_3 \, j_3'} \, \delta_{m_3 \, m_3'} \, \{j_1 \, j_2 \, j_3\}$$
$$\tag{10.64}$$

and orthogonality rule (10.63) will reduce to

$$\sum_{j_3} [j_3^2] \begin{pmatrix} j_1 & j_2 & j_3 \\ m_1 & m_2 & m_3 \end{pmatrix} \begin{pmatrix} m_1' & m_2' & m_3 \\ j_1 & j_2 & j_3 \end{pmatrix} = \delta_{m_1\, m_1'}\, \delta_{m_2\, m_2'}\,. \quad (10.65)$$

The summation over the magnetic momenta has been omitted by virtue of Einstein's summation convention over covariant-contravariant indices that are repeated during a contraction, that is during the summation over these indices.

4. Coupling of n Angular Momenta

4.1 Generalized Clebsch–Gordon Coefficients

4.1.1 Example

Consider the case of two particles with orbital momenta $\vec{\ell}_1$ and $\vec{\ell}_2$ and spins $\vec{s}_1$ and $\vec{s}_2$. The total angular momentum $\vec{J}$ will be equal to the sum $\vec{J} = \vec{\ell}_1 + \vec{s}_1 + \vec{\ell}_2 + \vec{s}_2$.

We can obtain the quantum state $|\,JM\rangle$, eigenstate of J^2 and J_z by first coupling $\vec{\ell}_1 + \vec{s}_1 = \vec{j}_1$, then $\vec{\ell}_2 + \vec{s}_2 = \vec{j}_2$ and finally $\vec{j}_1 + \vec{j}_2 = \vec{J}$. This mode of coupling will be noted A. Each of the intermediate stages will be obtained by using the Clebsch–Gordan series

$$|JM\rangle_A = |(\ell_1\, s_1)\, j_1\, (\ell_2\, s_2)\, j_2, JM\rangle = \sum_{m_1\, \sigma_1\, m_2\, \sigma_2 \mu_1\, \mu_2} \langle \ell_1\, m_1\, s_1\, \sigma_1 | j_1\, \mu_1 \rangle$$

$$\times \langle \ell_2\, m_2\, s_2\, \sigma_2 | j_2\, \mu_2 \rangle \langle j_1\, \mu_1\, j_2\, \mu_2\, | JM \rangle | \ell_1\, m_1 \rangle | \ell_2\, m_2 \rangle | s_1\, \sigma_1 \rangle | s_2\, \sigma_2 \rangle\,. \quad (10.66)$$

We can use the B coupling mode, first coupling numbers of the same type: $\big((\vec{\ell}_1 + \vec{\ell}_2) + (\vec{s}_1 + \vec{s}_2)\big)$, which leads to the following result:

$$|JM\rangle_B = |(\ell_1\, \ell_2)\, L\, (s_1\, s_2)\, S; JM\rangle = \sum_{m_1\, m_2\, \mu\, \sigma\, \sigma_1\, \sigma_2} \langle \ell_1\, m_1\, \ell_2\, m_2\, |\, LM \rangle$$

$$\times \langle s_1\, \sigma_1\, s_2\, \sigma_2 | S\, \sigma \rangle \langle LM\, S\, \sigma\, |\, JM \rangle | \ell_1\, m_1 \rangle | \ell_2\, m_2 \rangle | s_1\, \sigma_1 \rangle | s_2\, \sigma_2 \rangle\,. \quad (10.67)$$

We thus see that the result depends on the coupling mode used and also on the intermediate momenta introduced. By analogy with the Clebsch–Gordan series (10.13), we then introduce a generalized CG coefficient

$$|\,(\ell_1\, s_1)\, j_1\, (\ell_2\, s_2)\, j_2, \, JM\rangle\ =\ |\,JM\rangle_A$$

$$=\ \sum_{m_1\, m_2\, \sigma_1\, \sigma_2} \langle \ell_1\, m_1\, \ell_2\, m_2\, s_1\, \sigma_1\, s_2\, \sigma_2\, |\, (\ell_1\, s_1)\, j_1\, (\ell_2\, s_2)\, j_2, \, JM \rangle$$

$$\times |\, \ell_1\, m_1\, \ell_2\, m_2\, s_1\, \sigma_1\, s_2\, \sigma_2 \rangle\,. \quad (10.68)$$

Using the other coupling mode, we obtain

$$
\begin{aligned}
\mid (\ell_1 \ell_2) \, L \, (s_1 \, s_2) S, \, JM \rangle &= \mid JM \rangle_B \\
&= \sum_{m_1 \, m_2 \, \sigma_1 \, \sigma_2} \langle \ell_1 \, m_1 \, \ell_2 \, m_2 \, s_1 \, \sigma_1 \, s_2 \, \sigma_2 \mid (\ell_1 \ell_2) \, L \, (s_1 \, s_2) \, S, \, JM \rangle \\
&\quad \times \mid \ell_1 \, m_1 \, \ell_2 \, m_2 \, s_1 \, \sigma_1 \, s_2 \, \sigma_2 \rangle .
\end{aligned}
\tag{10.69}
$$

4.1.2 General Definition

Let $\langle j_1 \, m_1 \cdots j_{n-1} \, m_{n-1} \mid (j_1 \cdots j_{n-1})_A \, X_s ; j_n \, m_n \rangle$ be a generalized Clebsch–Gordan coefficient for a coupling mode A involving X_s (with $s = n - 3$) intermediate momenta. This leads to the series

$$
\begin{aligned}
\mid j_n \, m_n \rangle_A &= \sum_{m_1 \ldots m_{n-1}} \langle j_1 \, m_1 \cdots j_{n-1} \, m_{n-1} \mid (j_1 \cdots j_{n-1})_A \, X_s ; j_n \, m_n \rangle \\
&\quad \times \mid j_1 \, m_1 \cdots j_{n-1} \, m_{n-1} \rangle ,
\end{aligned}
\tag{10.70}
$$

and, in order to go from the uncoupled basis to the coupled basis

$$
\begin{aligned}
\mid j_1 \, m_1 \cdots j_{n-1} \, m_{n-1} \rangle &= \sum_{j_n \, m_n \, X_s} \langle j_1 \, m_1 \cdots j_{n-1} \, m_{n-1} \mid (j_1 \cdots j_{n-1})_A \, X_s ; j_n \, m_n \rangle \\
&\quad \times \mid (j_1 \cdots j_{n-1})_A \, X_s ; j_n \, m_n \rangle .
\end{aligned}
\tag{10.71}
$$

4.1.3 Symmetries and Orthogonality

The generalized CG coefficients will have all the properties of the constituent CG coefficients, in particular,

i)   the generalized CG coefficients are real numbers
ii) $m_1 + m_2 + \ldots m_{n-1} = m_n$
iii)   the sum $j_1 + j_2 + \ldots j_{n-1} + j_n$ is an integer
iv)   they satisfy the orthogonality relations:

$$
\begin{aligned}
\sum_{m_1 \ldots m_{n-1}} &\langle (j_1 \cdots j_{n-1})_A \, X_s, j_n \, m_n \mid j_1 \, m_1 \cdots j_{n-1} \, m_{n-1} \rangle \\
&\times \langle j_1 \, m_1 \cdots j_{n-1} \, m_{n-1} \mid (j_1 \cdots j_{n-1})_A \, X'_s, j'_n \, m'_n \rangle \\
&= \delta_{j_n \, j'_n} \, \delta_{m_n \, m'_n} \prod_{s=1}^{n-3} \delta_{X_s \, X'_s}
\end{aligned}
\tag{10.72}
$$

The second orthogonality rule will, for its part, be written as follows:

$$
\begin{aligned}
\sum_{X_s \, j_n \, m_n} &\langle j_1 \, m_1 \cdots j_{n-1} \, m_{n-1} \mid (j_1 \cdots j_{n-1}) \, X_s, j_n \, m_n \rangle \\
&\times \langle (j_1 \cdots j_{n-1})_A \, X_s, j_n \, m_n \mid j_1 \, m'_1 \cdots j_{n-1} \, m'_{n-1} \rangle = \prod_{i=1}^{n1} \delta_{m_i \, m'_i} .
\end{aligned}
\tag{10.73}
$$

4.2 njm Coefficient

4.2.1 Definition and Notation

As has already been mentioned in connection with the preceding example, the generalized Clebsch–Gordan (GCG) coefficient is a sum of the products of Clebsch–Gordan coefficients:

$$\langle j_1 \, m_1 \cdots j_{n-1} \, m_{n-1} \mid (j_1 \cdots j_{n-1})_A \, X_s; j_n \, m_n \rangle$$
$$= \sum_{m_{X_s}} \langle j_1 \, m_1 \, j_2 \, m_2 \mid X_1 \, m_{X_1} \rangle \, \langle j_3 \, m_3 \, j_4 \, m_4 \mid X_2 \, m_{X_2} \rangle \cdots$$
$$\langle j_{n-2} \, m_{n-2} \, j_{n-1} \, m_{n-1} \mid X_s \, m_{X_s} \rangle \cdots \langle \mid j_n m_n \rangle. \tag{10.74}$$

Each CG coefficient can therefore be transformed into a $3jm$ coefficient to obtain a generalized $3jm$ or njm coefficient noted as follows:

$$\begin{pmatrix} j_1 & j_2 & \cdots & j_n \\ & & & \\ m_1 & m_2 & & m_n \end{pmatrix} \; | X_s \Big) . \tag{10.75}$$

If necessary, the index A designating the coupling mode corresponding to the intermediate momenta X_s may be added.

4.2.2 Symmetries and Orthogonality

All the angular momenta $j_1 \ldots j_n$ now play all the same role since $\vec{j}_1 + \vec{j}_2 + \ldots \vec{j}_n = 0$.

The symmetry properties of the njm coefficient result from symmetry properties of the constituent $3jm$ coefficients:

i) $m_1 + m_2 + \ldots m_n = 0$
ii) invariant under circular permutation of the columns
iii) multiplied by $(-)^{j_1 + j_2 + \cdots j_n}$ for the change of sign of all the magnetic momenta
iv) they satisfy the orthogonality relations:

$$\begin{pmatrix} j_1 & \cdots & j_n \\ & & \\ m_1 & \cdots & m_n \end{pmatrix} \; | X_s \Big) \begin{pmatrix} m_1 & \cdots & m_{n-1} & m'_n \\ & & & \\ j_1 & \cdots & j_{n-1} & j'_n \end{pmatrix} \; | X_s \Big)$$

$$= [j_n^{-2}] \, \delta_{j_n \, j'_n} \, \delta_{m_n \, m'_n} \, \{j_1 \cdots j_n\} \tag{10.76}$$

The polygon delta is defined by the generalization of the triangular delta:

$$\{j_1 \cdots j_n\} = \begin{cases} 1 & \text{if} \quad \vec{j}_1 + \vec{j}_2 + \ldots \vec{j}_n = 0 \\ 0 & \text{otherwise} . \end{cases} \tag{10.77}$$

We obtain a second orthogonality rule by summing over j_n, m_n and X_s:

$$\sum_{j_n \, X_s} [j_n^2 \, X_s^2] \begin{pmatrix} j_1 & \cdots & j_n \\ & & \\ m_1 & \cdots & m_n \end{pmatrix} X_s \begin{pmatrix} m_1' & \cdots & m_{n-1}' & m_n \\ & & & \\ j_1 & \cdots & j_{n-1} & j_n \end{pmatrix} X_s$$

$$= \delta_{m_1 \, m_1'} \, \delta_{m_2 \, m_2'} \cdots \delta_{m_{n-1} \, m_{n-1}'} \tag{10.78}$$

The repeated covariant-contravariant indices in these relations are summed over according to Einstein's convention.

5. Change of Coupling Mode

Let us return to example (10.68) and try to define the change of basis matrix for going from one coupling mode to the other. If we project the state $|\, JM \rangle_A$ onto the state $|\, JM \rangle_B$, we write

$$|\, (\ell_1 \, s_1 \, \ell_2 \, s_2)_A \, j_1 \, j_2; JM \rangle = \sum_{LS} |\, (\ell_1 \, s_1 \, \ell_2 \, s_2)_B \, LS, J'M' \rangle$$

$$\langle (\ell_1 \, s_1 \, \ell_2 \, s_2)_B \, LS; J'M' \,|\, (\ell_1 \, s_1 \, \ell_2 \, s_2)_A \, j_1 \, j_2, JM \rangle . \tag{10.79}$$

The change of basis matrix is therefore the inner product of the quantum states written with the coupling modes A and B. This is represented with the Kronecker symbols $\delta_{J \, J'}$ and $\delta_{M \, M'}$. We notice then that the coupled states (irrespective of the coupling mode used to obtain them) are orthogonal, thus yielding

$$\langle (\ell_1 \, s_1 \, \ell_2 \, s_2)_B \, LS; J'M' \,|\, (\ell_1 \, s_1 \, \ell_2 \, s_2)_A \, j_1 \, j_2; JM \rangle$$

$$= \delta_{J \, J'} \, \delta_{M \, M'} \, [J^{-2}] \left\{ \begin{array}{l} \text{coefficient dependent upon} \\ \ell_1 \, \ell_2 \, s_1 \, s_2 \, j_1 \, j_2 \, LS \text{ and } J \end{array} \right\} . \tag{10.80}$$

The change of coupling matrix is generally of the form

$$\langle (j_1 \cdots j_{n-1})_A \, X_s; j_n \, m_n \,|\, (j_1 \cdots j_{n-1})_B \, Y_s; j_n' \, m_n' \rangle$$

$$= [j_n^{-2}] \, \delta_{j_n \, j_n'} \, \delta_{m_n \, m_n'} \text{ ``}3nj \text{ coefficient''} . \tag{10.81}$$

The coefficient thus obtained is generally a $3nj$ coefficient that is dependent upon the $3n$ angular momenta. The preceding example leads, for example, to a $9j$ coefficient as we will see in connection with the graphical method.

6. Introduction to the Graphical Method

The graphical method for the treatment of the algebra of angular momenta was first introduced in 1962 by Yutsis, Levinson, and Vanagas (Yutsis et al., 1960) It was further developed and improved upon in Lyon (Elbaz and Castel, 1972; Elbaz, 1985) and laboratories elsewhere. It is commonly referred to by the acronym GSA (for Graphical Spin Algebra). We will be content here to introduce the basics and refer the interested reader to more specialized textbooks on the subject.

6.1 Basic Graphical Representations

The conventional graphical representation of kets and bras of the Hilbert space subspace ξ_j is defined as:

$$| j\, m \rangle \; = \; \overset{jm}{\longmapsto\!\!\to\!-} \quad \text{and} \quad \langle j\, m\, | \; = \; \overset{jm}{\longmapsto\!\!\twoheadleftarrow\!-} \tag{10.82}$$

$$| j-m \rangle \; = \; \overset{jm}{\longmapsto\!\!\leftarrow\!-} \quad \text{and} \quad \langle j-m\,| = \; \overset{jm}{\longmapsto\!\!\twoheadrightarrow\!-}. \tag{10.83}$$

In the most common cases, only representations (10.82) are called into play:

$$\langle j\, m\, |\, j'\, m' \rangle \; = \delta_{j\,j'}\, \delta_{m\,m'} = \; \overset{jm \quad\; j'm'}{\twoheadrightarrow\!\!-\!\!\longrightarrow} \; = \; \overset{jm \quad\; j'm'}{\twoheadrightarrow\!\!\longrightarrow}. \tag{10.84}$$

Any diagram with just the double arrow (such as a bra) and the single arrow (such as a ket) is proportional to a Kronecker symbol. Considering that the proportionality coefficient can be determined using GSA rules, we will then write

$$\overset{jm \quad\;\; j'm'}{\twoheadrightarrow\!-\!(\alpha)\!-\!\to} \; = \; \alpha_j\, \delta_{j\,j'}\, \delta_{m\,m'}. \tag{10.85}$$

6.2 Rule for Summation Over a Magnetic Momentum

To sum over a magnetic momentum, we normally bring together the free extremities of the corresponding arrows, which boils down to applying Einstein's convention for summation over covariant indices (such as a bra) and contravariant indices (such as a ket):

$$\sum_m | j\, m \rangle \langle j\, m\, | \; = \sum_m \; \overset{jm \quad\;\; jm}{\longmapsto\!\to\;\; \twoheadrightarrow\!\dashv} \; = \; \overset{j}{\longmapsto\!\dashv} \tag{10.86}$$

If we assume that the magnetic momentum is always written in a natural way, that is in the form $| j\, m \rangle$ and not in the form $| j-m \rangle$, a related line can only be broken from (10.86).

We then show that the change in the direction of the related line j introduces a phase coefficient $(-)^{2j}$.

The completeness rule is expressed graphically by the relation

$$\sum_{j\,m} |j\,m\rangle \langle j\,m| = \sum_{j} \overset{j}{\longmapsto\!\!\!\bullet\!\!\!\longmapsto} = \mathbb{1} \qquad (10.87)$$

6.3 Clebsch–Gordan Coefficients

Let us introduce the graphical representation of Clebsch–Gordan coefficients by defining an a priori reading order $(-)$ for the clockwise direction and $(+)$ for the anticlockwise direction using the coupled angular momentum as departure point. Hence, we set

$$\langle j_1\,m_1\,j_2\,m_2\,|\,j_3\,m_3\rangle = \qquad = \qquad (10.88)$$

$$\langle j_3\,m_3\,|\,j_1\,m_1\,j_2\,m_2\rangle = \qquad = \qquad . \qquad (10.89)$$

The two Clebsch–Gordan series (10.13) and (10.14) can be obtained from these graphical representations and the procedure for summation over m:

$$\qquad = \qquad = \sum_{m_1 m_2} \qquad \qquad (10.90)$$

or by coupling the spaces ξ_{j_1} and ξ_{j_2}:

$$\qquad = \sum_{j} \qquad = \sum_{jm} \qquad . \qquad (10.91)$$

6.4 Orthogonality Relations

The second orthogonality rule is also easy to visualize with the aid of the graphical method by using the completeness relation (10.87):

$$\sum_{j} \qquad = \qquad . \qquad (10.92)$$

Because this relation is valid even when the arrows j_1 j_2 are linked, it constitutes a graphical rule for summation over j for any diagram whatsoever, as we can see by decomposing the process using relation (10.92) and definition (10.86):

$$\sum_j \left[\alpha + \!\!\!\!\!\!\!\!\!\!\!\!\!\!\!\!\! \begin{matrix} j_1 & & j_1 \\ & j & \\ j_2 & & j_2 \end{matrix} \!\!\!\!\!\!\!\!\!\!\!\!\!\!\!\!\! - \beta \right] = \sum_{\substack{m_1 m_1' \\ m_2 m_2'}} \alpha \;\left[\begin{matrix} j_1 m_1 \\ \\ j_2 m_2 \end{matrix}\right]\; \sum_j \left[+ \!\!\!\!\! \begin{matrix} j_1 m_1 & & j_1 m_1' \\ & j & \\ j_2 m_2 & & j_2 m_2' \end{matrix} \!\!\!\!\! - \right]\; \left[\begin{matrix} j_1 m_1' \\ \\ j_2 m_2' \end{matrix}\; \beta \right]$$

$$= \sum_{\substack{m_1 m_1' \\ m_2 m_2'}} \alpha \;\left[\begin{matrix} j_1 & m_1 & m_1' & j_1 \\ & & & \\ j_2 & m_2 & m_2' & j_2 \end{matrix}\right]\; \beta \;=\; \alpha \;\left[\begin{matrix} j_1 \\ \\ j_2 \end{matrix}\right]\; \beta$$

which finally gives the summation rule over j:

$$\sum_j \; \alpha \left[+ \!\!\!\!\! \begin{matrix} j_1 & & j_1 \\ & j & \\ j_2 & & j_2 \end{matrix} \!\!\!\!\! - \right] \beta \;=\; \alpha \left[\begin{matrix} j_1 \\ \\ j_2 \end{matrix}\right] \beta \,. \tag{10.93}$$

The first orthogonality relation is a direct application of relation (10.85) in which $\alpha_j = 1$

$$jm \;\left(+ \!\!\!\!\! \begin{matrix} j_1 \\ \\ j_2 \end{matrix} \!\!\!\!\! - \right)\; j'm' \;=\; \alpha_j \quad jm \;\longrightarrow\; j'm' \;=\; jm \;\longrightarrow\; j'm' \;=\; \delta_{jj'}\,\delta_{mm'}\,. \tag{10.94}$$

6.5 Generalized Clebsch–Gordan Coefficient

The graphical representation of a Generalized Clebsch–Gordan coefficient exhibits the coupled and uncoupled angular momentum:

$$\langle j_1\, m_1 \ldots j_{n-1}\, m_{n-1} \mid (j_1 \ldots j_{n-1})_A\, X_s; j_n\, m_n \rangle = \begin{matrix} j_1 m_1 \!\!\! & \boxed{A} & \\ & \vartriangleleft & \!\!\! j_n m_n \\ j_{n-1} m_{n-1} \!\!\! & \boxed{X_s} & \end{matrix} \,. \tag{10.95}$$

The box contains the coupling mode and the $X_s(s = n-3)$ intermediate momenta. Because the CG coefficients are real numbers, it is also possible to use the Hermitian conjugate to obtain

$$\langle (j_1 \ldots j_{n-1})_A \, X_s; j_n \, m_n \mid j_1 \, m_1 \ldots j_{n-1} \, m_{n-1} \rangle$$

$$= \qquad\qquad\qquad\qquad . \tag{10.96}$$

6.6 Example of Application

We will consider the the graphical representation of the coupling of four angular momenta

$$\vec{J} = \vec{\ell}_1 + \vec{s}_1 + \vec{s}_2 + \vec{\ell}_2 .$$

An X-ray of the black box will show the coupling mode and the intermediate momenta.

a) LS Coupling

$$\tag{10.97}$$

$$A = ((\ell_1 + s_1) + (\ell_2 + s_2)) \qquad X_s = j_1, j_2 \qquad n = 5 \qquad s = n - 3 = 2$$

b) $j\, j$ Coupling

$$\tag{10.98}$$

$$A = ((\ell_1 + \ell_2) + (s_1 + s_2)) \qquad X_s = L, S .$$

c) Orthogonality

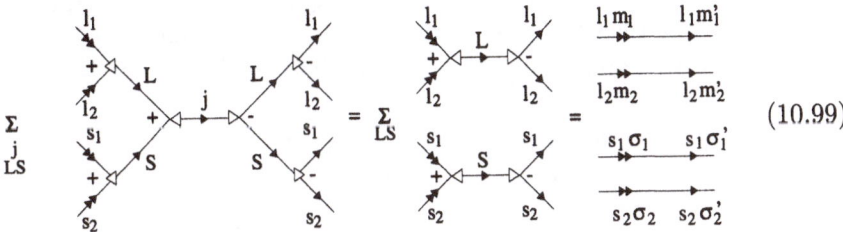

$$\underset{\underset{LS}{j}}{\Sigma} \quad = \quad \underset{LS}{\Sigma} \qquad = \tag{10.99}$$

6.7 Graphical Representation of the $3jm$ Coefficients

$$\begin{pmatrix} j_1 & j_2 & j_3 \\ m_1 & m_2 & m_3 \end{pmatrix} = \quad = \quad . \tag{10.100}$$

A negative magnetic momentum is represented by an inwards-pointing arrow, for example

$$\begin{pmatrix} j_1 & j_2 & j_3 \\ -m_1 & m_2 & m_3 \end{pmatrix} = \quad . \tag{10.101}$$

Next, we introduce the notation of the covariant angular momentum and the corresponding graphical representation

$$(-)^{j_3-m_3} \begin{pmatrix} j_1 & j_2 & j_3 \\ m_1 & m_2 & -m_3 \end{pmatrix} = \begin{pmatrix} j_1 & j_2 & m_3 \\ m_1 & m_2 & j_3 \end{pmatrix} = \quad . \tag{10.102}$$

Symmetries of the $3jm$ Coefficients

a) Change of sign $(-)^{j_1+j_2+j_3}$ for a change in the reading order, or for a change in the direction of the three arrows.

b) Absence of accumulation of magnetic momenta at pole $m_1+m_2+m_3 = 0$.

6.8 Rules for the Transition from a CG to a $3jm$ Coefficient

The graphical representation of relation (10.60) will give

$$\longrightarrow [1](-)^{2(2^-)} \quad . \tag{10.103}$$

We use 1 to designate the angular momentum, sum of the two other angular momenta in the CG coefficient, 1^+ if it is contravariant, 1^- if it is covariant. Similarly, the summed moments will be designated 2 or 3 depending on the reading order in the CG, 2^+ 3^+ if they are contravariant and 2^- 3^- if they are covariant. Expressed graphically, relation (10.61) will then take the following form:

$$\qquad\qquad\qquad [1]\,(\text{-})^{2(3^+)} \qquad\qquad\qquad\qquad\qquad (10.104)$$

To go from a CG to a $3jm$ coefficient we will therefore have to:

i)   introduce a $[1^\pm]$-dimensional coefficient
ii)   introduce a phase coefficient $(-)^{2(2^-\text{ or }3^+)}$
iii)   substitute the triangle pole with a simple one
iv)   change the reading order of the simple pole with respect to the triangle pole

To find out how the rule for summation over J (10.93) is written with $3jm$ coefficient by applying the transition rule, see (11.112) below.

6.9 njm Coefficients

The diagram representing a njm coefficient exhibits all the coupled angular momenta:

$$\begin{pmatrix} j_1 & j_2 & \cdots & j_n \\ m_1 & m_2 & \cdots & m_n \end{pmatrix} |X_s\rangle = \qquad\qquad\qquad (10.105)$$

The box contains the coupling mode and the X_s intermediate momenta.

6.10 $3nj$ Coefficients

These coefficients are not dependent on the magnetic momenta and express the change of coupling mode. They cannot be factorized into simpler coefficients.

$n = 1$ triad

$$\{j_1\ j_2\ j_3\} = \begin{cases} 1 & \text{if} \quad \vec{j}_1 + \vec{j}_2 + \vec{j}_3 = 0 \\ 0 & \text{otherwise}. \end{cases}$$

Its analytical statement uses two completely contracted $3jm$ coefficients.

$$\{j_1 \, j_2 \, j_3\} = \sum_{m_1 \, m_2 \, m_3} \begin{pmatrix} j_1 & j_2 & j_3 \\ m_1 & m_2 & m_3 \end{pmatrix} \begin{pmatrix} m_1 & m_2 & m_3 \\ j_1 & j_2 & j_3 \end{pmatrix}$$

$$= + \qquad (10.106)$$

$n = 2$ $6j$ coefficient

$$\begin{Bmatrix} j_1 & j_2 & j_3 \\ \ell_1 & \ell_2 & \ell_3 \end{Bmatrix} = \sum_{m_1 \, m_2 \, m_3 \, n_1 \, n_2 \, n_3} \begin{pmatrix} j_1 & j_2 & j_3 \\ m_1 & m_2 & m_3 \end{pmatrix} \begin{pmatrix} m_1 & n_2 & \ell_3 \\ j_1 & \ell_2 & n_3 \end{pmatrix}$$
$$\begin{pmatrix} n_3 & \ell_1 & m_2 \\ \ell_3 & n_1 & j_2 \end{pmatrix} \begin{pmatrix} n_1 & \ell_2 & m_3 \\ \ell_1 & n_2 & j_3 \end{pmatrix}$$

$$= + \qquad + = \qquad = \qquad (10.107)$$

The 6_j coefficient has the symmetry properties:

i)   invariant under circular permutation of columns,
ii) invariant under permutation of two elements of the first line with two elements of the second.

$n = 3$ $9j$ coefficient

$$\begin{Bmatrix} j_1 & j_2 & j_3 \\ \ell_1 & \ell_2 & \ell_3 \\ k_1 & k_2 & k_3 \end{Bmatrix} = +$$

$$\sum_{m_i \, n_i \, x_i} \begin{pmatrix} j_1 & j_2 & j_3 \\ m_1 & m_2 & m_3 \end{pmatrix} \begin{pmatrix} m_2 & n_2 & x_2 \\ j_2 & \ell_2 & k_2 \end{pmatrix} \begin{pmatrix} k_2 & k_3 & k_1 \\ x_2 & x_3 & x_1 \end{pmatrix}$$
$$\begin{pmatrix} x_3 & m_3 & n_3 \\ k_3 & j_3 & \ell_3 \end{pmatrix} \begin{pmatrix} \ell_1 & \ell_2 & \ell_3 \\ n_1 & n_2 & n_3 \end{pmatrix} \begin{pmatrix} n_1 & x_1 & m_1 \\ \ell_1 & k_1 & j_1 \end{pmatrix} . \qquad (10.108)$$

The properties of symmetry are apparent in the diagram:

i) invariance under reflection in a diagonal,

ii) multiplied by $(-)^{\sum_i^3 j_i + \ell_i + k_i}$ for an exchange of two rows or columns.

Owing to the latter property, a $9j$ coefficient with two equal rows or columns will be zero if the sum of the remaining moments is odd:

$$\left\{ \begin{array}{ccc} j_1 & j_2 & j_3 \\ j_1 & j_2 & j_3 \\ k_1 & k_2 & k_3 \end{array} \right\} = 0 \quad \text{if} \quad k_1 + k_2 + k_3 \quad \text{is odd}. \tag{10.109}$$

The $9j$ coefficient can be expressed as the sum of the products of three $6j$ coefficients as can be seen by using the rule for summation over j (10.93).

The transformation matrix for making the transition from the coupling mode $(\vec{\ell}_1 + \vec{\ell}_2) + (\vec{s}_1 + \vec{s}_2)$ to $(\vec{\ell}_1 + \vec{s}_1) + (\vec{\ell}_2 + \vec{s}_2)$ is a $9j$ coefficient, the graphical representation of (10.80):

$$|\,(\ell_1\,\ell_2)\,L\,(s_1\,s_2)|S, JM\rangle = \sum_{j_1\,j_2} [LS\,j_1\,j_2] \left\{ \begin{array}{ccc} j_1 & j_2 & J \\ \ell_1 & \ell_2 & L \\ s_1 & s_2 & S \end{array} \right\}$$
$$|\,(\ell_1\,s_1)\,(\ell_2\,s_2)\,j_2, JM\rangle. \tag{10.110}$$

Beyond $n = 3$, there are several possible determinations of the $3nj$ coefficients and we are thus obliged to introduce two "types" of $12j$ coefficients and three "types" of $15j$ coefficients. There is a graphical procedure (Elbaz and Castel, 1972; Elbaz, 1985) for simplifying the definition and manipulation of these coefficients. Let us also draw attention to the existence of $6j$ and $9j$ coefficient tables and the fact that other $3nj$ coefficients are worked out as the sum of the products of the $6j$ and $9j$ coefficients.

To identify a $3nj$ coefficient in its standard form, we need to:

i) verify the sign of the poles and return one $(j_1\,j_2\,j_3)$ pole to its standard sign by multiplying the result by the phase $(-)^{j_1 + j_2 + j_3}$,

ii) verify the reading order of the kinetic lines j, and return one j line to its standard form by multiplying the result by the phase $(-)^{2j}$.

Finally, we notice that a zero angular momentum $j = 0$ is expanded graphically by simply removing the corresponding j line .

6.11 Example

Analytical expression of a diagram.

$$= (-)^{j_2+\ell_2+\ell_3} \; (-)^{j_1+\ell_1+\ell_3+}$$

$$= (-)^{j_1+j_2+\ell_1+\ell_2+2\ell_3} \; (-)^{2\ell_2} \; (-)^{2\ell_1} +$$

$$= (-)^{-j_1+j_2+\ell_1-\ell_2} \begin{Bmatrix} j_2 & j_1 & j_3 \\ \ell_1 & \ell_2 & \ell_3 \end{Bmatrix}$$

Appendix 1: Table of Selected $3jm$ and $6j$ Coefficients

$$\begin{pmatrix} j_1 & j_2 & j_3 \\ 0 & 0 & 0 \end{pmatrix} = (-1)^{1/2\,J} \left[\frac{(j_1+j_2-j_3)!(j_1+j_3-j_2)!(j_2+j_3-j_1)!}{(j_1+j_2+j_3 1)!} \right]^{1/2}$$

$$\times \; \frac{(\tfrac{1}{2}J)!}{(\tfrac{1}{2}J-j_1)!(\tfrac{1}{2}J-j_2)!(\tfrac{1}{2}J-j_3)!} \quad \text{if } J \begin{cases} = j_1+j_2+j_3 & \text{is even} \\ = 0 & \text{otherwise} \end{cases}$$

$$\begin{pmatrix} J+\tfrac{1}{2} & J & \tfrac{1}{2} \\ M & -M-\tfrac{1}{2} & \tfrac{1}{2} \end{pmatrix} = (-1)^{J-M-\tfrac{1}{2}} \left[\frac{J-M+\tfrac{1}{2}}{(2J+2)(2J+1)} \right]^{1/2}$$

$$\begin{pmatrix} J+1 & J & 1 \\ M & -M-1 & 1 \end{pmatrix} = (-1)^{J-M-1} \left[\frac{(J-M)(J-M+1)}{(2J+3)(2J+2)(2J+1)} \right]^{1/2}$$

$$\begin{pmatrix} J+1 & J & 1 \\ M & -M & 0 \end{pmatrix} = (-1)^{J-M-1} \left[\frac{(J+M+1)(J-M+1)2}{(2J+3)(2J+2)(2J+1)} \right]^{1/2}$$

$$\begin{pmatrix} J & J & 1 \\ M & -M-1 & 1 \end{pmatrix} = (-1)^{J-M} \left[\frac{(J-M)(J+M+1)2}{(2J+2)(2J+1)(2J)} \right]^{1/2}$$

$$\begin{pmatrix} J & J & 1 \\ M & -M & 0 \end{pmatrix} = (-1)^{J-M} \frac{M}{(2J+1)(J+1)J]^{1/2}}$$

$$\begin{Bmatrix} a & b & c \\ 1 & c-1 & b-1 \end{Bmatrix} = (-)s \left[\frac{s(s+1)(s-2a-1)(s-2a)}{(2b-1)2b(2b+1)(2c-1)2c(2c+1)} \right]^{1/2}$$

$$\begin{Bmatrix} a & b & c \\ 1 & c-1 & b \end{Bmatrix} = (-)s \left[\frac{2(s+1)(s-2a)(s-2b)(s-2c+1)}{2b(2b+1)(2b+2)(2c-1)2c(2c+1)} \right]^{1/2}$$

$$\begin{Bmatrix} a & b & c \\ 1 & c-1 & b+1 \end{Bmatrix} = (-)s \left[\frac{(s-2b-1)(s-2b)(s-2c+1)(s-2c+2)}{(2b+1)(2b+2)(2b+3)(2c-1)2c(2c+1)} \right]^{1/2}$$

$$\begin{Bmatrix} a & b & c \\ 1 & c & b \end{Bmatrix} = (-)s+1 \frac{2[b(b+1)+c(c+1)-a(a+1)]}{[2b(2b+1)(2b+2)(2c(2c+1)(2c+2)]^{1/2}}$$

$$s = a+b+c$$

Appendix 2: Diquark

B. Silvestre-Brac has computed the probability of the formation of a diquark in a baryon comprising three colored quarks (Silvestre-Brac, 1987; Badhuri et al., 1981). The wave function with the $C = (1, 2, 3)$ color is an antisymmetric function of the form

$$C = \frac{1}{\sqrt{6}} (r\,g\,y + g\,y\,r + y\,r\,g - r\,y\,g - y\,g\,r - g\,r\,y).$$

By coupling the spin $\frac{1}{2}$ of each of the constituents, we obtain spin 0 or 1 for the diquark, which corresponds to a wave function with symmetric or antisymmetric spin for the exchange of particles 2 and 3:

$$\chi^{1/2}_{S_{12}=0} = |\,(1/2\ 1/2)_0\ 1/2\rangle_{1/2} = \frac{1}{\sqrt{2}} (\uparrow\uparrow\downarrow - \uparrow\downarrow\uparrow)\quad \text{antisymmetric}$$

$$\chi^{1/2}_{S_{12}=1} = |\,(1/2\ 1/2)_1\ 1/2\rangle_{1/2} = \frac{1}{\sqrt{6}} (\uparrow\uparrow\downarrow + \uparrow\downarrow\uparrow - 2\downarrow\uparrow\uparrow)\quad \text{symmetric}$$

or symmetric for the exchange of the three particles:

$$\chi^{3/2}_{S_{12}=1} = |\,(1/2\ 1/2)_1\ 1/2\rangle_{3/2} = \frac{1}{\sqrt{3}} (\uparrow\uparrow\downarrow + \uparrow\downarrow\uparrow + \downarrow\uparrow\uparrow)\quad \text{symmetric}.$$

Each quark is characterized by a position $\vec{r}_i$. The Hamiltonian is invariant under space translation, making it possible to separate the motion of the center of mass $\vec{R} = M^{-1} \sum_i m_i \vec{r}_i$ with $M = \sum_i m_i$ in the form of a plane wave $\exp\left(\frac{i}{\hbar} \vec{P}. \vec{R}\right)$ with $\vec{P} = \sum_i \vec{p}_i$, from the relative motion by setting:

$$\vec{p} = [2m_2 \, m_3 \, (m_2 + m_3)]^{-\frac{1}{2}} \, (m_3 \, \vec{p}_2 - m_2 \, \vec{p}_3) ,$$

$$\vec{q} = [2m_1 \, (m_2 + m_3) \, M]^{-\frac{1}{2}} \, (m_1 \, (\vec{p}_2 + \vec{p}_3) - (m_2 + m_3) \, \vec{p}_1) ,$$

$$\vec{x} = [2m_2 \, m_3/(m_2 + m_3)]^{\frac{1}{2}} \, (\vec{r}_2 - \vec{r}_3) ,$$

$$\vec{y} = [2m_1 \, (m_2 + m_3)/M]^{\frac{1}{2}} \left(\frac{(m_2 \, \vec{r}_2 + m_3 \, \vec{r}_3)}{m_2 + m_3} - \vec{r}_1 \right) .$$

The interaction potential between a diquark, which behaves like an antiquark from the point of view of color, and a quark can be modeled as follows using the potential of Badhuri et al. which depends on the color $\vec{\lambda}_i$ and the spin $\vec{\sigma}_i$:

$$V = -\frac{16}{3} \sum_{i \langle j} \vec{\lambda}_i \, \vec{\lambda}_j \left[-\frac{\chi}{r_{ij}} + \frac{r_{ij}}{a^2} - D + \frac{\hbar^2 \, \chi}{m_i \, m_j} \frac{\exp\left(-r_{ij}/r_0\right)}{r_0^2 \, r_{ij}} \vec{\sigma}_i \cdot \vec{\sigma}_j \right]$$

with the following parameters:

$$\chi = 102.67 \text{ MeV fm} \qquad D = 913.5 \text{ MeV}$$
$$a = 0.0326 \text{ (MeV}^{-1} \text{ fm)}^{1/2} \qquad r_0 = 112.2 \text{ fm} .$$

We are thus led to the solution of the Schrödinger equation:

$$\left[\frac{1}{2} \, (p^2 + q^2) + \sum_{i \langle j} V_{ij} \, (x, \, y) \right] \psi \, (1, \, 2, \, 3) = E \, \psi \, (1, \, 2, \, 3) .$$

The conclusion to be drawn from the above is that in a baryon $(q \, Q \, Q)$ for weak angular momenta, a diquark QQ in the state $\ell = 0$ is formed in a state with spin 1, whereas a diquark with two different quarks is for three quarters of the time $S = 0$ and for the remaining quarter $S = 1$.

Chapter 11
Rotations

1. Introduction

The rotation around a point O is defined as an overall movement of every point in the space during which the point O remains fixed. Each point P will in turn occupy a new position P' and the correspondence between points P and P' is a one-to-one relation. In this correspondence, point O corresponds to itself: lengths and angles as well as the direction of the trihedrons are all conserved.

Let $Oxyz$ be a given cartesian coordinate system and $OXYZ$ an another cartesian coordinate system obtained by performing three successive rotations in a specific sequence:

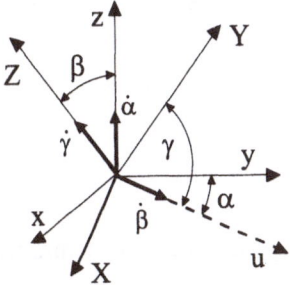

We define the rotations

$R_z(\alpha)$ through an angle α around Oz leading to Oy in Ou,
$R_u(\beta)$ through an angle β around Ou leading to Oz in OZ,
$R_Z(\gamma)$ through an angle γ around OZ leading to Ou in OY.

The overall rotation of the $Oxyz$ coordinate system will therefore be defined by the operator:

$$R(\alpha,\ \beta,\ \gamma) = R_Z(\gamma)\ R_u(\beta)\ R_z(\alpha). \tag{11.1}$$

2. Infinitesimal Rotations

A rotation of the system of axes around the Oz axis through an angle α will transform a function $\Psi_a(x,\ y,\ z)$ into a function $\Psi'_a(x,\ y,\ z)$, which can be written thus:

$$\Psi'_a\ (\vec{r}) = [\mathcal{R}\ \Psi_a]\ (\vec{r}) = \Psi_a\ (R^{-1}\ \vec{r})\,. \tag{11.2}$$

There is a one-to-one linear correspondence between the functions Ψ_a and Ψ'_a, making it possible to write

$$\Psi'_a = R\ \Psi_a\,. \tag{11.3}$$

The operator R must be unitary because the norm of the functions Ψ_a and Ψ'_a must be identical, that is, conserved during the rotation.

In the rotation of the system of axes through the angle α around the Oz axis, we will therefore write

$$R_z(\alpha)\ [\Psi_a\ (x,\ y,\ z)] = \Psi_a\ (x',\ y',\ z') \tag{11.4}$$

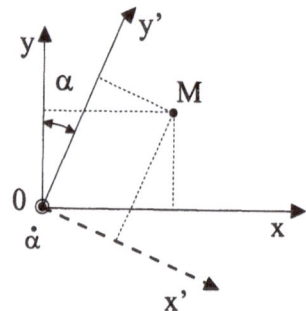

with transformations of the Cartesian coordinates:

$$\begin{aligned}
x' &= x\ \cos\ \alpha + y\ \sin\ \alpha\,, \\
y' &= -x\ \sin\ \alpha + y\ \cos\ \alpha\,, \\
z' &= z\,.
\end{aligned} \tag{11.5}$$

For an infinitesimal rotation ε, we therefore obtain:

$$R_z(\alpha)\ \Psi_a(x,\ y,\ z) = \Psi_a(x + y\ \varepsilon, -x\ \varepsilon + y, z) \tag{11.6}$$

and, by carrying out a Taylor expansion around the point $(x,\ y,\ z)$

$$R_z(\varepsilon)\ \Psi_a(x,\ y,\ z) = \Psi_a(x,\ y,\ z) + \varepsilon\ \left(y\ \frac{\partial\ \Psi}{\partial\ x} - x\ \frac{\partial\ \Psi}{\partial\ y} \right)$$

$$= \left[\mathbb{1} + \varepsilon\ \left(y\ \frac{\partial}{\partial\ x} - x\ \frac{\partial}{\partial\ y} \right) \right]\ \Psi_a\ (x,\ y,\ z)\,. \tag{11.7}$$

We can introduce the operator L_z, the projection of the orbital angular momentum onto the Oz axis, by applying the correspondence principle of quantum mechanics:

$$x \frac{\partial}{\partial y} - y \frac{\partial}{\partial x} = i L_z , \qquad (11.8)$$

which leads us to write statement (11.7) as follows:

$$R_z(\varepsilon) \, \Psi_a(x, \, y, \, z) = (1 - i \, \varepsilon \, L_z) \, \Psi_a(x, \, y, \, z) . \qquad (11.9)$$

We can infer from this the expression for the operator $R_z(\varepsilon)$:

$$R_z(\varepsilon) = 1 - i \, \varepsilon \, L_z . \qquad (11.10)$$

If an infinitesimal rotation ε had been made around the axis Ou of the unitary vector $\vec{u}$, we would have similarly obtained

$$R_u(\varepsilon) = 1 - i \, \varepsilon \, \vec{L} \cdot \vec{u} . \qquad (11.11)$$

We generalize the relation to make allowance for the case where the particle under consideration possesses an angular momentum $\vec{J}$ that is not necessarily orbital:

$$R_u(\varepsilon) = 1 - i \, \varepsilon \, \vec{J} \cdot \vec{u} \qquad (11.12)$$

3. Finite Rotations

3.1 Finite Rotations of the System of Coordinates

If we make a rotation around the Ou axis through an angle $\varphi + \Delta\varphi$ or successively two rotations around the same axis with angle $\Delta\varphi$ and angle φ, we obtain by construction the same result. We can therefore write

$$R_u(\varphi + \Delta\,\varphi) = R_u(\varphi) \, R_u(\Delta\,\varphi) = R_u(\Delta\,\varphi) \, R_u(\varphi) \qquad (11.13)$$

and, by using (11.8) for the infinitesimal rotation with angle $\Delta\,\varphi$,

$$R_u(\varphi + \Delta\,\varphi) = R_u(\varphi) \, (1 - i \, J_u \, \Delta\,\varphi) \qquad (11.14)$$

which can also be written as follows:

$$\frac{R_u(\varphi + \Delta\,\varphi) - R_u(\varphi)}{\Delta\,\varphi} = -i \, J_u \, R_u(\varphi) .$$

If $\Delta\,\varphi \to 0$, the first term of this equation will define the derivative with respect to φ of the operator $R_u\,(\varphi)$:

$$\frac{d \, R_u(\varphi)}{d \, \varphi} = -i \, J_u \, R_u(\varphi) . \qquad (11.15)$$

This is a first-order differential equation with separated variables and which directly leads to the result

$$R_u(\varphi) = R_u(0) \exp\left(-i\,\varphi\,J_u\right) \tag{11.16}$$

and, because a rotation with angle zero is the same as no rotation at all,

$$R_u(0) = 1 \quad \text{and} \quad R_u(\varphi) = \exp\left(-i\,\varphi\,J_u\right). \tag{11.17}$$

We can further calculate the operator $R(\alpha,\ \beta,\ \gamma)$ which makes the transition from the $Oxyz$ system to the $OXYZ$ system:

$$
\begin{aligned}
R(\alpha,\ \beta,\ \gamma) &= R_Z(\gamma)\,R_u(\beta)\,R_z(\alpha) \\
&= \exp\left(-i\,\gamma\,J_Z\right)\,\exp\left(-i\,\beta\,J_u\right)\,\exp\left(-i\,\alpha\,J_z\right). \tag{11.18}
\end{aligned}
$$

We can only know the components of the angular momentum via the $Oxyz$ axes; it is thus necessary to transform J_Z and J_u and to return to J_x, J_y or J_z.

3.2 Rotation of the Physical System. Rotation of Observables

A rotation can be described either by causing the axes of the system of coordinates to rotate while holding the physical objects fixed or by holding the axes fixed and causing the inverse rotation of the physical system.

Let us use $g = (\alpha,\ \beta,\ \gamma)$ to represent a rotation parametrized with the Euler angles $(\alpha,\ \beta,\ \gamma)$. If we make the inverse rotation $R^{-1}(g)$, this will amount to making a rotation R but with the Euler angles $g^{-1} = (-\gamma, -\beta, -\alpha)$:

$$R^{-1}(g) = R\left(g^{-1}\right) = R(\bar{g}). \tag{11.19}$$

The application of the rotation $R(g)$ to a dynamic state $|\,a\rangle$ will yield a state $|\,a'\rangle$:

$$|\,a'\rangle = R(g)\,|\,a\rangle, \tag{11.20}$$

and by taking the Hermitian conjugate

$$\langle a'\,| = \langle a\,|\,R^+(g). \tag{11.21}$$

The operator R is unitary since it conserves the norm of the vectors as we can see in (11.20) and (11.21):

$$\boxed{R(g)\,R^+(g) = R^+(g)\,R(g) = 1} . \tag{11.22}$$

The mean value of an observable Q (that is, the result of the measurement of the observable Q) is independent of the system of reference adopted and, hence, identical in the two systems $Oxyz$ and $OXYZ$. Thus

$$\langle a \mid Q \mid a \rangle \; = \; \langle a' \mid Q' \mid a' \rangle. \tag{11.23}$$

By using (11.19) and (11.20), we obtain,

$$\langle a \mid Q \mid a \rangle \; = \; \langle a \mid R^+(g)\, Q'\, R(g) \mid a \rangle \tag{11.24}$$

irrespective of the value of $\mid a \rangle$, and, hence, the relation between operators Q and Q':

$$Q = R^+(g)\, Q'\, R(g) \qquad \text{or} \quad Q' = R(g)\, Q\, R^+(g). \tag{11.25}$$

By fundamental rules of the rotation of physical systems we shall be referring to the two relations that allow the rotation of a quantum state or an operator:

$$\boxed{\begin{aligned} \mid a' \rangle &= R(g) \mid a \rangle \\ Q' &= R(g)\, Q\, R^+(g) \end{aligned}} \tag{11.26}$$

An observable is invariant under rotation if $Q' = Q$.

By right-multiplying by R we have $Q\,R(g) = R(g)\,Q$, that is $[Q, R(g)] = 0$.

Any observable represented by a Hermitian operator commuting with the rotation operator R is said to be invariant under rotation. This is particularly true of scalar operators.

3.3 Determination of $R\,(\alpha,\,\beta,\,\gamma)$

We can now evaluate the operator of the overall rotation of the system of coordinates:

$$\begin{aligned} R(\alpha,\,\beta,\,\gamma) &= R_Z(\gamma)\, R_u(\beta)\, R_z(\alpha) \\ &= \exp(-i\,\gamma\, J_Z)\, \exp\left(-i\,\beta\, J_u\right)\, \exp\left(-i\,\alpha\, J_z\right). \end{aligned} \tag{11.27}$$

The operator J_u is obtained by rotation around Oz through an angle α of the Hermitian operator J_y, that is of the system of coordinates defining the projection J_y of the operator $\vec{J}$, yielding (11.26):

$$J_u = R_z(\alpha)\, J_y\, R_z^+(\alpha), \tag{11.28}$$

or by direct application on the rotation operator:

$$R_u(\beta) = R_z(\alpha)\, R_y(\beta)\, R_z^+(\alpha). \tag{11.29}$$

The operator $R_Z(\gamma)$ is obtained from $R_z(\gamma)$ by first applying a rotation $R_z(\alpha)$ and then a rotation $R_u(\beta)$, that is,

$$R_Z(\gamma) = \left[R_z(\alpha)\ R_y(\beta)\ R_z^+(\alpha)\right] \left[R_z(\alpha)\ R_z(\gamma)\ R_z^+(\alpha)\right] \left[R_z(\alpha)\ R_y^+(\beta)\ R_z^+(\alpha)\right]$$
$$= R_z(\alpha)\ R_y(\beta)\ R_z(\gamma)\ R_y^+(\beta)\ R_z^+(\alpha). \tag{11.30}$$

By inserting over (11.29) and (11.30) into $R(\alpha,\ \beta,\ \gamma)$ defined with relation (11.27), we obtain the following result:

$$R(\alpha,\beta,\gamma) = \left[R_z(\alpha)\ R_y(\beta)\ R_z(\gamma)\ R_y^+(\beta)\ R_z^+(\alpha)\right] \left[R_z(\alpha)\ R_y(\beta)\ R_z^+(\alpha)\right] R_z(\alpha)$$
$$= R_z(\alpha)\ R_y(\beta)\ R_z(\gamma). \tag{11.31}$$

The operator for the overall rotation of the system of coordinates denoted $R(\bar{g})$ since the rotation of the physical system is denoted $R(g)$ in (11.20) can be written as follows:

$$R(\bar{g}) = R^+(g) = \exp\left(-i\ \alpha\ J_z\right)\ \exp\left(-i\ \beta\ J_y\right)\ \exp\left(-i\ \gamma\ J_z\right). \tag{11.32}$$

4. Standard Representation

4.1 Rotation Matrix

The form (11.32) of the rotation operator shows that the squared length J^2 of the angular momentum is invariant under rotation, since

$$[J^2, R(g)] = 0. \tag{11.33}$$

Hence, a rotation $R(g)$ does not remove a vector $|\ j\ m\rangle$ from the subspace ξ_j:

$$J^2\ R(g)\ |\ jm\rangle\ =\ R(g)\ J^2\ |\ j\ m\rangle\ =\ R(g)\ j\ (j+1)\ |\ j\ m\rangle$$
$$=\ j\ (j+1)\ R(g)\ |\ j\ m\rangle \tag{11.34}$$

and $R(g)\ |\ j\ m\rangle$, the eigenvector of J^2 (corresponding to the eigenvalue $j\ (j+1)$) is a vector in ξ_j, a linear combination of vectors with the same j eigenvalue but with different m eigenvalues. We therefore write

$$R(g)\ |\ j\ m\rangle\ =\ \sum_{m'} D_{m'\ m}^j\ (g)\ |\ j\ m'\rangle. \tag{11.35}$$

The projection of $R(g)\ |\ j\ m\rangle$ on ξ_j will immediately yield the expression

$$R(g)\ |\ j\ m\rangle = \sum_{m'}\ |\ j\ m'\rangle\ \langle j\ m'\ |\ R(g)\ |\ j\ m\rangle. \tag{11.36}$$

By identifying (11.35) and (11.36), we therefore obtain

$$D^j_{m'\,m}(g) = \langle j\,m' \mid R(g) \mid j\,m \rangle . \qquad (11.37)$$

Replacing the operator $R(g)$ with its value (11.32), we will write

$$D^j_{m'\,m}(\alpha,\,\beta,\,\gamma) = \langle j\,m' \mid \exp(i\gamma\,J_z)\,\exp(i\,\beta\,J_y)\,\exp(i\,\alpha\,J_z) \mid j\,m \rangle$$
$$= \exp(i\,\gamma\,m')\,\exp(i\,\alpha\,m)\,\langle j\,m' \mid \exp(i\,\beta\,J_y) \mid j\,m \rangle . \qquad (11.38)$$

4.2 Reduced Rotation Matrix

By reduced rotation matrix we shall be referring to the matrix element

$$\langle j\,m' \mid \exp(i\,\beta\,J_y) \mid j\,m \rangle = d^j_{m'\,m}(\beta) . \qquad (11.39)$$

This means that the matrix $D^j_{m'\,m}(\alpha,\,\beta,\,\gamma)$ will be known as soon as we are able to determine $d^j_{m'\,m}(\beta)$, since

$$D^j_{m'\,m}(\alpha,\,\beta,\,\gamma) = \exp(i\,\gamma\,m')\,d^j_{m'\,m}(\beta)\,\exp(i\,\alpha\,m) . \qquad (11.40)$$

It is easy to calculate $d^j_{m'\,m}(\beta)$ with the value $(1/2i)\,(J_+ - J_-)$ of J_y and by causing these operators to act on the states of the angular momentum $(j\,m)$ and $(j\,m')$.

For a given value j, the most general form of $d^j_{m'\,m}(\beta)$ will be

$$d^j_{m'\,m}(\beta) = \sum_{t=0} (-)^t \frac{[(j+m')\,!\,(j-m')\,!\,(j+m)\,!\,(j-m)\,!\,]^{1/2}}{(j+m'-t)\,!\,(j-m-t)\,!\,(t+m-m')\,!\,t\,!}$$
$$\times (\cos \frac{\beta}{2})^{2j+m'-m-2t}\,(\sin \frac{\beta}{2})^{2t+m-m'} . \qquad (11.41)$$

As an example, we will use the calculation of the matrix elements $d^{1/2}_{m'\,m}(\beta)$. If $m = 1/2$, then $m' = 1/2$:

$$\begin{array}{ll}
(j+m')\,! = 1\,! = 1 & (j+m)\,! = 1 \\
(j-m')\,! = 0\,! = 1 & (j-m)\,! = 0\,! = 1 \\
(j-m-t)\,! = (-t)\,! \ \text{implies} \ t = 0 \ \text{and} \ t\,! = 1 \\
(j+m-t)\,! = (1-t)\,! = 1 \\
(t+m-m')\,! = t\,! = 1 .
\end{array} \qquad (11.42)$$

We thus obtain the matrix element

$$d^{1/2}_{1/2\,1/2}(\beta) = \cos \frac{\beta}{2}, \qquad (11.43)$$

and similarly

$$d^{1/2}_{-1/2\,-1/2}(\beta) = \cos\frac{\beta}{2}$$

$$d^{1/2}_{1/2\,-1/2}(\beta) = -d^{1/2}_{-1/2\,1/2}(\beta) = \sin\frac{\beta}{2}.$$

(11.44)

The reduced rotation matrix for a particle with spin $1/2$ will therefore be of the form

$$d^{1/2}_{m'\,m}(\beta) = \begin{pmatrix} \cos\beta/2 & \sin\beta/2 \\ -\sin\beta/2 & \cos\beta/2 \end{pmatrix}.$$

(11.45)

The following table gives a few examples of reduced rotation matrices:

	$d^{1/2}_{1/2,1/2} = \cos\frac{\theta}{2}$	$d^{1/2}_{1/2,-1/2} = \sin\frac{\theta}{2}$
$d^{3/2}_{3/2,3/2} = \frac{1+\cos\theta}{2}\cos\frac{\theta}{2}$	$d^{1}_{1,1} = \frac{1+\cos\theta}{2}$	$d^{1}_{1,0} = -\frac{\sin\theta}{\sqrt{2}}$
$d^{3/2}_{3/2,1/2} = -\sqrt{3}\frac{1+\cos\theta}{2}\sin\frac{\theta}{2}$	$d^{2}_{2,2} = \left(\frac{1+\cos\theta}{2}\right)^2$	$d^{1}_{1,-1} = \frac{1-\cos\theta}{2}$
$d^{3/2}_{3/2,-1/2} = \sqrt{3}\frac{1-\cos\theta}{2}\cos\frac{\theta}{2}$	$d^{2}_{2,1} = -\frac{1+\cos\theta}{2}\sin\theta$	$d^{1}_{0,0} = \cos\theta$
$d^{3/2}_{3/2,-3/2} = -\frac{1-\cos\theta}{2}\sin\frac{\theta}{2}$	$d^{2}_{2,0} = \frac{\sqrt{6}}{4}\sin^2\theta$	$d^{2}_{1,1} = \frac{1+\cos\theta}{2}(2\cos\theta - 1)$
$d^{3/2}_{1/2,1/2} = \frac{3\cos\theta-1}{2}\cos\frac{\theta}{2}$	$d^{2}_{2,-1} = -\frac{1-\cos\theta}{2}\sin\theta$	$d^{2}_{1,0} = \sqrt{\frac{3}{2}}\sin\theta\cos\theta$
$d^{3/2}_{1/2,-1/2} = -\frac{3\cos\theta+1}{2}\sin\frac{\theta}{2}$	$d^{2}_{2,-2} = \left(\frac{1-\cos\theta}{2}\right)^2$	$d^{2}_{1,-1} = \frac{1\cos\theta}{2}(2\cos\theta + 1)$
		$d^{2}_{0,0} = (\frac{3}{2}\cos^2\theta - \frac{1}{2})$

4.3 Symmetry Properties

The analytical form (11.40) of the reduced matrix enables us to infer certain properties:

i) $d^j_{m'\,m}(\beta) = d^j_{m'\,m}(\beta)^*$ reduced matrices are real,

ii) $d^j_{m'\,m}(\beta) = d^j_{-m-m'}(\beta)$ (11.46)

iii) $d^j_{m'\,m}(\beta) = (-)^{j-m+j-m'}\, d^j_{m\,m'}(\beta) = (-)^{m-m'}\, d^j_{m\,m'}(\beta)$.

These properties yield the conjugate of a rotation matrix:

$$D^j_{m'\,m}{}^*(\alpha,\,\beta,\,\gamma) = (-)^{j-m+j-m'}\, D^j_{-m'-m}(\alpha,\,\beta,\,\gamma). \qquad (11.47)$$

In effect, the definition of the complex conjugate gives:

$$D^j_{m'\,m}{}^*(\alpha,\,\beta,\,\gamma) = \left(\exp\left(+i\,\gamma\,m'\right) d^j_{m'\,m}(\beta)\, \exp\left(+i\,\alpha\,m\right)\right)^*$$

$$= \exp\left(-i\,\gamma\,m'\right)\, \exp\left(-i\,\alpha\,m\right)\, d^j_{m'\,m}{}^*(\beta)$$

and property iii) then enables us to obtain the result (11.47):

$$D^j_{m'm}{}^*(\alpha,\beta,\gamma) = \Big[\exp\left(+i\,\gamma(-m')\right)\, d^j_{-m'-m}(\beta)\, \exp\left(+i\,\alpha(-m)\right)\Big]$$

$$\times\,(-)^{j-m+j-m'}.$$

4.4 Inverse Rotation

Because the operator R is unitary, it is possible to calculate the matrix elements of R^{-1} on the basis of the matrix elements of R:

$$\langle j\,m' \mid R^{-1}(g) \mid j\,m\rangle \;=\; \langle j\,m' \mid R^+(g) \mid j\,m\rangle \;=\; \langle j\,m \mid R(g) \mid j\,m'\rangle^*$$

$$= D^j_{m\,m'}{}^*(g)$$

$$= \langle j\,m' \mid R(g^{-1}) \mid j\,m\rangle \;=\; D^j_{m'\,m}(g^{-1}). \quad (11.48)$$

We therefore notice the following equivalence:

$$D^j_{mm'}{}^*(g) = D^j_{m'm}(g^{-1}) = D^j_{m'm}(\bar g). \qquad (11.49)$$

4.5 Closure Relation

The fact that R is unitary can be used to obtain a completeness relation for the rotation matrices:

$$\langle\,j\,m' \mid R(g)\,R^+(g) \mid j\,m\rangle \;=\; \langle j\,m' \mid j\,m\rangle \;=\; \delta_{m\,m'}$$

$$= \sum_{m''}\langle jm'|R(g)|\,j\,m''\rangle\langle j\,m''|R^+(g)|j\,m\rangle.$$

$$\qquad (11.50)$$

We only have to introduce the rotation matrices to obtain the completeness relation

$$\delta_{mm'} = \sum_{m''} D^j_{m'm''}(g)\, D^j_{mm''}(g)^* = \sum_{m''} D^j_{m'm''}(g)\, D^j_{m''m}(\bar{g}). \quad (11.51)$$

4.6 Orthonormalization Relation

The Euler angles $g = (\alpha,\ \beta,\ \gamma)$ have the following limit:

$$0 \le \alpha \le 2\pi, \qquad 0 \le \beta \le \pi \quad \text{and} \quad 0 \le \gamma \le 2\pi, \qquad (11.52)$$

and we write dg, the volume element of the rotation group, (Wigner, 1959), $dg = d\alpha \, \sin\beta \, d\beta \, d\gamma$. The integration of dg over the entire space defines what is generally termed the volume G of the rotation group, that is,

$$G = \int dg = \int d\alpha \, \sin\beta \, d\beta \, d\gamma = 8\pi^2. \qquad (11.53)$$

The integration of the product of the two rotation matrices over the group will yield the orthonormalization relation

$$\int D^j_{m\, m'}(g)\, D^{j'}_{n\, n'}(g)^* dg = \frac{8\pi^2}{[j^2]}\, \delta_{j\, j'}\, \delta_{m\, n}\, \delta_{m'\, n'}. \qquad (11.54)$$

4.7 Direct Product of Rotations

The $(2j_1 + 1)\,(2j_2 + 1)$ states in which j_1 and j_2 are independent form the irreducible representation base of the direct product of the rotation operators $R_{j_1}(g)$ and $R_{j_2}(g)$ of the rotation group, that is

$$(R_{j_1} \otimes R_{j_2})\ |\, j_1\, m_1\, j_2\, m_2\rangle = \sum_{m'_1\, m'_2} |\, j_1\, m'_1\, j_2\, m'_2\rangle$$
$$\times \langle j_1\, m'_1\, j_2\, m'_2 |\, R_{j_1}(g) \otimes R_{j_2}(g)\, |\, j_1\, m_1\, j_2\, m_2\rangle \qquad (11.55)$$

or by separating the subspaces of the direct product:

$$\langle j_1\, m'_1\, j_2\, m'_2 |\, R_{j_1} \otimes R_{j_2}\, |\, j_1\, m_1\, j_2\, m_2\rangle$$
$$= \langle j_1\, m'_1 |\, R_{j_1}\, |\, j_1\, m_1\rangle\, \langle j_2\, m'_2 |\, R_2\, |\, j_2\, m_2\rangle$$
$$= D^{j_1}_{m'_1\, m_1}(g)\, D^{j_2}_{m'_2\, m_2}(g). \qquad (11.56)$$

Relation (11.55) will therefore take the following form:

$$(R_{j_1} \otimes R_{j_2})\ |\, j_1\, m_1\, j_2\, m_2\rangle = \sum_{m'_1\, m'_2} D^{j_1}_{m'_1\, m_1}(g)\, D^{j_2}_{m'_2\, m_2}(g)\, |\, j_1\, m'_1\, j_2\, m'_2\rangle$$

$$(11.57)$$

and, by multiplying by the conjugate bra

$$\langle j_1\, m'_1\, j_2\, m'_2 \mid R_{j_1} \otimes R_{j_2} \mid j_1\, m_1\, j_2\, m_2 \rangle = D^{j_1}_{m'_1\, m_1}\, D^{j_2}_{m'_2\, m_2} \,. \qquad (11.58)$$

The direct product $R_{j_1} \otimes R_{j_2}$ of two rotation operators acting in the subspaces ξ_{j_1} and ξ_{j_2} is reducible, that is there exists a unitary matrix $C_{j_1 j_2}$ transforming the product into a direct sum, which reads

$$C^{-1}_{j_1 j_2}\, (R_{j_1} \otimes R_{j_2})\, C_{j_1 j_2} = R_{j_1+j_2} \oplus R_{j_1+j_2-1} \oplus \ldots \oplus R_{\mid j_1-j_2 \mid}\,. \qquad (11.59)$$

The Clebsch–Gordan coefficients may be considered as the elements of the unitary transformation $C_{j_1 j_2}$ and thus written $\langle m_1\, m_2 \mid C_{j_1 j_2} \mid j\, m \rangle$.

The decomposition of the direct product for a fixed angular momentum j will therefore be written

$$R_j(g) = R_{j_1}(g) \otimes R_{j_2}(g)\,. \qquad (11.60)$$

We apply this relation to a ket of the coupled base:

$$R_j \mid j\, m \rangle = R_{j_1} \otimes R_{j_2} \mid j\, m \rangle \qquad (11.61)$$

and then we decouple the ket $\mid j\, m \rangle$ with the Clebsch–Gordan series

$$R_j \mid j\, m \rangle = \sum_{m_1\, m_2} (R_{j_1} \otimes R_{j_2}) \mid j_1\, m_1\, j_2\, m_2 \rangle \langle j_1\, m_1\, j_2\, m_2 \mid j\, m \rangle\,. \qquad (11.62)$$

We multiply by the conjugate bra $\langle j\, m' \mid$ which we subsequently decouple:

$$\langle j\, m' \mid R_j \mid j\, m \rangle = \sum \langle j\, m' \mid j_1\, m'_1\, j_2\, m'_2 \rangle$$
$$\langle j_1\, m'_1\, j_2\, m'_2 \mid R_{j_1} \otimes R_{j_2} \mid j_1\, m_1\, j_2\, m_2 \rangle \langle j_1\, m_1\, j_2\, m_2 \mid j\, m \rangle\,. \qquad (11.63)$$

The first term represents the rotation matrix element

$$\langle j\, m' \mid R_j \mid j\, m \rangle = D^j_{m'\, m}(g)\,.$$

We thus obtain with (11.58) the fundamental decoupling relation

$$\boxed{\begin{aligned} D^j_{m'\, m}(g) = &\sum_{m'_1 m_1\, m'_2 m_2} \langle j\, m' \mid j_1\, m'_1\, j_2\, m'_2 \rangle \langle j_1\, m_1\, j_2\, m_2 \mid j\, m \rangle \\ &\times D^{j_1}_{m'_1\, m_1}(g)\, D^{j_2}_{m'_2\, m_2}(g) \end{aligned}}$$

$$(11.64)$$

This fundamental relation can be written in $3jm$ coefficients:

$$D^j_{m'\, m}(g) = \sum_{m'_1\, m'_2\, m_1\, m_2} [j^2]\, D^{j_1}_{m'_1\, m_1}(g)\, D^{j_2}_{m'_2\, m_2}(g)$$

$$\times \begin{pmatrix} j_1 & j_2 & j \\ m'_1 & m'_2 & m' \end{pmatrix} \begin{pmatrix} m_1 & m_2 & m \\ j_1 & j_2 & j \end{pmatrix}\,. \qquad (11.65)$$

4.8 Contraction of Two Rotation Matrices

By using the orthogonality relations for the Clebsch–Gordan coefficients and relation (11.64), we establish the fundamental rule for obtaining the product of two rotation matrices with the same Euler angle $g = (\alpha, \beta, \gamma)$. This is what has been termed the theorem of the contraction of two rotation matrices:

$$
\begin{aligned}
D^{j_1}_{m'_1 \, m_1}(g) \, D^{j_2}_{m'_2 \, m_2}(g) = \sum_{j \, m \, m'} & \langle j \, m \mid j_1 \, m_1 \, j_2 \, m_2 \rangle \\
& \times \langle j_1 \, m'_1 \, j_2 \, m'_2 \mid j \, m' \rangle \, D^{j}_{m' \, m}(g)
\end{aligned}
\tag{11.66}
$$

or by adopting Wigner's $3jm$ coefficients:

$$
\begin{aligned}
D^{j_1}_{m'_1 \, m_1}(g) \, D^{j_2}_{m'_2 \, m_2}(g) = \sum_{j \, m \, m'} [j^2] & \begin{pmatrix} j_1 & j_2 & j \\ m_1 & m_2 & m \end{pmatrix} \\
& \times \begin{pmatrix} m'_1 & m'_2 & m' \\ j_1 & j_2 & j \end{pmatrix} D^{j \, *}_{m' \, m}(g).
\end{aligned}
\tag{11.67}
$$

Let us note here that the transition from CG coefficients in (11.66) to $3jm$ coefficients in (11.67) introduces a conjugate rotation matrix or the inverse rotation $D^{j}_{m \, m'}(\bar{g})$.

We can use the $3jm$ orthogonality rules to obtain other formulations of the contraction rule, for example,

$$
\begin{aligned}
\sum_{m_1 \, m_2 \, m_3 \, m'_1 \, m'_2 \, m'_3} & D^{j_1}_{m'_1 \, m_1} \, D^{j_2}_{m'_2 \, m_2} \, D^{j_3}_{m'_3 \, m_3} \begin{pmatrix} j_1 & j_2 & j_3 \\ m'_1 & m'_2 & m'_3 \end{pmatrix} \\
& \times \begin{pmatrix} m_1 & m_2 & m_3 \\ j_1 & j_2 & j_3 \end{pmatrix} = 1.
\end{aligned}
\tag{11.68}
$$

4.9 Integration of a Product of Rotation Matrices

By combining the orthonormalization relation (11.54) for rotation matrices and the contraction theorem (11.67) giving the product of two rotation matrices, we obtain the integral over three rotation matrices:

$$
\begin{aligned}
I_3 &= \int D^{j_1}_{m'_1 \, m_1}(g) \, D^{j_2}_{m'_2 \, m_2}(g) \, D^{j_3}_{m'_3 \, m_3}(g) \, dg \\
&= 8\pi^2 \begin{pmatrix} j_1 & j_2 & j_3 \\ m_1 & m_2 & m_3 \end{pmatrix} \begin{pmatrix} m'_1 & m'_2 & m'_3 \\ j_1 & j_2 & j_3 \end{pmatrix}.
\end{aligned}
\tag{11.69}
$$

Furthermore, we can generalize this result to the integration of n rotation matrices by working out the product of the rotation matrices two-by-two, until we obtain the value I_n. We thus obtain the result

$$
I_n = \int D^{j_1}_{m'_1\,m_1}(g) \ldots D^{j_n}_{m'_n\,m_n}(g)\, dg
$$

$$
= 8\pi^2 \sum_{X_s} [X_s]^2 \begin{pmatrix} m'_1 & \cdots & m'_n \\ & & \\ j_1 & \cdots & j_n \end{pmatrix} X_s \begin{pmatrix} j_1 & \cdots & j_n \\ & & \\ m_1 & \cdots & m_n \end{pmatrix} X_s .
$$

$$\tag{11.70}$$

4.10 Product of Two Rotations

The matrices $R_j(g)$ constitute an irreducible representation of the rotation group. We can therefore write

$$
R_j(g_1)\, R_j(g_2) = R_j(g_2\, g_1) , \tag{11.71}
$$

and if we switch to the matrix elements

$$
\sum_{m''} \langle j\, m' \mid R_j(g_2) \mid j\, m'' \rangle \, \langle j\, m'' \mid R_j(g_1) \mid j\, m \rangle
$$

$$
= \langle j\, m' \mid R_j(g_2\, g_1) \mid j\, m \rangle , \tag{11.72}
$$

giving, with the notation (11.36) for the rotation matrices,

$$
\sum_{m''} D^j_{m'\,m''}(g_2)\, D^j_{m''\,m}(g_1) = D^j_{m'\,m}(g_2\, g_1) . \tag{11.73}
$$

We can also write more explicitly

$$
\sum_{m''} D^j_{m'\,m''}(\alpha_2, \beta_2, \gamma_2)\, D^j_{m''\,m}(\alpha_1, \beta_1, \gamma_1) = D^j_{m'\,m}(\alpha, \beta, \gamma) , \tag{11.74}
$$

where (α, β, γ) is the result of the rotation $(\alpha_1, \beta_1, \gamma_1)$, then $(\alpha_2, \beta_2, \gamma_2)$. We obtain, in particular, for the reduced matrices

$$
\sum_{m''} d^j_{m'\,m''}(\beta_2)\, d^j_{m''\,m}(\beta_1) = d^j_{m'\,m}(\beta_1 + \beta_2) . \tag{11.75}
$$

4.11 Rotation of Kets and Bras

If a system with axis $(Ox'y'z')$ is obtained by the rotation $R(g)$ of a system $(Oxyz)$, then we can define the rotation of the kets and the bras quantum states of the configuration space:

$$
\begin{aligned}
R(g) \mid x\, y\, z \rangle &= \mid x'\, y'\, z' \rangle \\
\langle x\, y\, z \mid R(\bar{g}) &= \langle x'\, y'\, z' \mid
\end{aligned}
\tag{11.76}
$$

since, operator $R(g)$ being unitary, $R^+(g) = R(\bar{g})$. Because the quantum system is not affected by the rotation, we have

$$| a \rangle \xrightarrow{\;R(g)\;} | a \rangle,$$
$$| x\, y\, z \rangle \longrightarrow R(g)\, | x\, y\, z \rangle = | x'\, y'\, z' \rangle. \tag{11.77}$$

From the definition of the wave function, we can write

$$\psi_a(x, y, z) = \langle x\, y\, z \,|\, a \rangle \text{ and } R(\bar{g})\langle x\, y\, z \,|\, a \rangle = \langle x'\, y'\, z' \,|\, a \rangle = \psi_a(x', y', z').$$

We infer from this the transformation of a wave function by rotation of the system of coordinate axes, returning us to relation (11.4):

$$\boxed{R(\bar{g})\, \Psi_a(x,\ y,\ z) = \Psi_a(x',\ y',\ z')} \ . \tag{11.78}$$

Comparing with (11.4) and (11.31) boils down to setting

$$R(\bar{g}) = \exp\left(-i\alpha J_z\right) \exp\left(-i\beta J_y\right) \exp\left(-i\gamma J_z\right). \tag{11.79}$$

If we cause the quantum system (and not the coordinate axes) to rotate, then

$$R(g)\, | a \rangle = | a' \rangle \tag{11.80}$$

and the effect of the rotation on the wave function is written as follows:

$$R(g)\, \Psi_a(x,\ y,\ z) = \langle x\, y\, z \,|\, R(g)\, |\, a \rangle = \langle x\, y\, z \,|\, a' \rangle = \Psi_{a'}(x,\ y,\ z).$$

We then obtain the following result for the rotation of the physical system:

$$\boxed{R(g)\, \Psi_a(x,\ y,\ z) = \Psi_{a'}(x,\ y,\ z)} \ . \tag{11.81}$$

Let us apply this result to the rotation of the ket $| j\, \mu \rangle$, eigenvector of J^2 and J_z: The eigenvectors $| j\, \mu \rangle'$ of J_z' will be obtained from the eigenstates of J_z with the aid of the rotation operator:

$$| j\, \mu \rangle' = R(g)\, | j\, \mu \rangle$$
$$= \sum_m | j\, m \rangle \, \langle j\, m \,|\, R(g)\, |\, j\, \mu \rangle. \tag{11.82}$$

This leads to a (11.80)-type relation:

$$\boxed{| j\mu \rangle' = \sum_m D^j_{m\,\mu}(g)\, | j\, m \rangle} \ . \tag{11.83}$$

The conjugate states $'\langle j\ \mu\ |$ of $|\ j\ \mu\rangle'$ are obtained by transposing (11.83):

$$'\langle j\ \mu\ | = \sum_m D^j_{m\ \mu}{}^*(g)\ \langle j\ m\ |\,. \qquad (11.84)$$

By expressing the conjugate rotation matrix using form (11.46) and by grouping terms dependent on m and those dependent on μ, Eq. (11.83) takes the form

$$(-)^{j-\mu'}\langle j\ \mu\ | = \sum_m D^j_{-m\ -\mu}\ \langle j\ m\ |\ (-)^{j-m}\,. \qquad (11.85)$$

Let us make a rotation transformation of the ket $|\ j - \mu\rangle$ with (11.83):

$$|\ j - \mu\rangle' = \sum_m D^j_{-m\ -\mu}\ |\ j - m\rangle\,. \qquad (11.86)$$

By comparing the last two equations, we notice that the ket $|\ j - m\rangle$ is transformed by rotation in the same manner as the bra $(-)^{j-m}\ \langle j\ m\ |$. We thus write formally the following important equivalence:

$$\boxed{\langle j\ m\ | = (-)^{j-m}\ |\ j - m\rangle} \,. \qquad (11.87)$$

If we use the graphical representations of each of the foregoing elements, we obtain the following result:

$$\langle j\ m\ | = \overset{jm}{\longleftarrow} = (-)^{j-m}\ |\ j - m\rangle = (-)^{j-m}\ \overset{jm}{\longleftarrow}\,. \qquad (11.88)$$

The second arrow therefore introduces the phase $(-)^{j-m}$. This is in fact what was obtained in the Wigner definition of $3j\ m$ coefficients:

$$\begin{pmatrix} m_3 & j_1 & j_2 \\ j_3 & m_1 & m_2 \end{pmatrix} = (-)^{j_3-m_3} \begin{pmatrix} j_3 & j_1 & j_2 \\ -m_3 & m_1 & m_2 \end{pmatrix} \qquad (11.89)$$

4.12 Graphical Representations

4.12.1 Rotation Matrices

$$D^j_{m\,m'}(g) = \;\;\begin{array}{c}\downarrow jm \\ \cdots\cdots g \\ \uparrow jm'\end{array}$$

$$D^j_{m\,m'}(g^{-1}) = \;\;\begin{array}{c}\downarrow jm \\ \cdots\cdots g \\ \uparrow jm'\end{array}$$

$$(11.90)$$

4.12.2 Completeness Relation

$$\sum_{m''} D^j_{m'\,m''}(g)\, D^{j}_{m\,m''}{}^{*}(g) = \sum_{m''} D^j_{m'\,m''}(g)\, D^j_{m''\,m}(\bar g) \qquad (11.91)$$

$$\begin{array}{cc}jm' \quad\quad jm \\ \xrightarrow{\quad\;\;\;}\;\xrightarrow{\quad\;\;\;} \\ \;\downarrow g \;\uparrow g\end{array} = \begin{array}{c}jm' \quad\quad jm \\ \xrightarrow{\quad\quad\quad\quad}\end{array} = \delta_{mm'}$$

4.12.3 Orthonormalization

$$\int D^j_{m\,m'}(g)\, D^{j'}_{n\,n'}(g)^{*}\, dg = \int D^j_{m\,m'}(g)\, D^j_{n'\,n}(\bar g)\, dg$$

$$= \frac{8\pi^2}{[j^2]}\,\delta_{j\,j'}\,\delta_{m\,n}\,\delta_{m'\,n'} \qquad (11.92)$$

$$\begin{array}{cc}\downarrow jm \\ \cdots\cdots\; dg \cdots\cdots \\ \uparrow jm'\end{array}\;\begin{array}{c}\uparrow j'n \quad jm \downarrow \\ \\ \uparrow j'n' \quad jm'\uparrow\end{array} = \begin{array}{c}\uparrow j'n \quad jm \\ \\ \uparrow j'n' \end{array} = 8\pi^2\; \begin{array}{c}jm \searrow \nearrow j'n \\ \\ jm'\nearrow \searrow j'n'\end{array}\;.$$

We can infer from this, by comparing the analytical and graphical results, that a pole on a kinetic line J makes a contribution $[J^{-1}]$ to the result.

4.12.4 Contraction

$$D^{j_1}_{m_1\,m'_1}(g)\, D^{j_2}_{m_2\,m'_2}(g) = \begin{array}{c}j_1 m_1 \\ \searrow \cdots g \cdots \\ j_1 m'_1 \\ j_2 m_2 \\ \searrow \cdots g \cdots \\ j_2 m'_2\end{array} = \begin{array}{c}j_1 m_1 \\ \searrow + \rhd \\ j_2 m_2 \nearrow\;\; jm \cdots g \cdots \\ \\ j_1 m'_1 \searrow\;\; jm' \\ + \rhd \\ j_2 m'_2\end{array}\;. \qquad (11.93)$$

The last diagram implies a summation over angular momenta that do not appear in the first term.

4.13 Rotation of the Spherical Harmonics

If we specifically consider the orbital momentum vector and identify the eigenvector of L_z with the angles θ, φ in the system $Oxyz$ and with θ', φ' in the system $Ox'y'z'$, we obtain with (11.78) and (11.76):

$$(\langle \theta\,\varphi \mid \ell\,n \rangle)' = \langle \theta'\,\varphi' \mid \ell\,n \rangle$$
$$= \sum_m D^\ell_{m\,n}(\bar{g})\,\langle \theta\,\varphi \mid \ell\,m \rangle. \qquad (11.94)$$

By introducing the spherical harmonic functions $\langle \theta\,\varphi \mid \ell\,m \rangle = Y_{\ell\,m}(\theta\,\varphi)$, we obtain the relation

$$Y_{\ell\,n}(\theta'\,\varphi') = \sum_m D^\ell_{m\,n}(\bar{g})\,Y_{\ell\,m}(\theta\,\varphi) \qquad (11.95)$$

describing the rotation of the spherical harmonics for the rotation of the system of coordinates.

4.14 Special Rotation Matrices

The Racah tensor is linked to the special value:

$$D^\ell_{m\,0}(\alpha\,\beta\,\gamma) = C^*_{\ell\,m}(\beta\,\alpha) = \frac{\sqrt{4\pi}}{[\ell]}\,Y^*_{\ell\,m}(\beta\,\alpha) \qquad (11.96)$$

whereas for $m = 0$ we obtain a Legendre polynomial

$$D^\ell_{00}(\alpha,\ \beta,\ \gamma) = C^*_{\ell\,0}(\beta,\ \alpha) = \frac{\sqrt{4\pi}}{[\ell]}\,Y^*_{\ell\,0}(\beta,\ \alpha) = P_\ell(\cos\,\beta). \qquad (11.97)$$

4.15 Integration of Three Spherical Harmonics

The integral over three rotation matrices defined in (11.69) is easily obtained in the special case $m'_1 = m'_2 = m'_3 = 0$ with (11.96):

$$\int D^{\ell_1}_{m_1 0}(g)\,D^{\ell_2}_{m_2 0}(g)\,D^{\ell_3}_{m_3 0}(g)\,dg = 8\pi^2 \begin{pmatrix} \ell_1 & \ell_2 & \ell_3 \\ 0 & 0 & 0 \end{pmatrix} \begin{pmatrix} \ell_1 & \ell_2 & \ell_3 \\ m_1 & m_2 & m_3 \end{pmatrix}. \qquad (11.98)$$

Relation (11.96) introduces the spherical harmonics of the angles β and α, the integration over the angle γ introduces a factor 2π, which by setting $\Omega = (\beta,\ \alpha)$ and transforming the conjugate spherical harmonics, gives

$$\int Y_{\ell_1 m_1}(\Omega)\,Y_{\ell_2 m_2}(\Omega)\,Y_{\ell_3 m_3}(\Omega)\,d\Omega$$
$$= \frac{[\ell_1\,\ell_2\,\ell_3]}{\sqrt{4\pi}} \begin{pmatrix} \ell_1 & \ell_2 & \ell_3 \\ 0 & 0 & 0 \end{pmatrix} \begin{pmatrix} \ell_1 & \ell_2 & \ell_3 \\ m_1 & m_2 & m_3 \end{pmatrix}. \qquad (11.99)$$

5. Fundamental Theorem of Pinch

Consider an expression F dependent on a contravariant magnetic momentum m_1 and a covariant magnetic momentum m_2, and solely on these two
magnetic momenta. The first momentum will necessarily come from a ket (in
a matrix element or in a spherical harmonic, for example) and the second
from a bra (Hermitian conjugate of a ket). This expression can be written
symbolically as follows:

$$F \left(\bar{\alpha}, \bar{\beta} \mid \begin{matrix} j_1 & m_2 \\ m_1 & j_2 \end{matrix} \right) = \langle \bar{\alpha} \mid j_1 \, m_1 \rangle \, \langle j_2 \, m_2 \mid \bar{\beta} \rangle. \tag{11.100}$$

With the exception of the magnetic momenta, we can include all quantum
numbers in $\bar{\alpha}$ and $\bar{\beta}$.

We apply the identity rotation $R \, R^+ = \mathbb{1}$ to this expression:

$$\langle \bar{\alpha} \mid j_1 \, m_1 \rangle \, \langle j_2 \, m_2 \mid \bar{\beta} \rangle = \langle \bar{\alpha} \mid R \mid j_1 \, m_1 \rangle \, \langle j_2 \, m_2 \mid R^+ \mid \bar{\beta} \rangle, \tag{11.101}$$

but the transformation of the ket

$$R(g) \mid j_1 \, m_1 \rangle = \sum_{m_1'} D^{j_1}_{m_1' \, m_1} \mid j_1 \, m_1' \rangle \tag{11.102}$$

and that of the bra

$$\langle j_2 \, m_2 \mid R^+(g) = \sum_{m_2'} D^{j_2*}_{m_2' \, m_2} \langle j_2 \, m_2' \mid \tag{11.103}$$

lead to the following expression for F transformed by rotation:

$$\langle \bar{\alpha} \mid j_1 \, m_1 \rangle \, \langle j_2 \, m_2 \mid \bar{\beta} \rangle = \sum_{m_1' \, m_2'} D^{j_1}_{m_1' \, m_1}(g) \, D^{j_2*}_{m_2' \, m_2}(g)$$
$$\times \, \langle \bar{\alpha} \mid j_1 \, m_1' \rangle \, \langle j_2 \, m_2' \mid \bar{\beta} \rangle. \tag{11.104}$$

Let us use the orthonormalization relation (11.54) and the rotation over rotation group (11.53) to integrate over the whole space of the two terms:

$$8\pi^2 \, \langle \bar{\alpha} \mid j_1 \, m_1 \rangle \, \langle j_2 \, m_2 \mid \bar{\beta} \rangle = \sum_{m_1' \, m_2'} \langle \bar{\alpha} \mid j_1 \, m_1' \rangle \, \langle j_2 \, m_2' \mid \bar{\beta} \rangle$$
$$\times \, \frac{8\pi^2}{[j_1^2]} \delta_{j_1 \, j_2} \, \delta_{m_1 \, m_2} \, \delta_{m_1' \, m_2'}. \tag{11.105}$$

We divide the two members by $8\pi^2$, which leaves us with

$$\langle \bar{\alpha} \mid j_1 \, m_1 \rangle \, \langle j_2 \, m_2 \mid \bar{\beta} \rangle = \left[\sum_{m_1' \, m_2'} \langle \bar{\alpha} \mid j_1 \, m_1' \rangle \, \langle j_2 \, m_2' \mid \bar{\beta} \rangle \, \delta_{m_1' \, m_2'} \right]$$
$$\times \, \frac{1}{[j_1^2]} \delta_{j_1 \, j_2} \, \delta_{m_1 \, m_2}. \tag{11.106}$$

The expression in square brackets, summed up over m_1' and m_2', is no longer dependent on the indices but rather solely on j_1 and j_2. We will denote

$$\langle \bar\alpha \mid j_1 \rangle \langle j_2 \mid \bar\beta \rangle = \sum_{m_1' \, m_2'} \langle \bar\alpha \mid j_1 \, m_1' \rangle \, \delta_{m_1' \, m_2'} \, \langle j_2 \, m_2' \mid \bar\beta \rangle \qquad (11.107)$$

which, upon inserting into (11.106), gives

$$\langle \bar\alpha \mid j_1 \, m_1 \rangle \langle j_2 \, m_2 \mid \bar\beta \rangle = \langle \bar\alpha \mid j_1 \rangle \langle j_2 \mid \bar\beta \rangle \, \frac{1}{[j_1^2]} \, \delta_{j_1 \, j_2} \, \delta_{m_1 \, m_2} . \qquad (11.108)$$

It would then appear that we necessarily have $j_1 = j_2$ and $m_1 = m_2$ since expression F is, in the final analysis, equivalent (to within $\langle \bar\alpha|j_1 \rangle \langle j_2|\bar\beta \rangle (1/\,[j_1^2])$ coefficient) to the Kronecker symbol $\delta_{j_1 \, j_2} \, \delta_{m_1 \, m_2}$. This important result is the foundation stone of the pinch theorem in the graphical treatment of the algebra of angular momenta (Elbaz and Castel, 1972 ; Elbaz, 1985).

$$(11.109)$$

6. Transformation Rules of Graphical Spin Algebra

Let us begin with a brief reminder of the results obtained with the graphical method, results we have already seen and which make for the instantaneous simplification of certain diagrams.

1) Zero angular momentum $j = 0$
   We eliminate the corresponding kinetic line j.
2) Change of sign of a pole $(j_1 \, j_2 \, j_3)$
   We multiply the result by the phase $(-)^{j_1+j_2+j_3}$.
3) Change of direction of a linked arrow j
   We multiply the result by the phase $(-)^{2j}$.
4) Summation over a magnetic momentum m
   We link the corresponding free extremities (contravariant-covariant)

$$\sum_m \overset{jm}{\underset{}{\longmapsto}} \overset{jm}{\underset{}{\longmapsto\!\!\!\!\!>}} = \overset{j}{\underset{}{\longmapsto}} . \qquad (11.110)$$

5) Summation over an angular momentum j

We withdraw the line j corresponding to the completeness relation,

$$\sum_{j\,m} |\,j\,m\rangle\,\langle j\,m\,| = \sum_{j} \;\overset{j}{\longmapsto}\; = \mathbb{1}\,, \qquad (11.111)$$

and link the corresponding lines (the same j and read in the same order), which leads to one of the following forms of the rule:

$$\sum_{j} \;\left[\;\alpha\;\right] \cdots = \sum_{j} [j^2]\;\left[\;\alpha\;\right]\cdots = \left[\;\alpha\;\right]\cdots\left[\;\beta\;\right]. \qquad (11.112)$$

6) Transition from a CG coefficient to a $3jm$ coefficient

We use the following procedure:

$$\cdots \;\longrightarrow\; [1]\,(-)^{2(2^-)}\;\cdots \qquad (11.113)$$

$$\cdots \;\longrightarrow\; [1]\,(-)^{2(3^+)}\;\cdots \qquad (11.114)$$

7) Pinch over two free lines

We use the result demonstrated in (11.108), which can be graphically expressed as follows (see (11.109)):

$$\left[\;\bar{\alpha}\;\right]\cdots = \left[\;\bar{\alpha}\;\right]\cdots = \left[\;\bar{\alpha}\;\right][j_1^{-2}]\cdots . \qquad (11.115)$$

8) Pole located on a kinetic line j introduces a weighting coefficient $[j^{-1}] = 1/\sqrt{2j+1}$.

The foregoing eight rules make it possible to transform some extremely complex expressions and to separate them into a first part that is completely independent of the magnetic momenta, sum of the products of $3nj$ coefficients and a second part comprising all the magnetic momenta in an njm coefficient with the well-known transformation and symmetry properties.

Let us examine some typical applications of these seven transformation rules.

Example 1: Value of a $3nj$ coefficient for zero angular momentum

$$\begin{Bmatrix} j_1 & j_2 & j_3 \\ \ell_1 & \ell_2 & 0 \end{Bmatrix} = +\begin{array}{c} \overset{-}{} \\ \text{(diagram)} \end{array} + = +\begin{array}{c} \text{(diagram)} \end{array} +$$

$$= (-)^{\ell_1 + \ell_2 + j_3}\, \delta_{\ell_1 j_2}\, \delta_{j_1 \ell_2}\, [j_1^{-1}\, j_2^{-1}]\, \{j_1\, j_2\, j_3\}. \quad (11.116)$$

Example 2: Summation over j for a closed diagram

$$\sum_{j=0}^{2j_1} (-)^j\, \{j\, j_1\, j_1\}\, [j^2] = \sum_j (-)^j\, [j^2]\quad +\begin{array}{c} \text{(diagram)} \end{array}.$$

$$= (-)^{2j_1} \sum_j [j^2]\quad +\begin{array}{c} \text{(diagram)} \end{array} +$$

$$= (-)^{2j_1}\, [j_1^2]. \quad (11.117)$$

We have used the $[j_1^2]$ value of the loop on j_1.

Example 3: Second orthogonality rule of the $3jm$ coefficient

$$\begin{array}{c} \text{(diagram)} \end{array} = +\begin{array}{c} \text{(diagram)} \end{array} \cdot [j_3^{-2}]\, \begin{array}{c} \text{(diagram)} \end{array}$$

$$= \sum_{m_1\, m_2} \begin{pmatrix} m_1 & m_2 & m_3 \\ j_1 & j_2 & j_3 \end{pmatrix} \begin{pmatrix} j_1 & j_2 & j_3' \\ m_1 & m_2 & m_3' \end{pmatrix}$$

$$= \{j_1\, j_2\, j_3\}\, [j_3^{-2}]\, \delta_{j_3\, j_3'}\, \delta_{m_3\, m_3'}. \quad (11.118)$$

Example 4: Pinch rule over three lines

$$(11.119)$$

The variances of angular momenta (1,2,3) are conserved at each of the CG poles.

By using transition rules (11.113) and (11.114) for each CG, it is easily shown that one can also write:

$$(11.120)$$

It should be noted that in the open diagram the variances are conserved whereas in the closed diagram the three kinetic lines point in the same direction.

Example 5: Simplification of a given expression

$$F = \sum_{n_2\, m_2\, m_3} \begin{pmatrix} j_1 & j_2 & m_3 \\ m_1 & m_2 & j_3 \end{pmatrix} \begin{pmatrix} m_2 & \ell_3 & \ell_2 \\ j_2 & n_3 & n_2 \end{pmatrix} \begin{pmatrix} \ell_1 & j_3 & n_2 \\ n_1 & m_3 & \ell_2 \end{pmatrix}$$

$$= - \quad \text{} \quad (-)^{2j_3+2\ell_3} \begin{Bmatrix} j_2 & j_1 & j_3 \\ \ell_1 & \ell_2 & \ell_3 \end{Bmatrix} \begin{pmatrix} j_1 & \ell_1 & \ell_3 \\ m_1 & n_1 & n_3 \end{pmatrix} \cdot \quad (11.121)$$

Chapter 12
Spherical Harmonics

In the last chapter, we defined spherical harmonics (SH) as the special values taken by rotation matrices when one of the magnetic momentum is zero.

In addition, SH are the eigenfunctions of the projections of the orbital momentum onto configuration space:

$$\langle \theta, \varphi \mid \ell\, m \rangle \;=\; \langle \Omega \mid \ell\, m \rangle \;=\; Y_{\ell\, m}(\Omega).$$

The differential equations corresponding to the eigenvalue equations of L^2 and L_z can be resolved in order to determine the analytical form of the spherical harmonic function $Y_{\ell\, m}(\Omega)$.

1. Orbital Angular Momentum

By writing the usual eigenvalue equations for the orbital angular momentum L, we obtain (by setting $\hbar = 1$):

$$L^2 \mid \ell\, m \rangle \;=\; \ell(\ell+1) \mid \ell\, m \rangle, \tag{12.1}$$

$$L_z \mid \ell\, m \rangle \;=\; m \mid \ell\, m \rangle, \tag{12.2}$$

with operators L^2 and L_z expanded in configuration space:

$$L^2 = -\left[\frac{1}{\sin\theta} \frac{\partial}{\partial\theta} \sin\theta \frac{\partial}{\partial\theta} + \frac{1}{\sin^2\theta} \frac{\partial^2}{\partial\varphi^2} \right], \tag{12.3}$$

$$L_z = -i \frac{\partial}{\partial\varphi}. \tag{12.4}$$

By projecting Eq. (12.1) and (12.2) onto configuration space, we obtain

$$L^2 \langle \theta\,\varphi \mid \ell\, m \rangle \;=\; L^2\, Y_{\ell\, m}(\theta\,\varphi) = \ell(\ell+1)\, Y_{\ell\, m}(\theta,\,\varphi), \tag{12.5}$$

$$L_z \langle \theta\,\varphi \mid \ell\, m \rangle \;=\; L_z\, Y_{\ell\, m}(\theta\,\varphi) = m\, Y_{\ell\, m}(\theta,\,\varphi). \tag{12.6}$$

If we use definitiion (12.4) of L_z, the last relation becomes

$$-i \frac{\partial}{\partial\varphi}\, Y_{\ell\, m}(\theta,\,\varphi) = m\, Y_{\ell\, m}(\theta,\,\varphi). \tag{12.7}$$

This is a first-order differential equation with separable variables. Its solution is of the form

$$Y_{\ell\, m}(\theta,\ \varphi) = Z_{\ell\, m}(\theta)\ e^{i\, m\, \varphi}. \tag{12.8}$$

The function $Y_{\ell\, m}(\theta,\ \varphi)$ is uniform if m is an integer by the very presence of the term $e^{i\, m\, \varphi}$. The relation

$$Y_{\ell\, m}(\theta, \varphi + 2\pi) = Y_{\ell\, m}(\theta\ \varphi)$$

in fact implies

$$Z_{\ell\, m}(\theta)\ e^{i\, m\, \varphi}\ e^{2i\, m\, \pi} = Z_{\ell\, m}(\theta)\ e^{i\, m\, \varphi}$$

and, hence, $e^{2i\, m\, \pi} = 1$, giving m integer.

In the discussion on angular momentum we saw that $\ell + m$ and $\ell - m$ needed to be integers. If m is an integer, then ℓ is also necessarily an integer, and since $\ell \geq 0$, the only possible values of ℓ and m are

$$\boxed{\begin{array}{l} \ell = 0, 1, 2, \ldots \\ -\ell \leq m \leq \ell \end{array}} \tag{12.9}$$

2. Spherical Harmonic $Y_{\ell 0}$

Regardless of the value of ℓ, the value $m = 0$ will always be an acceptable value and, hence, $Y_{\ell\, 0}(\theta,\ \varphi) = Z_{\ell\, 0}(\theta)$ is an eigenfunction of L^2

$$L^2\ Z_{\ell\, 0}(\theta) = \ell(\ell+1)\ Z_{\ell\, 0}(\theta). \tag{12.10}$$

By using relation (12.3) leading to the expression for the operator L^2 in configuration space, we obtain the following differential equation:

$$\left[-\frac{1}{\sin\theta} \frac{\partial}{\partial\theta}\left(\sin\theta\ \frac{\partial}{\partial\theta} \right) \right] Z_{\ell\, 0} = \ell(\ell+1)\ Z_{\ell\, 0}. \tag{12.11}$$

We set $\cos\theta = u$ to determine the first derivative:

$$\frac{d}{d\theta} = \frac{d}{du}\cdot\frac{d\, u}{d\, \theta} = -\sin\theta\ \frac{d}{du} \qquad \text{since} \qquad \frac{d\, u}{d\, \theta} = \sin\theta$$

and the second derivative with respect to θ:

$$\frac{d^2}{d\,\theta^2} = \frac{d}{d\theta}\left(\frac{d}{d\theta} \right) = \frac{d}{d\theta}\left(-\sin\theta\ \frac{d}{d\, u} \right)$$

$$= -\cos\theta\ \frac{d}{d\, u} - \sin\theta\ \frac{d}{d\, \theta}\frac{d}{d\, u}$$

$$= -\cos\theta\ \frac{d}{d\, u} - \sin\theta\ \frac{d}{d\, u}\frac{d}{d\, u}\frac{d\, u}{d\, \theta}$$

$$= -\cos\theta\ \frac{d}{d\, u} + \sin^2\theta\ \frac{d^2}{d\, u^2}$$

$$= (1-u^2)\ \frac{d^2}{d\, u^2} - u\ \frac{d}{d\, u}. \tag{12.12}$$

We thus obtain without great difficulty the differential operator in (12.11):

$$\frac{1}{\sin\theta}\frac{d}{d\theta}\sin\theta\frac{d}{d\theta} = \frac{d^2}{d\theta^2} + \cot\theta\frac{d}{d\theta}$$

$$= (1-u^2)\frac{d^2}{du^2} - u\frac{d}{du} + \frac{\cos\theta}{\sin\theta}\left(-\sin\theta\frac{d}{du}\right)$$

$$= (1-u^2)\frac{d^2}{du^2} - 2u\frac{d}{du}. \tag{12.13}$$

The change of function from θ to u:

$$Z_{\ell 0}(\theta) = a_\ell\, P_\ell(u) \tag{12.14}$$

in eigenvalue Eq. (12.10) leads to the second-order differential equation characteristic of Legendre polynomials $P_\ell(u)$:

$$(1-u^2)\frac{d^2\,P_\ell(u)}{d\,u^2} - 2u\frac{d\,P_\ell(u)}{d\,u} + \ell(\ell+1)\,P_\ell(u) = 0. \tag{12.15}$$

Legendre polynomials are orthogonal, that is, they verify the following orthogonality relation:

$$\int_{-1}^{+1} P_\ell(u)\,P'_\ell(u)\,d\,u = \frac{2}{2\ell+1}\,\delta_{\ell\,\ell'}. \tag{12.16}$$

Spherical harmonics, which are the eigenfunctions of the operators L^2 and L_z, are orthonormal functions since

$$\int \langle \ell\,m\mid\Omega\rangle\,d\,\Omega\,\langle\Omega\mid\ell'\,m'\rangle = \langle\ell\,m\mid\ell'\,m'\rangle = \delta_{\ell\,\ell'}\,\delta_{m\,m'}, \tag{12.17}$$

yielding with the spherical harmonics:

$$\int Y^*_{\ell\,m}(\Omega)\,Y_{\ell'\,m'}(\Omega)\,d\,\Omega = \delta_{\ell\,\ell'}\,\delta_{m\,m'}, \tag{12.18}$$

a relation holding for the value $m = m' = 0$:

$$\int Y^*_{\ell 0}(\Omega)\,Y_{\ell'\,0}(\Omega)\,d\,\Omega = \delta_{\ell\,\ell'} \quad\text{with}\quad d\,\Omega = \sin\theta\,d\,\theta\,d\,\varphi. \tag{12.19}$$

By replacing $Y_{\ell 0}$ by $a_\ell\,P_\ell(u)$, we obtain the following orthogonality relation:

$$|\,a_\ell\,|^2 \int_{-1}^{+1} P_\ell(u)\,P_{\ell'}(u)\,d\,u\,d\,\varphi = \delta_{\ell\,\ell'}.$$

By integrating over the angle φ and using the orthogonality of Legendre polynomials (12.16), we obtain

$$|\,a_\ell\,|^2 = (2\ell+1)/4\pi.$$

If we take the positive determination of a_ℓ, we obtain the spherical harmonic $Y_{\ell 0}\,(\theta,\,\varphi)$ in terms of the ℓth-order Legendre polynomial (12.16):

$$Y_{\ell 0}(\theta,\,\varphi) = \left[\frac{2\ell+1}{4\pi}\right]^{1/2} P_\ell(\cos\theta). \tag{12.20}$$

3. General Statement

We can construct $Y_{\ell m}(\theta \; \varphi)$ by applying the operator L_+ m times since

$$L_+ \, Y_{\ell m} = [(\ell - m) \; (\ell + m + 1)]^{1/2} \; Y_{\ell m+1} \, . \tag{12.21}$$

Using the starting value $Y_{\ell 0}$, the iteration will yield

$$L_+ \, Y_{\ell 0} \quad = \quad [\ell \, (\ell + 1)]^{1/2} \; Y_{\ell 1}$$
$$\vdots$$
$$L_+ \, Y_{\ell m-1} \quad = \quad [(\ell + m) \; (\ell - m + 1)]^{1/2} \; Y_{\ell m} \, .$$

Following iterated substitution, we obtain

$$Y_{\ell m}(\theta, \; \varphi) = \left[\frac{(\ell - m)!}{(\ell + m)!} \right]^{1/2} \; L_+^m \, Y_{\ell 0}(\theta) \; e^{i \, m \, \varphi} \, . \tag{12.22}$$

If we replace L_+ by the expression in terms of L_x and L_y, we obtain the following analytical statement:

---

for $m > 0$:

$$Y_{\ell m}(\theta, \; \varphi) = (-)^m \left[\frac{(\ell - m)!}{(\ell + m)!} \; \frac{(2\ell + 1)}{4\pi} \right]^{1/2} e^{i \, m \, \varphi} \, P_{\ell m}(\cos \, \theta) \tag{12.23}$$

---

where $P_{\ell m}$ is an associated Legendre polynomial.

For $m < 0$, and noting that $L_+^* = L_-$, we obtain the conjugation relation

$$Y_{\ell m}^*(\theta, \; \varphi) = (-)^m \; Y_{\ell - m}(\theta, \; \varphi) \, . \tag{12.24}$$

Special values	
$Y_{00} = \frac{1}{\sqrt{4\pi}}$	$Y_{11} = -\sqrt{\frac{3}{8\pi}} \; \sin \, \theta \; e^{i \, \varphi}$
$Y_{10} = \sqrt{\frac{3}{4\pi}} \; \cos \, \theta$	$Y_{21} = -\sqrt{\frac{15}{8\pi}} \; \sin \, \theta \; \cos \, \theta \; e^{i \, \varphi}$
$Y_{20} = \sqrt{\frac{5}{16\pi}} \; (3 \cos^2 \, \theta - 1)$	$Y_{31} = -\sqrt{\frac{21}{64\pi}} \; \sin \, \theta \; (5 \cos^2 \, \theta - 1) \; e^{i \, \varphi}$
$Y_{30} = \sqrt{\frac{7}{16\pi}} \; (5 \, \cos^3 \, \theta - 3 \cos \, \theta)$	

4. Parity

We introduce the parity operator P for transforming x into $-x$, y into $-y$, and z into $-z$ in a wave function Ψ (see Chap. 8):

$$P\,\Psi(x, y, z) = \Psi\,(-x,\ -y,\ -z)\,. \tag{12.25}$$

Expressed in spherical coordinates, this amounts to changing r into r, θ into $\pi - \theta$ and φ into $\varphi + \pi$:

$$P\,\Psi(r,\ \theta,\ \varphi) = \Psi\,(r,\ \pi - \theta,\ \varphi + \pi)\,. \tag{12.26}$$

This parity operator satisfies the condition

$$P^2 = \mathbb{1}\,. \tag{12.27}$$

It is easily verified that $L_x\, L_y\, L_z$ commute with P and the eigenfunctions of L^2 and L_z can therefore be considered as even and odd functions.

Replacing θ by $\pi - \theta$ and φ by $\varphi + \pi$ will yield

$$Y_{\ell\,m}(\pi - \theta,\ \varphi + \pi) = (-)^m \left[\frac{(\ell - m)!}{(\ell + m)!}\,\frac{2\ell + 1}{4\pi}\right]^{1/2} e^{i\,m\,\varphi}\,e^{i\,m\,\pi}\,P_{\ell\,m}(-\cos\theta)\,, \tag{12.28}$$

with the special values:

$$e^{i\,m\,\pi} = (-)^m \quad \text{and} \quad P_{\ell\,m}(-\cos\theta) = (-)^{\ell - m}\,P_{\ell\,m}(\cos\theta)\,.$$

We infer from this the parity transformation of the spherical harmonic that

$$P\,Y_{\ell\,m}(\theta\ \varphi) = Y_{\ell\,m}(\pi - \theta, \varphi + \pi) = Y_{\ell\,m}(-\Omega) = (-)^\ell\,Y_{\ell\,m}(\Omega)\,. \tag{12.29}$$

Spherical harmonics are functions with parity $(-)^\ell$.

5. Orthonormalization and Closure

As already noted in (12.18), spherical harmonics are orthonormal and obey the relation

$$\int Y_{\ell\,m}^*(\Omega)\,Y_{\ell'\,m'}(\Omega)\,d\Omega = \delta_{\ell\,\ell'}\,\delta_{m\,m'}\,. \tag{12.30}$$

By introducing the angles θ and φ with respect to the Oz and Ox axes, the orthonormalization relation can also take the form

$$\int_0^{2\pi} d\varphi \int_0^\pi \sin\theta\,d\theta\,Y_{\ell\,m}^*(\theta,\ \varphi)\,Y_{\ell'\,m'}(\theta,\ \varphi) = \delta_{\ell\,\ell'}\,\delta_{m\,m'}\,. \tag{12.31}$$

Spherical harmonics form a complete system since

$$\sum_{\ell\, m} |\,\ell\, m\rangle\, \langle\ell\, m\,| = \mathbb{1} \qquad (12.32)$$

and, by projection over the solid angles Ω and Ω', this will yield

$$\sum_{\ell\, m} \langle\,\Omega\,|\,\ell\, m\rangle\, \langle\ell\, m\,|\,\Omega'\rangle = \langle\Omega\,|\,\Omega'\rangle = \delta(\Omega - \Omega'). \qquad (12.33)$$

which can be written with spherical harmonics

$$\sum_{\ell\, m} Y_{\ell\, m}(\Omega)\, Y_{\ell\, m}^*(\Omega') = \delta(\Omega - \Omega') \qquad (12.34)$$

or, by introducing the angles θ and φ,

$$\sum_{\ell\, m} Y_{\ell\, m}(\theta,\ \varphi)\, Y_{\ell\, m}^*(\theta',\ \varphi') = \frac{\delta\,(\theta - \theta')}{\sin\theta}\, \delta(\varphi - \varphi'). \qquad (12.35)$$

The closure relation on the spherical harmonics shows that these functions constitute a complete system, that is, that any function of $\Omega = (\theta, \varphi)$ can be expanded in this basis. We can then write very generally

$$f(\Omega) = \sum_{\ell\, m} a_{\ell\, m}\, Y_{\ell\, m}(\Omega), \qquad (12.36)$$

and the expansion coefficient $a_{\ell\, m}$ will be obtained by multiplying the two terms by $Y_{\ell'\, m'}^*$, and then integrating over the direction Ω:

$$\int Y_{\ell'\, m'}^*(\Omega)\, f(\Omega)\, d\,\Omega = \sum_{\ell\, m} a_{\ell\, m} \int Y_{\ell'\, m'}^*(\Omega)\, Y_{\ell\, m}(\Omega)\, d\,\Omega$$

$$= \sum_{\ell\, m} a_{\ell\, m}\, \delta_{\ell\, \ell'}\, \delta_{m\, m'} = a_{\ell'\, m'}.$$

This determines the expansion coefficients $a_{\ell\, m}$:

$$a_{\ell\, m} = \int f(\Omega)\, Y_{\ell\, m}^*(\Omega)\, d\,\Omega. \qquad (12.37)$$

6. Graphical Representation

Spherical harmonics are the projections of the states of the orbital angular momentum $(\ell\, m)$ onto the configuration space. We therefore need to introduce the graphical representation of the vectors in this space. They obey the usual orthonormalization and closure relations for continuous variables:

$$\langle \Omega \mid \Omega' \rangle = \delta(\Omega - \Omega'),$$

$$\int \mid \Omega \rangle \, d\Omega \, \langle \Omega \mid = 1. \tag{12.38}$$

The vector $\mid \Omega \rangle$ is represented graphically by a dashed line with a single arrow for the ket and a double arrow for the bra:

$$\langle \, \Omega \mid = \overset{\hat{\Omega}}{\cdots\!\!\blacktriangleleft\!\!\blacktriangleleft\!\cdots} \quad \text{and} \quad \mid \Omega \, \rangle = \overset{\hat{\Omega}}{\cdots\!\!\blacktriangleright\!\cdots} \tag{12.39}$$

yielding the graphical representations of the spherical harmonics

$$Y_{\ell\,m}(\Omega) = \langle \Omega \mid \ell \, m \rangle = \overset{\hat{\Omega}}{\cdots\!\!\blacktriangleright\!\!\blacktriangleright}\overset{\text{lm}}{\!\!\!\!+\!\!\longrightarrow} = \overset{\hat{\Omega}}{\cdots}\overset{\text{lm}}{\!\!\!\!\!\longrightarrow} \tag{12.40}$$

and for the complex conjugate

$$Y_{\ell\,m}^{*}(\Omega) = \langle \ell \, m \mid \Omega \rangle = \overset{\text{lm}}{\longrightarrow\!\!\blacktriangleright}\overset{\hat{\Omega}}{+\!\cdots\!\!\blacktriangleright\!\cdots}. \tag{12.41}$$

This clarifies the graphical representations of orthonormalization relation (12.30) and closure relation (12.34):

$$\overset{\text{lm}}{\longrightarrow\!\!\blacktriangleright}\overset{\text{l'm'}}{+\!\cdots\cdots\!+\!\!\longrightarrow} = \overset{\text{lm}}{\longrightarrow\!\!\blacktriangleright}\overset{\text{l'm'}}{+\!\!\longrightarrow} = \delta_{\ell\,\ell'}\,\delta_{m\,m'} \tag{12.42}$$

$$\underset{i}{\sum} = \overset{\hat{\Omega}}{\cdots\!\!\blacktriangleright\!\!\blacktriangleright}\overset{\text{i}}{+\!\cdots\!+}\overset{\hat{\Omega}'}{\cdots\!\!\blacktriangleright\!\cdots} = \overset{\hat{\Omega}}{\cdots\!\!\blacktriangleright\!\!\blacktriangleright}\overset{\hat{\Omega}'}{\cdots\!\!\blacktriangleright\!\cdots} = \delta(\Omega - \Omega'). \tag{12.43}$$

7. Integration of Three Spherical Harmonics

Because spherical harmonics are special rotation matrices, the application of relation (11.96) to the integral (11.69) will yield the integral (11.99):

$$I_3 = \int Y_{\ell_1\,m_1}(\Omega) \, Y_{\ell_2\,m_2}(\Omega) \, Y_{\ell_3\,m_3}(\Omega) \, d\Omega$$

$$= \frac{[\ell_1 \, \ell_2 \, \ell_3]}{\sqrt{4\pi}} \begin{pmatrix} \ell_1 & \ell_2 & \ell_3 \\ 0 & 0 & 0 \end{pmatrix} \begin{pmatrix} \ell_1 & \ell_2 & \ell_3 \\ m_1 & m_2 & m_3 \end{pmatrix}.$$

Now, let us represent the first member graphically by assuming that the integration of the three angle lines is obtained by their union with a simple pole:

$$I_3 = \qquad\qquad \equiv \qquad\qquad . \tag{12.44}$$

The integration has been enclosed in a marking circle ($\ell_1\,\ell_2\,\ell_3$) (Elbaz et al., 1966, 1967).

Comparing to the second term of I_3, we notice that the marking circle is linked to the value $\dfrac{[\ell_1\,\ell_2\,\ell_3]}{\sqrt{4\pi}}\begin{pmatrix}\ell_1 & \ell_2 & \ell_3 \\ 0 & 0 & 0\end{pmatrix}$.

It should be noted that the variances in the marking circle are exactly the same as those of the delimited $3jm$ coefficients. For example, the integral

$$\int Y^*_{\ell_1\,m_1}(\Omega)\,Y_{\ell_2\,m_2}(\Omega)\,Y^*_{\ell_3\,m_3}(\Omega)\,d\,\Omega = \quad\text{(diagram)}\qquad (12.45)$$

leads to the analytical value

$$\int Y^*_{\ell_1\,m_1}\,Y_{\ell_2\,m_2}\,Y^*_{\ell_3\,m_3}\,d\Omega = \frac{[\ell_1\,\ell_2\,\ell_3]}{\sqrt{4\pi}}\begin{pmatrix}0 & \ell_2 & 0 \\ \ell_1 & 0 & \ell_3\end{pmatrix}\begin{pmatrix}m_1 & \ell_2 & m_3 \\ \ell_1 & m_2 & \ell_3\end{pmatrix}.$$
$$(12.46)$$

The $3j0$ coefficient is zero if the sum of the angular momenta is an odd number (see Chap. 10, Appendix 1):

$$\begin{pmatrix}\ell_1 & \ell_2 & \ell_3 \\ 0 & 0 & 0\end{pmatrix} = 0 \quad\text{if}\quad \ell_1+\ell_2+\ell_3 \text{ odd}. \qquad (12.47)$$

This shows that $\ell_1+\ell_2+\ell_3$ in a marking circle is necessarily an even number and that $(-)^{\ell_1+\ell_2+\ell_3} = +1$. A change of sign of a marking pole resulting in the introduction of a phase $(-)^{\ell_1+\ell_2+\ell_3} = +1$ can be made without need for a change of phase. It is therefore unnecessary to impose a $(+)$ or $(-)$ reading direction on a spherical harmonic marking circle. (We will have occasion to show that this is no longer the case with irreducible tensor operators.) In the final analysis, we will write

$$I_3 = \quad\text{(diagram)}\qquad (12.48)$$

irrespective of the variance of the SH in the integral.

8. Contraction

8.1 Contraction of Two Spherical Harmonics

From the closure relation in conjunction with the rule for integration over three SH, we immediately obtain the contraction relation

$$Y_{\ell_1\,m_1}(\Omega)\,Y_{\ell_2\,m_2}(\Omega) = \int Y_{\ell_1\,m_1}(\Omega')\,Y_{\ell_2\,m_2}(\Omega')\,\delta(\Omega'-\Omega)\,d\,\Omega'$$

$$\delta(\Omega'-\Omega) = \sum_{L\,M} Y^*_{L\,M}(\Omega')\,Y_{L\,M}(\Omega),$$

(12.49)

yielding graphically the following result:

(12.50)

This result can be obtained by expanding the product of the SH over the complete basis of the same harmonics and following (12.36):

$$Y_{\ell_1\,m_1}(\Omega)\,Y_{\ell_2\,m_2}(\Omega) = \sum_{\ell\,m} a_{\ell\,m}\,Y_{\ell\,m}(\Omega),$$

(12.51)

yielding with expansion coefficient (12.37)

$$a_{\ell\,m} = \int Y_{\ell_1\,m}(\Omega)\,Y_{\ell_2\,m_2}(\Omega)\,Y^*_{\ell\,m}(\Omega)\,d\,\Omega =$$

(12.52)

By substituting over the graphical statement of $a_{\ell\,m}$ in the expansion (12.51) and by summing over m, we again obtain (12.49). Of course, we can change the variance of the spherical harmonic in (12.49), expressing in this way the contraction of the two SH irrespective of their variances:

(12.53)

The analytical statement of (12.50), for example, is easily obtained:

$$Y_{\ell_1\,m_1}(\Omega)\,Y_{\ell_2\,m_2}(\Omega) = \sum_{L\,M} \frac{[\ell_1\,\ell_2\,L]}{\sqrt{4\pi}} \begin{pmatrix} \ell_1 & \ell_2 & L \\ 0 & 0 & 0 \end{pmatrix} \begin{pmatrix} \ell_1 & \ell_2 & L \\ m_1 & m_2 & M \end{pmatrix} Y^*_{L\,M}(\Omega).$$

(12.54)

By using the value $Y_{\ell\,0} = [\ell]/(\sqrt{4\pi})\,P_\ell(\cos\theta)$, we likewise infer the contraction rule for Legendre polynomials:

$$P_{\ell_1}(\cos\theta)\,P_{\ell_2}(\cos\theta) = \sum_{L} [L^2] \begin{pmatrix} \ell_1 & \ell_2 & L \\ 0 & 0 & 0 \end{pmatrix}^2 P_L(\cos\theta),$$

(12.55)

and in the special case in which $\theta = \frac{\pi}{2}$, $P_L(0) = 1$, we obtain the addition rule

$$\sum_{L} [L^2] \begin{pmatrix} \ell_1 & \ell_2 & L \\ 0 & 0 & 0 \end{pmatrix}^2 = 1,$$

(12.56)

which is simply a special case of the $3jm$ orthogonality rule.

8.2 Contraction and Integration of n Spherical Harmonics

The contraction rule for two spherical harmonics can be used and applied step-by-step with the SH taken two at a time in an A coupling mode. This will yield:

$$\begin{array}{ll}
l_1\,m_1 \quad \hat{\Omega} \\
\overline{}\cdots\cdots \\
\\
\overline{}\cdots\cdots \\
l_n m_n \quad \hat{\Omega}
\end{array} = \quad \boxed{\begin{array}{c} l_1\,m_1 \;\; A \\ \odot \quad l \quad \hat{\Omega} \\ L_s \\ l_n m_n \end{array}} \,. \tag{12.57}$$

The summation concerns all orbital magnetic momenta that do not appear in the first term. The marking circle in the box expresses that all nodes of the "njm" are marked. The integration is easily made with (12.49) or (12.44):

$$I_1 = \int Y_{\ell_1\,m_1}(\Omega)\, d\,\Omega = \int Y^*_{\ell_1\,m_1}\, Y_{00}\, \sqrt{4\pi} = \sqrt{4\pi}\, \delta_{\ell_1\,0}\, \delta_{m_1\,0}$$

$$I_2 = \int Y^*_{\ell_1\,m_1}\, Y_{\ell_2\,m_2}\, d\,\Omega = \; = \; \delta_{\ell_1\,\ell_2}\, \delta_{m_1\,m_2}$$

$$I_3 = \int Y_{\ell_1\,m_1}\, Y_{\ell_2\,m_2}\, Y_{\ell_3\,m_3} =$$

$$I_4 = \sum_L \qquad = \qquad \qquad \qquad \qquad (12.58)$$

$$\vdots$$

$$I_n = \sum_{L_1\ldots L_{n-3}} \boxed{\begin{array}{c} l_1 \;\; A \\ \odot \\ L_s \\ l_n \end{array}}\,.$$

We thus obtain in a general manner the analytical result of the integration of n spherical harmonics

$$I_n = \int Y_{\ell_1\,m_1}(\Omega)\ldots Y_{\ell_n\,m_n}(\Omega)\, d\,\Omega = \frac{[\ell_1\,\ell_2\ldots\ell_n]}{\left(\sqrt{4\pi}\right)^{n-2}} \sum_{X\,s} [X_s^2]$$

$$\times \begin{pmatrix} \ell_1 & \cdots & \ell_n \\ 0 & \cdots & 0 \end{pmatrix} |X_s \left.\right) \begin{pmatrix} \ell_1 & \cdots & \ell_n \\ m_1 & \cdots & m_n \end{pmatrix} |X_s \left.\right)\,. \tag{12.59}$$

8.3 Applications

8.3.1 Marked Orbital Momentum Line

By setting $\ell_3 = m_3 = 0$ in (12.48), we obtain the diagram

$$E = \qquad \qquad = \qquad \qquad . \qquad (12.60)$$

Its analytical statement stems from (12.46) and

$$E = \frac{[\ell_1 \, \ell_2 \, 0]}{\sqrt{4\pi}} \begin{pmatrix} \ell_1 & 0 & 0 \\ 0 & \ell_2 & 0 \end{pmatrix} \begin{pmatrix} \ell_1 & m_2 & 0 \\ m_1 & \ell_2 & 0 \end{pmatrix} = \frac{1}{\sqrt{4\pi}} \, \delta_{\ell_1 \, \ell_2} \, \delta_{m_1 \, m_2} \, . \tag{12.61}$$

8.3.2 Change of Coupling Mode and Summation of a Marked Line

Let us use (12.58) to expand the following diagram:

$$\sum_X \; \bar{\alpha} \quad X \; = \quad \bar{\alpha} \quad \sum_X \quad X \, . \tag{12.62}$$

The above last diagram is none other than I_4, giving

$$I_4 = \int Y_{\ell_1 \, m_1} \, Y_{\ell_2 \, m_2} \, Y_{\ell_3 \, m_3} \, Y_{\ell_4 \, m_4}(\Omega) \, d\Omega \, . \tag{12.63}$$

We can therefore change the coupling mode and make a fresh summation over the magnetic momenta to obtain

$$\sum_X \; \bar{\alpha} \quad X \; = \; \sum_Y \; \bar{\alpha} \quad Y \; = \; \sum_Z \; \bar{\alpha} \quad Z \, . \tag{12.64}$$

The change of coupling mode can sometimes save us an additional summation. For example,

$$\sum_X \quad = \; \sum_Y \quad = \qquad \qquad (12.65)$$

which analytically leads to the following statement:

$$\sum_X [X^2] \begin{Bmatrix} 1 & 2 & L \\ 4 & 3 & X \end{Bmatrix} \begin{pmatrix} 1 & 3 & X \\ 0 & 0 & 0 \end{pmatrix} \begin{pmatrix} 2 & 4 & 0 \\ 0 & 0 & X \end{pmatrix}$$

$$= \begin{pmatrix} 1 & 2 & L \\ 0 & 0 & 0 \end{pmatrix} \begin{pmatrix} 3 & 4 & 0 \\ 0 & 0 & L \end{pmatrix}. \tag{12.66}$$

9. Addition Theorem

Two SH with different directions but of the same order can be summed over the magnetic momenta to give a Legendre polynomial:

$$\sum_m Y^*_{\ell m}(\Omega_1) \, Y_{\ell m}(\Omega_2) = \frac{[\ell^2]}{4\pi} \, P_\ell(\cos \theta), \tag{12.67}$$

where θ is the relative direction of Ω_1 and Ω_2. The summation over both m and ℓ, on the other hand, will lead to the closure relation

$$\sum_{m \, \ell} Y^*_{\ell m}(\Omega_1) \, Y_{\ell m}(\Omega_2) = \sum_\ell \frac{[\ell^2]}{4\pi} \, P_\ell (\cos(\Omega_1, \, \Omega_2))$$

$$= \delta(\Omega_1 - \Omega_2). \tag{12.68}$$

Legendre polynomials therefore also constitute a complete basis for the expansion of any function whatsoever of the cosine of an angle.

Consider, for example, the function $\Psi(\vec{k}.\vec{r}) = \Psi(kr \cos \theta)$. It can be expanded in terms of Legendre polynomials by writing

$$\Psi(\vec{k} \cdot \vec{r}) = \Psi(kr \cos \theta) = \sum_\ell \Psi_\ell(k \, r) \, P_\ell(\cos \theta), \tag{12.69}$$

with the partial wave function for weighting Legendre polynomials:

$$\Psi_\ell(k \, r) = \frac{[\ell^2]}{2} \int_{-1}^{+1} \Psi(k \, r \, u) \, P_\ell(u) \, du. \tag{12.70}$$

We thus obtain the expansion of a plane wave:

$$e^{i \, \vec{k} \cdot \vec{r}} = \sum_\ell [\ell^2] \, i^\ell \, j_\ell(k \, r) \, P_\ell(\cos \theta), \tag{12.71}$$

where $j_\ell(k \, r)$ is the ℓth-order spherical Bessel function.

Another example here pertains to the expansion of the inverse of the position vector in many integrals:

$$\frac{1}{|\vec{r}_1 - \vec{r}_2|} = \sum_\ell \frac{r_<^\ell}{r_>^{\ell+1}} \, P_\ell(\cos \theta) \tag{12.72}$$

with the following definition of $r_>$ and $r_<$:

$$r_< = r_1 \quad \text{if} \quad r_1 < r_2 \quad \text{then} \quad r_> = r_2$$
$$= r_2 \quad \text{if} \quad r_2 < r_1 \qquad\qquad r_> = r_1$$

This result will prove very useful in the study of the hydrogen atom, for example.

By using the expansion of the Legendre polynomial in spherical harmonics (12.67), it is possible to introduce the direction of each vector with respect to an arbitrary quantization axis in the expansion (12.69):

$$\Psi(\vec{k} \cdot \vec{r}) = \sum_{\ell\, m} 4\pi [\ell^{-2}]\, \Psi_\ell(k\, r)\, Y^*_{\ell\, m}(\hat{k})\, Y_{\ell\, m}(\hat{r})\,. \tag{12.73}$$

We thus obtain the expansion of a plane wave in terms of spherical harmonics:

$$e^{i\, \vec{k} \cdot \vec{r}} = \sum_{\ell\, m} 4\pi\, i^\ell\, j_\ell(k\, r)\, Y^*_{\ell\, m}(\Omega_k)\, Y_{\ell\, m}(\Omega_r) \tag{12.74}$$

10. Coupled Spherical Harmonics

If the Hamiltonian H of a quantum system commutes with operators L^2 and L_z, then the quantum numbers $(\ell\ m)$ will be constant and conserved over time. These are good quantum numbers. The basis $Y_{\ell\, m}(\Omega)$ therefore constitutes a complete orthonormal basis over which can be developed any wave function dependent on Ω. If there exist a spin $\vec{S}$ and a total momentum $\vec{J} = \vec{L} + \vec{S}$, then the Hamiltonian H will commute with J^2, J_z, L^2, and S^2, that is ℓ, s, j, μ now become the good quantum numbers. This is the case for Hamiltonian H that comprise a spin–orbit coupling term:

$$H = \frac{p^2}{2m} + V(r) + f(r)\, \vec{L} \cdot \vec{S}\,. \tag{12.75}$$

By introducing the total angular momentum $\vec{J} = \vec{L} + \vec{S}$, and by squaring the corresponding operator, we obtain:

$$J^2 = L^2 + S^2 + 2\, \vec{L} \cdot \vec{S} \tag{12.76}$$

In other words, the inner product is expressed with the operators L^2, S^2 and J^2

$$\vec{L} \cdot \vec{S} = \frac{1}{2}\, \left(J^2 - L^2 - S^2\right)\,. \tag{12.77}$$

The eigenvectors of $\vec{L} \cdot \vec{S}$ are the coupled vectors $|\ (\ell\ s)\ j\ \mu\rangle$, since

$$\vec{L} \cdot \vec{S}\, |\ (\ell\ s)\ j\ \mu\rangle = \frac{1}{2}\, \left(J^2 - L^2 - S^2\right)\, |\ (\ell\ s)\ j\ \mu\rangle\,, \tag{12.78}$$

which leads to the eigenvalues $\frac{1}{2}\,[j\,(j+1) - \ell\,(\ell+1) - s\,(s+1)]$. The Hamiltonian (12.75) therefore has the following eigenvalues:

$$H \mid (\ell\ s)\ j\ \mu \rangle = \left[\frac{p^2}{2m} + V(r)\right.$$

$$\left. + \frac{1}{2}\ f(r)\ [j(j+1) - \ell\,(\ell+1) - s\,(s+1)]\right] \mid (\ell\ s)\ j\ \mu \rangle$$

(12.79)

By projecting the coupled states $\mid (\ell\ s)\ j\ \mu \rangle$ onto configuration space, we obtain coupled spherical harmonics:

$$\langle \Omega \mid (\ell\ s)\ j\mu \rangle = Y_{(\ell\ s)\ j\ \mu}(\Omega).$$

(12.80)

These coupled spherical harmonics can be expressed in terms of the usual spherical harmonics by using the Clebsch–Gordan series for decomposing $\mid j\ \mu \rangle$ over $\mid \ell\ m \rangle$ and $\mid s\sigma \rangle$:

$$Y_{(\ell\ s)\ j\ \mu}(\Omega) = \sum_{m\ \sigma} Y_{\ell\ m}(\Omega)\mid s\ \sigma \rangle\ \langle \ell\ m\ s\ \sigma \mid j\ \mu \rangle$$

(12.81)

It is easily shown (in particular, graphically) that these coupled spherical harmonics satisfy a closure relation of the same type as the ordinary spherical harmonics (the summation will then be over $\ell sj\mu$):

$$\sum_{\ell\ s\ j\ \mu} Y^*_{(\ell\ s)j\ \mu}(\Omega)\ Y_{(\ell\ s)\ j\ \mu}(\Omega') = \delta\,(\Omega' - \Omega),$$

(12.82)

whereas the orthonormalization relation will take the usual form:

$$\int Y^*_{(\ell\ s)\ j\ \mu}(\Omega)\ Y_{(\ell'\ s')\ j'\ \mu'}(\Omega)\ d\,\Omega = \delta_{jj'}\ \delta_{\mu\ \mu'}\ \delta_{\ell\ell'}\ \delta_{ss'}.$$

(12.83)

The Hamiltonian (12.75) can therefore be decomposed over the complete base of the coupled SH, which boils down to projecting (12.79) onto configuration space.

11. Solid Spherical Harmonics

By harmonic polynomial or ℓth-order solid spherical harmonic we mean the function

$$Y_{\ell\, m}(\vec{r}) = r^\ell \, Y_{\ell\, m}(\Omega) = \begin{array}{c} \vec{r} \quad \mathrm{lm} \\ \text{\textbf{→→}—\textbf{→}} \end{array} . \tag{12.84}$$

This leads, in particular, to the fact that the first-order harmonic polynomial is proportional to the vector $\vec{r}$ expressed in standard spherical coordinates. We shall return to this result in the next chapter (see (13.217)):

$$Y_{1\mu}(\vec{r}) = \frac{i}{c} \left[\frac{3}{4\pi}\right]^{1/2} r_{1\mu} \text{ with } |c|^2 = 1. \tag{12.85}$$

The decomposition relation for a solid spherical harmonic of the sum of two vectors is written

$$Y_{\ell\, m}(s\vec{r}_1 + t\vec{r}_2) = \sum_{m_1} B_{\ell\ell_1} \langle \ell_1\, m_1\, (\ell - \ell_1)(m - m_1)|\ell m\rangle$$
$$\times Y_{\ell_1\, m_1}(s\vec{r}_1)\, Y_{(\ell-\ell_1)(m-m_1)}(t\,\vec{r}_2). \tag{12.86}$$

with the expansion coefficient

$$B_{\ell\ell_1} = \sum_{\ell_1 = 0}^{\ell} \left[\frac{4\pi\,(2\ell + 1)!}{(2\ell_1 + 1)!(2\ell - 2\ell_1 + 1)!}\right]^{1/2}, \tag{12.87}$$

which in graphical representation gives

$$Y_{\ell\, m}(s\vec{r}_1 + t\vec{r}_2) = B_{\ell\ell_1} \quad \begin{array}{c} \vec{sr}_1 \quad l_1 \\ \text{→→} \searrow \quad \nearrow \quad \text{\scriptsize 1} \\ + \!\!\!\times\!\!\!—\text{→} \\ \text{→→} \nearrow \quad \searrow_{(l\text{-}l_1)} \\ \vec{tr}_2 \end{array} . \tag{12.88}$$

12. Schrödinger Equation in a Central Potential

The motion of a particle with reduced mass μ in a central potential $V(r)$ is given by the stationary Schrödinger Eq. (3.12) which we recall here:

$$\left[-\frac{\hbar^2}{2\mu}\,\nabla^2 + V(r)\right] \psi(\vec{r}) = E\,\psi(\vec{r}). \tag{12.89}$$

Let us express the kinetic part of the Hamiltonian in spherical coordinates:

$$-\frac{\hbar^2}{2\mu}\,\nabla^2 = -\frac{\hbar^2}{2\mu}\,\frac{1}{r^2}\,\frac{\partial}{\partial r}\,r^2\,\frac{\partial}{\partial r}$$
$$-\frac{\hbar^2}{2\mu}\,\frac{1}{r^2}\left[\frac{1}{\sin\theta}\,\frac{\partial}{\partial\theta}\,\sin\theta\,\frac{\partial}{\partial\theta} + \frac{1}{\sin^2\theta}\,\frac{\partial^2}{\partial\varphi^2}\right], \tag{12.90}$$

and introduce once more the square of the angular momentum vector (12.3):

$$L^2 = -\hbar^2 \left[\frac{1}{\sin\theta} \frac{\partial}{\partial\theta} \sin\theta \frac{\partial}{\partial\theta} + \frac{1}{\sin^2\theta} \frac{\partial^2}{\partial\varphi^2} \right],$$

which gives the Laplacean in spherical coordinates:

$$-\frac{\hbar^2}{2\mu} \nabla^2 = \frac{1}{2\mu} P_r^2 + \frac{L^2}{2\mu\,r^2}, \tag{12.91}$$

with the radial component of momentum $\vec{P}$:

$$P_r = -i\,\hbar\,\frac{1}{r}\frac{\partial}{\partial r}\,r\,. \tag{12.92}$$

It should be noted here that the correspondence principle is only applicable in cartesian coordinates as we can see in the preceding form of P_r.

The Schrödinger equation in a central potential therefore takes the following form:

$$\left[\frac{P_r^2}{2\mu} + \frac{L^2}{2\mu\,r^2} + V(r) \right] \psi(\vec{r}) = E\,\psi(\vec{r})\,. \tag{12.93}$$

We once more have the "effective" potential of classical mechanics:

$$V_{eff}(r) = V(r) + \frac{L^2}{2\mu\,r^2}\,.$$

The spherical harmonics $Y_{\ell\,m}(\hat{r})$ are eigenfunctions of L^2. We can therefore look for solutions of the form

$$\psi(\vec{r}) = \sum_{\ell m} \psi_\ell(r)\,Y_{\ell m}(\hat{r})\,. \tag{12.94}$$

Inserting this into the Schrödinger Eq. (12.93), we obtain for orbital momentum ℓ

$$\left[\frac{P_r^2}{2\mu} + \frac{L^2}{2\mu\,r^2} + V(r) \right] \psi_\ell(r)\,Y_{\ell m}(\hat{r}) = E_{n\,\ell}\,\psi_\ell(r)\,Y_{\ell m}(\hat{r}) \tag{12.95}$$

and we further use the eigenvalues $\hbar^2\,\ell(\ell+1)$ of the operator L^2 to obtain the equation of the radial wave function:

$$\left[\frac{P_r^2}{2\mu} + \frac{\hbar^2}{2\mu\,r^2}\,\ell(\ell+1) + V(r) - E_{n\,\ell} \right] \psi_\ell(r) = 0\,, \tag{12.96}$$

or, by expanding P_r with (12.92),

$$-\frac{\hbar^2}{2\mu}\frac{1}{r}\frac{d^2}{d\,r^2}(r\,\psi_\ell(r)) + \frac{\hbar^2}{2\mu}\frac{\ell(\ell+1)}{r^2}\,\psi_\ell(r) + (V - E_{n\,\ell})\,\psi_\ell(r) = 0\,. \tag{12.97}$$

Now, let us change functions by setting

$$r\,\psi_\ell(r) = \varphi_\ell(r)\,. \tag{12.98}$$

By simply rearranging terms, the above Schrödinger equation becomes

$$\left[-\frac{\hbar^2}{2\mu} \frac{d^2}{d\,r^2} + V\,(r) + \frac{\hbar^2}{2\mu} \frac{\ell\,(\ell+1)}{r^2} - E_{n\,\ell} \right] \varphi_\ell\,(r) = 0\,. \qquad (12.99)$$

This equation is formally the same as a one-dimensional Schrödinger equation if we add the centrifugal term

$$\frac{\hbar^2}{2\mu} \frac{\ell\,(\ell+1)}{r^2} \quad \text{to the potential } V(r)$$

that is, if we replace $V(r)$ with $V_{eff}\,(r)$.

The functions $\psi_{\ell m}\,(\vec{r}) = \psi_\ell\,(r)\,Y_{\ell m}\,(\hat{r})$ are orthonormal:

$$\int \psi_{\ell m}^*\,(\vec{r})\,\psi_{\ell' m'}\,(\vec{r})\,\vec{dr} = \int Y_{\ell m}^*\,(\hat{r})\,Y_{\ell' m'}\,(\hat{r})\,d\hat{r}$$
$$\times \int \psi_\ell^*\,(r)\,\psi_{\ell'}\,(r)\,r^2\,dr = \delta_{\ell\ell'}\,\delta_{mm'} \qquad (12.100)$$

hence yielding, with relations (12.30) and (12.98), the normalization

$$\int_0^\infty |\,\varphi_\ell\,(r)\,|^2\,dr = 1\,. \qquad (12.101)$$

The radial functions $\varphi_\ell\,(r)$ satisfy a one-dimensional normalization relation. In addition, they show that the operator P_r is Hermitian only if the square integrable functions $\psi_\ell\,(r)$ satisfy the boundary conditions:

$$\lim_{r \to 0} [r\,\psi\,(r)] = 0\,. \qquad (12.102)$$

In other words, the radial function $\varphi_\ell\,(r)$ should be regular at the origin and satisfy the condition:

$$\varphi_\ell\,(0) = 0 \qquad (12.103)$$

The solution $\varphi_\ell\,(r)$ should, in addition, be bounded in the whole space.

12.1 Free Particle

If the interaction potential is zero, then statement (12.99) will lead to the radial Schrödinger equation

$$-\frac{\hbar^2}{2\mu} \frac{d^2}{d\,r^2} \varphi_\ell\,(r) + \frac{\hbar^2}{2\mu} \frac{\ell\,(\ell+1)}{r^2} \varphi_\ell\,(r) = E\,\varphi_\ell\,(r)\,. \qquad (12.104)$$

We set $k^2 = 2\mu E/\hbar^2$ to simplify this equation as follows:

$$\left[\frac{d^2}{d\,r^2} - \frac{\ell\,(\ell+1)}{r^2} + k^2 \right] \varphi_\ell\,(r) = 0\,. \tag{12.105}$$

The fundamental state $\ell = 0$ is easily obtained and is written thus:

$$\varphi_0\,(r) = A\,\sin\,k\,r + B\,\cos\,k\,r$$

and, as a result of initial regularity condition (12.103) at the origin,

$$\varphi_0\,(0) = 0 \qquad \text{gives} \qquad \varphi_0\,(r) = A\,\sin\,k\,r\,. \tag{12.106}$$

Following normalization, we obtain the solution representing the fundamental state of orbital angular momentum $\ell = 0$:

$$\varphi_0\,(r) = \sqrt{\frac{2}{\pi}}\,\sin\,k\,r\,. \tag{12.107}$$

In the general case, the solution obtained for the radial part is of the spherical Bessel function type:

$$\psi_\ell\,(r) = A\,j_\ell\,(k\,r) + B\,\eta_\ell\,(k\,r)\,, \tag{12.108}$$

where $j_\ell\,(k\,r)$ and $\eta_\ell\,(k\,r)$ are first- and second-order spherical Bessel functions. If the particle is moving over the whole space, the solution being finite for $r = 0$, this will impose $B = 0$ and we obtain a solution of the type $\psi_\ell\,(r) = A\,j_\ell\,(k\,r)$.

The solutions of the Schrödinger equation for a free particle in a three-dimensional space will therefore take the form

$$\psi_{\ell m}\,(\vec{r}) = A\,j_\ell\,(k\,r)\,Y_{\ell m}\,(\hat{r})\,. \tag{12.109}$$

This will lead to the expansion (12.74) of the plane wave in terms of spherical harmonics.

12.2 Infinitely Deep Spherical Wells

Consider the potential

$$V\,(r) = \begin{cases} 0 & \text{if} \quad r \le a\,, \\ \infty & \text{if} \quad r \ge a\,. \end{cases} \tag{12.110}$$

Because the potential is zero when the particle is in the well, the solutions are of the Bessel function (12.109) type:

$$\psi_{\ell m}\,(k\,\vec{r}) = A\,j_\ell\,(k\,r)\,Y_{\ell m}\,(\hat{r})\,. \tag{12.111}$$

Because the potential outside is infinite, here the solutions are zero. The matching of internal and external wave functions at the edge of the potential well will lead to the following constraint:

$$j_\ell (k\,a) = 0. \tag{12.112}$$

If we denote the roots of the spherical Bessel functions $X_{n\,\ell}$,

$$k\,a = X_{n\,\ell}, \tag{12.113}$$

then the energy levels of the stationary states can be written

$$E_{n\,\ell} = \frac{\hbar^2\,k^2}{2\mu} = \frac{\hbar^2}{2\mu}\frac{X_{n\,\ell}^2}{a^2} \qquad n = 1,2,3\dots \quad \ell = 0,1,2. \tag{12.114}$$

It is by knowing the roots $X_{n\,\ell}$ of the spherical Bessel functions $j_\ell (k\,r)$ that we can define the value of the energy levels $E_{n\,\ell}$ for a given well of width a:

State $(n\,\ell)$	$X_{n\,\ell}$
$(10) = 1S$	3.142
$(11) = 1P$	4.493
$(12) = 1D$	5.763
$(20) = 2S$	6.283
$(13) = 1F$	6.988
$(21) = 2P$	7.725
—— —	—— —

12.3 Harmonic Well

The Schrödinger equation with potential $V(r) = \frac{1}{2}\,\mu\,\omega^2\,r^2$ is written

$$\left(-\frac{\hbar^2}{2\mu}\frac{d^2}{d\,r^2} + \frac{1}{2}\,\mu\,\omega^2\,r^2 + \frac{\hbar^2\,\ell\,(\ell+1)}{2\mu\,r^2} - E_{n\,\ell}\right)\varphi_{n\,\ell}(r) = 0. \tag{12.115}$$

By making the following changes of variable:

$$a = \left(\frac{\hbar}{\mu\,\omega}\right)^{1/2}, \quad \xi = \frac{r}{a}, \quad \varepsilon = \frac{E_{n\,\ell}}{\hbar\,\omega} = 2\left(n+s+\frac{1}{4}\right), \tag{12.116}$$

$$\ell\,(\ell+1) = 4s\left(s - \frac{1}{2}\right), \qquad \text{and} \qquad Z = \xi^2, \tag{12.117}$$

we are led to a solution of the type

$$\varphi_{n\,\ell}\,(\xi) = \exp\left(-\frac{Z}{2}\right) W\,(Z)\,. \tag{12.118}$$

The function $W\,(Z)$ satisfies the second-order differential equation

$$\left[Z\,\frac{d^2}{d\,Z^2} + \left(2s + \frac{1}{2} - Z\right)\frac{d}{d\,Z} + n\right]W\,(Z) = 0\,, \tag{12.119}$$

corresponding to a confluent hypergeometric series and the limit at infinity of the variable Z leading to the limit $\varphi_{n\,\ell}(\xi) \;\to\; 0$ imposes the conditions

$$n = 0, 1, 2, \ldots \qquad \text{and} \qquad s = \frac{1}{2}\,(\ell + 1)\,. \tag{12.120}$$

This defines the energy levels of the three-dimensional harmonic oscillator

$$E_{n\,\ell} = \hbar\,\omega\left(2n + \ell + \frac{3}{2}\right) \qquad n, \ell = 0, 1, 2, \ldots \tag{12.121}$$

and the corresponding wave functions.

The energy $E_{n\,\ell}$ depends solely on the combination $2n + \ell = \lambda = 0, 1, 2\ldots$, termed principal quantum number. Each λ value can be obtained with different combinations of n and ℓ; for $\lambda \geq 2$, the states are therefore degenerate.

The states of the harmonic well are denoted by the letter $S\ P\ D\ F\ \ldots$ for $\ell = 0, 1, 2, \ldots$ and by the number $n + 1$, where n is the maximum power of the polynomial defining the hypergeometric series.

For example, state $1S$ corresponds to $n = 0$, $\ell = 0$ and $1P$ to the quantum numbers $n = 0$ and $\ell = 1$.

The energy levels of a harmonic well and the corresponding wave functions will then be

$E_\lambda/\hbar\,\omega$	$\lambda = 2n + \ell$	$(n+1)\,\ell$	$\pi^{1/4}\,\frac{1}{\xi}\,\varphi_{n\,\ell}\,(\xi)$
$3/2$	0	$1S$	$2\exp\left(-\xi^2/2\right)$
$5/2$	1	$1P$	
$7/2$	2	$2S, 1D$	$\sqrt{\frac{8}{3}}\,\xi\,\exp\left(-\xi^2/2\right)$
$9/2$	3	$2P, 1F$	
$11/2$	4	$3S, 2D, 1G$	$\sqrt{\frac{8}{3}}\,\left(\xi^2 - \frac{3}{2}\right)\exp\left(-\xi^2/2\right)$

The computation of the energy levels of a particle with reduced mass μ in a potential well is the basis of the shell model in nuclear physics. A comparison of the results obtained with an infinitely deep spherical well and with a harmonic well shows that the experimental values lie between these two solutions. The introduction of a spin–orbit potential $f(r)\,\vec{L}\cdot\vec{S}$, developed in Sect. 10 of this chapter makes it possible to obtain the correct value for the energy levels and to explain the magic numbers $2, 8, 20, \ldots$ giving greater stability to the nuclei constituted by these numbers of nucleons (see Fig. 1).

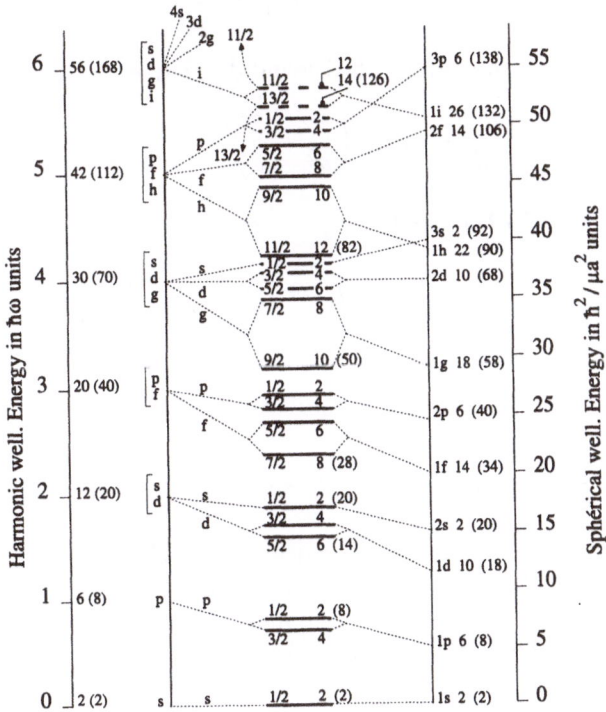

Figure 1. Shell model with two kinds of potential wells. The spectroscopic notation gives the number of angular momentum states $2(2L+1)$. In bracket the total number of occupied states.

12.4 The Hydrogenic Systems

The radial Schrödinger equation can be written as follows:

$$-\frac{d^2\,\varphi_\ell}{d\,r^2} + \left[K\,V(r) + \frac{\ell\,(\ell+1)}{r^2}\right]\varphi_\ell(r) = K\,E_{n\,\ell}\,\varphi_\ell(r) \qquad (12.122)$$

with $K = 2\mu/\hbar^2$ and $V(r) = -Ze^2/r$ for a hydrogenic atom with Z protons. It must be recalled that $\alpha = e^2/4\pi\,\varepsilon_0\hbar\,c = 1/137$.

Fundamental State S ($\ell = 0$)

The Schrödinger equation can be simplified as follows:

$$\frac{d^2\,\varphi_0}{d\,r^2} + K\left(\frac{Z\,e^2}{r} + E_{n0}\right)\varphi_0 = 0.$$ (12.123)

We are interested in decreasing exponential solutions

$$\varphi_0 = C\,\frac{r}{a}\,\exp\left(-\frac{r}{a}\right),$$ (12.124)

where C and a are constants. The second derivative is

$$\frac{d^2\,\varphi_0}{d\,r^2} = \frac{C}{a}\left(\frac{r}{a^2} - \frac{2}{a}\right)\exp\left(-\frac{r}{a}\right)$$

which we insert into Eq. (12.123):

$$\frac{r}{a^2} - \frac{2}{a} + Z\,K\,e^2 + K\,E_{n0}\,r = 0.$$ (12.125)

By cancelling out the terms in r, we obtain a relation for determining the parameter a and the Bohr radius a_0 obtained for $Z = 1$:

$$a = \frac{2}{K\,Z\,e^2} \simeq \frac{\hbar^2}{m\,e^2\,Z} \simeq \frac{a_0}{Z} \simeq \frac{0.5}{Z}\,10^{-10}\ \mathrm{m}$$ (12.126)

The following physical constants are usually introduced:
$a_0 = 4\pi\varepsilon_0\hbar^2/m_e e^2 = r_e/\alpha^2 = 0.529\ 177\ 249\ 10^{-10}\,\mathrm{m}$. The classical electron radius $r_e = e^2/4\pi\varepsilon_0\,m_e c^2$ is equal to $2.817\ 940\ 92\ 10^{-15}\,\mathrm{m}$
When r is large, the terms r/a^2 and $K\,E_{n0}\,r$ become overwhelming and we are left with

$$E_0^{(Z)} = -\frac{1}{K\,a^2} = -\frac{Z^2}{K\,a_0^2}$$

$$\simeq -\frac{m\,Z^2\,e^4}{2\hbar^2} = -13.6\,Z^2\ (\mathrm{eV}).$$ (12.127)

The normalization of φ_0 further yields

$$\int \varphi_0^2\,dr = \frac{C^2}{a^2}\int r^2\,e^{-\frac{2r}{a}}\,dr = \frac{C^2}{a^2}\frac{1}{4}\,a^3 = 1,$$ (12.128)

which determines the value of the parameter C:

$$C = \frac{2}{\sqrt{a}},$$ (12.129)

thus giving the fundamental state in the form:

$$\varphi_0 \left(r\right) = 2ra^{-3/2}\ \exp\ \left(-r/a\right) \ . \tag{12.130}$$

If for $\ell \neq 0$ we seek a solution of the form

$$\varphi_\ell \left(r\right) = \exp\left(-\frac{r}{a}\right) G_\ell(r), \tag{12.131}$$

then the function G_ℓ will satisy the differential equation

$$\frac{d^2\,G_\ell}{d\,r^2} - \frac{2}{a}\frac{d\,G_\ell}{d\,r} + \frac{1}{a^2}\,G_\ell + \left[K\,E_{n\,\ell} + \frac{K\,Z\,e^2}{r} - \frac{\ell\,(\ell+1)}{r^2}\right]G_\ell = 0. \tag{12.132}$$

Now, let us define function $G_\ell \left(r\right)$ in the form of a series:

$$G_\ell \left(r\right) = \sum_k A_k^\ell\,r^k\,. \tag{12.133}$$

By equating terms of the same order, we obtain

$$A_k^\ell\,r^{k-2}\,\left[k(k-1) - \ell(\ell+1) + \left(-\frac{2k}{a} + KZ\,e^2\right)r + \left(\frac{1}{a^2} + KE_{n\ell}\right)r^2\right] = 0 \tag{12.134}$$

which, after cancelling out coefficients of the power r^{k-2}, yields

$$k\,(k-1) - \ell\,(\ell+1) = 0 \quad\text{and}\quad k = \ell+1\,. \tag{12.135}$$

We determine A_k by recursion to obtain

$$\frac{A_{k+1}}{A_k} = \frac{(\ell+1 - Z\,K\,e^2\,a + k)}{(2\ell+2+k)(k+1)}\,\frac{2}{a} \tag{12.136}$$

which shows that the solution is a confluent hypergeometric series:

$$_1F_1\left(\ell+1-n\ ;\ 2\ell+2\ ;\ \frac{2r}{a}\right) \tag{12.137}$$

with the parameters

$$n = \frac{1}{2}\,K\,Z\,e^2\,a \quad\text{where}\quad a = \frac{2n}{K\,Z\,e^2} = \frac{n}{Z}\,a_0 \quad\text{and}\quad n \geq \ell+1\,. \tag{12.138}$$

The energy levels of the hydrogenic atoms are obtained by cancelling out the term in r^k in (12.134), which yields $E_{n\,\ell} = -\frac{1}{ka^2}$ and leads to the following form:

$$E_n^{(Z)} = -KZ^2e^4/4n^2 = E_0^{(1)}\,Z^2/n^2 \ . \tag{12.139}$$

We are left with the following normalized solution:

$$\varphi_{n\,\ell}(r) = N_{n\,\ell}\,\exp\left(-r/a\right)\left(\frac{2r}{a}\right)^{\ell+1}\,_1F_1\left(\ell+1-n\;;\;2\ell+2\;;\;\frac{2r}{a}\right).$$

(12.140)

The behavior of the radial solution at the origin is of the form

$$r \to 0 \qquad \varphi_{n\,\ell}(r) \sim \left(\frac{2r}{a}\right)^{\ell+1}$$

(12.141)

whereas, at infinity, it corresponds to

$$r \to \infty \qquad \varphi_{n\,\ell}(r) \sim \exp\left(\frac{-r}{a}\right).$$

(12.142)

The normalization constant is, for its part,

$$N_{n\,\ell} = \frac{\sqrt{Z}}{n\,(2\ell+1)!}\left[\frac{(n+\ell)!}{(n-\ell-1)!}\right]^{1/2}.$$

(12.143)

In most cases it is the integral corresponding to the expectation value of r^k that is used:

$$\langle r^k \rangle = I^k\,(n\,\ell, n'\,\ell') = \int \varphi_{n\,\ell}(r)\,r^k\,\varphi_{n'\,\ell'}(r)\,dr.$$

(12.144)

The normalization relation yields

$$\langle r^0 \rangle = I^0\,(n\,\ell, n\,\ell) = \int \varphi_{n\,\ell}(r)\,\varphi_{n\,\ell}(r)\,dr = 1.$$

(12.145)

The moments for a hydrogenic atom with electric charge Ze are thus

$$\langle r \rangle = \frac{1}{2Z}\left[3n^2 - \ell\,(\ell+1)\right]a_0,$$

$$\langle r^2 \rangle = \frac{n^2}{2Z^2}\left[5n^2 + 1 - 3\ell\,(\ell+1)\right]a_0^2,$$

$$\left\langle\frac{1}{r}\right\rangle = \frac{Z}{n^2}\frac{1}{a_0},$$

$$\left\langle\frac{1}{r^2}\right\rangle = \frac{Z^2}{n^3\,(\ell+\frac{1}{2})}\frac{1}{a_0^2},$$

$$\left\langle\frac{1}{r^3}\right\rangle = \frac{Z^3}{n^3\,(\ell+1)\,(\ell+\frac{1}{2})\,\ell}\frac{1}{a_0^3}.$$

(12.146)

This leads to the mean square deviation

$$\Delta r^2 = \langle(r - \langle r \rangle)^2\rangle = \langle r^2 \rangle - \langle r \rangle^2$$

$$= \frac{a^2}{4Z^2}\left(n^4 + 2n^2 - \ell^2\,(\ell+1)^2\right)$$

(12.147)

measuring the eccentricity of the electron orbit around the nucleus.

Chapter 13
Irreducible Tensor Operators

1. Definitions

The kth-order irreducible tensor operator (ITO) is a tensor T_{kq} with $(2k+1)$ components $(-k \leq q \leq +k)$ that are transformed by rotation of the coordinate system into a linear combination of the same components:

$$T'_{k\,q} = R\,T_{k\,q}\,R^+ = \sum_p D^k_{p\,q}\,(\bar{g})\,T_{k\,p}. \tag{13.1}$$

From the expression of the infinitesimal rotation operator $R_v(\varepsilon)$ through an angle ε around the Ov axis, the Israeli physicist G. Racah was able to give an equivalent definition of ITO's using the following procedure (Racah, 1942)

We use the generator of infinitesimal rotations through the ε angle around Ov axis (11.8),

$$R_v\,(\varepsilon) = 1 - i\,\varepsilon\,J_v, \tag{13.2}$$

to write the element of the rotation matrix

$$\begin{aligned} D^k_{p\,q}\,(\varepsilon) &= \langle k\,p \mid R_v\,(\varepsilon) \mid k\,q \rangle = \langle k\,p \mid 1 - i\,\varepsilon\,J_v \mid k\,q \rangle \\ &= \delta_{p\,q} - i\,\varepsilon\,\langle k\,p \mid J_v \mid k\,q \rangle. \end{aligned} \tag{13.3}$$

If we insert this into the definition (13.1) of ITOs, we obtain the following statement:

$$(1 - i\,\varepsilon\,J_v)\,T_{k\,q}\,(1 + i\,\varepsilon\,J_v) = \sum_p \left[\delta_{p\,q} - i\,\varepsilon\,\langle k\,p \mid J_v \mid k\,q \rangle\right]\,T_{k\,p}, \tag{13.4}$$

or, by making a first-order expansion in ε:

$$\boxed{[J_v, T_{k\,q}] = \sum_p \langle k\,p \mid J_v \mid k\,q \rangle\,T_{k\,p}} . \tag{13.5}$$

i) The choice of the Ov axis as Oz, defines the matrix element

$$\langle k\,p \mid J_z \mid k\,q \rangle = q\,\delta_{p\,q} \tag{13.6}$$

and (13.5) gives the commutator

$$[J_z, T_{k\,q}] = q\,T_{k\,q}$$

(13.7)

ii) By choosing the Ov axis to be Ox, and then Oy, and by forming the linear combination $J_x \pm i\,J_y = J_\pm$, the matrix element (13.5) becomes:

$$\langle k\,p\,|\,J_\pm\,|\,k\,p\rangle = [(k \pm q + 1)\,(k \mp q)]^{1/2}\,\delta_{p,q\pm 1}.$$

(13.8)

This transforms (13.5) into

$$[J_\pm, T_{k\,q}] = [(k \pm q + 1)\,(k \mp q)]^{1/2}\,T_{k,q\pm 1}$$

(13.9)

Relations (13.7) and (13.9) are termed the Racah definitions of ITOs.

2. Graphical Representation

The transformation by rotation of the spherical harmonics:

$$Y_{\ell\,m}\,(\theta',\,\varphi') = Y_{\ell\,m}\,(\theta,\,\varphi)' = \sum_n D^\ell_{n\,m}\,(\bar{g})\,Y_{\ell\,m}\,(\theta,\,\varphi)$$

(13.10)

shows that spherical harmonics are in fact ℓth-order ITOs projected onto the configuration space $\{\Omega\}$. They are diagonal in this space:

$$\begin{aligned} \langle \Omega\,|\,\Omega'\rangle &= \delta\,(\Omega - \Omega') \\ \int |\,\Omega\rangle\,d\Omega\langle\Omega\,| &= \mathbb{1} \\ \langle \Omega\,|\,Y_{\ell\,m}\,|\,\Omega'\rangle &= Y_{\ell\,m}(\Omega)\,\delta\,(\Omega - \Omega') \end{aligned}$$

(13.11)

The graphical representation of the spherical harmonics is (see (12.40))

$$Y_{\ell\,m}\,(\Omega) = \langle \Omega\,|\,\ell\,m\rangle = \begin{array}{c}\Omega \quad \text{lm} \\ \text{---}\!\!\blacktriangleright\!\text{---}\!\!\text{I}\!\text{---}\!\!\blacktriangleright\end{array}.$$

(13.12)

Now let us use the same procedure to represent the ITOs graphically, that is, by introducing a fictitious space $\{U\}$ (except for the spherical harmonics where the considered space is the configuration space), determined by or-thonormalization and completeness relations (13.11) and in which the ITO $T_{k\,q}$ is diagonal:

$$\langle u \mid u' \rangle \; = \delta \, (u - u')$$

$$\int \mid u \rangle \, du \rangle u \mid \; = \mathbb{1}$$

$$\langle u \mid T_{k\,q} \mid u' \rangle \; = \delta \, (u - u') \, T_{k\,q} \, (u)$$

(13.13)

The graphical representation of the ITOs will be as follows:

$$T_{k\,q} \, (u) = \; \langle u \mid k \, q \rangle \; = \quad \overset{u}{\underset{}{\bullet}}\!\!\!\!\!\!\!\rightarrow\!\!\!\!\!\overset{kq}{\underset{}{\vdash}}\!\!\!\!\rightarrow .$$

(13.14)

3. Hermitian Conjugate

Let us return to spherical harmonics by considering the Hermitian conjugate:

$$[\langle \Omega \mid \ell \, m \rangle]^{+} = \; \langle \ell \, m \mid \Omega \rangle$$
$$= [Y_{\ell\,m} \, (\Omega)]^{+} = Y^{+}_{\ell\,m} \, (\bar{\Omega}) = (-)^{\ell-m} \, Y_{\ell m} \, (\bar{\Omega}). \quad (13.15)$$

The variables $\bar{\Omega}$ are obtained by parity transformation:

$$Y_{\ell-m} \, (\bar{\Omega}) = P \, Y_{\ell-m} \, (\Omega) = (-)^{\ell} \, Y_{\ell-m} \, (\Omega), \quad (13.16)$$

which yields the relation between the Hermitian conjugate and the complex conjugate of the spherical harmonic:

$$Y^{+}_{\ell\,m} \, (\bar{\Omega}) = (-)^{\ell-m} \, Y_{\ell-m} \, (\bar{\Omega})$$
$$= (-)^{m} \, Y_{\ell-m} \, (\Omega) = Y^{*}_{\ell\,m} \, (\Omega) = \; \langle \ell \, m \mid \Omega \rangle. \quad (13.17)$$

or by the transformation of the parity of the space variable:

$$Y^{+}_{\ell\,m} \, (\Omega) = (-)^{\ell} \, Y^{*}_{\ell\,m} \, (\Omega). \quad (13.18)$$

By applying a second Hermitian conjugation to (13.15), we obtain:

$$[Y^{+}_{\ell\,m} \, (\bar{\Omega})]^{+} = Y^{++}_{\ell\,m} \, (\Omega) = (-)^{2\ell} \, Y_{\ell\,m} \, (\Omega) = Y_{\ell\,m} \, (\Omega), \quad (13.19)$$

because the orbital momentum is always an integer. The graphical representation of (13.17) is

$$\langle \ell \, m \mid \Omega \rangle \; = Y^{+}_{\ell\,m} \, (\bar{\Omega}) = Y^{*}_{\ell\,m} \, (\Omega) = \quad \overset{\Omega}{\underset{}{\bullet}}\!\!\!\!\leftarrow\!\!\!\!\overset{lm}{\underset{}{\vdash}}\!\!\!\!\leftarrow . \quad (13.20)$$

We use the kth-order ITOs in a similar manner:

$$(R \, T_{k\,q} \, R^{+})^{+} = R \, T^{+}_{k\,q} \, R^{+} = \left[\sum_{p} D^{k}_{p\,q} \, (\bar{g}) \, T_{k\,p} \right]^{+}$$

$$= \sum_{p} D^{k}_{p\,q} \, (\bar{g})^{*} \, T^{+}_{k\,p}. \quad (13.21)$$

Let us further use the complex conjugation of the rotation matrix to obtain the form

$$R\left((-)^{k-q} T_{kq}^+\right) R^+ = \sum_p D^k_{-p-q}(\bar{g}) \left((-)^{k-p} T_{kp}^+\right). \tag{13.22}$$

If we compare this with the transformation by rotation of T_{k-p} :

$$R\left(T_{k-q}\right) R^+ = \sum_p D^k_{-p-q}(\bar{g}) T_{k-p}, \tag{13.23}$$

we infer that $(-)^{k-q} T_{kq}^+$ is transformed by rotation just like T_{k-q}. Interestingly, this is the result obtained for the transformation by rotation of a ket. We therefore write a relation analogous to (13.15):

$$\langle k\,q \mid u \rangle = \left[T_{kq}(u)\right]^+ = T_{kq}^+(\bar{u}) = (-)^{k-q} T_{k-q}(\bar{u}). \tag{13.24}$$

The generalization of the parity transformation of the spherical harmonic will further yield

$$T_{k-q}(\bar{u}) = P\,T_{k-q}(u) = (-)^k\,T_{k-q}(u), \tag{13.25}$$

and, just like G. Racah, we introduce the complex conjugate of the ITO:

$$T_{kq}^*(u) = (-)^q\,T_{k-q}(u) = (-)^k\,T_{kq}^+(u) \tag{13.26}$$

which finally generalizes (13.18) as

$$T_{kq}^+(\bar{u}) = (-)^{k-q}\,T_{k-q}(\bar{u}) = (-)^q\,T_{k-q}(u) = T_{kq}^*(u)$$

$$= \langle k\,q \mid u \rangle = \overset{kq \quad u}{\xrightarrow{\hspace{1cm}}}. \tag{13.27}$$

By applying a second Hermitian conjugation, we obtain

$$\left[T_{kq}^+(\bar{u})\right]^+ = T_{kq}^{++}(u) = (-)^{2k}\,T_{kq}(u) = T_{kq}(u). \tag{13.28}$$

It is therefore necessary for k to be a positive integer.

4. Scalar Product

The kth-order scalar product of two ITOs is defined by the relation:

$$(T_k(u_1) \cdot U_k(u_2)) = \sum_q T_{kq}(u_1)\,U_{kq}^+(\bar{u}_2) = \sum_q T_{kq}(u_1)\,U_{kq}^*(u_2). \tag{13.29}$$

The graphical representation derives from (13.27) and (13.14):

$$(T_k\,(u_1)\cdot U_k\,(u_2)) = \sum_q \langle u_1\mid k\,q\rangle\,\langle k\,q\mid u_2\rangle \;=\; \overset{u_1\quad k\quad u_2}{\longrightarrow\!\!\!\longrightarrow\!\!-\!\!\!\longrightarrow\!\!\cdots\!\!\gg}. \qquad (13.30)$$

We notice in this way that the ℓth-order Legendre polynomials are, to within a normalization coefficient, the scalar products of the ℓth-order spherical harmonics of the directions Ω_1 and Ω_2:

$$(Y_\ell\,(\Omega_1).Y_\ell\,(\Omega_2)) = \sum_m \langle \Omega_1\mid \ell\,m\rangle\,\langle \ell\,m\mid \Omega_2\rangle = \sum_m Y_{\ell\,m}\,(\Omega_1)\,Y^*_{\ell\,m}\,(\Omega_2)$$

$$= \frac{[\ell^2]}{4\pi}\,P_\ell\,(\cos\,(\Omega_1,\Omega_2)). \qquad (13.31)$$

5. Tensor Product

5.1 Definition

The kth-order tensor product generalizes the scalar product and corresponds to the definition

$$\pi_{k\,q}\,(u_1,\,u_2) = \sum_{q_1\,q_2} \langle k_1\,q_1\,k_2\,q_2\mid k\,q\rangle\,T_{k_1\,q_1}\,(u_1)\,T_{k_2\,q_2}\,(u_2)$$

$$= \quad \cdots = [k] \qquad (13.32)$$

or, by inverting this relation using Clebsch–Gordan series one gets

$$T_{k_1\,q_1}\,(u_1)\,T_{k_2\,q_2}\,(u_2)$$

$$= \quad = \sum_k \quad = \sum_k [k^2] \qquad (13.33)$$

The analytical transcription of this statement further yields:

$$T_{k_1\,q_1}\,(u_1)\,T_{k_2\,q_2}\,(u_2) = \sum_{k\,q} \pi_{k\,q}\,(u_1,\,u_2)\langle k\,q\mid k_1\,q_1\,k_2\,q_2\rangle$$

$$= \sum_{k\,q} \pi_{k\,q}\,(u_1,\,u_2)\,[k] \begin{pmatrix} q & k_2 & k_1 \\ k & q_2 & q_1 \end{pmatrix}. \qquad (13.34)$$

If we consider the kth-order tensor product constructed from an ITO and a transpose, we obtain

$$\pi_{k\,q}\,(\bar{u}_1, u_2) =$$

$$= \sum_{q_1\,q_2} T^+_{k_1\,q_1}\,(\bar{u}_1)\,T_{k_2\,q_2}\,(u_2)\,(-)^{k_1+q_1}\,\langle k_1 - q_1\,k_2\,q_2 \mid k\,q\rangle,$$

(13.35)

which yields, through the application of the transition rule,

$$\pi_{k\,q}\,(\bar{u}_1, u_2) = [k]$$

$$= \sum_{q_1\,q_2} T^*_{k_1\,q_1}\,(u_1)\,T_{k_2\,q_2}\,(u_2)\,[k]\,\begin{pmatrix} k_1 & k & q_2 \\ q_1 & q & k_2 \end{pmatrix}.$$
(13.36)

5.2 Zero-Order Tensor Product

The zeroth-order tensor is easily calculated by eliminating the line $k = 0$, $q = 0$ in diagram (13.36):

$$\pi_{00}\,(\bar{u}_1, u_2) =$$

$$= [k_1^{-1}]$$

(13.37)

and, by comparing with definition (13.30) of the inner product,

$$\pi_{00}\,(\bar{u}_1\,u_2) = [k_1^{-1}]\,(T_{k_1}\,(u_1)\cdot T_{k_1}\,(u_2)).$$
(13.38)

The zeroth-order tensor product is, to within a normalization coefficient, the scalar product of the tensors of the product.

5.3 nth-Order Tensor Product

The structure of $\pi_{k\,q}\,(u_1, u_2)$ shows that we are dealing with an ITO although we can still verify this analytically by transforming by rotation each of the terms of (13.32) and by summing the result over q_1 and q_2.

By applying relation (13.32) step-by-step, it is possible to construct a k_nth-order ITO from the irreducible tensors $T_{k_1\,q_1}\cdots T_{k_{n-1}\,q_{n-1}}$.

$$\pi_{k_n\,q_n}\,(u_1,\ldots u_{n-1}) = \begin{array}{c}\text{(diagram)}\end{array} \quad s=1,2\ldots n-3$$

$$= \sum_{q_1\cdots q_{n-1}} \langle k_1\,q_1\ldots k_{n-1}\,q_{n-1}\mid (k_1\ldots k_{n-1})_A\,k_s\,;\,k_n\,q_n\rangle$$

$$\times T_{k_1\,q_1}\,(u_1)\ldots T_{k_{n-1}\,q_{n-1}}\,(u_{n-1}). \tag{13.39}$$

The definition of the Hermitian conjugate of an ITO is compatible with that of the tensor product of the two ITOs, that is,

$$\left[\pi_{k\,q}\,(u_1,\,u_2)\right]^{+} = \pi^{+}_{k\,q}\,(\bar{u}_1,\,\bar{u}_2) = (-)^{k-q}\,\pi_{k-q}\,(\bar{u}_1,\,\bar{u}_2). \tag{13.40}$$

The parity transformation of the variables u_1 and u_2 further yields

$$\pi^{+}_{k\,q}\,(\bar{u}_1,\,\bar{u}_2) = (-)^{k-q}\,(-)^{k_1+k_2}\,\pi_{k-q}\,(u_1,\,u_2). \tag{13.41}$$

If we set, as for the spherical harmonic,

$$(-)^{q}\,\pi_{k-q}\,(u_1,\,u_2) = \pi^{*}_{k\,q}\,(u_1,\,u_2), \tag{13.42}$$

we obtain the following significant result:

$$\pi^{+}_{k\,q}\,(\bar{u}_1,\,\bar{u}_2) = (-)^{k+k_1+k_2}\,\pi^{*}_{k\,q}\,(u_1,\,u_2) = \begin{array}{c}\text{(diagram)}\end{array}. \tag{13.43}$$

By changing variables $\bar{u}_1$ into u_1 and $\bar{u}_2$ into u_2, the coefficient $(-)^{k_1+k_2}$ disappears and relation (13.26) is generalized as follows:

$$\pi^{+}_{k\,q}\,(u_1,\,u_2) = (-)^{k}\,\pi^{*}_{k\,q}\,(u_1,\,u_2). \tag{13.44}$$

6. Standardization of a Vector Operator

6.1 General Procedure for Standardization

A vector operator $\vec{A}$ is an operator transformed by rotation R of the coordinate system according to the usual law:

$$\vec{A}' = R\,\vec{A}\,R^{+}, \tag{13.45}$$

yielding, for an infinitesimal rotation ε around the Oz axis:

$$\vec{A}\,' = R_z\,(\varepsilon)\,\vec{A}\,R_z^+(\varepsilon) = (1 - i\,\varepsilon\,J_z)\,\vec{A}\,(1 + i\,\varepsilon\,J_z)$$
$$= \vec{A} - i\,\varepsilon\,[J_z,\,\vec{A}]. \tag{13.46}$$

The vector operator $\vec{A}$ is also a vector whose elements are tranformed by infinitesimal rotation around the Oz axis according to the usual law (11.5):

$$A_x' = A_x + \varepsilon\,A_y\,,$$
$$A_y' = -\varepsilon\,A_x + A_y\,, \tag{13.47}$$
$$A_z' = A_z\,.$$

The term-by-term comparison of Eqs. (13.46) and (13.47) hence defines the commutators

$$[J_z,\,A_x] = i\,A_y\,,$$
$$[J_z,\,A_y] = -i\,A_x\,, \tag{13.48}$$
$$[J_z,\,A_z] = 0\,.$$

A first-order ITO will have three elements, A_{11}, A_{10} and A_{1-1}, and the Racah definition will lead to the commutation relation

$$[J_z,\,A_{1q}] = q\,A_{1q} \quad \text{with} \quad q = (1, -1, 0)\,. \tag{13.49}$$

The element A_{10} in this ITO obeys the commutation relation

$$[J_z,\,A_{10}] = 0\,. \tag{13.50}$$

It is then only too natural to consider that the elements A_z of the vector operator $\vec{A}$ and A_{10} of the irreducible tensor operator A_{1q} are proportional. We will therefore set

$$A_{10} = c\,(-i\,A_z)\,, \tag{13.51}$$

where c is an arbitrary coefficient. Some authors (e.g., Fano and Racah, 1959) have set $c = 1$ and others (e.g., Edmonds, 1959; Messiah, 1959; Wigner, 1959; Rose, 1955) $c = i$, which amounts to setting $A_{10} = A_z$. For our part, we will continue to use the coefficient c for the time being and determine the constraints that need to be satisfied to make the best of this choice of coefficient.

Associated with relation (13.51), the Racah definition (13.9) leads us to write

$$[J_\pm,\,A_{10}] = [J_\pm, -i\,c\,A_z] = -i\,c\,[J_\pm,\,A_z] = \sqrt{2}\,A_{1\pm1}\,. \tag{13.52}$$

By replacing $J_\pm$ by its statement in term sof J_x and J_y, we obtain the value of the elements A_{11} and A_{1-1} of the ITO A_{1q}, that is,

$$A_{1\pm1} = -\frac{i\,c}{\sqrt{2}}\,\left([J_x,\,A_z] \pm i\,[J_y,\,A_z]\right)\,. \tag{13.53}$$

We can proceed to evaluate the commutators by generalizing the formula introduced at the beginning of this section. Consider a vector operator $\vec{A}$, and a unitary vector $\vec{a}$. The projection of this operator will be transformed through the infinitesimal rotation ε around the (Ou) axis in a form similar to (13.46):

$$A_a = \vec{A}.\vec{a}$$
$$A'_a = R_u\,(\varepsilon)\,A_a\,R_u^+\,(\varepsilon) = A_a - i\,\varepsilon\,[J_u,\,A_a]\,. \tag{13.54}$$

The unitary vector $\vec{a}$ is transformed by rotation according to the relation

$$\vec{a}' = \vec{a} + \varepsilon\,(\vec{u} \wedge \vec{a})\,, \tag{13.55}$$

leading to the projection of the vector operator A'_a:

$$A'_a = \vec{A} \cdot \vec{a}' = \vec{A} \cdot (\vec{a} + \varepsilon\,(\vec{u} \wedge \vec{a})) = A_a + \varepsilon\,\vec{A} \cdot (\vec{u} \wedge \vec{a})\,. \tag{13.56}$$

By identification with (13.55), we obtain the generalization of the commutation relations (13.48), that is,

$$\boxed{[J_u,\,A_a] = i\,\vec{A} \cdot (\vec{u} \wedge \vec{a})} \;. \tag{13.57}$$

Special Case No. 1: $Ou = Oz$ and $Oa = Oz$:

$$[J_z,\,A_z] = i\,\vec{A} \cdot (\vec{e}_3 \wedge \vec{e}_3) = 0\,. \tag{13.58}$$

Special Case No. 2: $Ou = Ox$ and $Oa = Oz$:

$$[J_x,\,A_z] = i\,\vec{A} \cdot (\vec{e}_1 \wedge \vec{e}_3) = -i\,\vec{A} \cdot \vec{e}_2 = -i\,A_y\,. \tag{13.59}$$

Special Case No. 3: $Ou = Oy$ and $Oa = Oz$:

$$[J_y,\,A_z] = i\,\vec{A} \cdot (\vec{e}_2 \wedge \vec{e}_3) = i\,\vec{A} \cdot \vec{e}_1 = i\,A_x\,. \tag{13.60}$$

By substituting these results into (13.53), we obtain with (13.51) the standardization of the vector operator $\vec{A}$:

$$A_{11} = \frac{i\,c}{\sqrt{2}}\,(A_x + i\,A_y)\,,$$
$$A_{10} = -i\,c\,A_z\,, \tag{13.61}$$
$$A_{1-1} = -\frac{i\,c}{\sqrt{2}}\,(A_x - i\,A_y)\,.$$

It is easy to extract the Cartesian components with respect to the standard spherical components:

$$A_x = \frac{i}{c\,\sqrt{2}}\,(-A_{11} + A_{1-1})\,,$$
$$A_y = -\frac{1}{c\,\sqrt{2}}\,(A_{11} + A_{1-1})\,, \tag{13.62}$$
$$A_z = \frac{i}{c}\,A_{10}\,.$$

6.2 Standard Spherical Basis

If we introduce the unit vectors $\vec{e}_r$ $(r = 1, 2, 3 = x, y, z)$ of the Cartesian basis:

$$\vec{e}_r \cdot \vec{e}_s = \delta_{r\,s} \qquad (r, s) = (1, 2, 3) \tag{13.63}$$

and the unit vectors $\vec{e}_{1\,\mu}$ $(\mu = 1,\ 0,\ -1)$ of the standard spherical basis:

$$\vec{e}_{1\,\mu}^{+} \cdot \vec{e}_{1\,\mu'} = \delta_{\mu\,\mu'} \qquad (\mu, \mu') = (1,\ 0,\ -1) \tag{13.64}$$

relations (13.61) and (13.62) will give directly the change of basis:

$$\vec{e}_{1\,\mu} \begin{cases} \vec{e}_{11} = \dfrac{i\,c}{\sqrt{2}}\,(\vec{e}_1 + i\,\vec{e}_2) \\[2mm] \vec{e}_{10} = -i\,c\,\vec{e}_3 \\[2mm] \vec{e}_{1-1} = -\dfrac{i\,c}{\sqrt{2}}\,(\vec{e}_1 - i\,\vec{e}_2) \end{cases} \quad \begin{cases} \vec{e}_1 = \dfrac{i}{c\,\sqrt{2}}\,(-\vec{e}_{11} + \vec{e}_{1-1}) \\[2mm] \vec{e}_2 = -\dfrac{1}{c\,\sqrt{2}}\,(\vec{e}_{11} + \vec{e}_{1-1}) \\[2mm] \vec{e}_3 = \dfrac{i}{c}\,\vec{e}_{10} \end{cases} \tag{13.65}$$

The unit vectors of the standard spherical basis may be considered as the projections of the state of the angular momentum $(1,\ \mu)$ onto the configuration space $\{\hat{e}\}$ defined by the usual orthonormalization and completeness relations:

$$\boxed{\begin{aligned} \langle \hat{e} \mid \hat{e}' \rangle &= \delta\,(\hat{e} - \hat{e}') \\[2mm] \int \mid \hat{e} \rangle\, d\,\hat{e}\, \langle\, \hat{e} \mid &= \mathbb{1} \end{aligned}} \tag{13.66}$$

We therefore define by projection onto the configuration space $\{\hat{e}\}$:

$$\vec{e}_{1\,\mu}\,(\hat{e}) = \langle \hat{e} \mid 1\,\mu \rangle \quad \text{and} \quad \vec{e}_{1\,\mu}^{+}\,(\hat{e}) = \langle 1\,\mu \mid \hat{e} \rangle = \vec{e}_{1\,\mu}^{*}\,(\hat{e}). \tag{13.67}$$

The generalization of relation (13.27) between Hermitian conjugate and complex conjugate leads us to set

$$\vec{e}_{1\,\mu}^{*}\,(\hat{e}) = (-)^{\mu}\,\vec{e}_{1-\mu}\,(\hat{e}) \tag{13.68}$$

and the orthonormalization relation (13.69) thus reduces to

$$\begin{aligned} \int \langle 1\,\mu \mid \hat{e} \rangle\, d\,\hat{e}\, \langle \hat{e} \mid 1\,\mu' \rangle &= \int \vec{e}_{1\,\mu}^{*}\,(\hat{e})\, d\hat{e}\, \vec{e}_{1\,\mu'}\,(\hat{e}) \\[2mm] &= \langle 1\,\mu \mid 1\,\mu' \rangle = \delta_{\mu\,\mu'}. \end{aligned} \tag{13.69}$$

We may also consider that the unit vectors $\vec{e}_r$ along the Cartesian axes are the projections of the vector $\mid r \rangle$ onto the configuration space:

$$\vec{e}_r = \langle \hat{e} \mid r \rangle \quad \text{and} \quad \vec{e}_r^{+} = \langle r \mid \hat{e} \rangle. \tag{13.70}$$

In a three-dimensional Cartesian space, the covariant components are not distinguishable from the contravariant components such that

$$\vec{e}_r^{\,+} = \langle r \mid \hat{e} \rangle = \langle \hat{e} \mid r \rangle = \vec{e}_r \tag{13.71}$$

and the orthonormalization relation (13.63) can also take the form

$$\int \langle r \mid \hat{e} \rangle \, d\,\hat{e} \, \langle \hat{e} \mid s \rangle = \int \vec{e}_r \, (\hat{e}) \cdot \vec{e}_s \, (\hat{e}) \, d\,\hat{e} = \vec{e}_r \cdot \vec{e}_s = \delta_{r\,s}. \tag{13.72}$$

The change of basis (13.65) can therefore be written with a change of basis matrix:

$$\vec{e}_{1\,\mu} = \langle \hat{e} \mid 1\,\mu \rangle = \sum_r \langle \hat{e} \mid r \rangle \, \langle r \mid 1\,\mu \rangle = \sum_r \vec{e}_r \, \langle r \mid 1\,\mu \rangle,$$

$$\vec{e}_r = \langle \hat{e} \mid r \rangle = \sum_\mu \langle \hat{e} \mid 1\,\mu \rangle \, \langle 1\,\mu \mid r \rangle = \sum_\mu \vec{e}_{1\,\mu} \, \langle 1\,\mu \mid r \rangle. \tag{13.73}$$

By using standardization (13.65), we define the Λ matrix of change of basis element $\langle r \mid 1\,\mu \rangle$:

$r \backslash \mu$	1	0	-1
1	$ic/\sqrt{2}$	0	$-ic/\sqrt{2}$
2	$-c/\sqrt{2}$	0	$-c/\sqrt{2}$
3	0	$-ic$	0

$\mu \backslash r$	1	2	3
1	$-i/c\sqrt{2}$	$-1/c\sqrt{2}$	0
0	0	0	i/c
-1	$i/c\sqrt{2}$	$-1/c\sqrt{2}$	0

Matrix elements $\Lambda_r^\mu = \langle r \mid 1\,\mu \rangle$.   Matrix elements $\Lambda_\mu^r = \langle 1\,\mu \mid r \rangle$.

Generally speaking, a vector operator $\vec{A}$ may be considered as the projection of the action space $\{\hat{A}\}$ of the vector operator onto the configuration space $\{\hat{e}\}$:

$$\vec{A} = \langle \hat{e} \mid \hat{A} \rangle = \langle \hat{A} \mid \hat{e} \rangle \tag{13.74}$$

yielding, in Cartesian coordinates,

$$\vec{A} = \sum_r \langle \hat{e} \mid r \rangle \, \langle r \mid \hat{A} \rangle = \sum_r \vec{e}_r \, A_r, \tag{13.75}$$

and, in standard spherical coordinates, the contravariant and covariant components

$$\vec{A} = \sum_{\mu} \langle \hat{e} \mid 1 \, \mu \rangle \, \langle 1 \, \mu \mid \hat{A} \rangle \; = \sum_{\mu} \vec{e}_{1\mu}(\hat{e}) \, A_{1\mu}^{+}(\hat{e})$$

$$= \sum_{\mu} \langle \hat{A} \mid 1 \, \mu \rangle \, \langle 1 \, \mu \mid \hat{e} \rangle \; = \sum_{\mu} \vec{e}_{1\mu}^{+}(\hat{e}) \, A_{1\mu}(\hat{e}) \,. \tag{13.76}$$

The scalar product of two vector operators $\vec{A}$ and $\vec{B}$ thus becomes

$$\langle \hat{A} \mid \hat{B} \rangle \; = \; \langle \hat{B} \mid \hat{A} \rangle \; = \sum_{r} \langle \hat{A} \mid r \rangle \langle r \mid \hat{B} \rangle \; = \sum_{r} A_r \, B_r = \vec{A} \cdot \vec{B}$$

$$= \sum_{\mu} \langle \hat{A} \mid 1 \, \mu \rangle \langle 1 \, \mu \mid \hat{B} \rangle$$

$$= \sum_{r} A_{1\,\mu} \, B_{1\,\mu}^{+} = \sum_{\mu} A_{1\,\mu}^{+} \, B_{1\,\mu} \,. \tag{13.77}$$

By replacing the standard spherical components with Cartesian components (13.61), we obtain another expression for the scalar product:

$$\langle \hat{A} \mid \hat{B} \rangle \; = A_{11}^{+} \, B_{11} + A_{1-1}^{+} \, B_{1-1} + A_{10}^{+} \, B_{10} = \; |c \mid^2 \vec{A} \cdot \vec{B} \tag{13.78}$$

If we compare forms (13.77) and (13.78) of the inner product, we notice the constraint on the parameter c of the change (13.56) of coordinates:

$$|c \mid^2 \, = 1 \,. \tag{13.79}$$

6.3 Graphical Representations

Let us use the usual graphical representation of the basis vectors of the standard spherical representation and the configuration space $\{\hat{e}\}$:

$$\vec{e}_{1\,\mu} = \langle \hat{e} \mid 1 \, \mu \rangle = \quad \overset{\hat{e} \quad 1\mu}{\rule{0pt}{0pt}} \quad \text{and} \quad \vec{e}_{1\,\mu}^{+} = \langle 1 \, \mu \mid \hat{e} \rangle = \quad \overset{\hat{e} \quad 1\mu}{\rule{0pt}{0pt}} \tag{13.80}$$

By introducing the ITO action space $\{\hat{A}\}$ in the standard representation of the vector operator $\vec{A}$, we obtain

$$A_{1\,\mu} = \langle \hat{A} \mid 1 \, \mu \rangle = \quad \overset{\hat{A} \quad 1\mu}{\rule{0pt}{0pt}} \quad \text{and} \quad A_{1\,\mu}^{+} = \langle 1 \, \mu \mid \hat{A} \rangle = \quad \overset{1\mu \quad \hat{A}}{\rule{0pt}{0pt}} \,. \tag{13.81}$$

The graphical representation of a vector operator is therefore obtained from (13.76) by summing over the μ indices:

$$\vec{A} = \sum_{\mu} \vec{e}_{1\,\mu} \, A_{1\,\mu}^{+} = \quad \overset{\hat{e} \quad 1 \quad \hat{A}}{\rule{0pt}{0pt}} \quad = \; \hat{e} \overset{1}{\rule{0pt}{0pt}} \hat{A}$$

$$= \sum_{\mu} \vec{e}_{1\,\mu}^{+} \, A_{1\,\mu} = \quad \overset{\hat{e} \quad 1 \quad \hat{A}}{\rule{0pt}{0pt}} \quad = \; \hat{e} \overset{1}{\rule{0pt}{0pt}} \hat{A} \,. \tag{13.82}$$

There is therefore no need to indicate the direction of the kinetic arrow, thus simplifying the graphical representation:

$$\vec{A} = \hat{e} \xrightarrow{\;\;1\;\;} \hat{A}. \tag{13.83}$$

The scalar product of the two vector operators thus becomes

$$\langle \hat{A} \mid \hat{B} \rangle = \hat{A} \xrightarrow{\;\;1\;\;} \hat{B} = \vec{A} \cdot \vec{B}. \tag{13.84}$$

We notice that by graphically representing the unit vectors $\vec{e}_r$ of the Cartesian basis in a form that is analogous to (13.80), we are able to completely generalize the graphical representation of the vector operators:

$$\vec{e}_r = \langle \hat{e} \mid r \rangle = \langle r \mid \hat{e} \rangle = \hat{e} \xrightarrow{\;\;r\;\;} = \hat{e} \xrightarrow{\;\;r\;\;}, \tag{13.85}$$

$$A_r = \langle \hat{A} \mid r \rangle = \langle r \mid \hat{A} \rangle = \hat{A} \xrightarrow{\;\;r\;\;} = \hat{A} \xrightarrow{\;\;r\;\;}. \tag{13.86}$$

By summing over the Cartesian index $r = 1, 2, 3 = x, y, z$ we obtain the graphical representation of the vector operator:

$$\vec{A} = \sum_r \vec{e}_r \, A_r = \hat{e} \xrightarrow{\quad} \hat{A} = \hat{e} \xrightarrow{\quad} \hat{A}. \tag{13.87}$$

We can therefore very generally represent a vector operator in the form of (13.87) and, depending on the cut of the diagram made on $r = 1, 2, 3$ or $\mu = 1, 0, -1$, we obtain the Cartesian or standard spherical components:

$$
\begin{aligned}
\vec{A} &= \hat{A} \xrightarrow{\quad} \hat{e} \\
&= \sum_\mu \hat{A} \xrightarrow{\;1\mu\;} \xrightarrow{\;1\mu\;} \hat{e} \\
&= \sum_r \hat{A} \xrightarrow{\;r\;} \xrightarrow{\;r\;} \hat{e}
\end{aligned}
\tag{13.88}
$$

The scalar product (13.84) is also expressed in Cartesian or standard spherical coordinates by introducing constraint (13.79) on c:

$$
\begin{aligned}
\langle \hat{A} \mid \hat{B} \rangle &= \vec{A} \cdot \vec{B} = \hat{A} \xrightarrow{\quad} \hat{B} \\
&= \sum_\mu \hat{A} \xrightarrow{\;1\mu\;} \xrightarrow{\;1\mu\;} \hat{B} = \sum_r \hat{A} \xrightarrow{\;r\;} \xrightarrow{\;r\;} \hat{B} \\
&= \sum_\mu A_{1\,\mu} \, B^+_{1\,\mu} = \sum_r A_r \, B_r.
\end{aligned}
\tag{13.89}
$$

6.4 Tensor Product of Vector Operators

Let us consider the tensor product of two vector operators standardized using procedure (13.61):

$$\pi_{k\,q}(\hat{A},\hat{B}) = \quad \overset{\hat{A}\diagdown 1}{\underset{\hat{B}\diagup 1}{+}}\!\!\xrightarrow{kq} \;=\; \sum_{mn} \langle 1m\,1n \mid k\,q\rangle\, A_{1\,n}(\hat{A})\, B_{1\,n}(\hat{B})$$

$$ = [k] \quad \overset{\hat{A}\diagdown 1}{\underset{\hat{B}\diagup 1}{\longrightarrow}}\!\!\xrightarrow{kq} \;=\; \sum_{mn} [k] \begin{pmatrix} n & m & k \\ 1 & 1 & q \end{pmatrix} A_{1\,m}(\hat{A})\, B_{1\,n}(\hat{B}).$$

$$(13.90)$$

The quantum number k takes the values 0, 1, or 2, whereas q lies between k and $-k$ following the usual selection rule

i) Scalar Product

$$\pi_{00}(\hat{A},\hat{B}) = \quad \overset{\hat{A}\diagdown 1}{\underset{\hat{B}\diagup 1}{\Big\rangle}} \;=\; \frac{1}{\sqrt{3}} \; \hat{A}\!\xrightarrow{\ 1\ }\!\hat{B} \;=\; \frac{1}{\sqrt{3}} \,\vec{A}\cdot\vec{B}. \qquad (13.91)$$

ii) Vector Product

$$\pi_{1\,q}(\hat{A},\hat{B}) = \sqrt{3}\quad \overset{\hat{A}\diagdown 1}{\underset{\hat{B}\diagup 1}{\longrightarrow}}\!\!\xrightarrow{1q} \;=\sqrt{3}\sum_{mn} \begin{pmatrix} n & m & 1 \\ 1 & 1 & q \end{pmatrix} A_{1\,m}(\hat{A})\, B_{1\,n}(\hat{B}).$$

$$(13.92)$$

Let us evaluate one of the components $\pi_{10}(\hat{A},\hat{B})$, for example

$$\pi_{10}(\hat{A},\hat{B}) = \sqrt{3}\sum_{n} \begin{pmatrix} 1 & 1 & 1 \\ n & -n & 0 \end{pmatrix} A_{1-n}(\hat{A})\, B_{1\,n}(\hat{B}). \qquad (13.93)$$

The $3jm$ coefficient equals $(-)^{1-n}\, n/\sqrt{6}$ (see Chap. 10, Appendix 1). We thus obtain:

$$\pi_{10}(\hat{A},\hat{B}) = \frac{1}{\sqrt{2}}\,(A_{11}\,B_{1-1} - A_{1-1}\,B_{11}). \qquad (13.94)$$

The standardization procedure (13.61) yields the Cartesian form of this same component:

$$\pi_{10}(\hat{A},\hat{B}) = \frac{-i\,c^2}{\sqrt{2}}\,(A_x\,B_y - A_y\,B_x) = \frac{-i\,c^2}{\sqrt{2}}\,(\vec{A}\wedge\vec{B})_z. \qquad (13.95)$$

If we standardize the vector product with (13.62), we obtain

$$(\vec{A} \wedge \vec{B})_z = \frac{i}{c} (\vec{A} \wedge \vec{B})_{10}, \tag{13.96}$$

which gives, after inserting into $\pi_{10} (\hat{A}, \hat{B})$,

$$\pi_{10} (\hat{A}, \hat{B}) = \frac{c}{\sqrt{2}} (\vec{A} \wedge \vec{B})_{10}. \tag{13.97}$$

A similar, but lengthier, calculation of π_{11} and π_{1-1} finally leads to the general result

$$\pi_{1q} (\hat{A}, \hat{B}) = \frac{c}{\sqrt{2}} (\vec{A} \wedge \vec{B})_{1q} = -\sqrt{3} \sum_{mn} \begin{pmatrix} m & n & 1 \\ 1 & 1 & q \end{pmatrix} A_{1m} (\hat{A}) B_{1n} (\hat{B}). \tag{13.98}$$

We infer the expression for the standardized vector product:

$$(\vec{A} \wedge \vec{B})_{1q} = -\frac{\sqrt{6}}{c} \sum_{mn} \begin{pmatrix} m & n & 1 \\ 1 & 1 & q \end{pmatrix} A_{1m} B_{1n}. \tag{13.99}$$

By opting for $c = -1$ we avoid the sign $(-)$, which finally gives the $3jm$ value as a function of the Levi–Civita coefficient:

$$\begin{pmatrix} m & n & 1 \\ 1 & 1 & q \end{pmatrix} = \frac{1}{\sqrt{6}} \varepsilon^{m\,n}{}_{q} \qquad (m, n, q) = (1, 0, -1). \tag{13.100}$$

If we graphically represent the Levi–Civita tensor in Cartesian coordinates using a diagram similar to a $3jm$ but with a pole specific to Cartesian coordinates:

$$\varepsilon_{rst} = \quad \text{(diagram)} \quad = \sqrt{6} \quad \text{(diagram)}, \tag{13.101}$$

we obtain the graphical representations of the vector product in any basis whatsoever, be it standard Cartesian or spherical:

$$\vec{A} \wedge \vec{B} = \sum_{rst} \varepsilon_{rst} A_r B_s \vec{e}_t = \quad \text{(diagram)} \tag{13.102}$$

$$= \sum_{\mu\nu\lambda} \varepsilon^{\mu\nu\lambda} A_{1\mu} B_{1\nu} \vec{e}_{1\lambda} = \sqrt{6} \quad \text{(diagram)}. \tag{13.103}$$

To go then from Cartesian coordinates to spherical standard coordinates, we replace the Cartesian pole with a standard pole assigned the coefficient $\sqrt{6}$

using the following standardization procedure:

$$\vec{e}_{1\,\mu}\begin{cases} \vec{e}_{11} = \dfrac{-i}{\sqrt{2}}\,(\vec{e}_1 + i\,\vec{e}_2) \\[2mm] \vec{e}_{10} = i\,\vec{e}_3 \\[2mm] \vec{e}_{1-1} = \dfrac{i}{\sqrt{2}}\,(\vec{e}_1 - i\,\vec{e}_2) \end{cases} \qquad \vec{e}_r\begin{cases} \vec{e}_1 = \dfrac{i}{\sqrt{2}}\,(\vec{e}_{11} - \vec{e}_{1-1}) \\[2mm] \vec{e}_2 = \dfrac{1}{\sqrt{2}}\,(\vec{e}_{11} + \vec{e}_{1-1}) \\[2mm] \vec{e}_3 = -i\,\vec{e}_{10}\,. \end{cases} \qquad (13.104)$$

The foregoing results lead to the graphical vector analysis (GVA) presented in Appendix 2, at the end of this chapter. Let us simply add that the commutation relations of the angular momentum $\vec{J}$, that is,

$$\vec{J} \wedge \vec{J} = i\,\vec{J} \qquad (13.105)$$

will be graphically expressed as follows:

$$(13.106)$$

For the Pauli matrices $\vec{S} = \tfrac{1}{2}\,\vec{\sigma}$ we obtain

$$(13.107)$$

iii) Tensor Product

For $k = 2$, we obtain the tensor product of two vector operators:

$$(13.108)$$

whose components in the standard spherical basis are:

$$\pi_{2\pm2} = -\frac{1}{2}\,(A_{1\pm1} + B_{1\pm1})$$

$$\pi_{2\pm1} = \frac{1}{\sqrt{2}}\,(A_{10}\,B_{1\pm1} + A_{1\pm1}\,B_{10}) \qquad (13.109)$$

$$\pi_{20} = \frac{1}{\sqrt{6}}\,(A_{11}\,B_{1-1} + A_{1-1}\,B_{11} + 2A_{10}\,B_{10})\,.$$

An important special case here concerns the second-order tensor obtained from the Pauli matrices: it is identically equal to zero:

$$(13.110)$$

iv) Application

Now let us use the graphical representations of the vector operators to evaluate the product:

$$E = (\vec{\sigma} \cdot \vec{A})\,(\vec{\sigma} \cdot \vec{B}) \;=\; \begin{array}{c} \hat{\sigma} \overset{1}{\longrightarrow} \hat{A} \\ \hat{\sigma} \overset{1}{\longrightarrow} \hat{B} \end{array} . \tag{13.111}$$

Let us introduce the intermediate X momentum with the summation rule over j:

$$E = \begin{array}{c} \hat{\sigma} \overset{1}{\longrightarrow} \hat{A} \\ \hat{\sigma} \overset{1}{\longrightarrow} \hat{B} \end{array} = \sum_X [X^2]\; \begin{array}{c} \hat{\sigma} \;1\; X \;1\; \hat{A} \\ \hat{\sigma} \;1\;\;\;\;1\; \hat{B} \end{array} . \tag{13.112}$$

The splitting up of the sum over $X = 0, 1, 2$ will yield:

$$E = \quad\rangle\!\langle \quad + 3\;\rangle\!\langle \quad + 5\;\rangle\!\langle \quad . \tag{13.113}$$

The first term is easily calculated to have the value $\frac{1}{\sqrt{3}}\sigma^2\,\frac{1}{\sqrt{3}}\,\vec{A}\cdot\vec{B}$ and because $\sigma^2 = 3$, we are left with the scalar product $\vec{A}\cdot\vec{B}$. The last term is, as follows from (13.110), zero. The second is transformed with (13.107):

$$3i\,\sqrt{\frac{2}{3}}\;\hat{\sigma}\overset{1}{\longrightarrow}\!\!\begin{array}{c}1\;\hat{A}\\1\;\hat{B}\end{array} = i\sqrt{6}\;\;\hat{\sigma}\overset{1}{\longrightarrow}\!\!\begin{array}{c}1\;\hat{A}\\1\;\hat{B}\end{array} = i\,\vec{\sigma}\cdot(\vec{A}\wedge\vec{B}). \tag{13.114}$$

We are therefore left with the well-known relation introduced in (9.59):

$$(\vec{\sigma}\cdot\vec{A})\,(\vec{\sigma}\cdot\vec{B}) = \vec{A}\cdot\vec{B} + i\,\vec{\sigma}\cdot(\vec{A}\wedge\vec{B}), \tag{13.115}$$

in which the vector operators $\vec{A}$ and $\vec{B}$ do not necessarily commute with one another but do commute with the Pauli operator $\vec{\sigma}$, and do not act in the spin space.

7. Wigner–Eckart Theorem

7.1 Projection Space

To every observable one can associate a Hermitian operator that in turn may be associated with a kth-order ITO. We are thus inevitably led to evaluate the matrix element

$$\langle \alpha\, j\, m \mid T_{k\,q} \mid \alpha'\, j'\, m' \rangle, \tag{13.116}$$

where the quantum numbers α and α' complete the basis spanned by the angular momentum vectors:

$$\langle \alpha\, j\, m \mid \alpha'\, j'\, m' \rangle = \delta_{\alpha\,\alpha'}\, \delta_{j\,j'}\, \delta_{m\,m'}$$

$$\sum_{\alpha\, j\, m} \mid \alpha\, j\, m \rangle \langle \alpha\, j\, m \mid = \mathbb{1}$$

. (13.117)

If we place ourselves in the context of the eigenrepresentation of the ITO to evaluate the matrix element, that is if we introduce the projection space $\{U\}$ defined by the usual closure and orthonormalization relations, then

$$\langle u' \mid u' \rangle = \delta\,(u - u')$$

$$\int \mid u \rangle\, du\, \langle u \mid = \mathbb{1}$$

. (13.118)

Projecting the orthonormalization and completeness relations (13.117) onto the $\{U\}$ space they are transformed into

$$\int \langle \alpha\, j\, m \mid u \rangle\, du\, \langle u \mid \alpha'\, j'\, m' \rangle = \delta_{\alpha\,\alpha'}\, \delta_{j\,j'}\, \delta_{m\,m'}\,,$$

$$\sum_{\alpha\, j\, m} \langle u \mid \alpha\, j\, m \rangle \langle \alpha\, j\, m \mid u' \rangle = \delta\,(u - u'),$$

(13.119)

whereas $T_{k\,q}$ is diagonal in the projection space:

$$\langle u \mid T_{k\,q} \mid u' \rangle = \delta\,(u - u')\, T_{k\,q}(u)$$

. (13.120)

The projection of matrix element (13.116) on the space $\{U\}$ will then give

$$\langle \alpha\, j\, m \mid T_{k\,q} \mid \alpha'\, j'\, m' \rangle = \int \langle \alpha\, j\, m \mid u \rangle\, T_{k\,q}(u)\, \langle u \mid \alpha'\, j'\, m' \rangle\, du.$$ (13.121)

7.2 Reduced Matrix Element

Let us represent the foregoing result graphically and, just like in the integration of the three spherical harmonics, we will introduce a marking circle so as to recall the integration over lines u:

$$\langle \alpha\, j\, m \mid T_{k\,q} \mid \alpha'\, j'\, m' \rangle = \underset{kq}{\overset{\alpha jm \quad \alpha'j'm'}{\longrightarrow}} = \underset{kq}{\overset{+}{\underset{jm}{\bigodot}}} .$$ (13.122)

We analytically associate a reduced matrix element $\langle \alpha\, j \,||\, T_k \,||\, \alpha'\, j' \rangle$ with the marking circle, which leads to the Wigner–Eckart theorem:

$$\langle \alpha\, j\, m \,|\, T_{k\,q} \,|\, \alpha'\, j'\, m' \rangle = \underset{jm}{\overset{+\quad j'm'}{\longrightarrow\!\!\bigcirc\!\!\longleftarrow}}\Big\downarrow{}_{kq} \quad ,$$

$$\langle \alpha\, j\, m \,|\, T_{k\,q} \,|\, \alpha'\, j'\, m' \rangle = \begin{pmatrix} m & k & j' \\ j & q & m' \end{pmatrix} \langle \alpha\, j \,||\, T_{k\,q} \,||\, \alpha'\, j' \rangle. \tag{13.123}$$

The graphical expression for the reduced matrix element or marking circle is given, by comparing these statements, as

$$\langle \alpha\, j \,||\, T_k \,||\, \alpha'\, j' \rangle = \quad \quad . \tag{13.124}$$

A number of points need emphasizing here:

1. The pinching rule over three kinetic lines leads to three kinetic arrows pointing in the same direction in the marking circle whereas the variance is conserved in the $3jm$.
2. The reading order of the pole $(j\ k\ j')$ is indispensable.
3. In the case of spherical harmonics for which

$$\langle \alpha\, \ell \,||\, Y_k \,||\, \alpha'\, \ell' \rangle = \delta_{\alpha\,\alpha'} \frac{[\ell\ k\ \ell']}{\sqrt{4\pi}} \begin{pmatrix} 0 & k & \ell' \\ \ell & 0 & 0 \end{pmatrix} \tag{13.125}$$

the condition of existence of the $3j0$ forces $\ell + k + \ell'$ to be even and therefore renders the pole signature unnecessary.

4. The calculation of the reduced matrix element of a tensor product $\pi_{kq}(u_1, \ldots u_n)$ of the irreducibble tensor operators $T_{k_1 q_1}(u_1) \ldots T_{k_n q_n}(u_n)$ is made in a similar manner, by integrating the angle lines $u_1 \ldots u_n$, three at a time, and by marking the corresponding pole:

$$\langle \alpha\, j \,||\, \pi_k \,||\, \alpha'\, j' \rangle = \quad \quad . \tag{13.126}$$

7.3 Examples of Elementary Reduced Matrix Elements

An example of an elementary reduced matrix element (RME) has already been given in (13.125) above using the spherical harmonic operator. Examples are manifold but the underlying principle remains to directly use the Wigner–Eckart theorem to evaluate the matrix element.

i) Identity Operator
We have

$$\langle \alpha \, j \, m \mid \mathbf{1} \mid \alpha' \, j' \, m' \rangle = \langle \alpha \, j \, m \mid \alpha' \, j' \, m' \rangle = \delta_{\alpha \, \alpha'} \, \delta_{j \, j'} \, \delta_{m \, m'}$$

$$= \begin{pmatrix} m & 0 & j' \\ j & 0 & m' \end{pmatrix} \langle \alpha \, j \, \| \, \mathbf{1} \, \| \, \alpha' \, j' \rangle$$

$$= [j^{-1}] \, \delta_{j \, j'} \, \delta_{m \, m'} \, \langle \alpha \, j \, \| \, \mathbf{1} \, \| \, \alpha' \, j' \rangle, \quad (13.127)$$

from which we infer the reduced matrix element

$$\boxed{\langle \alpha \, j \, \| \, \mathbf{1} \, \| \, \alpha' \, j \rangle = \delta_{\alpha \, \alpha'} \, [j]} \, . \qquad (13.128)$$

ii) Angular Momentum Operator
Let us evaluate the RME for the component J_z using standardization (13.104), that is,

$$J_z = \frac{i}{c} \, J_{10} \, . \qquad (13.129)$$

This will lead to the matrix element:

$$\langle \alpha \, j \, m \mid J_z \mid \alpha' \, j' \, m' \rangle = m' \, \langle \alpha \, j \, m \mid \alpha' \, j' \, m' \rangle = m \, \delta_{\alpha \, \alpha'} \, \delta_{j \, j'} \, \delta_{m \, m'} \, . \qquad (13.130)$$

If we replace J_z with $\frac{i}{c} \, J_{10}$ and apply the Wigner–Eckart theorem

$$\langle \alpha \, j \, m \mid \frac{i}{c} \, J_{10} \mid \alpha' \, j' \, m' \rangle = \frac{i}{c} \begin{pmatrix} m & 1 & j' \\ j & 0 & m' \end{pmatrix} \langle \alpha \, j \, \| \, J_1 \, \| \, \alpha' \, j' \rangle, \quad (13.131)$$

the $3jm$ coefficient will have the value $m \, \delta_{j \, j'}/[j] \, \sqrt{j \, (j+1)} \, \delta_{m \, m'}$, giving

$$\boxed{\langle \alpha \, j \, \| \, J_1 \, \| \, \alpha' \, j' \rangle = -ic \, [j] \, \sqrt{j \, (j+1)} \, \delta_{\alpha \, \alpha'} \, \delta_{j \, j'}} \, . \qquad (13.132)$$

For Pauli matrices with spin $\frac{1}{2}$ for which $\vec{S} = \frac{1}{2} \, \vec{\sigma}$,

$$\boxed{\langle \alpha \, \frac{1}{2} \, \| \, \sigma_1 \, \| \, \alpha' \, \frac{1}{2} \rangle = -ic \, \sqrt{6} \, \delta_{\alpha \, \alpha'}} \, . \qquad (13.133)$$

7.4 Examples of RME Calculation

7.4.1 Coupled Angular Momentum State

By applying relation (13.126), we will develop the diagrams corresponding to
a few special reduced matrix elements, leaving it to the reader to verify the
analytical transcription of some of the results.

$$\langle (j_{12}\, j_3)\, j \parallel T_k \parallel (j'_{12}\, j'_3)\, j' \rangle = - \quad \text{(13.134)}$$

$$= \delta_{j_{12}\, j'_{12}}\, [j\, j']\, (-)^{j_{12}+j'_3+j+k} \begin{Bmatrix} j_{12} & j_3 & j \\ k & j' & j'_3 \end{Bmatrix} \langle j_3 \parallel T_k \parallel j'_3 \rangle . \quad \text{(13.135)}$$

For a spherical harmonic, we obtain

$$\left\langle \left(\ell\, \tfrac{1}{2}\right) j \parallel Y_L \parallel \left(\ell'\, \tfrac{1}{2}\right) j' \right\rangle = \frac{[j\, j'\, \ell\, \ell'\, L]}{\sqrt{4\pi}} \begin{pmatrix} 0 & L & \ell' \\ \ell & 0 & 0 \end{pmatrix}$$

$$\times (-)^{j'+\ell'+\frac{1}{2}} \begin{Bmatrix} \ell & j & \tfrac{1}{2} \\ j' & \ell' & L \end{Bmatrix} . \quad \text{(13.136)}$$

7.4.2 Two-ITO Tensor Product

i) If the ITOs are acting in the same space:

$$\langle \alpha\, j \parallel \pi_k \parallel \alpha'\, j' \rangle = - \quad \text{(13.137)}$$

ii) If the ITOs are acting in different spaces:

$$\langle (j_1\, j_2)\, j \parallel \pi_k \parallel (j'_1\, j'_2)\, j' \rangle = - \quad \text{(13.138)}$$

The analytical transcription of these results is given in Appendix 1, at the end of this chapter.

7.4.3 Two-ITO Scalar Product

The reduced matrix element of the scalar product of two k-order ITOs acting in different spaces $\{u_1\}$ and $\{u_2\}$ is written with (13.126) as follows:

$$\langle (j_1\ j_2)\ j \parallel (T_k.U_k) \parallel (j_1'\ j_2')\ j' \rangle \; = \;$$ 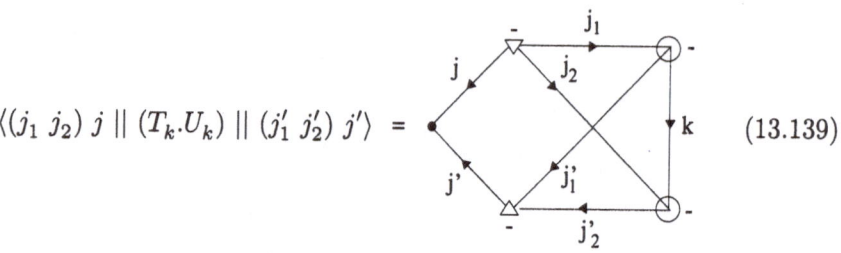 $$(13.139)$$

7.4.4 Vector Operators Scalar Product

The scalar product $\vec{J}\cdot\vec{A}$ expressed with (13.91) and (13.107) leads to the corresponding reduced matrix element:

$$(\vec{J}\cdot\vec{A}) = \sqrt{3}\ \pi_{00}\ (\hat{J},\ \hat{A})\ .$$ (13.140)

We obtain the reduced matrix element diagram:

$$\langle \alpha\ j \parallel \vec{J}\cdot\vec{A} \parallel \alpha'\ j' \rangle \; = \; \sum_{\alpha''j''}$$ (13.141)

$$= \sum_{\alpha''\ j''} (-)^{j+j''+1}\ [j^{-1}]\ \delta_{j\,j'}\ \langle \alpha\ j \parallel J_1 \parallel \alpha''\ j'' \rangle\ \langle \alpha''\ j'' \parallel A_1 \parallel \alpha'\ j' \rangle$$

(13.142)

By using the elementary RME (13.132), we obtain the following result:

$$\langle \alpha\ j \parallel \vec{J}\cdot\vec{A} \parallel \alpha'\ j' \rangle \; = \; \sum_{\alpha''\ j''} (-)^{j+j''+1}\ [j^{-1}]\ \delta_{j\,j'}\ (-ic)\ [j]\ \sqrt{j\ (j+1)}$$

$$\delta_{\alpha\,\alpha''}\ \delta_{j\,j''}\ \langle \alpha''\ j'' \parallel A_1 \parallel \alpha'\ j' \rangle$$

which finally leaves us with the important reduced matrix element:

$$\boxed{\langle \alpha\ j \parallel \vec{J}\cdot\vec{A} \parallel \alpha'\ j' \rangle \; = -ic\ \sqrt{j\ (j+1)}\ \langle \alpha\ j \parallel A_1 \parallel \alpha'\ j' \rangle\ \delta_{j\,j'}}\ .$$

(13.143)

8. Magnetic Moments

8.1 The Landé Formula

The ratio of the matrix elements of the ITOs $A_{1\,\mu}$ and $J_{1\,\mu}$ expressed with the Wigner–Eckart theorem in (13.143) and (13.132) yields:

$$\frac{\langle \alpha\,j\,m \mid A_{1\,\mu} \mid \alpha'\,j\,m' \rangle}{\langle \alpha\,j\,m \mid J_{1\,\mu} \mid \alpha\,j\,m' \rangle} = \frac{\langle \alpha\,j \| A_1 \| \alpha'\,j \rangle}{\langle \alpha\,j \| J_1 \| \alpha\,j \rangle}$$

$$= \frac{\langle \alpha\,j \| \vec{J}.\vec{A} \| \alpha'\,j \rangle}{\sqrt{j\,(j+1)}\,[j]\,\sqrt{j\,(j+1)}} \qquad (13.144)$$

This leads to the matrix element of any vector operator whatsoever as a function of the angular momentum operator:

$$\langle \alpha\,j\,m \mid A_{1\,\mu} \mid \alpha'\,j\,m' \rangle = [j]\,j(j+1)\,\langle \alpha\,j \| \vec{J}.\vec{A} \| \alpha'\,j \rangle$$
$$\times \langle \alpha\,j\,m \mid J_{1\,\mu} \mid \alpha\,j\,m' \rangle \qquad (13.145)$$

The Wigner–Eckart theorem enables us to replace the reduced matrix element by the matrix element of the scalar product:

$$\langle \alpha\,j\,m \mid \vec{J}\cdot\vec{A} \mid \alpha'\,j\,m \rangle = \begin{pmatrix} m & 0 & j \\ j & 0 & m \end{pmatrix}\langle \alpha\,j \| \vec{J}\cdot\vec{A} \| \alpha'\,j \rangle$$
$$= [j^{-1}]\,\langle \alpha\,j \| \vec{J}.\vec{A} \| \alpha'\,j \rangle, \qquad (13.146)$$

which finally leaves us with what has been termed the Landé formula:

$$\langle \alpha\,j\,m \mid A_{1\,\mu} \mid \alpha'\,j\,m' \rangle = \frac{1}{j\,(j+1)}\,\langle \alpha\,j\,m \mid \vec{J}\cdot\vec{A} \mid \alpha'\,j\,m \rangle$$
$$\times \langle \alpha\,j\,m \mid J_{1\,\mu} \mid \alpha'\,j\,m' \rangle. \qquad (13.147)$$

8.2 Magnetic Moment and the Landé Factor

Let us evaluate the expectation value of the magnetic moment $\vec{\mu}$ of a fermion with orbital angular momentum $\vec{\ell}$ and gyromagnetic ratio g_ℓ, with spin $\vec{s}$ and gyromagnetic ratio g_s. This magnetic moment is written with the Bohr magneton $\mu_B = Q\,e\,\hbar/(2m\,c)$, where Q is the number of charges,

$$\vec{\mu} = \mu_B\left(g_e\,\vec{\ell} + g_s\,\vec{s}\right). \qquad (13.148)$$

If a first-order ITO is associated with this magnetic moment, then it becomes possible with the Landé formula, to write

$$\langle(\ell\,s)\,jm \mid \mu_{1m'} \mid (\ell\,s)\,jm \rangle = \frac{1}{j\,(j+1)}\,\langle(\ell\,s)\,jm \mid \vec{J}.\vec{\mu} \mid (\ell\,s)\,jm \rangle$$
$$\times \langle(\ell\,s)\,jm \mid J_{1m'} \mid (\ell\,s)\,jm \rangle. \qquad (13.149)$$

Let us determine the scalar product with $\vec{J} = \vec{\ell} + \vec{s}$ of the magnetic moment $\vec{\mu}$:

$$\vec{J} \cdot \vec{\mu} = \mu_B \, (\vec{\ell} + \vec{s}) \, (g_e \, \vec{\ell} + g_s \, \vec{s})$$
$$= \mu_B \left[g_\ell \, \ell^2 + g_s \, s^2 + (g_\ell + g_s) \, \vec{\ell} \cdot \vec{s} \right] \tag{13.150}$$

and because $\vec{\ell} \cdot \vec{s} = \frac{1}{2} \, (J^2 - \ell^2 - s^2)$, the scalar product becomes

$$\vec{J} \cdot \vec{\mu} = \mu_B \left[\ell^2 (g_\ell - \frac{1}{2} g_\ell - \frac{1}{2} g_s) + s^2 (g_s - \frac{1}{2} g_\ell - \frac{1}{2} g_s) + J^2 \frac{1}{2} (g_\ell + g_s) \right] , \tag{13.151}$$

which can be simplified as follows:

$$\vec{J} \cdot \vec{\mu} = \frac{1}{2} \, \mu_B \left[\ell^2 \, (g_\ell - g_s) + s^2 \, (g_s - g_\ell) + J^2 \, (g_\ell + g_s) \right] . \tag{13.152}$$

The matrix element of $J_{1m'}$ is calculated with the Wigner–Eckart theorem:

$$\langle (\ell \, s) \, jm \mid J_{1m'} \mid (\ell \, s) \, jm \rangle = \begin{pmatrix} m & 1 & j \\ j & m' & m \end{pmatrix} \langle (\ell \, s) \, j \parallel J_1 \parallel (\ell \, s) \, j \rangle \tag{13.153}$$

The selection rule on the $3jm$ magnetic moments imposes the value $m' = 0$, whereas the $3jm$ coefficient itself is expressed as

$$\begin{pmatrix} m & 1 & j \\ j & 0 & m \end{pmatrix} = (-)^{j-m} \begin{pmatrix} j & 1 & j \\ -m & 0 & m \end{pmatrix} = \frac{m}{[j] \, \sqrt{j(j+1)}} . \tag{13.154}$$

Because, following (13.132), the reduced matrix element for $c = -1$ equals $i[j] \sqrt{j(j+1)}$, we are left with

$$\langle (\ell \, s) \, jm \mid J_{1m'} \mid (\ell \, s) \, jm \rangle = i \, m \, \delta_{m' \, 0} . \tag{13.155}$$

Inserting the various results in the matrix element of the magnetic moment, we obtain the expectation value (13.149):

$$\langle \mu_{1m'} \rangle = \langle (\ell \, s) \, jm \mid \mu_{1m'} \mid (\ell \, s) \, jm \rangle = i \, \mu_B \, m \, \delta_{m' \, 0} \, f_L . \tag{13.156}$$

The Landé factor f_L has the following value:

$$f_L = \left[(g_\ell - g_s) \, \frac{\ell(\ell+1) - s \, (s+1)}{2j \, (j+1)} + \frac{1}{2} \, (g_\ell + g_s) \right] . \tag{13.157}$$

8.3 Magnetic Moment of Baryons

The magnetic moment $\vec{\mu}$ of a particle with spin $\vec{J}$ is the mean value of a first-order ITO on the quantum state with the highest weight and by applying the Wicker–Eckart theorem, we obtain

$$\mu = \langle J\,J \mid T_{1q} \mid J\,J\rangle = \begin{pmatrix} J & 1 & J \\ J & q & J \end{pmatrix} \langle J \parallel T_1 \parallel J\rangle. \qquad (13.158)$$

The $3jm$ coefficient has the value

$$(-)^{J-M} \begin{pmatrix} J & 1 & J \\ -M & q & M \end{pmatrix} = \frac{M}{\sqrt{J\,(J+1)\,(2J+1)}}\, \delta_{q0} \quad \text{for} \quad M = J. \qquad (13.159)$$

The magnetic moment (13.158) is hence written with the RME of T_{1q}:

$$\boxed{\langle J \parallel T_1 \parallel J\rangle = \left[\frac{(J+1)\,(2J+1)}{J}\right]^{1/2} \mu}. \qquad (13.160)$$

We only have to evaluate the RME of the irreducible tensor operator T_{1q} to obtain the desired magnetic moment.

Consider a baryon formed by a diquark with spin $\vec{S}$ and a third quark with spin $\vec{s}_3$:

$$\begin{aligned} \vec{S} &= \vec{s}_1 + \vec{s}_2\,, \\ \vec{J} &= \vec{S} + \vec{s}_3\,. \end{aligned} \qquad (13.161)$$

The reduced matrix element of T_{1q} when the latter is acting only on $\vec{s}_3$ is calculated as in (13.134) and gives

$$\langle (S\,s_3)\,J \parallel T_1 \parallel (S\,s_3)\,J\rangle = -$$

$$= (-)^{S+s_3+J+1}[J^2] \begin{Bmatrix} S & s_3 & J \\ 1 & J & s_3 \end{Bmatrix} \langle s_3 \parallel T_1 \parallel s_3\rangle. \qquad (13.162)$$

The reduced matrix element of the action of T_{1q} on the diquark with spin S will also give

$$\langle (S\,s_3)\,J \parallel T_1 \parallel (S\,s_3)\,J\rangle = -$$

$$= (-)^{J+s_3+S+1}[J^2] \begin{Bmatrix} J & J & 1 \\ S & S & s_3 \end{Bmatrix} \langle S \parallel T_1 \parallel S\rangle. \qquad (13.163)$$

The irreducible tensor T_{1q} acting on the spin of the diquark is the sum of the irreducible tensors acting on spins s_1 and s_2 of the diquark $T_{1q}(\hat{s}) = T_{1q}(\hat{s}_1) + T_{1q}(\hat{s}_2)$. By using the foregoing result, we obtain

$$
\langle S \parallel T_1 \parallel S \rangle = [S^2] (-)^{s_1+s_2+S+1} \left[\left\{ \begin{matrix} S & S & 1 \\ s_1 & s_1 & s_2 \end{matrix} \right\} \langle s_1 \parallel T_1 \parallel s_1 \rangle \right.
$$
$$
\left. + \left\{ \begin{matrix} S & S & 1 \\ s_2 & s_2 & s_1 \end{matrix} \right\} \langle s_2 \parallel T_1 \parallel s_2 \rangle \right]. \tag{13.164}
$$

Similarly,

$$
T_{1q}(\hat{J}) = T_{1q}(\hat{S}) + T_{1q}(\hat{s}_3), \tag{13.165}
$$

giving for $s_1 = s_2 = s_3$ the reduced matrix element

$$
\langle J \parallel T_1 \parallel J \rangle = [J^2 \, S^2] (-)^{J-s} \left\{ \begin{matrix} J & J & 1 \\ S & S & s \end{matrix} \right\} \left\{ \begin{matrix} S & S & 1 \\ s & s & s \end{matrix} \right\}
$$
$$
\times [\langle s_1 \parallel T_1 \parallel s_1 \rangle + \langle s_2 \parallel T_1 \parallel s_2 \rangle]
$$
$$
+ [J^2] (-)^{J+S+s+1} \left\{ \begin{matrix} S & s & J \\ 1 & J & s \end{matrix} \right\} \langle s_3 \parallel T_1 \parallel s_3 \rangle. \tag{13.166}
$$

For baryons formed by quarks assigned spin $s_1 = s_2 = s_3 = \frac{1}{2}$, we obtain the reduced matrix element

$$
\langle J \parallel T_1 \parallel J \rangle = [J^2 \, S^2] (-)^{J-\frac{1}{2}} \left\{ \begin{matrix} J & J & 1 \\ S & S & \frac{1}{2} \end{matrix} \right\} \left\{ \begin{matrix} S & S & 1 \\ \frac{1}{2} & \frac{1}{2} & \frac{1}{2} \end{matrix} \right\}
$$
$$
\times [\langle s_1 \parallel T_1 \parallel s_1 \rangle + \langle s_2 \parallel T_1 \parallel s_2 \rangle]
$$
$$
+ [J^2] (-)^{J+S-\frac{1}{2}} \left\{ \begin{matrix} S & \frac{1}{2} & J \\ 1 & J & \frac{1}{2} \end{matrix} \right\} \langle s_3 \parallel T_1 \parallel s_3 \rangle. \tag{13.167}
$$

For baryons with spin $J = \frac{1}{2}$, the result is simplified and depends only on the spin S of the diquark:

$$
\langle J \parallel T_1 \parallel J \rangle = 2 \, [S^2] \left\{ \begin{matrix} S & S & 1 \\ \frac{1}{2} & \frac{1}{2} & \frac{1}{2} \end{matrix} \right\}^2 [\langle s_1 \parallel T_1 \parallel s_1 \rangle + \langle s_2 \parallel T_1 \parallel s_2 \rangle]
$$
$$
+ 2(-)^S \left\{ \begin{matrix} S & \frac{1}{2} & \frac{1}{2} \\ 1 & \frac{1}{2} & \frac{1}{2} \end{matrix} \right\} \langle s_3 \parallel T_1 \parallel s_3 \rangle \tag{13.168}
$$

i) Diquark with Zero Spin

Let us use the special values of the $6j$ coefficients:

$$
\left\{ \begin{matrix} 0 & 1 & 0 \\ \frac{1}{2} & \frac{1}{2} & \frac{1}{2} \end{matrix} \right\} = 0 \quad \text{and} \quad \left\{ \begin{matrix} \frac{1}{2} & \frac{1}{2} & \frac{1}{2} \\ \frac{1}{2} & \frac{1}{2} & \frac{1}{2} \end{matrix} \right\} = \frac{1}{2} \tag{13.169}
$$

to obtain the RME

$$
\langle J \parallel T_1 \parallel J \rangle = \langle s_3 \parallel T_1 \parallel s_3 \rangle. \tag{13.170}
$$

The magnetic moment μ of a baryon with spin 1/2 and comprising three quarks, one of which is a diquark with zero spin is the same as the magnetic moment of the unmatched quark $\mu = \mu_3$.

ii) Diquark with Spin 1

We use the value of the $6j$ coefficients:

$$\begin{Bmatrix} 1 & 1 & 1 \\ \frac{1}{2} & \frac{1}{2} & \frac{1}{2} \end{Bmatrix} = -\frac{1}{3} \quad \text{and} \quad \begin{Bmatrix} \frac{1}{2} & \frac{1}{2} & 1 \\ \frac{1}{2} & \frac{1}{2} & \frac{1}{2} \end{Bmatrix} = \frac{1}{6} \tag{13.171}$$

with statement (13.168) of the RME to obtain

$$\langle J \| T_1 \| J \rangle = \frac{2}{3} \left[\langle s_1 \| T_1 \| s_1 \rangle + \langle s_2 \| T_1 \| s_2 \rangle \right] - \frac{1}{3} \langle s_3 \| T_1 \| s_3 \rangle. \tag{13.172}$$

The magnetic moment μ of a baryon with spin $\frac{1}{2}$ comprising three quarks, one of which is a diquark with spin 1 is equal to

$$\mu = \frac{2}{3} (\mu_1 + \mu_2) - \frac{1}{3} \mu_3.$$

Table 1. Magnetic moments of the baryons $\frac{1}{2}^+$ in $e\, \hbar/(2m_N c)$ units (the used mass values of the model are: $m_u = m_d \simeq 337\,\text{MeV}$, $m_s = 569\,\text{MeV}$, $m_c = 1732\,\text{MeV}$).

Baryons	$f_1 \cdot (f_2 \wedge f_3)_S$	Theory	Experiment
$N \begin{cases} p \\ n \end{cases}$	$d \cdot (u \wedge u)_1$ $u \cdot (d \wedge d)_1$	2.773 -1.843	2.79 -1.86
Λ^0	$s \cdot (u \wedge d)_0$	-0.544	-0.605
$\Sigma \begin{cases} \Sigma^0 \\ \Sigma^+ \\ \Sigma^- \end{cases}$	$s \cdot (u \wedge d)_1$ $s \cdot (u \wedge u)_1$ $s \cdot (d \wedge d)_1$	0.801 2.648 -1.045	0.822 2.682 -1.039
$\Xi \begin{cases} \Xi^0 \\ \Xi^- \end{cases}$	$u \cdot (s \wedge s)_1$ $d \cdot (s \wedge s)_1$	-1.342 -0.418	-1.426 -0.496
Λ_c^+	$c \cdot (u \wedge d)_0$	0.362	0.391

These results have already been alluded to in Chap. 10. If a quark with the flavor f (equal to u, d, s, c, b, t) is represented as a vector in color space, the baryons will be the scalar triple products of the vectors f_1, f_2 and f_3 of the constituent quarks, and the evaluation of the magnetic moments of the baryons with spin $\frac{1}{2}$ will yield results (see Table 1) close to that of the experiment designed to test the foregoing theory (Elbaz and Meyer, 1981), or at least of its underlying principles. It would appear, in particular, that quarks with identical flavors tend to form a diquark with spin 1 but we should perhaps allow a mixture of configurations in order to fine-tune the results obtained.

9. Density Operator

9.1 Measurement of an Observable

It has been demonstrated in Chap. 5 that it might be appropriate to introduce a Hermitian density of states operator

$$\varrho = \varrho^+ \qquad \text{Tr } \varrho = 1, \tag{13.173}$$

such that the expectation value of an observable, the result of a measurement of the observable in question, is defined by a trace that is independent of the representation of the quantum states:

$$\langle A \rangle = \text{Tr } (\varrho\, A). \tag{13.174}$$

If we introduce the density of states with angular momentum $(j\mu)$, Eqs. (13.173) and (13.174) become

$$\text{Tr } \varrho = \sum_{\mu} \langle j\,\mu \,|\, \varrho \,|\, j\,\mu \rangle = 1, \tag{13.175}$$

$$\langle A \rangle = \sum_{\mu\,\mu'} \langle j\,\mu \,|\, \varrho \,|\, j\,\mu' \rangle \langle j\,\mu' \,|\, A \,|\, j\,\mu \rangle. \tag{13.176}$$

The matrix elements of ϱ and A will be represented graphically by black boxes that are dependent on the system under consideration:

$$\begin{array}{c} j\mu \qquad j\mu' \\ \longrightarrow\!\!\!\!\!\rightarrow\!(P)\!\rightarrow\!\longrightarrow \end{array} = \langle j\,\mu \,|\, \varrho \,|\, j\,\mu' \rangle, \tag{13.177}$$

$$\begin{array}{c} j\mu' \qquad j\mu \\ \longrightarrow\!\!\!\!\!\rightarrow\!(A)\!\rightarrow\!\longrightarrow \end{array} = \langle j\,\mu' \,|\, A \,|\, j\,\mu \rangle. \tag{13.178}$$

We have added poles to the kinetic lines $(j\ \mu)$ and $(j\ \mu')$ of the matrix element of ϱ to take into account condition (13.173) since the value of a loop is $\sum_\mu 1 = [j^2]$ and a pole on a kinetic line introduces a weighting coefficient $[j^{-1}]$, which yields, by linking the poles of the box,

$$\mathrm{Tr}\, \varrho = \;\bigcirc\hskip-0.4em\triangleright_j\; = 1. \qquad (13.179)$$

The graphical representation of the expectation value thus reduces to

$$\langle A \rangle \;=\; Tr\,(\varrho\, A) = \quad \overset{j}{\underset{j}{\fbox{}}} \quad . \qquad (13.180)$$

9.2 ITO Expansion

Let us introduce a summation over an angular momentum k with the rule for summation over j used in the graphical method (see (10.93)):

$$\langle A \rangle = \sum_k \; \text{(graph)} \; = \sum_{k\,\chi} \varrho^j_{k\,\chi} \, A^{j\,*}_{k\,\chi}. \qquad (13.181)$$

Using term-by-term identification, this will give the multipole (ITO) expansion of the density matrix:

$$\varrho^j_{k\,\chi} \;=\; \text{(graph)}$$

$$= \sum_{\mu\,\mu'} (-)^{j+\mu} \, \langle j - \mu\, j\, \mu' \mid k\, \chi \rangle \, \langle j\, \mu \mid \varrho \mid j\, \mu' \rangle, \qquad (13.182)$$

or, by using Clebsch–Gordan series to invert the relation,

$$\text{(graph)} \;=\; \text{(graph)}$$

$$= \sum_{k\,\chi} \varrho^j_{k\,\chi} \, (-)^{j-\mu} \, \langle\, k\chi \mid j - \mu\, j\, \mu' \,\rangle. \qquad (13.183)$$

The ITO expansion of $A_{k\,\chi}^*$ is obtained in a similar manner:

$$A_{k\,\chi}^{j\,*} = \quad \text{[diagram]} \quad =$$

$$= \sum_{\mu\,\mu'} \langle k\,\chi \mid j - \mu\, j\,\mu' \rangle\, (-)^{j-\mu}\, \langle j\,\mu' \mid A \mid j\,\mu \rangle. \qquad (13.184)$$

The reader will notice that $\varrho_{k\,\chi}^{j}$ has exactly the structure (13.35) of the tensor product $\pi_{k\,q}\,(\bar{u}_1\,u_2)$ in which $T_{k_1\,q_1}^{+}\,(\bar{u}_1)\,T_{k_2\,q_2}\,(u_2)$ has been replaced with

$$\langle k_1\,q_1 \mid u_1 \rangle\, \langle u_2 \mid k_2\,q_2 \rangle \;=\; \langle j\,\mu \mid \varrho \mid j\,\mu' \rangle. \qquad (13.185)$$

9.3 Density of States of Orbital Angular Momentum

By introducing the Hermitian operator $\varrho = \mid \Omega \rangle\,\langle \Omega \mid$ defining the the density of states in the direction $\Omega = (\theta, \varphi)$, condition (13.174) becomes

$$\mathrm{Tr}\,\varrho = \sum_{m} \langle \ell\,m \mid \Omega \rangle\,\langle \Omega \mid \ell\,m \rangle \;=\; \sum_{m} Y_{\ell\,m}^{*}\,(\Omega)\, Y_{\ell\,m}\,(\Omega) = \frac{[\ell^2]}{4\pi}. \qquad (13.186)$$

The orbital angular momentum density of states operator ℓ is therefore of the form

$$\varrho^{\ell} = \frac{4\pi}{[\ell^2]}\, \mid \Omega \rangle\,\langle \Omega \mid. \qquad (13.187)$$

The corresponding matrix element is expressed with spherical harmonics in the Ω direction:

$$\langle \ell\,m \mid \varrho^{\ell} \mid \ell\,m' \rangle = \quad \text{[diagram]} \quad = \langle \ell\,m \mid \Omega \rangle\,\langle \Omega \mid \ell\,m' \rangle\, \frac{4\pi}{[\ell^2]}$$

$$= \frac{4\pi}{[\ell^2]} \quad \text{[diagram]} \quad = \sum_{L} \frac{4\pi}{[\ell^2]} \quad \text{[diagram]}\, \Omega. \qquad (13.188)$$

The rule for contraction over spherical harmonics (12.53) enables us to introduce only one spherical harmonic with free variance. The calculation of the foregoing matrix element is an example in which the content of the black box is visualized:

$$\langle \ell \; m \mid \varrho \mid \ell \; m' \rangle \quad = \quad \text{(diagram)} \quad = \quad \text{(diagram)} . \tag{13.189}$$

The multipole expansion (in ITOs) of the density matrix is easily written:

$$\varrho^\ell_{k\;\chi} = \text{(diagram)} = \sum_L \frac{[4\pi]}{[l^2]} \; \text{(diagram)} . \tag{13.190}$$

By pinching on the kinetic lines $(k \; L)$, we obtain

$$\varrho^\ell_{k\;\chi} = \frac{4\pi}{[\ell^2 \; L^2]} \; \delta_{k\;L} \; \text{(diagram)} + \text{(diagram)} . \tag{13.191}$$

The $\varrho^\ell_{k\;\chi}$ is, to within one coefficient, a kth-order spherical harmonic since

$$\varrho^\ell_{k\;\chi} (\Omega) = \sqrt{4\pi} \begin{pmatrix} 0 & 0 & \ell \\ \ell & k & 0 \end{pmatrix} Y_{k\;\chi} (\Omega). \tag{13.192}$$

Let us draw attention here to the emphasis laid on the projection space of the even kth-order irreducible tensor operator, since the $3j0$ will be zero if $2\ell + k$ is odd.

9.4 Density of States with Spin 1/2

It has been shown that any 2×2 matrix can be expressed as a linear combination of Pauli matrices and the identity matrix. Condition (13.174) involving the trace leads to the following expression for the density of states operator with spin 1/2:

$$\varrho^{1/2} = \frac{1}{2} (\mathbb{1} + \vec{P} \cdot \vec{\sigma}) = \frac{1}{2} \begin{pmatrix} 1 + P_z & P_x - i \, P_y \\ P_x + i \, P_y & 1 - P_z \end{pmatrix} . \tag{13.193}$$

The vector $\vec{P}$ is the polarization vector of the particles, that is, the mean value of the Pauli operator $\vec{\sigma}$. In effect

$$\langle \vec{\sigma} \rangle \;\; = \text{Tr} \, (\varrho \, \vec{\sigma}) = \text{Tr} \left[\frac{1}{2} (\mathbb{1} + \vec{P} \cdot \vec{\sigma} \,) \, \vec{\sigma} \right] , \tag{13.194}$$

and, because Pauli matrices verify the relations

$$\text{Tr } \vec{\sigma} = 0 \quad \text{and} \quad \text{Tr } \sigma_j \, \sigma_k = 2 \, \delta_{jk}, \tag{13.195}$$

we easily infer that

$$\langle \vec{\sigma} \rangle = \vec{P}. \tag{13.196}$$

Using the procedure defined in (13.61) and (13.104), the standardization of the vector operator $\vec{P}$ yields the standard components

$$P_{11} = \frac{-i}{\sqrt{2}} \, (P_x + i \, P_y),$$

$$P_{10} = i \, P_z, \tag{13.197}$$

$$P_{1-1} = \frac{i}{\sqrt{2}} \, (P_x - i \, P_y).$$

Let us use definitions (13.182) and (13.193) to evaluate $\varrho_{k\,\chi}^{1/2}$:

$$\varrho_{k\,\chi}^{1/2} = \sum_{\mu\,\mu'} (-)^{1/2+\mu} \, \langle \tfrac{1}{2} -\mu \, \tfrac{1}{2} \, \mu' \mid k \, \chi \rangle \, \langle \tfrac{1}{2} \, \mu \mid \tfrac{1}{2} \, (\mathbb{1} + \vec{P}.\vec{\sigma}) \mid \tfrac{1}{2} \, \mu' \rangle$$

$$= \sum_{\mu\,\mu'} [k] \begin{pmatrix} \mu' & k & \tfrac{1}{2} \\ \tfrac{1}{2} & \chi & \mu \end{pmatrix} \, \langle \tfrac{1}{2} \, \mu \mid \tfrac{1}{2} \, (\mathbb{1} + \vec{P}.\vec{\sigma}) \mid \tfrac{1}{2} \, \mu' \rangle. \tag{13.198}$$

In particular, for $k = 0$ we obtain the value $1/\sqrt{2}$ whereas for $k = 1$ and $\chi = 0$ we obtain $\varrho_{10}^{1/2} = P_z/\sqrt{2}$. There is therefore a proportionality relation between $\varrho_{1\chi}$ and $P_{1\chi}$:

$$\varrho_{1\chi}^{1/2} = \frac{-i}{\sqrt{2}} \, P_{1\chi} \qquad \chi = (1, \, 0, \, -1) \tag{13.199}$$

which amounts more generally to setting

$$\varrho_{00}^{1/2} = \frac{1}{\sqrt{2}} \quad \text{and} \quad \begin{cases} \varrho_{11}^{1/2} = -\dfrac{1}{2} \, (P_x + i \, P_y) \\[2mm] \varrho_{10}^{1/2} = \dfrac{1}{\sqrt{2}} \, P_z \\[2mm] \varrho_{1-1}^{1/2} = \dfrac{1}{2} \, (P_x - i \, P_y). \end{cases} \tag{13.200}$$

In an experiment designed to measure the polarization of particles with spin 1/2, we determine a number N_+ of particles with spin $(\tfrac{1}{2} \, \tfrac{1}{2})$ and a number N_- of particles with spin $(\tfrac{1}{2} - \tfrac{1}{2})$ in a given direction. The angular distribution of particles, which is proportional to $N_+ + N_-$, is expressed by $\varrho_{00}^{1/2}$ whereas the polarization along the Oz axis is, for example, proportional to $(N_+ - N_-)/(N_+ + N_-)$ and will be expressed by $\varrho_{10}^{1/2}$, the statistical tensor along the Oz axis.

We have just seen the densities of states of particles with spin $\frac{1}{2}$. Let us use the same procedure to introduce the densities of states of particles with spin s, that is, $\varrho^s_{k\,\chi}$ and k take the values $0, 1, \ldots, 2s$. There are therefore tensor polarizations associated with the values of $k\rangle 1$. This point will not be treated any further in this text. The interested reader should consult more specialized textbooks, e.g. (Elbaz, 1972, 1985).

9.5 Density of States with Angular Momentum j

Particles with spin s and orbital angular momentum ℓ will be described by a density matrix ϱ^j, tensor product of ϱ^ℓ and ϱ^s with $\vec{J} = \vec{\ell} + \vec{s}$. The matrix element is easily determined as

$$\langle j\,\mu \mid \varrho^j \mid j\,\mu' \rangle = \quad \text{(13.201)}$$

The statistical density of these states is obtained by contracting the above matrix element with a Clebsch–Gordan coefficient:

$$\varrho^j_{k\,\chi} = \quad \text{(13.202)}$$

or by exhibiting the statistical tensors $\varrho^\ell_{k_1\,\chi_1}$ and $\varrho^s_{k_2\,\chi_2}$:

$$[k^2]. \quad \text{(13.203)}$$

We are brought back to the tensor product $\varrho^j_{k\,\chi}$ of $\varrho^{\ell_1}_{k_1\,\chi_1}$ and $\varrho^s_{k_2\,\chi_2}$, whereas the rearrangement of the closed diagram will lead to a $9j$ change-of-coupling mode coefficient (see Appendix 1):

$$\varrho^j_{k\,\chi} = \pi_{k\,\chi}\,(\hat{\ell},\,\hat{s})\,[j^2\,k_1\,k_2]\begin{Bmatrix} \ell & \ell & k_1 \\ s & s & k_2 \\ j & j & k \end{Bmatrix}(-)^{k+2s}. \quad \text{(13.204)}$$

We will return to this tensor product in connection with the polarization and distribution of emerging particles during a scattering process (Chap. 16).

10. Differential Operators

The standardization of the vector operators makes it possible to express the gradient operator $\vec{\nabla}$ in the Cartesian basis $r = (1, 2, 3) = (x, y, z)$ or in the standard spherical basis $\mu = (1,\ 0, -1)$:

$$\vec{\nabla} = \sum_e \vec{e}_r\, \partial_r = \sum_\mu \vec{e}^{\,\mu}\, \partial_\mu = \sum_\mu (-)^{1-\mu}\, \vec{e}_{-\mu}\, \partial_\mu\,. \tag{13.205}$$

We infer the usual vector operators:

$$\operatorname{div} \vec{A} = \vec{\nabla} \cdot \vec{A} = \sum_r \partial_r\, A_r = \sum_\mu \partial^\mu\, A_\mu = \sum_\mu (-)^{1-\mu}\, \partial_{-\mu}\, A_\mu\,, \tag{13.206}$$

$$\operatorname{curl} \vec{A} = \vec{\nabla} \wedge \vec{A} = \sum_{rst} \varepsilon_{rst}\, \partial_r\, A_s\, \vec{e}_t = \sum_{\mu\nu\tau} \varepsilon^{\mu\nu\tau}\, \vec{e}_\mu\, \partial_\nu\, A_\tau\,, \tag{13.207}$$

$$\nabla^2 = \vec{\nabla} \cdot \vec{\nabla} = \sum_r \partial_r\, \partial_r = \sum_\mu \partial_\mu\, \partial^\mu = \sum_\mu (-)^{1-\mu}\, \partial_\mu\, \partial_{-\mu}\,. \tag{13.208}$$

10.1 Reduced Matrix Element

Using a diagonal vector operator $A_{1\,\mu}$ in configuration space which acts on the functions $\phi^p_{\ell\,m}(\vec{r})$, it is possible to write the matrix element

$$\langle \phi^{p'}_{\ell'\,m'} \mid A_{1\,\mu} \mid \phi^p_{\ell\,m} \rangle = \int \phi^{p'\,*}_{\ell'\,m'}(\vec{r})\, A_{1\,\mu}(\vec{r})\, \phi^p_{\ell\,m}(\vec{r})\, d\vec{r} \tag{13.209}$$

and, by separating the angular and radial parts of the wave function

$$\phi^p_{\ell\,m}(\vec{r}) = \langle \vec{r} \mid \phi^p_{\ell\,m} \rangle = \langle r \mid f^p_\ell \rangle\, \langle \hat{r} \mid \ell\, m \rangle\,, \tag{13.210}$$

the matrix element becomes:

$$\langle \phi^{p'}_{\ell'\,m'} \mid A_{1\,\mu} \mid \phi^p_{\ell\,m} \rangle = \int r^2\, dr\, f^{p'}_{\ell'}{}^*(r)\, \langle \ell'\, m' \mid A_{1\,\mu} \mid \ell\, m \rangle\, f^p_\ell(r)\,, \tag{13.211}$$

when the Wigner–Eckart theorem is applied to the matrix element:

$$\langle \ell'\, m' \mid A_{1\,\mu} \mid \ell\, m \rangle = \begin{pmatrix} m' & 1 & \ell \\ \ell' & \mu & m \end{pmatrix} \langle \ell' \| A_1 \| \ell \rangle\,, \tag{13.212}$$

then we can write (13.211) in the usual form:

$$\langle \phi^{p'}_{\ell'\,m'} \mid A_{1\,\mu} \mid \phi^p_{\ell\,m} \rangle = \begin{pmatrix} m' & 1 & \ell \\ \ell' & \mu & m \end{pmatrix} \langle \phi^{p'}_{\ell'} \| A_1 \| \phi^p_\ell \rangle \tag{13.213}$$

with the reduced matrix element

$$\langle \phi^{p'}_{\ell'} \| A_1 \| \phi^p_\ell \rangle = \int r^2\, dr\, f^{p'}_{\ell'}{}^*(r)\, A_1(r)\, f^p_\ell(r)\, \langle \ell' \| Y_1 \| \ell \rangle\,. \tag{13.214}$$

We term $\langle \phi^{p'}_{\ell'} \| A_1 \| \phi^p_\ell \rangle$ the radial reduced matrix element, since the angular part has been integrated into the reduced matrix element of the spherical harmonic.

10.2 Example: The Position Operator

Let us introduce the the first-order ITO associated with the position vector by setting

$$r_{1\mu} = \lambda \, Y_{1\mu} \, (\vec{r}) = \lambda \, r \, Y_{1\mu} \, (\hat{r}) \,. \qquad (13.215)$$

The proportionality coefficient λ is determined upon the element r_{10}, that is,

$$r_{10} = \lambda \, r \, Y_{10} \, (\hat{r}) = \lambda \, r \, \sqrt{\frac{3}{4\pi}} \, \cos \theta = \lambda \, \sqrt{\frac{3}{4\pi}} \, r \, \cos \theta$$

$$= \lambda \, \sqrt{\frac{3}{4\pi}} \, z \qquad (13.216)$$

The standardization procedure (13.61) yields the value $r_{10} = -icz$. We infer from this the value of the coefficient λ which we insert into (13.215) to obtain

$$\boxed{r_{1\mu} = -ic \, \sqrt{\frac{4\pi}{3}} \, Y_{1\mu} \, (\vec{r})} \,. \qquad (13.217)$$

The calculation of the matrix element (13.213) yields, for example, with the coefficient $c = -1$:

$$\langle \phi^{p'}_{\ell' \, 0} | \, r_{10} | \, \phi^p_{\ell \, 0} \rangle = \begin{pmatrix} 0 & 1 & \ell \\ \ell' & 0 & 0 \end{pmatrix} \langle \phi^{p'}_{\ell'} \| \, r_1 \| \, \phi^p_\ell \rangle$$

$$= i \, \sqrt{\frac{4\pi}{3}} \int f^{p'}_{\ell'}{}^* (r) \, r^3 \, f^p_\ell \, (r) \, dr \, [\ell \ell'] \, \sqrt{\frac{3}{4\pi}} \begin{pmatrix} 0 & 1 & \ell \\ \ell' & 0 & 0 \end{pmatrix}^2 \,. \qquad (13.218)$$

By comparing term-by-term the two expressions for the matrix element, we obtain the radial matrix element of the position vector:

$$\langle \phi^{p'}_{\ell'} \| \, r_1 \| \, \phi^p_\ell \rangle = i \, \langle f^{p'}_{\ell'} \, | \, r \, | \, f^p_\ell \rangle \, B^\ell_{\ell'} \,, \qquad (13.219)$$

with angular and radial parts:

$$B^\ell_{\ell'} = [\ell \, \ell'] \begin{pmatrix} 0 & 1 & \ell \\ \ell' & 0 & 0 \end{pmatrix} = \left(\sqrt{\ell'} \, \delta_{\ell' \, \ell+1} - \sqrt{\ell} \, \delta_{\ell' \, \ell-1} \right) \,, \qquad (13.220)$$

$$\langle f^{p'}_{\ell'} \, | \, r \, | \, f^p_\ell \rangle = \int f^{p'}_{\ell'}{}^* (r) \, r^3 \, f^p_\ell \, (r) \, dr \,. \qquad (13.221)$$

10.3 Formula for the Gradient

Let us use the Wigner–Eckart theorem to evaluate the matrix element of the gradient operator over its component along the Oz axis:

$$\langle \ell' \, 0 \, | \, \nabla_{10} \, | \, \ell \, 0 \rangle = \begin{pmatrix} 0 & 1 & \ell \\ \ell' & 0 & 0 \end{pmatrix} \langle \ell' \| \, \nabla_1 \| \ell \rangle \,. \qquad (13.222)$$

Let us first show that

$$\nabla_{10} = \partial_0 = i\,\partial_z = i\,(\cos\theta\,\partial_r - \frac{\sin\theta}{r}\,\partial_\theta). \tag{13.223}$$

We express the partial derivative $\partial r/\partial z$ from $r = (x^2 + y^2 + z^2)^{1/2}$ to get $\partial r/\partial z = z/r = \cos\theta$

Differentiating z/r then gives

$$\frac{1}{r}\,dz - \frac{1}{r^2}\,z\,dr = \frac{1}{r}\,(1 - \frac{z^2}{r^2})\,dz = \frac{1}{r}\sin^2\theta\,dz,$$

and equating to the derivative of $\cos\theta$ yields $d\theta = -\sin\theta/r\,dz$

The total derivative of any function $f(r,\theta)$ is written

$$df = \frac{\partial f}{\partial r}\,dr + \frac{\partial f}{\partial\theta}\,d\theta = \frac{\partial f}{\partial r}\,\frac{z}{r}\,dz + \frac{\partial f}{\partial\theta}\,\left(-\frac{\sin\theta}{r}\right)\,dz$$

$$= (\cos\theta\,\partial f/\partial r - \sin\theta/r\,\partial f/\partial\theta)\,dz \equiv \frac{\partial f}{\partial z}\,dz$$

It is thus apparent that

$$\partial_z = \cos\theta\,\partial_r - \frac{\sin\theta}{r}\,\partial_\theta.$$

Inserting this expression into the matrix element (13.222) then gives

$$\langle \ell'\,0 \mid \nabla_{10} \mid \ell\,0 \rangle = i\int Y^*_{\ell'\,0}\,(\hat{r})\left[\cos\theta\,\partial_r - \frac{\sin\theta}{r}\,\partial_\theta\right]Y_{\ell\,0}\,(\hat{r})\,d\hat{r}. \tag{13.224}$$

The contraction relation for spherical harmonics, in conjunction with the fact that $\cos\theta = \sqrt{\frac{4\pi}{3}}\,Y_{10}\,(\hat{r})$, further gives the element $\cos\theta\,Y_{\ell\,0}\,(\hat{r})$:

$$\cos\theta\,Y_{\ell\,0}\,(\hat{r}) = \sum_L [\ell\,L]\begin{pmatrix}1 & \ell & L \\ 0 & 0 & 0\end{pmatrix}^2 Y_{L\,0}(\hat{r}) \tag{13.225}$$

and, by expanding the $3j0$ coefficients (see Chap. 10, Appendix 1),

$$\cos\theta\,Y_{\ell\,0} = \frac{\ell+1}{[\ell\,(\ell+1)]}\,Y_{\ell+1,0} + \frac{\ell}{[\ell\,(\ell-1)]}\,Y_{\ell-1,0}. \tag{13.226}$$

Let us use the expression for $Y_{\ell\,0}$ to evaluate $\sin\theta\,\partial_\theta\,Y_{\ell\,0}$ with respect to the associated Legendre polynomial:

$$Y_{\ell\,0}\,(\hat{r}) = \frac{[\ell]}{\sqrt{4\pi}}\,P^0_\ell\,(\cos\theta) \tag{13.227}$$

with the recursion relations

$$(\ell+1)\, P_{\ell+1}^0(x) - (2\ell+1)\, x\, P_\ell^0(x) + \ell\, P_{\ell-1}^0(x) = 0$$

$$(1-x^2)\, \partial_x\, P_\ell^0(x) = (\ell+1)\, x\, P_\ell^0(x) - (\ell+1)\, P_{\ell+1}^0(x). \qquad (13.228)$$

This yields, upon switching to spherical harmonics,

$$\sin\theta\, \partial_\theta\, Y_{\ell\,0}(\hat{r}) = \frac{\ell\,(\ell+1)}{[\ell\,(\ell+1)]}\, Y_{\ell+1,0} - \frac{\ell\,(\ell+1)}{[\ell\,(\ell-1)]}\, Y_{\ell-1,0}. \qquad (13.229)$$

By inserting these results into the direct calculation of the integral, we obtain two types of non-zero matrix elements

$$\langle \ell+1\ 0\ |\ \nabla_{10}\ |\ \ell\ 0\rangle = \frac{i(\ell+1)}{[\ell\,(\ell+1)]}\left[\partial_r - \frac{\ell}{r}\right]$$

$$= \frac{i(\ell+1)}{[\ell\,(\ell+1)]}\, r^{\ell-1}\, \partial_r\,\left(r^{1-\ell}\right),$$

$$\langle \ell-1\ 0\ |\ \nabla_{10}\ |\ \ell\ 0\rangle = \frac{i\ell}{[\ell(\ell-1)]}\left[\partial_r + \frac{\ell+1}{r}\right] \qquad (13.230)$$

$$= \frac{i\ell}{[\ell(\ell-1)]}\, r^{-\ell-1}\, \partial_r\,\left(r^{1+\ell}\right),$$

which amounts more generally to writing

$$\langle \ell'\ 0\ |\ \nabla_{10}\ |\ \ell\ 0\rangle = \frac{i}{[\ell\,\ell']}\left\{\ell'\,\delta_{\ell'\,\ell+1}\left[\partial_r - \frac{\ell}{r}\right] + \ell\,\delta_{\ell'\,\ell-1}\left[\partial_r + \frac{\ell+1}{r}\right]\right\}. \qquad (13.231)$$

We infer from this the reduced matrix element in the form

$$\langle \ell'||\,\nabla_1||\,\ell\rangle = \frac{i}{B_{\ell'}^\ell}\left\{\ell'\,\delta_{\ell'\,\ell+1}\left[\partial_r - \frac{\ell}{r}\right] + \ell\,\delta_{\ell'\,\ell-1}\left[\partial_r + \frac{\ell+1}{r}\right]\right\}$$

$$= i\left\{\sqrt{\ell'}\,\delta_{\ell'\,\ell+1}\left[\partial_r - \frac{\ell}{r}\right] + \sqrt{\ell}\,\delta_{\ell'\,\ell-1}\left[\partial_r + \frac{\ell+1}{r}\right]\right\}. \qquad (13.232)$$

The radial reduced matrix element of the gradient operator will then take the form

$$\langle \phi_{\ell'}^{p'}\,||\,\nabla_1\,||\,\phi_\ell^p\rangle = \langle f_{\ell'}^{p'}\,|\,\langle \ell'\,||\,\nabla_1\,||\,\ell\rangle\,|\,f_\ell^p\rangle$$

$$= \int_0^\infty f_{\ell'}^{p'\,*}(r)\,\langle \ell'\,||\,\nabla_1\,||\,\ell\rangle\, f_\ell^p(r)\, r^2\, dr. \qquad (13.233)$$

10.4 Special Case of Bessel Functions

The recursion relations for spherical Bessel functions:

$$\left(\partial_r - \frac{\ell}{r}\right) j_\ell\,(pr) = -p\,j_{\ell+1}\,(pr),$$
$$\left(\partial_r + \frac{\ell+1}{r}\right) j_\ell\,(pr) = p\,j_{\ell-1}\,(pr), \tag{13.234}$$

simplify the formula of the gradient for this type of function, and by setting

$$\alpha_{\ell'}^{\ell} = \sqrt{\ell'}\,\delta_{\ell'\,\ell+1} + \sqrt{\ell}\,\delta_{\ell'\,\ell-1}, \tag{13.235}$$

we obtain one of the following forms of the reduced matrix element:

$$\langle \phi_{\ell'}^{p'} \parallel \nabla_1 \parallel j_\ell^p \rangle = i\,p\,\alpha_{\ell'}^{\ell}\,\langle f_{\ell'}^{p'} \mid j_\ell^p \rangle, \tag{13.236}$$
$$\langle j_{\ell'}^{p'} \parallel \nabla_1 \parallel \phi_\ell^p \rangle = i\,p\,\alpha_{\ell'}^{\ell}\,\langle j_{\ell'}^{p'} \mid f_\ell^p \rangle, \tag{13.237}$$

whereas with two spherical Bessel functions we obtain

$$\langle j_{\ell'}^{p'} \parallel \nabla_1 \parallel j_\ell^p \rangle = -i\,p\,\alpha_{\ell'}^{\ell}\,\langle j_{\ell'}^{p'} \mid j_\ell^p \rangle$$
$$= -i\,p\alpha_{\ell'}^{\ell}\,\frac{\pi}{2p^2}\,\delta\,(p - p'). \tag{13.238}$$

The reader should bear in mind here that normalized solid spherical Bessel functions form a complete orthonormal basis. In effect, if we set

$$\langle \vec{r} \mid j_{\ell\,m}^p \rangle = \sqrt{\frac{2}{\pi}}\,j_{\ell\,m}\,(p\vec{r}) = \sqrt{\frac{2}{\pi}}\,j_\ell\,(pr)\,Y_{\ell\,m}\,(\hat{r}), \tag{13.239}$$

the orthonormalization relation for the angular parts will yield

$$\langle \ell'\,m' \mid \ell\,m \rangle = \int Y_{\ell'\,m'}^*\,(\hat{r})\,Y_{\ell\,m}\,(\hat{r})\,d\hat{r} = \delta_{\ell\,\ell'}\,\delta_{m\,m'}, \tag{13.240}$$

and for the radial parts,

$$\langle j_\ell^{p'} \mid j_\ell^p \rangle = \frac{2}{\pi}\int_0^\infty j_\ell\,(p'r)\,j_\ell\,(pr)\,r^2\,dr = \frac{1}{p^2}\,\delta\,(p - p'), \tag{13.241}$$

which finally gives the orthonormalization relation

$$\langle j_{\ell\,m}^p \mid j_{\ell'\,m'}^{p'} \rangle = \delta_{\ell\,\ell'}\,\delta_{m\,m'}\,\frac{1}{p^2}\,\delta\,(p - p'). \tag{13.242}$$

The completeness relation is written, with (13.239) and (13.241),

$$\sum_{\ell\,m}\int_0^\infty \langle \vec{r} \mid j_{\ell\,m}^q \rangle\,q^2\,dq\,\langle j_{\ell\,m}^q \mid \vec{r}' \rangle = \delta\,(\vec{r} - \vec{r}'). \tag{13.243}$$

10.5 Example: RME of the Laplacian

The RME can be calculated with the gradient theorem and expression (13.208) for the Laplacian (Elbaz, 1985). However, it is more expedient to note that the $j_{\ell\,m}\,(p\vec{r})$ are solutions to the Helmoltz equation

$$\nabla^2\,j_{\ell\,m}\,(p\vec{r}) + p^2\,j_{\ell\,m}\,(p\vec{r}) = 0, \tag{13.244}$$

because the plane waves $e^{i\,\vec{p}\cdot\vec{r}}$ are eigenfunctions of the operator ∇^2 corresponding to the eigenvalue $-p^2$:

$$\langle j^{p'}_{\ell'\,m'}\,|\,\nabla^2\,|\,j^{p}_{\ell\,m}\rangle = -p^2\,\langle j^{p'}_{\ell'\,m'}\,|\,j^{p}_{\ell\,m}\rangle$$

$$= -p^2\,\delta_{\ell\,\ell'}\,\delta_{m\,m'}\,\frac{1}{p^2}\,\delta\,(p-p'). \tag{13.245}$$

Using the Wigner–Eckart theorem, we obtain another expression for this matrix element:

$$\langle j^{p'}_{\ell'\,m'}\,|\,\nabla^2\,|\,j^{p}_{\ell\,m}\rangle = \begin{pmatrix} m' & 0 & \ell \\ \ell' & 0 & m \end{pmatrix} \langle j^{p'}_{\ell'}\,||\nabla^2\,||\,j^{p}_{\ell}\rangle$$

$$= [\ell^{-1}]\,\delta_{\ell\,\ell'}\,\delta_{m\,m'}\,\langle j^{p'}_{\ell'}\,||\,\nabla^2\,||\,j^{p}_{\ell}\rangle. \tag{13.246}$$

By comparing these matrix elements we obtain the desired RME:

$$\langle j^{p'}_{\ell'}\,||\,\nabla^2\,||\,j^{p}_{\ell}\rangle = -[\ell]\,\delta(p-p')\,\delta_{\ell\ell'}. \tag{13.247}$$

Knowing the RME of the gradient operator associated with the RME of the position operator enables us to calculate the reduced matrix elements of all operators in the configuration space.

11. Irreducible Tensor Operators –
Schematic Summary

Definition

$$R\,T_{kq}\,R^+ = \sum_p D^k_{pq}\,T_{kp}$$

$$[J_v, T_{kq}] = \sum_p \langle kp \mid J_v \mid kq \rangle\,T_{kp}$$

Tensor product

$$\pi_{kq} = \sum \langle k_1 q_1 k_2 q_2 \mid kq \rangle\,T_{k_1 q_1} T_{k_2 q_2}$$

Standardization of vector operators

$$A_{11} = \frac{i\,c}{\sqrt{2}}\,(A_x + i\,A_y)$$

$$A_{10} = -i\,c\,A_z$$

$$A_{1-1} = -\frac{i\,c}{\sqrt{2}}\,(A_x - i\,A_y)$$

$$A_x = \frac{i}{c\,\sqrt{2}}\,(-A_{11} + A_{1-1})$$

$$A_y = -\frac{1}{c\,\sqrt{2}}\,(A_{11} + A_{1-1})$$

$$A_z = \frac{i}{c}\,A_{10}$$

Wigner–Eckart theorem

$$\langle \alpha\,j\,m \mid T_{k\,q} \mid \alpha'\,j'\,m' \rangle = \begin{pmatrix} m & k & j' \\ j & q & m' \end{pmatrix} \langle \alpha\,j \parallel T_{k\,q} \parallel \alpha'\,j' \rangle$$

$$\langle \alpha\,j \parallel \mathbf{1} \parallel \alpha'\,j \rangle = \delta_{\alpha\,\alpha'}\,[j]$$

$$\langle \alpha\,\ell \parallel Y_k \parallel \alpha'\,\ell' \rangle = \delta_{\alpha\,\alpha'}\,\frac{[\ell\,k\,\ell']}{\sqrt{4\pi}}\begin{pmatrix} 0 & k & \ell' \\ \ell & 0 & 0 \end{pmatrix}$$

$$\langle \alpha\,j \parallel J_1 \parallel \alpha'\,j' \rangle = -ic\,[j]\,\sqrt{j\,(j+1)}\,\delta_{\alpha\,\alpha'}\,\delta_{j\,j'}$$

Appendix 1:   Reduced Matrix Elements

The following expressions give the analytical form of the RMEs that were
graphically determined in Sect. 8 of this chapter:

$$\langle \alpha\, j \parallel \pi_k \parallel \alpha'\, j' \rangle = \sum_{\alpha'' j''} [k] \begin{Bmatrix} k_1 & k & k_2 \\ j' & j'' & j \end{Bmatrix} (-)^{j+k+j'}$$
$$\langle \alpha\, j \parallel T_{k_1} \parallel \alpha''\, j'' \rangle \langle \alpha''\, j'' \parallel T_{k_2} \parallel \alpha'\, j' \rangle \qquad (13.137)$$

$$\langle (j_1\, j_2)\, j \parallel \pi_k \parallel (j_1'\, j_2')\, j' \rangle = \begin{Bmatrix} j_1 & j_2 & j \\ j_1' & j_2' & j' \\ k_1 & k_2 & k \end{Bmatrix}$$
$$\times \langle \alpha_1\, j_1 \parallel T_{k_1} \parallel \alpha_1'\, j_1' \rangle \langle \alpha_2\, j_2 \parallel T_{k_2} \parallel \alpha_2'\, j_2' \rangle\ [j\, j'\, k] \quad (13.138)$$

$$\langle (j_1\, j_2)\, j \parallel (T_k \cdot U_k) \parallel (j_1'\, j_2')\, j' \rangle = \delta_{j\, j'}\ \langle j_1 \parallel T_k \parallel j_1' \rangle \langle j_2 \parallel U_k \parallel j_2' \rangle$$
$$(-)^{k-j+j_1'+j_2} \begin{Bmatrix} j_1 & j & j_2 \\ j_2' & k & j_1' \end{Bmatrix} (13.139)$$

$$\langle \alpha\, j \parallel \vec{J} \cdot \vec{A} \parallel \alpha'\, j' \rangle = \sum_{\alpha'' j''} (-)^{j+j''+1}\ [j^{-1}]\ \delta_{j\, j'}\ \langle \alpha\, j \parallel J_1 \parallel \alpha''\, j'' \rangle$$
$$\langle \alpha''\, j'' \parallel A_1 \parallel \alpha'\, j' \rangle \qquad (13.142)$$

$$\langle j\, \mu | \varrho^j | j\, \mu' \rangle = \langle j\, \mu | \ell\, m\, s\, \sigma \rangle \langle \ell\, m | \varrho^\ell | \ell\, m' \rangle \langle s\, \sigma | \varrho^s | s\, \sigma' \rangle$$
$$\times \langle \ell\, m'\, s\, \sigma' | j\, \mu' \rangle \qquad (13.201)$$

$$\varrho^j_{k\,\chi} = \sum \langle j\, \mu | \ell\, m\, s\, \sigma \rangle \langle \ell\, m'\, s\, \sigma' | j\, \mu' \rangle (-)^{j+\mu} \langle j - \mu\, j\, \mu' | k\, \chi \rangle$$
$$\times \langle \ell\, m | \varrho^\ell | \ell\, m' \rangle \langle s\, \sigma | \varrho^s | s\, \sigma' \rangle \qquad (13.202)$$

$$\pi_{k\,\chi}\,(\hat{\ell},\, \hat{s}) = \sum_{\chi_1\, \chi_2} \langle k_1\, \chi_1\, k_2\, \chi_2 | k\, \chi \rangle\ \varrho^\ell_{k_1\, \chi_1}\ \varrho^s_{k_2\, \chi_2} \qquad (13.203)$$

$$\varrho^j_{k\,\chi} = \pi_{k\,\chi}\,(\hat{\ell},\, \hat{s})\ [j^2\, k_1\, k_2] \begin{Bmatrix} \ell & \ell & k_1 \\ s & s & k_2 \\ j & j & k \end{Bmatrix} (-)^{k+2s}. \qquad (13.204)$$

Appendix 2:   Graphical Vector Analysis

The graphical representation of the vectors of a Cartesian space is

$$\vec{e}_r = \hat{e}\!-\!\!-\!r$$

$$A_r = \hat{A}\!-\!\!-\!r \quad \text{and} \quad \vec{A} = \hat{A}\!-\!\!-\!\hat{e}$$

The graphical representation of the Levi–Civita tensor is

$$\varepsilon_{ijk} = \begin{cases} 1 & \text{if} & (i\,j\,k) = (1\,2\,3) \\ -1 & \text{if} & (i\,j\,k) = (2\,1\,3) \\ 0 & \text{otherwise}. \end{cases}$$

This makes it possible to obtain a diagrammatic representation of the elements of the vector analysis and to make some calculations or simplifications by inspection:

i) Scalar Product $\vec{A}\cdot\vec{B} = \hat{A}\!-\!\!-\!\hat{B}$

ii) Vector Product $\vec{A}\wedge\vec{B} = \hat{e}\!-\!\!-\!\langle$ (with $\hat{A}$, $\hat{B}$)

iii) Scalar Triple Product $(\vec{A}\wedge\vec{B})\cdot\vec{C} =$ (diagram with $\hat{A}$, $\hat{B}$, $\hat{C}$)

iv) Vector Crossing Rule

$$\sum_k \varepsilon_{ijk}\,\varepsilon_{i'j'k} = \text{(diagram)}$$

$$= \delta_{i\,i'}\,\delta_{j\,j'} - \delta_{i\,j'}\,\delta_{j\,i'} = \text{(diagram)}\;-\;\text{(diagram)}.$$

This simple rule is used to simplify by inspection vector products, for example:

$$(\vec{A}\wedge\vec{B})\cdot(\vec{C}\wedge\vec{D}) = \text{(diagram)} = \text{(diagram)} - \text{(diagram)}$$

$$= (\vec{A}\cdot\vec{C})\,(\vec{B}\cdot\vec{D}) - (\vec{A}\cdot\vec{D})\,(\vec{B}\cdot\vec{C})$$

If we multiply by $\sqrt{6}$ the special $3jm$ coefficient with angular momenta $J_1 = J_2 = J_3 = 1$, we obtain the graphical equivalent of the Levi–Civita tensor either in Cartesian or in spherical coordinates:

$$\begin{pmatrix} 1 & 1 & 1 \\ \alpha & \beta & \gamma \end{pmatrix} = \quad \text{[diagram]} = \sqrt{6} \quad \text{[diagram]}$$

$\alpha,\ \beta,\ \gamma = (1,\ 2,\ 3) = x,\ y,\ z$ Cartesian coordinates

$\alpha,\ \beta,\ \gamma = (1,\ 0,\ -1)$ and $\alpha + \beta + \gamma = 0$ spherical coordinates.

i) If $\vec{A}, \vec{B}$ are vectors:

$$(\vec{A} \wedge \vec{A}) = \quad \text{[diagram]} = 0.$$

ii) If $\vec{A}, \vec{B}$ are angular momentum vector operators:

$$\vec{J} \wedge \vec{J} = \quad \text{[diagram]} = \sqrt{6} \quad \text{[diagram]} = i\ \vec{J} \longrightarrow \hat{e},$$

which gives the following equivalences:

$$\text{[diagram]} = \frac{i}{\sqrt{6}}\ \vec{J} \longrightarrow \hat{e} \quad \text{and} \quad \text{[diagram]} = i\ \sqrt{2/3}\ \hat{\sigma} \longrightarrow \hat{e}$$

Let us recall result (13.110):

$$\text{[diagram]} = 0.$$

v) Tensor Crossing Rule

$$(\vec{A} \cdot \vec{B})\,(\vec{C} \cdot \vec{D}) = \quad \text{[diagram]} = \sum_X [X^2] \quad \text{[diagram]}.$$

We expand the values $X = 0$ and 1, as well as the resulting scalar products and vector multiplication rule:

$$6 \quad \text{[diagram]} = (\vec{A} \cdot \vec{B})\,(\vec{C} \cdot \vec{D}) - (\vec{A} \cdot \vec{D})\,(\vec{B} \cdot \vec{C})$$

which finally leaves us with the expression for the scalar product of the second-order tensors:

$$(\vec{A} \cdot \vec{B})(\vec{C} \cdot \vec{D}) = \frac{1}{3}(\vec{A} \cdot \vec{C})(\vec{B} \cdot \vec{D})$$

$$+ \frac{1}{2}\left[(\vec{A} \cdot \vec{B})(\vec{C} \cdot \vec{D}) - (\vec{A} \cdot \vec{D})(\vec{B} \cdot \vec{C})\right]$$

$$+ T_2(\hat{A}, \hat{C}) \cdot T_2(\hat{B}, \hat{D}).$$

Following simplification of like terms, we obtain the following equivalence:

$$= \frac{1}{2}(\vec{A} \cdot \vec{B})(\vec{C} \cdot \vec{D}) + \frac{1}{2}(\vec{A} \cdot \vec{D})(\vec{B} \cdot \vec{C}) - \frac{1}{3}(\vec{A} \cdot \vec{C})(\vec{B} \cdot \vec{D}).$$

Because this relation holds true irrespective of the vector operator, we infer the tensor crossing rule:

Example: Graphical representation of a tensor force:

$$\vec{A} = \vec{S}_1 \qquad \vec{B} = \vec{D} = \vec{r} \qquad \vec{C} = \vec{S}_2$$

We introduce the tensor:

$$S_{12} = \frac{T_2(\hat{S}_1, \hat{S}_2) \cdot T_2(\hat{r}, \hat{r})}{r^2}$$

to obtain the graphical expression for a tensor force in the form in which it is used in nuclear physics:

$$S_{12} = \frac{(\vec{S}_1 \cdot \vec{r})(\vec{S}_2 \cdot \vec{r})}{r^2} - \frac{1}{3}(\vec{S}_1 \cdot \vec{S}_2).$$

By using the tensor notation $r_{1\mu}$ of the position vector (13.217) with definition (12.84) we get

$$r_{1\mu} = -ic \sqrt{\frac{4\pi}{3}} \, Y_{1\mu}(\vec{r}) = -icr \sqrt{\frac{4\pi}{3}} \, Y_{1\mu}(\Omega)$$

The graphical expression for of the tensor force S_{12} can be written with the use of the contraction rule (12.50) of spherical harmonics as

$$S_{12} = -c^2 \sqrt{\frac{8\pi}{3}} \, \left(\vec{S}_1\right)_{1q_1} \left(\vec{S}_2\right)_{1q_2} \begin{pmatrix} 1 & 1 & q \\ q_1 & q_2 & 2 \end{pmatrix} Y_{2q}(\Omega),$$

where $\mid c \mid^2 = 1$. The diagram associated with S_{12} facilitates the derivation of the Wigner–Eckart theorem (Elbaz, 1985).

Chapter 14
Perturbations

Measurements of the energy states of a quantum system, whether a nucleus or an atom, are impossible without a thorough knowledge of the eigenvalues of the associated Hamiltonian. Some of the simple systems described in Chaps. 3 and 12 lead to analytical solutions, which, despite their elegance, are far from representing the general case. The usual thing to do is to make use of the results for known quantum systems by separating the Hamiltonian H into a part H_0, an eigenvalue equation of known (generally analytical) solution, and a part $V(x, t)$, termed perturbation. The perturbation will be a true one if the modification made on the eigenstates and eigenvalues of H_0 is negligible. If this is not the case, then the Hamiltonian H_0 was not properly chosen or, worse still, the perturbation method used does not converge well or perhaps even diverges. We will describe typical cases for which the perturbation method yields acceptable results before going on to show how to improve its use. First, we will consider the case of a time-independent perturbation $V(x)$ and then examine time-dependent perturbations and introduce the Feynman diagram method generally used in atomic physics, nuclear physics and, in particular, particle physics where it has become a very important tool.

1. Stationary Perturbations

Consider a Hamiltonian $H = H_0 + \lambda V$. We wish to determine its eigenstates and eigenvalues:

$$H \mid \Psi_n \rangle = E_n \mid \Psi_n \rangle \tag{14.1}$$

from the non-degenerate eigenvalues and eigenstates of the Hamiltonian H_0:

$$H_0 \mid n \rangle = \varepsilon_n \mid n \rangle . \tag{14.2}$$

1.1 The Rayleigh–Schrödinger Method

Let us make a λ power series expansion of the perturbed energy:

$$E_n = E_n^{(0)} + \lambda \, E_n^{(1)} + \lambda^2 \, E_n^{(2)} + \dots , \tag{14.3}$$

with the corresponding eigenstate

$$| \Psi_n \rangle = | \Psi_n^0 \rangle + \lambda | \Psi_n^{(1)} \rangle + \lambda^2 | \Psi_n^{(2)} \rangle + \dots , \qquad (14.4)$$

where we use the definition of the unperturbed states and energies

$$| \Psi_n^0 \rangle = | n \rangle \quad \text{and} \quad E_n^{(0)} = \varepsilon_n . \qquad (14.5)$$

We assume in the meantime that the level $| n \rangle$ corresponding to the eigenvalue ε_n of H_0 is not degenerate, that is, that there is only one level $| n \rangle$ corresponding to the eigenvalue ε_n.

The eigenstates of H_0 are normalized:

$$\langle n | n \rangle = 1 \qquad (14.6)$$

and orthogonal to every state $| \Psi_n^{(i)} \rangle$ of the different orders of the perturbation:

$$\langle n | \Psi_n \rangle = 1 \quad \text{and} \quad \langle n | \Psi_n^{(i)} \rangle = 0 \quad \forall i . \qquad (14.7)$$

The final step consists in inserting the expansion (14.3) and (14.4) into the eigenvalue Eq. (14.1) of the perturbed system to obtain

$$(H_0 + \lambda V) \left[| n \rangle + \lambda | \Psi_n^{(1)} \rangle + \dots \right]$$
$$= \left[\varepsilon_n + \lambda E_n^{(1)} + \dots \right] \left[| n \rangle + \lambda | \Psi_n^{(1)} \rangle + \dots \right] . \qquad (14.8)$$

By equating terms with the same power in λ, we obtain the following series of equations:

$$H_0 | n \rangle = \varepsilon_n | n \rangle ,$$
$$H_0 | \Psi_n^{(1)} \rangle + V | n \rangle = \varepsilon_n | \Psi_n^{(1)} \rangle + E_n^{(1)} | n \rangle ,$$
$$H_0 | \Psi_n^{(k)} \rangle + V | \Psi_n^{(k-1)} \rangle = \varepsilon_n | \Psi_n^{(k)} \rangle + E^{(1)} | \Psi_n^{(k-1)} \rangle + \dots E_n^{(k)} | n \rangle .$$
$$(14.9)$$

a) Perturbed Energies

If we multiply each of the foregoing equations from the left by the bra $\langle n |$, bearing in mind contraints (14.6) and (14.7), we are led to the following expressions

$$\varepsilon_n = \langle n | H_0 | n \rangle \qquad \text{unperturbed energy} , \qquad (14.10)$$
$$E_n^{(1)} = \langle n | V | n \rangle \qquad \text{first-order perturbed energy} , \qquad (14.11)$$
$$E_n^{(2)} = \langle n | V | \Psi_n^{(1)} \rangle \qquad \text{second-order perturbed energy} , \qquad (14.12)$$
$$E_n^{(k)} = \langle n | V | \Psi_n^{(k-1)} \rangle \qquad \text{kth-order perturbed energy} . \qquad (14.13)$$

Attention is drawn here to the fact that, by resumming series (14.3), the energy E_n becomes

$$E_n = \varepsilon_n + \lambda\, E_n^{(1)} + \lambda^2\, E_n^{(2)} + \dots$$
$$= \varepsilon_n + \lambda\, \langle n \mid V \mid n \rangle + \lambda^2\, \langle n \mid V \mid \Psi_n^{(1)} \rangle + \dots$$
$$= \varepsilon_n + \lambda \langle n \mid V \left[\mid n \rangle + \lambda \mid \Psi_n^{(1)} \rangle + \dots \right]$$
$$= \varepsilon_n + \langle n \mid \lambda\, V \mid \Psi_n \rangle .$$

By replacing ε_n with its value (14.10) and by using (14.1), we obtain

$$E_n = \langle n \mid H_0 \mid \Psi_n \rangle + \langle n \mid \lambda V \mid \Psi_n \rangle = \langle n \mid H \mid \Psi_n \rangle , \tag{14.14}$$

which is none other than eigenvalue Eq. (14.1) projected onto the unperturbed states $\mid n \rangle$ with the aid of orthonormalization relations (14.6) and (14.7).

This is a purely formal statement since the perturbed state $\mid \Psi_n \rangle$ is unknown.

b) Perturbed States

Let us multiply each of Eqs. (14.9) by the bra $\langle m \mid$, eigenbra of H_0 corresponding to the eigenvalue ε_m which is different from ε_n:

$$\langle m \mid H_0 \mid \Psi_n^{(k)} \rangle + \langle m \mid V \mid \Psi_n^{(k-1)} \rangle$$
$$= \varepsilon_n \langle m \mid \Psi_n^{(k)} \rangle + \dots + E_n^{(k-1)} \langle m \mid \Psi_n^{(1)} \rangle , \tag{14.15}$$

since $E^{(k)} \langle m \mid n \rangle = 0$ because $\mid m \rangle$ and $\mid n \rangle$ are orthogonal.

We note that the eigenvalue equation of H_0 yields the relation

$$\langle m \mid H_0 \mid \Psi_n^{(k)} \rangle = \varepsilon_m \langle m \mid \Psi_n^{(k)} \rangle . \tag{14.16}$$

By inserting this into (14.15), we infer the following inequality:

$$\langle m \mid \Psi_n^{(k)} \rangle = \frac{1}{\varepsilon_n - \varepsilon_m} \Big[\langle m \mid V \mid \Psi_n^{(k-1)} \rangle$$
$$- E_n^{(1)} \langle m \mid \Psi_n^{(k-1)} \rangle \dots - E_n^{(k-1)} \langle m \mid \Psi_n^{(1)} \rangle \Big] . \tag{14.17}$$

Following summation over all the values of m different from n, we obtain the kth-order perturbed state

$$\mid \Psi_n^{(k)} \rangle = \sum_m \mid m \rangle \langle m \mid \Psi_n^{(k)} \rangle = \sum_{m \neq n} \mid m \rangle \langle m \mid \Psi_n^{(k)} \rangle , \tag{14.18}$$

because hypothesis (14.7) implies

$$\langle n \mid \Psi_n^{(k)} \rangle = 0 \quad \forall\, k .$$

This gives with Eq. (14.17) the kth-order perturbed state

$$
\begin{aligned}
| \Psi_n^{(k)} \rangle = \sum_{m \neq n} \frac{1}{\varepsilon_n - \varepsilon_m} \; | m \rangle \; \langle m | \; \Big[V | \Psi_n^{(k-1)} \rangle \\
- E_n^{(1)} | \Psi_n^{(k-1)} \rangle \ldots - E_n^{(k-1)} | \Psi_n^{(1)} \rangle \Big] .
\end{aligned}
\tag{14.19}
$$

c) Example: Second-Order Perturbed State

Relation (14.19) leads us to write the first-order perturbed state as

$$
| \Psi_n^{(1)} \rangle = \sum_{m \neq n} \frac{1}{\varepsilon_n - \varepsilon_m} \; | m \rangle \; \langle m | V | n \rangle .
\tag{14.20}
$$

We obtain the second-order energy perturbation by inserting this expression into the definition (14.12):

$$
\begin{aligned}
E_n^{(2)} = \langle n | V | \Psi_n^{(1)} \rangle = \sum_{m \neq n} \frac{\langle n | V | m \rangle \; \langle m | V | n \rangle}{\varepsilon_n - \varepsilon_m} \\
= \sum_{m \neq n} \frac{| \langle n | V | m \rangle |^2}{\varepsilon_n - \varepsilon_m} .
\end{aligned}
\tag{14.21}
$$

The second-order perturbed state will be given by (14.38).

d) Renormalization Constant

We have assumed the unperturbed state $| n \rangle$ to be normalized, but not the state $| \Psi_n \rangle$. Let us now introduce a renormalization coefficient Z:

$$
\langle \Psi_n | \Psi_n \rangle^{-1/2} = Z^{1/2} \quad \text{and} \quad | \tilde{\Psi}_n \rangle = Z^{1/2} | \Psi_n \rangle .
\tag{14.22}
$$

The perturbed state $| \tilde{\Psi}_n \rangle$ is thus normalized by construction:

$$
\langle \tilde{\Psi}_n | \tilde{\Psi}_n \rangle = 1 .
\tag{14.23}
$$

Now let us calculate, for example, the second-order renormalization constant:

$$
\begin{aligned}
\langle \Psi_n | \Psi_n \rangle = \Big(\langle n | + \lambda \, \langle \Psi_n^{(1)} | + \lambda^2 \, \langle \Psi_n^{(2)} | \Big) \\
\times \Big(| n \rangle + \lambda | \Psi_n^{(1)} \rangle + \lambda^2 | \Psi_n^{(2)} \rangle \Big)
\end{aligned}
\tag{14.24}
$$

Because orthonormalization relations (14.6) and (14.7) impose

$$
\langle n | \Psi_n^{(i)} \rangle = 0 \quad \forall \, i \quad \text{and} \quad \langle n | n \rangle = 1 ,
$$

the normalization of the first-order truncated quantum state will be

$$\langle \Psi_n \mid \Psi_n \rangle = 1 + \lambda^2 \, \langle \Psi_n^{(1)} \mid \Psi_n^{(1)} \rangle . \qquad (14.25)$$

We then use statement (14.20) of the first-order perturbed state:

$$\mid \Psi_n^{(1)} \rangle = \sum_{m \neq n} \frac{\mid m \rangle \, \langle m \mid V \mid n \rangle}{\varepsilon_n - \varepsilon_m} , \qquad (14.26)$$

which leads to the value of the second term in (14.25):

$$\langle \Psi_n^{(1)} \mid \Psi_n^{(1)} \rangle = \sum_{m \neq n} \frac{\mid \langle m \mid V \mid n \rangle \mid^2}{\left(\varepsilon_n - \varepsilon_m \right)^2} \qquad (14.27)$$

and the perturbed state will have the following norm:

$$\langle \Psi_n \mid \Psi_n \rangle = 1 + \lambda^2 \sum_{m \neq n} \frac{\mid \langle m \mid V \mid n \rangle \mid^2}{\left(\varepsilon_n - \varepsilon_m \right)^2} . \qquad (14.28)$$

Since λ is small, the renormalization coefficient Z can be approximated as

$$Z = \langle \Psi_n \mid \Psi_n \rangle^{-1} \simeq 1 - \lambda^2 \sum_{m} \frac{\mid \langle m \mid V \mid n \rangle \mid^2}{\left(\varepsilon_n - \varepsilon_m \right)^2} . \qquad (14.29)$$

This expression is the differential of the real (perturbed) energy with respect to the unperturbed energy since

$$\frac{\partial E_n}{\partial \varepsilon_n} = \frac{\partial}{\partial \varepsilon_n} \left(\varepsilon_n + \lambda \, \langle n \mid V \mid n \rangle + \lambda^2 \sum_{m} \frac{\mid \langle m \mid V \mid n \rangle \mid^2}{\varepsilon_n - \varepsilon_m} \right)$$

$$= 1 - \lambda^2 \sum_{m} \frac{\mid \langle m \mid V \mid n \rangle \mid^2}{\left(\varepsilon_n - \varepsilon_m \right)^2} . \qquad (14.30)$$

This result can be generalized to all orders of perturbation:

$$\boxed{Z = \partial E_n / \partial \varepsilon_n} . \qquad (14.31)$$

The renormalization constant of the kth-order perturbed states is the partial derivative of the kth-order perturbed energy with respect to the unperturbed energy.

1.2 Wigner–Brillouin Method

Consider the eigenvalue equation of the Hamiltonian H:

$$(H_0 + \lambda V) \, | \, \Psi_n \rangle = E_n \, | \, \Psi_n \rangle. \tag{14.32}$$

Now let us isolate the perturbation λV, which gives

$$(E_n - H_0) \, | \, \Psi_n \rangle = \lambda V \, | \, \Psi_n \rangle. \tag{14.33}$$

Multiplying from the left by $\langle m \, |$ and using the fact that $\langle m \, | \, H_0 = \langle m \, | \, \varepsilon_m$, we obtain the relation

$$(E_n - \varepsilon_m) \, \langle m \, | \, \Psi_n \rangle = \lambda \, \langle m \, | \, V \, | \, \Psi_n \rangle. \tag{14.34}$$

The projection of the perturbed state onto the unperturbed states will further yield

$$
\begin{aligned}
| \, \Psi_n \rangle &= \sum_m | \, m \rangle \, \langle m \, | \, \Psi_n \rangle \\
&= | \, n \rangle \, \langle n \, | \, \Psi_n \rangle + \sum_{m \neq n} | \, m \rangle \, \langle m \, | \, \Psi_n \rangle,
\end{aligned} \tag{14.35}
$$

or, by noting that $\langle n \, | \, \Psi_n \rangle = 1$ following relation (14.7):

$$| \, \Psi_n \rangle = | \, n \rangle + \sum_{m \neq n} | \, m \rangle \, \langle m \, | \, \Psi_n \rangle. \tag{14.36}$$

We insert over $\langle m \, | \, \Psi_n \rangle$ from (14.34) into (14.36) to obtain the Wigner–Brillouin perturbed state:

$$| \, \Psi_n \rangle = | \, n \rangle + \lambda \sum_{m \neq n} | \, m \rangle \, \frac{1}{E_n - \varepsilon_m} \, \langle m \, | \, V \, | \, \Psi_n \rangle. \tag{14.37}$$

Higher orders of the perturbed wave function are obtained by iterating the last equation:

$$
\begin{aligned}
| \, \Psi_n \rangle = {} & | \, n \rangle + \lambda \sum_{m \neq n} | \, m \rangle \, \frac{1}{E_n - \varepsilon_m} \, \langle m \, | \, V \, | \, n \rangle \\
& + \lambda^2 \sum_{m,j \neq n} | \, j \rangle \, \frac{1}{E_n - \varepsilon_j} \, \langle j \, | \, V \, | \, m \rangle \, \frac{1}{E_n - \varepsilon_m} \, \langle m \, | \, V \, | \, n \rangle + \dots.
\end{aligned} \tag{14.38}
$$

Let us bear in mind that the energy E_n is also dependent on λ since, from (14.3),

$$E_n = \varepsilon_n + \lambda \, E_n^{(1)} + \lambda^2 \, E_n^{(2)} + \dots.$$

If we expand $1/E_n - \varepsilon_m$ to increasingly higher powers of λ, we arrive back at the Rayleigh–Schrödinger perturbation expansion.

It is possible to use the Wigner–Brillouin expansion (14.37) to calculate the energy perturbations since

$$E_n = \varepsilon_n + \lambda \langle n \mid V \mid \Psi_n \rangle \qquad (14.39)$$

yields, for example, the first terms of the perturbation series

$$E_n = \varepsilon_n + \lambda \langle n \mid V \mid n \rangle + \lambda^2 \sum_{m \neq n} \langle n \mid V \mid m \rangle \frac{1}{E_n - \varepsilon_m} \langle m \mid V \mid n \rangle + \dots .$$

$$(14.40)$$

This series (in which E_n is dependent on λ) converges more rapidly than the Rayleigh–Schrödinger series.

The Wigner–Brillouin method may seem too theoretical considering that the energy E_n in the denominator of (14.40) is not known. Its advantage lies in the fact that it leads, after recursion, to a better approximation of the perturbed energy.

1.3 Ground State of the Helium Atom

The Hamiltonian governing the motion of the two orbital electrons of the helium atom is written, with units such that $e^2 = 1/(4\pi \, \varepsilon_0) \, q^2$ for an electron charge q expressed in coulombs in the SI system,

$$H = \frac{p_1^2}{2m_e} + \frac{p_2^2}{2m_e} - \frac{2e^2}{r_1} - \frac{2e^2}{r_2} + \frac{e^2}{\mid \vec{r}_1 - \vec{r}_2 \mid} = H_0 + V . \qquad (14.41)$$

The two-body potential is considered as a perturbation:

$$V = \frac{e^2}{r_{12}} = \frac{e^2}{\mid \vec{r}_1 - \vec{r}_2 \mid} . \qquad (14.42)$$

The Hamiltonian of the two electrons in orbit leads to the wave function of the ground state which is written from that of the hydrogen atom (see (12.130) above), bearing in mind that $\Psi_\ell^{(r)} = \varphi_\ell^{(r)}/r$ and $Y_{00} \, (\hat{r}) = 1/\sqrt{4\pi}$:

$$\Psi_0 \, (\vec{r}_1, \vec{r}_2) = \langle \vec{r}_1, \, \vec{r}_2 \mid \Psi_0 \rangle = \frac{1}{\pi \, a^3} \exp \left(-\frac{r_1 + r_2}{a} \right) , \qquad (14.43)$$

with the Bohr radius a_0 and the energy of the ground state E_0 of the two electrons (see (12.126) and (12.127)):

$$a = \frac{1}{2} \, a_0 = \frac{\hbar^2}{2m \, e^2} \quad \text{and} \quad E_0 = -\frac{4m \, e^4}{\hbar^2} = -\frac{2e^2}{a} . \qquad (14.44)$$

The first-order perturbation of the ground-state energy is therefore written

$$E^{(1)} = \langle \Psi_0 \mid \frac{e^2}{r_{12}} \mid \Psi_0 \rangle$$

$$= \int \langle \Psi_0 | \vec{r}_1, \vec{r}_2 \rangle d\vec{r}_1 \, d\vec{r}_2 \langle \vec{r}_1, \vec{r}_2 | \frac{e^2}{r_{12}} | \vec{r}\,'_1, \vec{r}\,'_2 \rangle d\vec{r}\,'_1 \, d\vec{r}\,'_2 \langle \vec{r}\,'_1, \vec{r}\,'_2 | \Psi_0 \rangle$$

$$(14.45)$$

The matrix element is diagonal in configuration space:

$$\langle \vec{r}_1, \vec{r}_2 \mid \frac{e^2}{r_{12}} \mid \vec{r}\,'_1, \vec{r}\,'_2 \rangle = \frac{e^2}{r_{12}} \, \delta(\vec{r}_1 - \vec{r}\,'_1) \; \delta(\vec{r}_2 - \vec{r}\,'_2), \qquad (14.46)$$

whereas, because the function $\Psi_0 \, (\vec{r}_1, \vec{r}_2)$ is a real function, we obtain the value of the first-order perturbed energy

$$E^{(1)} = \frac{e^2}{\pi^2 \, a^6} \int \frac{e^{-\frac{2}{a}(r_1 + r_2)}}{|\, \vec{r}_1 - \vec{r}_2 \,|} \, d\vec{r}_1 \, d\vec{r}_2. \qquad (14.47)$$

We can calculate the integral (14.47) with the Fourier transformation of the denominator of the integrand:

$$\frac{1}{|\, \vec{r}_1 - \vec{r}_2 \,|} = \frac{1}{(2\pi)^3} \int d\vec{k} \; e^{i \, \vec{k} \cdot (\vec{r}_1 - \vec{r}_2)} \, \frac{4\pi}{k^2}, \qquad (14.48)$$

which enables us to express the perturbed energy as follows:

$$E^{(1)} = \frac{e^2}{\pi^2 \, a^6} \frac{1}{(2\pi)^3} \int d\vec{k} \; e^{i \, \vec{k} \cdot (\vec{r}_1 - \vec{r}_2)} \, \frac{4\pi}{k^2} \, d\vec{r}_1 \, d\vec{r}_2 \, \exp\left(-\frac{2}{a}(r_1 + r_2)\right)$$

$$= \frac{e^2}{a^6} \frac{1}{2\pi^6} \int d\vec{k} \; \frac{1}{k^2} \left| \int d\vec{r}_1 \, \exp\left(i \, \vec{k} \cdot \vec{r}_1 - \frac{2}{a} r_1\right) \right|^2. \qquad (14.49)$$

We recognize the Fourier transform of $\exp\left(-\frac{2}{a} r_1\right)$ in the square modulus:

$$\int d\vec{r} \; e^{i \, \vec{k} \cdot \vec{r}} \, e^{-\frac{2}{a} r} = \frac{16\pi}{a} \frac{1}{\left[k^2 + (2/a)^2\right]^2}. \qquad (14.50)$$

In this way, following integration over the direction of the vector $\vec{k}$, we obtain

$$E^{(1)} = \frac{128 \, e^2 \times 2\pi}{\pi^4 \, a^8} \int_0^\infty dk \; \frac{1}{\left[k^2 + (2/a)^2\right]^4}, \qquad (14.51)$$

which yields, following integration, the first-order perturbed energy:

$$\boxed{E^{(1)} = \frac{5}{8} \, e^2/a} \, . \qquad (14.52)$$

1.4 Stark Effect on a Rigid Rotator

The unperturbed Hamiltonian of a rigid rotator can be written as $H_0 = L^2/2I$ where I is the moment of inertia $\mu\, r_0^2$ and $\vec{L}$ the orbital angular momentum of the rotator. The eigenvalue equation

$$\frac{L^2}{2I}\, Y_{\ell\, m}(\Omega) = \frac{\hbar^2\, \ell(\ell+1)}{2I}\, Y_{\ell\, m}(\Omega) \tag{14.53}$$

will define the unperturbed energy of a state with angular momentum $(\ell\, m)$:

$$E_\ell^0 = \frac{\hbar^2\, \ell(\ell+1)}{2I} \tag{14.54}$$

and the corresponding unperturbed state $|\,\Psi_0\rangle \ = \ |\,\ell\, m\rangle$.

If we place this rotator in a uniform electric field, directed along the axis O_z, and introduce the dipole moment d of the rotator, we obtain the new Hamiltonian

$$H = \frac{L^2}{2I} - \mathcal{E}\, d\, \cos\Theta = H_0 + V. \tag{14.55}$$

The first-order perturbed energy will then become

$$E_\ell^{(1)} = \langle\Psi_0\,|\,V\,|\,\Psi_0\rangle = -\mathcal{E}\, d\, \langle\ell\, m\,|\,\cos\Theta\,|\,\ell\, m\rangle. \tag{14.56}$$

The calculation of the matrix element with the Wigner-Eckart theorem shows that $E_\ell^{(1)}$ is identically zero, since $\cos\Theta$ is proportional to $Y_{10}\,(\Theta,\varphi)$.

The first-order perturbed energy is zero.

The second-order perturbed energy, written

$$E_\ell^{(2)} = \sum_{E_0' \neq E_0} \frac{|\,\langle\Psi_0\,|\,V\,|\,\Psi_0'\rangle\,|^2}{E_0 - E_0'}, \tag{14.57}$$

will take the following form in terms of the states and energies of the ground state:

$$E_\ell^{(2)} = \frac{2I}{\hbar^2}\, \mathcal{E}^2\, d^2 \sum_{\ell' \neq \ell} \frac{|\,\langle\ell\, m\,|\,\cos\Theta\,|\,\ell'\, m\rangle\,|^2}{\ell(\ell+1) - \ell'(\ell'+1)}. \tag{14.58}$$

The matrix element is calculated with the Wigner–Eckart theorem as

$$\langle\ell\, m\,|\,\cos\Theta\,|\,\ell'\, m\rangle = A_{\ell'\, m}\, \delta_{\ell\, \ell'+1} + B_{\ell'\, m}\, \delta_{\ell\, \ell'-1} \tag{14.59}$$

using coefficients $A_{\ell\, m}$ and $B_{\ell\, m}$ with the values:

$$A_{\ell\, m} = \left[\frac{(\ell+1+m)\,(\ell+1-m)}{(2\ell+1)\,(2\ell+3)}\right]^{1/2},$$

$$B_{\ell\, m} = \left[\frac{(\ell+m)\,(\ell-m)}{(2\ell+1)\,(2\ell-1)}\right]^{1/2}. \tag{14.60}$$

By inserting matrix element (14.59) into the second-order perturbed energy (14.58), we obtain:

$$E_{\ell\,m}^{(2)} = \frac{2I}{\hbar^2}\,\mathcal{E}^2\,d^2\left[-\frac{1}{2(\ell+1)}\,A_{\ell\,m}^2 + \frac{1}{2\ell}\,B_{\ell\,m}^2\right],$$

and by writing out the coefficients $A_{\ell\,m}$ and $B_{\ell\,m}$, we further obtain

$$E_{\ell\,m}^{(2)} = \frac{\mathcal{E}^2\,d^2}{E_\ell^0}\,\frac{1}{2(2\ell-1)\,(2\ell+3)}\,[\ell(\ell+1) - 3m^2]\,. \tag{14.61}$$

The Stark effect separates the different magnetic momenta levels into absolute values. It should be noted that the quantum state $\mid \ell\ m\rangle$ is $(2\ell+1)$-fold degenerate for a fixed value of ℓ. We have fixed the value of ℓ and that of m, making it possible for us to treat the ket $\mid\ell m\rangle$ as non-degenerate.

1.5 Perturbation of a Degenerate System

a) Perturbed State and Energy

Let us assume now that the unperturbed state is degenerate, that is that there are r eigenstates corresponding to the energy ε_n of H_0:

$$H_0 \mid n\ r\rangle = \varepsilon_n \mid n\ r\rangle\,. \tag{14.62}$$

The normalization of these states will be written by choosing an unperturbed level ε_n corresponding to a given eigenstate $\mid n\ r\rangle$:

$$\langle n'\ r' \mid n\ r\rangle = \delta_{n\,n'}\,\delta_{r'\,r} \tag{14.63}$$

The eigenvalue equation of the perturbed Hamiltonian will then take the following form:

$$(H_0 + \lambda\,V) \mid \Psi_{nr}\rangle = E_{nr} \mid \Psi_{nr}\rangle \tag{14.64}$$

with the perturbation expansions of the energy and eigenstate:

$$\begin{aligned} E_{nr} &= \varepsilon_n + \lambda\,E_{nr}^{(1)} + \lambda^2\,E_{nr}^{(2)} + \dots\,, \\ \mid \Psi_{nr}\rangle &= \mid \Psi_{nr}^0\rangle + \lambda \mid \Psi_{nr}^{(1)}\rangle + \dots\,. \end{aligned} \tag{14.65}$$

The state $\mid \Psi_{nr}^{(0)}\rangle$ is not necessarily an eigenstate $\mid nr\rangle$ of H_0 but rather a linear combination of these states:

$$\mid \Psi_{nr}^{(0)}\rangle = \sum_{r'} \mid nr'\rangle\,\langle nr' \mid \Psi_{nr}^{(0)}\rangle\,. \tag{14.66}$$

We thus obtain, for the first-order perturbation,

$$(H_0 - \varepsilon_n) \mid \Psi_{nr}^{(1)}\rangle + \left(V - E_{nr}^{(1)}\right) \mid \Psi_{nr}^{(0)}\rangle = 0\,. \tag{14.67}$$

We multiply by the eigenbra $\langle nr' \mid$ of H_0, eliminating in this way the first term of the above equation:

$$\langle nr' \mid V \mid \Psi_{nr}^{(0)} \rangle = E_{nr}^{(1)} \langle nr' \mid \Psi_{nr}^{(0)} \rangle. \tag{14.68}$$

If the states $\mid \Psi_{nr}^{(0)} \rangle$ are the basis states of H_0 diagonalizing V in the subspace $\{\mid n\,r\rangle\}$ of $\{\mid n\rangle\}$, Eq. (14.68) will yield the following first-order perturbation:

$$\boxed{E_{nr}^{(1)} = \langle nr \mid V \mid nr \rangle} . \tag{14.69}$$

In the more general case, by projecting (14.68) onto the complete orthonormal basis of the eigenstates $\mid nr\rangle$ of H_0, we obtain the fundamental equation

$$\sum_{r''} \langle nr' \mid V \mid nr'' \rangle \langle nr'' \mid \Psi_{nr}^{(0)} \rangle = E_{nr}^{(1)} \langle nr' \mid \Psi_{nr}^{(0)} \rangle. \tag{14.70}$$

As a way of simplifying the notation, we will set:

$$\begin{aligned}
V_{rr'} &= \langle nr \mid V \mid nr' \rangle, \\
\langle r' \mid r \rangle &= \langle nr' \mid \Psi_{nr}^{(0)} \rangle, \\
e_r &= E_{nr}^{(1)}.
\end{aligned} \tag{14.71}$$

This will transform the fundamental Eq. (14.69) as follows:

$$\sum_{r''} V_{r'\,r''} \langle r'' \mid r \rangle = e_r \langle r' \mid r \rangle, \tag{14.72}$$

or, by introducing the overlap factor $\langle r'' \mid r \rangle$:

$$\sum_{r''} (V_{r'\,r''} - e_r\, \delta_{r''\,r'}) \langle r'' \mid r \rangle = 0. \tag{14.73}$$

We thus obtain a system of N coupled equations:

$$\begin{aligned}
(V_{11} - e_r) \langle 1 \mid r \rangle + V_{12} \langle 2 \mid r \rangle + \ldots V_{1N} \langle N \mid r \rangle &= 0, \\
V_{21} \langle 1 \mid r \rangle + (V_{22} - e_r) \langle 2 \mid r \rangle + \ldots V_{2N} \langle N \mid r \rangle &= 0, \\
&\quad \vdots \\
V_{N1} \langle 1 \mid r \rangle + \ldots (V_{NN} - e_r) \langle N \mid r \rangle &= 0.
\end{aligned} \tag{14.74}$$

A solution is only possible if the secular determinant is zero:

$$0 = \det\,(V_{r'\,r''} - e_r\, \delta_{r'\,r''}) = \det \begin{pmatrix} V_{11} - e_r & V_{12} & \cdots \\ V_{21} & V_{22} - e_r & \cdots \\ \vdots & & \end{pmatrix}. \tag{14.75}$$

The first-order perturbed states are obtained by multiplying Eq. (14.67) by the eigenstate $\langle n'r' |$ of H_0:

$$\langle n'r' | \Psi_{nr}^{(1)} \rangle = \sum_{r''} \frac{\langle n'r' | V | nr'' \rangle}{\varepsilon_n - \varepsilon_{n'}} \langle nr'' | \Psi_{nr}^{(0)} \rangle . \tag{14.76}$$

If the states $| \Psi_{nr}^{(0)} \rangle$ are the basis states of H_0, then they will satisfy the relation

$$\langle nr'' | \Psi_{nr}^{(o)} \rangle = \langle nr'' | nr \rangle = \delta_{rr''} , \tag{14.77}$$

and, finally, the projection of the first-order perturbed state onto the unperturbed states will yield

$$\langle n'r' | \Psi_{nr}^{(1)} \rangle = \frac{\langle n'r' | V | nr \rangle}{\varepsilon_n - \varepsilon_{n'}} . \tag{14.78}$$

This shows that to the first order of perturbation, we can write

$$| \Psi_{nr}^{(1)} \rangle = | nr \rangle + \lambda \sum_{r'n' \neq n} \frac{\langle n'r' | V | nr \rangle}{\varepsilon_n - \varepsilon_{n'}} | n'r' \rangle . \tag{14.79}$$

Let us assume for the sake of simplicity that $| \Psi_{nr}^0 \rangle = | nr \rangle$ such that

$$\langle nr' | \Psi_{nr}^0 \rangle = 0 . \tag{14.80}$$

The pth-order equation in λ will then give

$$(H_0 - E_{nr}^0) | \Psi_{nr}^{(p)} \rangle + (V - E_{nr}^1) | \Psi_{nr}^{(p-1)} \rangle = E_{nr}^2 | \Psi_{nr}^{(p-2)} \rangle + \dots E_{nr}^p | nr \rangle , \tag{14.81}$$

or, by multiplying by $\langle n'r' |$, the unperturbed degenerate state

$$\langle n'r' | \Psi_{nr}^{(p)} \rangle = \frac{\langle n'r' | V | \Psi_{nr}^{(p-1)} \rangle}{\varepsilon_n - \varepsilon_{n'}} . \tag{14.82}$$

By successive iterations, we obtain the perturbed state

$$| \Psi_{nr} \rangle = | nr \rangle + \lambda \sum_{n' \neq n \ r'} \frac{\langle n'r' | V | nr \rangle}{\varepsilon_n - \varepsilon_{n'}} | n'r' \rangle$$

$$+ \lambda^2 \sum_{n' \ n'' \ r' \ r''} \frac{\langle n'r' | V | n''r'' \rangle \langle n''r'' | V | nr \rangle}{\varepsilon_n - \varepsilon_{n'} \qquad \varepsilon_n - \varepsilon_{n'}} | n'r' \rangle + \dots \tag{14.83}$$

and the corresponding perturbed energy:

$$E_{nr} = \varepsilon_n + \lambda \langle nr | V | nr \rangle + \lambda^2 \sum_{n' \neq n \ r'} \frac{| \langle n'r' | V | nr \rangle |^2}{\varepsilon_n - \varepsilon_{n'}} + \dots . \tag{14.84}$$

b) Example: Perturbation of a Two-Fold Degenerate Level

Equation (14.74) takes the following form in a second-order degeneracy:

$$\det \begin{pmatrix} V_{11} - x & V_{12} \\ V_{21} & V_{22} - x \end{pmatrix} = 0 \tag{14.85}$$

which leads to the second-degree equation

$$x^2 - x \ (V_{11} + V_{22}) + V_{11}V_{22} - V_{12}V_{21} = 0 , \tag{14.86}$$

with the two first-order perturbed energy levels as solutions:

$$E_{11}^{(1)} = \bar{V} - \frac{1}{2} \left[(V_{22} - V_{11})^2 + 4V_{12} \ V_{21} \right]^{1/2}$$

$$E_{12}^{(1)} = \bar{V} + \frac{1}{2} \left[(V_{22} - V_{11})^2 + 4V_{12} \ V_{21} \right]^{1/2} \tag{14.87}$$

and the expectation value of the potential

$$\bar{V} = \frac{1}{2} \ (V_{11} + V_{22}) = \frac{1}{2} \ [\langle n1 \mid V \mid n1 \rangle + \langle n2 \mid V \mid n2 \rangle] . \tag{14.88}$$

By inserting $E_{1r}^{(1)}$ into the coupled Eq. (14.74), we obtain

$$\left(V_{11} - E_{1r}^{(1)} \right) \langle 1 \mid r \rangle + V_{12} \ \langle 2 \mid r \rangle = 0$$

$$V_{21} \ \langle 1|r \rangle + (V_{22} - E_{1r}^{(1)}) \ \langle 2|r \rangle = 0 . \tag{14.89}$$

We introduce the normalization condition for the degenerate states:

$$1 = \langle \Psi_{nr}^{(0)} \mid \Psi_{nr}^{(0)} \rangle = \sum_{r'} \langle \Psi_{nr}^{(0)} \mid nr' \rangle \langle nr' \mid \Psi_{nr}^{(0)} \rangle \tag{14.90}$$

expressed in the form (14.71) as

$$\sum_{r'} \langle r' \mid r \rangle^2 = 1 . \tag{14.91}$$

Following expansion, the perturbed states will give for $r = 1$ and then $r = 2$:

$$\langle 1 \mid r \rangle^2 + \langle 2 \mid r \rangle^2 = 1 . \tag{14.92}$$

Statements (14.87) and (14.89) hence determine the overlap

$$\langle 1 \mid r \rangle^2 = \frac{V_{12}^2}{\left(V_{11} - E_{1r}^{(1)} \right)^2 + V_{12}^2} ,$$

$$\langle 2 \mid r \rangle^2 = \frac{\left(V_{11} - E_{1r}^{(1)} \right)^2}{\left(V_{11} - E_{1}^{(1)} \right)^2 + V_{12}^2} , \tag{14.93}$$

$$\langle 2 \mid r \rangle = \frac{E_{1r}^{(1)} - V_{22}}{V_{21}} \ \langle 1 \mid r \rangle .$$

In the special case in which $V_{11} = V_{22} = V$ and $\bar{V} = V$ and $V_{21} = V_{12}$, we obtain the very simple result

$$E_{1r}^{(1)} = V \mp V_{12}, \tag{14.94}$$

whereas overlaps (14.93) of the quantum states become

$$\langle 1 \mid 1 \rangle^2 = \langle 2 \mid 1 \rangle^2 = \frac{1}{2} \quad \text{and} \quad \langle 2 \mid 1 \rangle = - \langle 1 \mid 1 \rangle. \tag{14.95}$$

If $E_{11}^{(1)} = V - V_{12}$ and $E_{12}^{(1)} = V + V_{12}$ while $\langle 1 \mid 1 \rangle = 1$ then $\langle 2 \mid 1 \rangle = -1/\sqrt{2}$ and $\langle 2 \mid 2 \rangle = \langle 1 \mid 2 \rangle = 1/\sqrt{2}$. We infer the degenerate perturbed states

$$\mid \Psi_{nr} \rangle = \sum_{r'} \mid nr' \rangle \langle nr' \mid \Psi_{nr}^{(o)} \rangle = \sum_{r'} \mid nr' \rangle \langle r' \mid r \rangle. \tag{14.96}$$

This leads to two quantum states, with the first written as follows:

$$\mid \Psi_{11} \rangle = \langle 1 \mid 1 \rangle \mid 11 \rangle + \langle 2 \mid 1 \rangle \mid 12 \rangle = \frac{1}{\sqrt{2}} [\mid 11 \rangle - \mid 12 \rangle]. \tag{14.97}$$

This is an antisymmetric state corresponding to the energy

$$E_1 = \varepsilon + \lambda \left(\bar{V} - V_{12} \right), \tag{14.98}$$

and the second quantum state will be of the form:

$$\mid \Psi_{12} \rangle = \mid 11 \rangle \langle 1 \mid 2 \rangle + \mid 12 \rangle \langle 2 \mid 2 \rangle = \frac{1}{\sqrt{2}} [\mid 11 \rangle + \mid 12 \rangle], \tag{14.99}$$

a symmetrical state corresponding to the energy

$$E_2 = \varepsilon + \lambda \left(\bar{V} + V_{12} \right). \tag{14.100}$$

We notice here that the antisymmetrical state has the lowest energy. It will therefore be filled first.

2. Variational Method

2.1 Underlying Principle

Let us assume that the equation examined has as solutions the functions of a functional space $\mathcal{F}$, which renders a function $E(\Psi)$ stationary, that is, such that:

$$\delta E = 0. \tag{14.101}$$

We will look for the solutions to Eq. (14.101) among functions belonging to a domain $\mathcal{F}'$ that is more restricted than $\mathcal{F}$.

The success of the method will depend on the choice of the trial functions of the space $\mathcal{F}'$.

2.2 Determination of the Eigenvalues

Let us choose the function of $E(\Psi)$ in the following form:

$$E(\Psi) = \frac{\langle \Psi \mid H \mid \Psi \rangle}{\langle \Psi \mid \Psi \rangle} \tag{14.102}$$

in the space $\mathcal{E}$ of the $|\Psi\rangle$ states. Any vector capable of rendering the functional stationary is an eigenvector of H and vice-versa. The corresponding eigenvalue is the stationary value of the functional $E(\Psi)$.

The eigensolutions of H in $\mathcal{E}$ are therefore solutions to the variational equation

$$\delta\, E(\Psi) = 0. \tag{14.103}$$

The functional $E(\Psi)$ defined in (14.102) can, in effect, be written as follows:

$$E(\Psi)\, \langle \Psi \mid \Psi \rangle \; = \; \langle \Psi \mid H \mid \Psi \rangle. \tag{14.104}$$

The successive variation of each element of this equation will give:

$$\langle \Psi \mid \Psi \rangle\, \delta\, E = \delta\, \langle \Psi \mid H \mid \Psi \rangle \; - E\, \delta\, \langle \Psi \mid \Psi \rangle. \tag{14.105}$$

We cause the ket and bra to vary in the right-hand term:

$$\langle \Psi \mid \Psi \rangle\, \delta\, E = \langle \delta\, \Psi \mid (H - E) \mid \Psi \rangle \; + \; \langle \Psi \mid (H - E) \mid \delta\, \Psi \rangle. \tag{14.106}$$

If $\langle \Psi \mid \Psi \rangle$ remains non-zero and finite, then $\delta\, E = 0$ will imply

$$\langle \delta\, \Psi \mid H - E \mid \Psi \rangle \; + \; \langle \Psi \mid H - E \mid \delta\, \Psi \rangle \; = 0. \tag{14.107}$$

Notice that variations $\mid \delta\, \Psi\rangle$ and $\langle \delta\, \Psi \mid$ belonging to dual spaces are not independent. Yet we can consider them as such since (14.107) is satisfied for every arbitrary infinitely small $\mid \delta\, \Psi\rangle$. If we replace $\mid \delta\, \Psi\rangle$ with $\mid i\, \delta\, \Psi\rangle = i \mid \delta\, \Psi\rangle$, we get

$$\langle i\, \delta\, \Psi \mid \; = [\, \mid i\delta\, \Psi\rangle\,]^{+} = -i\, \langle \delta\, \Psi \mid,$$

and, by inserting this into (14.107),

$$-i\, \langle \delta\, \Psi \mid (H - E) \mid \Psi \rangle \; + i\, \langle \Psi \mid (H - E) \mid \delta\, \Psi \rangle \; = 0. \tag{14.108}$$

We are then left with the two variational equations:

$$\begin{aligned}
\langle \Psi \mid (H - E) \mid \delta\, \Psi \rangle \; + \; \langle \delta\, \Psi \mid (H - E) \mid \Psi \rangle \; = 0, \\
\langle \Psi \mid (H - E) \mid \delta\, \Psi \rangle \; - \; \langle \delta\, \Psi \mid (H - E) \mid \Psi \rangle \; = 0.
\end{aligned} \tag{14.109}$$

Because H is Hermitian, the matrix element of $(H - E)$ will take the form

$$\langle \delta\, \Psi \mid (H - E) \mid \Psi \rangle \; = [\langle \Psi \mid H - E \mid \delta\, \Psi \rangle\,]^{+} \tag{14.110}$$

and, irrespective of $\langle \delta \Psi \mid$, we should therefore obtain

$$\langle \delta \Psi \mid H - E \mid \Psi \rangle + \langle \Psi \mid H - E \mid \delta \Psi \rangle = 2R_e \langle \delta \Psi \mid H - E \mid \Psi \rangle = 0. \quad (14.111)$$

The result is the following eigenvalue equation:

$$(H - E) \mid \Psi \rangle = 0,$$

which proves the foregoing proposition:

$$\boxed{\delta E(\Psi) = 0 \implies (H - E) \mid \Psi \rangle = 0} \ . \qquad (14.112)$$

Any ket $\mid \Psi_1 \rangle$ capable of rendering E stationary is therefore an eigenket of H corresponding to the eigenvalue $E = \langle \Psi_1 \mid H \mid \Psi_1 \rangle / (\langle \Psi_1 \mid \Psi_1 \rangle)$ and vice-versa.

This theorem implies that the expectation value of the energy of a system is higher than or equal to the eigenvalue of its ground state.

In effect, let us assume that H has a discrete spectrum, and let $E_0, E_1, \ldots, E_n$ be the energy levels arranged in the ascending order $E_0 < E_1 < E_2 < E_3 \ldots$ and $P_0, P_1, \ldots, P_n$ the projectors onto their respective subspaces, satisfying the completeness relation $\sum_n P_n = \sum_n \mid n \rangle \langle n \mid = \mathbb{1}$.

Now, let us evaluate the difference between the energies E and E_0:

$$E - E_0 = \frac{\langle \Psi \mid H \mid \Psi \rangle}{\langle \Psi \mid \Psi \rangle} - E_0 \frac{\langle \Psi \mid \Psi \rangle}{\langle \Psi \mid \Psi \rangle} = \frac{\langle \Psi \mid H - E_0 \mid \Psi \rangle}{\langle \Psi \mid \Psi \rangle} . \qquad (14.113)$$

By projecting onto the subspace of the eigenvalue E_n we obtain

$$H P_n \mid \Psi \rangle = E_n P_n \mid \Psi \rangle, \qquad (14.114)$$

and substituting this into (14.113), we further obtain

$$E - E_0 = \sum_n \frac{\langle \Psi \mid H P_n - P_n E_0 \mid \Psi \rangle}{\langle \Psi \mid \Psi \rangle} = \sum_n (E_n - E_0) \frac{\langle \Psi \mid P_n \mid \Psi \rangle}{\langle \Psi \mid \Psi \rangle}$$

$$= \sum_n (E_n - E_0) \frac{\mid \langle \Psi \mid n \rangle \mid^2}{\langle \Psi \mid \Psi \rangle} . \qquad (14.115)$$

Each term in this sum is either positive or zero, since $E_0 < E_1 < \ldots < E_n \ldots$. Hence, $E - E_0$ is either positive or zero, proving that $E_0 \leq E = \langle \Psi \mid H \mid \Psi \rangle / (\langle \Psi \mid \Psi \rangle)$, which is the proposition formulated above.

2.3 Example: Ground State of the Helium Atom

We will take the wave function used in perturbation theory as the trial wave function:

$$\Phi_a\,(\vec{r}_1,\ \vec{r}_2) = \left(\frac{1}{\pi\,a^3}\right)\exp\left(-\frac{r_1+r_2}{a}\right)\quad \text{with}\quad a = \frac{\hbar^2}{2m\,e^2}\,. \qquad (14.116)$$

The function is normalized to unity:

$$\int \Phi_a^*\,(\vec{r}_1,\ \vec{r}_2)\ \Phi_a\,(\vec{r}_1,\ \vec{r}_2)\ d\vec{r}_1\ d\vec{r}_2 = 1\,, \qquad (14.117)$$

giving the following functional:

$$E(a) = \frac{\langle\Phi_a\,|\,H\,|\,\Phi_a\rangle}{\langle\Phi_a\,|\,\Phi_a\rangle} = \int \Phi_a^*\,(\vec{r}_1,\ \vec{r}_2)\ H\ \Phi_a\,(\vec{r}_1,\ \vec{r}_2)\ d\vec{r}_1\ d\vec{r}_2\,. \quad (14.118)$$

The Hamiltonian H in the integral will be written

$$H = t_1 + t_2 + v_1 + v_2 + V_{12} \qquad (14.119)$$

t_i kinetic energy of the electron $i = -\hbar^2/(2m_i)\,\nabla_i^2$
v_i interaction of ith electron with the nucleus $= -2e^2/r_i$
V_{ij} interaction of the ith electron with the jth electron $= e^2/r_{ij}$

We therefore obtain kinetic terms of the following form:

$$\langle\Phi_a\,|\,t_1\,|\,\Phi_a\rangle = \int \Phi_a^*\,(\vec{r}_1,\ \vec{r}_2)\ d\vec{r}_1\ d\vec{r}_2 \left(-\frac{\hbar^2}{2m_1}\,\nabla_1^2\right)\Phi_a\,(\vec{r}_1,\ \vec{r}_2)\,, \tag{14.120}$$

with the radial Laplacian in spherical coordinates:

$$\nabla_1^2 = \frac{1}{r_1^2}\,\frac{\partial}{\partial\,r_1}\,r_1^2\,\frac{\partial}{\partial\,r_1}\,. \qquad (14.121)$$

Because the function $\Phi_a\,(\vec{r}_1,\ \vec{r}_2)$ is dependent neither on θ nor on φ, only the term in r_1 will have an effect on the wave function:

$$\begin{aligned}
\nabla_1^2\,\Phi_a\,(\vec{r}_1,\ \vec{r}_2) &= \frac{1}{\pi\,a^3}\left(\frac{d^2}{d\,r_1^2} + \frac{2}{r_1}\,\frac{d}{d\,r_1}\right)\exp\left(-\frac{r_1+r_2}{a}\right)\\
&= \frac{1}{\pi\,a^4}\left(\frac{1}{a} - \frac{2}{r_1}\right)\exp\left(-\frac{r_1+r_2}{a}\right),
\end{aligned} \qquad (14.122)$$

and we are left with the matrix element of the kinetic term

$$\langle\Phi_a|t_1\,|\,\Phi_a\rangle = -\frac{\hbar^2}{2m_1}\left(\frac{1}{\pi\,a^3}\right)^2\frac{1}{a}\int\left(\frac{1}{a} - \frac{2}{r_1}\right)\exp\left(-\frac{2(r_1+r_2)}{a}\right)d\vec{r}_1\ d\vec{r}_2\,. \tag{14.123}$$

The integration over the angular variables will yield 4π. The definite integral

$$\int_0^\infty x^n \, e^{-s\,x} \, dx = \frac{n!}{s^{n+1}} \qquad (14.124)$$

makes it possible to calculate the remaining radial integrals, that is

$$\int_0^\infty \exp\left(-\frac{2r_2}{a}\right) r_2^2 \, dr_2 = \frac{a^3}{4} \qquad (14.125)$$

$$\int_0^\infty \left(\frac{1}{a} - \frac{2}{r_1}\right) \exp\left(-\frac{2r_1}{a}\right) r_1^2 \, dr_1 = \frac{a^2}{4} - \frac{2a^2}{4} = -\frac{a^2}{4} \qquad (14.126)$$

Hence, we obtain the matrix element of the kinetic energy:

$$\langle \Phi_a \mid t_1 \mid \Phi_a \rangle = \left(-\frac{\hbar^2}{2m_1}\right) \left(\frac{16\,\pi^2}{a^6\,\pi^2}\right) \left[\frac{1}{a}\left(\frac{a^3}{4}\right)\left(-\frac{a^2}{4}\right)\right] \qquad (14.127)$$

which yields, after the necessary simplifications

$$\langle \Phi_a \mid t_1 \mid \Phi_a \rangle = \frac{\hbar^2}{2m_1} \frac{1}{a^2}. \qquad (14.128)$$

The second interaction element is easily calculated:

$$\langle \Phi_a \mid v_1 \mid \Phi_a \rangle = \left(\frac{1}{\pi\,a^3}\right)^2 \int d\vec{r}_1 \, d\vec{r}_2 \, \frac{2e^2}{r_2} \exp\left[-\frac{2}{a}(r_1 + r_2)\right]$$

$$= (4\pi)^2 \, 2e^2 \left(\frac{1}{\pi\,a^3}\right)^2 \frac{a^3}{4} \left(-\frac{a^2}{4}\right), \qquad (14.129)$$

leaving us with the matrix element

$$\langle \Phi_a \mid v_1 \mid \Phi_a \rangle = -\frac{2e^2}{a}. \qquad (14.130)$$

The third two-body interaction term will then be written as follows:

$$\langle \Phi_a \mid V_{12} \mid \Phi_a \rangle = \frac{e^2}{(\pi\,a^3)^2} \int \exp\left(-\frac{2}{a}(r_1 + r_2)\right) \frac{1}{|\,\vec{r}_1 - \vec{r}_2\,|} \, d\vec{r}_1 \, d\vec{r}_2. \qquad (14.131)$$

This integral was calculated within the framework of the theory of perturbations. We now extract the result (see (14.47) and (14.52)).

$$\langle \Phi_a \mid V_{12} \mid \Phi_a \rangle = \frac{5}{8} \frac{e^2}{a}. \qquad (14.132)$$

We can therefore determine the functional $E(a)$ with (14.119), (14.127), (14.128) and (14.129):

$$E(a) = \frac{\hbar^2}{2m_1} \frac{1}{a^2} + \frac{\hbar^2}{2m_2} \frac{1}{a^2} - \frac{4e^2}{a} + \frac{5}{8} \frac{e^2}{a}, \qquad (14.133)$$

and since $m_1 = m_2 = m_e = m$, we are led to the expression

$$E(a) = \frac{\hbar^2}{m} \frac{1}{a^2} - \frac{27}{8} \frac{e^2}{a}. \tag{14.134}$$

Now, let us apply the variational principle by calculating $\delta E = 0$ for a variation of the Bohr radius a:

$$\delta \frac{\hbar^2}{m} \frac{1}{a^2} - \delta \frac{27}{8} \frac{e^2}{a} = -\frac{\delta a}{a^3} \left(\frac{2\hbar^2}{m} - \frac{27}{8} e^2 a \right) = 0. \tag{14.135}$$

The non-trivial solution will lead to the value of a:

$$\frac{2\hbar^2}{m} = \frac{27}{8} e^2 a_{var} \quad \text{and} \quad a_{var} = \frac{16}{27} \frac{\hbar^2}{m e^2} = \frac{32a}{27} \rangle a. \tag{14.136}$$

The inequality $a_{var} \rangle a$ derives from the fact that the variational method describes the effect of the residual interaction between electrons, reducing the attraction by the nucleus of the electron cloud.

If we introduce the ground-state energy E_H of the hydrogen atom, then

$$E_H = \frac{1}{2} \frac{m e^4}{\hbar^2} \quad \text{and} \quad a_{var} = \frac{8}{27} \frac{e^2}{E_H}. \tag{14.137}$$

The functional $E(a)$ goes through an extremum. By inserting this value (14.137) in $E(a)$ given by (14.134), we obtain the value of the variational energy:

$$E_{var} = \frac{\hbar^2}{m} \left(\frac{1}{a_{var}} \right)^2 - \frac{27}{8} \frac{e^2}{a_{var}} = -2E_H \left(\frac{27}{16} \right)^2 = -5.70 E_H. \tag{14.138}$$

First-order perturbation theory leads to the value

$$E_{pert} = E_0 + E_1 = -\frac{2e^2}{a} + \frac{5}{8} \frac{e^2}{a} = -\frac{11}{8} \frac{e^2}{a} \quad \text{with} \quad a = \frac{\hbar^2}{2m e^2}, \tag{14.139}$$

that is, by introducing E_H,

$$E_{pert} = -8E_H + 2.5E_H = -5.50 E_H, \tag{14.140}$$

a value that is quite close to the variational result:

$$E_{vart} - E_{pert} = -0.20 E_H. \tag{14.141}$$

3. Time-Dependent Perturbations

Consider a quantum state initially in the eigenstate $|\varphi_i\rangle$ of the Hamiltonian H_0, and a subsequent evolution of the system caused by the application of an explicitly time-dependent potential

$$i\hbar\,\partial_t\,|\,\Psi(t)\rangle = (H_0 + \lambda\,V(t))\,|\,\Psi(t)\rangle \qquad (14.142)$$

with the initial condition

$$|\,\Psi(0)\rangle = |\,\varphi_i\rangle\,.$$

The system returns to an eigenstate $|\,\varphi_f\rangle$ of the Hamiltonian H_0 when the action of the perturbing potential becomes negligible. As a result, we want to determine the transition probability of the initial state $|\,\varphi_i\rangle$ to the final state $|\,\varphi_f\rangle$:

$$P_r\,(\varphi_i\ \rightarrow\ \Psi(t)) = P_r\,(i\ \rightarrow\ f) = |\,\langle\varphi_i\,|\,\Psi(t)\rangle\,|^2\,. \qquad (14.143)$$

This problem already came up in the discussion on the intermediate representation of quantum dynamics as we will see in Sect. 3.2 below.

3.1 Analytical Method

The projection of the quantum state $|\,\Psi(t)\rangle$ onto the complete orthonormal system of the eigenfunctions $|\,\varphi_n\rangle$ of the Hamiltonian H_0 will give

$$|\,\Psi(t)\rangle = \sum_n C_n\,|\varphi_n\rangle \quad \text{with} \quad C_n(t) = \langle\varphi_n\,|\,\Psi(t)\rangle\,, \qquad (14.144)$$

whereas evolution Eq. (7.142) will become

$$i\hbar\,\partial_t\,C_n(t) = E_n\,C_n(t) + \sum_k \lambda\,V_{n\,k}(t)\,C_k(t)\,, \qquad (14.145)$$

with the matrix element of the interaction potential

$$V_{nk}(t) = \langle\varphi_n\,|\,V(t)\,|\,\varphi_k\rangle\,. \qquad (14.146)$$

If $\lambda = 0$, then we obtain from (14.142) the unperturbed equation with the solution

$$C_n(t) = b_n\,\exp\left(-\frac{i}{\hbar}\,E_n\,t\right)\,, \qquad (14.147)$$

where b_n is a constant dependent on the initial conditions.

If $\lambda \neq 0$, then we can look for solutions of a similar form but with b_n time dependent:

$$C_n(t) = b_n(t)\,\exp\left(-\frac{i}{\hbar}\,E_n\,t\right)\,. \qquad (14.148)$$

Inserting this into (14.145), we obtain the perfectly equivalent differential equation:

$$i \, \hbar \, \partial_t \, b_n(t) = \lambda \sum_k \left[\exp \left(i \, \omega_{nk} \, t \right) \right] V_{nk}(t) \, b_k(t) \,, \tag{14.149}$$

with Bohr's angular velocity defined through the equation

$$\omega_{nk} = \frac{1}{\hbar} \left(E_n - E_k \right). \tag{14.150}$$

Now, let us look for an approximate solution by making a series expansion of the coefficient $b_n(t)$ appearing in (14.148):

$$b_n(t) = b_n^{(0)}(t) + \lambda \, b_n^{(1)}(t) + \ldots = \sum_{r=0} \lambda^r \, b_n^{(r)}(t). \tag{14.151}$$

Inserting this into (14.149), we obtain, by identification of terms of the same order for $r \neq 0$,

$$i \, \hbar \, \partial_t \, b_n^{(r)}(t) = \sum_k e^{i \, \omega_{nk} t} \, V_{nk}(t) \, b_k^{(r-1)}(t), \tag{14.152}$$

whereas for $r = 0$ we regain the initial value (14.142):

$$i \, \hbar \, \partial_t \, b_n^{(0)}(t) = 0 \quad \text{and} \quad b_n^{(0)}(t) = C_n(0) = \delta_{ni}. \tag{14.153}$$

The first-order term can be evaluated from this with (14.152):

$$i \, \hbar \, \partial_t \, b_n^{(1)}(t) = \sum_k e^{i \, \omega_{nk} t} \, V_{nk}(t) \, \delta_{ki} = e^{i \, \omega_{ni} t} \, V_{ni}(t). \tag{14.154}$$

By integration over the variable t this will yield

$$b_n^{(1)}(t) = \frac{1}{i \, \hbar} \int_0^t e^{i \, \omega_{ni} t'} \, V_{ni}(t') \, dt'. \tag{14.155}$$

We infer from this the first-order coefficients (14.147)

$$C_n^{(1)}(t) = b_n^{(1)}(t) \, \exp \left(-\frac{i}{\hbar} \, E_n \, t \right), \tag{14.156}$$

and the first-order transition probability

$$P_r \, (i \rightarrow f) = |\, C_f^{(1)} \,|^2 \; = \; |\, b_f^{(1)} \,|^2. \tag{14.157}$$

3.2 Intermediate Representation

Let us recall the results obtained with quantum dynamics (Chap. 6) for the intermediate representation:

$$V_I(t) = U_0^+(t,t_0)\, V(t)\, U_0(t,t_0)\,, \tag{14.158}$$

$$i\,\hbar\,\partial_t \mid I,t\rangle = V_I(t) \mid I,t\rangle\,, \tag{14.159}$$

$$\mid I,t\rangle = U_I(t,t_0) \mid I,t_0\rangle\,. \tag{14.160}$$

The evolution operator $U_0(t,t_0)$ is determined from the time-independent Hamiltonian H_0 and, hence, is written as follows:

$$U_0(t,t_0) = \exp\left(-\frac{i}{\hbar}\, H_0(t-t_0)\right). \tag{14.161}$$

We assumed in Chap. 6 that the intermediate states constitute a complete orthonormal discrete system. Now, let us examine the case in which the intermediate states constitute a complete orthonormal continuous basis, that is, satisfy the following conditions at a given point in time:

$$\int \mid I,t\rangle\, dI \langle I,t \mid\, = \mathbb{1}$$

$$\langle I,t \mid I',t\rangle = \langle I,t_0 \mid I',t_0\rangle = \delta(I - I')\,. \tag{14.162}$$

The evolution operator $U_I(t,t_0)$ will therefore be written

$$U_I(t,t_0) = \int \mid I,t\rangle\, dI \langle I,t_0 \mid \tag{14.163}$$

and we can easily verify that $U_I(t,t_0)$ is a unitary operator permitting the transition from $\mid I,t_0\rangle$ to $\mid I,t\rangle$, that is,

$$U_I(t,t_0) \mid I,t_0\rangle = \mid I,t\rangle\,. \tag{14.164}$$

The probability of transition from one state $\mid \alpha\rangle = \mid a,\, t_0\rangle$ to another $\mid \beta\rangle = \mid b,t\rangle$ can be obtained with the intermediate states (14.160):

$$\mid I,t_0\rangle = U_I(t_0,t_0) \mid a,t_0\rangle = \mid \varphi_i,t_0\rangle = \mid \varphi_i\rangle\,,$$
$$\mid I',t_0\rangle = U_{I'}(t_0,t) \mid b,t\rangle = \mid \varphi_f,t_0\rangle = \mid \varphi_f\rangle\,. \tag{14.165}$$

We infer from this and (14.164) the matrix element

$$\langle I',t_0 \mid U_I(t,t_0) \mid I,t_0\rangle = \langle I',t_0 \mid I,t\rangle\,, \tag{14.166}$$

and the 4th postulate of quantum mechanics will then give the probability of transition from the state $\mid \varphi_i\rangle$ to $\mid \varphi_f\rangle$, as the square modulus of the above matrix element:

$$P_r\,(i \rightarrow f) = \mid \langle I',t_0 \mid I,t\rangle \mid^2 = \mid\langle \varphi_f \mid U_I(t,t_0) \mid \varphi_i\rangle \mid^2\,. \tag{14.167}$$

The operator $U_I(t, t_0)$ is determined from (14.158) in the form

$$U_I(t, t_0) = \tau \, \exp\left[-\frac{i}{\hbar} \int_{t_0}^{t} V_I(t') \, dt'\right], \qquad (14.168)$$

where the time-ordering operator τ is responsible for ordering the intermediate times of the series expansion of the exponential.

The matrix element of $U_I(t, t_0)$ will cause the matrix element of $V_I(t)$ to be introduced; the latter must therefore be fully determined with (14.158) and (14.161), which gives

$$\langle \varphi_f | V_I(t) | \varphi_i \rangle = \langle \varphi_f | U_0^+(t, t_0) V(t) U_0(t, t_0) | \varphi_i \rangle$$
$$= \langle \varphi_f | \exp\left[\frac{i}{\hbar}(t - t_0)H_0\right] V(t) \exp\left[-\frac{i}{\hbar}(t - t_0)H_0\right] | \varphi_i \rangle .$$
$$(14.170)$$

The action of H_0 on the states $| \varphi_f \rangle$ or $| \varphi_i \rangle$ will lead to the eigenvalues E_f and E_i of the energy, leaving us with the matrix element

$$\langle \varphi_f | V_I(t) | \varphi_i \rangle = \exp\left[\frac{i}{\hbar}(t - t_0)E_f\right] \langle \varphi_f | V | \varphi_i \rangle \exp\left[-\frac{i}{\hbar}(t - t_0) E_i\right]$$
$$= \exp\left[\frac{i}{\hbar}(t - t_0)(E_f - E_i)\right] \langle \varphi_f | V(t) | \varphi_i \rangle . \qquad (14.171)$$

We can then introduce Bohr's frequency

$$(E_f - E_i) = h(\nu_f - \nu_i) = h \, \nu_{fi} = \hbar \, \omega_{fi} \qquad (14.172)$$

to obtain the matrix element of the interaction in the intermediate representation:

$$\langle \varphi_f | V_I(t) | \varphi_i \rangle = \exp\left[i \, \omega_{fi}(t - t_0)\right] \langle \varphi_f | V(t) | \varphi_i \rangle . \qquad (14.173)$$

The matrix element of the first-order evolution operator, for example, is obtained by the series expansion of the exponential (14.168), which leaves us with

$$\langle \varphi_f | U_I^{(1)}(t, t_0) | \varphi_i \rangle = -\frac{i}{\hbar} \int_{t_0}^{t} \langle \varphi_f | V_I(t') | \varphi_i \rangle dt'$$
$$= -\frac{i}{\hbar} \int_{t_0}^{t} \exp\left(i \, \omega_{fi}(t' - t_0)\right) \langle \varphi_f | V(t') | \varphi_i \rangle \, dt' ,$$
$$(14.174)$$

giving with (14.167) the first-order transition probability:

$$P_r \, (i \rightarrow f) = \frac{1}{\hbar^2} \, | \int_{t_0}^{t} \langle \varphi_f | V(t') | \varphi_i \rangle \, e^{i \, \omega_{fi} \, (t' - t_0)} \, dt' |^2 . \qquad (14.175)$$

3.3 Example: Sinusoidal Perturbation

Let us introduce a sinusoidal perturbation $V(t')$:

$$V(t') = V^0 \cos \omega\, t' \tag{14.176}$$

This will determine the matrix elements between the intial and final states:

$$\langle \varphi_f \mid V(t') \mid \varphi_i \rangle = \langle \varphi_f \mid V^0 \mid \varphi_i \rangle \cos \omega\, t'$$

$$= V_{fi}^0 \frac{1}{2} \left(e^{i\omega t'} + e^{-i\omega t'} \right). \tag{14.177}$$

We use the initial time $t_0 = 0$ as the origin, thus simplifying the integration over the variable t' in statement (14.175), and making it easy to obtain the first-order transition

$$P_r\,(i \rightarrow f) = \frac{1}{4\hbar^2}\, |V_{fi}^0|^2\, \left| \frac{e^{i(\omega_{fi}+\omega)t} - 1}{\omega_{fi} + \omega} + \frac{e^{i(\omega_{fi}-\omega)t} - 1}{\omega_{fi} - \omega} \right|^2. \tag{14.178}$$

If $\omega = 0$, then the perturbation applied at instant $t_0 = 0$ is a constant equal to V^0 and leads to the transition probability (Fig. 1):

$$P_r(i \rightarrow f) = \frac{1}{\hbar^2}\, |V_{fi}^0|^2\, \left[\frac{\sin \omega_{fi}\, t/2}{\omega_{fi}/2} \right]^2$$

$$= \frac{4\, |V_{fi}^0|^2}{|E_f - E_i|^2}\, \sin^2 \left[(E_f - E_i)\, \frac{t}{2\hbar} \right]. \tag{14.179}$$

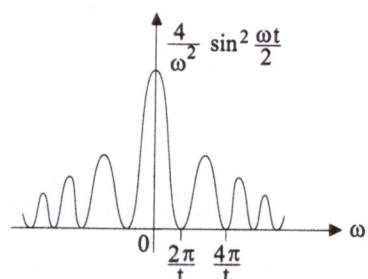

Figure 1

The half-maximum height of the curve centered around the value $\omega = 0$ for a fixed interval t is t^2, and the full width at half-maximum height proportional to $\frac{1}{t}$. When t becomes large, $P_r(i \rightarrow f)$ is only appreciably large for final states such as

$$t \sim \frac{2\pi}{\omega_{fi}} = \frac{2\pi\, \hbar}{E_f - E_i}. \tag{14.180}$$

If we term Δt the time duration of the action of the perturbation V^0, then the transition probability will only yield appreciable values for

$$\Delta t \, \Delta E \sim \hbar. \tag{14.181}$$

Here we arrive once more at Heisenberg's uncertainty relation although time t in non-relativistic quantum mechanics is not an observable. This is a point we have already alluded to in Chap. 2 dedicated to wave mechanics.

For $\omega \neq 0$, we can neglect the first term of (14.178) for $\omega \ll \omega_{fi}$ and $t \gg 1/\omega_{fi} \simeq \frac{1}{\omega}$ that is if the sinusoidal perturbation acts for a given length of time t significantly greater than $1/\omega$. We obtain a response similar to the one obtained for a constant perturbing potential:

$$F\left(t, \omega_{fi} - \omega\right) = \left[\frac{\sin \left(\omega_{fi} - \omega\right)t/2}{\left(\omega_{fi} - \omega\right)/2}\right]^2, \tag{14.182}$$

$$P_r(i \rightarrow f) = \frac{1}{4\hbar^2} \mid V_{fi}^0 \mid^2 F(t, \omega_{fi} - \omega). \tag{14.183}$$

We also note from the relation $\delta(x) = \lim_{\varepsilon \to 0} \frac{\varepsilon}{\pi} \left(\frac{\sin x/\varepsilon}{x}\right)^2$ that the response function yields, at the limit $t \to \infty$, a Dirac function centered on the energy $E_{fi} = \hbar \omega_{fi} = E_f - E_i$:

$$F\left(t, \omega_{fi} - \omega\right) \equiv F\left(t, \frac{E_{fi} - E}{\hbar}\right) \xrightarrow[t \to \infty]{} 2\pi \hbar t \, \delta(E_{fi} - E). \tag{14.184}$$

This leads to Fermi's golden rule:

$$P_r(i \rightarrow f) \xrightarrow[t \to \infty]{} \frac{\pi^2}{h} (t - t_0) \, \delta(E_{fi} - E) \mid V_{fi}^0 \mid^2. \tag{14.185}$$

The rate of transition, or transition probability per unit time,

$$\frac{d \, P_r(i \rightarrow f)}{dt} = \Gamma(i \rightarrow f) = \frac{\pi^2}{h} \, \delta\left(E - E_{fi}\right) \mid V_{fi}^0 \mid^2 \tag{14.186}$$

is therefore constant, centered on the value $E_f = E_i$ of the unperturbed Hamiltonian.

3.4 Feynman Diagrams

Let us use the form (14.168) of the evolution operator,

$$U_I(t, t_0) = \tau \, \exp\left(-\frac{i}{\hbar} \int_{t_0}^{t} V_I(t') \, dt'\right), \tag{14.187}$$

and, following a series expansion with the time-ordering operator,

$$U_I\,(t,\,t_0) = 1 - \frac{i}{\hbar} \int_{t_0}^{t} V_I\,(t_1)\,dt_1$$

$$+ \left(-\frac{i}{\hbar}\right)^2 \int_{t_0}^{t} V_I(t_1) \int_{t_0}^{t_1} V_I(t_2)\,dt_1\,dt_2 + \ldots$$

$$+ \left(\frac{-i}{\hbar}\right)^n \int_{t_0}^{t} dt_1 \int_{t_0}^{t_1} dt_2 \ldots \int_{t_0}^{t_{n-1}} dt_n\, V_I(t_1)\,V_I(t_2)\ldots V_I(t_n)$$

(14.188)

with $t_1 < t_2 < t_3 \ldots < t_n$. This is the Dyson series.
Since $|\,\alpha\,t_0\rangle = \exp\left[-(i/\hbar)\,E_\alpha\,t_0\right]\,|\,\alpha\rangle$ and $|\,\beta\,t\rangle = \exp[-(i/\hbar)\,E_\beta t]\,|\,\beta\rangle$ the
matrix element of the $U_I(t,t_0)$ operator takes the form

$$\langle\beta\,t\mid U_I(t,t_0)\mid\alpha\,t_0\rangle = \exp[\frac{i}{\hbar}\,(E_\beta t - E_\alpha t_0)]\,\langle\beta\mid U_I(t,t_0)\mid\alpha\,\rangle.\quad(14.189)$$

This leads to the matrix element of the evolution operator:

$$\langle\beta\,t\mid U_I(t,\,t_0)\mid\alpha t_0\rangle = \langle\beta\mid\alpha\rangle\,\exp(\frac{i}{\hbar}\,E_\alpha(t-t_0)$$

$$+\frac{1}{i\,\hbar}\int \exp\left[\frac{i}{\hbar}\,E_\beta\,(t-t_1)\right]\,\langle\beta|V|\alpha\rangle\exp\left[\frac{i}{\hbar}E_\alpha(t_1-t_0)\right]dt_1 + \ldots$$

$$+\left(\frac{1}{i\hbar}\right)^n \sum_{\gamma_i}\int_{t_0}^{t} dt_1 \int_{t_0}^{t_1} dt_2 \ldots \int_{t_0}^{t_{n-1}} dt_n\,\exp\left[\frac{i}{\hbar}E_\beta(t-t_1)\right]$$

$$\times\,\langle\beta|V|\gamma_1\rangle\exp\left[+\frac{i}{\hbar}E_{\gamma_1}(t_1-t_2)\right]\,\langle\gamma_1|V|\gamma_2\rangle\ldots\exp\left[\frac{i}{\hbar}E_\alpha(t_n-t_0)\right].$$

(14.190)

We further visualize this result with a Feynman diagram by choosing a vertical time axis and by representing each interaction at time t_k with a vertex,
that is, with a junction point of two lines (see Fig. 2).

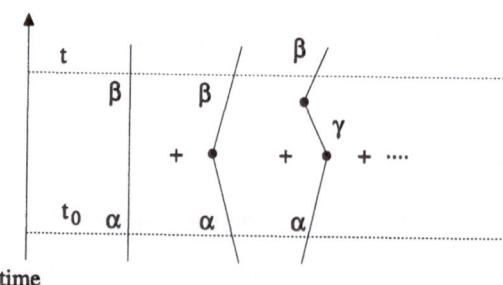

Figure 2

Notation Rules for Feynman Diagrams

i) each line λ beginning at time t_1 and ending at time t_2 contributes $K_\lambda(t_2 - t_1)$ with:

$$K_\lambda(t_2 - t_1) = \begin{cases} \exp\left[\frac{i}{\hbar} E_\lambda (t_2 - t_1)\right] & \text{if } t_2 - t_1 > 0 \\ 0 & \text{if } t_2 - t_1 < 0. \end{cases} \qquad (14.191)$$

It is thus possible to change the integration limits to $-\infty$ and $+\infty$, thus facilitating the calculation.

ii) Each vertex linking lines β and α at time t_k contributes

$$\frac{1}{i\hbar} \langle \beta \mid V(t_k) \mid \alpha \rangle . \qquad (14.192)$$

iii) We then integrate over all the intermediate times from $-\infty$ to $+\infty$ and sum up over all the intermediate states.

The evolution operator is defined from $U_I(t, t_0)$ through

$$\mathcal{K}(t, t_0) = \begin{cases} U_I(t, t_0) & \text{if } t - t_0 > 0 \\ 0 & \text{if } t - t_0 < 0. \end{cases} \qquad (14.193)$$

The series (14.190) will then take the following form:

$$\langle \beta t \mid \mathcal{K}(t, t_0) \mid \alpha t_0 \rangle = \langle \beta \mid \alpha \rangle K_\alpha(t - t_0)$$
$$+ \frac{1}{i\hbar} \int dt_1 \, K_\beta(t - t_1) \langle \beta \mid V(t_1) \mid \alpha \rangle K_\alpha(t_1 - t_0) \cdots$$
$$+ \left(\frac{1}{i\hbar}\right)^n \int dt_1 \cdots \int dt_n \, K_\beta(t - t_1) \langle \beta \mid V(t_1) \mid \gamma \rangle \cdots \langle \mid V(t_n) \mid \alpha \rangle K_\alpha(t_n - t_0) .$$
$$(14.194)$$

Now let us introduce the Fourier transform of the functions $\mathcal{K}$ and K_α:

$$\mathcal{K}(\omega) = \int e^{-\frac{i}{\hbar}\omega t} \mathcal{K}(t) \, dt \quad \text{and} \quad \mathcal{K}(t) = \frac{1}{2\pi\hbar} \int \mathcal{K}(\omega) \, e^{+\frac{i}{\hbar}\omega t} \, d\omega , \qquad (14.195)$$

$$K_\alpha(\omega) = \int e^{-\frac{i}{\hbar}\omega t} K_\alpha(t) \, dt \quad \text{and} \quad K_\alpha(t) = \frac{1}{2\pi\hbar} \int K_\alpha(\omega) \, e^{+\frac{i}{\hbar}\omega t} \, d\omega . \qquad (14.196)$$

By using the step function (14.191) defining $K_\alpha(t)$, we obtain

$$K_\alpha(\omega) = \int e^{-\frac{i}{\hbar}\omega t} e^{+\frac{i}{\hbar} E_\alpha t} \, dt = \frac{i\hbar}{\omega - E_\alpha + i\eta} \qquad \eta \to 0. \qquad (14.197)$$

The series (14.194) then reduces to

$$-\frac{i}{\hbar}\langle\beta t|\mathcal{K}(\omega)|\alpha t_0\rangle = \frac{\langle\beta|\alpha\rangle}{\omega - E_\alpha + i\eta} + \frac{1}{\omega - E_\beta + i\eta}\langle\beta|V|\alpha\rangle\frac{1}{\omega - E_\alpha + i\eta}$$

$$+\sum_\gamma \frac{1}{\omega - E_\beta + i\eta}\langle\beta|V|\gamma\rangle\frac{1}{\omega - E_\gamma + i\eta}\langle\gamma|V|\alpha\rangle\frac{1}{\omega - E_\alpha + i\eta} + \cdots.$$

$$(14.198)$$

This result can be written directly in terms of operators:

$$-\frac{i}{\hbar}\mathcal{K}(\omega) = \frac{\mathbb{1}}{\omega - H_0 + i\eta} + \frac{1}{\omega - H_0 + i\eta} V \frac{1}{\omega - H_0 + i\eta} + \cdots = \frac{1}{\omega - H + i\eta}.$$

$$(14.199)$$

To demonstrate the foregoing result, we first establish the identity:

$$\frac{1}{A+B} = \frac{1}{A} - \frac{1}{A} B \frac{1}{A} + \frac{1}{A} B \frac{1}{A} B \frac{1}{A} + \cdots \qquad (14.200)$$

Working from the identity between operators:

$$X(A+B) = XA + XB = \mathbb{1} \qquad (14.201)$$

we obtain, by isolating the term XA,

$$XA = \mathbb{1} - XB. \qquad (14.202)$$

By multiplying each of the right-hand members by $A^{-1} = \frac{1}{A}$, we obtain:

$$X = \frac{\mathbb{1}}{A} - XB\frac{1}{A}, \qquad (14.203)$$

and get back to (14.200) after iteration. We next set $A = \omega - H_0$ and $B = -V$, which gives

$$-\frac{i}{\hbar}\mathcal{K}(\omega) = \frac{1}{\omega - H_0 - V + i\eta} = \frac{1}{\omega - H + i\eta}. \qquad (14.204)$$

Each line in the Feynman diagram therefore represents the energy propagation of the quantum state $1/(\omega - H_0 + i\eta)$ and each vertex the interaction operator V in the total Hamiltonian $H = H_0 + V$. This boils down to saying that the Feynman expression for the evolution operator U_I is none other than a perturbation expansion of this operator.

Let us now write the eigenvalue equation of the total Hamiltonian

$$H|\Psi\rangle = (H_0 + V)|\Psi\rangle = E|\Psi\rangle, \qquad (14.205)$$

in a form that isolates the perturbation potential V:

$$(E - H_0)|\Psi\rangle = V|\psi\rangle. \qquad (14.206)$$

This formally leads to a solution of the form

$$| \Psi \rangle \ = \ \frac{1}{E - H_0} \, V \, | \Psi \rangle + \text{constant} \, . \qquad (14.207)$$

When $V = 0$, $H = H_0$ and $| \Psi \rangle$ becomes the unperturbed solution $| \varphi \rangle$, an eigenfunction of H_0. This gives the following form to solution (14.207):

$$| \Psi \rangle \ = \ | \varphi \rangle + \frac{1}{E - H_0} \, V \, | \Psi \rangle \, . \qquad (14.208)$$

By iterating this solution, we obtain the expansion for the perturbed wave function:

$$| \Psi \rangle \ = \ | \varphi \rangle + \frac{1}{E - H_0} \, V \, | \varphi \rangle + \frac{1}{E - H_0} \, V \, \frac{1}{E - H_0} \, V \, | \varphi \rangle + \dots . \quad (14.209)$$

Because the matrix element of the operator U_I only translates the crossover of the state $| \varphi_i \rangle$ with $| \Psi \rangle$, we arrive back at the matrix element (14.147) of the transition probability:

$$\langle \varphi_i \, | \, \Psi \rangle \ = \ \langle \varphi_i \, | \, \varphi_f \rangle + \langle \varphi_i \, | \, \frac{1}{E - E_i} \, V \, | \varphi_f \rangle$$

$$+ \, \langle \varphi_i \, | \, \frac{1}{E - E_i} \, V \, \frac{1}{E - H_0} \, V \, | \varphi_f \rangle + \dots . \qquad (14.210)$$

This is the result we obtained with Feynman diagrams and to which we shall return in the discussion on the scattering of particles in the next chapter and in the study of Fermion systems Chap. 18.

Chapter 15
Scattering of Particles

1. Scattering Operator

A quantum system is prepared at an instant t_0 to enable the measurement of certain observables: energy, angular momentum, and q linear momentum. For a very brief period, the system enters into interaction with another quantum system (we are not interested in how this interaction is produced). At a later instant t_1 we again measure the system's observables. It goes without saying that for most of the duration of the experiment (i.e., between instants t_1 and t_0) the different constituents of the physical system under observation are almost free (Fig. 1).

This is precisely what happens in classical mechanics when we observe the asymptotic values of particles in interaction.

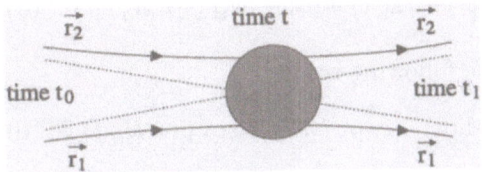

Figure 1

The free particles are characterized by their vector positions

$$\vec{r}(t) = \vec{r}_0 + \vec{v}\, t.$$ (15.1)

The position vector of a particle k with momentum $\vec{k}$ will thus be:

$$\begin{aligned}
\vec{r}_k(t) &= \vec{r}_k^{\,in} + \vec{v}_k^{\,in}\, t & t \simeq t_0 \\
&= \vec{r}_k^{\,out} + \vec{v}_k^{\,out}\, t & t \simeq t_1.
\end{aligned}$$ (15.2)

Mathematically expressed, we say that the limit of the vector length between a current point of the trajectory and a point on the asymptote tends to zero:

$$\lim_{t \to \pm\infty}\; |\, \vec{r}_k - (\vec{r}_k^{\,ex} + \vec{v}_k^{\,ex}\, t)\, | = 0,$$ (15.3)

with the asymptotic conditions:

$$t \rightarrow +\infty \quad ex = out \quad \text{outgoing particle},$$
$$t \rightarrow -\infty \quad ex = in \quad \text{incoming particle}. \tag{15.4}$$

In quantum mechanics, we will similarly write

$$\lim_{t \rightarrow \pm \infty} | \Psi(t) - \Psi^{ex}(t) | = 0, \tag{15.5}$$

where $\Psi^{ex}(t)$ is a free incoming $(t \rightarrow -\infty)$ or outgoing $(t \rightarrow +\infty)$ asymptotic state.

If we introduce the Hamiltonian of the system,

$$H = \left(-\frac{\hbar^2}{2m_1} \Delta_1 - \frac{\hbar^2}{2m_2} \Delta_2 \right) + V(r) = H_0 + V(r), \tag{15.6}$$

the free states at instant t are obtained from the initial free states $\Psi_f(0)$ through

$$\Psi_f(t) = \exp\left(-i H_0 t\right) \Psi_f(0), \tag{15.7}$$

whereas the actual state $\Psi(t)$ will require the total Hamiltonian to be introduced at instant t:

$$\Psi(t) = \exp\left(-i H t\right) \Psi(0). \tag{15.8}$$

Condition (15.4) will therefore take the following form:

$$\lim_{t \rightarrow \pm \infty} | \exp\left(-i H t\right) \Psi(0) - \exp\left(-i H_0 t\right) \Psi^{ex}(0) | = 0. \tag{15.9}$$

We then prove the following result:

$$\Psi(0) = \lim_{t \rightarrow \mp \infty} \left(\exp\left(i H t\right) \exp\left(-i H_0 t\right) \Psi^{ex}(0) \right). \tag{15.10}$$

A state $\Psi(t)$ with an incoming asymptotic state and an outgoing asymptotic state is termed a a scattering state.

Using Eqs. (15.9) and (15.10), the uniformly bounded Moller operator (Prugovescki, 1971)

$$\boxed{\Omega = \exp\left(i H t\right) \exp\left(-i H_0 t\right)} \tag{15.11}$$

enables us to write

$$\lim_{t \rightarrow \mp \infty} | \Psi(0) - \Omega(t) \Psi^{ex}(0) | = 0. \tag{15.12}$$

There exists an incoming Moller operator Ω_- corresponding to the incoming scattering waves and responsible for the transition from $\Psi^{in}_-(0)$ to $\Psi_-(0)$:

$$\Psi_-(0) = \lim_{t \rightarrow -\infty} \exp\left(i H t\right) \exp\left(-i H_0 t\right) \Psi^{in}_-(0) = \Omega_- \; \Psi^{in}_-(0) \tag{15.13}$$

as well as an outgoing Moller operator Ω_+ corresponding to the outgoing scattering waves:

$$\Psi_+(0) = \lim_{t \to +\infty} \exp\left(i\,H\,t\right) \exp\left(-i\,H_0\,t\right) \Psi_+^{out}(0) = \Omega_+\,\Psi_+^{out}(0). \quad (15.14)$$

A scattering operator is the operator product:

$$\boxed{S = \Omega_-^*\,\Omega_+} \quad (15.15)$$

and we can determine the scattering probability of a state Ψ_- to another state Ψ_+ using this operator:

$$P_r\,(\Psi_- \to \Psi_+) = |\ <\Psi_i\ |\ S\ |\ \Psi_f\rangle\ |^2\,. \quad (15.16)$$

The scattering probability of a state Ψ_- to another state Ψ_+ is equal to the square modulus of the matrix element of the scattering operator considered between the initial $\Psi_-^{in}(0) = \Psi_i$ and the final $\Psi_+^{out}(0) = \Psi_f$ scattering states.

2. Transition Operator

2.1 Definition

Let us introduce the transition operator T with the scattering operator S using the relation:

$$S = \mathbb{1} - 2\pi\,i\,T\,. \quad (15.17)$$

This operator is zero if there is no interaction since in this case:

$$H = H_0 \quad \text{and} \quad \Omega = \mathbb{1} \quad \text{hence} \quad S = \mathbb{1} \quad \text{implies} \quad T = 0\,.$$

The matrix element $\langle \Psi_f\ |\ S\ |\ \Psi_i \rangle$ will be written with the transition operator:

$$\begin{aligned}
\langle \Psi_f\ |\ S\ |\ \Psi_i \rangle &= \langle \Psi_f\ |\ \mathbb{1} - 2\pi\,i\,T\ |\ \Psi_i \rangle \\
&= \langle \Psi_f\ |\ \Psi_i \rangle - 2\pi\,i\langle \Psi_f\ |\ T\ |\ \Psi_i \rangle\,. \quad (15.18)
\end{aligned}$$

2.2 Optical Theorem

Now let us evaluate S^*S and SS^* from the definition of the transition operator:

$$\begin{aligned}
S^*S &= (\mathbb{1} + 2\pi\,i\,T^+)\,(\mathbb{1} - 2\pi\,i\,T) = \mathbb{1} - 2\pi\,i(T - T^+) + 4\pi^2\,T^+T\,, \\
SS^* &= (\mathbb{1} - 2\pi\,i\,T)\,(\mathbb{1} + 2\pi\,i\,T^+) = \mathbb{1} - 2\pi\,i(T - T^+) + 4\pi^2\,T\,T^+\,.
\end{aligned}$$
$$(15.19)$$

The operator S is therefore unitary, that is, it satisfies the condition $S^*S = SS^* = \mathbb{1}$ if and only if

$$T^+ - T = 2\pi\, i\, TT^+ = 2\pi\, i\, T^+T = 2\pi\, i \mid T \mid^2. \qquad (15.20)$$

This can also be written with the imaginary part of the transition operator:

$$\boxed{\text{Im}\,\{T\} = -\pi \mid T \mid^2} \qquad (15.21)$$

a result termed the optical theorem, by analogy with the relation obtained for the elastic scattering of electromagnetic waves. Some authors define the operator T from $S = \mathbb{1} + 2\pi i\, T$, leading to the sign $(+)$ in relation (15.21).

3. Differential Cross Section

3.1 Definition

A beam of particles with a well-defined direction and energy impinges on a target (Fig. 2)

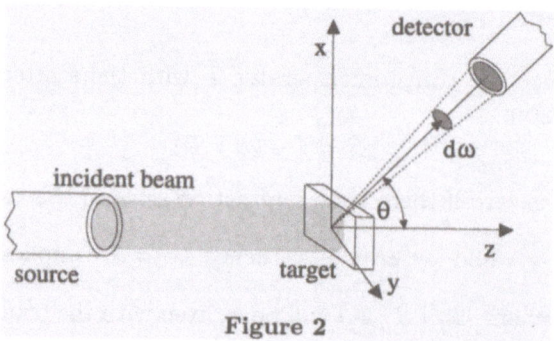

Figure 2

Let Φ_i be the incident flux, that is, the number of incident particles with velocity v_i crossing a unit-area perpendicular to the incident beam per unit time. If N_i is the number of incident particles,

$$\Phi_i = N_i\, v_i. \qquad (15.22)$$

The number of detected particles N_d in the direction $w = (\Theta,\, \varphi)$ will be proportional to the incident flux and to the aperture dw of the detector:

$$N_d \propto \Phi_i\, dw. \qquad (15.23)$$

By differential cross section we mean the proportionality coefficient for just one particle of the target. If N_T particles in the target are in interaction, we will therefore write

$$N_d = \sigma(\omega) \, N_T \, \Phi_i \, d\omega \,. \qquad (15.24)$$

The differential cross section $\sigma(\omega)$ has the dimensions of a surface and is expressed in barns (1 barn $= 10^{-24}$ cm^2). The total cross section is obtained by integration over the solid angle:

$$\sigma = \int \sigma(\omega) \, d\omega = \int \frac{d\sigma}{d\omega} \, d\omega = \int d\sigma \,. \qquad (15.25)$$

3.2 Scattering Amplitude

When two particles with mass m_1 and m_2 are in interaction, it is possible to isolate the motion of the center of mass R from the relative motion (see (3.8)) by setting $M = m_1 + m_2$ and $1/\mu = 1/m_1 + 1/m_2$, and if we introduce the total wave function of the system $\Psi\,(\vec{r}_1, \, \vec{r}_2) = f(\vec{R}) \, \Psi(\vec{r})$, we obtain the system of differential equations:

$$-\frac{\hbar^2}{2M} \, \nabla_R^2 \, f(\vec{R}) = E_{CM} \, f(\vec{R}) \,,$$
$$\left(-\frac{\hbar^2}{2\,\mu} \, \nabla_r^2 + V(\vec{r}) \right) \, \Psi(\vec{r}) = E \, \Psi(\vec{r}) \,. \qquad (15.26)$$

The total energy is the sum of the kinetic energy of the center of mass and the relative energy:

$$E_{tot} = E_{CM} + E \,. \qquad (15.27)$$

The incident particle with mass m_i is a free particle, described by a plane wave

$$e^{i \, \vec{k}_i \cdot \vec{r}} = A \, \langle \vec{r} \mid \vec{k}_i \rangle \,. \qquad (15.28)$$

During the interaction, the wave $\Psi(\vec{r})$ will be described by the Schrödinger equation of relative motion,

$$-\frac{\hbar^2}{2\mu} \, \nabla_r^2 \, \Psi(\vec{r}) + V(r) \, \Psi(\vec{r}) = E \, \Psi(\vec{r}) \,. \qquad (15.29)$$

If we eliminate the reduced mass in E and V by setting:

$$E = \frac{p^2}{2\mu} = \frac{\hbar^2 \, k^2}{2\mu} \quad \text{and} \quad V(r) = \frac{\hbar^2 \, U(r)}{2\mu} \,, \qquad (15.30)$$

the Schrödinger Eq. (15.29) becomes

$$[\nabla^2 + k^2 - U(r)] \, \Psi(\vec{r}) = 0 \,. \qquad (15.31)$$

Now let us seek a solution $\Psi^+_{k_i}(\vec{r})$ to the equation with an incident energy E_i, a solution corresponding to an outgoing wave.

At infinity, this $\Psi^+_{k_i}(\vec{r})$ will behave like a plane wave $e^{i\,\vec{k}_i \cdot \vec{r}}$ added to a scattered spherical wave $e^{i\,k_i\,r}/r$ with amplitude $f(\omega)$.

The asymptotic form of the outgoing wave $\Psi^+_{k_i}(\vec{r})$ is written as follows:

$$\Psi^+_{k_i}(r) \xrightarrow[r \to \infty]{} A\left(e^{i\,\vec{k}_i \cdot \vec{r}} + f(\omega)\,\frac{e^{i\,k_i\,r}}{r} \right). \qquad (15.32)$$

By scattering amplitude $f(\omega)$, we mean the amplitude of the outgoing wave function. If the interaction potential $V(r)$ is spherically symmetrical, then the scattering amplitude will exhibit the same symmetry and will depend solely on the scattering angle Θ.

3.3 Relation Between $\sigma(\omega)$ and $f(\omega)$

The radial flux corresponding to a wave function $\Psi(\vec{r})$ is defined from the probabilty current (see (2.30)) by projection onto the unit vector in the direction r, that is $\Phi = \vec{J}.\vec{e}_r$, with the probability current

$$\vec{J} = \frac{\hbar}{2\mu\,i}\,(\Psi^*\,\vec{\nabla}\,\Psi - \Psi\,\vec{\nabla}\,\Psi^*) = \mathrm{Re}\left\{ \frac{\hbar}{\mu\,i}\,\Psi^*\,\vec{\nabla}\,\Psi \right\}. \qquad (15.33)$$

An incoming plane wave $A\,\exp\,(-i\,\vec{k}_i \cdot \vec{r}) = A\,\exp\,(-i\,k_i\,r\,\cos\,\Theta)$ leads to an incoming radial flux $\Phi_{in} = \vec{J}_{i\,n} \cdot \vec{e}_r$, which is easily evaluated as

$$\Phi_{in} = \mathrm{Re}\left[\frac{\hbar}{\mu\,i}\,A^*A\,\exp\,(+i\,k_i\,r\,\cos\,\Theta)\,\partial_r\,\exp\,(-i\,k_i\,r\,\cos\,\Theta) \right]_{\Theta=0}$$

$$= \frac{\hbar\,k_i}{\mu}\,|\,A\,|^2. \qquad (15.34)$$

The flux of the outgoing spherical wave $\frac{1}{r}\,\exp\,(i\,k_i\,r)$ will further yield

$$\Phi_{out} = \mathrm{Re}\left\{ \frac{\hbar}{\mu\,i}\,A^*A\,f^*\,\frac{e^{i\,k_i\,r}}{r}\,\partial_r\,f\,\frac{e^{i\,k_i\,r}}{r} \right\} \qquad (15.35)$$

$$= \mathrm{Re}\left\{ \frac{\hbar}{\mu\,i}\,|\,A\,|^2\,|\,f\,|^2\,\left(\frac{i\,k_i}{r^2} - \frac{1}{r^3} \right) \right\}. \qquad (15.36)$$

The terms in $\frac{1}{r^3}$ are not taken into account because, far from the target, r is big and $1/r^3$ negligible compared to $1/r^2$. The interference term between the plane wave and the spherical wave is zero for $\Theta = 0$ but oscillates rapidly with respect to r for $\Theta \neq 0$, thus leading to a zero mean value which permits us to neglect it.

The flux of outgoing particles through a surface element $r^2 \, d\omega$ away from the interaction zone (r large) is therefore

$$\Phi_{out} = \mid A \mid^2 \frac{\hbar \, k_i}{\mu} \mid f(\omega) \mid^2 , \qquad (15.37)$$

and, by dividing by the flux determined earlier on in (15.34),

$$\Phi_{in} = \mid A \mid^2 \frac{\hbar \, k_i}{\mu} , \qquad (15.38)$$

we obtain the differential cross section with respect to the scattering coefficient:

$$\sigma(\omega) = \frac{d\,\sigma}{d\,\omega} = \mid f(\omega) \mid^2 . \qquad (15.39)$$

3.4 Elastic Scattering by a Coulomb Potential

The Coulomb potential is of the form $V\,(r) = Z_1 \, Z_2 \, e^2/r$, leading to the Schrödinger equation

$$\left[-\frac{\hbar^2}{2\mu} \nabla^2 + \frac{Z_1 \, Z_2 \, e^2}{r} \right] \psi_c\,(\vec{r}) = E \, \psi_c\,(\vec{r}) . \qquad (15.40)$$

We introduce the velocity of the particles through their relative kinetic energy

$$E = \frac{\hbar^2 \, k^2}{2\mu} = \frac{1}{2} \, \mu v^2 \qquad (15.41)$$

and a parameter that is inversely proportional to the velocity,

$$\gamma = \frac{Z_1 \, Z_2 \, e^2}{\hbar \, v} . \qquad (15.42)$$

This will enable us to write the Schrödinger equation in the following simple form:

$$\left(\nabla^2 + k^2 - \frac{2\gamma \, k}{r} \right) \psi_c\,(\vec{r}) = 0 . \qquad (15.43)$$

By setting $z = r \cos \Theta$, let us seek scattered solutions along the direction Θ of the form

$$\psi_c\,(\vec{r}) = e^{i\,k\,z} \, f\,(r - z) = e^{i\,k\,z} \, f\,(u) \qquad \text{with} \qquad r - z = u . \qquad (15.44)$$

The differential equation satisfied by $f\,(u)$ then reduces to

$$\left[u \frac{d^2}{d\,u^2} + (1 - i\,k\,u) \frac{d}{d\,u} - \gamma \, k \right] f\,(u) = 0 . \qquad (15.45)$$

By finally setting

$$w = i\,k\,u = i\,k\,(r - z),$$ (15.46)

we obtain a second-order differential Laplace equation

$$\left(w\,\frac{d^2}{d\,w^2} + (1 - w)\,\frac{d}{d\,w} + i\,\gamma\right) f\,(w) = 0.$$ (15.47)

The regular solution at the origin is the confluent hypergeometric series $F\,(-i\,\gamma,\ 1\ ;w)$:

$$\psi_c\,(k\,r) = A\,e^{i\,k\,z}\,F\,(-i\,\gamma,\ 1\ ;\ i\,k\,(r - z)).$$ (15.48)

The hypergeometric series behaves at infinity like the sum of a plane wave and a spherical wave:

$$\psi_c\,(\vec{r}) = \psi_{in}\,(\vec{r}) + \psi_{out}\,(\vec{r}),$$ (15.49)

whose asymptotic behavior for $|\,r - z\,| \rightarrow \infty$ is given by

$$\psi_{in}\,(\vec{r}) = \exp i\,((kz + \gamma\,\log\,k\,(r - z))\left[1 + \frac{\gamma^2}{i\,k\,(r - z)} + \dots\right]$$ (15.50)

$$\psi_{out}\,(\vec{r}) = -\frac{\gamma\,\Gamma\,(1 + i\,\gamma)}{k\,(r - z)\,\Gamma\,(1 - i\,\gamma)}\,\exp i\,((kr - \gamma\,\log\,k\,(r - z))$$

$$\times \left[1 + \frac{(1 + i\,\gamma)^2}{i\,k\,(r - z)} + \dots\right].$$ (15.51)

The function Γ is the factorial function $\Gamma(x)$.

Since $z = r\cos\Theta$, the first term of the expansion of the outgoing wave is written

$$\psi_{out} = \frac{1}{r}\,\exp\,[i\,(k\,r - \gamma\,\log\,2k\,r]\,f_c\,(\Theta)$$ (15.52)

with the resulting Coulomb scattering amplitude $f_c\,(\Theta)$:

$$f_c\,(\Theta) = -\frac{\gamma}{2k\,\sin^2\frac{\theta}{2}}\,\exp\left[-i\,\gamma\,\log\,\sin^2\frac{\Theta}{2} + 2i\,\sigma_0\right],$$

$$\exp\,[2i\,\sigma_0] = \frac{\Gamma\,(1 + i\,\gamma)}{\Gamma\,(1 - i\,\gamma)}.$$ (15.53)

We can easily infer the Coulomb scattering or Rutherford scattering differential cross section:

$$\sigma_c\,(\Theta) = |\,f_c\,(\Theta)\,|^2$$

$$= \frac{\gamma^2}{4k^2\,\sin^4\frac{\Theta}{2}} = \left(\frac{Z_1\,Z_2\,e^2}{4E}\right)^2\,\frac{1}{\sin^4\frac{\Theta}{2}}.$$ (15.54)

This is precisely the result obtained in classical mechanics by using Kepler's laws to evaluate the scattering of an electrically charged particle with charge $Z_1\,e$ on a nucleus with charge $Z_2\,e$.

We notice in the foregoing result that:

(i) $\sigma_c\,(\Theta)$ depends on the absolute value of the potential and not on the sign,
(ii) at a given angle Θ, $\sigma_c\,(\Theta)$ decreases with $\frac{1}{E^2}$,
(iii) the total cross section is infinite since

$$\int \sigma_c\,(\Theta)\,d\,\omega = \left(\frac{Z_1\,Z_2\,e^2}{4E}\right)^2 2\pi \int \frac{\sin\Theta\,d\Theta}{\sin^4\frac{\Theta}{2}} \tag{15.55}$$

diverges at small angles. This is as a result of the presence of a screening effect at small angles and of the fact that the effect of the Coulomb electronic cloud becomes zero at distances that are sufficiently large compared to the radius of the atom. Rutherford scattering of charged particles has provided evidence in favor of the atomic nucleus being made up of electrical charges borne by particles (protons) and not being a charged fluid compensating the electric charge of the electrons. This type of scattering is today a general tool used by atomic, nuclear, and particle physicists.

4. Scattering Amplitude and Interaction Potential

If we normalize the plane wave (15.28) propagating in the positive r direction:

$$\langle \vec{r}\,|\,\vec{k}\rangle = \frac{1}{(2\pi)^{3/2}}\,e^{i\,\vec{k}\cdot\vec{r}}, \tag{15.56}$$

the scattered wave will take the asymptotic form

$$\Psi_{k_i}^+(\vec{r}) \xrightarrow[r\to\infty]{} \frac{1}{(2\pi)^{3/2}}\left[e^{i\,\vec{k}_i\cdot\vec{r}} + f(\omega)\,\frac{e^{i\,k_i\,r}}{r}\right], \tag{15.57}$$

and the formal solution of Schrödinger Eq. (15.31) can be written:

$$\Psi_{k_i}^+(\vec{r}) = \frac{1}{(2\pi)^{3/2}}\left[e^{i\,\vec{k}_i\cdot\vec{r}} - \frac{\mu}{2\pi\,\hbar^2}\int d\vec{r}\,'\,\frac{e^{i\,k_i\,|\vec{r}-\vec{r}\,'|}}{|\vec{r}-\vec{r}\,'|}\right.$$

$$\left. \times\,V(r')\,\Psi_{k_i}^+(\vec{r}\,')\right]. \tag{15.58}$$

Beyond the range of the interaction, that is for $r\rangle\rangle r'$, we can take an approximate value of the denominator:

$$|\vec{r}-\vec{r}\,'| = \left[r^2 + r'^2 - 2\vec{r}.\vec{r}\,'\right]^{1/2} = r\left[1 - \frac{2\vec{r}.\vec{r}\,'}{r^2} + \frac{r'^2}{r^2}\right]^{1/2}$$

$$\simeq r\left(1 - \frac{\vec{r}.\vec{r}\,'}{r^2}\right). \tag{15.59}$$

The spherical wave of the integrand thus becomes

$$\frac{e^{i\,k_i\,|\,\vec{r}-\vec{r}\,'\,|}}{|\,\vec{r}-\vec{r}\,'\,|} = \frac{e^{i\,k_i\,r\left(1-\frac{\vec{r}\cdot\vec{r}\,'}{r^2}\right)}}{r\left(1-\frac{\vec{r}\cdot\vec{r}\,'}{r^2}\right)} \simeq \frac{e^{i\,k_i\,r}}{r}\,\exp\left(-i\,k_i\,\frac{\vec{r}\cdot\vec{r}\,'}{r}\right). \qquad (15.60)$$

The emerging momentum vector is defined by setting

$$\vec{k}_f = k_i\,\frac{\vec{r}}{r}. \qquad (15.61)$$

The emerging wave will thus take the form

$$\Psi_{k_i}^+(\vec{r}) \xrightarrow[r\to\infty]{} \frac{1}{(2\pi)^{3/2}}\left[e^{i\,\vec{k}_i\cdot\vec{r}} - \frac{\mu}{2\pi\,\hbar^2}\,\frac{e^{i\,k_i\,r}}{r}\int e^{-i\,\vec{k}_f\cdot\vec{r}\,'}\right.$$
$$\left.\times V(r')\,\Psi_{k_i}^+(\vec{r}\,')\,d\vec{r}\,'\right]. \qquad (15.62)$$

By comparing this with the expression (15.57) for $\Psi_{k_i}^+(\vec{r})$, we obtain an important relation defining the scattering amplitude in terms of the interaction potential:

$$\boxed{f(\omega) = -\frac{\mu}{2\pi\,\hbar^2}\int e^{-i\,\vec{k}_f\cdot\vec{r}\,'}\,V(r')\,\Psi_{k_i}^+(\vec{r}\,')\,d\vec{r}\,'}\,. \qquad (15.63)$$

We can take advantage of the fact that the interaction potential is diagonal in its eigenrepresentation, that is, in configuration space,

$$\langle\vec{r}\,|\,V\,|\,\vec{r}\,'\rangle = V(\vec{r})\,\delta\,(\vec{r}-\vec{r}\,'), \qquad (15.64)$$

to write the scattering amplitude with the momentum $\vec{k}_f$ of the plane wave:

$$f(\omega) = -\frac{\mu}{2\pi\,\hbar^2}\,(2\pi)^{3/2}\int \langle\vec{k}_f\,|\,\vec{r}\,'\rangle\,d\vec{r}\,'\,\langle\vec{r}\,'\,|\,V\,|\,\vec{r}\,''\rangle\,d\vec{r}\,''$$
$$\times \langle\vec{r}\,''\,|\,\Psi_{k_i}^+\rangle\,(2\pi)^{3/2}, \qquad (15.65)$$

which yields, following the extraction of the completeness relations for $\vec{r}\,'$ and $\vec{r}\,''$:

$$\boxed{f(\omega) = -\frac{4\pi^2\,\mu}{\hbar^2}\,\langle\vec{k}_f\,|\,V\,|\,\Psi_{k_i}^+\rangle}\,. \qquad (15.66)$$

The transition operator is the operator obtained by identifying the above matrix element with the matrix element between plane waves:

$$\boxed{\langle \vec{k}_f \mid V \mid \Psi_{\vec{k}_i}^+ \rangle \; = \; \langle \vec{k}_f \mid T \mid \vec{k}_i \rangle} \; . \tag{15.67}$$

We can next determine the scattering operator S with relation (15.18):

$$\langle \vec{k}_f \mid S \mid \vec{k}_i \rangle \; = \delta \, (\vec{k}_f - \vec{k}_i) - 2\pi \, i \, \langle \vec{k}_f \mid T \mid \vec{k}_i \rangle ,$$
$$= \delta \, (\vec{k}_f - \vec{k}_i) - 2\pi \, i \, \langle \vec{k}_f \mid V \mid \Psi_{\vec{k}_i}^+ \rangle . \tag{15.68}$$

5. Lippmann–Schwinger Equations

5.1 Transition Amplitude and Interaction Potential

The resolvants of the operators H and H_0 with $H = H_0 + V$ are defined by the following relations:

$$G^\pm = \frac{1}{E - H \pm i\,\varepsilon} , \tag{15.69}$$

$$G_0^\pm = \frac{1}{E - H_0 \pm i\,\varepsilon} . \tag{15.70}$$

The easily verified identity between operators A and B,

$$\frac{1}{A} - \frac{1}{B} = \frac{1}{B} \, (B - A) \, \frac{1}{A} , \tag{15.71}$$

yields with the Hamiltonian operators, that is, by setting:

$$A = E - H_0 \quad \text{and} \quad B = E - H = E - H_0 - V , \tag{15.72}$$

the formal relation linking resolvants G_0 and G to the potential V:

$$\boxed{G = G_0 + G \, V \, G_0} \; . \tag{15.73}$$

Similarly, by setting $A = E - H$ and $B = E - H_0$, we obtain:

$$\boxed{G = G_0 + G_0 \, V \, G} \; . \tag{15.74}$$

Now, consider the eigenvalue equation of the total Hamiltonian operator written with $H = H_0 + V$ as follows:

$$(E - H_0) \mid \Psi \rangle \; = V \mid \Psi \rangle . \tag{15.75}$$

For a zero interaction potential $V = 0$, the eigenvalue equation of H_0 will introduce an eigenfunction $\mid \varphi \rangle$ such that:

$$(E - H_0) \mid \varphi \rangle \; = 0 . \tag{15.76}$$

The formal solution of (15.75) can be written by allowing the $\pm i\varepsilon$ to tend to zero and omitting it:

$$| \Psi \rangle = | \varphi \rangle + \frac{1}{E - H_0} V | \Psi \rangle, \tag{15.77}$$

and by introducing the resolvant G_0:

$$| \Psi \rangle = | \varphi \rangle + G_0 V | \Psi \rangle. \tag{15.78}$$

We thus obtain the scattering states with the formal relation:

$$\boxed{| \Psi \rangle = | \varphi \rangle + G_0 V | \Psi \rangle} . \tag{15.79}$$

It is interesting to note that the expansion of this relation by iteration yields exactly the result obtained from Feynman diagrams with energy propagators:

$$| \Psi \rangle = | \varphi \rangle + G_0 V | \varphi \rangle + G_0 V G_0 V | \varphi \rangle + \dots . \tag{15.80}$$

Since $G_0 = G - G V G_0$ from relation (15.73), we obtain with (15.79)

$$| \Psi \rangle = | \varphi \rangle + GV | \Psi \rangle - G V G_0 V | \Psi \rangle. \tag{15.81}$$

We obtain from (15.79) the value of the vector $G_0 V | \Psi \rangle$, that is,

$$G_0 V | \Psi \rangle = | \Psi \rangle - | \varphi \rangle, \tag{15.82}$$

which we insert into (15.81) to finally obtain

$$| \Psi \rangle = | \varphi \rangle + G V | \Psi \rangle - G V | \Psi \rangle + G V | \varphi \rangle$$

$$\boxed{| \Psi \rangle = | \varphi \rangle + G V | \varphi \rangle} . \tag{15.83}$$

This is the perturbation expansion of the state $| \Psi \rangle$ with respect to the interaction V and the resolvant $G = 1/(E - H)$.

Now, let us compare the matrix elements of V and those of T using definition (15.67):

$$\langle \varphi | T | \varphi \rangle = \langle \varphi | V | \Psi \rangle = \langle \varphi | V | \varphi \rangle + \langle \varphi | V G V | \varphi \rangle, \tag{15.84}$$

which leads us to the Lippmann–Schwinger equation, relation between the transition operator T and the interaction operator V:

$$\boxed{T = V + V G V = V + V \frac{1}{E - H} V} . \tag{15.85}$$

The resolvant G, also called energy propagator or Green's function, written with relation (15.74) as

$$\frac{1}{E-H} = \frac{1}{E-H_0-V} = \frac{1}{E-H_0} + \frac{1}{E-H_0} V \frac{1}{E-H}, \qquad (15.86)$$

enables us to iterate the statement of the transition amplitude:

$$T = V + V \frac{1}{E-H_0} V + V \frac{1}{E-H_0} V \frac{1}{E-H_0} V + \dots . \qquad (15.87)$$

5.2 The Born Approximation

By truncating the expansion of the scattered wave Ψ after the first order, we obtain an identity between the matrix element of the transition operator T and that of the potential V:

$$\langle \varphi \mid T \mid \varphi \rangle = \langle \varphi \mid V \mid \Psi \rangle = \langle \varphi \mid V \mid \varphi \rangle + \langle \varphi \mid V\, G\, V \mid \varphi \rangle . \qquad (15.88)$$

The Born approximation takes into account only the first term:

$$\boxed{\langle \varphi \mid T \mid \varphi \rangle = \langle \varphi \mid V \mid \varphi \rangle} . \qquad (15.89)$$

Let us calculate the matrix element of V between plane waves:

$$\langle \varphi_f \mid V \mid \varphi_i \rangle = \langle \vec{k}_f \mid V \mid \vec{k}_i \rangle$$

$$= \int \langle \vec{k}_f \mid \vec{r} \rangle \, d\vec{r} \, \langle \vec{r} \mid V \mid \vec{r}\,' \rangle \, d\vec{r}\,' \, \langle \vec{r}\,' \mid \vec{k}_i \rangle$$

$$= (2\pi)^{-3} \int e^{i\,(\vec{k}_i - \vec{k}_f)\cdot\vec{r}} \, V(\vec{r}) \, d\vec{r}. \qquad (15.90)$$

If we introduce the Fourier transform of the potential,

$$v(k) = (2\pi)^{-3/2} \int e^{i\,\vec{k}\cdot\vec{r}} \, V(r) \, d\vec{r}, \qquad (15.91)$$

the matrix element (15.90) reduces to

$$\langle \varphi_f \mid V \mid \varphi_i \rangle = (2\pi)^{-3/2} \, v(q) \quad \text{with} \quad \vec{q} = \vec{k}_f - \vec{k}_i . \qquad (15.92)$$

In elastic scattering $\mid \vec{k}_f \mid = \mid \vec{k}_i \mid$ such that the transferred momentum is expressed by

$$q^2 = k_i^2 + k_2^2 - 2\,\vec{k}_i \cdot \vec{k}_f = 2\,k_i^2\,(1 - \cos\Theta) = 4\,k_i^2\,\sin^2\Theta/2 \qquad (15.93)$$

where Θ is the scattering angle. The scattering amplitude is therefore written in the Born approximation as follows:

$$f(\omega) = -\frac{4\pi^2\,\mu}{\hbar^2} \, \langle \varphi_f \mid V \mid \varphi_i \rangle = -\sqrt{2\pi} \, \frac{\mu}{\hbar^2} \, v(q) . \qquad (15.94)$$

Some forms of the interaction potential $V(r)$ lead to an analytical expression for the scattering amplitude.

6. Phase-Shift Method

Expressed in spherical coordinates, the Schrödinger equation for a central potential is written with relation (12.99)

$$\left[\frac{d^2}{dr^2} + \varepsilon - u(r) - \frac{\ell\,(\ell+1)}{r^2}\right] \varphi_\ell(r) = 0 \,, \tag{15.95}$$

with the reduced forms of energy and potential

$$\varepsilon = \frac{2m\,E}{\hbar^2} = k^2, \quad u(r) = \frac{2m\,V(r)}{\hbar^2} \,.$$

The solution $\varphi_\ell(r)$ is the regular solution of the radial Eq. (15.95), and its asymptotic form is

$$\varphi_\ell(r) \xrightarrow[r\to\infty]{} a_\ell \, \sin\left(k\,r - \frac{\ell\,\pi}{2} + \delta_\ell\right) . \tag{15.96}$$

The phase shift δ_ℓ will be determined by the radial equation whereas a_ℓ will be adjusted so that the wave function

$$\Psi\,(r,\,\Theta) = \sum_\ell \frac{\varphi_\ell(r)}{r}\,P_\ell\,(\cos\,\Theta) \tag{15.97}$$

has the desired asymptotic behavior.

Let us expand $f(\Theta)$ over the Legendre polynomial basis by setting

$$f(\Theta) = \sum_\ell f_\ell\,P_\ell\,(\cos\,\Theta) . \tag{15.98}$$

The scattered wave $\Psi\,(r,\,\Theta)$ is written with (15.57) as

$$\psi\,(r,\,\Theta) = e^{i\,\vec{k}.\vec{r}} + f(\Theta)\,\frac{e^{i\,k\,r}}{r} \,. \tag{15.99}$$

With Legendre polynomial expansion (12.71) of the plane wave,

$$e^{i\,\vec{k}.\vec{r}} = \sum_\ell [\ell^2]\,i^\ell\,j_\ell(k\,r)\,P_\ell\,(\cos\,\Theta) , \tag{15.100}$$

and expansion (15.98) of the spherical wave, we obtain

$$\Psi\,(r,\,\Theta) = \sum_\ell \left([\ell^2]\,i^\ell\,j_\ell(k\,r) + f_\ell\,\frac{e^{i\,k\,r}}{r}\right)\,P_\ell\,(\cos\,\Theta) . \tag{15.101}$$

Let us use the asymptotic form of the spherical Bessel function:

$$j_\ell\,(k\,r)\xrightarrow[r\to\infty]{}\frac{1}{k\,r}\,\sin\left(k\,r-\frac{\ell\,\pi}{2}\right)$$

$$=\frac{1}{2i\,k\,r}\left[e^{i\,(k\,r-\frac{\ell\pi}{2})}-e^{-i\,(k\,r-\frac{\ell\pi}{2})}\right].\qquad(15.102)$$

Inserting this into expression (15.101) for the Legendre polynomial expansion of the scattered wave, we obtain

$$\Psi\,(r,\,\Theta)\xrightarrow[r\to\infty]{}\sum_\ell\left[\frac{(2\ell+1)\,i^\ell}{2i\,k\,r}\left(e^{i\,(k\,r-\frac{\ell\pi}{2})}\,e^{-i\,(k\,r-\frac{\ell\pi}{2})}\right)\right.$$

$$\left.+f_\ell\,\frac{e^{i\,k\,r}}{r}\right]\,P_\ell\,(\cos\,\Theta).\qquad(15.103)$$

With the phases

$$e^{-i\,\frac{\ell\pi}{2}}=(-i)^\ell\quad\text{and}\quad e^{i\,\frac{\ell\pi}{2}}=i^\ell,\qquad(15.104)$$

the asymptotic form of the scattered wave is written as follows:

$$r\,\Psi\,(r,\,\Theta)\xrightarrow[r\to\infty]{}\sum_\ell\left[(-)^{\ell+1}\,\frac{2\ell+1}{2i\,k}\,e^{-i\,k\,r}+\left(\frac{2\ell+1}{2i\,k}+f_\ell\right)\,e^{i\,k\,r}\right]\,P_\ell.$$

$$(15.105)$$

Let us identify this result with (15.96), the expression for the asymptotic behavior of the regular radial solution at the origin:

$$a_\ell=i^\ell\,\frac{2\ell+1}{k}\,e^{i\,\delta_\ell},$$

$$f_\ell=\frac{2\ell+1}{k}\,e^{i\,\delta_\ell}\,\sin\,\delta_\ell.\qquad(15.106)$$

By reconstructing the scattering amplitude (15.98), we obtain

$$f(\Theta)=\sum_{\ell=0}^{\infty}\frac{(2\ell+1)}{k}\,e^{i\,\delta_\ell}\,\sin\,\delta_\ell\,P_\ell\,(\cos\,\Theta)$$

$$=\sum_{\ell=0}^{\infty}\frac{2\ell+1}{2i\,k}\,\left[e^{2i\,\delta_\ell}-1\right]\,P_\ell\,(\cos\,\Theta).\qquad(15.107)$$

The differential cross section is the square modulus of the scattering amplitude:

$$\sigma(\omega)=\mid f(\omega)\mid^2.\qquad(15.108)$$

It is written with expansion (15.107):

$$\sigma(\omega)=\frac{1}{k^2}\sum_{\ell\,\ell'}\left[\ell^2\,\ell'^2\right]\,e^{i\,(\delta_\ell-\delta'_\ell)}\,\sin\,\delta_\ell\,\sin\,\delta_{\ell'}\,P_{\ell'}\,(\cos\,\Theta)P_\ell\,(\cos\,\Theta).$$

$$(15.109)$$

The total cross section will be obtained by integration over the solid angle (Θ, φ) and with the orthogonal Legendre polynomials,

$$\int_{-1}^{+1} P_\ell(u) \, P_{\ell'}(u) \, du = \frac{2}{2\ell+1} \, \delta_{\ell\,\ell'} , \qquad (15.110)$$

we obtain the following result:

$$\boxed{\sigma_{tot} = \frac{4\pi}{k^2} \sum_\ell (2\ell+1) \, \sin^2 \delta_\ell} \, . \qquad (15.111)$$

For $\Theta = 0$, Eq. (15.107) coupled with the fact that $P_\ell(1) = 1$, will give the scattering amplitude

$$f(0) = \frac{1}{k} \sum_\ell (2\ell+1) \, e^{i\,\delta_\ell} \, \sin \delta_\ell , \qquad (15.112)$$

with an imaginary part equal to

$$\text{Im} \, \{f(0)\} = \frac{1}{k} \sum_\ell (2\ell+1) \, \sin^2 \delta_\ell . \qquad (15.113)$$

This leads to the optical theorem that we have already had occasion to discuss in Sect. 2.2 of this chapter.

$$\boxed{\sigma_{tot} = \frac{4\pi}{k} \, \text{Im} \, \{f(0)\}} \, . \qquad (15.114)$$

7. Angular Distribution and Polarization

7.1 Transition Amplitude

In a nuclear interaction process, a nucleus A comprising Z protons and N neutrons may be in direct interaction with another nucleus a to produce a residual nucleus B and an emerging particle b. Suppose that B and b are identical to A and a. We then call the scattering elastic if the energy of the emerging particles is equal to those of the incident particles, and inelastic if the contrary is the case. If nuclei $(A\ a)$ and $(B\ b)$ are different, we say that there has been a nuclear reaction since there is a rearrangment of the nucleons involved in the entrance channel $(A\ a)$ and the exit channel $(B\ b)$.

The transition amplitude is the matrix element of the interaction potential between the incoming and the outgoing waves, and we can write it in the following form:

$$T = \langle \Psi_{B\,b}^- \, | \, V \, | \, \Psi_{A\,a}^+ \rangle . \qquad (15.115)$$

The particles involved in this reaction possess a spin or a total angular momentum defined as: $\vec{j}_A$ for the target nucleus, $\vec{j}_a = \vec{\ell}_a + \vec{s}_a$ for the incident nucleus, $\vec{j}_b$ for the particle b and $\vec{j}_b = \vec{\ell}_b + \vec{s}_b$, and $\vec{j}_B$ for the residual nucleus. The transition amplitude can be fully expressed by projecting the matrix element onto configuration space (Elbaz, 1985) but because we are only interested in the geometry inherent in the angular distribution or in the polarization of the emerging particles, we will represent the transition amplitude with a black box, with the incoming and outgoing angular momenta, derived from the kets or bras of the transition amplitude (15.115):

$$T = \boxed{T} \begin{array}{l} \text{s}_a \\ \text{j}_A \\ \text{j}_B \\ \text{s}_b \end{array} . \tag{15.116}$$

7.2 Detection of Emerging Particles

There is a statistical angular momentum density of states ϱ^{sa} and ϱ^{jA} in the entrance channel and a statistical detection efficiency of the angular momentum states ε^{sb} and ε^{jB} in the exit channel. The observation of the emerging and residual particles is proportional to the mean value of the detection efficiency tensor ε. This can be written as:

$$\langle \varepsilon \rangle = W = \mathrm{Tr}(\varrho\,\varepsilon). \tag{15.117}$$

By introducing the statistical tensors of the entrance channel (initial density of states) and those of the exit channel (detection efficiency of the final state), we obtain the mean value of the detection efficiency tensor W:

$$W = \boxed{T}\begin{array}{c} \text{s}_a \text{\textcircled{ρ}} \text{s}_a \\ \text{j}_A \text{\textcircled{ρ}} \text{j}_A \\ \text{s}_b \text{\textcircled{ε}} \text{s}_b \\ \text{j}_B \text{\textcircled{ε}} \text{j}_B \end{array}\boxed{T^+} . \tag{15.118}$$

7.3 Distribution of Emerging Particles

If we do not measure the polarization of emerging and residual particles, then there is no need to introduce the ITOs associated with ε^{sb} and ε^{jB}. We therefore have to link the extremities of corresponding kinetic lines.

If the incident and target particles are not polarized, then there is no need either to introduce the irreducible tensor operators (ITOs) related to ϱ^{sa} and ϱ^{jA}. This means linking corresponding kinetic lines. This particular case, the simplest imaginable, therefore corresponds to the following W_0 diagram:

$$W_0 = \boxed{T \begin{array}{c} s_a \\ \dot{j_A} \\ s_b \\ \dot{j_B} \end{array} T^+} \qquad (15.119)$$

$$W_0 = \frac{1}{[s_a^2 \, j_A^2]} \sum_{\sigma_a \, \mu_A \, \sigma_b \, \mu_B} | \, T_{\mu_A \, \sigma_a \, \mu_B \, \sigma_b} \, |^2 . \qquad (15.120)$$

Let μ_i and μ_f be the reduced masses of the incident and emerging channels,

$$\frac{1}{\mu_i} = \frac{1}{m_a} + \frac{1}{m_A} \quad \text{and} \quad \frac{1}{\mu_f} = \frac{1}{m_b} + \frac{1}{m_B} , \qquad (15.121)$$

and k_a and k_b the moduli of the wave vectors of the incident and emerging particles,

$$E_a = \frac{\hbar^2}{2m_a} k_a^2 \quad \text{and} \quad E_b = \frac{\hbar^2}{2m_b} k_b^2 . \qquad (15.122)$$

The angular distribution of the emerging particles is written with W_0

$$\frac{d\sigma}{d\Omega} = \frac{\mu_i \, \mu_f}{(2\pi \, \hbar^2)^2} \frac{k_b}{k_a} W_0 . \qquad (15.123)$$

Notice that the angular distribution defines the relationship between the number of particles detected in a given direction and the number of incident particles. It is therefore a magnitude that is invariant under rotation of the system of axes considering that only the relative direction of incident and emerging particles are relevant. This is clearly visible in the diagram representing W_0. Rather than try to detail the foregoing diagram, we will be content here to show that, in the transition amplitude, the angular dependence of the incident particles $\hat{k}_a$ and the emerging particles $\hat{k}_b$, associated with the spherical harmonics $Y_{\ell_a m_a}(\hat{k}_a)$ and $Y_{\ell_b \, m_b}^*(\hat{k}_b)$, give:

$$T = \boxed{\begin{array}{c} \hat{k}_a \xrightarrow{l_a} \\[4pt] T \\[4pt] \hat{k}_b \xrightarrow{l_b} \end{array} \begin{array}{c} s_a \\ \dot{j_A} \\ \dot{j_B} \\ s_b \end{array}} . \qquad (15.124)$$

The W_0 diagram associated with the angular distribution becomes, with (15.119),

$$W_0 = \boxed{\begin{array}{c} \hat{k}_a \xrightarrow{l_a} \\[4pt] T \\[4pt] \hat{k}_b \xrightarrow{l_b} \end{array} \begin{array}{c} s_a \\ \dot{j_A} \\ s_b \\ \dot{j_B} \end{array} \begin{array}{c} T^+ \end{array} \begin{array}{c} l_a' \quad \hat{k}_a \\[4pt] \\[4pt] l_b' \quad \hat{k}_b \end{array}} . \qquad (15.125)$$

The contraction (12.53) of the spherical harmonics of the same angle introduces the intermediate orbital momenta L_a and L_b:

$$W_0 = \qquad (15.126)$$

The pinch (11.109) over lines L_a and L_b introduces the Legendre polynomial (13.31) of the relative direction $(\hat{k}_a, \hat{k}_b)$:

$$W_0 = \qquad (15.127)$$

We only have to express W_0 analytically to obtain the theoretical value of the angular distribution of the emerging particles for each type of nuclear reaction, or for elastic or inelastic scattering of particles (Elbaz, 1985).

7.4 Polarization of Emerging Particles

Consider the simplest case in which the incident particles a on nuclei A are unpolarized and in which only the polarization of emerging particles b is detected. It has been shown in Chap. 13 that the mean value of ε can be expressed with ITOs associated with the statistical densities,

$$W = \langle \varepsilon \rangle = \sum_{k \, \chi} \varrho_{k \, \chi} \, \varepsilon^*_{k \, \chi}, \qquad (15.128)$$

and that the statistical density of states $\varrho^j_{k \, \chi}$ associated with a detection efficiency $\varepsilon^{j \, *}_{k \, \chi}$ defined the components of the polarization vector (for particles with spin $\frac{1}{2}$), or more generally speaking, of the polarization tensor (for particles with greater spins):

$$W = \langle \varepsilon \rangle = \text{Tr}(\varrho\, \varepsilon) =$$

$$= \sum_{k\,\chi} \varrho^j_{k\,\chi}\, \varepsilon^{j\,*}_{k\,\chi} = \sum_k$$

. $\qquad(15.129)$

The statistical polarization tensor of emerging particles $\varrho^{s_b}_{k\,\chi}$ is therefore obtained by introducing the intermediate momentum k and the statistical efficiency tensor $\varepsilon^*_{k\chi}$ in W, giving:

$$W =$$

$\qquad(15.130)$

By comparison with the foregoing diagram (15.129) and by normalizing the diagram (15.129) at the value W_0 in the absence of any polarization measurement (we should in this case obtain W_0), we infer the polarization tensor:

$$\varrho^{s_b}_{k\,\chi} = \frac{1}{[s_a^2\, j_A^2]}\, \frac{1}{W_0}$$

. $\qquad(15.131)$

We must recall that, for particles with spin $\frac{1}{2}$,

$$\varrho^{1/2}_{00} = \frac{1}{\sqrt{2}} \qquad \varrho^{1/2}_{10} = \frac{1}{\sqrt{2}}\, P_z \qquad \text{and}$$
$$\varrho^{1/2}_{1\pm1} = \mp\, (P_x \pm i\, P_y) \qquad\qquad (15.132)$$

whereas for particles with spin 1, we obtain (Elbaz, 1985)

$$\varrho_{00}^1 = \frac{1}{\sqrt{3}} \qquad \varrho_{10}^1 = \frac{1}{\sqrt{2}} P_z \qquad \varrho_{1\pm1}^1 = \mp \frac{1}{2} (P_x \pm i\, P_y),$$

$$\varrho_{20}^1 = \frac{1}{\sqrt{6}} P_{zz} \qquad \varrho_{2\pm1}^1 = \mp \frac{1}{3} (P_{x\,z} \mp i\, P_{y\,z}), \quad (15.133)$$

$$\varrho_{2\pm2}^1 = \frac{1}{2} (P_{xx} - P_{yy} \pm 2i\, P_{xy}).$$

Just like with the angular distribution, it is possible to exhibit the angular dependence of the polarization. The only difference here is that the pinch of the orbital lines L_a and L_b will no longer be possible because of the kinetic line $(k\,\chi)$. We thus obtain, by expressing the angular dependences of diagram (15.131),

$$\varrho_{k\,\chi}^{s_b} = \frac{1}{W_0} \qquad \text{} \qquad (15.134)$$

The pinch (11.120) on lines $(L_a\, L_b\, k)$ introduces a function of $(\hat{k}_a,\, \hat{k}_b)$ that is no longer a Legendre polynomial:

$$M_{k\,\chi} (\hat{k}_a, \hat{k}_b) = - \quad \text{} \qquad (15.135)$$

$$= \sum_{M_a\, M_b} \begin{pmatrix} M_a & k & L_b \\ L_a & \chi & M_b \end{pmatrix} Y_{L_a M_b} (\hat{k}_a)\, Y_{L_b\, M_b}^* (\hat{k}_b). \quad (15.136)$$

Notice here that the special value of this function:

$$M_{00} (\hat{k}_a, \hat{k}_b) = \frac{[L_a]}{4\pi} \delta_{L_a\, L_b}\, P_{L_a} (\cos (\hat{k}_a, \hat{k}_b)) \qquad (15.137)$$

is the value used in the calculation of the angular distribution in (15.129).

The remaining closed diagram has the following graphical structure:

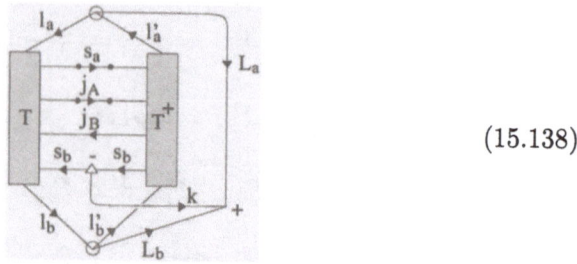

$$(15.138)$$

As an exercise, let us make a complete evaluation of the angular distribution and the polarization of the emerging particles for a nuclear reaction involving the direct transfer of a nucleon x of a nucleus to another nucleus A:

$$A + a \ (= b + x) \ \rightarrow \ B \ (= A + x) + b. \qquad (15.139)$$

This gives a transition amplitude of the following form:

$$T_{\sigma_b \ \sigma_a \ \mu_A \ \mu_B} = \mathcal{R} \ (k_a, \ k_b) \ G \ (\hat{k}_a, \ \hat{k}_b), \qquad (15.140)$$

in which the radial integral will be of the form

$$\mathcal{R} \ (k_A, \ k_b) = \frac{(4\pi)^{3/2}}{k_a \ k_b} \ i^{\ell_a - \ell_b} \int dr \ \chi^*_{\ell_b} \ (k_b \ r) \ \varphi^*_{\ell_1}(r) \ \chi_{\ell_a} \ (k_a \ r), \quad (15.141)$$

whereas the angular part (hidden in the black box of the transition amplitude) has the following graphical representation with an interaction potential of zero range:

$$(15.142)$$

The analytical transcription of this result is not hard to find although, for the time being of no use. In the meantime, we will note that, following relations (15.120) and (15.123), the angular distribution of emerging particles is written

$$\frac{d\sigma}{d\Omega} = \frac{\mu_i \ \mu_f}{(2\pi \ \hbar^2)^2} \frac{k_b}{k_a} \frac{1}{[s_a^2 \ j_A^2]} \sum_{\sigma_a \ \sigma_b \ \mu_A \ \mu_B} | \ T_{\sigma_b \ \sigma_a \ \mu_A \ \mu_B} \ |^2, \qquad (15.143)$$

and by extracting the square modulus of the radial integral of the transition amplitude (15.141) we obtain

$$\frac{d\sigma}{d\Omega} = \frac{\mu_i \, \mu_f}{(2\pi \, \hbar^2)^2} \frac{k_b}{k_a} \frac{1}{[s_a^2 \, j_A^2]} \, |\, \mathcal{R}\,(k_a, \, k_b)|^2 \sum_{\sigma_a \, \sigma_b \, \mu_A \, \mu_B} |\, G(\hat{k}_a, \, \hat{k}_b)\,|^2 .$$

(15.144)

The graphical statement of the sum is easily obtained with diagram (15.142):

$$E_1 = \sum_{\sigma_b \, \sigma_a \, \mu_B \, \mu_A} |\, G(\hat{k}_a, \, \hat{k}_b)\,|^2$$

(15.145)

Following the extraction of the triangular deltas by pinch on the two kinetic lines (j_1, j_1'), (j_2, j_2') and then (LL'), we are left with a diagram denoted E_2:

$$E_1 = \{j_a \, s_b \, j_2\} \, \delta_{j_2 \, j_2'} \, \{s_a \, j_B \, j_1\} \, \delta_{j_1 \, j_1'} \, \{j_1 \, j_2 \, L\} \, \delta_{LL'}$$
$$\times [j_2^{-2} \, j_1^{-2} \, L^{-2}] \, E_2 .$$

(15.146)

The contraction of the spherical harmonics will lead to the angular dependence of E_2:

(15.147)

The analytical transcription of this diagram is all the more easy since all the poles are marked, making it possible to avoid introducing the reading orders at the poles or the direction of the kinetic lines and leaving the value outside the triangular deltas:

$$E_2 = \frac{[L_a \, L_a' \, \mathcal{L}]}{\sqrt{4\pi}} \frac{[L_b \, L_b' \, \mathcal{L}]}{\sqrt{4\pi}} \frac{[L_a \, L_b \, L]}{\sqrt{4\pi}} \frac{[L_a' \, L_b' \, L]}{\sqrt{4\pi}} \frac{1}{4\pi} \, P_{\mathcal{L}} \, (\cos(\hat{k}_a, \, \hat{k}_b))$$

$$\times \begin{pmatrix} 0 & L_a' & \mathcal{L} \\ L_a & 0 & 0 \end{pmatrix} \begin{pmatrix} 0 & L & L_b' \\ L_a' & 0 & 0 \end{pmatrix} \begin{pmatrix} 0 & 0 & 0 \\ L_b' & \mathcal{L} & L_b \end{pmatrix} \begin{pmatrix} L_a & L_b & 0 \\ 0 & 0 & L \end{pmatrix} \begin{Bmatrix} L_a' & L_b' & L \\ L_b & L_a & \mathcal{L} \end{Bmatrix} .$$

(15.148)

It is also possible to rearrange the summation over the intermediate orbital angular momentum L in order to replace E_2 by E_3 which has the graphical expression

$$E_3 = \qquad\qquad\qquad\qquad\qquad (15.149)$$

leading to the simplest analytical form:

$$E_3 = \frac{1}{4\pi} P_{\mathcal{L}} \left(\cos(\hat{k}_a, \hat{k}_b)\right)[\mathcal{L}^{-2}] \frac{[L_a^2 \, L_a'^2 \, \mathcal{L}^2]}{4\pi} \frac{[L_b^2 \, L_b'^2 \, \mathcal{L}^2]}{4\pi}$$

$$\times \begin{pmatrix} L_a & L_a' & \mathcal{L} \\ 0 & 0 & 0 \end{pmatrix}^2 \begin{pmatrix} L_b & L_b' & \mathcal{L} \\ 0 & 0 & 0 \end{pmatrix}^2 . \qquad (15.150)$$

The angular distribution (15.144) therefore depends on the relative direction of the vectors $\vec{k}_a$ and $\vec{k}_b$ through the Legendre polynomial as we can see in E_2 or in E_3, and is, in the final analysis, written with E_3:

$$\frac{d\sigma}{d\Omega} = \frac{\mu_i \, \mu_f}{(2\pi\hbar^2)^2} \frac{k_b}{k_a} \frac{1}{[s_a^2 \, j_A^2]} |\mathcal{R}(k_a, k_b)|^2 \{j_a \, s_b \, j_2\} \{s_a \, j_B \, j_1\} \frac{(4\pi)^{-3}}{[j_1^2 \, j_2^2]}$$

$$\times [L_a^2 \, L_b^2 \, L_a'^2 \, L_b'^2 \, l^2] \begin{pmatrix} L_a & L_a' & l \\ 0 & 0 & 0 \end{pmatrix}^2 \begin{pmatrix} L_b & L_b' & l \\ 0 & 0 & 0 \end{pmatrix}^2$$

$$P_l \left(\cos(\hat{k}_a, \hat{k}_b)\right) . \qquad (15.151)$$

The above example of the calculation of angular distributions of the emerging particles b in the direct nuclear reaction $A + a \rightarrow B + b$ with a zero-range interaction potential is the simplest example imaginable. It shows however the procedure for obtaining the analytical statement of the angular distribution (or by using (15.38) of the polarization) for any form whatsoever of the interaction potential.

8. Scattering of Particles – Schematic Summary

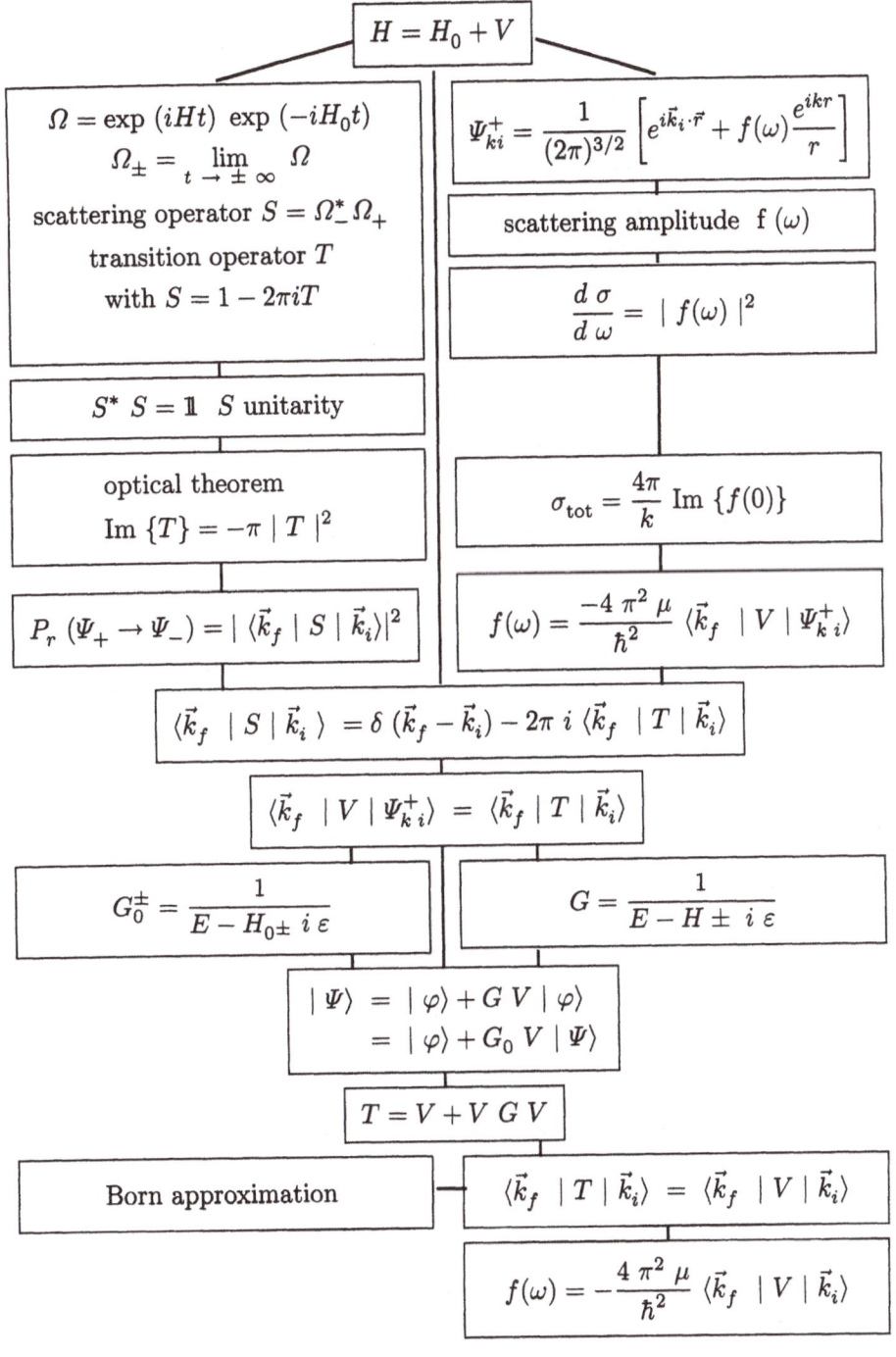

$$H = H_0 + V$$

$$\Omega = \exp\left(iHt\right)\exp\left(-iH_0 t\right)$$
$$\Omega_\pm = \lim_{t \to \pm\infty} \Omega$$
scattering operator $S = \Omega_-^* \Omega_+$
transition operator T
with $S = 1 - 2\pi i T$

$$\Psi_{ki}^+ = \frac{1}{(2\pi)^{3/2}}\left[e^{i\vec{k}_i \cdot \vec{r}} + f(\omega)\frac{e^{ikr}}{r}\right]$$

scattering amplitude $f(\omega)$

$$\frac{d\sigma}{d\omega} = |f(\omega)|^2$$

$$S^* S = \mathbb{1} \quad S \text{ unitarity}$$

optical theorem
$$\text{Im}\{T\} = -\pi\,|T|^2$$

$$\sigma_{\text{tot}} = \frac{4\pi}{k}\,\text{Im}\{f(0)\}$$

$$P_r\,(\Psi_+ \to \Psi_-) = |\langle \vec{k}_f\,|\,S\,|\,\vec{k}_i\rangle|^2$$

$$f(\omega) = \frac{-4\pi^2\mu}{\hbar^2}\,\langle \vec{k}_f\,|\,V\,|\,\Psi_{ki}^+\rangle$$

$$\langle \vec{k}_f\,|\,S\,|\,\vec{k}_i\rangle = \delta\,(\vec{k}_f - \vec{k}_i) - 2\pi\,i\,\langle \vec{k}_f\,|\,T\,|\,\vec{k}_i\rangle$$

$$\langle \vec{k}_f\,|\,V\,|\,\Psi_{ki}^+\rangle = \langle \vec{k}_f\,|\,T\,|\,\vec{k}_i\rangle$$

$$G_0^\pm = \frac{1}{E - H_0 \pm i\varepsilon}$$

$$G = \frac{1}{E - H \pm i\varepsilon}$$

$$|\Psi\rangle = |\varphi\rangle + G\,V\,|\varphi\rangle$$
$$= |\varphi\rangle + G_0\,V\,|\Psi\rangle$$

$$T = V + V\,G\,V$$

Born approximation

$$\langle \vec{k}_f\,|\,T\,|\,\vec{k}_i\rangle = \langle \vec{k}_f\,|\,V\,|\,\vec{k}_i\rangle$$

$$f(\omega) = -\frac{4\pi^2\mu}{\hbar^2}\,\langle \vec{k}_f\,|\,V\,|\,\vec{k}_i\rangle$$

Chapter 16
Second Quantization

The second quantization formalism is based on an assumed knowledge of the possible quantum states of a system and their occupation or otherwise by particles. Naturally, it takes into account all the postulates of quantum mechanics although it rests essentially on the algebra of the operators acting in the Hilbert space of states. Its use is indispensible in the description of quantum systems with a large number of particles: atoms, nuclei, as well as in quantum field theory. We will introduce the formalism with a discussion of the harmonic oscillator, then go on to generalize the method to boson and fermion systems, and finally return to the harmonic oscillator with field quantization.

1. One-Dimensional Harmonic Oscillator

The importance of the harmonic oscillator in quantum physics is two-fold. The first aspect has to do with the possibility of describing to a first approximation a bound system, such as a particle in a harmonic well. This yields precious information on energy levels, their spacing, and their degeneracy. This is the idea behind the nuclear shell model. The second aspect concerns the formalism actually used, based on the description of creation and annihilation operators, to construct a representation of quantum states in a discrete basis.

This method underlies the second quantization formalism as well as the factorization method for solving certain types of differential equation. Finally, it is important to be familiar with the harmonic oscillator if we want to get to grips with the quantum field theory.

1.1 Creation and Annihilation Operator

Consider a particle with mass m in a harmonic potential well. The motion of the particle will be defined by the classical Hamiltonian

$$ H = \frac{p^2}{2m} + \frac{1}{2} m \, \omega^2 \, x^2 , \qquad (16.1) $$

where ω is the angular velocity of the oscillator.

We introduce a quantum creation operator:

$$a^+ = \sqrt{\frac{m\,\omega}{2\hbar}}\, x - \frac{i}{\sqrt{2m\,\hbar\,\omega}}\, p \qquad (16.2)$$

and its transpose, the annihilation operator:

$$a = (a^+)^+ = \sqrt{\frac{m\,\omega}{2\hbar}}\, x + \frac{i}{\sqrt{2m\,\hbar\,\omega}}\, p\,. \qquad (16.3)$$

The operator $N = a^+ a$ (in this order) is the number of quanta operator:

$$N = a^+\,a = \left(\sqrt{\frac{m\,\omega}{2\hbar}}\, x - \frac{i}{\sqrt{2m\,\hbar\,\omega}}\, p \right) \left(\sqrt{\frac{m\,\omega}{2\hbar}}\, x + \frac{i}{\sqrt{2m\,\hbar\,\omega}}\, p \right).$$

The expansion of this operator product yields

$$N = \frac{m\,\omega}{2\hbar}\, x^2 + \frac{1}{2m\,\hbar\,\omega}\, p^2 + \frac{1}{2\hbar}\, i\,(xp - px)\,. \qquad (16.4)$$

The postulates of quantum mechanics provide the value of the operator commutator of x and p, that is $[x, p] = i\hbar$, making it possible to write:

$$N = \frac{1}{\hbar\,\omega} \left(\frac{1}{2}\, m\,\omega^2\, x^2 + \frac{p^2}{2m} \right) - \frac{1}{2}\,\mathbb{1}\,. \qquad (16.5)$$

We recognize the Hamiltonian H in the term in brackets, which leads to the important relation

$$N = H\,\frac{1}{\hbar\,\omega} - \frac{1}{2}\,\mathbb{1}\,, \qquad (16.6)$$

or, by expressing the Hamiltonian with respect to N,

$$\boxed{H = \left(N + \frac{1}{2}\,\mathbb{1} \right) \hbar\,\omega} \qquad (16.7)$$

The search for the eigenvalues of the Hamiltionian H therefore reduces to the search for the eigenvalues n and the eigenvectors $\mid n\rangle$, of N.

$$N \mid n\rangle = n \mid n\rangle\,. \qquad (16.8)$$

The operator is Hermitian since $N^+ = (a^+a)^+ = a^+\,a = N$, which defines the real eigenvalues n:

$$n^* = n = \langle n \mid N \mid n\rangle = \langle n \mid a^+\,a \mid n\rangle\,. \qquad (16.9)$$

The bra vector $\langle n \mid a^+$ is the transpose of $a \mid n\rangle$ and hence n represents the norm (positive or zero) of the vector $a \mid n\rangle$:

$$\boxed{n = \text{Norm}\,(a\mid n)) \geq 0} \, . \tag{16.10}$$

The eigenvalues n are therefore positive or zero and a zero value of the norm implies that the vector itself is zero.

Let us use definitions (16.2) and (16.3) and the commutator $[x, p]$ to calculate the commutation relation between a and a^+:

$$a\, a^+ - a^+ \, a = [a, a^+] = \mathbb{1} \, . \tag{16.11}$$

This yields the important relations

$$Na = a^+ \, aa = (aa^+ - \mathbb{1})a = aa^+ \, a - a = a(a^+ \, a - \mathbb{1}) = a\,(N - \mathbb{1}), \tag{16.12a}$$

which correspond to the commutation relation

$$[N, a] = -a \, , \tag{16.12b}$$

and, similarly, with the creation operator,

$$Na^+ = a^+ \, aa^+ = a^+ \, (a^+ \, a + \mathbb{1}) = a^+ \, (N + \mathbb{1}) \, . \tag{16.13a}$$

This is written in the form of a commutator as follows:

$$[N, a^+] = a^+ \, . \tag{16.13b}$$

Thus, by applying relation (16.12) to the vector $\mid n)$, we obtain

$$N\,(a\mid n)) = a\,(N - \mathbb{1})\mid n) \; = a\,(n - 1)\mid n) \; = (n - 1)\,(a\mid n)) \, . \tag{16.14}$$

The vector $A \mid n)$ is the eigenvector of N corresponding to the eigenvalue $(n-1)$. We can therefore note that the vector $a \mid n)$ is proportional to a vector $\mid n - 1)$, which is also an eigenvector of N corresponding to the eigenvalue $(n - 1)$, that is

$$\begin{aligned} N\,(a\mid n)) &= (n - 1)\,(a\mid n)) \, , \\ N\mid n - 1) &= (n - 1)\mid n - 1) \, . \end{aligned} \tag{16.15}$$

By comparing these results, we obtain

$$a\mid n) \; = c\mid n - 1) \, . \tag{16.16}$$

We may assume that the vectors $\mid n)$ are normalized and proceed to write

$$(n\mid n') \; = \delta_{n\,n'} \, . \tag{16.17}$$

Now let us evaluate the norm of the vector $a \mid n)$ using (16.16):

$$(n\mid a^+ \, a\mid n) \; = (n - 1\mid c^* \, c\mid n - 1) \; = \mid c\mid^2 (n - 1\mid n - 1) \; = \mid c\mid^2 \, ,$$

but following (16.10), the norm of $(a \mid n))$ is equal to n. We therefore choose

$$c = \sqrt{n}. \tag{16.18}$$

The annihilation operator a causes the eigenvalue to go from n of N to the eigenvalue $(n-1)$, which corresponds to the eigenvector $|\,n-1\rangle$ of N, which, with the aid of relation (16.16), then becomes:

$$\boxed{a\,|\,n\rangle \;=\; \sqrt{n}\,|\,n-1\rangle} \;. \tag{16.19}$$

It can further be shown that $a^2\,|\,n\rangle$ is an eigenvector of N, but corresponding to the eigenvalue $(n-2)$ and we can continue in this manner to the pth-order:

$$a\,|\,n\rangle \;=\; \sqrt{n}\,|\,n-1\rangle$$
$$a^2\,|\,n\rangle \;=\; a\,a\,|\,n\rangle \;=\; a\,\sqrt{n}\,|\,n-1\rangle \;=\; \sqrt{n\,(n-1)}\,|\,n-2\rangle$$
$$\vdots \qquad\qquad\qquad\qquad\qquad\qquad\qquad\qquad \vdots$$
$$a^p\,|\,n\rangle \;= \qquad\qquad\qquad\quad = \sqrt{n\,(n-1)\ldots(n-p+1)}\,|\,n-p\rangle. \tag{16.20}$$

If $n = 0$, that is if Norm $(a\,|\,n\rangle) = 0$, then the vector $a\,|\,n\rangle$ will itself be zero:

$$\boxed{a\,|\,0\rangle \;=\; 0} \;. \tag{16.21}$$

The value 0 belongs to the series of possible values of n and considering that these values vary by one-unit jumps at a time, we infer that n can take positive or zero integer values $n = 0, 1, 2 \ldots$.

We can also show in an analogous manner, using relation (16.13), that $a^+\,|\,n\rangle$ is an eigenvector of N corresponding to the eigenvalue $n+1$.

The normalization of the vector $a^+\,|\,n\rangle$ then leads us to write

$$\boxed{a^+\,|\,n\rangle \;=\; \sqrt{(n+1)}\,|\,n+1\rangle} \;. \tag{16.22}$$

We can construct the vector $|\,n+1\rangle$ by the action of the creation operator a^+, or the vector $|\,n-1\rangle$ by the action of the annihilation operator a (Fig. 1) from an eigenvalue n corresponding to the eigenvector $|\,n\rangle$ of N.

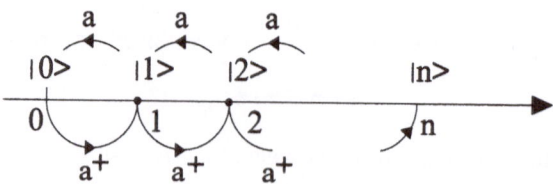

Figure 1

We thus formulate two theorems:

Theorem 1: The positive real eigenvalues of N are the integers 0, 1, 2, ...

$$N \mid n\rangle = n \mid n\rangle \qquad n = 0, 1, 2, \ldots \qquad (16.23)$$

This leads to the eigenvalue equation of the Hamiltonian operator:

$$H \mid n\rangle = \left(N + \frac{1}{2} \mathbb{1}\right) \hbar\,\omega \mid n\rangle = \left(n + \frac{1}{2}\right) \hbar\,\omega \mid n\rangle. \qquad (16.24)$$

The eigenvalues of the Hamiltonian H are therefore discrete and are expressed as follows:

$$E_n = \left(n + \frac{1}{2}\right) \hbar\,\omega = \left(n + \frac{1}{2}\right) h\,\nu. \qquad (16.25)$$

Theorem 2: To the eigenvalue n corresponds the eigenfunction $\mid n\rangle$ of N and any one of the eigenfunctions can be used to construct the others by application of the creation operator a^+ or the annihilation operator a:

$$
\begin{aligned}
\mid 0\rangle & \\
\mid 1\rangle &= a^+ \mid 0\rangle \\
\mid 2\rangle &= \frac{a^+}{\sqrt{2}} \mid 1\rangle = \frac{1}{\sqrt{2}} (a^+)^2 \mid 0\rangle \\
&\ \ \vdots \qquad\qquad\quad \vdots \\
\mid n\rangle &= \ldots \qquad\quad = \frac{1}{\sqrt{n!}} a^{+\,n} \mid 0\rangle.
\end{aligned}
\qquad (16.26)
$$

In the space $\{N\}$, or Fock space, determined by the vectors $\mid 0\rangle \mid 1\rangle \ldots \mid n\rangle$, a vector can be constructed from the "quantum" vacuum $\mid 0\rangle$ using the preceding relation:

$$\boxed{\mid n\rangle = (n!)^{-1/2} a^{+\,n} \mid 0\rangle} \ . \qquad (16.27)$$

We can also reduce the eigenvalue, one unit at a time, until we obtain a zero eigenvalue:

$$
\begin{aligned}
a \mid n\rangle &= \sqrt{n} \mid n - 1\rangle \\
a \mid n - 1\rangle &= \sqrt{n - 1} \mid n - 2\rangle \\
&\ \ \vdots \\
a \mid 0\rangle &= 0.
\end{aligned}
\qquad (16.28)
$$

The eigenvalues of N constitute a complete basis in which we can represent the harmonic oscillator; this is the "occupation number $\{N\}$ representation of energy quanta $h\nu$" or Fock space.

1.2 $\{N\}$ Representation

The Fock space in this representation is therefore spanned by the enumerable infinity of vectors $|\,n\,\rangle$, eigenvectors of the operator N:

$$
\boxed{
\begin{aligned}
N\,|\,n\,\rangle &= n\,|\,n\,\rangle \qquad\qquad n = 0, 1, 2, \ldots \\
\langle\,n\,|\,n'\,\rangle &= \delta_{n\,n'} \\
\sum_n |\,n\,\rangle\,\langle\,n\,| &= \mathbb{1}
\end{aligned}
}
\tag{16.29}
$$

The transition from one basis vector to another is made by applying the creation and annihilation operators.

In this representation, the matrix representing the operator N and that representing the Hamiltonian operator $H = \left(N + \frac{1}{2}\,\mathbb{1}\right)\hbar\,\omega$ are diagonal:

$$
H = \hbar\,\omega
\begin{bmatrix}
1/2 & 0 & 0 & 0 & \ldots \\
0 & 3/2 & 0 & 0 & \ldots \\
0 & 0 & 5/2 & 0 & \ldots \\
\cdot & 0 & 0 & 7/2 &
\end{bmatrix}.
\tag{16.30}
$$

The matrices representing the operators p and x are not diagonal and are constructed from a^+ and a using definitions (16.2) and (16.3):

$$
x = \sqrt{\frac{\hbar}{2m\,\omega}}\,(a^+ + a) = \sqrt{\frac{\hbar}{2m\,\omega}}
\begin{bmatrix}
0 & 1 & & \\
1 & 0 & \sqrt{2} & \\
& \sqrt{2} & 0 & \sqrt{3} \\
& & \sqrt{3} & 0 & \sqrt{4}
\end{bmatrix},
$$

$$
\tag{16.31}
$$

$$
p = i\,\sqrt{\frac{m\,\hbar\,\omega}{2}}\,(a^+ - a) = i\,\sqrt{\frac{m\,\hbar\,\omega}{2}}
\begin{bmatrix}
0 & -\sqrt{1} & & \\
+\sqrt{1} & 0 & -\sqrt{2} & \\
& +\sqrt{2} & 0 & -\sqrt{3}
\end{bmatrix}.
$$

1.3 $\{X\}$ Representation

We can seek the eigenfunctions of the Hamiltonian H in configuration space by projecting the eigenstates $|\,n\,\rangle$ onto $\{X\}$, introducing in this way the wave function

$$
\langle\,x\,|\,n\,\rangle = \Psi_n(x). \tag{16.32}
$$

The ground state is determined by definition (16.21) of the quantum vacuum $|\,0\,\rangle$, that is,

$$
a\,|\,0\,\rangle = 0
$$

and by projecting onto $\{X\}$, we obtain the equivalent equation

$$\langle x \mid a \mid 0 \rangle = \int \langle x \mid a \mid x' \rangle \, dx' \langle x' \mid 0 \rangle = \int a(x) \, \delta(x - x') \, \Psi_0(x') \, dx'$$

$$= a(x) \, \Psi_0(x)$$

This yields with (16.21) the definition of the ground state $\Psi_0(x)$ of the oscillator:

$$\boxed{a(x) \, \Psi_0(x) = 0} \, . \tag{16.33}$$

Expression (16.3) for the annihilation operator $a(x)$ in configuration space leads to the equation

$$\left(\sqrt{\frac{m \, \omega}{2\hbar}} \, x + \frac{i \, p}{\sqrt{2m \, \omega \, \hbar}} \right) \Psi_0(x) = 0 \, , \tag{16.34}$$

which, with the correspondence principle, yields the differential equation

$$\left(\sqrt{\frac{m \, \omega}{2\hbar}} \, x + \sqrt{\frac{\hbar}{2m \, \omega}} \, \frac{d}{dx} \right) \Psi_0(x) = 0 \, . \tag{16.35}$$

This is a separable first-order equation of the form

$$\frac{d \, \Psi_0(x)}{\Psi_0(x)} = -\frac{m \, \omega}{\hbar} \, x \, dx \, . \tag{16.36}$$

When normalized to unity, the solution is written as follows:

$$\Psi_0(x) = \left(\frac{m \, \omega}{\hbar \, \pi} \right)^{1/4} \exp \left(\frac{-m \, \omega}{2\hbar} \, x^2 \right) \, . \tag{16.37}$$

The first excited state is obtained with relation (16.26) projected onto the configuration space:

$$\Psi_1(x) = \langle x \mid 1 \rangle = \langle x \mid a^+ \mid 0 \rangle = \int \langle x \mid a^+ \mid x' \rangle \, dx' \, \langle x' \mid 0 \rangle \, ,$$

$$\Psi_1(x) = a^+(x) \, \Psi_0(x) \, .$$

With the correspondence principle, we are led to the differential equation

$$\Psi_1(x) = \left(\sqrt{\frac{m \, \omega}{2\hbar}} \, x - \sqrt{\frac{\hbar}{2m \, \omega}} \, \frac{d}{dx} \right) \Psi_0(x) \, . \tag{16.38}$$

By differentiating (16.37) and inserting th result into (16.38), we obtain the first excited state:

$$\Psi_1(x) = \sqrt{\frac{2m \, \omega}{\hbar}} \, x \, \Psi_0(x) \, . \tag{16.39}$$

We can thus construct, step-by-step, the wave functions of the excited states of the one-dimensional harmonic oscillator in configuration space.

To simplify the notation, we change variables and functions by setting

$$X = \sqrt{\frac{m\,\omega}{\hbar}}\,x, \qquad u_n(X) = \langle X \mid n\rangle. \tag{16.40}$$

The unit-normalized function $u_n(X)$, corresponding to the energy eigenvalue $E_n = \left(n + \frac{1}{2}\right)\hbar\,\omega$, is the solution of eigenvalue Eq. (16.24) in which H is expressed with (16.1) and we change coordinates according to (16.40):

$$\frac{1}{2}\left(X^2 - \frac{d^2}{d\,X^2}\right)u_n(X) = \left(n + \frac{1}{2}\right)\hbar\,\omega\,u_n(X). \tag{16.41}$$

The functions $u_n(x)$ constitute by construction a complete orthonormal basis:

$$\sum_{n=0}^{\infty} u_n^*(X)\,u_n(X') = \sum_n \langle X' \mid n\rangle \langle n \mid X\rangle = \langle X' \mid X\rangle = \delta(X'-X), \tag{16.42}$$

$$\int_{-\infty}^{+\infty} u_n^*(X)\,u_p(X)\,dX = \int_{-\infty}^{+\infty} \langle n \mid X\rangle\,dX\,\langle X \mid p\rangle = \langle n \mid p\rangle = \delta_{np}. \tag{16.43}$$

The projection onto configuration space $\{X\}$ of relations (16.19) and (16.22) defining the annihilation and creation operators gives the first-order differential equations constituting the recursion relations for the functions $u_n(X)$:

$$\frac{1}{\sqrt{2}}\left(X + \frac{d}{dX}\right)u_n(X) = \sqrt{n}\,u_{n-1}(X),$$
$$\frac{1}{\sqrt{2}}\left(X - \frac{d}{dX}\right)u_n(X) = \sqrt{n+1}\,u_{n+1}(X). \tag{16.44}$$

Calculating the term-by-term sum of these two equations, we have

$$X\,u_n(X) = \sqrt{\frac{n}{2}}\,u_{n-1}(X) + \sqrt{\frac{n+1}{2}}\,u_{n+1}(X). \tag{16.45}$$

We can use the recursion relations to construct the analytical form of the $u_n(X)$. This involves the introduction of the nth-order Hermite polynomials generated by the relation

$$H_n(X) = (-)^n\,e^{X^2}\,\frac{d^n}{dX^n}\left(-e^{X^2}\right) \quad n = 0, 1, 2 \ldots \infty. \tag{16.46}$$

The Hermite polynomials have n zeros and a $(-)^n$ parity. We note the following special values:

$$H_0(X) = 1,$$
$$H_1(X) = 2X,$$
$$H_2(X) = 4X^2 - 2. \tag{16.47}$$

The general form of the $u_n(X)$ with $(-)^n$ parity such as $H_n(X)$ is then written, after normalization,

$$u_n(X) = (\sqrt{\pi}\, 2^n\, n!)^{-1/2}\, e^{-X^2}\, H_n(X).$$

(16.48)

2. The p-Dimensional Harmonic Oscillator

2.1 General Formulation

In the one-dimensional problem, we constructed the eigenstates of the harmonic oscillator via the particle number operator $N = a^+\, a$. We will generalize this method to the case of a p-dimensional harmonic oscillator by setting

$$H = \sum_{i=1}^{p} H_i \qquad \text{with} \qquad H_i = \frac{1}{2m}\,(p_i^2 + m^2\,\omega^2\,q_i^2).$$

(16.49)

If H_i is defined in the space of states in terms of the variables p_i and q_i, then the space of the dynamical states envisaged will be the tensor product of all existing $\mathcal{E}_i$, that is,

$$\mathcal{E} = \mathcal{E}_1 \otimes \mathcal{E}_2 \otimes \ldots \otimes \mathcal{E}_p$$

and the vector $\mid n_1,\ldots,n_p\rangle = \mid n_1\rangle \ldots \mid n_p\rangle$ will be the eigenstate of the overall Hamiltonian, with, for each i state,

$$H_i \mid n_i\rangle = \left(n_i + \frac{1}{2}\right)\hbar\omega \mid n_i\rangle \qquad i = 1, 2, \ldots p.$$

(16.50)

This will lead, with (16.49), to the eigenvalue equation

$$H \mid n_1 \ldots n_p\rangle = \left(n_1 + \ldots n_p + \frac{p}{2}\right)\hbar\omega \mid n_1 \ldots n_p\rangle.$$

(16.51)

The corresponding energy $E_n = (n + p/2)\hbar\,\omega = (n_1 + \ldots n_p + p/2)\,\hbar\,\omega$ depends solely on the sum $n = n_1 + \ldots n_p$ of the p positive integers $n_1 \ldots n_p$. There exist $C_{n+p-1}^n = (n + p - 1)!/[n!\,(p-1)!]$ different possible values and, hence, the eigenvalue $(n + p/2)\,\hbar\,\omega$ of the p-dimensional oscillator will be C_{n+p-1}^n-fold degenerate.

We introduce the type i creation and annihilation operators. They satisfy the commutation relations

$$[a_i,\, a_j] = [a_i^+,\, a_j^+] = 0$$
$$[a_i,\, a_j^+] = \delta_{ij}\,\mathbf{1} \qquad i,j = 1,\ldots,p.$$

(16.52)

Let $\mid 0\rangle$ be the eigenvector of the ground state, while $\mid 0\rangle \equiv \mid \overset{p\ \text{times}}{0\ \ldots\ 0}\rangle$ being the expression for the quantum vacuum. When relations (16.26) are applied to each of the i states, we obtain an eigenvector of H:

$$a_1 \mid 0\rangle = a_2 \mid 0\rangle \ldots = a_p \mid 0\rangle = 0, \tag{16.53}$$

$$\mid n_1 \ldots n_p\rangle = (n_1! \ldots n_p!)^{-1/2} \, a_1^{+n_1} \ldots a_p^{+n_p} \mid 0\rangle. \tag{16.54}$$

The operator for the total number of quanta, n, eigenvalue of N, will be the sum of n_i eigenvalues of the number of type i quanta N_i:

$$N = \sum_{i=1}^{p} a_i^+ a_i = \sum_i N_i. \tag{16.55}$$

2.2 Two-Dimensional Harmonic Oscillator

We can use the above case study, that is, Eq. (16.51) for $p = 2$, to set up a table of the eigenvalues and eigenvectors of H (Table 1).

Table 1. Eigenvalues and eigenvectors of the two-dimensional harmonic oscillator.

Eigenvalues of H	Degeneracy	Eigenvectors common to N_1 and N_2
$\hbar\,\omega$	1	$\mid 00\rangle$
$2\hbar\,\omega$	2	$\mid 10\rangle \mid 01\rangle$
$3\hbar\,\omega$	3	$\mid 20\rangle \mid 11\rangle \mid 02\rangle$
$\ldots$	$\ldots$	$\ldots$
$(n+1)\,\hbar\,\omega$	$n+1 = C_{n+1}^n$	$\mid n\,0\rangle \mid n-1\,1\rangle \ldots \mid n-s\,s\rangle$ $\ldots \mid 0n\rangle$

A two-dimensional "angular momentum" is defined by the relation

$$L = \frac{1}{\hbar}\,(q_1\,p_2 - q_2\,p_1) = i\,(a_1\,a_2^+ - a_1^+\,a_2). \tag{16.56}$$

The operator L commutes with H; it is therefore a constant of the motion. We will show that N and L constitute another complete set of commuting (or compatible) observables.

We will introduce for the purpose two types of creation and annihilation operators:

$$A_{\pm} = \frac{1}{\sqrt{2}} (a_1 \mp i\, a_2),$$

$$A_{\pm}^+ = \frac{1}{\sqrt{2}} (a_1^+ \pm i\, a_2^+).$$

(16.57)

It is now easy to determine the commutation relations:

$$[A_r,\, A_s] = [A_r^+,\, A_s^+] = 0$$
$$[A_r,\, A_s^+] = \delta_{r\,s}\, \mathbb{1} \qquad\qquad r = \pm,\ s = \pm.$$

(16.58)

We also introduce the type $+$ and type $-$ $N_+ = A_+^+ A_+$ and $N_- = A_-^+ A_-$ "number of quanta" operators.

These eigenvalues can be studied in the same manner as above, in a one-dimensional space, and the study leads us to write

$$N_+ \mid n_+\, n_- \rangle = n_+ \mid n_+\, n_- \rangle,$$
$$N_- \mid n_+\, n_- \rangle = n_- \mid n_+\, n_- \rangle,$$

(16.59)

with the eigenvalues n_+ and n_- being real numbers and integers:

$$n_+ = 0, 1, 2, .. \qquad \text{and} \qquad n_- = 0, 1, 2, \dots.$$

(16.60)

Using the definition of the vacuum of the $+$ state or $-$ state:

$$A_+ \mid 00 \rangle = A_- \mid 00 \rangle = 0,$$

(16.61)

we obtain the eigenfunctions of N_+ and N_-:

$$\mid n_+\, n_- \rangle = (n_+ !\, n_- !)^{-\frac{1}{2}}\, A_+^{+n_+}\, A_-^{+n_-} \mid 00 \rangle.$$

(16.62)

The operators N, L and H can be expressed with A and A^+ using relations (16.56) and (16.57) and the definitions of N_+ and N_-:

$$L = i\, (a_1 a_2^+ - a_1^+ a_2) = N_+ - N_-$$
$$N = a_1^+ a_1 + a_2^+ a_2 = N_+ + N_-$$
$$H = (N_+ + \frac{1}{2}\, \mathbb{1} + N_- + \frac{1}{2}\, \mathbb{1})\, \hbar\omega$$
$$= (N_+ + N_- + \mathbb{1})\, \hbar\omega = (N + \mathbb{1})\, \hbar\omega.$$

(16.63)

The operators N_+ and N_- constitute a complete set of commuting observables. This is also true of the operators L and N:

$$[N_+, N_-] = 0 \quad \text{and} \quad [L, N] = 0 \quad \text{imply} \quad [H, N] = 0 \quad \text{and} \quad [H, L] = 0. \quad (16.64)$$

We can therefore proceed to write the eigenvalue equations

$$
\begin{aligned}
H \mid n_+ \, n_- \rangle &= (n_+ + n_- + 1) \, \hbar \, \omega \mid n_+ \, n_- \rangle, \\
L \mid n_+ \, n_- \rangle &= (n_+ - n_-) \mid n_+ \, n_- \rangle.
\end{aligned}
\tag{16.65}
$$

N_+ is interpreted as the number of particles with positive charge and N_- as the number of particles with negative charge, with operator L representing the total charge operator.

In the theory of crystal vibrations, the motions of lattices are represented with the two-dimensional isotropic oscillators and the oscillation quanta are termed phonons. Type $+$ or type $-$ phonons then represent traveling waves that are propagating in opposite directions.

2.3 Three-Dimensional Oscillator

We now write the Hamiltonian as follows:

$$
H = \frac{p^2}{2\mu} + \frac{1}{2} \, \mu \, \omega^2 \, r^2 = H_x + H_y + H_z.
\tag{16.66}
$$

The notation has been changed to avoid any confusion between the reduced mass μ and the orbital magnetic quantum number m.

The eigenvalues of H will be $\left(n + \frac{3}{2} \right) \hbar \, \omega$ and they are $\frac{1}{2} \, (n+1) \, (n+2)$-fold degenerate. The base vectors will be in the Cartesian basis:

$$
\mid n_x \, n_y \, n_z \rangle = (n_x! \, n_y! \, n_z!)^{-1/2} \, a_x^{+n_x} \, a_y^{+n_y} \, a_z^{+n_z} \mid 000 \rangle.
\tag{16.67}
$$

In the standard spherical basis, the radial and angular parts separate and we write them with the spherical harmonics as follows:

$$
\langle r \, \Theta \, \varphi \mid n \, \ell \, m \rangle = \Psi_{n \, \ell \, m} \, (r, \, \Theta, \, \varphi) = u_\ell^n(r) \, Y_{\ell \, m}(\Theta, \, \varphi).
\tag{16.68}
$$

The quantum numbers n, ℓ and m assume the values:

$$
\begin{aligned}
&n \geq \ell + 1 \quad \text{azimuthal or principal quantum number}, \\
&\ell = 0, 1, 2, \ldots \quad \text{orbital quantum number}, \\
&-\ell \leq m \leq +\ell \quad \text{magnetic quantum number}.
\end{aligned}
\tag{16.69}
$$

The functions $u_\ell^n(r)$ are obtained by solving the radial equation and using $\alpha = \mu \, \omega / \hbar$. One finds

$$
u_\ell^n(r) = \frac{\alpha^{\ell+3/2}}{\pi^{1/4}} \left(\frac{2^{n+\ell+1}(n-1)!}{(2n+2\ell-1)!!} \right)^{1/2} r^\ell \, \exp\left(\frac{-\alpha^2 \, r^2}{2} \right) L_{n-1}^{\ell+1/2} (\alpha^2 \, r^2).
\tag{16.70}
$$

These are unit-normalized functions

$$\int_0^\infty u_\ell^n(r) \, u_\ell^n(r) \, r^2 \, dr = 1 \,, \tag{16.71}$$

corresponding to the eigenvalue $E_n = \hbar \, \omega \, (2n + \ell - \frac{1}{2})$ of the oscillator energy.

The functions $L_{n-1}^{\ell+1/2} \, (\alpha^2 \, r^2)$ are ℓth-order Laguerre polynomials.

The three-dimensional harmonic oscillator plays an important role in the shell model of the nucleus (Ripka and Blaizot, 1978).

3. Bosons and Fermions

Let us define a particle as a well-defined quantum state, localized in space-time, and characterized by the elements that are conserved over time (invariants): restmass; charges [electric, weak (or neutral), leptonic, baryonic]; spin; parity .... .

Each species of particle i is defined by a statistical distribution function in a gas in equilibrium at the temperature T:

$$f \, (x \, z_i \, \xi_i; q) = \left[\exp \, (\sqrt{x^2 + z_i^2} - \xi_i) - q \right]^{-1}. \tag{16.72}$$

This function determines the numerical density n_i, the pressure P_i, and the energy density ϱ_i of these particles at a temperature T. The parameter q is a number equal to ± 1, which determines the type, boson or fermion, of the particles. In the statistical distribution function, we have set

$$x = \frac{p \, c}{k_B \, T} \qquad z_i = \frac{m_i \, c^2}{k_B \, T} \qquad \text{and} \qquad \xi_i = \frac{\mu_i}{k_B \, T}. \tag{16.73}$$

The parameter μ_i represents the chemical potential of a type i particle and k_B is Boltzmann's constant:

$$k_B = 8.617 \; 658 \; 10^{-5} \; \text{eV K}^{-1} = 1.3801 \; 10^{-23} \text{J K}^{-1}. \tag{16.74}$$

The restmass m_i of a particle is equivalent to an energy with mass $m_i \, c^2$ or to a formation temperature

$$T_i \, (\text{in K}) = \frac{m_i \, c^2}{k_B} = 1.16 \; 10^{10} \; m_i \, (\text{in MeV}/c^2). \tag{16.75}$$

We thus speak of the π-meson with mass $m_\pi \simeq 140 \; \text{MeV}/c^2$ or with formation temperature 10^{12} K. The number g_i of degrees of spin freedom for mass particles with spin S_i is equal to $(2 \, S_i + 1)$; for particles with zero mass (at rest), such as photons, the number of degrees of spin freedom is $g_i = 2$.

A knowledge of the distribution function $f(x\, z_i\, g_i; \varepsilon)$ completely determines the statistical characteristics of a gas of type i particles. By setting

$$\frac{g_i}{2\pi^2}\left(\frac{k_B\, T}{\hbar\, c}\right)^3 = g\,(g_i, T),\qquad (16.76)$$

we obtain the numerical density, number of type i particles per unit volume:

$$n_i = g\,(g_i, T)\int_0^\infty f\,(x\, z_i\, \xi_i; q)\, x^2\, dx,\qquad (16.77)$$

the energy density, energy of type i particles per unit volume:

$$\varrho_i = g(g_i, T)\int_0^\infty f(x\, z_i\, \xi; q)\, x^2\, \sqrt{x^2 + z_i^2}\, dx,\qquad (16.78)$$

and the pressure of the particles i in a volume element:

$$P_i = \frac{1}{3}\, g(g_i, T)\int_0^\infty f\,(x\, z_i\, \xi_i; q)\, \frac{x^4}{\sqrt{x^2 + z_i^2}}\, dx.\qquad (16.79)$$

Because the overall densities and pressures are arithmetic sums over particle type of the above observables:

$$n = \sum_i n_i\qquad \varrho = \sum_i \varrho_i\qquad P = \sum_i P_i.\qquad (16.80)$$

We will now examine the possible constituents of this gas of particles and their fundamental interactions.

3.1 Fermions

The matter constituting the Universe is made up of fermions. Fermions are particles with mass m_i (at rest), spin $\frac{1}{2}$, and charge Q (leptonic, baryonic, electric, weak, color or gravitational).

These particles obey Fermi–Dirac statistics for which the parameter q takes the value -1 defining the statistical distribution function $f\,(x\, z_i\, \xi_i; -1) = [\exp() + 1]^{-1}$. The wave function of a fermion system is antisymmetric with respect to the exchange of the position of two fermions, implying that fermions obey the Pauli Exclusion Principle.

In Chap. 9, we described the two main classes of fermions in nature, leptons and hadrons, the former being insensitive to strong interaction and the latter being sensitive to all forms of fundamental interactions (Table 2).

Electrically charged leptons (electrons, muons, tau) are massive whereas their corresponding neutrinos possess only a small mass and, maybe, even none at all.

Table 2. The standard model of fundamental interactions

The masses attributed in 1996 to leptons and quarks, in GeV/c^2 $(= 1.782\ 662\ 70 \times 10^{-24}\text{g})$ are as follows:

	1st generation	2nd generation	3rd generation
Leptons	$m_{\nu_e} \langle 7 \cdot 10^{-9}$ $m_e = 5.1 \cdot 10^{-4}$	$m_{\nu_\mu} \langle 3 \cdot 10^{-4}$ $m_\mu = 0.106$	$m_{\nu_\tau} \langle 4 \cdot 10^{-2}$ $m_\tau = 1.777$
Quarks	$m_u = 5 \cdot 10^{-3}$ $m_d = 9 \cdot 10^{-3}$	$m_c = 1.350$ $m_s = 0.175$	$m_t = 175 \pm 6$ $m_b = 4.5$

3.2 Bosons Mediating Fundamental Interactions

These are particles with integer spin obeying a Bose-Einstein statistical distribution in which the coefficient q is equal to $+1$, thus defining the statistical distribution function $f\ (x\ z_i\ \xi_i; 1) = [\exp() - 1]^{-1}$. The wave function of a boson system is symmetrical with respect to the exchange of the position of two of them, and bosons are not subjected to the Pauli exclusion principle.

By bringing together an even number of fermions, we can create a boson, for example, a deuteron comprising a neutron and a proton, or an α-particle comprising two protons and two neutrons. Perhaps even more interesting are the bosons that mediate the fundamental interactions, i.e. the gauge bosons. From the quantum-mechanical point of view, a fundamental interaction corresponds to the exchange of virtual bosons between fermions. If the restmass of these bosons is m_B, then the range of the interaction will be

$$\lambda_B = \frac{\hbar\, c}{m_B\, c^2} = \frac{197}{m_B\, c^2} \frac{(\text{MeV fm})}{(\text{MeV})} = \frac{3 \cdot 10^{-43}}{m_B} \frac{\text{kg m}}{(\text{kg})}. \tag{16.81}$$

Thus, a photon with zero mass confers an infinite range to the electromagnetic interaction, a meson with mass 140 MeV/c^2 corresponds to a range on the order of 1.4 fermi (1 fermi $= 10^{-13}$ cm), whereas an intermediate boson with mass 90 GeV/c^2 confers a range of about $2 \cdot 10^{-3}$ fm, that is, about 10^{-16} cm.

The strength of the interaction between a fermion with charge Q and another with charge Q' is measured by the dimensionless number:

$$\alpha_f = \frac{Q\, Q'}{\hbar\, c} = \frac{Q\, Q'}{197} \frac{(\text{MeVfm})}{(\text{MeVfm})}. \tag{16.82}$$

For the exchange of only one virtual boson (a first-order term in the Feynman perturbation expansion), a fundamental interaction can be expressed with the attractive or repulsive potential

$$V_f = \pm\, \alpha_f\, \frac{e^{-r/\lambda_B}}{r} = \pm\, \frac{Q\, Q'}{\hbar\, c} \frac{e^{-r/\lambda_B}}{r}. \tag{16.83}$$

Fundamental interactions with infinite range, carried by bosons with zero mass will thus be written as follows:

$$V_f = \pm\, \alpha_f\, \frac{1}{r}. \tag{16.84}$$

We can include in this category:

i) The gravitational interaction carried by gravitons, bosons with zero mass, with a non-zero gravitational charge, and spin 2. This interaction is always attractive and has a strength characterized by the number

$$\alpha_G = \frac{Q_G\, Q'_G}{\hbar\, c} = \frac{G\, m\, m'}{\hbar\, c}. \tag{16.85}$$

By evaluating the gravitational interaction between a proton with mass

$$m_p = 1.672\ 623\ 10^{-27}\ \text{kg} = 938.27231\ (\text{MeV}/c^2),$$

and an electron with mass

$$m_e = 9.109\ 389\ 10^{-31}\ \text{kg} = 0.510\ 999\ 06\ (\text{MeV}/c^2), \tag{16.86}$$

we obtain the strength of the gravitational interaction

$$\alpha_G \simeq 10^{-42}. \tag{16.87}$$

ii) The electromagnetic interaction borne by virtual photons, gauge bosons (see Chap. 8) of the electromagnetic field, particles with zero mass and spin 1, which can be attractive (when the electrical charges have opposite signs) or repulsive (electrical charges of the same sign). The intensity of the electromagnetic interaction between an electron and a proton (attractive force) is easily evaluated as

$$\alpha_{em} = \frac{Q\,Q'}{\hbar\,c} = \frac{e^2}{\hbar\,c} \simeq 10^{-2}\,. \tag{16.88}$$

This is the finestructure constant of the electromagnetic interaction, which is known to great precision; expressed in MKS units, it has the value

$$\alpha = \frac{e^2}{4\pi\,\varepsilon_0\,\hbar c} = 1/137.035\ 989\ 5\,.$$

Heisenberg's uncertainty relation enables us to understand why the gravitational and electromagnetic interactions that are propagated at the speed of light c lead to forces behaving as $1/r^2$. Within a domain of size $\Delta\,x \simeq r$, exchange bosons will have the following momentum:

$$\Delta\,x\,\Delta\,p_x \simeq \hbar \quad \text{that is} \quad \Delta\,p_x \simeq \frac{\hbar}{r}\,. \tag{16.89}$$

The time required for a boson to cover the distance $\Delta\,x = r$, at the speed c, will be $\Delta\,t = r/c$. We infer from this the force applied to the exchange boson:

$$F_x = \frac{\Delta\,p_x}{\Delta\,t} \simeq \frac{\hbar\,c}{r^2}\,. \tag{16.90}$$

iii) The third force carried by massless bosons is the strong interaction or color interaction, carried by gluons G_i^j, field gauge bosons that change the color j of a quark q to a color i:

$$G_i^j\,q_j = q_i \qquad i \neq j = 1, 2, 3\,. \tag{16.91}$$

The strength of the strong (color) interaction is

$$\alpha_S = 0.1134 \pm 0.0035\,. \tag{16.92}$$

The confinement force between the quarks, which is proportional to their separation, makes them unobservable outside the nuclei, in the free state. We notice that the series expansion of the fundamental interaction (16.83) yields $\pm\,\alpha_f\,\frac{1}{r}$ as the first term, but only by further introducing a constant term in α_f/λ_B, and then a term proportional to r in α_f/λ_B^2, responsible for the confinement. This is the approximate form of the strong color interaction usually chosen empirically, and which adequately accounts for the confinement of quarks inside the nuclei:

$$V_f = \frac{A}{r} + Br\,. \tag{16.93}$$

The coefficients A and B are determined in a purely phenomenological manner.

The strong nuclear interaction is carried by massive mesons. These particles represent a structure linked by the color exchange of a quark with flavor f (that is of the u, d, s, c, b, t type) and color i (r, y, g red, yellow, green) and of an antiquark f' (that is of the $\bar{u}, \bar{d}, \bar{s}, \bar{c}, \bar{b}, \bar{t}$ type) and an anticolor i (a complementary color of i):

$$M = \sum_i f_i \, \bar{f}'^{\,i} \tag{16.94}$$

Mesons are not colored (they may be considered as being white) and, hence, observable. Stable mesons possess spin 0 and isospin 0, 1, or $\frac{1}{2}$. In this way, we can observe the π-mesons with mass about 140 meV/c^2 and isospin 1:

$$\pi^+ = u\,\bar{d} \qquad \pi^0 = \frac{1}{\sqrt{2}}\left(u\,\bar{u} - d\,\bar{d}\right) \qquad \pi^- = \bar{u}\,d, \tag{16.95}$$

the η-mesons with mass 547.45 MeV/c^2 and isospin 0:

$$\eta = c_1 \left(u\,\bar{u} + d\,\bar{d}\right) + c_2 \left(s\,\bar{s}\right), \tag{16.96}$$

the K-mesons with isospin $\frac{1}{2}$ and mass about 490 MeV/c^2:

$$K^+ = u\,\bar{s} \qquad K^- = \bar{u}\,s \qquad K^0 = d\,\bar{s} \qquad \text{and} \qquad \bar{K}^0 = \bar{d}\,s, \tag{16.97}$$

and then the D-mesons with isospin $\frac{1}{2}$, constructed with first- and second-generation quarks and antiquarks and finally the B-mesons, also with isospin $\frac{1}{2}$, constructed with first- and third-generation quarks and antiquarks.

The strong nuclear interaction is an effective color interaction that is analogous to the van der Waals interaction between molecules. Its strength is about one.

iv) The weak nuclear interaction responsible for radioactive phenomena is carried by the intermediate bosons W^+, W^-, and Z^0, gauge bosons of the weak interaction, particles with spin 1 and mass:

$$\begin{aligned} m_W &= 80.356 \ \pm \ 0.125 \ \text{GeV}/c^2, \\ m_{Z^0} &= 91.1863 \ \pm \ 0.0020 \ \text{GeV}/c^2. \end{aligned} \tag{16.98}$$

Intermediate bosons were first characterized at CERN in Geneva in 1984. These particles are not directly observable, because their lifetime is about 10^{-24}s and their path of about 10^{-14} cm is smaller than the diameter of the electron. Their existence can only be inferred from the products of their disintegration. Since then, these particles have been abundantly produced in $e^+ e^-$ experiments with the LEP accelerator at CERN. The range of the weak interaction is about 10^{-16} cm, and its strength about 1/30. One is inclined to think that the weak nuclear interaction is that part of the interaction of a fundamental interaction that can be termed hypercolor and a number of theories (Elbaz, 1989) attributing a common sub-structure (termed rishon by H. Harari) to leptons and quarks have been developed. But in the absence of experimental evidence in favor of the existence of excited states of intermediate bosons, such theories cannot yet be accepted. Experimental developments with future high-energy particle accelerators will no doubt provide answers to these questions.

4. Fermion Systems

Although the Dirac formalism is well-adapted to the description of individual quantum states, it is too cumbersome for the description of a large number of particles: fermions in one nucleus, or exchange bosons between the fermions. A better suited formalism has therefore been introduced. It is based on creation and annihilation operators from a formal quantum vacuum and termed the second quantization formalism. It will be detailed in Sect. 4.3 below.

4.1 Pauli Exclusion Principle

Fermions, particles with half integer spin obeying Fermi–Dirac statistics, are described by antisymmetric wave functions and subject to an exclusion principle termed the Pauli exclusion principle:

$$\Psi\left(\vec{r}_1,\ \vec{r}_2\right) = -\Psi\left(\vec{r}_2,\ \vec{r}_1\right). \tag{16.99}$$

Consider a quantum system comprising a large number of fermions in interaction. This might, for example, be a nucleus comprising Z protons and N neutrons (with atomic number $A = Z + N$), or an atom in which the nucleus with charge Ze interacts with the Z electrons of the electron cloud. Taken individually, each fermion can be described by a wave function $\varphi_\alpha\left(x_i\right)$ of the quantum state φ_α at point x_i. The system formed by the individual wave functions $\varphi_\alpha\left(x_i\right)$ is a complete orthonormal system. We can therefore form, from these functions, an overall wave function of the n fermions which should be antisymmetric in order to satisfy the Pauli exclusion principle. We will, for example, write

$$\Psi_{\alpha\,\beta\,\gamma\,...}\left(x_1,\ x_2\,...\right) = \mathcal{A}\,\varphi_\alpha\left(x_1\right)\varphi_\beta\left(x_2\right)...$$

$$= \frac{1}{\sqrt{n\,!}}\sum_p \varepsilon_p\,P\left(i\,j\,k\,...\right)\varphi_\alpha\left(x_i\right)\varphi_\beta\left(x_j\right)... \tag{16.100}$$

where ε_p defines the parity ± 1 of the permutation $P(i\,j\,...)$ of the n particles, and $\mathcal{A}$ the antisymmetrization operator.

For a two-fermion system, for example, we have

$$\Psi_{\alpha\beta}(x_1, x_2) = \frac{1}{\sqrt{2!}} \left[\varphi_\alpha(x_1)\,\varphi_\beta(x_2) - \varphi_\alpha(x_2)\varphi_\beta(x_1) \right], \qquad (16.101a)$$

and for a three-fermion system,

$$\begin{aligned}
\Psi_{\alpha\beta\gamma}(x_1, x_2, x_3) = \frac{1}{\sqrt{3!}} \Big[&\varphi_\alpha(x_1)\,\varphi_\beta(x_2)\,\varphi_\gamma(x_3) \\
&+ \varphi_\alpha(x_2)\,\varphi_\beta(x_3)\,\varphi_\gamma(x_1) \\
&+ \varphi_\alpha(x_3)\,\varphi_\beta(x_1)\,\varphi_\gamma(x_2) \\
&- \varphi_\alpha(x_1)\,\varphi_\beta(x_3)\,\varphi_\gamma(x_2) \\
&- \varphi_\alpha(x_3)\,\varphi_\beta(x_2)\,\varphi_\gamma(x_1) \\
&- \varphi_\alpha(x_2)\,\varphi_\beta(x_1)\,\varphi_\gamma(x_3) \Big]. \qquad (16.102a)
\end{aligned}$$

We can also fix the positions x_1, x_2, and x_3 and then modify the position of the occupied states to obtain an equivalent form of these wave functions:

$$\Psi_{\alpha\beta}(x_1, x_2) = \frac{1}{\sqrt{2!}} \left[\varphi_\alpha(x_1)\,\varphi_\beta(x_2) - \varphi_\beta(x_1)\,\varphi_\alpha(x_2) \right] \qquad (16.101b)$$

and for a three-fermion system:

$$\begin{aligned}
\Psi_{\alpha\beta\gamma}(x_1, x_2, x_3) = \frac{1}{\sqrt{3!}} \Big[&\varphi_\alpha(x_1)\,\varphi_\beta(x_2)\,\varphi_\gamma(x_3) \\
&+ \varphi_\beta(x_1)\,\varphi_\gamma(x_2)\,\varphi_\alpha(x_3) \\
&+ \varphi_\gamma(x_1)\,\varphi_\alpha(x_2)\,\varphi_\beta(x_3) \\
&- \varphi_\alpha(x_1)\,\varphi_\gamma(x_2)\,\varphi_\beta(x_3) \\
&- \varphi_\gamma(x_1)\,\varphi_\beta(x_2)\,\varphi_\alpha(x_3) \\
&- \varphi_\beta(x_1)\,\varphi_\alpha(x_2)\,\varphi_\gamma(x_3) \Big] \qquad (16.102b)
\end{aligned}$$

It is easily noticed that the antisymmetrized wave function thus constructed can take the form of a determinant of the individual wave functions; it is called the Slater determinant:

$$\Psi_{\alpha\beta\gamma\ldots}(x_1, x_2 \ldots) = \frac{1}{\sqrt{n!}} \begin{vmatrix} \varphi_\alpha(x_1) & \varphi_\alpha(x_2) & \cdots \\ \varphi_\beta(x_1) & \cdots & \\ \varphi_\gamma(x_1) & \cdots & \end{vmatrix}. \qquad (16.103)$$

It follows from the above that Ψ will change sign if two states corresponding to two columns or two rows of the Slater determinant are interchanged, and that $\Psi = 0$ if two states are identical. This is the Pauli exclusion principle.

4.2 Occupied States

Let us write the wave function as follows:

$$\Psi_{\alpha\,\beta\,\gamma...}\,(x_1, x_2, x_3, ...) = |\,\alpha\,\beta\,\gamma...\rangle \qquad (16.104)$$

This simply means that the states α β $\gamma...$ are occupied in $x_1\ x_2\ x_3...$ (in that order) just like in (16.101 b) and (16.102b), for example. A more orthodox notation would in fact be $\Psi_{\alpha\,\beta\,\gamma\,...}\,(x_1, x_2, x_3, ...) = \langle x_1, x_2, x_3, ...\ |\,\alpha\,\beta\,\gamma\,...\rangle$ in which we specify the projection space.

Because the wave function is antisymmetric, the permutation of two individual states will change its sign

$$|\,\alpha\,\beta\,\gamma\,...\rangle = -\,|\,\beta\,\alpha\,\gamma\,...\rangle = -\,|\,\alpha\,\gamma\,\beta\,...\rangle. \qquad (16.105)$$

We will, more generally, write

$$|\,\alpha\,\beta\,\gamma\,...\rangle = \varepsilon_p\,|\,P\,\alpha\,\beta\,\gamma\,...\rangle. \qquad (16.106)$$

If we make an even number of permutations $\varepsilon_p = +1$ and $\varepsilon_p = -1$ for an odd number of permutations of the states $\alpha, \beta...$ at positions $x_1, x_2...$.

The problem at hand now is to expand the operators in this representation, in particular, the single- and two-body potentials $\sum_i v\,(r_i)$ and $\sum_{i<j} v\,(r_{ij})$, respectively.

4.3 Creation and Annihilation Operators

a) Occupation Number

Let us first define an "occupation number" operator n_λ expressing the fact that λ is occupied or not occupied:

$$\begin{aligned} n_\lambda\,|\,\alpha\,\beta...\,\lambda...\rangle &= |\,\alpha\,\beta...\,\lambda...\rangle, \\ n_\lambda\,|\,\alpha\,\beta...\ \text{not}\ \lambda...\rangle &= 0. \end{aligned} \qquad (16.107)$$

b) Creation Operators

The operators n_λ are diagonal in the above representation; they are not enough to constitute a complete set of commuting operators. We therefore introduce an operator c_α^+ creating a particle in the state α, that is, an operator such that when applied to a state in which α is not occupied, will transform the overall quantum state into a state in which α is occupied. c_α^+ is a creation operator.

To simplify the notation, we also introduce a vacuum state $|\ 0\rangle$ in which none of the states α, $\beta, ...$ is occupied. This therefore defines the creation operator through the relation $c_\alpha^+ \mid 0\rangle = \mid \alpha\rangle$, or, more generally speaking,

$$\mid \alpha \, \beta \, \gamma \ldots\rangle = c_\alpha^+ \, c_\beta^+ \, c_\gamma^+ \ldots \mid 0\rangle, \tag{16.108}$$

$$c_\lambda^+ \mid \alpha \, \beta \, \gamma \ldots\rangle = \mid \lambda \, \alpha \, \beta \, \gamma \ldots\rangle. \tag{16.109}$$

Definition (16.109) should be compatible with (16.101), antisymmetrical if the order of the states α in x_1 and β in x_2 is changed:

$$\mid \alpha \, \beta \, \gamma \ldots\rangle = c_\alpha^+ \, c_\beta^+ \mid \gamma \ldots\rangle = -c_\beta^+ \, c_\alpha^+ \mid \gamma \ldots\rangle.$$

We therefore obtain the relation between the creation operators:

$$c_\alpha^+ \, c_\beta^+ = -c_\beta^+ \, c_\alpha^+.$$

In this case, we say that the creation operators anticommute:

$$\boxed{\{c_\alpha^+, \, c_\beta^+\} = c_\alpha^+ \, c_\beta^+ + c_\beta^+ \, c_\alpha^+ = 0} \, . \tag{16.110}$$

The Pauli exclusion principle is reflected by the fact that two identical states cannot exist together:

$$\boxed{(c_\alpha^+)^2 = 0} \tag{16.111}$$

since (16.111) will be equivalent to (16.110) if $\alpha = \beta$. The Pauli exclusion principle is therefore contained in the anticommutation rule for creation operators.

c) Annihilation Operators

We define the Hermitian operator c_λ, conjugate of c_λ^+, that is,

$$(c_\alpha \mid 0\rangle)^+ = (\mid \alpha\rangle)^+ \quad \text{and} \quad \langle \alpha \mid = \langle 0 \mid c_\alpha. \tag{16.112}$$

This is represented by the more general form

$$\langle \alpha \, \beta \, \gamma \ldots \mid = \langle 0 \mid \ldots c_\gamma \, c_\beta \, c_\alpha, \tag{16.113}$$
$$\langle \alpha \, \beta \, \gamma \ldots \mid c_\lambda = \langle \lambda \, \alpha \, \beta \, \gamma \ldots \mid.$$

By taking the Hermitian conjugates of (16.110) and (16.111), we further obtain

$$\boxed{\begin{aligned} \{c_\alpha, \, c_\beta\} &= 0 \\ (c_\alpha)^2 &= 0 \end{aligned}} \, . \tag{16.114}$$

What does the state $c_\lambda \mid \alpha \, \beta \, \ldots \rangle$ then represent? This is easily found out by calculating its inner product with a state $\mid \alpha' \, \beta' \, \ldots \rangle$, that is, the matrix element $\langle \alpha' \, \beta' \, \ldots \mid c_\lambda \mid \alpha \, \beta \, \gamma \, \ldots \rangle$.

Following definition (16.113), $\langle \alpha' \, \beta' \, \ldots \mid c_\lambda = \langle \lambda \, \alpha' \, \beta' \, \ldots \mid$, which gives the value of the inner product as

$$\langle \alpha' \, \beta' \, \ldots \mid c_\lambda \mid \alpha \, \beta \, \gamma \, \ldots \rangle = \langle \lambda \, \alpha' \, \beta' \, \ldots \mid \alpha \, \beta \, \gamma \, \ldots \rangle .$$

The inner product will be zero except where the states $\lambda \, \alpha' \, \beta' \, \ldots$ are the same as the states $\alpha \, \beta \, \ldots$, in which case the inner product ± 1 will follow the parity of the permutation. Therefore, $\alpha \, \beta \, \gamma \, \ldots$ should contain λ.

If λ is moved to the first position, then we can write

$$c_\lambda \mid \lambda \, \alpha \, \beta \, \gamma \, \ldots \rangle = \mid \alpha \, \beta \, \gamma \, \ldots \rangle ,$$
$$c_\lambda \mid \alpha \, \beta \, \gamma \, \ldots \text{ not } \lambda \, \ldots \rangle = 0 . \tag{16.115}$$

Thus, the operator c_λ will act on a ket as an annihilation operator just like c_λ^+ acts on a bra as an annihilation operator.

Usually, we would observe the action of the operators on the kets alone and we would say:

---

c_λ^+ is a creation operator for the state λ of a ket

c_λ is an annihilation operator for the state λ of a ket

$\qquad$. $\tag{16.116}$

4.4 Operator Algebra

We can easily use the above definition to write the following relations involving the product of the operators c_λ and c_λ^+:

$$c_\lambda^+ c_\lambda \mid \lambda \, \alpha \, \beta \, \ldots \rangle = \mid \lambda \, \alpha \, \beta \, \ldots \rangle , \tag{16.117}$$

$$c_\lambda c_\lambda^+ \mid \lambda \, \alpha \, \beta \, \ldots \rangle = 0 , \tag{16.118}$$

$$c_\lambda^+ c_\lambda \mid \alpha \, \beta \, \ldots \text{ not } \lambda \, \ldots \rangle = 0 , \tag{16.119}$$

$$c_\lambda c_\lambda^+ \mid \alpha \, \beta \, \ldots \text{ not } \lambda \, \ldots \rangle = \mid \alpha \, \beta \, \ldots \text{ not } \lambda \, \ldots \rangle . \tag{16.120}$$

If we compare results (16.117) and (16.119) with definition relation (16.107) of n_λ, we infer the equivalence between the operators:

$$\boxed{c_\lambda^+ c_\lambda = n_\lambda} \quad . \tag{16.121}$$

We also note by action on a quantum state containing λ:

$$\left(c_\lambda c_\lambda^+ + c_\lambda^+ c_\lambda \right) \mid \lambda \, \alpha \, \beta \, \ldots \rangle = c_\lambda c_\lambda^+ \mid \lambda \, \alpha \, \ldots \rangle + \mid \lambda \, \alpha \, \beta \, \ldots \rangle$$
$$= 0 + \mid \lambda \, \alpha \, \beta \, \ldots \rangle ,$$

which yields the anticommutation relation

$$c_\lambda \, c_\lambda^+ + c_\lambda^+ \, c_\lambda = \{c_\lambda, \, c_\lambda^+\} = \mathbb{1}\,, \tag{16.122}$$

and with relation (16.121), the operator $c_\lambda \, c_\lambda^+$ expressed with n_λ:

$$c_\lambda \, c_\lambda^+ = \mathbb{1} - n_\lambda\,. \tag{16.123}$$

Now let us seek the commutation rule for the creation and annihilation operators for the different states λ and μ:

$$c_\lambda \, c_\mu^+ \mid \text{not } \lambda \, \ldots \, \text{not } \mu\rangle = 0 = c_\mu^+ \, c_\lambda \mid \text{not } \lambda \, \ldots \, \text{not } \mu\rangle\,,$$
$$c_\lambda \, c_\mu^+ \mid \mu \, \alpha \, \beta \, \gamma \, \ldots \, \text{not } \lambda \, \rangle = 0 = c_\mu^+ \, c_\lambda \mid \mu \, \alpha \, \beta \, \ldots \, \text{not } \lambda\rangle\,,$$
$$c_\lambda \, c_\mu^+ \mid \lambda \, \mu \, \alpha \, \beta \, \ldots\rangle = 0 = c_\mu^+ \, c_\lambda \mid \lambda \, \mu \, \alpha \, \beta \, \ldots\rangle\,,$$
$$c_\lambda \, c_\mu^+ \mid \lambda \, \alpha \, \beta \, \ldots \, \text{not } \mu\rangle = c_\lambda \mid \mu \, \lambda \, \alpha \, \beta \, \ldots\rangle = -\mid \mu \, \alpha \, \beta \, \ldots \, \text{not } \lambda\rangle\,,$$
$$c_\mu^+ \, c_\lambda \mid \lambda \, \alpha \, \beta \, \ldots \, \text{not } \mu \, \rangle = c_\mu^+ \mid \alpha \, \beta \, \gamma \, \ldots \, \text{not } \mu \, \ldots \, \text{not } \lambda\rangle\,,$$
$$= \mid \mu \, \alpha \, \beta \, \ldots \, \text{not } \lambda\rangle\,. \tag{16.124}$$

We notice that in all cases $c_\lambda \, c_\mu^+ + c_\mu^+ \, c_\lambda = 0$, giving in the final analysis, and by recalling (16.110), (16.114) and (16.122), the anticommutation relations

$$\{c_\lambda^+, \, c_\mu^+\} = 0\,, \tag{16.125}$$
$$\{c_\lambda, \, c_\mu\} = 0\,, \tag{16.126}$$
$$\{c_\lambda, \, c_\mu^+\} = \delta_{\lambda \, \mu} \, \mathbb{1}\,. \tag{16.127}$$

4.5 Creation and Annihilation Operators for an Arbitrary State

Consider a quantum state a that is not a member of the set $\alpha \, \beta \, \gamma \, \ldots$ of occupied states, the latter constituting a complete orthonormal system permitting the projection of the quantum state $\mid a\rangle$ onto such a basis.

We use the usual notation of occupied states with the antisymmetry operator $\mathcal{A}$ and the individual wave functions $\varphi_\alpha(x_i)$:

$$c_a^+ \mid \alpha \, \beta \, \ldots\rangle = \mid a \, \alpha \, \beta \, \ldots\rangle$$
$$= \mathcal{A} \, \varphi_a \, (x_1) \, \varphi_\alpha \, (x_2) \, \ldots \tag{16.128}$$

By using the Dirac notation for the individual wave functions, we can introduce the complete set λ of the occupied states:

$$\varphi_a \, (x) = \langle x \mid a\rangle = \sum_\lambda \langle x \mid \lambda\rangle \, \langle \lambda \mid a\rangle$$
$$= \sum_\lambda \varphi_\lambda \, (x) \, \langle \lambda \mid a\rangle\,. \tag{16.129}$$

Inserting this expression into (16.128), we obtain

$$| a\,\alpha\,\beta\,\ldots\rangle = \sum_{\lambda} | \lambda\,\alpha\,\beta\,\ldots\rangle \langle \lambda\,|\,a\rangle, \qquad (16.130)$$

or, better still, by introducing creation operators c_a^+ and c_λ^+:

$$c_a^+ | \alpha\,\beta\,\ldots\rangle = \sum_{\lambda} c_\lambda^+ | \alpha\,\beta\,\ldots\rangle \langle \lambda\,|\,a\rangle. \qquad (16.131)$$

Because this result is true irrrespective of the state $| \alpha\,\beta\,\ldots\rangle$, we are left with

$$\boxed{c_a^+ = \sum_{\lambda} c_\lambda^+ \langle \lambda\,|\,a\rangle} \; . \qquad (16.132)$$

The operator c_λ^+ will then cause a state λ among all the possible vacant states to be occupied.

Using the Hermitian conjugate of definition (16.132), we also obtain

$$\boxed{c_a = \sum_{\lambda} \langle a\,|\,\lambda\rangle\, c_\lambda} \; . \qquad (16.133)$$

The anticommutation relations for the creation and annihilation operators of any state whatsoever can be determined with the aid of (16.125), (16.126) and (16.127) and definitions (16.132) and (16.133):

$$\{c_a^+,\, c_b^+\} = \sum_{\lambda\lambda'} \{c_\lambda^+,\, c_{\lambda'}^+\} \langle \lambda\,|\,a\rangle \langle \lambda'\,|\,b\rangle = 0\,,$$

$$\{c_a,\, c_b\} = \sum_{\lambda\lambda'} \{c_\lambda,\, c_{\lambda'}\} \langle a\,|\,\lambda\rangle \langle b\,|\,\lambda'\rangle = 0\,,$$

$$\{c_a,\, c_b^+\} = \sum_{\lambda\lambda'} \{c_\lambda,\, c_{\lambda'}^+\} \langle a\,|\,\lambda\rangle \langle \lambda'\,|\,b\rangle \qquad (16.134)$$

$$= \sum_{\lambda\lambda'} \delta_{\lambda\lambda'} \langle a\,|\,\lambda\rangle \langle \lambda'\,|\,b\rangle$$

$$= \sum_{\lambda} \langle a\,|\,\lambda\rangle \langle \lambda\,|\,b\rangle = \langle a\,|\,b\rangle\,.$$

We thus obtain the anticommutation relations for the creation and annihilation operators of an arbitrary state:

$$\{c_a^+,\, c_b^+\} = 0\,, \qquad (16.135)$$
$$\{c_a,\, c_b\} = 0\,, \qquad (16.136)$$
$$\{c_a,\, c_b^+\} = \langle a\,|\,b\rangle\,\mathbb{1}\,. \qquad (16.137)$$

4.6 Single-Body Operator

Let V be a single operator describing the interaction of the particles with the force center:

$$V = \sum_i v_i. \tag{16.138}$$

We can examine the action of the potential v_i on a quantum state $\mid \alpha \rangle$ by projecting it onto the set of occupied states forming a complete orthonormal system:

$$v_i \mid \alpha \rangle = \sum_{\alpha'} \mid \alpha' \rangle \langle \alpha' \mid v_i \mid \alpha \rangle, \tag{16.139}$$

and by projectiong onto the configuration space:

$$v_i \langle x_1 \mid \alpha \rangle = \sum_{\alpha'} \langle x_1 \mid \alpha' \rangle \langle \alpha' \mid v_i \mid \alpha \rangle. \tag{16.140}$$

Because the quantum states labeled α and α' are not dependent on i, the matrix element is the same for every i. It is therefore denoted $\langle \alpha' \mid v \mid \alpha \rangle$, meaning that the single-body potential is the same irrespective of i:

$$v_i \langle x_1 \mid \alpha \rangle = \sum_{\alpha'} \langle x_1 \mid a' \rangle \langle \alpha' \mid v \mid \alpha \rangle,$$

$$v_i \, \varphi_\alpha (x_1) = \sum_{\alpha'} \varphi_{\alpha'} (x_1) \langle \alpha' \mid v \mid \alpha \rangle. \tag{16.141}$$

Let us determine $V \mid \alpha \, \beta \, \ldots \rangle$ by examining the action of V in the space $(x_1, x_2, \ldots)$ in which the quantum states $\alpha \, \beta \, \ldots$ may be located:

$$V \, \varphi_\alpha (x_1) \, \varphi_\beta (x_2) \, \ldots = \sum_i v_i \, \varphi_\alpha (x_1) \, \varphi_\beta (x_2) \, \ldots. \tag{16.142}$$

The action of v_i on an individual state has already been expanded in (16.141):

$$v_i \, \varphi_\alpha (x_1) = \sum_{\alpha'} \langle \alpha' \mid v \mid \alpha \rangle \, \varphi_{\alpha'} (x_1), \tag{16.143}$$

but if the v_i are acting on the state β, then we similarly obtain

$$v_i \, \varphi_\beta (x_2) = \sum_{\beta'} \langle \beta' \mid v \mid \beta \rangle \, \varphi_{\beta'} (x_2). \tag{16.144}$$

By using these results in (16.142), we obtain the following expression:

$$V \, \varphi_\alpha (x_1) \, \varphi_\beta (x_2) \, \ldots = \sum_{\alpha'} \langle \alpha' \mid v \mid \alpha \rangle \, \varphi_{\alpha'} (x_1) \, \varphi_\beta (x_2) \, \ldots$$

$$+ \sum_{\beta'} \langle \beta' \mid v \mid \beta \rangle \, \varphi_\alpha (x_1) \, \varphi_{\beta'} (x_2) + \ldots$$

$$+ \sum_{\lambda'} \langle \lambda' \mid v \mid \lambda \rangle \, \varphi_\alpha (x_1) \, \ldots \, \varphi_{\lambda'} (x_k) \, \ldots, \tag{16.145}$$

or, by returning to the notation based on occupied states,

$$V \mid \alpha \beta \gamma \ldots\rangle = \sum_{\alpha'} \langle \alpha' \mid v \mid \alpha \rangle \mid \alpha' \beta \gamma \ldots\rangle$$
$$+ \sum_{\beta'} \langle \beta' \mid v \mid \beta \rangle \mid \alpha \beta' \gamma \ldots\rangle + \ldots. \quad (16.146)$$

Each of the occupied states can be expanded from a reference state with the creation and annihilation operators:

$$\mid \alpha' \beta \gamma \ldots\rangle = c_{\alpha'}^+ c_\alpha \mid \alpha \beta \gamma \ldots\rangle,$$
$$\mid \alpha \beta' \gamma \ldots\rangle = c_{\beta'}^+ c_\beta \mid \alpha \beta \gamma \ldots\rangle, \quad (16.147)$$

thus transforming relation (16.146) as follows:

$$V \mid \alpha \beta \gamma \ldots\rangle = \left[\sum_{\alpha'} \langle \alpha' \mid v \mid \alpha \rangle c_{\alpha'}^+ c_\alpha \right.$$
$$\left. + \sum_{\beta'} \langle \beta' \mid v \mid \beta \rangle c_{\beta'}^+ c_\beta + \ldots \right] \mid \alpha \beta \ldots\rangle. \quad (16.148)$$

Because this result holds irrespective of the reference state $\mid \alpha \beta \gamma \ldots\rangle$, we infer the expression for the single-body potential in the representation of the occupied states:

$$\boxed{V = \sum_{\lambda \mu} \langle \lambda \mid v \mid \mu \rangle c_\lambda^+ c_\mu} \quad . \quad (16.149)$$

The matrix element $\langle \lambda \mid v \mid \mu \rangle$ can be evaluated by expanding the single-body operator in configuration space in which the interaction operator is diagonal, that is, local:

$$\langle \lambda \mid v \mid \mu \rangle = \int \langle \lambda \mid x \rangle \, dx \, \langle x \mid v \mid x' \rangle \, dx \, \langle x' \mid \mu \rangle$$
$$= \int \varphi_\lambda^* (x) \, v(x) \, \varphi_\mu (x) \, dx. \quad (16.150)$$

The functions $\varphi_\lambda^*(x)$ and $\varphi_\mu(x)$ are individual wave functions at point x, obtained by solving the nonrelativistic Schrödinger equation for a potential well (whether spherical or harmonic or Woods-Saxon, etc.).

4.7 Two-Body Operator

We will proceed in a similar manner to determine the two-body operator which is dependent on the relative position of the two particles i and j:

$$V = \sum_{i\langle j} v(ij) = \frac{1}{2} \sum_{ij \neq i} v_{ij} . \tag{16.151}$$

We can describe the action of v_{ij} on the two quantum states α and β at positions x_i and x_j by projecting onto the complete orthonormal system of the occupied states:

$$v_{ij} \, \varphi_\alpha \, (x_i) \, \varphi_\beta \, (x_j) = \sum_{\alpha' \, \beta'} \varphi_{\alpha'} \, (x_i) \, \varphi_{\beta'} \, (x_j) \, \langle \alpha' \, \beta' \mid v \mid \alpha \, \beta \rangle . \tag{16.152}$$

Here again, $\langle \alpha' \, \beta' \mid v \mid \alpha \, \beta \rangle$ is independent of the labels ij of the particles and has replaced the matrix element $\langle \alpha' \, \beta' \mid v_{ij} \mid \alpha \, \beta \rangle$. The sum extends over all the one-particle states that could be occupied.

We should perhaps add that the symmetrical role played by the coordinates i and j is represented by the equality of the matrix elements:

$$\langle \alpha' \, \beta' \mid v \mid \alpha \, \beta \rangle = \langle \beta' \, \alpha' \mid v \mid \beta \, \alpha \rangle . \tag{16.153}$$

With the preceding elements, the action of the interaction operator V on a two-particle state can be written as follows:

$$V \, \varphi_\alpha \, (x_i) \, \varphi_\beta \, (x_j) \, \ldots$$
$$= \sum_{\alpha' \, \beta'} \langle \alpha' \, \beta' \mid v \mid \alpha \, \beta \, \rangle \varphi_{\alpha'} \, (x_i) \, \varphi_{\beta'} \, (x_j) \, \varphi_\gamma \, (x_k) \, \ldots +$$
$$+ \sum_{\alpha' \, \gamma'} \langle \alpha' \, \gamma' \mid v \mid \alpha \, \gamma \rangle \, \varphi_{\alpha'} \, (x_i) \, \varphi_\beta \, (x_j) \, \varphi_{\gamma'} \, (x_k) \ldots + \ldots . \tag{16.154}$$

This amounts to using the following notation for the occupied states:

$$V \mid \alpha \, \beta \, \gamma \, \ldots \rangle = \sum_{\alpha' \, \beta'} \langle \alpha' \, \beta' \mid v \mid \alpha \, \beta \rangle \mid \alpha' \, \beta' \, \gamma \, \ldots \rangle$$
$$+ \sum_{\alpha' \, \gamma'} \langle \alpha' \, \gamma' \mid v \mid \alpha \, \gamma \rangle \mid \alpha' \, \beta \, \gamma' \, \ldots \rangle + \ldots , \tag{16.155}$$

or, by introducing a reference state and the creation and annihilation operators

$$\mid \alpha' \, \beta' \, \gamma \, \ldots \rangle = c_{\alpha'}^+ \, c_{\beta'}^+ \, c_\beta \, c_\alpha \mid \alpha \, \beta \, \gamma \, \ldots \rangle ,$$
$$\mid \alpha' \, \beta \, \gamma' \, \ldots \rangle = c_{\alpha'}^+ \, c_{\gamma'}^+ \, c_\gamma \, c_\alpha \mid \alpha \, \beta \, \gamma \, \ldots \rangle . \tag{16.156}$$

These results are used in (16.155) and, because the reference state appears in both terms of the equation, we obtain the expression for the two-body operator:

$$V = \sum_{\lambda \mu \nu \varrho} c_\lambda^+ c_\mu^+ \langle \lambda \mu \mid v \mid \nu \varrho \rangle c_\varrho c_\nu. \tag{16.157}$$

The sum $\lambda \mu$ is over all states capable of being occupied, whereas the sum over ν and ϱ includes all pairs of unordered states capable of being occupied such that

$$\sum_{\lambda \mu} c_\lambda^+ c_\mu^+ \langle \lambda \mu \mid v \mid \nu \varrho \rangle c_\varrho c_\nu$$
$$= \sum_{\lambda \mu} c_\lambda^+ c_\mu^+ \langle \mu \lambda \mid v \mid \varrho \nu \rangle c_\varrho c_\nu, \tag{16.158}$$

since the matrix element is invariant by simultaneous permutation of the two index pairs:

$$\langle \lambda \mu \mid v \mid \nu \varrho \rangle = \langle \mu \lambda \mid v \mid \varrho \nu \rangle. \tag{16.159}$$

We use the anticommutation relations between the creation and annihilation operators to obtain the equivalent form

$$\sum_{\lambda \mu} c_\lambda^+ c_\mu^+ \langle \lambda \mu \mid v \mid \nu \varrho \rangle c_\varrho c_\nu$$
$$= \sum_{\lambda \mu} c_\mu^+ c_\lambda^+ \langle \mu \lambda \mid v \mid \varrho \nu \rangle c_\nu c_\varrho, \tag{16.160}$$

or, by changing the name of the index λ to μ and vice versa, we obtain the result

$$\sum_{\lambda \mu} c_\lambda^+ c_\mu^+ \langle \lambda \mu \mid v \mid \nu \varrho \rangle c_\varrho c_\nu$$
$$= \sum_{\lambda \mu} c_\lambda^+ c_\mu^+ \langle \lambda \mu \mid v \mid \varrho \nu \rangle c_\nu c_\varrho. \tag{16.161}$$

If we swap indices ν and ϱ, the sum over μ and ν will remain the same. We can therefore sum up over all ν and ϱ and then divide the result by two:

$$\boxed{V = \frac{1}{2} \sum_{\lambda \mu \nu \varrho} c_\lambda^+ c_\mu^+ \langle \lambda \mu \mid v \mid \nu \varrho \rangle c_\varrho c_\nu} . \tag{16.162}$$

If we antisymmetrize the matrix element, that is, if we set

$$\langle \lambda \mu \mid \bar{v} \mid \nu \varrho \rangle = \langle \lambda \mu \mid v \mid \nu \varrho \rangle - \langle \lambda \mu \mid v \mid \varrho \nu \rangle, \tag{16.163}$$

this will confer the matrix element with the following symmetry properties:

$$\langle \lambda \mu \mid \bar{v} \mid \nu \varrho \rangle = \langle \mu \lambda \mid \bar{v} \mid \varrho \nu \rangle$$
$$= -\langle \lambda \mu \mid \bar{v} \mid \varrho \nu \rangle = -\langle \mu \lambda \mid \bar{v} \mid \nu \varrho \rangle, \tag{16.164}$$

and lead to another form of the two-body operator:

$$V = \frac{1}{4} \sum_{\lambda\,\mu\,\nu\,\varrho} c_\lambda^+ c_\mu^+ \langle \lambda\,\mu \mid \bar{v} \mid \nu\,\varrho \rangle c_\varrho\, c_\nu \,.$$

(16.165)

Single- and two-body potentials are of prime importance in the study of quantum systems. Consider, for example, an atom with two outer electrons (Fig. 2). The interaction of an electron of charge $-e$ with the nucleus of charge $+Ze$ will be described by a single-body potential that is dependent solely on the distance to the center of nucleus, whereas the residual interaction between the two outer electrons will depend on the position of each of the electrons, that is, on the relative distance $V(\mid \bar{r}_1 - \bar{r}_2 \mid)$. (See Chap. 14 on the perturbations of the energy levels of hydrogenic atoms.)

In most n-body problems, it is sufficient to introduce single- or two-body potentials. The effects of three-body interaction potentials are sometimes taken into account, but anything beyond that is simply discarded.

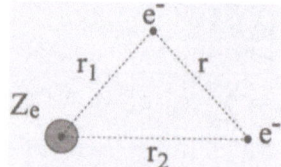

Figure 2

4.8 Wick's Theorem

In the second quantization formalism for fermion systems, a quantum state is always expanded from an arbitrary reference state termed the vacuum state. It is the state in which all possible states possible are empty $\mid 00\ldots 0\rangle$. It is simply denoted $\mid 0\rangle$. If, for example, we want to describe oxygen 17 from oxygen 16 by adding a neutron, the reference (or vacuum) state will be oxygen 16, and it will be denoted $\mid 0\rangle$. It is a "relative" vacuum filled with all the quantum states of oxygen 16's nucleons but which has no part to play in the structure of oxygen 17.

The calculation of single- or two-body potential matrix elements will therefore always lead to the manipulation of a large number of creation and annihilation operators, acting on the vacuum. We can use the anticommutation relations step-by-step. There is a theorem, the Wick's theorem, for obtaining the result more rapidly.

Let us use the calculation of the matrix element $\langle \alpha \mid V \mid \beta \rangle = \langle 0 \mid c_\alpha\, V\, c_\beta^+ \mid 0 \rangle$ as an example and replace V with the expression (16.49) for single-body potential in second quantization:

$$\langle 0 \mid c_\alpha \, V \, c_\beta^+ \mid 0 \rangle \;=\; \sum_{\lambda \, \mu} \langle \lambda \mid v \mid \mu \rangle \, \langle 0 \mid c_\alpha \, c_\lambda^+ \, c_\mu \, c_\beta^+ \mid 0 \rangle \, . \qquad (16.166)$$

The anticommutation relations make it possible for us to write

$$c_\alpha \, c_\lambda^+ \, c_\mu \, c_\beta^+ = c_\alpha \, c_\lambda^+ \, (\delta_{\mu \, \beta} - c_\beta^+ \, c_\mu) \, , \qquad (16.167)$$

which leads to the matrix element

$$\langle 0 \mid c_\alpha \, V \, c_\beta^+ \mid 0 \rangle \;=\; \sum_{\lambda \, \mu} \langle \lambda \mid v \mid \mu \rangle \, [\langle 0 \mid \delta_{\mu \, \beta} \, c_\alpha \, c_\lambda^+ \mid 0 \rangle$$
$$- \langle 0 \mid c_\alpha \, c_\lambda^+ \, c_\beta^+ \, c_\mu \mid 0 \rangle] \, . \qquad (16.168)$$

We further note that $c_\mu \mid 0 \rangle \; = 0$ since the state $\mid 0 \rangle$ represents a vacuum state $\mid 0000 \ldots \rangle$ in which no state is occupied. The second term in the brackets is therefore zero.

If we again use the anticommutation relation to express the first term in (16.168), we obtain the mean value over the vacuum:

$$\langle 0 \mid c_\alpha \, c_\lambda^+ \mid 0 \rangle \;=\; \delta_{\lambda \, \alpha} \, \langle 0 \mid 0 \rangle - \langle 0 \mid c_\lambda^+ \, c_\alpha \mid 0 \rangle$$
$$= \delta_{\lambda \, \alpha} \, . \qquad (16.169)$$

Inserting these result into the matrix element (16.168), we obtain

$$\langle 0 \mid c_\alpha \, V \, c_\beta^+ \mid 0 \rangle \;=\; \sum_{\lambda \, \mu} \langle \lambda \mid v \mid \mu \rangle \, \langle 0 \mid c_\alpha \, c_\lambda^+ \mid 0 \rangle \, \delta_{\mu \, \beta}$$
$$= \sum_{\lambda \, \mu} \langle \lambda \mid v \mid \mu \rangle \, \delta_{\mu \, \beta} \, \delta_{\alpha \, \lambda} \, , \qquad (16.170)$$

which, we must admit, is a forseeable result:

$$\langle \alpha \mid V \mid \beta \rangle \;=\; \langle \alpha \mid v \mid \beta \rangle \, . \qquad (16.171)$$

If we wish to calculate $\langle 0 \mid c_\alpha \, c_\beta \, V \, c_\gamma^+ \, c_\delta^+ \mid 0 \rangle$, we can proceed in the same manner to obtain the result. However, this procedure is long and tedious. So we will use Wick's theorem to show how such calculations can be significantly simplified.

a) Normal Product

Calculating the normal product of creation and annihilation operators boils down to simply all c^+ on the left and all c on the right. A $-$ sign is added if the rearrangement requires an odd number of permutations. A normal product is either denoted $N(\ )$ or enclosed between two colons.

Examples:

$$N\left(c_i^+ c_j^+ c_k c_\ell c_n^+\right) \equiv \ : c_i^+ c_j^+ c_k c_\ell c_n^+ : \ = c_i^+ c_j^+ c_n^+ c_k c_\ell \, ,$$
$$N\left(c_i^+ c_j c_k c_\ell c_m^+\right) \equiv \ : c_i^+ c_j c_k c_\ell c_m^+ : \ = -c_i^+ c_m^+ c_j c_k c_\ell \, .$$

Let us, in particular, note the normal products of two operators:

$$\begin{aligned}
N\left(c_1^+ c_2^+\right) &\equiv \ : c_1^+ c_2^+ : \ = c_1^+ c_2^+ \, , \\
N\left(c_1 c_2\right) &\equiv \ : c_1 c_2 : \ = c_1 c_2 \, , \\
N\left(c_1 c_2^+\right) &\equiv \ : c_1 c_2^+ : \ = -c_2^+ c_1 \, , \\
N\left(c_1^+ c_2\right) &\equiv \ : c_1^+ c_2 : \ = c_1^+ c_2 \, .
\end{aligned} \tag{16.172}$$

b) Properties of Normal Products

i) The mean value over the vacuum $| 0 \rangle$ of a normal product is zero.

ii) Anticommutative:

$$N\left(c_i^+ c_j^+ c_k\right) = -N\left(c_j^+ c_i^+ c_k\right) \, .$$

iii) Distributive with respect to the sum:

$$\begin{aligned}
N\left[\left(\lambda c_i + \mu c_j^+\right) c_k^+ c_\ell^+\right] &= N\left[\lambda c_i c_k^+ c_\ell^+\right] + N\left[\mu c_j^+ c_k^+ c_\ell^+\right] \\
&= \lambda N\left(c_i c_k^+ c_\ell^+\right) + \mu N\left(c_j^+ c_k^+ c_\ell^+\right)
\end{aligned}$$

c) Contraction

The contraction of two products is simply the mean value over the reference state of their product. A contraction of $c_i c_j$ is represented as $\overline{c_i c_j}$. It is easily noticed that:

$$\begin{aligned}
\overline{c_1^+ c_2^+} &= \langle \ | c_1^+ c_2^+ | \ \rangle = 0 \, , \\
\overline{c_1 c_2} &= \langle \ | c_1 c_2 | \ \rangle = 0 \, , \\
\overline{c_1^+ c_2} &= \langle \ | c_1^+ c_2 | \rangle = 0 \, , \\
\overline{c_1 c_2^+} &= \langle \ | c_1 c_2^+ | \rangle = \delta_{12} \, .
\end{aligned} \tag{16.173}$$

The contraction of two operators is an inner product that commutes with all other operators and the only non-zero contraction is that of $c_1 c_2^+$ (in this particular order) which equals δ_{12}.

d) Normal Contracted Product

The normal contracted product will be defined by the product of the contractions and the normal product of the remaining operators, assigned a parity coefficient $(-)^\nu$ where ν is the permutation parity that brings the contracted products next to one another. The contracted operators, which are mean values (and, hence, numbers), are therefore extracted from the normal product.

Exapmles:

$$\overline{c_i \ c_j \ c_k^+ \ c_\ell^+ \ c_m^+ \ c_n^+} = (-) \ \overline{c_j \ c_\ell^+ \ c_m^+ \ c_n^+} \ N \ (c_i \ c_k^+),$$

$$\overline{c_i^+ \ c_j \ c_k^+ \ c_\ell} = (-) \ \overline{c_j \ c_\ell} \ N \ (c_i^+ \ c_k^+).$$

e) Wick's Theorem

> The product of creation and annihilation products is equal
> to the sum of all of their normal contracted products
> (including the uncontracted term)   .

f) Example

What is the value of the matrix element of the operators $c_\alpha \ c_\lambda^+ \ c_\mu \ c_\beta^+$ taken between the vacuum states?

$$c_\alpha \ c_\lambda^+ \ c_\mu \ c_\beta^+ = \overline{c_\alpha \ c_\lambda^+} \ \overline{c_\mu \ c_\beta^+} - \overline{c_\alpha \ c_\mu} \ \overline{c_\lambda^+ \ c_\beta^+} + \overline{c_\alpha \ c_\beta^+} \ \overline{c_\lambda^+ \ c_\mu}$$
$$+ \overline{c_\alpha \ c_\lambda^+} \ N \ (c_\mu \ c_\beta^+) - \overline{c_\alpha \ c_\mu} \ N \ (c_\lambda^+ \ c_\beta^+) + \overline{c_\alpha \ c_\beta^+} \ N \ (c_\lambda^+ \ c_\mu)$$
$$+ \ N \ (c_\alpha \ c_\lambda^+) \ \overline{c_\mu \ c_\beta^+} - N \ (c_\alpha \ c_\mu) \ \overline{c_\lambda^+ \ c_\beta^+} + N \ (c_\alpha \ c_\beta^+) \ \overline{c_\lambda^+ \ c_\mu}$$
$$+ \ N \ (c_\alpha \ c_\lambda^+ \ c_\mu \ c_\beta^+). \tag{16.174}$$

We thus obtain, by evaluating the mean value of these operators:

$$\langle 0 \mid c_\alpha \ c_\lambda^+ \ c_\mu \ c_\beta^+ \mid 0 \rangle = \overline{c_\alpha \ c_\lambda^+} \ \overline{c_\mu \ c_\beta^+} - \overline{c_\alpha \ c_\mu} \ \overline{c_\lambda^+ \ c_\beta^+}$$
$$+ \overline{c_\alpha \ c_\beta^+} \ \overline{c_\lambda^+ \ c_\mu} + \overline{c_\alpha \ c_\lambda^+} \ \langle 0 \mid N \ (c_\mu \ c_\beta^+) \mid 0 \rangle$$
$$- \overline{c_\alpha \ c_\mu} \ \langle 0 \mid N \ (c_\lambda^+ \ c_\beta^+) \mid 0 \rangle + \overline{c_\alpha \ c_\beta^+} \ \langle 0 \mid N \ (c_\lambda^+ \ c_\mu) \mid 0 \rangle$$
$$+ \langle 0 \mid N \ (c_\alpha \ c_\lambda^+) \mid 0 \rangle \ \overline{c_\mu \ c_\beta^+} - \langle 0 \mid N \ (c_\alpha \ c_\mu) \mid 0 \rangle \ \overline{c_\lambda^+ \ c_\beta^+}$$
$$+ \langle 0 \mid N \ (c_\alpha \ c_\beta^+) \mid 0 \rangle \ \overline{c_\lambda^+ \ c_\mu} + \langle 0 \mid N \ (c_\alpha \ c_\lambda^+ \ c_\mu \ c_\beta^+) \mid 0 \rangle. \tag{16.175}$$

The mean value of a normal product is always zero and the only non-zero contraction is $\overline{c_1 \ c_2^+} = \delta_{12}$. Thus, the only contribution to the result comes from the first contraction which yields

$$\langle 0 \mid c_\alpha \ c_\lambda^+ \ c_\mu \ c_\beta^+ \mid 0 \rangle = \delta_{\alpha \lambda} \ \delta_{\mu \beta}. \tag{16.176}$$

In fact, to apply Wick's theorem to the calculation of a mean value, we simply look for contractions in the form of $\overline{c_1 \ c_2^+}$ with a non-zero δ_{12} contribution to the result.

g) Matrix Element of a Single-Body Operator Q

If we calculate the matrix element of the single-body operator Q using (16.149) of the operator, we will write

$$\langle \alpha \mid Q \mid \beta \rangle = \langle 0 \mid c_\alpha \, Q \, c_\beta^+ \mid 0 \rangle$$

$$= \sum_{\lambda \, \mu} \langle 0 \mid c_\alpha \, c_\lambda^+ \, c_\mu \, c_\beta^+ \mid 0 \rangle \, \langle \lambda \mid q \mid \mu \rangle , \qquad (16.177)$$

and by applying Wick's theorem to the results obtained in (16.176):

$$\langle \alpha \mid Q \mid \beta \rangle = \sum_{\lambda \, \mu} \delta_{\alpha \, \lambda} \, \delta_{\mu \, \beta} \, \langle \lambda \mid q \mid \mu \rangle ,$$

$$\boxed{\langle 0 \mid c_\alpha \, Q \, c_\beta \mid 0 \rangle = \langle \alpha \mid q \mid \beta \rangle} . \qquad (16.178)$$

h) Matrix Element of a Two-Body Operator V

To calculate $\langle \alpha \, \beta \mid V \mid \gamma \, \delta \rangle$, we will write in second quantization (see (16.165))

$$\langle 0 \mid c_\alpha \, c_\beta \, V \, c_\gamma^+ \, c_\delta^+ \mid 0 \rangle = \frac{1}{4} \sum_{\lambda \, \mu \, \nu \, \varrho} \langle \lambda \, \mu \mid \bar{v} \mid \nu \, \varrho \rangle$$

$$\langle 0 \mid c_\alpha \, c_\beta \, c_\lambda^+ \, c_\mu^+ \, c_\varrho \, c_\nu \, c_\gamma^+ \, c_\delta^+ \mid 0 \rangle . \, (16.179)$$

We apply Wick's theorem to this operator product by conserving only terms like $c_1 \, c_2^+ = \delta_{12}$ in order to obtain the non-zero contracted products

$$c_\alpha \, c_\beta \, c_\lambda^+ \, c_\mu^+ \, c_\varrho \, c_\nu \, c_\gamma^+ \, c_\delta^+ = \delta_{\alpha \, \lambda} \, \delta_{\beta \, \mu} \, \delta_{\varrho \, \gamma} \, \delta_{\nu \, \delta} - \delta_{\alpha \, \lambda} \, \delta_{\beta \, \mu} \, \delta_{\varrho \, \delta} \, \delta_{\nu \, \gamma}$$

$$- \delta_{\alpha \, \mu} \, \delta_{\beta \, \lambda} \, \delta_{\varrho \, \gamma} \, \delta_{\nu \, \delta}$$

$$+ \delta_{\alpha \, \mu} \, \delta_{\beta \, \lambda} \, \delta_{\varrho \, \delta} \, \delta_{\nu \, \gamma} . \qquad (16.180)$$

Substituting into the two-body matrix element (16.179), we obtain the four antisymmetrized matrix elements

$$\frac{1}{4} \left(\langle \alpha \, \beta \mid \bar{v} \mid \delta \, \gamma \rangle - \langle \alpha \, \beta \mid \bar{v} \mid \gamma \, \delta \rangle - \langle \beta \, \alpha \mid \bar{v} \mid \delta \, \gamma \rangle \right.$$

$$\left. + \langle \beta \, \alpha \mid \bar{v} \mid \gamma \, \delta \rangle \right) \qquad (16.181)$$

and by using the symmetry properties of these matrix elements, we are left with the mean value of a two-body operator:

$$\boxed{\langle 0 \mid c_\alpha \, c_\beta \, V \, c_\gamma^+ \, c_\delta^+ \mid 0 \rangle = \langle \alpha \, \beta \mid \bar{v} \mid \delta \, \gamma \rangle} . \qquad (16.182)$$

5. Boson Systems

Bosons are particles with integer spin that are not subject to the Pauli exclusion principle; they are therefore represented by symmetrical wave functions. This replaces the anticommutation relations with commutation relations. The results obtained, in fact, generalize the conclusions of the study of the harmonic oscillator whose "associated particles" constitute a typical boson system.

5.1 Occupied States

The occupation number n_i of the state λ_i is no longer limited to 0 or 1 by the Pauli exclusion principle:

$$n_i = 0, 1, 2, \ldots \leq N. \tag{16.183}$$

The vectors $| n_1 \, n_2 \ldots n_i \ldots \rangle$ for all the possible sets such that $\sum_i n_i = N$ form a complete basis spanning the Hilbert space:

$$\langle n_1 \ldots n_i \ldots | n_1' \ldots n_i' \ldots \rangle = \delta_{n_1 \, n_1'} \cdots \delta_{n_i \, n_i'} \cdots ,$$

$$\sum_{n_1 \ldots n_i \ldots} | n_1 \ldots n_i \ldots \rangle \langle n_1 \ldots n_i \ldots | = \mathbb{1}. \tag{16.184}$$

5.2 Creation and Annihilation Operators

We introduce the annihilation operator a_i which destroys a particle occupying the quantum state $| n_i \rangle$.

$$a_i | n_1 \ldots n_i \ldots (N) \rangle = \sqrt{n_i} \, | n_1 \ldots n_i - 1 \ldots (N-1) \rangle. \tag{16.185}$$

Because the operator a_i only acts upon the number n_i of states $| n_i \rangle$:

$$[a_i, a_j] = 0. \tag{16.186}$$

The matrix element of the operator a_i is easily obtained with (16.185) where n_i represents the column index and $n_i - 1$ the row index:

$$\langle n_i - 1 | a_i | n_i \rangle = (a_i)^{n_i}_{n_i - 1} = \sqrt{n_i}. \tag{16.187}$$

By returning to (16.185), this can also be written as

$$\langle n_1' \, n_2' \ldots n_i' \ldots (N') | a_i | n_1 \, n_2 \ldots n_i \ldots (N) \rangle$$
$$= \sqrt{n_i} \, \delta_{n_1' \, n_1} \, \delta_{n_2' \, n_2} \cdots \delta_{n_i' \, n_i - 1} \cdots \delta_{N'(N-1)}. \tag{16.188}$$

We define the boson creation operator, Hermitian conjugate of a_i, using its matrix element

$$\langle n_i - 1 \mid a_i \mid n_i \rangle^+ = \left[(a_i)^{n_i}_{n_i - 1} \right]^* = (a_i^+)^{n_i - 1}_{n_i} = \sqrt{n_i} = \langle n_i \mid a_i^+ \mid n_i - 1 \rangle.$$
(16.189)

The action of the operator a_i^+ on the state $\mid n_i - 1 \rangle$ can also be written directly, that is

$$a_i^+ \mid n_i - 1 \rangle = \sqrt{n_i} \mid n_i \rangle,$$
(16.190)

or by changing the label n_i into $n_i + 1$:

$$a_i^+ \mid n_i \rangle = \sqrt{n_i + 1} \mid n_i + 1 \rangle.$$
(16.191)

This will imply the definition of the operator a_i^+ acting on a (N)-occupied state ket:

$$a_i^+ \mid n_1 \ldots n_i \ldots (N) \rangle = \sqrt{n_i + 1} \mid n_1 \ldots n_i + 1 \ldots (N+1) \rangle.$$
(16.192)

The commutation relations between the creation and annihilation operators can be derived from relations (16.185) and (16.192):

$$\boxed{\begin{aligned} [a_i^+, \, a_j^+] &= 0 \\ [a_i, \, a_j^+] &= \delta_{ij} \, \mathbb{1} \end{aligned}}.$$
(16.193)

5.3 Particle-Number Operator

Consider the Hermitian operator N_i, product of the operators a_i^+ and a_i in this order:

$$N_i = a_i^+ \, a_i.$$
(16.194)

Its action on a vector $\mid n_1 \ldots n_i \ldots \rangle$ will give, with the foregoing definitions, the eigenvalue n_i the number of particles in the state λ_i. It therefore represents the occupation number operator for the state λ_i.

$$\begin{aligned} a_i^+ a_i \mid n_1 \ldots n_i \ldots \rangle &= a_i^+ \sqrt{n_i} \mid n_1 \ldots n_i - 1 \ldots \rangle \\ &= \sqrt{n_i} \sqrt{n_i} \mid n_1 \ldots n_i \ldots \rangle \\ &= n_i \mid n_1 \ldots n_i \ldots \rangle. \end{aligned}$$
(16.195)

We infer from this the eigenvalue equation

$$\boxed{N_i \mid n_1 \ldots n_i \ldots \rangle = n_i \mid n_1 \ldots n_i \ldots \rangle}.$$
(16.196)

The operator for the total number of particles will be obtained by summing the operator N_i over all possible states i:

$$N = \sum_i N_i .$$ (16.197)

It is easy to use definition (16.194) of N_i and commutation relations (16.193) to derive the commutation relations satisfied by the operator N_i:

$$
\begin{aligned}
[N_i, \ N_j] &= 0 \\
[N_i, \ a_j] = [N_i, \ a_j^+] &= 0 \\
[N_i, \ a_i^+] &= a_i^+ \\
[N_i, \ a_i] &= -a_i
\end{aligned}
$$ (16.198)

5.4 Occupation Number of the State i

If we apply (16.196) to an eigenstate $\mid n_i \rangle$ of N_i:

$$N_i \mid n_i \rangle \ = n_i \mid n_i \rangle ,$$ (16.199)

we obtain, by raplacing a_i^+ with the commutator $[N_i, \ a_i^+]$, the following vector:

$$
\begin{aligned}
a_i^+ \mid n_i \rangle \ = [N_i, \ a_i^+] \mid n_i \rangle \ &= (N_i \, a_i^+ - a_i^+ \, N_i) \mid n_i \rangle \\
&= N_i \, a_i^+ \mid n_i \rangle \ - a_i^+ \, n_i \mid n_i \rangle .
\end{aligned}
$$ (16.200)

Collecting the terms $a_i^+ \mid n_i \rangle$ will further yield:

$$N_i \, (a_i^+ \mid n_i \rangle) \ = (n_i + 1) \, (a_i^+ \mid n_i \rangle)$$ (16.201)

In other words, $a_i^+ \mid n_i \rangle$ is an eigenvector of N_i corresponding to the eigenvalue $(n_i + 1)$. We will also show that $a_i \mid n_i \rangle$ is an eigenvector of N_i corresponding to the eigenvalue $(n_i - 1)$:

$$N_i \, (a_i \mid n_i \rangle) \ = (n_i - 1) \, (a_i \mid n_i \rangle)$$ (16.202)

The norm of the vector $a_i \mid n_i \rangle$ can be determined as

$$\text{Norm} \, (a_i \mid n_i \rangle) \ = \langle n_i \mid a_i^+ \, a_i \mid n_i \rangle \ = \ \langle n_i \mid N_i \mid n_i \rangle \ = n_i .$$ (16.203)

The norm of a vector of the Hilbert space is always positive or zero and when it is zero this implies the nullity of the vector itself, thus leaving us with

$$n_i \geq 0$$
$$n_i = 0 \iff a_i \mid n_i \rangle = 0. \tag{16.204}$$

The eigenvalues of N_i are therefore positive integers and zero.

The eigenvalue zero corresponds to the zero norm of the vector $a_i \mid 0 \rangle$, hence to the vector $a_i \mid 0 \rangle$ which is itself zero. This constitutes the definition of the quantum vacuum or zero-particle state irrespective of the value of i:

$$\boxed{a_i \mid 0 \rangle = 0 \ \forall i} \ . \tag{16.205}$$

By the repeated action of a_i or a_i^+ on one of the eigenvectors of N_i, we obtain a set of vectors each corresponding to an eigenvalue of the spectrum:

$$a_i \mid 0 \rangle = 0,$$
$$a_i \mid n_i \rangle = \sqrt{n_i} \mid n_i - 1 \rangle,$$
$$a_i^+ \mid n_i \rangle = \sqrt{n_i + 1} \mid n_i + 1 \rangle,$$
$$N_i \mid n_i \rangle = n_i \mid n_i \rangle \ \text{with} \ N_i = a_i^+ a_i, \tag{16.206}$$
$$\langle n_i \mid n_i' \rangle = \delta_{n_i \, n_i'} \ \text{and} \ \sum_{n_i} \mid n_i \rangle \langle n_i \mid = \mathbb{1}.$$

These results generalize the relations obtained in the framework of the harmonic oscillator. The creation and annihilation operators are no longer directly linked to the position and momentum operators. It is therefore impossible to obtain the corresponding wave functions in the configuration space from the foregoing relations.

5.5 Creation and Annihilation Operators at Point $\vec{r}$

a) Definitions

Let us introduce the annihilation operator of a boson at point $\vec{r}$ at a given instant t.

$$\Psi(t, \vec{r}) \equiv \Psi(\vec{r}) = \sum_i \Psi_{\lambda_i}(\vec{r}) \, a_i = \sum_i \langle \vec{r} \mid \lambda_i \rangle \, a_i. \tag{16.207}$$

The Hermitian conjugate operator will be written as follows

$$\Psi^+(\vec{r}) = \sum_i \Psi_{\lambda_i}^*(\vec{r}) \, a_i^+ = \sum_i a_i^+ \, \Psi_{\lambda_i}^*(\vec{r}) = \sum_i a_i^+ \langle \lambda_i \mid \vec{r} \rangle. \tag{16.208}$$

This operator creates a boson at point $\vec{r}$. (Particular attention should be paid to the notation: $\Psi(\vec{r})$ and $\Psi^+(\vec{r})$ are now operators and no longer wave functions.)

These operators satisfy the commutation relations

$$[\Psi(\vec{r}),\ \Psi^+(\vec{r}\,')] = \delta(\vec{r} - \vec{r}\,')\ \mathbb{1}\,, \tag{16.209}$$

$$[\Psi(\vec{r}),\ \Psi(\vec{r}\,')] = [\Psi^+(\vec{r}),\ \Psi^+(\vec{r}\,')] = 0\,. \tag{16.210}$$

Let us demonstrate, for example, the first of these relations using definitions (16.207) and (16.208) of these operators

$$[\Psi(\vec{r}),\ \Psi^+(\vec{r}\,')] = \sum_{ij} \Psi_{\lambda_i}(\vec{r})\, [a_i,\ a_j^+]\, \Psi_{\lambda_j}^*(\vec{r}\,')\,. \tag{16.211}$$

Using commutation relation (16.193) and definitions (16.207) and (16.208), we obtain

$$[\Psi(\vec{r}),\ \Psi^+(\vec{r}\,')] = \sum_{ij} \Psi_{\lambda_i}(\vec{r})\, \delta_{ij}\, \mathbb{1}\, \Psi_{\lambda_j}^*(\vec{r}\,') = \sum_{i} \Psi_{\lambda_i}(\vec{r})\, \Psi_{\lambda_i}^*(\vec{r}\,')\, \mathbb{1}$$

$$= \sum_{i} \langle \vec{r} \mid \lambda_i \rangle \langle \lambda_i \mid \vec{r}\,' \rangle = \langle \vec{r} \mid \vec{r}\,' \rangle \mathbb{1}$$

$$= \delta(\vec{r} - \vec{r}\,')\mathbb{1}\,. \tag{16.212}$$

The total number of bosons (16.147) can be expressed with the boson creation and annihilation operators at point $\vec{r}$. We artificially introduce a Kronecker symbol of the form

$$\delta_{ij} = \langle \lambda_j \mid \lambda_i \rangle\,, \tag{16.213}$$

in the equation for the operator expressing the number of bosons occupying the state $\mid \lambda_i \rangle$:

$$N_i = a_i^+ a_i = \sum_j a_j^+ a_i\, \delta_{ij} = \sum_j a_j^+ a_i\, \langle \lambda_j \mid \lambda_i \rangle\,. \tag{16.214}$$

By using the projector $\int \mid \vec{r} \rangle\, d\vec{r}\, \langle \vec{r} \mid = \mathbb{1}$ over configuration space, the total number of bosons $N = \sum_i N_i$ becomes

$$N = \sum_{ij} a_j^+ \langle \lambda_j \mid \lambda_i \rangle a_i = \sum_{ij} \int a_j^+ \langle \lambda_j \mid \vec{r} \rangle\, d\vec{r}\, \langle \vec{r} \mid \lambda_i \rangle a_i$$

$$= \sum_{ij} \int a_i^+ \Psi_{\lambda_j}^*(\vec{r})\, d\vec{r}\, \Psi_{\lambda_i}(\vec{r})\, a_i\,. \tag{16.215}$$

By using definitions (16.207) of $\Psi(\vec{r})$ and (16.208) of $\Psi^+(\vec{r})$, we are left with

$$\boxed{N = \int \Psi^+(\vec{r})\, d\vec{r}\, \Psi(\vec{r})}\,. \tag{16.216}$$

The operator product $\Psi^+(\vec{r})\,\Psi(\vec{r})$ can therefore be interpreted as the particle density operator at point $\vec{r}$.

If we use definition (16.197) and the commutation relations, we also easily obtain

$$[N,\,\Psi^+(\vec{r})] = \Psi^+(\vec{r}). \qquad (16.217)$$

It is readily shown that $\Psi^+(\vec{r})\,|\,0\rangle$ is a one-boson state at point $\vec{r}$ since we obtain, with the foregoing relation

$$[N,\,\Psi^+(\vec{r})]\,|\,0\rangle = \Psi^+(\vec{r})\,|\,0\rangle, \qquad (16.218)$$

and by expanding the commutator

$$N\,\Psi^+(\vec{r})\,|\,0\rangle = \Psi^+(\vec{r})\,N\,|\,0\rangle + \Psi^+(\vec{r})\,|\,0\rangle.$$

The definition of the vacuum $N_i\,|\,0\rangle = 0$ shows that $\sum_i N_i\,|\,0\rangle = N\,|\,0\rangle = 0$, which finally leaves us with

$$N\left(\Psi^+(\vec{r})\,|\,0\rangle\right) = \left(\Psi^+(\vec{r})\,|\,0\rangle\right). \qquad (16.219)$$

As a result, $\Psi^+(\vec{r})\,|\,0\rangle$ is an eigenstate of N corresponding to the eigenvalue 1. The operator $\Psi^+(\vec{r})$ therefore creates a particle at point $\vec{r}$. Similarly, the operator $\Psi(\vec{r})$ will destroy a particle at point $\vec{r}$.

b) Fock Space

It is usual to introduce Fock space (or, rather, the Fock representation) in which the vacuum $|\,0\rangle$ is a quantum state with no bosons at any point $\vec{r}$:

$$\Psi(\vec{r})\,|\,0\rangle = 0. \qquad (16.220)$$

The base vectors of Fock space are:

$$
\begin{aligned}
&|\,0\rangle && \text{vacuum state}, \\
&|\,\vec{r}_1\rangle && \text{a particle at } \vec{r}_1, \\
&|\,\vec{r}_1\,\vec{r}_2\rangle && \text{two particles at } \vec{r}_1 \text{ and } \vec{r}_2 \text{ respectively}, \\
&|\,\vec{r}_1\ldots\vec{r}_N\rangle && N \text{ particles at } \vec{r}_1\ldots\vec{r}_N, \text{ respectively}.
\end{aligned}
\qquad (16.221)
$$

The basis S of Fock space is therefore the row of the base vectors:

$$S = [\,|\,0\rangle \,|\,\vec{r}_1\rangle \,|\,\vec{r}_1\,\vec{r}_2\rangle \ldots |\,\vec{r}_1\ldots\vec{r}_N\rangle]. \qquad (16.222)$$

The action of $\Psi(\vec{r})$ and $\Psi^+(\vec{r})$ on the Fock states is given by one of these relations:

$$\langle\vec{r}_1\ldots\vec{r}_{N-1}\,(N-1)\,|\,\Psi(\vec{r}) = \sqrt{N}\,\langle\vec{r}\,\vec{r}_1\ldots\vec{r}_{N-1}\,(N)\,| \qquad (16.223)$$

and if we transpose this equation,

$$\Psi^+ \left(\vec{r} \right) | \vec{r}_1 \ldots \vec{r}_{N-1} \left(N - 1 \right) \rangle = \sqrt{N} \, | \vec{r} \, \vec{r}_1 \ldots \vec{r}_{N-1} \left(N \right) \rangle . \qquad (16.224)$$

The action of the annihilation operator at point $\vec{r}$ of a Fock space ket is, in fact, written

$$\Psi \left(\vec{r} \right) | \vec{r}_1 \ldots \vec{r}_{N+1} \left(N + 1 \right) \rangle$$

$$= \frac{1}{\sqrt{N+1}} \sum_{\ell=1}^{N+1} \delta \left(\vec{r}_\ell - \vec{r} \right) | \vec{r}_1 \ldots \vec{r}_{\ell-1} \, \vec{r}_{\ell+1} \ldots \vec{r}_{N+1} \left(N \right) \rangle \qquad (16.225)$$

and, by transposing this equation,

$$\langle \vec{r}_1 \ldots \vec{r}_{N+1} \left(N + 1 \right) | \, \Psi^+ \left(\vec{r} \right) = \frac{1}{\sqrt{N+1}} \sum_{\ell=1}^{N+1} \delta \left(\vec{r}_\ell - \vec{r} \right)$$

$$\times \langle \vec{r}_1 \ldots \vec{r}_{\ell-1} \, \vec{r}_{\ell+1} \ldots \vec{r}_{N+1} \left(N \right) | . \qquad (16.226)$$

We infer from this the formal representation of the Fock space operator matrix elements.

Let V be an operator which does not change the number of particles and which is diagonal in configuration space:

$$\langle \vec{r}_1 \ldots \vec{r}_N | V | \vec{r}\,'_1 \ldots \vec{r}\,'_N \rangle$$

$$= V \left(\vec{r}_1, \ldots, \vec{r}_N \right) \delta \left(\vec{r}_1 - \vec{r}\,'_1 \right) \ldots \delta \left(\vec{r}_N - \vec{r}\,'_N \right) . \qquad (16.227)$$

This operator is represented in Fock space by the integral

$$V = \frac{1}{N!} \int d\vec{r}_1 \ldots d\vec{r}_N \, \Psi^+ \left(\vec{r}_N \right) \ldots \Psi^+ \left(\vec{r}_1 \right) V \left(\vec{r}_1, \ldots, \vec{r}_N \right)$$

$$\Psi \left(\vec{r}_1 \right) \ldots \Psi \left(\vec{r}_N \right) . \qquad (16.228)$$

Case No. 1: V is a sum of single-particle operators:

$$V \left(\vec{r}_1, \ldots, \vec{r}_N \right) = \sum_{i=1}^{N} v \left(\vec{r}_i \right) . \qquad (16.229)$$

Substituting into (16.228) and using (16.216), we obtain the simple form

$$V = \int \Psi^+ \left(\vec{r} \right) v \left(\vec{r} \right) \Psi \left(\vec{r} \right) d\vec{r} . \qquad (16.230)$$

Case No. 2: V is a sum of two-particle operators:

$$V\,(\vec{r}_1,\ \vec{r}_2\ldots,\vec{r}_N) = \sum_{i\langle j} V\,(\vec{r}_i,\ \vec{r}_j) = \frac{1}{2}\sum_{ij\neq i} V\,(\vec{r}_i,\ \vec{r}_j)\,. \qquad (16.231)$$

In this case, we obtain in Fock space:

$$V = \frac{1}{2}\int\,d\vec{r}\,'\,d\vec{r}\,\Psi^+\,(\vec{r}\,')\,\Psi^+\,(\vec{r})\,v\,(\vec{r},\ \vec{r}\,')\,\Psi\,(\vec{r})\,\Psi\,(\vec{r}\,')\,. \qquad (16.232)$$

The form of single- and two-body operators in Fock space will be very similar to those of single- and two-body operators for fermions and is obtained by replacing the creation and annihilation operators by the operators that create and annihilate a boson at a given point in space-time.

6. Simple Harmonic Oscillator – Schematic Summary

$$a^+ = \sqrt{\frac{m\,\omega}{2\hbar}}\; x - \frac{i}{\sqrt{2m\,\hbar\,\omega}}\; p$$

$$a = (a^+)^+$$

$$[a,\, a^+] = \mathbb{1}$$

$$N = a^+\, a$$

$$H = \left(N + \frac{1}{2}\,\mathbb{1}\right)\hbar\,\omega$$

$$[N,\, a] = -a$$

$$[N,\, a^+] = a^+$$

$$N\,|\,n\rangle = n\,|\,n\rangle$$

$$n = n^* = 0, 1, 2, \dots .$$

$$_n = (n + \frac{1}{2})\,\hbar\omega$$

$$a\,|\,n\rangle = \sqrt{n}\,|\,n-1\rangle$$

$$a^+\,|\,n\rangle = \sqrt{n+1}\,|\,n+1\rangle$$

$$a\,|\,0\rangle = 0$$

$$\langle x\,|\,n\rangle = \Psi_n(x)$$

$$X = \sqrt{\frac{m\,\omega}{\hbar}}\; x$$

$$\langle X\,|\,n\rangle = \Psi_n(X)$$

$$a^+ = \frac{1}{\sqrt{2}}\left(X - \frac{d}{dX}\right)$$

$$a = \frac{1}{\sqrt{2}}\left(X + \frac{d}{dX}\right)$$

$$\Psi_n(X) = (-)^n\, e^{+X^2}\, \frac{d^n}{dX^n}\, e^{-X^2}$$

$$= \left(\sqrt{\pi}\, 2^n\, n!\right)^{1/2}\, e^{-X^2}\, H_n(X)$$

7. Second Quantization – Schematic Summary

Fock space

$$\text{quantum vacuum } a_\lambda \,|\, 0 \rangle \;=0 \quad \forall \lambda$$
$$N_\lambda = a_\lambda^+ \, a_\lambda$$

$$[a_\lambda, \, a_{\lambda'}^+]_q = \delta_{\lambda \, \lambda'} \, \mathbb{1}$$

$q = -1$

$q = +1$

fermions

bosons

$$\{a_\lambda^+, \, a_\mu^+\} = 0$$
$$\{a_\lambda, \, a_\mu\} = 0$$
$$\{a_\lambda, \, a_\mu^+\} = \delta_{\lambda \, \mu} \, \mathbb{1}$$

$$[a_\lambda^+, \, a_\mu^+] = 0$$
$$[a_\lambda, \, a_\mu] = 0$$
$$[a_\lambda, a_\mu^+] = \delta_{\lambda \, \mu} \, \mathbb{1}$$

$$a_\lambda^+ \,|\, ..\text{not } \lambda ... \rangle \;=\; |\, ... \lambda ... \rangle$$
$$a_\lambda^+ \,|\, ... \lambda ... \rangle \;= 0$$
$$a_\lambda \,|\, ... \text{not } \lambda ... \rangle \;= 0$$
$$a_\lambda \,|\, ... \lambda ... \rangle \;=\; |\, ... \text{not } \lambda ... \rangle$$

$$a_\lambda \,|\, ... n_\lambda ... \rangle$$
$$=\sqrt{n_\lambda} \,|\, ... n_\lambda - 1 ... \rangle$$
$$a_\lambda^+ \,|\, ... n_\lambda ... \rangle$$
$$=\sqrt{n_\lambda + 1} \,|\, ... n_\lambda + 1 ... \rangle$$

Pauli exclusion principle
$$\left(a_\lambda^+\right)^2 = 0 \quad a_\lambda^2 = 0$$

$$N_\lambda \,|\, ... \lambda ... \rangle \;=\; |\, ... \lambda ... \rangle$$
$$N_\lambda \,|\, ... \text{not } \lambda ... \rangle \;= 0$$
$$n_\lambda = 0 \text{ or } 1$$

$$N_\lambda \,|\, n_\lambda \rangle \;= n_\lambda \,|\, n_\lambda \rangle$$
$$n_\lambda = 0, \, 1, \, 2, ...$$

$$V = \sum_{\lambda \, \mu} \langle \lambda \,|\, v \,|\, \mu \rangle \, a_\lambda^+ \, a_\mu$$
$$V = \frac{1}{2} \sum_{\lambda \, \mu \, \nu \, \varrho} a_\lambda^+ \, a_\mu^+ \, \langle \lambda_\mu \,|\, v \,|\, \nu \, \varrho \rangle \, a_\nu \, a_\varrho$$

Appendix: The q-Commutators

Since 1982, numerous authors (see Kibler, 1993) have developed a quantum algebra based on the introduction of deformed numbers (qp):

$$[x] = [x]_{qp} = [x]_{pq} = \frac{q^x - p^x}{q - p} \qquad q, p, x \in \mathbb{C}. \tag{A.1}$$

For $n \in \mathbb{N} - [0]$ we obtain the expression for the deformed number n (qp):

$$[n] = q^{n-1} + pq^{n-2} + p^2 q^{n-3} + \ldots + qp^{n-2} + p^{n-1} \tag{A.2}$$

with, in particular, the first three deformed numbers (qp):

$$[0] = 0 \quad [1] = 1 \quad [2] = q + p \quad [3] = q^2 + qp + p^2 . \tag{A.3}$$

Deformed operators (qp), creation operators a^+, annihilation operators a, and occupation number operators N are defined by their action on the Fock vector space $\mathcal{F} = \{\, |\, n\rangle : n \in \mathbb{N}\}$, that is by generalizing relations (16.206):

$$\begin{aligned}
a \,|\, 0\rangle &= 0, \\
a \,|\, n\rangle &= \sqrt{[n]} \,|\, n-1\rangle, \\
a^+ \,|\, n\rangle &= \sqrt{[n+1]} \,|\, n+1\rangle, \\
N \,|\, n\rangle &= [n] \,|\, n\rangle.
\end{aligned} \tag{A.4}$$

This will yield, by iteration, the deformed quantum state (qp) with $N = a^+ a$:

$$|\, n\rangle = \frac{1}{\sqrt{[n]!}} \, (a^+)^n \,|\, 0\rangle, \tag{A.5}$$

in which the deformed factorial (qp) is written:

$$[n]! = [n]\,[n-1]\ldots[1]. \tag{A.6}$$

The operators a, a^+, and N obey the usual commutation relations (16.11), (16.12) and (16.13):

$$[a, a^+] = \mathbb{1}, \qquad [N, a^+] = a^+, \qquad [N, a] = -a. \tag{A.7}$$

Relations (A.7) can be used to show an important property of these operators:

$$a\, a^+ = [N + \mathbb{1}] \qquad \text{and} \qquad a^+ a = [N]. \tag{A.8}$$

This will lead to the expression for the commutator of the quantum algebra annihilation and creation operators:

$$[a, a^+] = [N + \mathbb{1}] - [N]. \tag{A.9}$$

The general form of a Q-commutator can be written as:

$$[a, a^+]_Q = aa^+ - Q\, a^+a = \frac{1}{q-p}\left(q^N\,(q-Q) - p^N\,(p-Q)\right), \qquad (A.10)$$

which gives, for arbitrary parameters p and q,

$$[a, a^+]_q = p^N,\ [a, a^+]_p = q^N \text{ and } [a, a^+]_1 = \frac{1}{q-p}\left(q^N\,(q-1) - p^N\,(p-1)\right).$$
$$(A.11)$$

A particularly interesting case is obtained for $Q = q$ and $p = 1$ since

$$[a, a^+]_q = aa^+ - q\, a^+a = \mathbb{1} \qquad (A.12)$$

leads to the commutation relations for $q = 1$, and for $q = -1$ to the anti-commutation relations. We can use the (qp) deformed operators to introduce a position operator x_k and momentum operator p_k which are also (qp) deformed, by drawing the analogy between relations (16.2) and (16.3) of the one-dimensional harmonic oscillator:

$$p_k = i\sqrt{\frac{\hbar\,\mu\,\omega}{2}}\,(a_k^+ - a_k) \quad \text{and} \quad x_k = \sqrt{\frac{\hbar}{2\mu\,\omega}}\,(a_k^+ + a_k) \quad k = 1, 2, \ldots, d.$$
$$(A.13)$$

The parameters μ and ω represent respectively the reduced mass and angular velocity of the (qp)-deformed harmonic oscillator.

The commutation relation between these operators is obtained from (A.8) and (A.9):

$$[x_k, p_k] = i\hbar\,[a_k, a_k^+] = i\hbar\,([N_k + \mathbb{1}] - [N_k])$$
$$= \frac{i\hbar}{q-p}\left(q^{N_k}\,(q-1) - p^{N_k}\,(p-1)\right). \quad (A.14)$$

The generalization of Heisenberg's uncertainty relation to the (qp)-deformed algebras shows that, for $p = q^{-1}$,

$$[x_k, p_k] = i\hbar\,\frac{\cosh\left(\left(N_k + \frac{1}{2}\,\mathbb{1}\right)\log\,q\right)}{\cosh\left(\frac{1}{2}\log\,q\right)} \qquad (A.15)$$

and if $q \to 1$, we regain the usual expression for this commutator. The Hamiltonian of the (qp)-deformed oscillator is written, with (A.13), in the usual form

$$H_{def} = \sum_{k=1}^{d}\frac{1}{2\mu}\,p_k^2 + \frac{1}{2}\,\mu\,w^2\,x_k^2 = \frac{\hbar w}{2}\sum_k a_k^+\,a_k + \alpha_k\,a_k^+$$
$$= \frac{\hbar w}{2}\sum_{k=1}^{d}[N_k] + [N_k + \mathbb{1}] \quad (A.16)$$

and the eigenvalues, expressed with the deformed numbers (qp), reduce to

$$E_{def} = \frac{\hbar w}{2} \sum_{k=1}^{d} [n_k] + [n_k + 1] = \frac{\hbar w}{2(q-p)} \sum_{k=1}^{d} (q^{n_k}(q+1) - p^{n_k}(p+1)) .$$

(A.17)

In the special case $p = q^{-1}$, we obtain

$$E_{def} = \frac{\hbar w}{2} \sum_{k=1}^{d} \frac{\sinh((n_k + \frac{1}{2}) \ell_n q)}{\sinh(\frac{1}{2} \ell_n q)} .$$

(A.18)

Ever more attention (Goodison and Toms, 1993) is now being paid to the q-commutators generalizing the commutation and anticommutation relations with the definition

$$[a_i, a_j^+]_q = a_i a_j^+ - q a_j^+ a_i = \delta_{ij} \mathbb{1} ,$$

(A.19)

in which q is a real number between -1 and $+1$, with the boundary values corresponding to boson statistics ($q = +1$ defines a commutator) and to fermion statistics ($q = -1$ defines an anticommutator). For $|q| = 1$, the statistics are those of Bose–Einstein or Fermi–Dirac. For $|q| > 1$, the Fock space states have a negative norm, showing that the structure of the Hilbert space is no longer present whereas for $|q| < 1$ the Hilbertian structure of the Fock space persists, but the many-state particles no longer have a well-defined symmetry for the exchange of these particles.

We are therefore going to make a detailed study of the statistics for which $|q| < 1$ and develop a field operator $\Psi(x)$ on the basis of the creation and annihilation operators obeying a q-commutator:

$$\Psi(x) = \sum_i F_i(x) a_i + F_i^*(x) a_i^+ = \sum_j G_j(x) b_j + G_j^*(x) b_j^+$$

(A.20)

with the q-commutations

$$[a_i, a_j^+]_q = [b_i, b_j^+]_{q'} = \delta_{ij} \mathbb{1} \quad |q| \text{ and } |q'| < 1.$$

(A.21)

The functions $F_i(x)$ and $G_j(x)$ are complete sets of positive frequency functions that are solutions of the Klein–Gordon equation. The expansion of the functions $G_i(x)$ in the basis of the functions $F_i(x)$ will introduce Bogoliubov coefficients α_{ij} and β_{ij}:

$$G_i(x) = \sum_j \alpha_{ij} F_j(x) + \beta_{ij} F_j^*(x) .$$

(A.22)

Substituting this expansion into $\Psi(x)$, we obtain the annihilation operator a_i in terms of b_j and b_j^+:

$$a_i = \sum_j \alpha_{ji} b_j + \beta_{ji}^* b_j^+ .$$

(A.23)

Let us now make use of the Friedmann–Robertson–Walker space-time with a flat space metric until time t_1 and thereafter set (see Chap. 20)

$$ds^2 = dt^2 - R^2(t) \, (dx^2 + dy^2 + dz^2), \tag{A.24}$$

with the definition of the quantum vacuum, before and after the change of Universe-scale factor at time t_1:

$$
\begin{aligned}
&\text{for } t < t_1 \quad R(t) = R_1 \quad \text{and} \quad a_i \mid 0, \text{ in} \rangle = 0 \; \forall i, \\
&\text{for } t < t_1 \quad R(t) = R_2 \quad \text{and} \quad b_j \mid 0, \text{ out} \rangle = 0 \; \forall j.
\end{aligned}
\tag{A.25}
$$

The Bogoliubov coefficients become diagonal in such a hypothesis:

$$\alpha_{ij} = \alpha_i \, \delta_{ij} \quad \text{and} \quad \beta_{ij} = \beta_i \, \delta_{ij}. \tag{A.26}$$

The q-commutation relation for the operators a_i associated with their expression in terms of the operators b_j will lead to the fundamental relation

$$
\begin{aligned}
\delta_{ij} =\ & (\mid \alpha_i \mid^2 - q \mid \beta_i \mid^2) \, \delta_{ij} + \alpha_i \, \beta_i \, (b_i \, b_j - q \, b_j \, b_i) \\
& + \beta_i^* \alpha_j^* \, (b_i^+ \, b_j^+ - q \, b_j^+ \, b_i^+) + (1 - q'q) \, \beta_i^* \, \beta_j \, b_i^+ \, b_j \\
& + (q' - q) \, \alpha_i \, \alpha_j^* \, b_j^+ \, b_i.
\end{aligned}
\tag{A.27}
$$

The mean value over the vacuum $\mid 0, \text{out} \rangle$ of this expression eliminates all operator terms, leaving the constraint

$$1 = \mid \alpha_i \mid^2 - q \mid \beta_i \mid^2. \tag{A.28}$$

The quantum states for $t \rangle t_1$ are defined with the creation operators on the vacuum of the outgoing states:

$$b_k^+ \, b_\ell^+ \mid 0, \text{out} \rangle = \mid k \, \ell, \text{out} \rangle. \tag{A.29}$$

By using the q-commutator of operators b and b^+, the orthonormalization of the states $\mid k \, \ell, \text{out} \rangle$ will yield the following relation:

$$\langle k \, \ell, \text{ out} \mid mn, \text{out} \rangle = \delta_{km} \, \delta_{\ell n} + q' \, \delta_{\ell m} \, \delta_{kn}. \tag{A.30}$$

For symmetrical states, it is important that $q' = +1$ whereas antisymmetrical states impose $q' = -1$. States $\mid k \, \ell, \text{out} \rangle$ do not have a well-defined symmetry for $\mid q \mid < 1$.

By introducing relation (A.27), and using constraint (A.28) on the vacuum state $\mid 0 \text{ out} \rangle$ in order to eliminate constant terms, it is easily shown that we are left with

$$\beta_i^* \, \alpha_j^* \, (\mid ij, \text{out} \rangle - q \mid ji, \text{out} \rangle) = 0, \tag{A.31}$$

and because the quantum states $\mid ij, \text{out} \rangle$ and $\mid ji, \text{out} \rangle$ are linearly independent:

$$\beta_i^* \, \alpha_j^* = 0 \quad \text{and hence} \quad \alpha_i \, \beta_j = 0. \tag{A.32}$$

This will leave (A.27) with only the following non-zero terms:

$$(q' - q)\, \alpha_i\, \alpha_j^*\, b_j^+\, b_i + (1 - qq')\, \beta_i^*\, \beta_j\, b_i^+\, b_j = 0. \qquad (A.33)$$

This relation is satisfied irrespective of the values of i and j, and hence, in particular, for $i = j$:

$$\{(q' - q)\,|\,\alpha_i\,|^2 + (1 - qq')\,|\,\beta_i\,|^2\}\, b_i^+\, b_i = 0. \qquad (A.34)$$

The mean value taken over states $|\,i, \mathrm{out}\rangle$ leads to the expression

$$\{(q' - q)\,|\,\alpha_i\,|^2 + (1 - qq')\,|\,\beta_i\,|^2\}\,\langle i, \mathrm{out}\,|\, b_i^+\, b_i\,|\, i, \mathrm{out}\rangle = 0 \qquad (A.35)$$

and the non-zero norm of the vector $|\,0, \mathrm{out}\rangle$ leaves us with the condition:

$$(q' - q)\,|\,\alpha_i\,|^2 + (1 - qq')\,|\,\beta_i\,|^2 = 0. \qquad (A.36)$$

When associated with constraint (A.28), this fixes $|\,\alpha_i\,|^2$ and $|\,\beta_i\,|^2$ with respect to q and q':

$$|\,\alpha_i\,|^2 = \frac{1 - qq'}{1 - q^2} \quad \text{and} \quad |\,\beta_i\,|^2 = \frac{q - q'}{1 - q^2}, \qquad (A.37)$$

whereas the product $\alpha_i\,\beta_i$ must remain zero following relation (A.32).

- If $\alpha_i = 0$ and $\beta_i \neq 0$ then $qq' = 1$, which is incompatible with the initial hypothesis $|\,q\,| < 1$ and $|\,q'\,| < 1$.
- If $\alpha_i \neq 0$ and $\beta_i = 0$, then $q = q'$ although this leads to $|\,\alpha_i\,|^2 = 1$ and the positive frequency solutions for $t < t_1$ remain of positive frequency for $t > t_1$, even if there is a dynamical evolution such as the expansion of the Universe.
- The only possibility then is to have the bosons ($q = +1$) or fermions ($q = -1$) at time $t < t_1$ and to retain the same type of particles at time $t > t_1$, $q' = +1$ or $q' = -1$.

Statistics other than Bose–Einstein or Fermi–Dirac are incompatible with the Big Bang theory in its simple model as presented here.

Chapter 17
Boson Fields

A system with a large number of particles (or degrees of freedom to be more formal) can be described in its classical discontinuous (corpuscular) aspect with the equations of analytical mechanics (see Chap. 1): Lagrange's equations, the Hamiltonian formalism, the Poisson equation, etc., or in its quantum discontinuous aspect with the second quantization formalism (see Chap. 16). It can also be described globally in its classical continuous aspect using Hamilton–Jacobi equations or Maxwell equations for the electromagnetic field or, more generally, in terms of classical fields.

We begin the discussion here by developing these remarks further. Then we will quantize these fields using the second quantization formalism, and the creation and annihilation operators at point $\vec{r}$ thus become field operators. We shall also study boson systems, boson fields, which, in principle, are not subjected to the Pauli exclusion principle.

1. Classical Field Theory

To describe the continuous evolution of a physical system with an infinite number of degrees of freedom, it is often expedient to introduce a functional (a field) $\varphi\,(t,\,\vec{x})$ such that a value of the field is defined at each point in the space $\vec{x}$. The meaning of such a functional in physics will be adapted to suit each field treated, probability amplitude of presence, amplitude of an electric field, etc. Next, we introduce a fundamental scalar function in a flat Minkowski space characterized by the metric $\eta_{\mu\,\nu}$ that is dependent on the functional $\varphi(x^\mu)$ with $x^\mu = (x^0,\,x^i) = (t,\,\vec{x})$ and its first derivatives $\partial_\nu\,\varphi(x^\mu) = \partial\,\varphi(x^\mu)/\partial x^\nu$. To simplify the notation, the four-vector x^μ is denoted x and the functional associated with the field under study $\varphi(x)$. By analogy with the analytical mechanics of a system of discrete point particles, we will assume that there exists a Lagrangian density, a fundamental scalar function

$$\mathcal{L}\,(\varphi,\,\partial_\nu\,\varphi) = \mathcal{L}\,(\varphi(x^\mu),\,\partial_\nu\,\varphi(x^\mu)) \tag{17.1}$$

containing all the information on the physical state of the system and the associated observables.

In essence, this Lagrangian density, integrated over the whole space, should lead to the usual Lagrangian of the system:

$$L = \int \mathcal{L}\left(\varphi,\, \partial_\mu\, \varphi\right) d^3 x \tag{17.2}$$

and the functional $\varphi(x)$ will play the role of the generalized coordinates of analytical mechanics (cf. Chap. 1) in the Lagrangian density.

1.1 Variational Principle and Field Equations

Using relation (17.2), the Hamiltonian action takes the form

$$S = \int L\, dt = \int \mathcal{L}\, d^3 x\, dt = \int \mathcal{L}\, d^4 x. \tag{17.3}$$

We postulate the existence of the following variational principle: Field equations correspond to a stationary Hamiltonian action:

$$\delta\, S = 0 \tag{17.4}$$

for zero field variations at the boundary of the integration zone.

Now, let us evaluate the variation of the Hamiltonian action (17.3):

$$\delta\, S = \delta \int \mathcal{L}\, d^4 x = \int \delta\, \mathcal{L}\, d^4 x. \tag{17.5}$$

Because the Lagrangian density depends on the variables $\varphi(x)$ and $\partial_\mu\, \varphi$, we obtain

$$\delta\, \mathcal{L} = \frac{\partial\, \mathcal{L}}{\partial\, \varphi}\, \delta\, \varphi + \frac{\partial\, \mathcal{L}}{\partial\, \partial_\mu\, \varphi}\, \delta\, \partial_\mu\, \varphi. \tag{17.6}$$

The last term can also be written

$$\frac{\partial\, \mathcal{L}}{\partial\, \partial_\mu\, \varphi}\, \delta\, \partial_\mu\, \varphi = \frac{\partial\, \mathcal{L}}{\partial\, \partial_\mu\, \varphi}\, \partial_\mu\, \delta\, \varphi$$

$$= \partial_\mu\left(\frac{\partial\, \mathcal{L}}{\partial\, \partial_\mu\, \varphi}\, \delta\, \varphi\right) - \partial_\mu\left(\frac{\partial\, \mathcal{L}}{\partial\, \partial_\mu\, \varphi}\right) \delta\, \varphi. \tag{17.7}$$

Putting this form into the Hamiltonian actions (17.6) and (17.5), we obtain

$$\delta\, S = \int \left(\frac{\partial\, \mathcal{L}}{\partial\, \varphi} - \partial_\mu\, \frac{\partial\, \mathcal{L}}{\partial\, \partial_\mu\, \varphi}\right) \delta\, \varphi\, d^4\, x + \int \partial_\mu\left(\frac{\partial\, \mathcal{L}}{\partial\, \partial_\mu\, \varphi}\, \delta\, \varphi\right) d^4 x. \tag{17.8}$$

By setting

$$j^\mu\,(x) = \frac{\partial\, \mathcal{L}}{\partial\, \partial_\mu\, \varphi}\, \delta\, \varphi, \tag{17.9}$$

the integrand of the second integral will take the form of a current quadrivergence $\int \partial_\mu j^\mu (x) \, d^4 (x)$, and the introduction of Ostogradsky's theorem will transform this volume integral into a surface integral which cancels itself out on the boundary surface Σ of the integration zone. This corresponds to the conservation of the current $j^\mu (x)$. This is Noether's theorem associated with the invariance of the Lagrangian density during an overall transformation, as we will have occasion to see in detail in the next chapter. The variational principle (17.4) is therefore written as follows:

$$\delta S = \int \left[\frac{\partial \mathcal{L}}{\partial \varphi} - \partial_\mu \left(\frac{\partial \mathcal{L}}{\partial \partial_\mu \varphi} \right) \right] \delta \varphi \, d^4 x = 0 \quad \forall \, \delta \varphi. \tag{17.10}$$

Thus reformulated, the least-action principle implies that the integrand must be zero irrespective of $\delta \varphi$, which leads to the Euler–Lagrange field equations:

$$\boxed{ \frac{\partial \mathcal{L}}{\partial \varphi} - \partial_\mu \frac{\partial \mathcal{L}}{\partial \partial_\mu \varphi} = 0 } . \tag{17.11}$$

1.2 Hamiltonian Formalism

The canonical conjugate momentum of the field $\varphi(x)$ is defined by the relation

$$\pi(x) = \frac{\partial \mathcal{L}}{\partial \partial_0 \varphi} = \frac{\partial \mathcal{L}}{\partial \dot{\varphi}} \tag{17.12}$$

and the Hamiltonian density $\mathcal{H}$ takes the same form as the Hamiltonian H of analytical mechanics:

$$\mathcal{H} = \pi \, \dot{\varphi} - \mathcal{L}, \tag{17.13}$$

yielding, following integration over the entire space, the Hamiltonian

$$H = \int \mathcal{H} \, d^3 x. \tag{17.14}$$

Using an analogous procedure to that used in Chap. 1, we obtain Poisson-type field equations

$$\dot{\varphi} = \{\varphi, \, \mathcal{H}\} \qquad \dot{\pi}, = \{\pi, \, \mathcal{H}\}. \tag{17.15}$$

2. Massive Scalar Fields

2.1 Field Equation

A real scalar field $\varphi (x^\mu) \equiv \varphi (x)$ is associated to a spinless particle with mass m and the Lagrangian density of the field is written (using the notation

$\partial^\mu = \eta^{\mu\nu} \partial_\nu$) in the form

$$\mathcal{L} = \frac{1}{2} \left(\partial_\mu \varphi \, \partial^\mu \varphi - m^2 \, \varphi^2 \right) . \tag{17.16}$$

The Euler–Lagrange equation of this field or the propagation equation of the field associated with spinless mass particle is the Klein–Gordon equation

$$\left(\partial_\mu \, \partial^\mu + m^2 \right) \varphi \equiv (\Box + m^2) \, \varphi = 0 . \tag{17.17a}$$

Since the d'Alembertian $\Box = \frac{\partial^2}{\partial t^2} - \nabla^2$, we are led to the field equation

$$\left(\frac{\partial^2}{\partial t^2} - \nabla^2 + m^2 \right) \varphi = 0 . \tag{17.17b}$$

In a $(+ - - -)$ signature of the Minkowski space-time, usually chosen in field theory, the Lagrangian and Hamiltonian densities are written

$$\mathcal{L} = \frac{1}{2} \left(\dot\varphi^2 - (\vec\nabla \varphi)^2 - m^2 \, \varphi^2 \right) , \tag{17.18}$$

$$\mathcal{H} = \frac{1}{2} \left(\pi^2 + (\vec\nabla \varphi)^2 + m^2 \, \varphi^2 \right) . \tag{17.19}$$

If the scalar field $\varphi(x)$ is not restricted to the real field the Lagrangian density for a complex field will become

$$\mathcal{L} = \partial_\mu \varphi \, \partial^\mu \varphi^* - m^2 \, \varphi^* \, \varphi \tag{17.20}$$

and the variation with respect to φ or to φ^* will lead to the same Klein–Gordon equation (17.17) for the field $\varphi(x)$ or its complex conjugate $\varphi^*(x)$. By multiplying (17.17) by φ^* and the Hermitian conjugate of (17.17) by φ, we obtain through term-by-term subtraction:

$$\varphi^* \frac{\partial^2 \varphi}{\partial t^2} - \varphi \frac{\partial^2 \varphi^*}{\partial t^2} = \varphi^* \, \nabla^2 \varphi - \varphi \, \nabla^2 \varphi^* , \tag{17.21}$$

which may also be written in the form

$$\partial_t \left(\varphi^* \, \partial_t \varphi - \varphi \, \partial_t \varphi^* \right) = \vec\nabla \cdot \left(\varphi^* \, \vec\nabla \varphi - \varphi \, \vec\nabla \varphi^* \right) . \tag{17.22}$$

However, the condensed notation

$$\varphi^* \, \partial_t \varphi - \varphi \, \partial_t \varphi^* = \varphi^* \, \overleftrightarrow{\partial_t} \, \varphi , \tag{17.23}$$

is often used. With the arrows indicating which function is differentiated, Eq. (17.22) is also written as follows:

$$\partial_t \left(\varphi^* \, \overleftrightarrow{\partial_t} \, \varphi \right) = \vec\nabla \cdot \left(\varphi^* \, \overleftrightarrow{\vec\nabla} \, \varphi \right) . \tag{17.24}$$

There thus exists a complex function from which the following quantity can be constructed:

$$\varrho(x) \; \alpha \; \varphi^* \; \overset{\leftrightarrow}{\partial_t} \; \varphi, \tag{17.25}$$

whose integral over the whole space is conserved in time. Equation (17.22) can then, in fact, be written in the form of a hydrodynamic continuity equation:

$$\partial_t \; \varrho(x) + \vec{\nabla} \cdot \vec{J} = 0 \quad \text{with} \quad \vec{J} = \varphi \; \overset{\leftrightarrow}{\nabla} \; \varphi^*. \tag{17.26}$$

Unfortunately, the form taken by $\varrho(x)$ in (17.25) shows that the function is not necessarily positive definite. We will therefore choose among the possible solutions of (17.17) those that fulfill the condition:

$$\varrho(x) \geq 0. \tag{17.27}$$

2.2 Solutions of the Propagation Equation

Let us introduce the Fourier transform of the field function:

$$\varphi(x) = \int \exp(-i \; k.x) \; \varphi(k) \; d^4 \, k,$$

with the following notation for the scalar product:

$$k \cdot x = k_\mu \, x^\mu = k_0 \, x^0 - \mathbf{k} \cdot \mathbf{x} = k_0 \, t - \mathbf{k} \cdot \mathbf{x}. \tag{17.28}$$

Substituting into the Klein–Gordon equation (17.17), we obtain

$$0 = (\Box + m^2) \, \varphi(x) = \int \exp\left(-i \; k \cdot x\right) \left[-k^2 + m^2\right] \varphi(k) \; d^4 \, k. \tag{17.29}$$

This imposes the following constraint on the values of the four-wave-vector k^μ:

$$0 = -k^2 + m^2 = -k_\mu \, k^\mu + m^2 = -k_0^2 + \mathbf{k}^2 + m^2, \tag{17.30}$$

a constraint that can also be written in the following different form:

$$k_0^2 = \omega^2 = \mathbf{k}^2 + m^2 \quad \text{and} \quad k_0 = \omega = \pm \sqrt{m^2 + |\mathbf{k}|^2}. \tag{17.31}$$

There exist two possible solutions, corresponding to the positive and negative values of the frequency ω. Solution (17.27) is written thus:

$$\varphi(x) = \varphi_+(x) + \varphi_-(x),$$
$$\varphi_\pm(x) = \int \varphi(k) \; \exp\left[-i \left(\pm \, \omega \, t - \mathbf{k} \cdot \mathbf{x}\right)\right] \; d^4 \, k. \tag{17.32}$$

The function $\varrho(x)$ can be expressed with positive frequency solutions, that is, for $k_0 = \omega = +\sqrt{m^2 + |\mathbf{k}|^2} \,)0$ and definition (17.25) here becomes

$$\varrho(x) = i \; \varphi_+^* \; \overset{\leftrightarrow}{\partial_t} \; \varphi_+. \tag{17.33}$$

By integrating over the entire space using (17.32), we obtain

$$\int \varrho(x) \, d^3 \, x = (2\pi)^3 \int 2\omega \, \varphi^*(k) \, \varphi(k) \, d^3 \, k \geq 0. \tag{17.34}$$

Analogous reasoning will lead to the same result with the negative frequencies. We therefore impose the orthonormalization condition on the field functions of the wave vector $\mathbf{k}$:

$$\int d^3 \, x \, \varphi^*_{\mathbf{k}}(x) \, i \, \overleftrightarrow{\partial}_t \, \varphi_{\mathbf{k}'}(x) = \delta^3 \, (\mathbf{k} - \mathbf{k}') \tag{17.35}$$

and the orthogonality condition on the positive and negative energy field functions:

$$\int d^3 \, x \, \varphi^*_- \, i \, \overleftrightarrow{\partial}_t \, \varphi_+ = 0. \tag{17.36}$$

A free spinless particle with mass m is thus described by a plane wave

$$\varphi_+ \, (x) = \frac{1}{\sqrt{2\omega(\mathbf{k})}} \, \exp \left(-i \, k \cdot x\right). \tag{17.37}$$

The general solution of the Klein–Gordon equation can therefore be written with (17.32) and (17.37) and a unit-normalization of (17.35):

$$\varphi(x) = \frac{1}{(2\pi)^{3/2}} \int \frac{d^3 \, k}{\sqrt{2\omega(\mathbf{k})}} \, [a(\mathbf{k}) \, \exp \left(-i \, k \cdot x\right) + a^*(\mathbf{k}) \, \exp \left(i \, k \cdot x\right)] \, . \tag{17.38}$$

The conjugate momentum, calculated with definition (17.12), is of the form

$$\pi(x) = \dot{\varphi}(x)$$

$$= \frac{i}{(2\pi)^{3/2}} \int \frac{d^3 \, k}{\sqrt{2\omega(\mathbf{k})}} \, \omega(\mathbf{k}) \, [a^*(\mathbf{k}) \, \exp \left(i \, k \cdot x\right) - a(\mathbf{k}) \, \exp \left(-i \, k \cdot x\right)] \, . \tag{17.39}$$

These expressions will prove extremely useful in the definition of the quantization procedure. It should be noted here that some authors prefer to introduce the normalization $(2\pi)^{-3}$ in $\varphi(x)$ in (17.38), which leaves a $(2\pi)^3$ coefficient in equation (17.35). The coefficient $(2\pi)^{-3/2}$ allows one to normalize (17.35) to unity.

The form (17.38) of the functional is not invariant with respect to the Lorentz group, and it does not eliminate the negative energy contributions. It is therefore necessary to find a procedure that eliminates undesirable solutions.

Let us recall the fundamental property of the Dirac function:

$$\delta \, (x^2 - a^2) = \frac{1}{2 \, | \, a \, |} \, \{\delta \, (x - a) + \delta \, (x + a)\} \, . \tag{17.40}$$

By introducing Heaviside's step function:

$$\theta(x) = 0 \quad \text{if} \quad x < 0$$
$$= 1 \quad \text{if} \quad x > 0 \tag{17.41}$$

we eliminate the $a \langle 0$ contributions by noting that

$$\delta\,(x^2 - a^2)\,\theta(x) = \frac{1}{2a}\,\delta\,(x - a) \qquad \text{with} \qquad a > 0. \tag{17.42}$$

We can therefore write the following equality:

$$\int d^4\,k\,\delta\,(k^2 - m^2)\,\theta\,(\omega) = \int d^3\,k\,d\omega\,\delta\,(\omega^2 - \mathbf{k}^2 - m^2)\,\theta\,(\omega)$$
$$= \int d^3\,k \int \frac{d\,\omega}{2\,|\,\omega\,|}\,\left[\delta\,(\omega - \sqrt{\mathbf{k}^2 + m^2})\,\theta\,(\omega) + \delta\,(\omega + \sqrt{\mathbf{k}^2 + m^2})\,\theta\,(\omega)\right]. \tag{17.43}$$

Because the last term makes a zero contribution to the integral as a result of property (17.42), $\int d^4\,k\,\delta\,(k^2 - m^2)\,\theta(\omega)$ can be replaced with $\int d^3k/2\omega$. To eliminate negative energy contributions, we will have to replace the Fourier transform (17.38) with the functional $\varphi(x)$ by

$$\varphi(x) = \frac{1}{(2\pi)^{3/2}} \int \frac{d^3\,k}{2\omega(\mathbf{k})}\,[b\,(\mathbf{k})\,\exp\,(-i\,k\cdot x) + b^*(\mathbf{k})\,\exp\,(i\,k\cdot x)]\,, \tag{17.44}$$

which modifies relation (17.34), while leaving the definite integral positive. This boils down to replacing $a(\mathbf{k})$ by $(1/\sqrt{2\omega(\mathbf{k})})\,b(\mathbf{k})$ and $a^*(\mathbf{k})$ by $1/(\sqrt{2\omega(\mathbf{k})})\,b^*(\mathbf{k})$ in equation (17.38) of the functional $\varphi(x)$.

3. Higgs Fields

The Higgs scalar field was introduced to describe the process of acquisition of mass by some gauge bosons during the spontaneous symmetry breaking of gauge fields. Consider a complex scalar field with Lagrangian density

$$\mathcal{L}_H = \partial_\mu\,\varphi\,\partial^\mu\,\varphi^* - V\,(\varphi^*\,\varphi)\,, \tag{17.45}$$

in which the Higgs potential $V(\varphi^*\,\varphi)$ is chosen in the form

$$V\,(\varphi^*\,\varphi) = M^2\,(\varphi^*\,\varphi) + h\,(\varphi^*\,\varphi)^2 \qquad M^2 < 0 \quad h > 0. \tag{17.46}$$

Particular attention should be paid to the fact that M is not the mass of the boson associated with the Higgs complex scalar field, but rather a simple parameter.

If we introduce the density by setting

$$\varrho = \varphi^*\,\varphi = |\,\varphi\,|^2 \tag{17.47}$$

for $M^2 > 0$ the potential $V(\varrho)$ is symmetrical and has only one minimum for $\varrho = 0$ whereas for $M^2 < 0$, there is a second minimum:

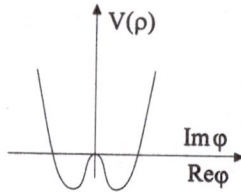

$$\varrho = \varphi^* \, \varphi = -M^2/2h = \frac{v^2}{2} > 0 \, . \tag{17.48}$$

The variation of $\mathcal{L}_H$ with respect to φ^* leads to the Euler–Lagrange equations of the Higgs scalar field:

$$0 = \frac{\delta \, \mathcal{L}_H}{\delta \, \varphi^*} \implies \partial_\mu \, \partial^\mu \, \varphi + \frac{\partial \, V}{\partial \, \varphi^*} = 0 \, , \tag{17.49}$$

and, by introducing the d'Alembertian of the Minkowski space and the partial derivative of the Higgs potential (17.46), we have

$$\Box \, \varphi + M^2 \, \varphi + 2h \, \varphi \, (\varphi^* \, \varphi) = 0 \, . \tag{17.50}$$

For a constant value $\varphi = \varphi_0$ of the scalar field, we obtain the obvious solution

$$0 = \frac{\partial \, V}{\partial \, \varphi^*} = \varphi \, (M^2 + 2h \, \varphi^* \, \varphi) \, . \tag{17.51}$$

- If $M^2 > 0$, only the solution $\varphi_0 = 0$ is possible, which corresponds to the only minimum of the Higgs potential,
- If $M^2 < 0$, the solution $\varphi_0 = 0$ is unstable and a second solution is possible:

$$\varrho_0 = \varphi_0 \, \varphi_0^* = -\frac{M^2}{2h} = \frac{v^2}{2} > 0 \, . \tag{17.52}$$

We can obtain from this the solution φ_0 describing this fundamental state:

$$\langle 0 \mid \varphi \mid 0 \rangle \; = \varphi_0 = \frac{v}{\sqrt{2}} \, \exp{(i \, \beta)} \, . \tag{17.53}$$

This solution breaks the global symmetry of the gauge field (see Chap. 8):

$$\varphi \; \rightarrow \; \varphi' = \exp{(i \, \wedge)} \, \varphi \, , \tag{17.54}$$

since, for the ground state, (17.53) becomes during the transformation:

$$\varphi_0' = \; \langle 0 \mid \varphi' \mid 0 \rangle \; = \; \langle 0 \mid \exp{(i \, \wedge)} \, \frac{v}{\sqrt{2}} \, \exp{(i \, \beta)} \mid 0 \rangle$$

$$= \exp{(i \, \wedge)} \, \exp{(i \, \beta)} \, \frac{v}{\sqrt{2}} \neq \varphi_0 \, . \tag{17.55}$$

The minimum φ_0 of φ has broken the initial symmetry, and φ_0 is no longer invariant under global gauge transformation. This process is termed spontaneous symmetry breaking and it confers a mass to gauge bosons.

Now let us replace φ with $\varphi + \varphi_0$ in the field equation (17.50). For simplicity we choose the phase β to be zero, and separate the real part φ_1 from the imaginary part φ_2 of the Higgs field:

$$\varphi = \varphi + \varphi_0 = \frac{1}{\sqrt{2}} \, (v + \varphi_1 + i \, \varphi_2). \qquad (17.56)$$

Inserting this into the Lagrangian density $\mathcal{L}_H$ of the Higgs field, we obtain

$$\partial_\mu \, \partial^\mu \, \varphi_1 + i \, \partial_\mu \, \partial^\mu \, \varphi_2 + M^2 \, (v + \varphi_1 + i \, \varphi_2) + h \, (v + \varphi_1 + i \, \varphi_2)^2 \, (v + \varphi_1 - i \, \varphi_2) = 0.$$

We separate the real and imaginary parts, which gives two coupled equations:

$$\partial_\mu \, \partial^\mu \, \varphi_1 + M^2 \, (v + \varphi_1) + h \, \left[(v + \varphi_1)^2 + \varphi_2^2 \right] \, (v + \varphi_1) = 0,$$
$$\partial_\mu \, \partial^\mu \, \varphi_2 + M^2 \, \varphi_2 + h \, \left[(v + \varphi_1)^2 + \varphi_2^2 \right] \, \varphi_2 = 0. \qquad (17.57)$$

If we use the value (17.48) of $h = -M^2/v^2$, these equations can also take the form

$$\partial_\mu \, \partial^\mu \, \varphi_1 - 2M^2 \, \varphi_1 = -h \, (3v \, \varphi_1^2 + v \, \varphi_2^2 + \varphi_1^3 + \varphi_1 \, \varphi_2^2),$$
$$\partial_\mu \, \partial^\mu \, \varphi_2 = -h \, (2v \, \varphi_1 \, \varphi_2 + \varphi_1^2 \, \varphi_2 + \varphi_2^3). \qquad (17.58)$$

The right-handed terms represent interaction terms of real scalar fields φ_1 and φ_2. The real field φ_1 has therefore acquired mass m_1 (the coefficient of φ_1 in the (17.16)-type field equation) whereas the field φ_2 remains a field with zero mass m_2:

$$m_1^2 = -2M^2 \quad > 0,$$
$$m_2^2 = 0. \qquad (17.59)$$

This Higgs process is extremely important in the definition of the mass of gauge bosons, and has received spectacular confirmation through the determination of the mass of the intermediate bosons W^+, W^- and Z^0. The Higgs boson, a neutral scalar boson with mass $m_1 = \sqrt{-2M^2}$ has yet to be experimentally confirmed and is currently the object of intense research activity at the major particle accelerators.

The most recent research findings of particle physicists assign an upper and a lower limit to the mass of this boson (assuming that it really exists and conform to the proposed process):

$$60 - 62 \text{ GeV} \; < m_H < 200 - 250 \text{ GeV}. \qquad (17.60)$$

4. The Electromagnetic Field

The functional usually chosen to describe the electromagnetic (e.m.) field is the potential four-vector:

$$A^\mu (x^\nu) \equiv A^\mu (x) = (A^0, \ A^i) = (\phi, \ \vec{A}). \qquad (17.61)$$

The antisymmetrical second-order Faraday tensor

$$F_{\mu \nu} = \partial_\mu A_\nu - \partial_\nu A_\mu \qquad (17.62)$$

next determines the electric field $\vec{E}$ with the components

$$F_{0\,i} = -E_i = \partial_0 A_i - \partial_i A_0 = -(\partial_t \vec{A} + (\vec{\nabla} \ \phi)_i) \qquad (17.63)$$

and the magnetic field $\vec{B}$ with the other components:

$$F_{i\,j} = B_k = \partial_i A_j - \partial_j A_i = -(\vec{\nabla} \wedge \vec{A})_k. \qquad (17.64)$$

The Faraday tensor is invariant under gauge transformation (see Chap. 8):

$$\begin{aligned} A_\mu &\rightarrow \tilde{A}_\mu = A_\mu + \partial_\mu \chi, \\ F_{\mu \nu} &\rightarrow \tilde{F}_{\mu \nu} = F_{\mu \nu}, \end{aligned} \qquad (17.65)$$

and the Jacobi relation applied to the antisymmetric Faraday tensor:

$$\partial_\mu F_{\lambda \nu} + \partial_\lambda F_{\nu \mu} + \partial_\nu F_{\mu \nu} = 0, \qquad (17.66)$$

leads to the first group of Maxwell equations:

$$\partial_t \vec{B} + \vec{\nabla} \wedge \vec{E} = 0 \quad \text{and} \quad \vec{\nabla} \cdot \vec{B} = 0. \qquad (17.67)$$

Because relation (17.65) does not determine the vector potential uniquely, we introduce a gauge condition, for example, the Lorentz gauge:

$$\partial_\mu A^\mu = \partial_0 A^0 + \partial_i A^i = \partial_t \phi + \vec{\nabla} \cdot \vec{A} = 0. \qquad (17.68)$$

The second group of Maxwell equations can be obtained from the Lagrangian density of the e.m. field in interaction with the existing electric charges represented by a four-current-density:

$$J^\mu = (J^0, \ J^i) = (\varrho, \ \vec{J}), \qquad (17.69)$$

$$\mathcal{L}_{e\,m} = -\frac{1}{4} F_{\mu \nu} F^{\mu \nu} - J^\mu A_\mu. \qquad (17.70)$$

The free electromagnetic field likewise corresponds to a zero four-current J^μ.

The variation of $\mathcal{L}_{e\,m}$ with respect to the vector potential A_μ leads to the Euler–Lagrange equation of the e.m. field:

$$\partial_\mu F^{\mu \nu} = J^\nu \qquad (17.71)$$

corresponding to the second group of Maxwell equations

$$\vec{\nabla} \cdot \vec{E} = \varrho \quad \text{and} \quad \partial_t \vec{E} + \vec{J} = \vec{\nabla} \wedge \vec{B}. \tag{17.72}$$

Relation (17.71), together with the Lorentz gauge condition (17.68), leads to the propagation equation of four-potential:

$$\Box A^\nu = J^\nu. \tag{17.73}$$

In vacuum, the propagation of e.m. waves is described by the d'Alembert equation

$$\Box A^\nu = 0. \tag{17.74}$$

The analysis made for the Klein–Gordon equation is therefore valid if one considers photons as particles with zero mass. Inserting over the Fourier transform of the vector potential

$$A_\mu(x) = \int d^3k \, a_\mu(\mathbf{k}) \exp(-i \, k \cdot x) \tag{17.75}$$

into the covariant form of the d'Alembert equation (17.74) imposes a four-wave-vector with zero length:

$$k^2 = k_\mu \, k^\mu = k_0^2 - \mathbf{k}^2 = 0 \quad \text{and} \quad k_0 = \omega = \pm \, |\, \mathbf{k} \,|. \tag{17.76}$$

The vector potential can therefore take the form of (17.38) for each component.

Normalized positive frequency waves will, for example, be written

$$A_\mu(x) = \frac{1}{(2\pi)^{3/2}} \int \frac{d^3k}{\sqrt{2\omega(\mathbf{k})}} \, a_\mu(\mathbf{k}) \exp(-i \, k \cdot x). \tag{17.77}$$

Let us introduce the polarization of the e.m. waves with the unit vector $\vec{e}_\mu^\lambda$, that is by expanding $a(\mathbf{k})$ in this base:

$$a_\mu(\mathbf{k}) = \sum_\lambda \vec{e}_\mu^\lambda(\mathbf{k}) \, a_\lambda(\mathbf{k}). \tag{17.78}$$

We demonstrate that the Lorentz gauge condition is satisfied $A_\mu(x)$ by writing, with (17.77) and (17.78),

$$0 = \partial_\mu A^\mu = \frac{1}{(2\pi)^{3/2}} \sum_\lambda \int \frac{d^3k}{\sqrt{2\omega(\mathbf{k})}} \, \vec{e}_\mu^\lambda(-i \, k^\mu) \, a_\lambda(\mathbf{k}) \exp(-i \, k \cdot x). \tag{17.79}$$

We obtain from this the constraint relation which augments (17.76) with the condition

$$k^\mu \, \vec{e}_\mu^\lambda = 0. \tag{17.80}$$

Of the four degrees of freedom λ of $\vec{e}_\mu^\lambda$, only three are therefore independent since the vectors $\vec{e}_\mu^\lambda$ are orthogonal to k^μ. Constraint (17.76) also imposes that one of the vectors orthogonal to k^μ should be the vector k_μ itself. There are therefore two linearly independent vectors $\vec{e}_\mu^\lambda$, with the third $\vec{e}_\mu^3$ being proportional to k_μ:

$$\vec{e}_\mu^3 = \lambda \, k_\mu \, . \tag{17.81}$$

The proportionality coefficient λ can be chosen zero by a suitable gauge transformation. We are then left with only two non-zero polarization vectors that are orthogonal to the e.m. wave propagation direction. This is the transversality property of e.m. waves (or the transversality of the mediating boson photon of e.m. waves).

The vector potential is therefore written in the general form

$$A_\mu\,(x) = \frac{1}{(2\pi)^{3/2}} \sum_{\lambda=1,2} \int \frac{d^3\,k}{\sqrt{2\omega\,(\mathbf{k})}}$$
$$\times \left[\vec{e}_\mu^\lambda\,(\mathbf{k})\,a_\lambda\,(\mathbf{k})\,\exp(-i\,k.x) + \vec{e}_\mu^\lambda\,(\mathbf{k})\,a_\lambda^*(\mathbf{k})\,\exp(i\,k.x) \right] \,. \tag{17.82}$$

5. Quantization of a Boson Field

5.1 Schrödinger Canonical Quantization

The state of a field at instant t is described by $\mid \psi(t) \rangle$ and $\varphi(x)$, and $\pi(x)$ become operators obeying commutation relations of type (17.85). The description of a field takes the form of a functional $\langle \varphi \mid \psi \rangle = \psi(\varphi, t)$.

Because the operators $\varphi(x)$ and $\pi(x)$ are acting on $\mid \varphi \rangle$ and $\mid \psi \rangle$, we obtain

$$\varphi(x) \mid \varphi \rangle = \varphi \mid \varphi \rangle \,,$$
$$\pi(x)\,\psi(\varphi, t) = -i\hbar\,\frac{\delta\,\psi}{\delta\,\varphi} \, . \tag{17.83}$$

The correspondence principle associates the functional derivative $-i\hbar\,\delta/\delta\,\varphi$ to $\pi(x)$. The evolution of ψ is given by the Schrödinger equation

$$\mathcal{H}\,\left(\varphi,\,-i\hbar\,\frac{\delta}{\delta\,\varphi}\right)\,\psi(\varphi, t) = i\hbar\,\frac{\partial\,\psi}{\partial\,t} \, . \tag{17.84}$$

The square modulus of $\psi(\varphi,\,t)$ represents the probability density that a field will have the configuration $\varphi(x)$ at instant t.

5.2 Heisenberg Canonical Quantization

We postulate that the field $\varphi\,(x^\mu)$ and its conjugate momentum $\pi\,(x^\mu)$ satisfy at a given instant the same types of commutation rules as the position and momentum operators:

$$
\begin{aligned}
[\varphi\,(x,t),\ \varphi\,(y,t)] &= 0\,, \\
[\pi\,(x,t),\ \pi\,(y,t)] &= 0\,, \\
[\varphi\,(x,t),\ \pi\,(y,t)] &= i\,\delta^3\,(x-y)\,.
\end{aligned}
\tag{17.85}
$$

If we consider that $a(\mathbf{k})$ is an operator in (17.38) and (17.39), and replace $a^*(\mathbf{k})$ with the operator $a^+(\mathbf{k})$, the foregoing relations will lead to the commutation relations

$$
\begin{aligned}
[a\,(\mathbf{k}),\ a\,(\mathbf{k}')] &= 0\,, \\
[a^+\,(\mathbf{k}),\ a^+\,(\mathbf{k}')] &= 0\,, \\
[a\,(\mathbf{k}),\ a^+\,(\mathbf{k}')] &= \delta^3\,(\mathbf{k}-\mathbf{k}')\,.
\end{aligned}
\tag{17.86}
$$

The Hamiltonian density (17.18), expanded with forms (17.38) and (17.39) of $\varphi\,(\mathbf{x})$ and $\pi\,(\mathbf{x})$, will yield, following integration over space variables, the Hamiltonian

$$
H = \frac{1}{2} \int d^3 k\,\omega\,(\mathbf{k})\,\left[a^+\,(\mathbf{k})\,a\,(\mathbf{k}) + a\,(\mathbf{k})\,a^+\,(\mathbf{k}) \right]\,.
\tag{17.87}
$$

The Hamiltonian is therefore in the form of a superposition of harmonic oscillators of wave vector $\mathbf{k}$, whereas the operators $a\,(\mathbf{k})$ and $a^+\,(\mathbf{k})$ are in the form of boson creation and annihilation operators of wave vector $\mathbf{k}$.

5.3 Vacuum Energy

The quantum vacuum is defined by the equation:

$$
a\,(\mathbf{k})\,|\,0\rangle = 0
$$

and a wave vector boson $\mathbf{k}$ will be obtained by the action of the operator $a^+\,(\mathbf{k})$ on the quantum vacuum:

$$
a^+\,(\mathbf{k})\,|\,0\rangle\ =\ |\,\mathbf{k}\rangle\,.
$$

An n-boson vector state $\mathbf{k}$ is obtained by the repeated action of the creation operator

$$
|\,n\,(\mathbf{k})\rangle\ =\ \frac{1}{\sqrt{n\,k!}}\,a^+\,(\mathbf{k})^{\,n\,(\mathbf{k})}\,|\,0\rangle
\tag{17.88}
$$

and we go from $\mid n\,(\mathbf{k})\rangle$ to $\mid n(\mathbf{k}+1)\rangle$ or $\mid n\,(\mathbf{k}-1)\rangle$ by applying the creation and annihilation operators:

$$a^+\,(\mathbf{k}')\mid n\,(\mathbf{k})\rangle = \delta^3\,(\mathbf{k}-\mathbf{k}')\,(n\,(\mathbf{k})+1)^{1/2}\mid n\,(\mathbf{k})+1\rangle\,,$$
$$a\,(\mathbf{k}')\mid n\,(\mathbf{k})\rangle = \delta^3\,(\mathbf{k}-\mathbf{k}')\,(n\,(\mathbf{k}))^{1/2}\mid n\,(\mathbf{k})-1\rangle\,. \qquad (17.89)$$

This simplifies the calculation of the matrix element of H:

$$\langle n'\,(\mathbf{k}')\mid H\mid n\,(\mathbf{k})\rangle = \delta^3\,(\mathbf{k}-\mathbf{k}')\,\{n\,(\mathbf{k})+\tfrac{1}{2}\}\,\omega\,(\mathbf{k})\,\delta_{n\,n'}\,. \qquad (17.90)$$

The trace of the Hamiltonian operator, the integral over all the $\mathbf{k}$ vectors of the above mean value is easily evaluated:

$$\mathrm{Tr}\,H = \int H\,(\mathbf{k})\,\frac{d^3\,k}{(2\pi)^{3/2}} = \int \left(n\,(\mathbf{k})+\frac{1}{2}\right)\,\omega\,(\mathbf{k})\,\frac{d^3\,k}{(2\pi)^{3/2}}\,. \qquad (17.91)$$

The integral is divergent since $n\,(\mathbf{k}) = 0$ for the vacuum, and the trace for a field with mass m becomes

$$\frac{1}{2}\int \omega\,(\mathbf{k})\,\frac{d^3\,k}{(2\pi)^{3/2}} = \frac{1}{2}\int_0^\infty \sqrt{k^2+m^2}\,\frac{d^3\,k}{(2\pi)^{3/2}} = \infty\,. \qquad (17.92)$$

The energy of the vacuum is infinite. To eliminate this discrepancy, the vacuum energy is chosen by convention to be zero:

$$H\mid 0\rangle = 0\,, \qquad (17.93)$$

and to obtain this result, we replace all products of creation and annihilation operators with normal products (see Chap. 16):

$$\frac{1}{2}:a^+\,(\mathbf{k})\,a\,(\mathbf{k})+a\,(\mathbf{k})\,a^+\,(\mathbf{k}):=a^+\,(\mathbf{k})\,a\,(\mathbf{k})\,,$$
$$:H:=\int \frac{d^3\,k}{(2\pi)^{3/2}}\,\omega\,(\mathbf{k})\,a^+(\mathbf{k})\,a(\mathbf{k})\,. \qquad (17.94)$$

The trace of H no longer has a constant term:

$$\mathrm{Tr}:H:=\int n\,(\mathbf{k})\,\omega\,(\mathbf{k})\,\frac{d^3\,k}{(2\pi)^{3/2}}\,. \qquad (17.95)$$

The normal product has therefore subtracted the vacuum energy, and the boson field energy is no longer infinite.

5.4 Normalization of Boson States

To eliminate the negative energy contributions, we replace the boson creation and annihilation operators as in (17.44):

$$a \ (\mathbf{k}) \ \text{by} \ b \ (\mathbf{k})/\sqrt{2\omega \ (\mathbf{k})},$$
$$a^+ \ (\mathbf{k}) \ \text{by} \ b^+ \ (\mathbf{k})/\sqrt{2\omega \ (\mathbf{k})}. \tag{17.96}$$

This transforms the commutation relations (17.84) as follows:

$$[b \ (\mathbf{k}), \ b \ (\mathbf{k}')] = [b^+ \ (\mathbf{k}), \ b^+ \ (\mathbf{k}')] = 0,$$
$$[b \ (\mathbf{k}), \ b^+ \ (\mathbf{k}')] = 2\omega \ (\mathbf{k}) \ \delta^3 \ (\mathbf{k} - \mathbf{k}'), \tag{17.97}$$

but not the definition of the quantum vacuum:

$$b \ (\mathbf{k}) \ | \ 0\rangle \ = 0 \quad \text{and} \quad \langle 0 \ | \ b^+ \ (\mathbf{k}) = 0. \tag{17.98}$$

A boson wave vector $\mathbf{k}$ is then written from the quantum vacuum

$$b^+ \ (\mathbf{k}) \ | \ 0\rangle \ = \ | \ \mathbf{k}\rangle. \tag{17.99}$$

We obtain from this the normalization of the $\mathbf{k}$ wave boson states in the form

$$\langle \mathbf{k} \ | \ \mathbf{k}'\rangle \ = \ \langle 0 \ | \ b \ (\mathbf{k}) \ b^+ \ (\mathbf{k}') \ | \ 0\rangle \ = \ \langle 0 \ | \ [b \ (\mathbf{k}), \ b^+ \ (\mathbf{k}')] \ | \ 0\rangle$$
$$= \ \langle 0 \ | \ 2\omega \ (\mathbf{k}) \ \delta^3 \ (\mathbf{k} - \mathbf{k}') \ | \ 0\rangle \tag{17.100}$$

and because the vacuum is a unit-noralized state,

$$\langle 0 \ | \ 0\rangle \ = 1, \tag{17.101}$$

we infer the desired relation:

$$\langle \ \mathbf{k} \ | \ \mathbf{k}'\rangle \ = 2\omega \ (\mathbf{k}) \ \delta^3 \ (\mathbf{k} - \mathbf{k}'). \tag{17.102}$$

5.5 Absorption of a Boson

The absorption probability amplitude of a boson can be evaluated from the expression (17.44) for the boson field function and the foregoing relations:

$$\langle 0 \ | \ \varphi \ (x) \ | \ \mathbf{k}\rangle \ = \ \frac{1}{(2\pi)^{3/2}} \ \int \ \frac{d^3 \ k'}{2\omega \ (k')} \Big\{ \langle 0 \ | \ b \ (\mathbf{k}') \ | \ \mathbf{k}\rangle \ \exp \ (-i \ k' \cdot x)$$
$$+ \ \langle 0 \ | \ b^+ \ (\mathbf{k}') \ | \ \mathbf{k}\rangle \ \exp \ (i \ k \cdot x) \Big\}. \tag{17.103}$$

The last matrix element is zero following (17.98) and the first matrix element is given by (17.99) and (17.100), which, in the final analysis, leaves us with the absorption amplitude

$$\langle 0 \mid \varphi\,(x) \mid \mathbf{k} \rangle \;=\; \frac{1}{(2\pi)^{3/2}}\,\exp\,(-i\,k\cdot x),\qquad (17.104)$$

and, using the Hermitian conjugate

$$\langle \mathbf{k} \mid \varphi\,(x) \mid 0 \rangle \;=\; \frac{1}{(2\pi)^{3/2}}\,\exp\,(i\,k\cdot x).\qquad (17.105)$$

In order to describe the e.m. field, we will have to introduce the polarization λ of the photon:

$$A^{\mu}\,(x) = \frac{1}{(2\pi)^{3/2}}\,\sum_{\lambda=1,2}\,\int \frac{d^3\,k}{2\omega\,(\mathbf{k})}\Big\{\vec{e}_{\mu}^{\,\lambda}\,(\mathbf{k})\,b_{\lambda}\,(\mathbf{k})\,\exp\,(-i\,k\cdot x)$$

$$+\,\vec{e}_{\mu}^{\,\lambda\,*}\,(\mathbf{k})\,b_{\lambda}^{+}\,(\mathbf{k})\,\exp\,(i\,k\cdot x)\Big\}.\qquad (17.106)$$

This leads to the absorption amplitude of a photon:

$$\langle 0 \mid A^{\mu}\,(x) \mid k\lambda \rangle \;=\; \frac{1}{(2\pi)^{3/2}}\,\vec{e}_{\lambda}^{\,\mu}\,(\mathbf{k})\,\exp\,(-i\,k\cdot x).\qquad (17.107)$$

The absorption of a photon in a volume V in which the field energy is ω will be obtained by replacing the normalization coefficient $(2\pi)^{3/2}$ with $\sqrt{2\omega\,(\mathbf{k})\,V}$.

The creation amplitude of a photon with momentum k and polarization λ is obtained from the Hermitian conjugate of (17.107), that is with $\langle \mathbf{k}\,\lambda \mid A^{\mu}(x) \mid 0 \rangle$.

Chapter 18
Fermion Fields

The Pauli–Schrödinger equation which replaces p^2 by $(\vec{\sigma} \cdot \vec{p}\,)^2$, or π^2 by $(\vec{\sigma} \cdot \vec{\pi})^2 - \vec{\mu} \cdot \vec{B}$, enables us to take into account the spin of a particle (see Chap. 9). Yet it is a nonrelativistic equation in which the mass energy is not taken into consideration, since $E = \frac{p^2}{2m}$, whereas the relativistic energy is written (in c units)

$$E^2 = p^2 + m^2 \,. \tag{18.1}$$

The correspondence principle replacing E by $i\,\partial\,t$, and $\vec{p}$ by $-i\,\vec{\nabla}$ (in $\hbar = 1$ natural units) leads to the Klein–Gordon equation (see Chaps. 5 and 17)

$$(i\,\partial_t)^2\,\Psi\,(\vec{x},t) = (-i\,\vec{\nabla})^2\,\Psi\,(\vec{x},t) + m^2\,\Psi\,(\vec{x},t)\,. \tag{18.2}$$

It is usually written in the form

$$(-\partial_t^2 + \nabla^2 - m^2)\,\Psi\,(\vec{x},t) = 0\,, \tag{18.3}$$

and by introducing the $(+ - --)$ signature in the Minkowski space with $x = (t,\vec{x}) = x^\mu = (x^0,\,\vec{x})$ and the d'Alembertian

$$\Box = \partial_t^2 - \nabla^2\,, \tag{18.4}$$

the Klein–Gordon equation describes the propagation of the presence probability amplitude of a spinless particle with mass m:

$$(\Box + m^2)\,\Psi\,(x) = 0\,. \tag{18.5}$$

The Dirac equation is a first-order differential equation with the primary aim of taking into account the relativistic covariance and including the spin of a particle with mass m. First, we examine the case of the free particle, then that of an electrically charged particle in interaction with the electromagnetic field and, finally, the quantization of a fermion field.

1. Dirac Equation of a Free Particle

The linearized relativistic energy (18.1) is written in natural units

$$E = \vec{v} \cdot \vec{p} + \sqrt{1 - v^2}\, m \,. \tag{18.6}$$

We therefore assume a priori the existence of a Hamiltonian with analogous form:

$$H = \vec{\alpha} \cdot \vec{p} + \beta\, m \,, \tag{18.7}$$

in which $\vec{\alpha}$ is a vector operator and β a scalar operator to be determined. The eigenvalue equation of the linearized Hamiltonian will then be

$$\begin{aligned} E\, \Psi(x) &= H\, \Psi(x) \\ &= (\vec{\alpha} \cdot \vec{p} + m\,\beta)\, \Psi(x) \,, \end{aligned} \tag{18.8}$$

or, using the correspondence principle (1.242) of wave mechanics,

$$i\, \partial_t\, \Psi(x) = (\vec{\alpha} \cdot \vec{p} + m\,\beta)\, \Psi(x) \,. \tag{18.9}$$

Let us once more apply the operator E to Eq. (18.8) to determine the form of the operators $\vec{\alpha}$ and β:

$$\begin{aligned} E^2\, \Psi(x) &= (\vec{\alpha} \cdot \vec{p} + m\,\beta)\,(\vec{\alpha} \cdot \vec{p} + m\,\beta)\, \Psi(x) \\ &= (\vec{\alpha} \cdot \vec{p})^2\, \Psi(x) + m\,\beta\,(\vec{\alpha} \cdot \vec{p})\, \Psi(x) \\ &\quad + m\,(\vec{\alpha} \cdot \vec{p})\,\beta\, \Psi(x) + m^2\,\beta^2\, \Psi(x) \,. \end{aligned} \tag{18.10}$$

By comparing term-by-term with the eigenvalue equation written with (18.1), we obtain two expressions for $E^2\, \Psi$:

$$\begin{aligned} E^2\, \Psi &= (p^2 + m^2)\, \Psi \\ &= (\vec{\alpha} \cdot \vec{p})^2\, \Psi + \beta^2\, m^2\, \Psi + (m\,\beta\,(\vec{\alpha} \cdot \vec{p}) + m\,(\vec{\alpha} \cdot \vec{p})\,\beta)\, \Psi \,. \end{aligned} \tag{18.11}$$

We infer the following relations:

$$\alpha_i^2 = \mathbb{1} \,, \qquad \beta^2 = \mathbb{1} \,, \tag{18.12}$$

$$\alpha_i\,\beta + \beta\,\alpha_i = 0 \,, \tag{18.13}$$

$$\alpha_i\,\alpha_j + \alpha_j\,\alpha_i = 2\,\delta_{ij}\,\mathbb{1} \,. \tag{18.14}$$

α_i and β are Hermitian matrices if H is Hermitian and $\Psi(x) = \Psi(x^\mu) = \Psi(x^0,\ \vec{x}) = \Psi\,(t,\ \vec{x})$ is an n-row column vector or spinor if the matrices $\vec{\alpha}$ and β are $n \times n$ matrices.

The right or left multiplication of relation (18.13) by β shows, with (18.12), that

$$\alpha_i = -\beta\,\alpha_i\,\beta \,. \tag{18.15}$$

The trace of each member further yields

$$\mathrm{Tr}\ \alpha_i = -\mathrm{Tr}\ \beta\ \alpha_i\ \beta = -\mathrm{Tr}\ \alpha_i\,. \qquad (18.16)$$

This forces the matrices $\vec{\alpha}$ to be traceless. The same reasoning with (18.13) and the multiplication by α_i shows that β is also a traceless matrix:

$$\begin{aligned} \mathrm{Tr}\ \alpha_i &= 0\,, \\ \mathrm{Tr}\ \beta &= 0\,. \end{aligned} \qquad (18.17)$$

The matrices α_i and β, whose square is the identity matrix following (18.12), will have $+1$ or -1 as eingenvalues (real values). Because the trace is the sum of the eigenvalues (diagonal elements of the matrix), there should be as many $+1$ eigenvalues as -1. The matrices α_i and β are therefore of even dimension. Four 2×2 matrices (that is the three α_i matrices and matrix β) cannot simultaneously satisfy relations (18.12) and (18.14). We therefore have to introduce at least 4×4 matrices and a 1×4 spinor. Dirac sets:

$$\boxed{\gamma^0 = \beta \qquad \text{and} \qquad \gamma^i = \beta\ \alpha_i}\,. \qquad (18.18)$$

The Dirac Eq. (18.9) is written with the correspondence principle

$$i\ (\partial_0 + \alpha_k\ \partial_k)\ \Psi(x) - m\ \beta\ \Psi(x) = 0\,. \qquad (18.19)$$

Following multiplication from the left by the matrix β, and by using relations (18.18) and (18.12), we introduce the Dirac γ matrices:

$$i\ (\gamma^0\ \partial_0 + \gamma^k\ \partial_k)\ \Psi(x) - m\ \Psi(x) = 0\,. \qquad (18.20)$$

Using the usual covariant notation, this equation reduces to

$$(i\ \gamma^\mu\ \partial_\mu - m)\ \Psi = 0 \qquad \mu = 0, 1, 2, 3\,. \qquad (18.21)$$

The following notation is often used:

$$\rlap{/}{A}\ (\text{we read } A \text{ slash}) = \gamma^\mu\ A_\mu = \gamma^0\ A_0 + \gamma^i\ A_i = \gamma^0\ A_0 - \vec{\gamma} \cdot \vec{A}. \qquad (18.22)$$

This gives the simpler condensed form of the Dirac equation of a free particle with mass m:

$$\boxed{(i\ \rlap{/}{\partial} - m)\ \Psi(x) = 0}\,. \qquad (18.23)$$

2. Dirac γ Matrices

Relations (18.12), (18.14), and (18.18) lead to the anti-commutator of Dirac matrices

$$\{\gamma^\mu, \gamma^\nu\} = \gamma^\mu \gamma^\nu + \gamma^\nu \gamma^\mu = 2\eta^{\mu\nu} \mathbb{1}. \tag{18.24}$$

The Minkowski metric tensor $\eta^{\mu\nu}$ has a $(+ - --)$, or (-2), signature. It is represented by the 4×4 matrix

$$\eta_{\mu\nu} = \eta^{\mu\nu} = \begin{pmatrix} 1 & 0 & 0 & 0 \\ 0 & -1 & 0 & 0 \\ 0 & 0 & -1 & 0 \\ 0 & 0 & 0 & -1 \end{pmatrix}. \tag{18.25}$$

This signature is different from that usually employed in general relativity theory, the $(+2)$ signature, but is more convenient because of the form of the d'Alembertian it introduces.

The γ^μ matrices satisfy the relations of:

i)   hermiticity $(\gamma^0)^+ = \gamma^0$ and $(\gamma^0)^2 = \mathbb{1}$ (18.26)

ii)   antihermiticity $(\gamma^k)^+ = -\gamma^k$ and $(\gamma^k)^2 = \mathbb{1}$ (18.27)

as can be seen in the definitions and by setting $\mu = \nu$ in anti-commutator (18.24). We condense these relations by writing:

$$\gamma^{\mu+} = \gamma^0 \gamma^\mu \gamma^0, \qquad \mu = 0, k. \tag{18.28}$$

We often introduce the right and left parts of a $\Psi(x)$ spinor with the γ_5 matrix:

$$\gamma_5 = \begin{pmatrix} 0 & \mathbb{1} \\ \mathbb{1} & 0 \end{pmatrix} = i\,\gamma^0 \gamma^1 \gamma^2 \gamma^3 = \gamma_5^+ = \gamma^5. \tag{18.29}$$

This matrix obeys the following relations:

$$\{\gamma_5, \gamma^\mu\} = 0, \qquad \gamma_5^2 = \mathbb{1}, \qquad \gamma_5^+ = -\gamma^0 \gamma_5 \gamma^0. \tag{18.30}$$

It is easily seen from definition (18.29) and relation (18.30) that

$$\left[\frac{1}{2} (\mathbb{1} \pm \gamma_5)\right]^2 = \frac{1}{2} (\mathbb{1} \pm \gamma_5). \tag{18.31}$$

A spinor can always be separated into a right part Ψ_R, and a left part Ψ_L by setting

$$\Psi_R = \frac{1}{2} (1 + \gamma_5) \Psi \quad \text{and} \quad \Psi_L = \frac{1}{2} (1 - \gamma_5) \Psi. \tag{18.32}$$

A possible representation of the γ^μ matrices, proposed by Dirac (other forms have been proposed by Weyl and Majorana), is as follows:

$$\gamma^k = \begin{pmatrix} 0 & \sigma^k \\ -\sigma^k & 0 \end{pmatrix} \quad \text{and} \quad \gamma^0 = \begin{pmatrix} \mathbb{1} & 0 \\ 0 & -\mathbb{1} \end{pmatrix}, \tag{18.33}$$

where σ^k are the 2×2 Pauli matrices and $\mathbb{1}$ the 2×2 identity matrix.

The Dirac Eq. (18.23), expanded as (18.20), is written with the correspondence principle, and by introducing the constants $\hbar$ and c,

$$i\,\hbar\,\gamma^0\,\partial_0\,\Psi = c\,(\gamma^k\,p^k + m\,c)\,\Psi,$$
$$(\gamma^0\,\frac{E}{c} - \gamma^k\,p^k - m\,c)\,\Psi = 0. \tag{18.34}$$

We introduce the covariant and contravariant four-vectors:

$$p^\mu = (p^0,\ p^k) = \left(\frac{E}{c},\ p^k\right) = \left(\frac{E}{c},\ \vec{p}\right),$$
$$p_\mu = \eta_{\mu\,\nu}\,p^\nu = (p_0,\ p_k) = \left(\frac{E}{c},\ -\vec{p}\right), \tag{18.35}$$

thus transforming the Dirac Eq. (18.23) into

$$\boxed{(\gamma^\mu\,p_\mu - m\,c)\,\Psi = (\not{p} - m\,c)\,\Psi = 0}. \tag{18.36}$$

In the chosen signature (18.25), that is the (- 2) signature, and for $\hbar = c = 1$,

$$p^\mu = i\,\partial^\mu \quad \text{and} \quad p_\mu = i\,\partial_\mu, \tag{18.37}$$

and with the expansion

$$\not{p} = \gamma^\mu\,p_\mu = \gamma^0\,p_0 + \gamma^k\,p_k = \gamma^0\,p^0 - \vec{\gamma}\cdot\vec{p}, \tag{18.38}$$

$$\partial_\mu = \frac{\partial}{\partial\,x^\mu} = (\partial_0,\ \partial_k) = (\partial_0,\ \vec{\nabla}),$$
$$\partial^\mu = \eta^{\mu\,\nu}\,\partial_\nu = \frac{\partial}{\partial\,x_\mu} = (\partial^0,\ \partial^k) = (\partial_0,\ -\vec{\nabla}), \tag{18.39}$$

we easily derive the Dirac equation.

3. The Adjoint Dirac Equation

The spinor $\Psi(x) = \begin{pmatrix} \Psi_1(x) \\ \Psi_2(x) \\ \Psi_3(x) \\ \Psi_4(x) \end{pmatrix}$ will have the row of conjugate components as its Hermitian conjugate:

$$\Psi^+(x) = (\Psi_1^*(x) \; \Psi_2^*(x) \; \Psi_3^*(x) \; \Psi_4^*(x)).$$

We introduce the adjoint vector

$$\overline{\Psi}(x) = \Psi^+(x) \, \gamma^0, \tag{18.40}$$

whose components are easily obtained with the expression for the γ^0 matrix:

$$\overline{\Psi}(x) = (\Psi_1^* \; \Psi_2^* \; \Psi_3^* \; \Psi_4^*) \begin{pmatrix} \mathbb{1} & 0 \\ 0 & -\mathbb{1} \end{pmatrix} = (\Psi_1^* \; \Psi_2^* - \Psi_3^* - \Psi_4^*). \tag{18.41}$$

To obtain the adjoint Dirac equation, we use the fundamental equation

$$i \, \partial_t \, \Psi = \vec{\alpha} \cdot (-i \, \vec{\nabla} \, \Psi) + m \, \beta \, \Psi. \tag{18.42}$$

We take the Hermitian conjugate:

$$-i \, \partial_t \, \Psi^+ = i \, \vec{\nabla} \, \Psi^+ \cdot \vec{\alpha} + m \, \Psi^+ \, \beta. \tag{18.43}$$

By multiplying from the right by $\gamma^0 = \beta$, with $(\gamma^0)^2 = \mathbb{1}$, we obtain

$$-i \, \partial_t \, \overline{\Psi} \, \gamma^0 = i \, \vec{\nabla} \, \overline{\Psi} \cdot \vec{\gamma} + m \, \overline{\Psi}. \tag{18.44}$$

The adjoint Dirac equation is written in its covariant form:

$$\boxed{ i \, (\partial_\mu \, \overline{\Psi}) \, \gamma^\mu + m \, \overline{\Psi} = \overline{\Psi} \, (i \, \overleftarrow{\partial\!\!\!/} + m) = 0 } \; . \tag{18.45}$$

4. Lagrangian Density and Field Equations

The least-action principle enables us to obtain the field equation, that is the Dirac equation, with the Lagrangian density:

$$\begin{aligned} \mathcal{L} &= i \, \overline{\Psi} \, \gamma^\mu \, \partial_\mu \, \Psi - m \, \overline{\Psi} \, \Psi \\ &= i \, \overline{\Psi} \, \partial\!\!\!/ \, \Psi - m \, \overline{\Psi} \, \Psi \\ &= \overline{\Psi} \, (i \, \overrightarrow{\partial\!\!\!/} - m) \, \Psi = -\overline{\Psi} \, (i \, \overleftarrow{\partial\!\!\!/} + m) \, \Psi. \end{aligned} \tag{18.46}$$

The Lagrangian density is often written in the symmetrical form

$$\mathcal{L} = \frac{1}{2}\,(i\,\overline{\Psi}\,\overrightarrow{\partial}\,\Psi - i\,\overline{\Psi}\,\overleftarrow{\partial}\,\Psi) - m\,\overline{\Psi}\,\Psi = \overline{\Psi}\Big(\frac{i}{2}\,\overleftrightarrow{\partial} - m\Big)\,\Psi$$

$$= \frac{1}{2}\,(i\,\overline{\Psi}\,\gamma^\mu\,\partial_\mu\,\Psi - i\,\partial_\mu\,\overline{\Psi}\,\gamma^\mu\,\Psi) - m\,\overline{\Psi}\,\Psi\,. \tag{18.47}$$

The variation with respect to $\overline{\Psi}$ leads to the Dirac equation:

$$\frac{\delta\,\mathcal{L}}{\delta\,\overline{\Psi}} = 0 \implies (i\,\partial\!\!\!/ - m)\,\Psi = (i\,\gamma^\mu\,\partial_\mu - m)\,\Psi = 0\,. \tag{18.48}$$

The variation with respect to Ψ leads to the adjoint equation:

$$\frac{\delta\,\mathcal{L}}{\delta\,\Psi} = 0 \implies \overline{\Psi}\,(i\,\overleftarrow{\partial\!\!\!/} + m) = i\,(\partial_\mu\,\overline{\Psi})\,\gamma^\mu + m\,\overline{\Psi} = 0\,. \tag{18.49}$$

The conjugate momentum is next defined in the usual way:

$$\pi = \frac{\partial\,\mathcal{L}}{\partial\,\partial_0\,\Psi} = \frac{\partial\,\mathcal{L}}{\partial\,\dot{\Psi}} = i\,\Psi^+\,. \tag{18.50}$$

This leads to a Hamiltonian density

$$\mathcal{H} = \pi\,\dot{\Psi} - \mathcal{L} = i\,\Psi^+\,(\partial_0\,\Psi) - \overline{\Psi}\,(i\,\partial\!\!\!/ - m)\,\Psi\,, \tag{18.51}$$

but the Dirac equations satisfied by the spinor Ψ cancel out the last term, finally leaving the Hamiltonian density

$$\mathcal{H} = i\,\Psi^+\,\partial_0\,\Psi = i\,\overline{\Psi}\,\gamma^0\,\partial_0\,\Psi\,. \tag{18.52}$$

The integration over the whole space leads to the Hamiltonian

$$H = \int \mathcal{H}\,d^3\,x = i\,\int \overline{\Psi}\,\gamma^0\,\partial_0\,\Psi\,d^3\,x\,. \tag{18.53}$$

5. Probability Density

The probability density, square modulus of the wave function, is written

$$j^0(x) = \varrho(x) = \Psi^+(x)\,\Psi(x)\,, \tag{18.54}$$

and by using the identity $(\gamma^0)^2 = \mathbb{1}$ and the adjoint $\overline{\Psi}(x)$ of the spinor, we also obtain

$$\varrho(x) = j^0(x) = \overline{\Psi}(x)\,\gamma^0\,\Psi(x)\,. \tag{18.55}$$

We introduce a probability current density in an analogous form:

$$j^k(x) = \Psi^+(x)\,\alpha_k\,\Psi(x) = \overline{\Psi}(x)\,\gamma^k\,\Psi(x)\,. \tag{18.56}$$

This amounts to introducing the probability four-current:

$$\boxed{j^\mu(x) = \overline{\Psi}(x)\,\gamma^\mu\,\Psi(x)}\,. \tag{18.57}$$

Now, let us directly evaluate the time evolution of the probability density:

$$\begin{aligned}
i\,\partial_0\,\varrho(x) &= i\,\partial_0\,j^0(x)\\
&= i\,\partial_0\,\Psi^+(x)\,\Psi(x)\\
&= i\,\Psi^+(x)\,\partial_0\,\Psi(x) + i\,(\partial_0\,\Psi^+(x))\,\Psi(x)\,.
\end{aligned} \tag{18.58}$$

We next evaluate the right-hand part with Dirac equation (18.43) and (18.42), which leaves us with

$$i\,\partial_0\,j^0(x) = -i\,\vec{\nabla}\cdot\Psi^+(x)\,\vec{\alpha}\,\Psi(x)\,. \tag{18.59}$$

This is the vector expression of the hydrodynamic continuity of the probability four-current $j^\mu(x)$:

$$\boxed{\partial_\mu\,j^\mu(x) = 0}\,. \tag{18.60}$$

It is also possible to obtain this relation by evaluating the term-by-term sum of equation (18.48) left-multiplied by $\overline{\Psi}$ and (18.49) right-multiplied by the spinor Ψ:

$$0 = \overline{\Psi}\,(\gamma^\mu\,\partial_\mu\,\Psi) + (\partial_\mu\,\overline{\Psi})\,\gamma^\mu\,\Psi = \partial_\mu\,(\overline{\Psi}\,\gamma^\mu\,\Psi) = \partial_\mu\,j^\mu(x)\,. \tag{18.61}$$

To identify the probability current $j^\mu(x)$ with the electric current, we simply multiply $j^\mu(x)$ by the electric charge e, which leads to an electric charge density:

$$j^0(x) = \varrho(x) = e\,\overline{\Psi}\,\gamma^0\,\Psi = e\,\Psi^+(x)\,\Psi(x)\,. \tag{18.62}$$

The electric charge is then obtained by integrating the density $j^0(x)$ over the whole space:

$$Q = \int d^3x\,j^0(x) = e\int d^3x\,\overline{\Psi}(x)\,\gamma^0\,\Psi(x)\,. \tag{18.63}$$

One should note here the very useful Pauli–Krofink relation applying to the matrices $M = \mathbb{1}, \gamma^\mu, \gamma^5, \gamma^\mu\gamma_5$,

$$(\overline{\Psi}\,M\,\gamma_\lambda\,\psi)\,\gamma^\lambda\,\psi = (\overline{\Psi}\,M\,\psi)\psi - (\overline{\Psi}\,M\,\gamma_5\,\psi)\,\gamma_5\,\psi\,. \tag{18.64}$$

6. Gauge Invariance and Noether's Theorem

The conservation property (18.60) is easily understood with Noether's theorem as the outcome of a global gauge invariance.

In effect, in a global gauge transformation (see Chap. 8), we have

$$\Psi_i(x) \;\rightarrow\; \Psi_i'(x) = \exp\left(-i\,\Lambda\,T_{ij}\right)\Psi_j(x)\,.$$

The first-order expansion in Λ will give the variation:

$$\psi_i'(x) - \Psi_i(x) = \delta\,\Psi_i(x) = -i\,\Lambda\,T_{ij}\,\Psi_j(x)\,. \tag{18.65}$$

Let us evaluate the Lagrangian density variation $\delta\,\mathcal{L}\,(\Psi_i,\,\partial_\mu\,\Psi_i)$ of the spinor field:

$$\delta\,\mathcal{L} = \frac{\partial\,\mathcal{L}}{\partial\,\Psi_i}\,\delta\,\Psi_i + \frac{\partial\,\mathcal{L}}{\partial\,\partial_\mu\,\Psi_i}\,\delta\,(\partial_\mu\,\Psi_i)$$

$$= \partial_\mu\left(\frac{\partial\,\mathcal{L}}{\partial\,\partial_\mu\,\Psi_i}\,\delta\,\Psi_i\right) - \left(\partial_\mu\,\frac{\partial\,\mathcal{L}}{\partial\,\partial_\mu\,\Psi} - \frac{\partial\,\mathcal{L}}{\partial\,\Psi_i}\right)\delta\,\Psi_i\,. \tag{18.66}$$

The Euler–Lagrange equations, after canceling the second term, lead to the field equations. The variational principle therefore also forces the first term to be zero. By replacing $\delta\,\Psi_i$ with its value (18.65), we obtain the expression for the conserved four-current:

$$\partial_\mu\,j^\mu = 0 \qquad \text{with} \qquad j^\mu(x) = -i\,\frac{\partial\,\mathcal{L}}{\partial\,\partial_\mu\,\Psi_i}\,T_{ij}\,\Psi_j\,. \tag{18.67}$$

The Dirac spinor field described by the Lagrangian density (18.47) is invariant under the global gauge transformation

$$\Psi(x) \;\rightarrow\; \Psi'(x) = \exp\left(-i\,q\,\Lambda\right)\Psi(x)\,,$$
$$\overline{\Psi}(x) \;\rightarrow\; \overline{\Psi}' = \overline{\Psi}(x)\,\exp\left(i\,q\,\Lambda\right)\,, \tag{18.68}$$
$$\mathcal{L} \;\rightarrow\; \mathcal{L}' = \mathcal{L}\,.$$

and the partial derivative with respect to $\partial_\mu\,\Psi$ is written

$$\frac{\partial\,\mathcal{L}}{\partial\,\partial_\mu\,\Psi} = i\,\overline{\Psi}\,\gamma^\mu\,. \tag{18.69}$$

This effectively leads to the probability current (18.57), since (18.67) gives

$$-i\,(i\,\overline{\Psi}\,\gamma^\mu)\,\Psi = \overline{\Psi}\,\gamma^\mu\,\Psi = j^\mu(x)\,. \tag{18.70}$$

This current is conserved and relation (18.67) is termed Noether's theorem.

7. Interaction with the Electromagnetic Field

The interaction of a particle with mass m and electric charge e with the electromagnetic e.m. field with electric vector

$$\vec{E} = -\vec{\nabla}\,\Phi - \frac{\partial \vec{A}}{\partial t} \tag{18.71}$$

and magnetic vector

$$\vec{B} = \vec{\nabla} \wedge \vec{A} \tag{18.72}$$

is described in classical mechanics by replacing $\vec{p}$ with the kinematic momentum $\vec{\pi}$ (see Chap. 1, Sects. 12.3 and 16.3):

$$\vec{p} \;\rightarrow\; \vec{\pi} = \vec{p} - e\,\vec{A} \qquad H = \frac{p^2}{2m} \;\rightarrow\; H = \frac{\pi^2}{2m} + e\,\Phi\,. \tag{18.73}$$

The correspondence principle next leads to a nonrelativistic Schrödinger equation.

The relativistic approach consists in replacing the Einstein energy equation (18.1) with

$$(E - e\,\Phi)^2 = (\pi^2 + m^2)\,. \tag{18.74}$$

If we introduce the π^0 component of the kinematic four-momentum

$$\pi^0 = (E - e\,\Phi)\,, \tag{18.75}$$

the Einstein equation (18.74) becomes

$$(\pi^0)^2 = (\pi^2 + m^2)\,. \tag{18.76}$$

The Klein–Gordon equation for a spinless particle interacting with the e.m. field, expressed in $\hbar = c = 1$ natural units, will therefore be

$$(i\,\partial_t - e\,\Phi)^2\,\Psi = (-i\,\vec{\nabla} - e\,\vec{A})^2\,\Psi + m^2\,\Psi\,. \tag{18.77}$$

The procedure for writing the relativistic wave equation will be the same as that used in Sect. 1, if the linear form is used as the departure point:

$$(\vec{\alpha}\cdot\vec{\pi} + m\,\beta)\,\Psi = \pi^0\,\Psi\,, \tag{18.78}$$

which naturally leads to the same γ matrices and to the Dirac equation for a particle interacting with the e.m. field:

$$\begin{aligned}
0 = (\not{\pi} - m)\,\Psi &= (\gamma^\mu\,\pi_\mu - m)\,\Psi \\
&= [\gamma^\mu\,(i\,\partial_\mu - e\,A_\mu) - m]\,\Psi \\
&= (i\,\not{\partial} - e\,\not{A} - m)\,\Psi\,.
\end{aligned} \tag{18.79}$$

Let us recall the remark made in (5.91) in connection with the correspondence principle: The principle introduced in (18.37) should be used in a (-2)

signature:

$$p^{\mu} \rightarrow i \, \partial^{\mu} \qquad \text{and} \qquad p_{\mu} \rightarrow i \, \partial_{\mu} \, . \tag{18.80}$$

Similarly, if we introduce the operator D_{μ} associated with the kinematic momentum π_{μ}, the correspondence principle will yield

$$\pi_{\mu} \rightarrow i \, D_{\mu} \, . \tag{18.81}$$

The covariant differentiation operator D_{μ} will therefore have to be written in the presence of the e.m. field:

$$D_{\mu} = \partial_{\mu} + i \, e \, A_{\mu} \tag{18.82}$$

in order to obtain form (18.79) of the Dirac equation.

The Lagrangian density for a particle with spin $\frac{1}{2}$ and mass m in the presence of the e.m. field is written, as in (18.46), by replacing ∂_{μ} with D_{μ},

$$\mathcal{L}_{DEM} = \overline{\Psi} \left(i \, \not{D} - m \right) \Psi = \overline{\Psi} \left(i \, \not{\partial} - e \, \not{A} - m \right) \Psi \tag{18.83}$$

and by adding the Lagrangian density of the free e.m. field, that is,

$$\mathcal{L}_{EM} = -\frac{1}{4} \, F_{\mu \nu} \, F^{\mu \nu} \quad \text{with} \quad F_{\mu \nu} = \partial_{\mu} \, A_{\nu} - \partial_{\nu} \, A_{\mu}, \tag{18.84}$$

we obtain the Lagrangian density of the e.m. field interacting with a particle of mass m and electric charge e, the basis of quantum electrodynamics (QED),

$$\begin{aligned}
\mathcal{L}_{QED} &= \mathcal{L}_{EM} + \mathcal{L}_{DEM} \\
&= -\frac{1}{4} \, F_{\mu \nu} \, F^{\mu \nu} + \overline{\Psi} \left(i \, \not{\partial} - e \, \not{A} - m \right) \Psi \\
&= -\frac{1}{4} \, F_{\mu \nu} \, F^{\mu \nu} + \overline{\Psi} \left(i \, \not{D} - m \right) \Psi \, .
\end{aligned} \tag{18.85}$$

8. Local Gauge Invariance

We notice here that the Dirac Lagrangian density is not invariant under a local gauge transformation since, in such a case,

$$\begin{aligned}
\Psi &\rightarrow \Psi' = \exp \left(-i \, e \, \Lambda(x) \right) \Psi \, (x) \, , \\
\overline{\Psi} &\rightarrow \overline{\Psi}' = \overline{\Psi} \, (x) \, \exp \left(i \, e \, \Lambda(x) \right) ,
\end{aligned} \tag{18.86}$$

and the transformation of the Lagrangian density is written

$$\mathcal{L} = \overline{\Psi} \left(i \, \not{\partial} - m \right) \Psi \rightarrow \mathcal{L}' = \overline{\Psi}' \left(i \, \not{\partial} - m \right) \Psi' \, . \tag{18.87}$$

After expansion, we obtain with (18.86) and the definition (18.70) of the current $j^{\mu} \, (x)$

$$\mathcal{L} \rightarrow \mathcal{L}' = \mathcal{L} + e \, j^{\mu} \, \partial_{\mu} \, \Lambda(x) \, . \tag{18.88}$$

This clearly shows that the Dirac Lagrangian density is not conserved under local gauge transformation.

If the differentiation operator ∂_μ is replaced in (18.87) by the covariant differentiation operator in the presence of the e.m. field (18.82):

$$\partial_\mu \; \to \; D_\mu = \partial_\mu + i\,e\,A_\mu \,,$$

the Lagrangian density becomes invariant under local gauge transformation. Let us now use the Lagrangian density $\mathcal{L}_1$:

$$\begin{aligned} \mathcal{L}_1 &= \mathcal{L} - e\,j^\mu\,A_\mu = i\,\overline{\Psi}\,\gamma^\mu\,\partial_\mu\,\Psi - m\,\overline{\Psi}\,\Psi - e\,\overline{\Psi}\,\gamma^\mu\,A_\mu\,\Psi \\ &= i\,\overline{\Psi}\,\gamma^\mu\,(\partial_\mu + i\,e\,A_\mu)\,\Psi - m\,\overline{\Psi}\Psi \\ &= \overline{\Psi}\,(i\,\not{D} - m)\,\Psi \,. \end{aligned} \qquad (18.89)$$

In a local gauge transformation we have

$$\begin{aligned} \mathcal{L}_1 &\to \mathcal{L}_1' = \mathcal{L}' - e\,j'^\mu\,A_\mu' \,, \\ A_\mu &\to A_\mu' = A_\mu + \partial_\mu\,\Lambda(x) \,, \\ j^\mu &\to j'^\mu = j^\mu \,. \end{aligned} \qquad (18.90)$$

By using (18.87) and the above relations, we notice that

$$\mathcal{L}_1 \; \to \; \mathcal{L}_1' = \mathcal{L}_1 \,. \qquad (18.91)$$

The Lagrangian density

$$\mathcal{L}_{DEM} = \overline{\Psi}\,(i\,\not{D} - m)\,\Psi \qquad (18.92)$$

of a fermion interacting with the e.m. field is invariant under a local gauge transformation:

$$\Psi \; \to \; \Psi' = \exp\,(-i\,e\,\Lambda(x))\,\Psi \,, \qquad (18.93)$$

$$A_\mu \; \to \; A_\mu' = A_\mu + \partial_\mu\,\Lambda(x) \,, \qquad (18.94)$$

$$\mathcal{L}_{DEM} \; \to \; \mathcal{L}_{DEM}' = \mathcal{L}_{DEM} = \overline{\Psi}\,(i\,\not{D} - m)\,\Psi \,. \qquad (18.95)$$

It is interesting here to note that $D_\mu\,\Psi$ behaves like Ψ under the local gauge transformation:

$$\begin{aligned} D_\mu'\,\Psi' &= (\partial_\mu + i\,e\,A_\mu')\,\exp\,(-i\,e\,\Lambda)\,\Psi \qquad &(18.96) \\ &= (\partial_\mu + i\,e\,A_\mu + i\,e\,\partial_\mu\,\Lambda)\,\exp\,(-i\,e\,\Lambda)\,\Psi \qquad &(18.97) \\ &= \exp\,(-i\,e\,\Lambda)\,D_\mu\,\Psi \,. \qquad &(18.98) \end{aligned}$$

The comparison of relations (18.86) and (18.98) demonstrates the result postulated.

9. Electron Spin

Poisson's equation, in association with the correspondence principle, gives the time evolution of the observables (see Chaps. 1 and 6):

$$\frac{d\,A}{dt} = \dot{A} = \frac{1}{i\,\hbar}\,[A,\,H] + \frac{\partial\,A}{\partial\,t}\,. \tag{18.99}$$

Using the form (18.8) of H we infer the two equations of motion:

$$\dot{x}_k = \frac{1}{i\,\hbar}\,[x_k, H] = \frac{1}{i\,\hbar}\,[x_k, (\vec{\alpha}\cdot\vec{p})] = \alpha_k\,, \tag{18.100}$$

$$\dot{\pi}_k = \frac{1}{i\,\hbar}\,[\pi_k, H] + \frac{\partial\,\pi_k}{\partial\,t} = e\,(E_k + (\vec{\alpha}\wedge\vec{B})_k)\,. \tag{18.101}$$

By inserting the value $\vec{\dot{r}} = \vec{v} = \vec{\alpha}$ from the first of these equations into the second, we obtain the quantum expression for the Lorentz force (obtained directly in (1.175)):

$$\frac{d\,\vec{\pi}}{dt} = \frac{d}{dt}\,(\vec{p} - e\,\vec{A}) = e\,(\vec{E} + \vec{v}\wedge\vec{B})\,. \tag{18.102}$$

The angular momentum of the electron in the e.m. field is described by the operator $\vec{L}$, which is not explicitly time dependent,

$$\vec{L} = \vec{r}\wedge\vec{\pi}\,, \tag{18.103}$$

and its time evolution is obtained with (18.99) as

$$\vec{\dot{L}} = \frac{d\,\vec{L}}{dt} = \frac{1}{i\,\hbar}\,[\vec{L},\,H] = \vec{\alpha}\wedge\vec{\pi}\,. \tag{18.104}$$

We introduce the electron spin operator using the Dirac matrix commutator

$$\sigma^{\mu\,\nu} = \frac{i}{2}\,[\gamma^{\mu},\,\gamma^{\nu}]\,, \tag{18.105}$$

and the components associated with the Pauli matrices:

$$\sigma_i = \sigma^{j\,k}\quad i,j,k = 1,2,3\,. \tag{18.106}$$

By using the anti-commutation relations (18.24), equations (18.12) and (18.14) give, for example,

$$\sigma_3 = \sigma_z = -i\,\alpha_1\,\alpha_2 \tag{18.107}$$

whose time evolution is fixed by (18.99):

$$\dot{\sigma}_3 = \frac{1}{i\,\hbar}\,[\sigma_3, H] = \frac{1}{i\,\hbar}\,[-i\alpha_1\,\alpha_2,\,\vec{\alpha}.\vec{\pi}]$$

$$= -\frac{2}{\hbar}\,(\vec{\alpha}\wedge\vec{\pi})_3\,. \tag{18.108}$$

We infer from this the general vector relation

$$\frac{d}{dt} \left(\frac{\hbar}{2} \vec{\sigma} \right) = -(\vec{\alpha} \wedge \vec{\pi}) \,. \tag{18.109}$$

By comparing relations (18.104) and (18.109), we notice that the operator $\vec{J} = \vec{L} + \frac{\hbar}{2} \vec{\sigma}$ is a constant of motion.

The description of the relativistic motion of the electron in an e.m. field implies the existence of a spin $\vec{S} = \frac{1}{2} \hbar \vec{\sigma}$ for this particle.

Pauli matrices were introduced in the Dirac Hamiltonian (18.8) without any reference to the spin. The latter appears to be indispensable in the coherent description of the motion of the electron in an e.m. field. It is the same spin $\vec{\sigma}$ which we find in the nonrelativistic Pauli–Schrödinger equation, a nonrelativistic limit of the Dirac equation.

10. Solutions for a Free Electron

Let us look for the solutions of the Dirac equation

$$(i \not{\partial} - m) \Psi = (i \gamma^\mu \partial_\mu - m) \Psi = 0 \tag{18.110}$$

for a free electron, in the form of plane waves by setting

$$\Psi(x) = \Psi(x^\mu) = u(\mathbf{p}) \, \exp \left(-i \, p \cdot x \right) \tag{18.111}$$

with the scalar product

$$p \cdot x = p^\mu \, x_\mu = p^0 \, x^0 - \mathbf{p} \cdot \mathbf{x} = \frac{E}{c} \, t - \mathbf{p} \cdot \mathbf{x} \,. \tag{18.112}$$

The Dirac equation (18.110) thus leads to the formal equation

$$(\gamma^\mu \, p_\mu - m) \, u(\mathbf{p}) = 0 \,, \tag{18.113}$$

describing a system of homogeneous linear equations in $u_\beta(\mathbf{p})$ when the matrix element is introduced:

$$(\gamma^0_{\alpha \, \beta} \, p^0 - \vec{\gamma}_{\alpha \, \beta} \cdot \vec{p} - m \, \delta_{\alpha \, \beta}) \, u_\beta(\mathbf{p}) = 0 \,. \tag{18.114}$$

The condition for the existence of non-trivial solutions is that the determinant should be zero, which yields the relation

$$(p_0^2 - \mathbf{p}^2 - m^2)^2 = 0 \,, \tag{18.115}$$

leading to the double root

$$p_0 = \pm \, (\mathbf{p}^2 + m^2)^{1/2} \,. \tag{18.116}$$

We therefore obtain four possible solutions, two with positive eneregies $u(\mathbf{p})$ and the other two with negative energies $v(\mathbf{p})$, with each of the spinors naturally having four components. By setting $E = (\mathbf{p}^2 + m^2)^{1/2}$ and $c = 1$ we obtain:

$$
\begin{aligned}
\Psi_I(x) &= u_1\,(\mathbf{p}, E)\,\exp\,(-i\,p \cdot x) = u_1\,(\vec{p},\ E)\,\exp\,[-i\,(Et - \mathbf{p} \cdot \mathbf{x})]\,, \\
\Psi_{II}(x) &= u_2\,(\mathbf{p}, E)\,\exp\,(-i\,p \cdot x) = u_2\,(\vec{p},\ E)\,\exp\,[-i\,(Et - \mathbf{p} \cdot \mathbf{x})]\,, \\
\Psi_{III}(x) &= v_1\,(\mathbf{p}, -E)\,\exp\,(i\,p \cdot x) = v_1\,(\vec{p}, -E)\,\exp\,[i\,(Et + \mathbf{p} \cdot \mathbf{x})]\,, \\
\Psi_{IV}(x) &= v_2\,(\mathbf{p}, -E)\,\exp\,(i\,p \cdot x) = v_2\,(\vec{p}, -E)\,\exp\,[i\,(Et + \mathbf{p} \cdot \mathbf{x})]\,.
\end{aligned}
$$
$$(18.117)$$

The positive energy solutions $u(\mathbf{p})$ satisfy the Dirac equation

$$(\not{p} - m)\,u(\mathbf{p}) = 0\,, \tag{18.118}$$

and the negative energy solutions the equation

$$(\not{p} + m)\,v(\mathbf{p}) = 0\,. \tag{18.119}$$

The solutions corresponding to the free particle at rest are obtained by setting

$$\mathbf{p} = 0 \quad \text{and} \quad p^0 = \frac{E}{c} = \frac{m\,c^2}{c} = m\,c\,. \tag{18.120}$$

The Dirac equations (18.118) and (18.119) are then written

$$(\gamma^0\,p_0 - m)\,u(0) \equiv (\gamma^0 - 1)\,m\,u(0) = 0\,, \tag{18.121}$$
$$(\gamma^0\,p_0 + m)\,v(0) = (\gamma^0 + 1)\,m\,v(0) = 0\,. \tag{18.122}$$

The Dirac matrix γ^0 of the form (18.33) leads to the solutions

$$
u_1(0) = \begin{pmatrix} 1 \\ 0 \\ 0 \\ 0 \end{pmatrix} \quad
u_2(0) = \begin{pmatrix} 0 \\ 1 \\ 0 \\ 0 \end{pmatrix} \quad
v_1(0) = \begin{pmatrix} 0 \\ 0 \\ 1 \\ 0 \end{pmatrix} \quad
v_2(0) = \begin{pmatrix} 0 \\ 0 \\ 0 \\ 1 \end{pmatrix}\,. \tag{18.123}
$$

The solutions $\Psi_0(x)$ can therefore be written with (18.111). The solution

$$\Psi_0^I(x) = A_I \begin{pmatrix} 1 \\ 0 \\ 0 \\ 0 \end{pmatrix} \exp\,(-i\,m\,t)\,, \tag{18.124}$$

for example, describes a positive-energy state, with a spin projection $+\frac{1}{2}$. The solution

$$\Psi_0^{II} = A_{II} \begin{pmatrix} 0 \\ 1 \\ 0 \\ 0 \end{pmatrix} \exp\,(-i\,m\,t) \tag{18.125}$$

also describes a positive-energy state, with a spin projection $-\frac{1}{2}$ whereas the negative-energy states with spin projection $\pm\frac{1}{2}$ will be

$$\Psi_0^{III}(x) = A_{III} \begin{pmatrix} 0 \\ 0 \\ 1 \\ 0 \end{pmatrix} \exp{(i\, m\, t)} \quad \text{and} \quad \Psi_0^{IV}(x) = A_{IV} \begin{pmatrix} 0 \\ 0 \\ 0 \\ 1 \end{pmatrix} \exp{(i\, m\, t)}.$$

(18.126)

To obtain the solutions corresponding to an electron in motion, we note that (18.118) and (18.119) also permit us to write

$$(\not{p} + m)\,(\not{p} - m) = 0.$$

(18.127)

This shows that $u(\mathbf{p})$ must contain $(\not{p}+m)$, and $v(\mathbf{p})$ must contain $(\not{p}-m)$. The positive-energy normalized solutions should be written

$$u_r(\mathbf{p}) = \frac{(\not{p}+m)}{\sqrt{2m(E+m)}}\, u_r(0) \qquad r = 1, 2,$$

(18.128)

whereas the negative-energy solutions are

$$v_r(\mathbf{p}) = \frac{-\not{p}+m}{\sqrt{2m(E+m)}}\, v_r(0) \qquad r = 1, 2.$$

(18.129)

The adjoint solutions will be $\bar{u}_r(\mathbf{p})$ and $\bar{v}_r(\mathbf{p})$. The positive-energy states and the negative-energy states are normalized:

$$\bar{u}_r(\mathbf{p})\, u_{r'}(\mathbf{p}) = \delta_{rr'},$$
$$\bar{v}_r(\mathbf{p})\, v_{r'}(\mathbf{p}) = \delta_{rr'}.$$

(18.130)

(The reader's attention is drawn here to the order in which the states and adjoint states are written: $u_{r'}(\mathbf{p})\,\bar{u}_r(\mathbf{p})$ would define a matrix and not a number, as it should in an orthonormalization relation.) The positive-energy states are orthogonal to the negative-energy states:

$$\bar{u}_r(\mathbf{p})\, v_{r'}(\mathbf{p}) = 0,$$
$$\bar{v}_r(\mathbf{p})\, u_{r'}(\mathbf{p}) = 0.$$

(18.131)

In the Dirac representation, the solutions for a free electron moving in any direction whatsoever may be written, with $p_\pm = p_x \pm i\, p_y$ and $p_z = p_3$

$$u_1(\mathbf{p}) = \frac{1}{\sqrt{E+m}} \begin{pmatrix} E+m \\ 0 \\ p_3 \\ p_+ \end{pmatrix} \quad \text{and} \quad u_2(\mathbf{p}) = \frac{1}{\sqrt{E+m}} \begin{pmatrix} 0 \\ E+m \\ p_- \\ -p_3 \end{pmatrix},$$

(18.132)

whereas for negative-energy states we obtain

$$v_1(\mathbf{p}) = \frac{1}{\sqrt{E+m}} \begin{pmatrix} p_3 \\ p_+ \\ E+m \\ 0 \end{pmatrix} \quad \text{and} \quad v_2(\mathbf{p}) = \frac{1}{\sqrt{E+m}} \begin{pmatrix} p_- \\ -p_3 \\ 0 \\ E+m \end{pmatrix}. \tag{18.133}$$

11. Charge Conjugation

To interpret the negative-energy states that are inevitably present in the solution of his fundamental equation, Dirac assumed that all states of this type are already occupied, thus forming a "sea of negative-energy states". A positive-energy electron cannot make the transition to an already occupied negative-energy state (Pauli exclusion principle). Conversely, a negative-energy state can absorb energy (a photon), and move into a positive-energy state, leaving a hole in the sea of negative-energy states.

The vacuum $|0\rangle$ will be termed the fundamental state corresponding to all the negative-energy states being occupied and all the positive-energy states being unoccupied.

The holes in the sea of negative-energy states can be interpreted as quantum states characterized by quantum numbers that are all opposite to those of positive-energy quantum states. For a long time, Dirac thought these holes were simply protons, reducing in this way the number of particles known in his time to two. It required the discovery by Anderson of a particle with the same mass as the electron and with a opposite electric charge to establish the existence of the anti-electron or positron. Negative-energy states are therefore antiparticles with the same mass as particles but with opposite quantum numbers. The absorption of a photon by a negative-energy state with a transition to a positive-energy state is therefore interpreted as the annihilation of a photon and the creation of a particle–antiparticle pair.

The spinor Ψ describing a particle with electric charge e in an e.m. field obeys the Dirac equation

$$(i\,\partial\!\!\!/ - e\,A\!\!\!/ - m)\,\Psi = 0 \tag{18.134}$$

and the spinor ψ_c describing the antiparticle with the same mass, but with electric charge $-e$, has a Dirac equation of the same form

$$(i\,\partial\!\!\!/ + e\,A\!\!\!/ - m)\,\Psi_c = 0. \tag{18.135}$$

The transformation responsible for the transition from Ψ_c to Ψ is termed charge conjugation.

The Hermitian conjugation of the Dirac equation (18.134) gives

$$0 = \Psi^+ \left(-i\,\slashed{\partial}^+ - e\,\slashed{A}^+ - m\right) = \Psi^+ \left(-i\,(\gamma^\mu)^+\,\partial_\mu - e\,(\gamma^\mu)^+\,A_\mu - m\right). \quad (18.136)$$

We introduce $(\gamma_0)^2 = \mathbb{1}$ on the right-hand side of Ψ^+ and then use definition (18.40) of $\overline{\Psi}$ to write

$$0 = \overline{\Psi}\left(-i\,\gamma^0\,(\gamma^\mu)^+\,\partial_\mu - e\,\gamma^0\,(\gamma^\mu)^+\,A_\mu - \gamma^0\,m\right). \quad (18.137)$$

By multiplying from the right by γ^0 which commutes with ∂_μ, A_μ, and m, we are left with

$$0 = \overline{\Psi}\left(-i\,\gamma^0\,(\gamma^\mu)^+\,\gamma^0\,\partial_\mu - e\,\gamma^0\,(\gamma^\mu)^+\,\gamma^0\,A_\mu - m\right), \quad (18.138)$$

but, following relation (18.28), we can also write

$$\gamma^0\,(\gamma^\mu)^+\,\gamma^0 = \gamma^\mu, \quad (18.139)$$

thus expressing equation (18.138) in the following alternative form:

$$0 = \overline{\Psi}\left(-i\,\slashed{\partial} - e\,\slashed{A} - m\right). \quad (18.140)$$

By transposition (changing of lines into rows) it then becomes

$$\left(-i\,\slashed{\partial}^T - e\,\slashed{A}^T - m\right)\overline{\Psi}^T = 0. \quad (18.141)$$

Let us define the charge-conjugation operator C by

$$\boxed{C\,(\gamma^\mu)^T\,C^{-1} = -\gamma^\mu} \qquad \boxed{-(\gamma^\mu)^T = C^{-1}\,\gamma^\mu\,C} \quad (18.142)$$

By introducing $C^{-1}\,C = \mathbb{1}$ on the left-hand side of $\overline{\Psi}^T$ in (18.141), and left-multiplying the equation by C, we obtain with the above definition

$$\left(i\,\slashed{\partial} + e\,\slashed{A} - m\right)C\,\overline{\Psi}^T = 0. \quad (18.143)$$

Comparing with the Dirac equation (18.135) permits us to set

$$\Psi_c = \left(C\,\overline{\Psi}^T\right)\eta_c \quad (18.144)$$

with η_c a phase factor which may be chosen equal to 1:

$$\boxed{\Psi_c = C\,\overline{\Psi}^T}\;. \quad (18.145)$$

The charge conjugate of a positive-energy solution with spin up will be a negative-energy solution with spin down and vice versa:

$$v(\mathbf{p}) = C \, \bar{u}^T(\mathbf{p}) \,,$$
$$u(\mathbf{p}) = C \, \bar{v}^T(\mathbf{p}) \,. \tag{18.146}$$

In Chap. 8 on the invariances and symmetries of physical laws, mention was made of the invariance of PT. The invariance under charge conjugation should, in actual fact, be added to this invariance, thus making CPT one of the most important symmetry properties of elementary particles.

Spinors that are equal to their charge conjugates

$$\Psi = \Psi_c \tag{18.147}$$

are termed Majorana spinors.

12. Hydrogen Atom

Let us return to the discussion of the hydrogen atom as a means of testing and comparing the Schrödinger, Dirac, and Klein–Gordon equations for determining the probability amplitude and the energy levels of an electron in the e.m. field of a proton.

12.1 The Schrödinger Equation

By introducing the Coulomb field of the proton with the Coulomb potential α/r, with the structure constant of the electromagnetic interactions (see (16.89))

$$\alpha = \frac{e^2}{4\pi \, \varepsilon_0 \, \hbar \, c} \simeq 1/137 \,,$$

we obtain the usual Schrödinger equation (for $\hbar = 1$)

$$\left(-\frac{1}{2m} \, \nabla^2 - \frac{\alpha}{r} - E \right) \Psi = 0 \,, \tag{18.148}$$

which can also be written in spherical coordinates as follows:

$$\left[\left(\frac{\partial^2}{\partial r^2} + \frac{2}{r} \frac{\partial}{\partial r} - \frac{L^2}{r^2} \right) + \frac{2\alpha \, m}{r} + 2m \, E \right] \Psi(\vec{r}) = 0 \,. \tag{18.149}$$

The expression for $\Psi(\vec{r})$ in terms of spherical harmonics will yield the eigenvalues $\ell(\ell+1)$ of L^2 and the principal quantum number $0 \leq n \leq \ell - 1$, which leads to the energy levels $E_{n\ell}$ of the hydrogen atom (see Chap. 12):

$$E_{n\ell} = -m \, \frac{\alpha^2}{2n^2} \,. \tag{18.150}$$

These energy levels exhibit a degeneracy of the order:

$$\sum_{\ell=0}^{n-1} (2\ell + 1) = n^2 . \tag{18.151}$$

This is the fundamental result of atomic spectroscopy, the basis of Mendeleev's classification of chemical elements.

12.2 The Klein–Gordon Equation

The Schrödinger equation is not relativistic. The above results can be improved upon by using the Klein–Gordon equation, that is, by discarding the electron spin. We therefore replace the operator $i\,\partial_t$ in the Klein–Gordon equation (18.3) by $E + \alpha/r$, which defines the Klein–Gordon equation of a spinless particle with mass m and electric charge $-e$, in the e.m. field of a proton:

$$\left[\left(E + \frac{\alpha}{r} \right)^2 + \nabla^2 - m^2 \right] \Psi = 0 . \tag{18.152}$$

Expressing this in a spherical coordinate system thus gives

$$\left[\left(\frac{\partial^2}{\partial\, r^2} + \frac{2}{r} \frac{\partial}{\partial\, r} - \frac{L^2}{r^2} \right) + \frac{2\alpha\, E}{r} + \frac{\alpha^2}{r^2} + E^2 - m^2 \right] \Psi(\vec{r}) = 0 . \tag{18.153}$$

This equation is identical to (18.149) if we make the following changes:

$$L^2 \;\rightarrow\; L^2 - \alpha^2 ,$$
$$\alpha \;\rightarrow\; \alpha\,\frac{E}{m} , \tag{18.154}$$
$$E \;\rightarrow\; \frac{E^2 - m^2}{2m} .$$

The eigenvalues of the operator $\Lambda^2 = L^2 - \alpha^2$ are of the form $\lambda\,(\lambda + 1)$ with

$$\lambda = \ell - \delta\,\ell \quad\text{and}\quad \delta\,\ell = \left(\ell + \frac{1}{2} \right) - \left[\left(\ell + \frac{1}{2} \right)^2 - \alpha^2 \right]^{1/2} . \tag{18.155}$$

The principal quantum number ν replaces the quantum number n; the quantum number $\nu - (\lambda + 1) = \nu - (\ell - \delta\,\ell + 1)$ must be an integer, but, following what we saw in the preceding paragraph, the quantum numbers $n - (\ell + 1)$ should also be integers. This therefore fixes the possible values of $\nu - n + \delta\,\ell$ integers:

$$\nu = n - \delta\,\ell \quad \nu \text{ and } n \text{ integers.} \tag{18.156}$$

By replacing $E_{n\,\ell}$ by $(E_{\nu\,\lambda}^2 - m^2)/2m$ and α by $(E_{\nu\,\lambda})/m$ in (18.150), we obtain the corresponding energy levels $E_{\nu\,\lambda}$, which we write in the form $E_{n\,\ell}$

using the value (18.156) of the quantum number:

$$E_{n\,\ell} = \frac{m}{[1+\alpha^2/(n-\delta\,\ell)^2]^{1/2}} = m - \frac{m\,\alpha^2}{2n^2} - \frac{m\,\alpha^4}{n^3\,(2\ell+1)} + \ldots . \quad (18.157)$$

The first term is the mass energy, which is discarded in the nonrelativistic Schrödinger equation. The second term corresponds to the solution (18.150) of the Schrödinger equation and the remainder are correction terms.

12.3 The Dirac Equation

To include the relativistic effects and the spin of the electron, we use relation (18.127) in which we replace $\not{p} = i\,\not\partial$ by $\not\pi = i\,\not\partial - e\,\not{A}$. This leads to a Klein–Gordon equation modified by taking into account the spin of the electron in the e.m. field of the proton:

$$(\not\pi + m)\,(\not\pi - m)\,\Psi = (\not\pi^2 - m^2)\,\Psi$$
$$= [(i\,\not\partial - e\,\not{A})^2 - m^2]\,\Psi = 0 . \quad (18.158)$$

By using anticommutation and commutation relations (18.24) and (18.105) for the Dirac matrices γ, the Faraday tensor of the e.m. field appears in a contraction with the spin operator $\sigma^{\mu\,\nu}$, thus transforming (18.158) as follows:

$$\left[(i\,\partial^\mu - e\,A^\mu)\,(i\,\partial_\mu - e\,A_\mu) - \frac{e}{2}\,\sigma^{\mu\,\nu}\,F_{\mu\,\nu} - m^2\right]\Psi = 0 . \quad (18.159)$$

The only non-zero component of the vector potential is A_0:

$$A_0 = -\frac{\alpha}{e}\,\frac{1}{r} , \quad (18.160)$$

which leads to the expression for the corresponding electric field:

$$\vec{E} = -\vec{\nabla}\,A_0 = \frac{\alpha}{e}\,\frac{\vec{r}}{r^3} \quad \text{since} \quad -E_i = F_{0\,i} = -\partial_i\,A_0 . \quad (18.161)$$

If we use expression (18.105) for the tensor $\sigma^{0\,i}$, that is,

$$\sigma^{0\,i} = \frac{i}{2}\begin{pmatrix} \sigma_i & 0 \\ 0 & -\sigma_i \end{pmatrix} , \quad (18.162)$$

we obtain the spin term of the modified Klein–Gordon equation (18.159):

$$\frac{e}{2}\,\sigma^{\mu\,\nu}\,F_{\mu\,\nu} \equiv \frac{e}{2}\,\sigma^{0\,i}\,F_{0\,i} = i\,e\begin{pmatrix} \vec{\sigma}\cdot\vec{E} & 0 \\ 0 & -\vec{\sigma}\cdot\vec{E} \end{pmatrix} . \quad (18.163)$$

The eigenvalue equation (18.159) is therefore defined with components Ψ_+ and Ψ_- of the wave function in a two-dimensional space, the spin space of the particle. The component Ψ_+ will describe the wave function of a particle with mass m, charge e and energy E, and a spin $|\frac{1}{2}\frac{1}{2}\rangle$ in the e.m. field of the proton, and the component Ψ_- the wave function of a particle with spin $|\frac{1}{2}-\frac{1}{2}\rangle$. Using (18.161), the Klein–Gordon equation thus becomes

$$\left[\left(\frac{\partial^2}{\partial r^2} + \frac{2}{r}\frac{\partial}{\partial r} - \frac{L^2}{r^2}\right) + \frac{2\alpha E}{r} + \frac{\alpha^2}{r^2} + E^2 - m^2 \pm \frac{i\alpha}{r^3}\,\vec{\sigma}.\vec{r}\right]\Psi_\pm(\vec{r}) = 0.$$

(18.164)

This is equation (18.153) with an additional spin term

$$\pm\frac{i\alpha}{r^3}\,\vec{\sigma}\cdot\vec{r}.$$

(18.165)

The orbital angular momentum $\vec{L}$ is no longer a good quantum number, and the wave functions $\Psi_\pm$ will have to be projected onto the coupled spherical harmonics basis, since the total angular momentum $\vec{J} = \vec{L} + \frac{1}{2}\,\vec{\sigma}$ is a constant of motion (see Chap. 12, Sect. 10 or Chap. 18, Sect. 9). The application of the Wigner–Eckart theorem to the matrix element of the irreducible tensor operator $\vec{\sigma}.\vec{r}$ in the coupled basis readily yields:

$$\langle(\ell\,\tfrac{1}{2})\,j\mu\,|\,\vec{\sigma}\cdot\vec{r}\,|\,(\ell'\,\tfrac{1}{2})\,j'\mu'\rangle = [j^{-1}]\,\delta_{j\,j'}\,\delta_{\mu\,\mu'}\,\langle(\ell\tfrac{1}{2})\,j\,\|\,(\vec{\sigma}\cdot\vec{r})_0\,\|\,(\ell'\tfrac{1}{2})\,j\rangle,$$

(18.166)

with the analytical value of the reduced matrix element

$$\langle(\ell\,\tfrac{1}{2})\,j\,\|\,(\vec{\sigma}.\vec{r})_0\,\|\,(\ell'\,\tfrac{1}{2})\,j\rangle$$

$$= [j\,\ell\,\ell'](-)^{j+\frac{1}{2}}\,\sqrt{6}\,r\begin{pmatrix}\ell & 1 & \ell' \\ 0 & 0 & 0\end{pmatrix}\begin{Bmatrix}j & \frac{1}{2} & \ell' \\ 1 & \ell & \frac{1}{2}\end{Bmatrix}.$$

(18.167)

The orbital momentum ℓ' can only take the values $\ell \pm 1$ for the "$3j$ 0" to exist, whereas $j = \ell \pm \frac{1}{2}$ are the only possible values of j. We notice then that the only non-zero reduced matrix elements couple the orbital momenta $\ell = j - \frac{1}{2}$ and $\ell = j + \frac{1}{2}$. The direct calculation of matrix elements (18.167) or the replacement of ℓ with j in (18.153) and (18.154) lead to the same results. By defining the eigenvalues of the matrix representing the operator $L^2 - \alpha^2 \mp \frac{i\alpha}{r}\,\vec{\sigma}.\vec{r}$ in the form $\lambda\,(\lambda+1)$, we obtain

$$\lambda = \left(j \pm \frac{1}{2}\right) - \delta\,j \quad\text{and}\quad \delta\,j = \left(j+\frac{1}{2}\right) - \left[\left(j+\frac{1}{2}\right)^2 - \alpha^2\right]^{1/2}.$$

(18.168)

This simply generalizes (18.155), whereas, by using (18.157), the energy levels reduce to

$$E_{n\,j} = \frac{m}{[1+\alpha^2/(n-\delta\,j)^2]^{1/2}} \simeq m - \frac{m\,\alpha^2}{2n^2} - \frac{m\,\alpha^4}{n^3\,(2j+1)} + \frac{3}{8}\frac{m\,\alpha^4}{n^4} + \cdots.$$

(18.169)

The electron spin is only introduced then by the replacement for the orbital angular momentum $\ell = 0, 1, 2\ldots$ by the total angular momentum $j = \frac{1}{2}, \frac{3}{2}, \frac{5}{2}, \ldots$. Experimental results confirm this fine structure of the energy levels of the hydrogen atom.

13. Coulomb Elastic Scattering

The perturbation expansion of the transition amplitude (15.66) and the Feynman diagram method (see Chap. 14) are, of course, valid for the scattering of electrically charged fermions (electron, positron, or proton, for example) by an atom with electric charge Ze. The initial and final wave functions will be relativistic solutions satisfying the Dirac equation (18.79), which can be rewritten in the form

$$(\not{p} - m)\,\Psi = e\,\not{A}\,\Psi\,. \tag{18.170}$$

This defines the interaction potential as the operator $e\,\not{A} = e\,\gamma^\mu\,A_\mu$ acting on the wave function Ψ.

Let us evaluate the Fourier transform $\not{A}(p)$ of the four-vector-potential:

$$\not{A}(p) = \int d^4x\,\not{A}(x)\,\exp(i\,p\cdot x),$$
$$\not{A}(x) = \int \frac{d^4p}{(2\pi)^4}\,\not{A}(p)\,\exp(-i\,p\cdot x)\,. \tag{18.171}$$

The electrostatic potential responsible for the Coulomb scattering has the following non-zero component

$$A_0(x) = -\frac{Z\,e}{4\pi\,r} \quad\text{and}\quad \not{A}(x) = \gamma^0\,A_0 = -\frac{\gamma^0\,Ze}{4\pi\,r}\,, \tag{18.172}$$

such that the Fourier transform $\not{A}(p)$ is written with (18.171) in the form:

$$\not{A}(p) = -\int d^4x\,\frac{\gamma^0\,Ze}{4\pi\,r}\,\exp(i\,p\,x) \quad\text{with}\quad p\cdot x = E\,t - \mathbf{p}\cdot\mathbf{x}\,. \tag{18.173}$$

The integration over the variable t leads to the Dirac function $2\pi\,\delta(E)$, and the integration over the angles θ and φ in the relative direction of the vectors $\mathbf{p}$ and $\mathbf{x}$ finally leave us with

$$\not{A} = -2\pi\,\delta(E)\,\frac{\gamma^0\,Ze}{|\,\mathbf{p}\,|^2}\,. \tag{18.174}$$

The first-order transition amplitude between the initial and final states, $\Psi_i(x)$ and $\Psi_f(x)$, of expansion (15.66), that is, in the Born approximation, is written with the potential $e\,\not{A}$ in the form

$$T_{fi} = -i\,e\int d^4x\,\Psi_f^+(x)\,\not{A}(x)\,\Psi_i(x)\,. \tag{18.175}$$

Replacing $\Psi_i(x)$ by $u_i(\mathbf{p}_i)\exp(-i\cdot p_i\cdot x)$ and $\Psi_f^+(x)$ by $\bar{u}_f(\mathbf{p}_f)\exp(i\,p_f\cdot x)$, we obtain the first-order transition amplitude

$$T_{fi} = 2\pi\,\delta\,(E_f - E_i)\,\bar{u}_f\,(\mathbf{p}_f)\left(\frac{i\,\gamma^0\,Ze^2}{|\,\mathbf{p}_f - \mathbf{p}_i\,|^2}\right)u_i\,(\mathbf{p}_i)\,. \qquad (18.176)$$

The momentum transferred during the elastic scattering can be expressed with the scattering angle θ:

$$|\,\mathbf{p}_f\,| = |\,\mathbf{p}_i\,| = k\,,$$
$$|\,\mathbf{p}_f - \mathbf{p}_i\,|^2 = 4k^2\,\sin^2\frac{\theta}{2}\,, \qquad (18.177)$$
$$E_f = E_i = E\,.$$

We shall skip the details of the calculation of the term $|\,\bar{u}_f\,(p_f)\,\gamma^0\,u_i\,(p_i)\,|^2$ yielding the value $2\,(E_i\,E_f + \mathbf{p}_i\cdot\mathbf{p}_f + m^2)$, which leads to the relativistic Coulomb elastic scattering cross-section:

$$\frac{d\,\sigma}{d\,\Omega} = \frac{(Z\,\alpha)^2}{8k^4\,\sin^4\theta/2}\,[E^2 + m^2 + k^2\,\cos\theta]$$
$$= \frac{(Z\,\alpha)^2}{4}\,\frac{E^2}{k^4}\,\frac{1 - v^2\,\sin^2\theta/2}{\sin^4\theta/2}\,. \qquad (18.178)$$

For nonrelativistic particles, we have $v\to0$ or $k/m\to0$, which leads to the classical Rutherford scattering.

14. Quantization of a Fermion Field

A set of fermions with spin $\frac{1}{2}$ and mass m can be described as a spinor field $\Psi(x^\mu)\equiv\Psi(x)\equiv\Psi(t,\mathbf{x})$ associated with a Lagrangian density $\mathcal{L}$ (cf. (18.47)) and a Hamiltonian density $\mathcal{H}$ (cf. (18.52)).

In the second quantization formalism (see Chap. 16), one defines a quantum vacuum $|\,0\rangle$, and field operators $\Psi(t,\mathbf{x})$ and $\Psi^+(t,\mathbf{x})$ annihilating or creating particles occupying the possible quantum states.

The fermion creation and annihilation operators acting on a quantum state α obey the definitions

$$\begin{array}{ll} c_\alpha^+\,|\,0\rangle = |\,\alpha\rangle\,, & c_\alpha^+\,|\,\alpha\rangle = 0\,, \\ c_\alpha\,|\,0\rangle = 0\,, & c_\alpha\,|\,\alpha\rangle = |\,0\rangle\,, \end{array} \qquad (18.179)$$

whereas the operator $N_\alpha = c_\alpha^+\,c_\alpha$ represents the number of fermions in the state $|\,\alpha\rangle$, since the application of relations (18.179) leads to the eigenvalue equations:

$$N_\alpha \,|\, \alpha\rangle \;=\; |\, \alpha\rangle \,,$$
$$N_\alpha \,|\, \text{not } \alpha\rangle \;=\; 0 \,. \tag{18.180}$$

The operators c_α and c_α^+ obey the anticommutation rules that are specific to fermion states:

$$\{c_\alpha,\, c_\beta\} = \{c_\alpha^+,\, c_\beta^+\} = 0 \,,$$
$$\{c_\alpha,\, c_\beta^+\} = \delta_{\alpha\beta}\, \mathbb{1} \,. \tag{18.181}$$

The fermion field operators will be defined with the individual wave functions (solutions of the Dirac equation) and fermion creation and annihilation operators:

$$\Psi(t,\, \mathbf{x}) = \sum_\alpha c_\alpha\, \Psi_\alpha(t,\, \mathbf{x}) \,,$$
$$\Psi^+(t,\, \mathbf{x}) = \sum_\alpha \Psi_\alpha^*(t,\, \mathbf{x})\, c_\alpha^+ \,. \tag{18.182}$$

The use of relations (18.182) and the anticommutation relations (18.181) leads to anticommutation relations for the field operators themselves:

$$\{\Psi(t,\, \mathbf{x}),\, \Psi(t,\, \mathbf{x}')\} = \{\Psi^+(t,\, \mathbf{x}),\, \Psi^+(t,\, \mathbf{x}')\} = 0 \,,$$
$$\{\Psi(t,\, \mathbf{x}),\, \Psi^+(t,\, \mathbf{x}')\} = \delta^3(\mathbf{x} - \mathbf{x}') \,. \tag{18.183}$$

The quantization procedure applied to the field operators $\Psi(t,\, \mathbf{x})$ and $\Psi^+(t,\, \mathbf{x})$ and to the conjugate momentum (see (18.50)):

$$\pi(t,\, \mathbf{x}) = i\, \Psi^+(t,\, \mathbf{x}) \,, \tag{18.184}$$

therefore leads to anticommutation relations at a particular instant t:

$$\{\Psi(t,\, \mathbf{x}),\, \Psi^+(t,\, \mathbf{x}')\} = \delta^3(\mathbf{x} - \mathbf{x}') \,,$$
$$\{\Psi(t,\, \mathbf{x}),\, \pi(t,\, \mathbf{x}')\} = i\, \delta^3(\mathbf{x} - \mathbf{x}') \,. \tag{18.185}$$

Let us now examine how these spinor field operators can be constructed. There is one positive-energy (or negative-energy) spinor for each fermion (or antifermion). We therefore set

$$\Psi_n(t,\, \mathbf{x}) = w_n(\mathbf{p})\, \exp\left(-i\, \varepsilon_n\, (p \cdot x)\right) \tag{18.186}$$

with $p.x = p_\mu\, x^\mu$ and $\varepsilon_n = +1$ for the positive-energy solutions (fermions), $\varepsilon_n = -1$ for the antifermions (negative-energy solutions). The individual wave functions $w_n(\mathbf{p})$ are solutions of the Dirac equation (18.118) or (18.119):

$$(\not{p} - \varepsilon_n\, m)\, w_n(\mathbf{p}) = 0 \,. \tag{18.187}$$

The positive-energy fermion states are denoted $u^{(r)}(\mathbf{p})$ (where r is the projection index with spin projection $r = 1, 2$) and $v^{(r)}(\mathbf{p})$ are the negative-energy states.

Each fermion with energy E and spin $\frac{1}{2}$ or $-\frac{1}{2}$ will occupy an elementary volume dV in configuration space, and Pauli's principle states that only one fermion can exist in this volume dV. We may consider that there are $2E$ fermions per unit volume (or that a fermion with energy E occupies a volume $1/2E$). The normalization of the spinor field can therefore be written in configuration space as follows:

$$\int \Psi^+ (t,\ \mathbf{x})\ \Psi\ (t,\ \mathbf{x})\ dV = 2E\,. \qquad (18.188)$$

This changes both the positive- and the negative-energy orthonormalization relations, (18.130) and (18.131) respectively, which become

$$\bar{u}^{(r)}\ (\mathbf{p})\ u^{(s)}\ (\mathbf{p}) = 2E\ \delta_{rs}\,,$$
$$\bar{v}^{(r)}\ (\mathbf{p})\ v^{(s)}\ (\mathbf{p}) = -2E\ \delta_{rs}\,, \qquad (18.189)$$
$$\bar{u}^{(r)}\ (\mathbf{p})\ v^{(s)}\ (\mathbf{p}) = \bar{v}^{(r)}\ (\mathbf{p})\ u^{(s)}\ (\mathbf{p}) = 0\,.$$

The Fourier transform of the field operator of the spinor $\Psi\ (x^\mu)$ can be written in the form

$$\Psi\ (t,\ \mathbf{x}) = \sum_{r=1,2} \int \frac{d^3\ p}{(2\pi)^3}\ \frac{1}{(2E_p)}$$
$$\times \left[b_r\ (\mathbf{p})\ u^{(r)}\ (\mathbf{p})\ \exp\ (-i\ p.x) + d_r^*\ (\mathbf{p})\ v^{(r)}\ (\mathbf{p})\ \exp\ (i\ p.x) \right]\,.$$
$$(18.190)$$

The spinor $\bar{\Psi}$ is obtained by replacing the term in brackets with its complex conjugate, whereas u and v become $\bar{u}$ and $\bar{v}$.

The quantization of the Dirac field is obtained by replacing the coefficients b_r and d_r^* with the fermion or antifermion annihilation and creation operators $b_r(\mathbf{p})$ and $d_r^+(\mathbf{p})$ of spin projection $\pm \frac{1}{2}$, energy $2E$, and momentum $\mathbf{p}$.

By substituting for $\Psi\ (t,\ \mathbf{x})$ and $\Psi^+\ (t,\ \mathbf{x}')$ in (18.183), we obtain the anticommutation relations for the creation and annihilation operators of individual states of fermions (or antifermions):

$$\{b_r\ (\mathbf{p}),\ b_{r'}^+\ (\mathbf{p}')\} = (2\pi)^3\ (2E_p)\ \delta_{r\ r'}\ \delta^3\ (\mathbf{p} - \mathbf{p}')\,,$$
$$\{d_r\ (\mathbf{p}),\ d_{r'}^+\ (\mathbf{p}')\} = (2\pi)^3\ (2E_p)\ \delta_{r\ r'}\ \delta^3\ (\mathbf{p} - \mathbf{p}')\,. \qquad (18.191)$$

All the other anticommutators are zero.

The integration over configuration space can be replaced by an integration in phase space by using the Fourier transform (18.190) of the field operators:

$$\Psi\ (\mathbf{p}) = \sum_r \frac{1}{2E_p} \left[b_r\ (\mathbf{p}) u^{(r)}(\mathbf{p}) \exp\ (i\ E_p\ t) \right.$$
$$\left. + d_r^+\ (\mathbf{p})\ v^{(r)}\ (\mathbf{p})\ \exp(-i\ E_p\ t) \right]\,, \qquad (18.192)$$

whereas with the procedure described above the operator $\bar{\Psi}$ becomes

$$\bar{\Psi}\,(\mathbf{p}) = \sum_r \frac{1}{2E_p} \Big[d_r\,(\mathbf{p})\,\bar{v}^{(r)}\,(\mathbf{p})\,\exp\,(-i\,E_p\,t)$$

$$+\, b_r^+\,(-\mathbf{p})\,\bar{u}^r\,(-\mathbf{p})\,\exp(i\,E_p\,t)\Big]. \qquad (18.193)$$

The Hamiltonian density is represented with (18.52) using the operator

$$\mathcal{H} = i\,\Psi^+\,\partial_0\,\Psi = i\,\bar{\Psi}\,\gamma^0\,\partial_0\,\Psi, \qquad (18.194)$$

and the energy associated with the fermion field via the eigenvalue of the Hamiltonian

$$H = \int \mathcal{H}\,d^3\,x = i \int d^3\,x\,\bar{\Psi}\,(x)\,\gamma^0\,\partial_0\,\Psi. \qquad (18.195)$$

Parceval's relation allows us to write

$$\int d^3\,x\,\Psi\,(\mathbf{x})\,\varphi\,(\mathbf{x}) = \int \frac{d^3\,p}{(2\pi)^3}\,\Psi\,(\mathbf{p})\,\varphi\,(-\,\mathbf{p}), \qquad (18.196)$$

and by considering $\varphi(\mathbf{x})$ as $\partial_0\,\Psi\,(x)$, we obtain, after some transformations and a redefinition of the energy zero, the Hamiltonian operator

$$H = \sum_{r=1,2} \int \frac{d^3\,p}{(2\pi)^3}\,\frac{1}{(2E_p)}\,E_p\,\big[b_r^+\,(\mathbf{p})\,b_r\,(\mathbf{p}) + d_r^+\,(\mathbf{p})\,d_r\,(\mathbf{p})\big]. \qquad (18.197)$$

This is, to within a normalization constant, the form (17.65) of the Hamiltonian of a field in which the creation and annihilation operators are fermion creation and annihilation operators.

The electric charge can also be evaluated with the charge operator Q as defined in (18.63):

$$Q = e \sum_{r=1,2} \int \frac{d^3\,p}{(2\pi)^3\,(2E_p)}\,\big[b_r^+\,(\mathbf{p})\,b_r\,(\mathbf{p}) - d_r^+\,(\mathbf{p})\,dr\,(\mathbf{p})\big]. \qquad (18.198)$$

If we bear in mind that b and b^+ are the annihilation and creation operators associated with the particles, and d and d^+ those associated with the antiparticles, we obtain the expected result: the electric charge is equal to the charges of the particles, minus the charges of the antiparticles.

15. Dirac Equation – Schematic Summary

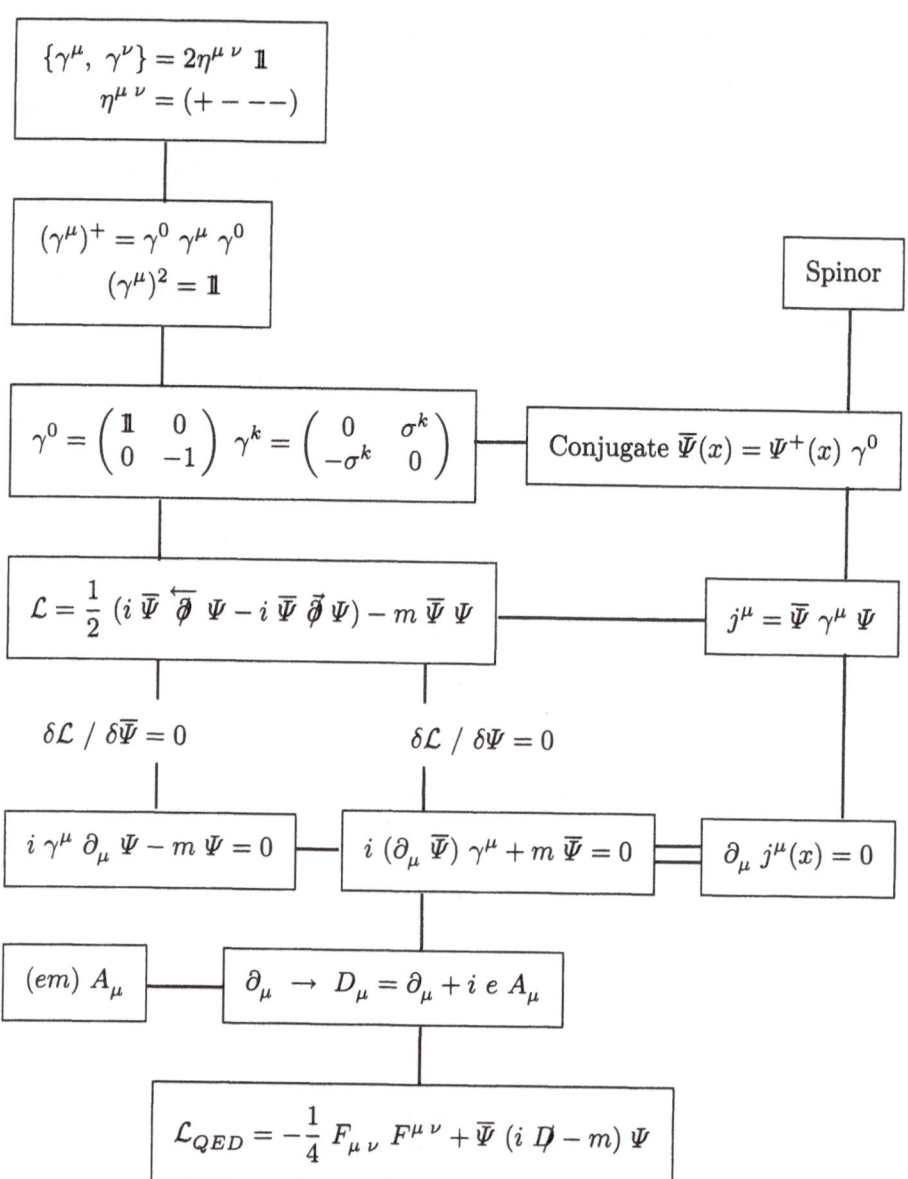

$$\{\gamma^\mu,\ \gamma^\nu\} = 2\eta^{\mu\nu}\ \mathbb{1}$$
$$\eta^{\mu\nu} = (+---)$$

$$(\gamma^\mu)^+ = \gamma^0\ \gamma^\mu\ \gamma^0$$
$$(\gamma^\mu)^2 = \mathbb{1}$$

Spinor

$$\gamma^0 = \begin{pmatrix} \mathbb{1} & 0 \\ 0 & -1 \end{pmatrix} \quad \gamma^k = \begin{pmatrix} 0 & \sigma^k \\ -\sigma^k & 0 \end{pmatrix}$$

Conjugate $\overline{\Psi}(x) = \Psi^+(x)\ \gamma^0$

$$\mathcal{L} = \frac{1}{2}\left(i\ \overline{\Psi}\ \overleftarrow{\partial}\!\!\!/\ \Psi - i\ \overline{\Psi}\ \vec{\partial}\!\!\!/\ \Psi\right) - m\ \overline{\Psi}\ \Psi$$

$$j^\mu = \overline{\Psi}\ \gamma^\mu\ \Psi$$

$$\delta\mathcal{L}\ /\ \delta\overline{\Psi} = 0 \qquad\qquad \delta\mathcal{L}\ /\ \delta\Psi = 0$$

$$i\ \gamma^\mu\ \partial_\mu\ \Psi - m\ \Psi = 0$$

$$i\ (\partial_\mu\ \overline{\Psi})\ \gamma^\mu + m\ \overline{\Psi} = 0$$

$$\partial_\mu\ j^\mu(x) = 0$$

$$(em)\ A_\mu$$

$$\partial_\mu\ \rightarrow\ D_\mu = \partial_\mu + i\,e\,A_\mu$$

$$\mathcal{L}_{QED} = -\frac{1}{4}\ F_{\mu\nu}\ F^{\mu\nu} + \overline{\Psi}\ (i\ D\!\!\!\!/ - m)\ \Psi$$

Chapter 19
Quantum X-Dynamics

The theories describing the interaction of fermions (quarks and/or leptons) with gauge boson fields are usually designated quantum X-dynamics where X is the relevant gauge field. It is thus that quantum electrodynamics (QED) describes the interaction of the electromagnetic field with electrons, and quantum chromodynamics (QCD) the interaction of quarks with the color gauge field or strong nuclear interaction. The GSW standard theory of electroweak interaction of Glashow–Salam–Weinberg ought to be described too as quantum electroweak-dynamics; however, in current usage the term (GSW) standard model of electromagnetic and weak interactions is preferred. It describes the interaction of quarks and leptons with the electromagnetic gauge field (Abelian gauge field with symmetry $U(1)$) and with the weak gauge field (non-Abelian gauge field with symmetry $SU(2)$) in the same procedure. This GSW standard model is one of the most significant advances of the last twenty years and although, properly speaking, it is a topic that is beyond a masters program, we thought it might be a good idea to take a cursory look at its most significant results and its use in demonstrating a Higgs process for the acquisition of mass by the intermediate bosons W that carry the weak interaction.

1. Quantum Electrodynamics

Quantum electrodynamics (QED) is a quantum and relativistic description of the interaction between the electromagnetic field, the Abelian gauge field with symmetry $U(1)$, carried by gauge bosons (photons), and electrons, first-generation electrically charged leptons.

QED is today the physical theory with the most thorough experimental verification. The magnetic moment, expressed in terms of spin, electric charge, and mass,

$$\vec{\mu} = g \, \frac{e \, \hbar}{2m \, c} \, \vec{S}, \tag{19.1}$$

yields experimental values that are in excellent agreement with the theory (see Chap. 9). The gyromagnetic ratio g is equal to 1 in classical theory and 2 in Dirac's theory and the latter has been confirmed by experiment (De Rafael, 1976). The first-order corrections of perturbation theory predict a modification in powers of the finestructure constant α of electromagnetic interactions (see Chap. 16):

$$\alpha = \frac{e^2}{4\pi\,\varepsilon_0\,\hbar\,c} = 1/137.035\ 989\ 5\,. \tag{19.2}$$

The evaluation of the corrections to the fourth-order (by summing about 900 Feynman diagrams), leads to the theoretical value of the gyromagnetic factor

$$\frac{g_{\mathrm{th}}-2}{2} = \frac{1}{2}\frac{\alpha}{\pi} - 0.328\ 48\ \left(\frac{\alpha}{\pi}\right)^2 + 1.49\ \left(\frac{\alpha}{\pi}\right)^3 + O\ \left(\frac{\alpha}{\pi}\right)^4 \tag{19.3}$$

yielding, with (19.2), the theoretical value

$$\frac{g_{\mathrm{th}}-2}{2} = (115\ 965\ 5 \pm 3.3)\ 10^{-9}\,, \tag{19.4}$$

whereas experiment gives the result

$$\frac{g_{\mathrm{exp}}-2}{2} = (115\ 965\ 7.7 \pm 3.5)\ 10^{-9}\,. \tag{19.5}$$

We will not go into the details of QED and instead refer the interested reader to more specialized textbooks (Elbaz, 1989; Le Bellac, 1988).

The Lagrangian density of QED is written in the form (cf. 18.85)

$$\mathcal{L}_{\mathrm{QED}} = -\frac{1}{4}\ F_{\mu\nu}\ F^{\mu\nu} + \bar{\Psi}\ (i\ \partial\!\!\!/ - m)\ \Psi - J^{\mu}\ A_{\mu}\,, \tag{19.6}$$

with expression (18.55) for the electromagnetic current. It is also written

$$\mathcal{L}_{\mathrm{QED}} = -\frac{1}{4}\ F_{\mu\nu}\ F^{\mu\nu} + \bar{\Psi}\ (i\ D\!\!\!\!/ - m)\ \Psi \tag{19.7}$$

by using the covariant derivative (18.82) of the electromagnetic field:

$$D_{\mu} = \partial_{\mu} + i\,e\,A_{\mu}\,. \tag{19.8}$$

It was shown in (18.95) that the second term in $\mathcal{L}_{\mathrm{QED}}$ is invariant for a local gauge transformation:

$$\mathcal{G}\ (x) = \ \exp\ \left(-\frac{i\,e}{\hbar\,c}\ \Lambda\ (x)\right)\,, \tag{19.9}$$

and since $F_{\mu\nu}$ is also invariant for this type of transformation, $\mathcal{L}_{\mathrm{QED}}$ is invariant under a local gauge transformation:

$$\begin{aligned}
\Psi\ &\rightarrow\ \Psi' = \mathcal{G}\ (x)\ \Psi\,,\\
A_{\mu}\ &\rightarrow\ A'_{\mu} = A_{\mu} + \partial_{\mu}\ \Lambda\ (x)\,,\\
D_{\mu}\ \Psi\ &\rightarrow\ D'_{\mu}\ \Psi' = \mathcal{G}\ (x)\ D_{\mu}\ \Psi\,,\\
\mathcal{L}_{\mathrm{QED}}\ &\rightarrow\ \mathcal{L}'_{\mathrm{QED}} = \mathcal{L}_{\mathrm{QED}}\,.
\end{aligned} \tag{19.10}$$

The canonical quantization rules (17.82) naturally apply to the e.m. field and if we consider the potentials $A_\mu (x)$ as generalized coordinates, then the conjugate momenta $\pi^\mu (x)$ will be the four-vectors

$$\pi^\mu \equiv \pi^{\mu\,0} = \frac{\partial \mathcal{L}}{\partial \left(\partial A_\mu / \partial x^0 \right)} = \frac{\partial \mathcal{L}}{\partial \left(\partial A_\mu / \partial t \right)} = F^{\mu\,0}. \tag{19.11}$$

The conservation of the momentum π^k is another way of stating Gauss' law:

$$\partial_k \pi^k = \partial_k F^{k\,0} = \partial_k E_k = \vec{\nabla} \cdot \vec{E} = 0. \tag{19.12}$$

The canonical quantization of the e.m. field raises the issue of the choice of the gauge appearing in the expression for the propagator of the e.m. field, which we write here in the following form, without any attempt at demonstration (Leader and Predazzi, 1982):

$$-\frac{i}{k^2} \left[\eta_{\mu\,\nu} - (1 - a) \frac{k_\mu\, k_\nu}{k^2} \right]. \tag{19.13}$$

Feynman's choice is $a = 1$, whereas Landau's is $a = 0$.

The Feynman rules for transcribing diagrams of the different expansion orders of the Lippman–Schwinger transition amplitude:

$$T = V + V\, G_0\, V + \dots \tag{19.14}$$

are next applied with a few modifications. They form the basis for the calculations of QED which have met with spectacular success. We have already had occasion to implicitly use the first-order statement of the transition amplitude in the calculation of the relativistic Coulomb scattering (Chap. 18).

An important aspect of QED concerns the renormalization procedure for eliminating divergences through the introduction of the concept of effective electric charge e_R, which is different from the naked electric charge, or equivalently of the e.m. interaction coupling constant $\alpha_E (Q^2)$, which is different from the e.m. finestructure constant α. In fact, $\alpha = \alpha_E (0)$ and the Q-dependence of $\alpha_E (Q^2)$ is written

$$\frac{1}{\alpha_E (Q^2)} = \frac{1}{\alpha (\mu^2)} - \frac{1}{3\pi} \log \left(\frac{Q^2}{\mu^2} \right). \tag{19.15}$$

The momentum μ is a reference momentum and Q is the transferred momentum.

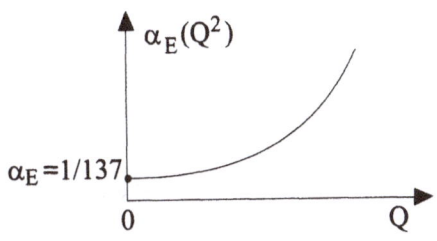

The coupling constant α_E (Q^2) describes the evolution of the effective electric charge with respect to the distance between two charged particles. The summation at every order of perturbation therefore describes the polarization of the vacuum and the screening effect experienced by the charged particle in interaction. With every successive increase in Q^2 comes an increase in the number of charges seen by the exchanged photon and, if the finite value of Q^2 is very high, the coupling constant will become infinite.

2. Elementary Introduction to Quantum Chromodynamics

Quantum chromodynamics (QCD) describes the interaction of quarks with flavor f (associated with the electric charge $\frac{2}{3}$ e or $-\frac{1}{3}$ e), mass m_f, color α ($= 1, 2, 3, = r, g, y$), and Lagrangian density

$$\mathcal{L}_q = \bar{\Psi}_f^\alpha \left(i\, \partial\!\!\!/ - m_f \right) \Psi_\alpha^f \tag{19.16}$$

with the gluon gauge field associated with the non-Abelian transformation group $SU(3)_c$ of covariant derivative (see Chap. 8):

$$D_\mu = \partial_\mu - i\, g\, W_\mu = \partial_\mu - i\, g\, T_a\, W_\mu^a \qquad a = 1, \ldots, 8. \tag{19.17}$$

The Lagrangian density of QCD is written in a totally analogous form to that of QED (with summation over a and α):

$$\mathcal{L}_{QCD} = -\frac{1}{4}\, F_{\mu\nu}^a\, F_a^{\mu\nu} + \sum_f \bar{\Psi}_f^\alpha \left(i\, D\!\!\!\!/ - m_f \right) \Psi_\alpha^f. \tag{19.18}$$

The problem of choosing a gauge is as crucial here as it is in QED, and following renormalization, the coupling constant of strong interactions α_s is also dependent on Q^2, although tending to zero for $Q \to \infty$:

$$\frac{1}{\alpha_s\,(Q^2)} = \frac{1}{\alpha_s\,(\mu^2)} + \frac{1}{12\pi}\,(33 - 2n_f)\, \log \frac{Q^2}{\mu^2} \tag{19.19}$$

$$n_f = \text{number of flavors} = 1, \ldots, 6.$$

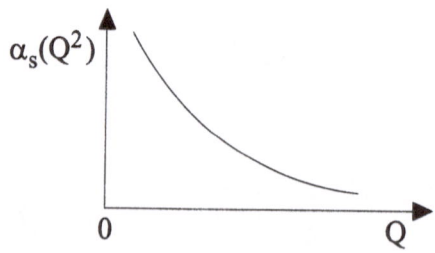

For $n_f > 16$, we once more obtain an increasing function of Q^2 just like in QED. The asymptotic freedom ($\alpha_s \rightarrow 0$ for $Q \rightarrow \infty$) exists only for a number $n_f < 16$ (that is for less than 8 generations of quarks). By introducing the scale parameter

$$\Lambda = \mu^2 \, \exp \left[\frac{-12\pi}{(33 - 2n_f) \, \alpha_s \, (\mu^2)} \right], \tag{19.20}$$

the coupling constant of strong interactions can take another form equivalent to (19.19):

$$\alpha_s \, (Q^2) = \frac{12\pi}{(33 - 2n_f) \, \log \frac{Q^2}{\Lambda^2}}. \tag{19.21}$$

For $Q^2 \gg \Lambda^2$, the perturbation expansion of QCD remains valid because $\alpha_s \, (Q^2)$ is small, which is no longer the case for $Q^2 \simeq \Lambda^2$.

The scale parameter Λ, which is experimentally determined, is of the order of 0.1 GeV, giving a coupling constant for strong interactions on the order of 0.1 for Q on the order of 10 GeV. This justifies the use of QCD in high-energy physics, at least in the energy regime covered by present-day accelerators.

3. The GSW Standard Model

The Glashow–Salam–Weinberg standard model of electroweak interactions represents one of the most significant advances of the last twenty years in our knowledge of the structure of matter. It describes the e.m. interaction by a gauge field with symmetry $U(1)$, and the weak interaction by a gauge field with symmetry $SU(2)$. However, the originality of the theory derives from its unified description of these interactions in a gauge field with symmetry $SU(2)_L \times U(1)$. Because the e.m. interaction has an infinite range and the weak interaction a finite range, these characteristics needed to be further associated with a process conferring mass to gauge bosons. Quite naturally, Higgs spontaneous symmetry breaking was used and the results obtained were experimentally confirmed beyond doubt. We will be content here with a very simplified description of the GSW model, referring the interested reader as usual to more specialized texts (Leader an Predazzi, 1982; Zee, 1982). For example, we shall be considering only first-generation fermions (leptons: electrons, neutrinos; hadrons: quarks u and d) and only boson exchanges between leptons will be taken into account (thus exluding exchanges between hadrons and leptons or between hadrons).

3.1 The Fermion Matter Field

The weak nuclear interaction, responsible in particular for the phenomenon of radioactivity, is sometimes purely leptonic (for example, in the disintegration of the muon $\mu^- \rightarrow e^- + \bar{\nu}_e + \nu_\mu$), or purely hadronic (as in the disintegration of the particle $\Lambda \rightarrow p + \mu^-$), or semi-leptonic in the typical example of neutron disintegration in natural β radioactivity $n \rightarrow p + e^- + \bar{\nu}_e$.

Fermi has described the weak interaction as a current-current interaction by using the Lagrangian density

$$\mathcal{L}_F = \frac{G_F}{\sqrt{2}} \, j_\alpha^+ \, (x) \, j^\alpha \, (x). \tag{19.22}$$

The Fermi coupling constant G_F takes the value

$$G_F = 1.16 \; 10^{-5} \; \text{GeV}^{-2} \tag{19.23}$$

and the current $j_\alpha \, (x)$ is, more generally, the sum of a leptonic current $\ell_\alpha \, (x)$ and a hadronic current $h_\alpha \, (x)$. We will limit ourselves here to first-generation leptons (e^-, ν) by using the spinors associated with these fermions to describe the leptonic current:

$$j^\lambda \equiv \ell^\lambda = \bar{\Psi}_e \, \gamma^\lambda \, (1 - \gamma_5) \, \Psi_\nu \,. \tag{19.24}$$

This current is of the axial vector type. We note here that any spinor can be separated into a right and a left part with the matrix γ_5, by writing (cf. (18.32)):

$$\psi = \frac{1}{2} \, (1 + \gamma_5) \, \Psi + \frac{1}{2} \, (1 - \gamma_5) \, \Psi = \Psi_R + \Psi_L \,. \tag{19.25}$$

From a strictly formal point of view, the two left-hand leptons, ν_L and e_L^-, are grouped together in a doublet Ψ_1 with weak isospin $1/2$, and we assign the projection $I_3 = +\frac{1}{2}$ to the neutrino, and $I_3 = -\frac{1}{2}$ to the electrically charged lepton, that is to the left electron:

$$\Psi_1 = \begin{pmatrix} \nu_L \\ e_L^- \end{pmatrix} = \begin{pmatrix} \frac{1}{2}(1-\gamma_5)\Psi_\nu \\ \frac{1}{2}(1-\gamma_5)\Psi_{e^-} \end{pmatrix} = \begin{pmatrix} f_L \\ f_L' \end{pmatrix} \,. \tag{19.26}$$

The electric charge is linked to the weak isospin and the hypercharge Y through the Gell–Mann–Nishijima (cf. (9.76)) relation

$$Q = I_3 + \frac{1}{2} \, Y \,, \tag{19.27}$$

which confers the doublet of left leptons with $Y_L = -1$, the right electron with hypercharge $Y_R = -2$, and the right neutrino for which $I_3 \equiv 0$ with $Y_R = 0$. The right-hand sides of the lepton that do not take part in the weak interactions are kept in singlets denoted Ψ_2 and Ψ_3, that is,

$$\Psi_2 = \nu_R \qquad \text{and} \qquad \Psi_3 = e_R^- . \tag{19.28}$$

This allows the introduction of the spinor Ψ_j ($j = 1, 2, 3$) for describing the left doublet and the two right singlets of the lepton (cf. (19.26) and (19.28)). We will note that the Pauli matrices $\vec{\tau}$ are acting on the doublet, and not on the singlets, such that

$$\vec{\tau}\,\Psi_2 = \vec{\tau}\,\Psi_3 = 0 . \tag{19.29}$$

Fermi's theory of weak interactions assumes that the neutrino is a particle with spin $\frac{1}{2}$ and zero mass with left chirality, and that the right-hand part does not exist. The major weakness of the theory is that it is not renormalizable and efforts are underway to replace it with a renormalizable theory in which the weak interaction (with a very short range) is carried by an electrically charged or neutral massive vector boson.

At the outset, leptons, just like gauge fields, are assumed to have zero mass and it is through the process of Higgs spontaneous symmetry breaking that the bosons and fermions in interaction, in the GSW standard model, acquire mass.

3.2 The GSW Boson Gauge Field

The GSW standard model of e.m. and weak interactions is based on the symmetry gauge field $SU(2)_L \times U(1)$. The local gauge transformation operator of this model is written in the form

$$\mathcal{G}\,(x) = \mathcal{G}_{SU_2}\,(x) \times \mathcal{G}_{U_1}\,(x)$$

$$= \exp\,\left(\frac{1}{2}\, g\, \vec{\tau} \cdot \vec{\alpha}\,(x) \right)\, \exp\,(i\, g'\, Y_j\, \beta\,(x)) \tag{19.30}$$

where the index $j = 1, 2, 3$ refers to the type-(19.26) or type-(19.28) spinor on which the operator should act, $\vec{\tau}$ are the Pauli matrices, generators of the group $SU(2)_L$, whereas g and g' are the coupling constants of the leptons in the gauge fields $B_\mu\,(x)$ associated with the Abelian group $U(1)$ and $W_\mu^a\,(x)$, $a = 1, 2, 3$ is associated with the non-Abelian gauge field $SU(2)_L$ (see Sect. 3.4 of Chap. 8). Thus, the covariant differentiation operator will take the form

$$D_\mu = \partial_\mu - i\, \frac{g}{2}\, \vec{\tau} \cdot \vec{W}_\mu\,(x) - i\, g'\, Y_j\, B_\mu\,(x) \tag{19.31}$$

and the tensors associated with the gauge fields will be denoted

$$\begin{aligned} B_{\mu\,\nu}\,(x) &= \partial_\mu\, B_\nu\,(x) - \partial_\nu\, B_\mu\,(x) , \\ W_{\mu\,\nu}\,(x) &= \partial_\mu\, W_\nu\,(x) - \partial_\nu\, W_\mu\,(x) + g\, \vec{W}_\mu\,(x) \wedge \vec{W}_\nu\,(x) . \end{aligned} \tag{19.32}$$

The Lagrangian density of the gauge field will be written in a form analogous to that of the QED:

$$\mathcal{L}_{\text{GSW}} = \mathcal{L}_B + \mathcal{L}_W + \mathcal{L}_{int}$$
$$= -\frac{1}{4} B_{\mu\nu} B^{\mu\nu} - \frac{1}{4} \vec{W}_{\mu\nu} \cdot \vec{W}^{\mu\nu} + \sum_j i \bar{\Psi}_j \not{D} \Psi_j . \quad (19.33)$$

The important term here is the interaction Lagrangian of the gauge field with the leptons:

$$\mathcal{L}_{int} = \sum_j i \bar{\Psi}_j \gamma^\mu D_\mu \Psi_j \qquad j = 1, 2, 3 . \quad (19.34)$$

It is written with the right and left parts of the spinor

$$\mathcal{L}_{int} = \bar{\Psi}_R \, i \, \gamma^\mu \, (\partial_\mu - i \, g' \, Y_R \, B_\mu) \, \Psi_R$$
$$+ \bar{\Psi}_L \, i \, \gamma^\mu \, (\partial_\mu - i \, g' \, Y_L \, B_\mu - \frac{i}{2} \, g \, \vec{\tau} \cdot \vec{W}_\mu) \, \Psi_L . \quad (19.35)$$

The right-hand parts comprise two singlets e_R^- and ν_R, and the left the doublet ν_L and e_L^-. The compact form (19.34) will be used more often here than the expanded form (19.35) for the interaction Lagrangian density.

3.3 Electrically Charged Weak Currents

Electrically charged weak currents are carried by the gauge bosons W_μ^1 and W_μ^2 and described with the Lagrangian density

$$\mathcal{L}_{cc} = \frac{g}{2} \, \bar{\Psi}_1 \, \gamma^\mu \, (\tau_1 \, W_\mu^1 + \tau_2 \, W_\mu^2) \, \Psi_1 . \quad (19.36)$$

We introduce the linear combination of the fields W_μ^1 and W_μ^2 as follows:

$$W_\mu^\pm = \frac{1}{\sqrt{2}} \, (W_\mu^1 \mp i \, W_\mu^2), \quad (19.37)$$

yielding, with the Pauli matrices τ_1 and τ_2,

$$\tau_1 \, W_\mu^1 + \tau_2 \, W_\mu^2 = \begin{pmatrix} 0 & W_\mu^1 + i \, W_\mu^2 \\ W_\mu^1 - i \, W_\mu^2 & 0 \end{pmatrix} = \sqrt{2} \begin{pmatrix} 0 & W_\mu^- \\ W_\mu^+ & 0 \end{pmatrix} . \quad (19.38)$$

Inserting this into (19.36) with the left doublet of fermions (19.26), we obtain

$$\mathcal{L}_{cc} = \frac{g}{\sqrt{2}} \, (\bar{f}_L \, \bar{f}'_L) \, \gamma^\mu \begin{pmatrix} 0 & W_\mu^- \\ W_\mu^+ & 0 \end{pmatrix} \begin{pmatrix} f_L \\ f'_L \end{pmatrix}$$
$$= \frac{g}{\sqrt{2}} \, [\bar{f}_L \, \gamma^\mu \, W_\mu^- \, f'_L + \bar{f}'_L \, \gamma^\mu \, W_\mu^+ \, f_L] . \quad (19.39)$$

We introduce the Pauli matrix $\tau^- = \frac{1}{2}(\tau_1 - i\,\tau_2)$ and the charged weak current

$$j_{cc}^{-\mu} = \bar{f}_L\,\gamma^\mu\,f'_L = \bar{f}\,\gamma^\mu\,\frac{1}{2}\,(1 - \gamma_5)\,f'$$

$$= \sum_j \bar{\Psi}_j\,\gamma^\mu\,\tau^-\,\Psi_j\,. \tag{19.40}$$

The Lagrangian density $\mathcal{L}_{cc}$ is then written as the interaction of a current j^μ with a gauge field W_μ with electric charge $(+)$ or $(-)$ as follows:

$$\boxed{\begin{aligned} \mathcal{L}_{cc} &= \frac{g}{\sqrt{2}}\,[j_{cc}^{-\mu}\,W_\mu^- + j_{cc}^{+\mu}\,W_\mu^+] \\ &= \frac{g}{\sqrt{2}}\,[j_{cc}^{-\mu}\,W_\mu^- + \text{hermitian conjugate}] \end{aligned}} \tag{19.41}$$

3.4 Neutral Current

After extraction of the charged weak current, we are left with the neutral current Lagrangian density (not electrically charged):

$$\mathcal{L}_{NC} = \bar{\Psi}_j\,\gamma^\mu\left(g\,\frac{\tau_3}{2}\,W_\mu^3 + g'\,Y_j\,B_\mu\right)\Psi_j \quad j = 1,2,3\,. \tag{19.42}$$

Let us make a canonical transformation of the fields B_μ and W_μ^3 by introducing two fields A_μ and Z_μ, such that A_μ is not coupled to the neutrino but solely to the electron, with parity conservation (characteristics associated with the e.m. field):

$$\begin{pmatrix} W_\mu^3 \\ B_\mu \end{pmatrix} = \begin{pmatrix} \cos\theta & \sin\theta \\ -\sin\theta & \cos\theta \end{pmatrix} \begin{pmatrix} Z_\mu \\ A_\mu \end{pmatrix}. \tag{19.43}$$

The rotation angle is termed the Weinberg angle, and thus often denoted θ_W. By using (19.43) in $\mathcal{L}_{NC}$, we obtain two terms, with the first dependent only upon A_μ (for simplicity we write θ for the Weinberg angle):

$$\mathcal{L}_{NC}^A = \bar{\Psi}_j\,\gamma^\mu\left(g\,\frac{\tau_3}{2}\,\sin\theta + g'\,Y_j\,\cos\theta\right)\Psi_j\,A_\mu \quad j = 1,2,3 \tag{19.44}$$

and the second term depending only on Z_μ:

$$\mathcal{L}_{NC}^Z = \bar{\Psi}_j\,\gamma^\mu\left(g\,\frac{\tau_3}{2}\,\cos\theta - g'\,Y_j\,\sin\theta\right)\Psi_j\,Z_\mu\,. \tag{19.45}$$

Let us write the Lagrangian density of the e.m. field in interaction with the leptons f with electric charge eQ_f, and f' with electric charge $eQ_{f'}$ (see Chap. 18):

$$\mathcal{L}_{em} = e\,(\bar{f}\,\gamma^\mu\,Q_f\,f + \bar{f}'\,\gamma^\mu\,Q_{f'}\,f')\,A_\mu\,. \tag{19.46}$$

The number of electric charges Q_j ($j = 1, 2, 3$) will be $Q_2 = Q_f$ for the neutrino, $Q_3 = Q_{f'}$ for the electron. The number of electric charges Q_1 will be associated with the left lepton doublet:

$$Q_1 = \begin{pmatrix} Q_f & 0 \\ 0 & Q_{f'} \end{pmatrix} = \frac{\tau_3}{2} + (Q_f - \frac{1}{2})\, \mathbb{1} \equiv \frac{\tau_3}{2} + Y_1\, \mathbb{1}. \tag{19.47}$$

This defines the hypercharge Y_1 of the left lepton:

$$Y_1 = Q_f - \frac{1}{2}. \tag{19.48}$$

The Lagrangian density $\mathcal{L}_{em}$ can now be written in the form

$$\mathcal{L}_{em} = \bar{\Psi}_j\, \gamma^\mu\, (e\, Q_j)\, \Psi_j\, A_\mu, \tag{19.49}$$

and, by comparing with (19.44), we infer

$$e\, Q_j = g\, \frac{\tau_3}{2}\, \sin\theta + g'\, Y_j\, \cos\theta. \tag{19.50}$$

This leads to three equations corresponding to the values of the index j, with the values (19.47) of Q_1, Q_2 and Q_3:

$$\begin{aligned} e\, Q_1 &= e\, \left(\frac{\tau_3}{2} + \left(Q_f - \frac{1}{2} \right)\, \mathbb{1} \right) = g\, \frac{\tau_3}{2}\, \sin\theta + g'\, Y_1\, \cos\theta, \\ e\, Q_2 &= e\, Q_f = g'\, Y_2\, \cos\theta, \\ e\, Q_3 &= e\, Q_{f'} = g'\, Y_3\, \cos\theta. \end{aligned} \tag{19.51}$$

These equations determine the hypercharge Y_j:

$$Y_1 = Q_f - \frac{1}{2} \qquad Y_2 = Q_f \qquad Y_3 = Q_{f'}, \tag{19.52}$$

and a relation between the coupling constants g, g' and e and the Weinberg angle:

$$\boxed{e = g\, \sin\theta = g'\, \cos\theta}. \tag{19.53}$$

We can eliminate the Weinberg angle θ from these equations, yielding

$$\frac{1}{e^2} = \frac{1}{g^2} + \frac{1}{g'^2} \qquad \text{and} \qquad e = \frac{g\, g'}{\sqrt{g^2 + g'^2}}, \tag{19.54}$$

or eliminate the electric charge e, while retaining the Weinberg angle:

$$\tan\theta = \frac{g'}{g} \qquad \cos\theta = \frac{g}{\sqrt{g^2 + g'^2}} \qquad \text{and} \qquad \sin\theta = \frac{g'}{\sqrt{g^2 + g'^2}}. \tag{19.55}$$

It might be useful to evaluate $g\cos\theta$ and $g'\sin\theta$, which appear in $\mathcal{L}_{NC}^Z$ in (19.45) by using (19.53):

$$g\,\cos\theta = \frac{e\,\cos\theta}{\sin\theta} = \frac{e\,\cos^2\theta}{\sin\theta\,\cos\theta} \qquad \text{and} \qquad g'\,\sin\theta = \frac{e\,\sin^2\theta}{\sin\theta\,\cos\theta}. \tag{19.56}$$

This leads to the following form of $\mathcal{L}_{NC}^Z$ taken from expression (19.45):

$$\mathcal{L}_{NC}^Z = \bar{\Psi}_j\,\gamma^\mu\left(\frac{T_3}{2}\frac{e\,\cos^2\theta}{\sin\theta\,\cos\theta} - \frac{e\,\sin^2\theta}{\sin\theta\,\cos\theta}\,Y_j\right)\Psi_j\,Z_\mu, \tag{19.57}$$

which may also be written in the form of a matrix:

$$\mathcal{L}_{NC}^Z = \frac{2e}{\sin 2\theta}\,\{(\bar{f}_L\ \bar{f}'_L)\,\gamma^\mu$$
$$\times\begin{pmatrix} \frac{1}{2}\cos^2\theta - (Q_f - \frac{1}{2})\sin^2\theta & 0 \\ 0 & -\frac{1}{2}\cos^2\theta - (Q_f - \frac{1}{2})\sin^2\theta) \end{pmatrix}\begin{pmatrix} f_L \\ f'_L \end{pmatrix}$$
$$-\bar{f}_R\,\gamma^\mu\,Q_f\,\sin^2\theta\,f_R - \bar{f}'_R\,\gamma^\mu\,Q_{f'}\,\sin^2\theta\,f'_R\}\,Z_\mu. \tag{19.58}$$

Expanding the matrix product leaves us with

$$\mathcal{L}_{NC}^Z = \frac{2e}{\sin 2\theta}\,\Big\{\bar{f}_L\,\gamma^\mu\left(\frac{1}{2} - Q_f\,\sin^2\theta\right)f_L$$
$$+ \bar{f}'_L\,\gamma^\mu\left(-\frac{1}{2} - (Q_f - 1)\,\sin^2\theta)\right)f'_L$$
$$-\bar{f}_R\,\gamma^\mu\,Q_f\,\sin^2\theta\,f_R - \bar{f}'_R\,\gamma^\mu\,Q_{f'}\,\sin^2\theta\,f'_R\}\,Z_\mu. \tag{19.59}$$

If we introduce the matrices γ_5 by writing out the left and right parts of the fermions with (19.25) and then group together similar terms, we have

$$\mathcal{L}_{NC}^Z = \frac{2e}{\sin 2\theta}\,\Big\{\bar{f}\,\gamma^\mu\left(\frac{1-\gamma_5}{2}\right)f + \bar{f}'\left(-\frac{1}{4} + \frac{\gamma_5}{4} - \sin^2\theta\right)f'\Big\}\,Z_\mu. \tag{19.60}$$

As can be seen in the first term, only the left neutrino appears in the weak interaction through the exchange of neutral current.

By using the form (19.57) of the Lagrangian density and hypercharges (19.51), we can further write

$$\mathcal{L}_{NC}^Z = \frac{2e}{\sin 2\theta}\,\Big[\bar{\Psi}_j\,\gamma^\mu\left(\frac{T_3}{2}\cos^2\theta - Y_j\,\sin^2\theta\right)\Psi_j\Big]\,Z_\mu, \tag{19.61}$$

and by replacing $\cos^2\theta$ with $1 - \sin^2\theta$, and Q_1 with $\frac{T_3}{2} + Y_1\,\mathbb{1}$, we obtain the relation

$$\mathcal{L}_{NC}^Z = \frac{2e}{\sin 2\theta}\,\Big[\bar{\Psi}_j\,\gamma^\mu\left(\frac{T_3}{2} - Q_j\,\sin^2\theta\right)\Psi_j\Big]\,Z_\mu. \tag{19.62}$$

This is the interaction of the field Z_μ with the neutral current

$$
\begin{aligned}
j^\mu_{NC} &= \bar{\Psi}_j \, \gamma^\mu \, \left(\tau_3 - 2 \, \sin^2 \theta \, Q_j \right) \, \Psi_j \\
&= \bar{\Psi}_j \, \gamma^\mu \, \tau_3 \, \Psi_j - 2 \, \sin^2 \theta \, \bar{\Psi}_j \, Q_j \, \Psi_j \, .
\end{aligned}
\tag{19.63}
$$

If we introduce the electromagnetic current, we obtain

$$
\boxed{
\begin{aligned}
j^\mu_{NC} &= j^\mu_3 - 2 \, \sin^2 \theta \, j^\mu_{em} \\
\mathcal{L}^Z_{NC} &= e/\sin 2\theta \, j^\mu_{NC} \, Z_\mu
\end{aligned}
}
\tag{19.64}
$$

The neutral weak current is the difference between the current j^μ_3 and the electromagnetic current, weighted with the Weinberg angle. It corresponds to an exchange of photons γ with neutral intermediate bosons Z^0, mixed with the Weinberg angle.

3.5 Interaction of the Electroweak Field with the Higgs Field

Let us introduce a scalar field φ in the form of a doublet $\Phi = \begin{pmatrix} \varphi^+ \\ \varphi^0 \end{pmatrix}$ in which the first element φ^+ can be coupled with electrically charged gauge bosons $W^\pm$, and the second element φ^0 with a photon and Z^0, which are electrically neutral. The interaction Lagrangian density of the GSW gauge field with the scalar field, characterized by the Higgs potential $V (\Phi^+ \, \Phi)$:

$$
V (\Phi^+ \, \Phi) = M^2 \, \Phi^+ \, \Phi + h \, (\Phi^+ \, \Phi)^2 \quad h > 0 \quad M^2 < 0 \, ,
\tag{19.65}
$$

will be written with the covariant derivative D_μ defined in (19.31) in the form:

$$
\mathcal{L}_{GSW-H} = (D_\mu \, \Phi)^+ \, (D_\mu \, \Phi) - V (\Phi^+ \, \Phi) \, .
\tag{19.66}
$$

To evaluate this Lagrangian density, we expand the matrix form of the covariant derivative:

$$
D_\mu = \partial_\mu - \frac{i}{2} \, g \, \vec{\tau} \cdot \vec{W}_\mu - i \, g' \, Y \, B_\mu \, ,
\tag{19.67}
$$

by fixing the constant Y such that the e.m. field is not coupled to the neutral Higgs field φ^0.

Let us begin by rewriting the GSW Lagrangian density associated with the neutral currents

$$
\mathcal{L}_{NC} = \bar{\Psi}_j \, \gamma^\mu \, \left(\frac{g}{2} \, \tau_3 \, W^3_\mu + g' \, Y \, B_\mu \right) \, \Psi_j
\tag{19.68}
$$

with the Weinberg mixing angle of Z_μ and A_μ; we obtain from relation (19.43)

$$
\begin{aligned}
W^3_\mu &= \cos \theta \, Z_\mu + \sin \theta \, A_\mu \, , \\
B_\mu &= - \sin \theta \, Z_\mu + \cos \theta \, A_\mu \, .
\end{aligned}
\tag{19.69}
$$

Recalling the relation between the coupling constants

$$g \sin \theta = g' \cos \theta = e,$$
(19.70)

the neutral current operator in (19.68) can finally be written

$$\frac{g}{2} \tau_3 W_\mu^3 + g' Y B_\mu = \left[\left(\frac{g}{2} \cos \theta \, \tau_3 - g' Y \sin \theta \right) Z_\mu + e \left(\frac{\tau_3}{2} + Y \right) A_\mu \right].$$
(19.71)

Expressed in the form of a matrix, the term in A_μ will therefore be written

$$\begin{pmatrix} \left(\frac{1}{2} + Y \right) e A_\mu & 0 \\ 0 & \left(-\frac{1}{2} + Y \right) e A_\mu \end{pmatrix}.$$
(19.72)

If we apply this operator to the Higgs field doublet Φ, we obtain the doublet

$$\begin{pmatrix} \left(\frac{1}{2} + Y \right) & e \varphi^+ A_\mu \\ \left(-\frac{1}{2} + Y \right) & e \varphi_0 A_\mu \end{pmatrix}.$$
(19.73)

We force the electromagnetic field into not coupling to the electrically neutral Higgs field φ_0. This defines the hypercharge $Y = \frac{1}{2}$ and the covariant derivative (19.67) thus becomes, for the doublet part, i.e., for the part acting in the two-dimensional space with isospin,

$$D_\mu = \left(\partial_\mu - \frac{i}{2} g' B_\mu \right) \mathbb{1} - \frac{i}{2} g \, \vec{\tau} \cdot \vec{W}_\mu.$$
(19.74)

This corresponds to the matrix form of D_μ:

$$D_\mu = \begin{pmatrix} \partial_\mu - \frac{i}{2} \frac{g^2 - g'^2}{\sqrt{g^2 + g'^2}} Z_\mu - i e A_\mu & -\frac{i g}{\sqrt{2}} W_\mu^+ \\ -\frac{i g}{\sqrt{2}} W_\mu^- & \partial_\mu + \frac{i}{2} \sqrt{g^2 + g'^2} \, Z_\mu \end{pmatrix}.$$
(19.75)

By applying this operator to the doublet of the Higgs complex scalar field $\Phi = \begin{pmatrix} \varphi^+ \\ \varphi_0 \end{pmatrix}$, we obtain the term $D_\mu \Phi$ of the Lagrangian density $\mathcal{L}_{GSW-H}$:

$$D_\mu \Phi = \begin{pmatrix} \left(\partial_\mu - \frac{i}{2} \frac{g^2 - g'^2}{\sqrt{g^2 + g'^2}} Z_\mu - i e A_\mu \right) \varphi^+ - \frac{i g}{\sqrt{2}} - C W_\mu^+ \varphi_0 \\ -\frac{i g}{\sqrt{2}} W_\mu^- \varphi^+ + \left(\partial_\mu + \frac{i}{2} \sqrt{g^2 + g'^2} \, Z_\mu \right) \varphi_0 \end{pmatrix}.$$
(19.76)

3.6 Spontaneous Symmetry Breaking of the Higgs Field

The spontaneous symmetry breaking of the Higgs field (see Chap. 17) will be described by writing the doublet Φ as

$$\Phi = \begin{pmatrix} 0 \\ \frac{1}{\sqrt{2}} \left(\chi \left(x \right) + v \right) \end{pmatrix}.$$
(19.77)

This is equivalent to introducing the special values of the Higgs field:

$$\varphi^+ = 0 \qquad \text{and} \qquad \varphi^0 = \frac{1}{\sqrt{2}} (\chi + v). \qquad (19.78)$$

The matrix expression (19.76) for $D_\mu \phi$ then becomes

$$D_\mu \Phi = \begin{pmatrix} -\frac{i\,g}{2} \, W_\mu^+ \, (\chi + v) \\ \frac{1}{\sqrt{2}} \partial_\mu \, \chi + \frac{i}{2\sqrt{2}} \, \sqrt{g^2 + g'^2} \, Z_\mu \, (\chi + v) \end{pmatrix} . \qquad (19.79)$$

It is then easy to evaluate the Lagrangian density $\mathcal{L}_{GSW-H}$, following the spontaneous symmetry breaking, by inserting this result and (19.78) into (19.66):

$$\mathcal{L}_{GSW-H} = (D_\mu \, \Phi)^+ \, (D_\mu \, \Phi) - \frac{M^2}{2} \, (\chi + v)^2 + \frac{h}{4} \, (\chi + v)^4 . \qquad (19.80)$$

By using the Weinberg angle, expressed in terms of the coupling constants:

$$\cos \theta = \frac{g}{\sqrt{g^2 + g'^2}} \,,, \qquad (19.81)$$

this Lagrangian density can be rewritten in the form

$$\begin{aligned} \mathcal{L}_{GSW-H} = & \frac{1}{4} \, g^2 \, v^2 \, (W_\mu^+ \, W_\mu^- + \frac{1}{2 \cos^2 \theta} \, Z_\mu \, Z^{\mu+}) \\ & + \frac{g^2}{4} \, (2v \, \chi + \chi^2) \, (W_\mu^+ \, W_\mu^- + \frac{1}{2 \, \cos^2 \theta} \, Z_\mu \, Z^{\mu+}) \\ & + \frac{1}{2} \, (\partial_\mu \, \chi \, \partial^\mu \, \chi + 2 M^2 \, \chi^2) \\ & + \frac{M^2}{v} \, \chi^3 + \frac{M^2}{4 v^2} \, \chi^4 - \frac{1}{4} \, M^2 \, v^2 . \end{aligned} \qquad (19.82)$$

The reader is reminded here that the Lagrangian of a complex scalar field with mass is written $\mathcal{L} = \frac{1}{2} \, \partial_\mu \, \varphi \, \partial^\mu \, \varphi^+ - m^2 \, | \, \varphi \, |^2$ whereas for a real field it is $\mathcal{L} = \frac{1}{2} \, (\partial_\mu \, \varphi \, \partial^\mu \, \varphi - m^2 \, \varphi^2)$. The boson field is complex since we have set $W_\mu^\pm = (1/\sqrt{2}) \, (W_\mu^1 \pm i \, W_\mu^2)$ whereas the boson field Z_μ is a real field. The first line of (19.82) therefore shows that the mass acquired by the intermediate bosons in the spontaneous symmetry breaking process is

$$M_W = \frac{1}{2} \, g \, v \,, \qquad (19.83)$$

$$M_Z = \frac{1}{2} \, \frac{g \, v}{\cos \theta} = \frac{M_W}{\cos \theta} . \qquad (19.84)$$

Furthermore, the terms of the second line of (19.82) show that there is interaction between the massive bosons $W^\pm$ and Z and the Higgs scalar field χ. Finally, the brackets of the second line show that the Higgs bosons, real scalar bosons, have acquired a mass

$$M_H = \sqrt{-2\,M^2}\,. \qquad (19.85)$$

The parameter $M^2 \langle 0$ is a free parameter in the theory but the fact that such massive particles have never been observed suggests that Higgs bosons are extremely massive. Current and future experiments in the physics of elementary particles are striving to verify their existence and to determine their mass, that is the parameter M. This is one of the major objectives of the Large Hadron Collider (LHC) to be constructed at CERN in Geneva.

3.7 Mass of Intermediate Bosons

To determine a theoretical value of M_W and M_Z, the coupling constant g will have to be linked to the Fermi constant of weak interactions G_F. To this end, the phenomenology of weak interactions is used, by evaluating the differential cross-section of electron–neutrino scattering:

$$\nu_e + e^- \;\rightarrow\; \nu_e + e^-\,,$$

given to first-order in the center of mass system by

$$\frac{d\,\sigma}{d\,\Omega} = \frac{G_F^2\,k^2}{\pi^2}\,. \qquad (19.86)$$

Higher-order corrections lead to divergences which cannot be eliminated by the appropriate renormalizations. The first attempt to eliminate these divergences consisted in assuming that the weak interaction is due to the exchange of massive bosons. The Fermi Lagrangian (19.22)

$$\mathcal{L}_F = \frac{G_F}{\sqrt{2}}\; j^\mu\,(x)\, j_\mu^+\,(x) \qquad (19.87)$$

will then have to be replaced by an interaction Lagrangian with these massive bosons W_μ written in the form

$$\mathcal{L}_W = g_W\; j^\mu\,(x)\, W_\mu\,(x) + \mathrm{h}\cdot\mathrm{c}\,. \qquad (19.88)$$

This leads to the following value of the electron–neutrino differential scattering cross-section:

$$\frac{d\,\sigma}{d\,\Omega} = \frac{2\,g_W^4\,k^2}{\pi^2\,(Q^2 - M_W^2)^2}\,, \qquad (19.89)$$

where M_W is the mass of the intermediate boson W and Q the transferred momentum.

If $Q \rightarrow 0$, we obtain the Fermi differential cross-section, and by identifying (19.86) with the limit obtained with (19.89)

$$\frac{2 g_W^4}{\pi^2\, M_W^4} = \frac{G_F^2}{\pi^2} \quad \text{or} \quad \frac{g_W^2}{M_W^2} = \frac{G_F}{\sqrt{2}} . \tag{19.90}$$

It remains to find a relation between g_W and the coupling constant g of the GSW standard theory. The Lagrangian density $\mathcal{L}_{cc}$ of the charged weak currents (19.41) is written with (19.40) as

$$\mathcal{L}_{cc} = \frac{g}{2\sqrt{2}}\, \left(\bar{f}\, \gamma^\mu\, (1 - \gamma_5)\, f'\, W_\mu + \text{h} \cdot \text{c} \right) . \tag{19.91}$$

It is to be compared with Lagrangian density (19.88) of the weak interactions carried by massive bosons, which shows that

$$\boxed{\; g/2\sqrt{2} = g_W \quad \text{and} \quad g = 2\sqrt{2}\, M_W\, \left(\frac{G_F}{\sqrt{2}} \right)^{1/2} \;} . \tag{19.92}$$

By returning to the value (19.83) of the mass of the boson W, we finally obtain the Higgs parameter v (expressed in natural units $\hbar = c = 1$):

$$v^2 = \frac{4\, M_W^2}{g^2} = \frac{1}{\sqrt{2}\, G_F} . \tag{19.93}$$

With the experimental value of the Fermi constant G_F:

$$G_F/(\hbar\, c)^3 = 1.166\, 39\ 10^{-5}\ \text{GeV}^{-2} , \tag{19.94}$$

the Higgs parameter assumes the value $v \simeq 246$ GeV. We further use the value of the fine structure constant of the e.m. interactions $\alpha_{em} \simeq 1/137$, which boils down fixing e at the experimental value of the electric charge of the electron in order to determine the mass of the intermediate bosons:

$$M_W = \frac{1}{2}\, g\, v = \frac{e}{2\,\sin\theta}\, \frac{1}{(\sqrt{2}\, G_F)^{1/2}}$$

$$= \left(\frac{\pi\, \alpha_{em}}{\sqrt{2}\, G_F} \right)^{1/2} \frac{1}{\sin\theta} \simeq \frac{38}{\sin\theta}\ (\text{GeV}/c^2) , \tag{19.95}$$

$$M_Z = \frac{M_W}{\cos\theta} \simeq \frac{76}{\sin 2\theta}\ \ \text{GeV}/c^2$$

The Weinberg angle θ can be experimentally determined in neutrino–nucleon or electron–deuteron scattering and the value currently accepted by particle physicists* is

$$\sin^2 \theta = 0.23061 \pm 0.00047, \tag{19.96}$$

which gives a theoretical value for the masses of intermediate bosons

$$M_W \simeq 80 \text{ GeV}/c^2 \quad \text{and} \quad M_Z \simeq 90 \text{ GeV}/c^2. \tag{19.97}$$

The experimental value of these masses was determined for the first time at CERN in 1983, with lifetimes in the order of 10^{-25} seconds.

The accepted experimental value in 1995 was

$$\begin{aligned} M_W &= 80.356 \pm 0.125 \text{ GeV}/c^2, \\ M_Z &= 91.1863 \pm 0.0020 \text{ GeV}/c^2. \end{aligned} \tag{19.98}$$

The numerous experiments producing intermediate bosons which permit the continual improvement of these results, and the excellent agreement obtained with the theoretical determinations, constitute an eloquent verification of the standard model of electroweak interactions as well as of the Higgs spontaneous symmetry breaking. It remains, however, to unequivocally characterize Higgs bosons, neutral (with neither electric nor weak charge) scalar bosons (spin 0) with mass $M_H = \sqrt{-2\,M^2}$, assumed to lie between 60 and 250 GeV. Numerous experiments with high-energy particle accelerators (LHC at CERN, for example) are geared toward determining this mass and confirming the Higgs theory.

4. Grand Unification Theories

The success of the standard model encouraged particle physicists to unify strong interactions with symmetry $SU(3)$ and electroweak interactions with symmetry $SU(2) \times U(1)$ into one gauge theory, with a gauge field of symmetry $SU(3) \times SU(2) \times U(1)$. Many attempts have been made by theoreticians to find a symmetry group larger than $SU(3)$, capable of yielding $SU(3)$ after symmetry breaking, that is the gauge field associated with the strong nuclear interaction and quantum chromodynamics, and the electroweak gauge field with symmetry $SU(2) \times U(1)$, a symmetry breaking with lower energy separating the weak interactions $SU(2)$ from the electromagnetic ones $U(1)$. This is very important in understanding the production of matter in the Universe just after the Big Bang, i.e., in the current standard model of cosmology (see Chap. 20).

The Grand Unification Model with $SU(5)$ seemed to be able to meet this expectation, especially because the coupling constants of strong, electromagnetic, and weak interactions seemed to converge toward the same value at very high energy (10^{15} GeV) (Zee, 1992).

---

* Rochester Conference Warsaw (1996)

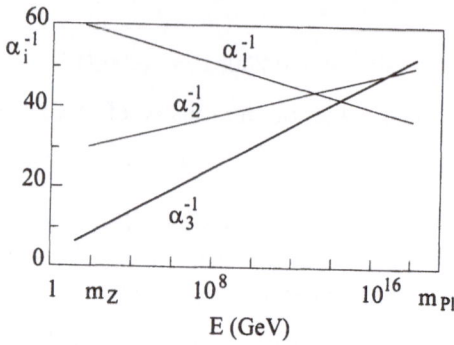

The grand unification model based on the gauge group $SU(5)$ was therefore developed. It predicts $5^2 - 1 = 24$ gauge bosons and, by extracting the 8 gluons, the three intermediate bosons and the photon, that is the 12 bosons carrying the $SU(3) \times SU(2) \times U(1)$ gauge fields, there remained the 12 gauge bosons of the grand unification model, termed X and Y bosons.

The X bosons have electric charge $\frac{4}{3}e$ and the Y bosons electric charge $+\frac{1}{3}e$, together with a color charge like the quarks. In this model, one thus has 3 X bosons, 3 Y bosons and their corresponding antiparticles, 3 $\bar{X}$ bosons and 3 $\bar{Y}$ bosons, making exactly the 12 expected bosons.

These bosons (also termed leptoquarks) can transform a quark into a lepton, for example, during the disintegration process

$$u \; \rightarrow \; e^+ + \bar{Y}. \tag{19.99}$$

The quark u with color i (red, blue, green) transfers its color to the $\bar{Y}$ boson with electric charge $-\frac{1}{3} e$. The electric charges are of course conserved. If we introduce a weak charge $N = -\frac{1}{3}$ for the quark u and 0 for the positron, the $\bar{Y}$ will also have a weak charge $-\frac{1}{3}$ and a total charge $Q + N = -\frac{2}{3}$. It should be noted that the disintegration process of the quark u does not conserve the baryon number $B = Y = Q + N = \frac{1}{3}$ of the quark or the lepton number $L = Q + N = 1$ of the positron. The $\bar{Y}$ leptoquark will have the quantum number $B - L = \frac{1}{3} - 1 = -\frac{2}{3}$ and the baryon and lepton numbers, B and L respectively, are not conserved individually during the disintegration process, although the global quantum number $B - L$ is conserved.

The $\bar{Y}$ leptoquark produced in disintegration (19.99) can subsequently recombine with a quark u to produce an antiquark $\bar{d}$:

$$\bar{Y} + u \; \rightarrow \; \bar{d}. \tag{19.100}$$

We can therefore imagine exchange processes of the following type:

$$\begin{aligned} (u,d) \; &\rightarrow \; e^+ + (\bar{Y}, \bar{X}), \\ (\bar{Y}, \bar{X}) + u \; &\rightarrow \; (\bar{d}, \bar{u}). \end{aligned} \tag{19.101}$$

This should result in the disintegration of the proton, for example, according to the following procedure:

$$p \rightarrow \pi^0 + e^+ . \qquad (19.102)$$

Because the disintegration time is proportional to the fourth power of the mass of the leptoquark,

$$\tau_p \simeq \frac{M_X^4}{m_p^5} \simeq \frac{M_Y^4}{m_p^5}, \qquad (19.103)$$

we obtain for $M_X \simeq M_Y \simeq 10^{15} \text{GeV}/c^2$ a disintegration time on the order of 10^{31} years.

This lifetime of the proton is enormous compared to the assumed age of the Universe which is about 1.510^{10} years. To better appreciate this lifetime, we could, for example, determine the frequency at which we will have to draw water from the ocean drop-by-drop to be able to empty all the oceans of the earth. A mean depth of 3000 m is equal to a density per m^2 of $\varrho_{ocean} \simeq 3 \times 10^6$ kg m^{-2}. The surface of the oceans (including seas, lakes, oceans, etc.) is about 2/3 of the surface of the earth, leading to a water mass of about 10^{21} kg, which is low compared to the mass of the earth, estimated at some 2.4×10^{26} kg by the law of universal gravitation. If we grant that a drop of liquid represents a mere cubic millimeter of water, we obtain:

$$\frac{10^{21}\text{kg}}{10^{31}\text{years}} = \frac{10^{27}}{10^{31}} \text{ mg/years} = \frac{1}{10000} \text{mg/years}. \qquad (19.104)$$

It is enough here to draw 1 drop (1mg of water) every 10 000 years to empty all the oceans of the earth in 10^{31} years. At the same rate of filling, the content of oceans since the creation of the earth (4×10^9 years ago) would be:

$$\frac{4 \cdot 10^9}{10^4} = 4 \cdot 10^5 = 400 \text{ g}$$

Different research laboratories have tried to experimentally demonstrate up the disintegration of the photon, for example, by placing a reservoir containing 8000 m^3 of very pure water (that is about 3×10^{34} protons) in a 600-meter deep salt mine. A Franco-German experiment of this type used the Modane highway tunnel for an experiment using solid detectors equivalent to a number of protons of the same magnitude.

The experiments indicated a lifetime greater than 10^{31} years, thus ruling out the $SU(5)$ model. The latter, in fact, did not gave a truly satisfactory relation between the intensities of the three interactions.

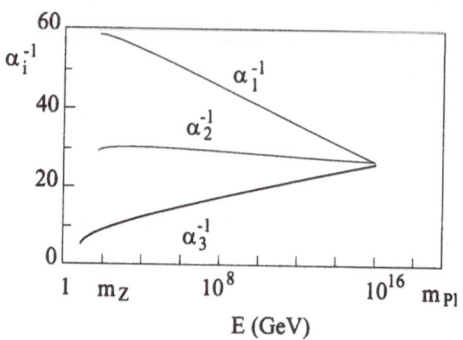

The grand unification model $SU(5)$ has, in the meantime, been abandoned or, at least, modified by the introduction of supersymmetric partners (see above figure). Supersymmetry is a special kind of symmetry; it associates boson partners to fermions and vice versa. Winos, zinos, photinos and gluinos, all particles with spin $\frac{1}{2}$, are thus associated with W, Z, γ and gluons, whereas squarks and selectrons with spin 0 are associated with quarks and electrons (see Appendix to this chapter).

Research on supersymmetric particles began as early as 1970 with high-energy accelerators but has yet to produce results. Theoreticians have also explored unification theories in many-dimensional spaces in which some dimensions are compact and are rolled up upon themselves in spheres, rings, etc. Such dimensions may be manifest on the Planck scale (10^{19} GeV or 10^{-33} cm), in which the fundamental interactions should also include gravitation. At this scale, elementary particles are no longer considered as point-like, but are rather represented as supersymmetric strings, etc. This is the Total Unification Theory being sought by some theoretical physicists.

Appendix: Introduction to Supersymmetry

This appendix presents a brief introduction to the supersymmetric theories being developed in particle physics. As a simple examle (Grosse, 1989) we consider a harmonic oscillator with mass m and spin $\vec{\frac{1}{2}}$.

A.1 The Supersymmetric Harmonic Oscillator

The eigenstates $\mid n\rangle$, which are boson states as already indicated in Chap. 17, are obtained from the creation and annihilation operators, b^{+} and b respectively, acting on the vacuum state $\mid 0\rangle$. First, a reminder of the definitions proposed for a one-dimensional oscillator with mass m, in a potential $\frac{1}{2} m \omega^2 x^2$ (we set $\hbar = 1$):

$$b = \frac{1}{\sqrt{2m\,\omega}} \left(\partial_x + m\,\omega\,x \right) \quad \text{and} \quad b^+ = \frac{1}{\sqrt{2m\,\omega}} \left(-\partial_x + m\,\omega\,x \right), \text{(A.1)}$$

$$H_B = \omega \left(b^+ b + \mathbb{1} \frac{1}{2} \right). \qquad \text{(A.2)}$$

The boson operators obey the commutation relations

$$[b, b] = [b^+, b^+] = 0,$$
$$[b, b^+] = \mathbb{1}, \qquad \text{(A.3)}$$
$$[H_B, b^+] = \omega\, b^+.$$

We use the fundamental state $| 0 \rangle$ with energy $\frac{1}{2}\omega$ to construct the excited states, or expressed in the language of occupied eigenstates peculiar to the second quantization theory, we construct the state with n bosons from the boson vacuum state:

$$\begin{array}{ll} b\,|\,0\rangle = 0 & b^+\,|\,0\rangle = |1\rangle \\ b\,|\,1\rangle = |\,0\rangle & b^+\,|\,1\rangle = \sqrt{2}\,|\,2\rangle = \sqrt{2}(b^+)^2\,|\,0\rangle \\ b\,|\,2\rangle = \sqrt{2}\,|\,1\rangle & \vdots \\ & (b^+)^n\,|\,0\rangle = \sqrt{n!}\,|\,n\rangle. \end{array} \qquad \text{(A.4)}$$

The eigenstates of H_B are defined with the operator $N_B = b^+ b$ whose eigenvalue gives the number of bosons:

$$H_B = \omega \left(N_B + \frac{1}{2}\mathbb{1} \right) \quad \text{and} \quad H_B\,|\,n\rangle = E_n\,|\,n\rangle \qquad \text{(A.5)}$$

with the eigenstates and eigenvalues:

$$|\,n\rangle = \frac{1}{\sqrt{n!}} \left(b^+ \right)^n\,|\,0\rangle \quad \text{and} \quad E_n = \omega \left(n + \frac{1}{2} \right). \qquad \text{(A.6)}$$

Now, consider the states with spin $|\frac{1}{2}\sigma\rangle$ of this particle. They can be obtained from the Pauli operators (cf. Chap. 9)

$$\sigma_+ = \frac{1}{2} \left(\sigma_x + i\,\sigma_y \right) \quad \text{and} \quad \sigma_- = (\sigma_+)^+ = \frac{1}{2} \left(\sigma_x - i\,\sigma_y \right), \qquad \text{(A.7)}$$

obeying the fundamental equations

$$\sigma_+\,|\,\tfrac{1}{2}\,\tfrac{1}{2}\rangle = 0 \qquad\qquad \sigma_-\,|\,\tfrac{1}{2}\,-\tfrac{1}{2}\rangle = 0$$
$$\sigma_+\,|\,\tfrac{1}{2}\,-\tfrac{1}{2}\rangle = |\,\tfrac{1}{2}\,\tfrac{1}{2}\rangle \qquad \sigma_-\,|\,\tfrac{1}{2}\,\tfrac{1}{2}\rangle = |\,\tfrac{1}{2}\,-\tfrac{1}{2}\rangle. \qquad \text{(A.8)}$$

We obtain from the anticommutation relations, cf. (9.58), satisfied by the Pauli matrices

$$\{\sigma_+, \sigma_+\} = \{\sigma_-, \sigma_-\} = 0 \quad \text{and} \quad (\sigma_+)^2 = (\sigma_-)^2 = 0$$
$$\{\sigma_-, \sigma_+\} = \mathbb{1}.$$
(A.9)

A representation of these Pauli operators can be obtained in the 2×2 matrix space by setting (see Chap. 9)

$$|\frac{1}{2} - \frac{1}{2}\rangle \equiv |0\rangle = \begin{pmatrix} 0 \\ 1 \end{pmatrix} \quad \text{and} \quad |\frac{1}{2}\frac{1}{2}\rangle \equiv |1\rangle = \begin{pmatrix} 1 \\ 0 \end{pmatrix}$$
$$f^+ = \sigma_+ = \begin{pmatrix} 0 & 1 \\ 0 & 0 \end{pmatrix} \quad \text{and} \quad f = \sigma_- = \begin{pmatrix} 0 & 0 \\ 1 & 0 \end{pmatrix}.$$
(A.10)

In the second quantization, the state $|\frac{1}{2} - \frac{1}{2}\rangle \equiv |0\rangle$ becomes a fermion vacuum state, whereas the state $|\frac{1}{2}\frac{1}{2}\rangle \equiv |1\rangle$ becomes a fermion-occupied space. The Pauli exclusion principle characteristic of fermion systems is then expressed through the fact that

$$f^2 = (f^+)^2 = 0.$$
(A.11)

If we assign the energy $-\frac{1}{2}\omega$ to the state $|\frac{1}{2} - \frac{1}{2}\rangle$, that is, to the fermion vacuum state (this reference energy is by convention assigned to a reference state termed vacuum), we can then write the fermion Hamiltonian

$$H_F = \omega \left(f^+ f - \mathbb{1}\frac{1}{2} \right) \quad \text{and} \quad H_F |0\rangle = -\frac{1}{2}\omega |0\rangle.$$
(A.12)

The commutation relations (A3) for the boson operators find their equivalent in fermion space when (A9) is rewritten in terms of fermion creation and annihilation operators:

$$\{f, f\} = \{f^+, f^+\} = 0,$$
$$\{f, f^+\} = \mathbb{1}.$$
(A.13)

It is possible to condense the commutation and anticommutation rules of the boson and fermion operators, b and f respectively, by introducing the graduated commutator $[.., ..]$ representing a commutator, except where the two are both fermion operators.

If $a = b, f$ we obtain

$$[a, a^+\} = \mathbb{1} \qquad [a, a\} = [a^+, a^+\} = 0,$$
(A.14)

whereas the operators b and f mutually commute:

$$[b, f] = [b, f^+] = [b^+, f] = [b^+, f^+] = 0.$$
(A.15)

The occupied or unoccupied fermion states are defined with respect to the vacuum $|\,0\rangle$ using the relations:

$$f\,|\,0\rangle = 0 \qquad\qquad f^+\,|\,0\rangle = |\,1\rangle$$
$$f\,|\,1\rangle = |\,0\rangle \qquad\qquad f^+\,|\,1\rangle = 0. \tag{A.16}$$

The matrix representation (A10) of the fermion operators and states makes the direct verification of these relations possible.

Let us add together the Hamiltonians H_B and H_F that describe the particle with mass m and spin $\frac{1}{2}$ in a harmonic well $\frac{1}{2}\,m\,\omega^2\,x^2$. The energy of the boson vacuum compensates the energy of the fermion vacuum, and the energy of the fundamental state $|\,0,0\rangle = |\,0\rangle \otimes \begin{pmatrix} 0 \\ 1 \end{pmatrix}$ of the total Hamiltonian

$$H = H_B + H_F = \omega\,(b^+ b + f^+ f) \tag{A.17}$$

becomes zero by construction:

$$H\,|\,0,0\rangle = 0, \tag{A.18}$$

$$b\,|\,0,0\rangle = f\,|\,0,0\rangle = 0. \tag{A.19}$$

The eigenstates of H are tensor products of the boson states $|\,n\rangle$ (given by (A.6)) and the fermion states $|\,m\rangle$ (given by (A.16)):

$$|\,n,m\rangle = \frac{1}{\sqrt{n!}}\,(b^+)^n\,(f^+)^m\,|\,0,0\rangle \qquad n = 0,1,2\ldots \quad m = 0,1. \tag{A.20}$$

By adding and subtracting the term $b^+ b\,f^+ f$ in H, that is, in (A.17), we obtain

$$H = \omega\,\left(b^+ b\,(\mathbb{1} - f^+ f) + (b^+ b + \mathbb{1})f^+ f\right). \tag{A.21}$$

The use of commutation and anticommutation relations, (A.3) and (A.13), in conjunction with the fact the operators f and b mutually commute, leads to a different form of the Hamiltonian:

$$H = \omega\,(bf^+\,fb^+ + fb^+\,bf^+). \tag{A.22}$$

Now let us introduce the supercharge operators Q and Q^+ using the matrix representation (A.10) of f and f^+:

$$Q = bf^+ = \begin{pmatrix} 0 & b \\ 0 & 0 \end{pmatrix},$$

$$Q^+ = fb^+ = \begin{pmatrix} 0 & 0 \\ b^+ & 0 \end{pmatrix}. \tag{A.23}$$

The Hamiltonian H thus takes the form of an anticommutator:

$$H = \omega\,\{Q, Q^+\} = \omega\begin{pmatrix} bb^+ & 0 \\ 0 & b^+ b \end{pmatrix}. \tag{A.24}$$

The matrix representations of the operators H, Q and Q^+ lead to the commutation relations:

$$[H, Q] = [H, Q^+] = 0.$$
(A.25)

This shows that if $\mid n, m \rangle$ is an eigenstate of H corresponding to the non-zero eigenvalue E of the energy, then $Q \mid n, m \rangle$ and $Q^+ \mid n, m \rangle$ are equally eigenstates of H corresponding to the same energy. In effect, we can use (A.25) to write

$$
\begin{aligned}
H \mid n, m \rangle &= E \mid n, m \rangle \text{ for } E \neq 0, \\
H Q \mid n, m \rangle &= Q H \mid n, m \rangle = E \left(Q \mid n, m \rangle \right), \\
H Q^+ \mid n, m \rangle &= Q^+ H \mid n, m \rangle = E \left(Q^+ \mid n, m \rangle \right).
\end{aligned}
$$
(A.26)

The supercharge operators Q and Q^+ therefore generate a new symmetry: the supersymmetry. They induce the transition from a boson state to a fermion state since

$$Q \mid n, 0 \rangle = \sqrt{\omega n} \mid n - 1, 1 \rangle,$$
(A.27)

or from a fermion state to a boson state:

$$Q^+ \mid n, 1 \rangle = \sqrt{\omega(n + 1)} \mid n + 1, 0 \rangle.$$
(A.28)

It is really here a question of language: The operator Q induces the transition from the energy state E_n of the oscillator with spin element $\mid \frac{1}{2} - \frac{1}{2} \rangle$ to the same energy state but with spin element $\mid \frac{1}{2} \frac{1}{2} \rangle$. The energy states E_n of the harmonic oscillator constitutes a doublet.

The Hilbert space $\mathcal{H}$ in which the operator H is defined can be separated into two complementary subspaces, $\mathcal{H}_0$ and $\mathcal{H}_1$, in which the operators N_0 and N_1 act, such that

$$
\begin{aligned}
N_0 + N_1 &= \mathbb{1}, \\
N_0 - N_1 &= K.
\end{aligned}
$$
(A.29)

The operator K is termed the Klein operator or involution operator. A representation of N_0, N_1 and K in the space of 2×2 matrices can be written:

$$N_0 = \begin{pmatrix} 1 & 0 \\ 0 & 0 \end{pmatrix} \quad N_1 = \begin{pmatrix} 0 & 0 \\ 0 & 1 \end{pmatrix} \quad K = \begin{pmatrix} -1 & 0 \\ 0 & 1 \end{pmatrix}.$$
(A.30)

We will note from these representations the following properties:

i) The operator K with ± 1 as eigenvalue, as can be seen on the first diagonal, satisfies the condition:

$$K^2 = \mathbb{1}.$$
(A.31)

ii) The operator N_0 represents the number of fermions, since, applied to the fermion states $\mid 0 \rangle = \begin{pmatrix} 0 \\ 1 \end{pmatrix}$ and $\mid 1 \rangle = \begin{pmatrix} 1 \\ 0 \end{pmatrix}$, it restores the eigenvalue equations

$$N_0 \,|\, 0\rangle \,= 0 \qquad N_0 \,|\, 1\rangle \,= \,|\, 1\rangle. \qquad (A.32)$$

iii) Any operator of the form $F = \begin{pmatrix} 0 & f_1 \\ f_2 & 0 \end{pmatrix}$ where f_1 and f_2 are operators, anticommutes with K:

$$\{K, F\} = 0. \qquad (A.33)$$

The operator F is said to be of the fermion type or odd or of graduation one.

iv) Any operator of the form $B = \begin{pmatrix} b_1 & 0 \\ 0 & b_2 \end{pmatrix}$ where b_1 and b_2 are operators, commutes with K:

$$[K, B] = 0. \qquad (A.34)$$

The operator B is said to be of the boson type or even or of zero graduation.

v) The form (A.23) of the operators Q and Q^+ shows that we are dealing with odd operators whereas H is an even operator as follows from expression (A.24), giving with (A.33) and (A.34) the relations

$$[K, H] = 0,$$
$$\{K, Q\} = \{K, Q^+\} = 0. \qquad (A.35)$$

vi) Because the operators b and f (or b^+ and f^+) mutually commute, the definition (A.21) of the supercharge operators leads to the relations

$$[Q, b^+] = f^+ \qquad [Q, b] = 0$$
$$\{Q, f\} = b \qquad \{Q, f^+\} = 0. \qquad (A.36)$$

This mirrors the fact that Q is an odd operator whereas b is an even operator and f an odd operator.

A.2 Supersymmetric Quantum Mechanics

The Hermitian (selfadjoint) Hamiltonian operators H, with supersymmetry Q and involution K constitute a supersymmetric system if they satisfy the conditions (Gierès, 1993)

$$H = Q^2,$$
$$K^2 = \mathbb{1},$$
$$\{K, Q\} = 0. \qquad (A.37)$$

from this we directly infer the two commutation relations

$$[H, Q] = 0 \qquad [H, K] = 0. \qquad (A.38)$$

Now let us examine the principal properties of this system:

i)   The Hamiltonian H is positive definite $H \geq 0$. Hence, the spectrum of its eigenvalues does not contain negative values.

ii) If $\mid \varphi \rangle$ is the eigenket of H corresponding to the eigenvalue $E \geq 0$, then $Q \mid \varphi \rangle$ is also an eigenket with the same eigenvalue because

$$H \mid \varphi \rangle = E \mid \varphi \rangle,$$
$$H \left(Q \mid \varphi \rangle \right) = QH \mid \varphi \rangle = E \left(Q \mid \varphi \rangle \right). \qquad (A.39)$$

The eigenvalues of H are therefore degenerate.

iii) By quantum vacuum, we mean the state $\mid 0 \rangle$, which is the lowest-energy or ground state. According to the foregoing property, this energy will be zero if and only if

$$Q \mid 0 \rangle = 0 \quad \forall\, Q,$$
$$H \mid 0 \rangle = 0. \qquad (A.40)$$

iv) The involution operator K has ± 1 as eigenvalues. We can therefore decompose the Hilbert space into an even and an odd part by setting

$$\begin{aligned}
\mid \varphi \rangle &= \frac{1}{2} \left(\mathbb{1} + K \right) \mid \varphi \rangle + \frac{1}{2} \left(\mathbb{1} - K \right) \mid \varphi \rangle \\
&= \mid \varphi_B \rangle + \mid \varphi_f \rangle \\
&= K_B \mid \varphi \rangle + K_F \mid \varphi \rangle. \qquad (A.41)
\end{aligned}$$

The vector $\mid \varphi_B \rangle = K_B \mid \varphi \rangle$ of the subspace $\mathcal{H}_B$ of $\mathcal{H}$ is different from zero for the eigenvalue $+1$ of K, since $K_B = \frac{1}{2} \left(\mathbb{1} + K \right)$; the vector $\mid \varphi_f \rangle = K_F \mid \varphi \rangle$ corresponds to the eigenvalue -1 of the operator K. The Klein involution operator can be represented with a 2×2 matrix and, if we assume that the vector $\mid \varphi \rangle$ is a 2×1 column, we obtain

$$K = \begin{pmatrix} -1 & 0 \\ 0 & 1 \end{pmatrix} \quad K_B = \frac{1}{2} \left(\mathbb{1} + K \right) = \begin{pmatrix} 0 & 0 \\ 0 & 1 \end{pmatrix} \quad K_F = \frac{1}{2} \left(\mathbb{1} - K \right) = \begin{pmatrix} 1 & 0 \\ 0 & 0 \end{pmatrix}. \tag{A.42}$$

Applied to $\mid \varphi \rangle = \begin{pmatrix} \varphi_1 \\ \varphi_2 \end{pmatrix}$, we easily obtain

$$\begin{aligned}
\mid \varphi_B \rangle &= K_B \mid \varphi \rangle = \begin{pmatrix} 0 & 0 \\ 0 & 1 \end{pmatrix} \begin{pmatrix} \varphi_1 \\ \varphi_2 \end{pmatrix} = \begin{pmatrix} 0 \\ \varphi_2 \end{pmatrix} \equiv \begin{pmatrix} 0 \\ \varphi_B \end{pmatrix}, \\
\mid \varphi_F \rangle &= K_F \mid \varphi \rangle = \begin{pmatrix} 1 & 0 \\ 0 & 0 \end{pmatrix} \begin{pmatrix} \varphi_1 \\ \varphi_2 \end{pmatrix} = \begin{pmatrix} \varphi_1 \\ 0 \end{pmatrix} \equiv \begin{pmatrix} \varphi_f \\ 0 \end{pmatrix}.
\end{aligned} \tag{A.43}$$

The part $\mid \varphi_B \rangle$ of $\mid \varphi \rangle$ is the boson part, while $\mid \varphi_f \rangle$ is the fermion part.

v) A 2×2 matrix may be decomposed into an even (boson) part and an odd (fermion) part and, if we set $M = \begin{pmatrix} A & B \\ C & D \end{pmatrix}$, the conditions

$$[K, M] = 0 \quad \Longrightarrow \quad M = \begin{pmatrix} A & 0 \\ 0 & D \end{pmatrix},$$

$$\{K, M\} = 0 \quad \Longrightarrow \quad M = \begin{pmatrix} 0 & B \\ C & 0 \end{pmatrix}, \tag{A.44}$$

which amounts to writing

$$M = \begin{pmatrix} A & 0 \\ 0 & D \end{pmatrix} + \begin{pmatrix} 0 & B \\ C & 0 \end{pmatrix} = M_B + M_F. \tag{A.45}$$

An even (boson) operator commutes with the Klein operator whereas an odd (fermion) operator anticommutes with it.

vi) The supersymmetric operator Q (selfadjoint $Q = Q^+$) anticommutes with K following definition (A.35). It is therefore an odd operator of the form

$$Q = \begin{pmatrix} 0 & A^* \\ A & 0 \end{pmatrix}. \tag{A.46}$$

Applied to an eigenstate $\mid \varphi \rangle = \begin{pmatrix} \varphi_f \\ \varphi_B \end{pmatrix}$ of H corresponding to the eigenvalue E, this operator transforms it into $Q \mid \varphi \rangle$, another eigenstate of H corresponding to the same eigenvalue:

$$Q \mid \varphi \rangle = \begin{pmatrix} 0 & A^* \\ A & 0 \end{pmatrix} \begin{pmatrix} \varphi_f \\ \varphi_B \end{pmatrix} = \begin{pmatrix} A^* \varphi_B \\ A \varphi_f \end{pmatrix}. \tag{A.47}$$

The fermion state φ_f has been transformed into a boson state $A^* \varphi_B$, whereas the boson state φ_B has been transformed into a fermion state $A \varphi_f$. We say in such a case that $A^* \varphi_B$ is the superpartner of φ_f and $A \varphi_f$ the superpartner of φ_B.

vii) Any eigenstate $\mid \varphi \rangle$ whatsoever of H represented by the matrix $H = \begin{pmatrix} A A^* & 0 \\ 0 & A A^* \end{pmatrix}$ has a superpartner $Q \mid \varphi \rangle$, for an eigenvalue different from zero.

To the one-particle state $\mid 1 \rangle$, corresponds the superpartner $Q \mid 1 \rangle$. The eigenstates $\mid 1 \rangle$ and $Q \mid 1 \rangle$ have the same energy but a spin differing by $\frac{1}{2}$.

The eigenstate $\mid 0 \rangle$ of the lowest-energy (positive) H is termed a vacuum. Following property ii the quantum vacuum will have zero energy if and only if

$$Q \mid 0 \rangle = 0 \quad \forall Q \quad \Longrightarrow \quad H \mid 0 \rangle = 0. \tag{A.48}$$

viii) By convention, the superpartners of bosons are bosinos (gravitinos, photinos, gluinos), whereas the superpartners of fermions are sfermions (squarks or sleptons). Supersymmetric minimal theories attribute a unique supersymmetric partner to all fundamental bosons and fermions.

ix) Supersymmetric field theories generalize the first hypothesis (A.37), by introducing the generators Q_α ($\alpha = 1, \ldots, 4$) of this superalgebra, odd operators that satisfy the anticommutation relation

$$\{Q_\alpha, Q_\beta\} = \sum_{\mu=0}^{3} f^\mu_{\alpha\beta} \, P_\mu, \tag{A.49}$$

where the $f^\mu_{\alpha\beta}$ are complex numbers termed structure constants and the operators P_μ ($\mu = 0, 1, 2, 3$) are generators of the time and space translations.

A field theory of space-dimension zero leads us to the form

$$\{Q_1, Q_1\} = f^0_{11} \, P_0 \tag{A.50}$$

and, since P_0 is the generator of the time translations, that is the Hamiltonian, we obtain a form that is analogous to (A.37), that is a Hamiltonian H proportional to Q^2.

x) All states capable of being transformed into one another by the action of the supersymmetric operator Q form a supersymmetric multiplet or supermultiplet, whose elements possess the same restmass but differ by a new quantum number, the R-charge.

xi) In the supersymmetric standard model, the partner of the photon γ, the photino, possesses a spin $\frac{1}{2}$ and the partner of the gluon, the gluino, a spin $\frac{1}{2}$. Sleptons and squarks are supersymmetric partners with spin 0. The Higgs boson with spin 0 has a supersymmetric partner with spin $\frac{1}{2}$, the higgsino, conferring mass to the supersymmetric partners of fundamental bosons.

As of today (1997), the lower limit of the mass of supersymmetric particles is 45 GeV, where the mass of the gluino and squarks should be greater than 100 GeV. Yet only the supersymmetric standard theory satisfactorily addresses the convergence of coupling constants in the Grand Unification Theories (Fayet, 1994).

Research on supersymmetrical particles is being actively pursued at the large particle accelerators and is the object of a program at the LHC accelerator at CERN in Geneva (around 2005 probably).

Chapter 20
Quantum Cosmology

1. Introduction

Gravitation is remarkably well described by Einstein's general relativity theory as demonstrated by a variety of observations: the advance of the perihelion of planets, the deviation of light rays due to stellar masses leading to gravitational lenses, radar or laser echoes from different planets, emission of gravitational waves by binary pulsars and, finally, various cosmological tests, in particular, the isotropic 3 K cosmic background radiation. Yet, some physicists think that the Einstein equation linking the geometric structure of space-time (described by a curvature tensor) to its physical matter or radiation content (described by an energy–momentum tensor) is simply an effective theory describing the macroscopic properties of matter. The general relativity theory has made it possible to develop a coherent cosmological model (standard model or the Big Bang theory), and the current state of knowledge in particle, nuclear, and atomic physics as well as in thermodynamics permits the description of the evolution of the Universe from soon after the first minutes of creation.

In the standard or Big Bang model, the initial conditions are approximately in the Planck domain and we may safely assume that quantum effects become overwhelming at this scale ($\ell_p \simeq 10^{-33}$ cm $t_p \simeq 10^{-44}$s $\varrho_p \simeq 10^{94}$gcm^{-3} $T_p \simeq 10^{32}$ K). Quantum cosmology, in which both matter and the gravitational field are quantized, is therefore the natural framework for addressing the issue of the initial conditions of the formation of the Universe.

The intention is to determine the Universe's wave function $\psi(h_{ij}(x), \phi(x), \Sigma)$ describing a closed Universe in which a three-dimensional metric $h_{ij}(x)$ and a matter field $\phi(x)$ are defined on a suface Σ. We use a dynamic theory of gravitation, deriving for example from the general relativity theory, to obtain a Hamiltonian formulation of Einstein's theory and then quantize the associated operators. This permits us to write a wave equation that is analogous to the Schrödinger equation: the Wheeler–DeWitt (WDW) equation. Such an equation leads to numerous possible solutions, and for it to be able to predict anything, we will need to complement it with initial and boundary conditions, leading to a unique solution. Next, we will need to interpret the wave function thus obtained and, in particular, to explain why the Universe is approximately classical in present-day observations.

We begin with a reminder of the fundamental equations of general relativity (Elbaz, 1992; Linde, 1990) before going on to examine the Hamiltonian formulation of the theory, using the same general plan as in Chap. 4: mathematical formalism, dynamical laws, quantization rules and the correspondence principle.

After separating the space-time curvature into an intrinsic curvature of a three-dimensional space and an extrinsic curvature describing the evolution of this space, we introduce the Einstein–Hilbert action using the scalar curvature of the space-time with the foregoing data. This action is subsequently expressed in a six-dimensional superspace, the space of all three-dimensional metrics, which leads to a Hamiltonian formulation of the Einstein equations. Quantization is thereafter introduced with the commutation relations of the operators associated with three-dimensional metrics (considered as the position variables of the superspace) and with their conjugate momenta. This will lead to an equation analogous to the Schrödinger equation, but which introduces functional derivatives, the WDW equation of Wheeler and DeWitt. By reducing the number of degrees of freedom of the superspace, this WDW equation will be written in a mini-superspace, making it possible to obtain the time-independent wave function of the Universe in a form that is in all respects analogous to the Schrödinger equation. Examples of applications will be given in particular for the the B.K.W.-type semi-classical solutions (see Chap. 3).

One of the most fundamental problems facing to theoreticians of quantum cosmology concerns the fact that the universal wave function is time independent, whereas the observable Universe, in the standard model, is naturally time-dependent. We will try to address this problem by quoting from A. Linde's *Particle Physics and Inflationary Cosmology* (Harwood Academic Press, 1990, page 271)

"The Universe *as a whole* does not depend on time because the very concept of such a change presumes the existence of some immutable reference that does not appertain to the universe, but relative to which the latter evolves. If by "the universe" we mean "everything", then there remains no *external observer* according to whose clocks the universe could develop. But in actuality, we are not asking why the universe is developing, we are inquiring as to why we *perceive* it to be developing. We have thereby separated the universe into two parts: a macroscopic observer with clocks, and all the rest. The latter can perfectly well evolve in time (according to the clocks of the observer), despite the fact that the wave function of the *entire* universe is time-independent".

In other words, we return to our familiar vision of the world evolving with time, but only after separating the Universe into two macroscopic parts, each of which evolves in a semi-classical fashion. The resulting situation reminds us of quantum mechanical tunneling through a barrier: The wave function is defined within the barrier, but it produces a non-zero probability amplitude of finding a particle propagating in real time, outside the barrier where a classical motion is allowed.

Thus, by the very fact of his existence, an observer reduces somewhat the total wave function of the Universe to that part describing the world he can observe. This is exactly the Copenhagen School's point of view for interpreting quantum mechanics: "The observer is not simply a passive spectator but, rather, a participant in the creation of the Universe".

2. Essentials of General Relativity

2.1 Mathematical Formalism

In classical mechanics, time is represented by a space-independent parameter and the interval between two infinitely close points is represented by the scalar

$$ds^2 \equiv d\ell^2 = \delta_{ij} \, dx^i dx^j \quad j, i = 1, 2, 3 \,. \tag{20.1}$$

Changes of coordinate system are made independently of time, and if the position x^i in a system is a function of the position $x^{j'}$ in the other system (we will use Einstein's convention for summation over repeated indices appearing in the lower and upper positions),

$$dx^i = \frac{\partial x^i}{\partial x^{j'}} \, dx^{j'} = \Lambda^i_{j'} \, dx^{j'} \,. \tag{20.2}$$

In relativistic mechanics, time is considered as an additional coordinate, and an event is represented by a point in space-time, a four-dimensional vector space in which the interval is represented by the scalar

$$ds^2 = \eta_{\mu\nu} \, dx^\mu \, dx^\nu \quad \mu, \nu = 0, 1, 2, 3 \,, \tag{20.3}$$

and in a Minkowski space with (+2) signature, with coordinate 0 being reserved for time:

$$ds^2 = -(dx^0)^2 + (dx^i)^2 \quad i = 1, 2, 3 \,. \tag{20.4}$$

If the interval is negative, then it is time-like, and the "proper time" τ will be the term for the time associated with the event being described, by setting

$$d\tau^2 = -ds^2 = -\eta_{\mu\nu} \, dx^\mu \, dx^\nu \quad \mu, \nu = 0, 1, 2, 3 \,. \tag{20.5}$$

During the change of coordinate system, we can write, just like in the three-dimensional space,

$$dx^\mu = \frac{\partial x^\mu}{\partial x^{\nu'}} \, dx^{\nu'} = \Lambda^\mu_{\nu'} \, dx^{\nu'} \,. \tag{20.6}$$

Note that by combining this transformation with the expression for the interval ds^2 which is invariant with respect to the coordinate changes, we obtain the transformation of the metric $\eta_{\mu\nu}$ of the Minkowski space:

$$ds^2 = \eta_{\mu\nu}\, dx^\mu\, dx^\nu = \eta_{\mu'\nu'}\, dx^{\mu'}\, dx^{\nu'}$$
$$= \eta_{\mu'\nu'}\, \Lambda_\mu^{\mu'}\, dx^\mu\, \Lambda_\nu^{\nu'}\, dx^\nu, \qquad (20.7)$$

which, on comparison with the foregoing statements, leaves us with

$$\eta_{\mu\nu} = \eta_{\mu'\nu'}\, \Lambda_\mu^{\mu'}\, \Lambda_\nu^{\nu'}. \qquad (20.8)$$

Any mathematical object V^μ that is transformed like dx^μ is termed a contravariant vector, or more precisely, a contravariant component of a vector V. By projecting this vector onto the base unit vectors $\vec{e}_\mu$ ($\mu = 0,1,2,3$), it is easy to see that

$$V = V^\mu\, \vec{e}_\mu \quad \text{and} \quad V^\mu = \Lambda_{\mu'}^\mu\, V^{\mu'} = \frac{\partial x^\mu}{\partial x^{\mu'}}\, V^{\mu'}. \qquad (20.9)$$

We define the dual space spanned by the unit vectors $\vec{e}^\mu$ such that the product of a base vector of each of these spaces will lead to a scalar:

$$\vec{e}^\mu \cdot \vec{e}_\nu = \delta_\nu^\mu. \qquad (20.10)$$

The covariant components of the vector V, that is the elements V_μ are obtained in a similar manner as (20.9), by projecting V onto the basis $\{\vec{e}^\mu\}$:

$$V = V_\mu\, \vec{e}^\mu. \qquad (20.11)$$

They transform as follows:

$$V_\mu = \Lambda_\mu^{\mu'}\, V_{\mu'} = \frac{\partial x^{\mu'}}{\partial x^\mu}\, V_{\mu'}. \qquad (20.12)$$

It is easily noticed here that the transformation matrices $\Lambda_\mu^{\mu'}$ and $\Lambda_{\mu'}^\mu$ are inverse with respect to each other:

$$\Lambda_\mu^{\mu'}\, \Lambda_{\mu'}^\nu = \frac{\partial x^{\mu'}}{\partial x^\mu}\, \frac{\partial x^\nu}{\partial x^{\mu'}} = \frac{\partial x^\nu}{\partial x^\mu} = \delta_\mu^\nu. \qquad (20.13)$$

The elements $\eta_{\mu\nu}$ of the Minkowski metric transform just like the product of the two vectors: They are covariant tensors of rank-two. We can, in a general fashion, define the covariant or contravariant or mixed elements, $T_{\mu\nu}$, $T^{\mu\nu}$ and T_ν^μ of a tensor T (in this case, a rank-two tensor, but nothing stops us from going beyond this; zero-order tensors are scalars and first-order tensors vectors of the space under consideration).

In general relativity, space-time can be described by a more general metric $g_{\mu\nu}$, and the space-time interval is the scalar

$$ds^2 = g_{\mu\nu} \, dx^\mu \, dx^\nu \qquad \mu, \nu = 0, 1, 2, 3 . \qquad (20.14)$$

The transformation laws for elements of the metric tensor, for a change of coordinate system are the same as those of the Minkowski space, although the elements of the metric $g_{\mu\nu}$ are not constants and may depend on the co-ordinates x^μ. The transformation of a covariant element into a contravariant element is obtained by contraction (summation over the repeated upper and lower indices) with the metric tensor $g_{\mu\nu}$:

$$V^\mu = g^{\mu\nu} V_\nu \text{ and } V_\mu = V^\nu g_{\mu\nu} , \qquad (20.15)$$

with the contraction rule over the metric tensor itself:

$$g^{\mu\nu} g_{\mu\nu'} = \delta^\nu_{\nu'} . \qquad (20.16)$$

The differentiation of a vector is straightforward in a "flat" space charcterized by unit vectors $\vec{e}_\mu$ with a constant direction:

$$dV = d(V^\mu \vec{e}_\mu) = (dV^\mu) \, \vec{e}_\mu = (\partial_\nu V^\mu) \, dx^\nu \, \vec{e}_\mu . \qquad (20.17)$$

In a "curved" space the unit vectors $\vec{e}_\mu$ change direction during the trans-lation from point x^μ to $x^\mu + dx^\mu$, and the derivative of the vector $\vec{e}_\mu$ can be expressed as a linear combination of the base vectors with coefficient $\Gamma^\sigma_{\mu\nu}$ (which is not a third-rank tensor), termed the connection, or Christoffel sym-bol:

$$d\vec{e}_\mu = \Gamma^\sigma_{\mu\nu} \, dx^\nu \, \vec{e}_\sigma . \qquad (20.18)$$

By using the last two results, the covariant derivative of a contravariant vector is defined by

$$\begin{aligned} dV &= (\partial_\nu V^\mu) \, dx^\nu \vec{e}_\mu + V^\mu \, d\vec{e}_\mu = (\partial_\nu V^\mu) \, dx^\nu \vec{e}_\mu + \Gamma^\mu_{\nu\sigma} \, dx^\nu \, V^\sigma \, \vec{e}_\mu \\ &\equiv (D_\nu V^\mu) \, dx^\nu \, \vec{e}_\mu = (DV^\mu) \, \vec{e}_\mu . \end{aligned} \qquad (20.19)$$

The covariant derivative (which transforms like a vector during the change of coordinate system) therefore takes the form

$$DV^\mu = dx^\nu (D_\nu V^\mu) = dx^\nu (\partial_\nu V^\mu + \Gamma^\mu_{\nu\sigma} V^\sigma) . \qquad (20.20)$$

An analogous approach leads to the covariant derivative of a covariant vector, with the only possible change being in the sign before the Christoffel symbol:

$$DV_\mu = dx^\nu (D_\nu V_\mu) = dx^\nu (\partial_\nu V_\mu - \Gamma^\sigma_{\nu\mu} V_\sigma) . \qquad (20.21)$$

The covariant derivative of a tensor is taken, just as in the covariant differenti-ation of a vector product, with the corresponding variances. The conservation of the energy–momentum tensor T^μ_ν, which plays an especially important role in general relativity, defined with its four-divergence $D_\mu\, T^\mu_\nu$ is written, for example,

$$D_\mu\, T^\mu_\nu \;=\; \partial_\mu\, T^\mu_\nu + \Gamma^\mu_{\mu\sigma}\, T^\sigma_\nu - \Gamma^\sigma_{\mu\nu}\, T^\mu_\sigma\,, \qquad (20.22)$$

and by using the special value of the first Christoffel symbol depending on the derivative of the determinant $g = |\det\, g_{\mu\nu}|$ of the metric, that is,

$$\Gamma^\mu_{\mu\sigma} \;=\; \frac{1}{\sqrt{g}}\, \partial_\sigma\, \sqrt{g}\,, \qquad (20.23)$$

we obtain the following statement of relation (20.22):

$$D_\mu\, T^\mu_\nu \;=\; \frac{1}{\sqrt{g}}\, \partial_\sigma\, (\sqrt{g}\, T^\sigma_\nu) - \Gamma^\sigma_{\mu\nu}\, T^\mu_\sigma\,. \qquad (20.24)$$

We call a Riemann vector space one in which the covariant derivative of the metric tensor is zero:

$$D_\mu\, g^{\lambda\sigma} \;=\; 0\,, \qquad (20.25)$$

which derives from the following expression for the Christoffel symbol:

$$\Gamma^\mu_{\nu\lambda} \;=\; \frac{1}{2}\, g^{\mu\sigma}\, (\partial_\lambda\, g_{\nu\sigma} + \partial_\nu\, g_{\lambda\sigma} - \partial_\sigma\, g_{\nu\lambda})\,. \qquad (20.26)$$

The covariant second derivative of a vector leads to the definition of the (fourth-rank) Riemann–Christoffel curvature tensor with the aid of the commutator

$$[D_\nu,\, D_\mu]\, V^\alpha \;=\; R^\alpha_{\varepsilon\mu\nu}\, V^\varepsilon\,. \qquad (20.27)$$

The contraction of this tensor over its indices $\alpha = \mu$ defines the (second-rank) Ricci curvature tensor:

$$R^\alpha_{\varepsilon\alpha\nu} \;=\; R_{\varepsilon\nu}\,, \qquad (20.28)$$

and by setting, in a general manner,

$$(A\,B)_{\mu\nu} \;=\; A^\tau_{\mu\varepsilon}\, B^\varepsilon_{\nu\tau} - A^\tau_{\mu\nu}\, B^\varepsilon_{\tau\varepsilon}\,, \qquad (20.29)$$

we obtain the expression for the Ricci curvature tensor:

$$R_{\mu\nu} \;=\; \partial_\nu\, \Gamma^\alpha_{\mu\alpha} - \partial_\alpha\, \Gamma^\alpha_{\mu\nu} + (\Gamma\Gamma)_{\mu\nu}\,. \qquad (20.30)$$

The contraction of the Ricci curvature symbol over its indices finally defines the scalar curvature (zero-rank tensor) of the space-time:

$$R^\mu_\mu \;\equiv\; R\,. \qquad (20.31)$$

Einstein's curvature tensor $\mathcal{G}_{\mu\nu}$ used in his gravitation equations,

$$\mathcal{G}_{\mu\nu} \;=\; R_{\mu\nu} - \frac{1}{2}\, g_{\mu\nu}\, R\,, \qquad (20.32)$$

has the fundamental property of possessing a zero four-divergence

$$D_\mu\, \mathcal{G}^{\mu\nu} \;=\; 0\,. \qquad (20.33)$$

2.2 Dynamical Laws

The laws of motion of a free particle can be inferred from Hamilton's principle of least action by introducing a Hamiltonian action integral. For a free particle, this action is written

$$S = -mc^2 \int ds,$$ (20.34)

and in the most general case, the least-action principle $\delta S = 0$ leads to trajectories termed geodesics, which may be expressed with the four-velocity $dx^\mu/d\tau = u^\mu$ as follows:

$$Du^\mu = dx^\nu \, D_\nu u^\mu = \frac{du^\mu}{d\tau} + \Gamma^\mu_{\nu\sigma} u^\nu u^\sigma = 0$$

$$= \frac{d^2 x^\mu}{d\tau^2} + \Gamma^\mu_{\nu\sigma} \frac{dx^\nu}{d\tau} \frac{dx^\sigma}{d\tau} = 0.$$ (20.35)

When geodesics are straight lines in space-time (rectilinear and uniform trajectories), the space-time is said to be flat, in a Cartesian coordinate system in which all Christoffel symbols are zero.

If there is no such coordinate system, then the space-time will be said to be curved and the Christoffel symbols will no longer all be zero.

As we saw above (Chaps. 17 and 18), in field theory, the Hamiltonian action is defined by a Lagrangian density $\mathcal{L}(\varphi, \partial_\mu \varphi)$, and the Hamiltonian action for a scalar field in a curved space-time is written

$$S = \int \tilde{\mathcal{L}}(\varphi, \partial_\mu \varphi) \, d^\mu x = \int \sqrt{g} \left(\frac{1}{2} \partial^\mu \varphi \, \partial_\mu \varphi - V(\varphi) \right) d^4 x$$ (20.36)

with $\partial^\mu \varphi = g^{\mu\nu} \partial_\nu \varphi$. The principe of least action leads to the field equation (of the Klein–Gordon type) in a curved space-time

$$\delta S = 0 \quad \longrightarrow \quad -D^\mu D_\mu \varphi + \frac{\partial V}{\partial \varphi} = 0.$$ (20.37)

The energy–momentum tensor is defined with the Lagrangian density of the matter or radiation field $\tilde{\mathcal{L}}_{mat} = \sqrt{g} \, \mathcal{L}_{mat}$ by the relations

$$T_{\mu\nu} = \frac{2c}{\sqrt{g}} \frac{\delta \tilde{\mathcal{L}}m}{\delta g^{\mu\nu}} = \frac{2c}{\sqrt{g}} \left[\partial_\sigma \frac{\partial \tilde{\mathcal{L}}_{mat}}{\partial \partial_\sigma g^{\mu\nu}} - \frac{\partial \tilde{\mathcal{L}}_{mat}}{\partial g^{\mu\nu}} \right],$$ (20.38)

whereas $D_\mu T^{\mu\nu} = 0$.

For a scalar field with (20.36)-type Lagrangian density, this definition leads to an energy–momentum tensor

$$T_{\mu\nu} = \partial_\nu \varphi \, \partial_\nu \varphi - g_{\mu\nu} \left(\frac{1}{2} \partial^\sigma \varphi \, \partial_\sigma \varphi + V(\varphi) \right), \qquad (20.39)$$

whereas for a perfect fluid with pressure P and energy density ϱ, the energy–momentum tensor is written directly in the form (Elbaz, 1992)

$$T_{\mu\nu} = P \, g_{\mu\nu} + (P + p) \, u_\mu \, u_\nu. \qquad (20.40)$$

3. The Correspondence Principle

Einstein's gravitational theory expressed with general relativity is based on the principle of equivalence between inertial mass and gravitational mass (the equality between these two masses has been tested in various older experiments – that of Eötvos, for example – and or in more recent ones, to within 10^{-11}). This implies that it is always possible to compensate the gravitational force with an inertial force, and that it is therefore possible to locally eliminate the effect of gravitation by expressing all physical laws in a locally flat reference system (Minkowski Cartesian coordinate system with metric $\eta_{\mu\nu}$).

The effect of gravitation is subsequently introduced by replacing the Minkowski space-time of special relativity with the Riemann space-time of general relativity, that is, by replacing $\eta_{\mu\nu}$ with $g_{\mu\nu}$, the ordinary derivatives ∂_μ with the covariant derivatives D_μ, and in the integrals, the volume element d^4x with $\sqrt{g} \, d^4x$.

It is in this way that, in the absence of gravitation, the trajectory of a free particle, represented by the generalized Newton equation,

$$du^\mu = 0, \qquad (20.41)$$

is replaced using the correspondence principle (which mathematicians term the covariance principle) by the geodesics equation:

$$Du^\mu = dx^\nu \, (D_\nu u^\mu) = dx^\nu \, (\partial_\nu u^\mu + \Gamma^\mu_{\nu\sigma} u^\sigma u^\nu) = 0. \qquad (20.42)$$

We notice that the application of the correspondence, or covariance principle to the probability scalar field Ψ, leading to the Klein–Gordon equation in the Minkowski space,

$$(-\Box + m^2) \, \Psi = 0, \qquad (20.43)$$

remains the same if we replace the definition of the d'Alembertian of Minkowski space by the operator defined in Riemann space:

$$\Box \Psi = D_\mu D^\mu \Psi = \frac{1}{\sqrt{g}} \partial_\mu \partial^\mu (\sqrt{g} \, \Psi). \qquad (20.44)$$

We are naturally brought back to the usual form for $g = 1$, that is for a flat Minkowski-type metric in Cartesian coordinates.

Einstein's fundamental hypothesis is to further assume that the gravitational field is described by the metric tensor $g_{\mu\nu}$ of the space-time. E. Cartan has shown that, for the equations of the field to be second-order, the Hamiltonian action of the gravitational force should be written with the scalar curvature of the space-time and the constant Λ (the cosmological constant):

$$S_g = -\frac{1}{2\chi} \int (R - 2\Lambda)\sqrt{g}\, d^4 x. \tag{20.45}$$

The presence of matter and/or radiation is next described by a Lagrangian density $\mathcal{L}_m$ and the corresponding Hamiltonian action:

$$S_m = \int \mathcal{L}_m \sqrt{g}\, d^4 x. \tag{20.46}$$

The application of the principle of least action to the total Hamiltonian action $S = S_g + S_m$ leads to the Einstein equations of the gravitational theory of general relativity (Levy–Leblond, 1988)

$$\mathcal{G}_{\mu\nu} = R_{\mu\nu} - \frac{1}{2} g_{\mu\nu} R = -\chi (T_{\mu\nu} - \Lambda\, g_{\mu\nu}). \tag{20.47}$$

It should be noted here that relations (20.25) (20.33) and (20.38) permit us to intuitively obtain the Einstein equations by writing the conservation of four-divergence

$$D_\mu (\mathcal{G}^{\mu\nu} - \chi\,\Lambda\, g^{\mu\nu} + \chi\, T^{\mu\nu}) = 0, \tag{20.48}$$

with each of these terms possessing a zero four-divergence. The cosmology constant Λ seems to play an important role in current theories of cosmology. The constant χ determined from the Newtonian limit of the Einstein equations is given by

$$\chi = \frac{8\pi\, G}{c^4}, \tag{20.49}$$

where G is the gravitational constant.

4. Essentials of the Standard Model of Cosmology

The current standard model of cosmology, or Big Bang model, has been receiving wider and wider attention since the discovery of cosmic background radiation at 2.73 K. The observable facts upon which the standard model is based are, in fact, very few: the realization that the sky is black at night, the realization that at a higher scale the distribution of stars in the sky appears to be homogeneous and isotropic, the discovery of a spectral red shift of the light emitted by stars, the experimental evidence for a cosmic background electromagnetic radiation, and the agreement between the abundance of light elements determined from experimental measurements and the result deduced from the standard theory.

The standard cosmological model is based on three main ideas. The first assumes the existence of a cosmological principle following which the large-scale homogeneous and isotropic Universe may be represented by a cosmic fluid that behaves like a perfect gas with energy density $\varrho(t)$ and pressure $p(t)$, and which is only time dependent. The second assumes that the physical laws obeyed in the laboratory are transposable to cosmic level, and the third, which assumes that general relativity is applicable on the cosmic scale and is capable of accounting for the evolution of the Universe since creation (or almost, since quantum cosmology is intended to account for the initial conditions at the creation of the Universe and during the transition from a quantum state to a classical state).

To obtain a statement of physical laws that is not subordinate to the observer-dependent time, but to a universal cosmic time, cosmology uses a special comoving, coordinate system in which the space and time coordinates of an event are chosen in the same manner as those of a free-falling particle at a particular point with the proper time linked to the particle. The cosmological principle therefore allows us to write the metric in the so-called Friedmann–Robertson–Walker (FRW) form:

$$ds^2 = -d\tau^2 = -dt^2 + R^2(t)\left(\frac{dr^2}{1-kr^2} + r^2\,d\Omega^2\right), \qquad (20.50)$$

with $d\Omega^2 = d\theta^2 + \sin^2\theta\,d\varphi^2$.

The coefficient $R(t)$, often erroneously termed the "radius" of the Universe, is the expansion scale factor of the Universe. (Care should be taken here not confuse $R(t)$ with the space-time scalar curvature R^μ_μ.) The coefficient k is termed the curvature constant, because it represents the sign of intrinsic curvature of three-dimensional space. If $k = 1$, then the three-dimensional space is finite, an elliptical model, and $R(t)$ can be interpreted as the radius of curvature of the Universe. If $k = 0$ or -1, the scale function $R(t)$ can only yield a scale measurement of the geometry in a three-dimensional space; for $k = 0$ we obtain an Euclidean model and, finally, a hyperbolic model for $k = -1$.

The use of the Einstein Eq. (20.47) for a cosmic fluid described by the energy–momentum tensor (20.40), in an FRW metric (20.50), leads to the fundamental equations of the standard model of cosmology:

$$\frac{\ddot{R}}{R} = -\frac{4\pi G}{3}\,(\varrho + 3P), \qquad \left(\frac{\dot{R}}{R}\right)^2 + \frac{k}{R^2} = \frac{8\pi G}{3}\,\varrho,$$

$$\frac{d}{dR}\,(\varrho\,R^3) = -3PR^2. \qquad (20.51)$$

We will need to add to these equations a state equation $P = f(\varrho)$ for the cosmic fluid, for example, an equation of the type

$$P = (\gamma - 1)\,\varrho, \tag{20.52}$$

yielding the state-of-the-vacuum equation for $\gamma = 0$, that of an ultra-relativistic fluid for $\gamma = 4/3$, that of an incoherent fluid with zero pressure for $\gamma = 1$ and, finally, that of a completely rigid fluid for $\gamma = 2$.

Using the most recent data made available by particle and nuclear physics, cosmologists have been able to coherently explain the evolution of the Universe, from a few minutes after its creation (Big Bang) to the present (Elbaz, 1992). The simplest mathematical model of the Big Bang implies the existence of an initial singularity at time t_0, for which the scale factor $R(t_0)$ is zero, whereas the energy density and the equivalent temperature of the cosmic fluid are infinite. This means that cosmological time introduced with the FRW comoving coordinates is decribed on the open time interval $]\,t_0\ +\infty\,[$ excluding this initial time to which can be approximated without ever being reached (Levy-Leblond, 1988). Physical models of the Big Bang are saved from this singularity by introducing a Planck domain spanning the initial time to 10^{-44} seconds, a period from which physical, quantum, then semi-classical, and, finally, classical laws apply. Quantum cosmology seeks to interpret the transition from this purely quantum period to the semi-classical one by determining the wave function of the Universe obeying to adequate initial or boundary conditions to render this fundamental transition possible.

5. Hamiltonian Statement of Einstein Equations

To be able to use the usual quantization method of field theory, we will need to construct a Hamiltonian from the Lagrangian density $\tilde{\mathcal{L}}_g = \frac{1}{2\chi}\,\tilde{R}$ of general relativity. Space-time is therefore considered as a time-dependent three-dimensional space, with the dynamical variables constituting the metric elements.

5.1 Intrinsic and Extrinsic Curvature

Let us introduce a space-time folding by time-constant surfaces Σ. At time t, a point of the space Σ is defined by the length element

$$d\,\sigma'^2 = g_{ij}\,(t, x, y, z)\,dx^i\,dx^j\,.$$

At time $t + dt$, the surface Σ has moved to Σ' on which a point is defined by the length element

$$d\sigma^2 = g_{ij}\,(t + dt,\ x, y, z)\,dx'^i\,dx'^j\,.$$

During the transition from Σ to Σ', there is a time lapse $N(t, x, y, z)\, dt$, whereas dx^i is transformed into

$$dx'^i = dx^i + N^i\, dt.$$

The space-time interval therefore takes the form

$$
\begin{aligned}
ds^2 &= d\sigma'^2 - (N\, dt)^2 \\
&= g_{ij}\, dx'^i\, dx'^j - (N\, dt)^2 \\
&= g_{ij}\, (dx^i + N^i\, dt)(dx^j + N^j\, dt) - (N\, dt)^2 \\
&= g_{ij}\, dx^i\, dx^j + g_{ij}\, N^j\, dx^i\, dt + g_{ij}\, N^i\, dx^j\, dt \\
&\quad + (g_{ij}\, N^i\, N^j - N^2)\, dt^2,
\end{aligned}
$$

which finally becomes

$$\boxed{ds^2 = -(N^2 - N_i\, N^i)\, dt^2 + 2\, N_i\, dx^i\, dt + g_{ij}\, dx^i\, dx^j} \qquad (20.53)$$

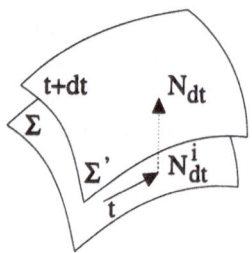

5.1.1 Extrinsic Curvature

Let η^β be the field of vectors normal to the surface Σ. Now, let us choose the direction of the unitary vector η_β by setting

$$\eta_\beta = (-1, 0, 0, 0). \qquad (20.54)$$

By the fundamental second form we mean the symmetric tensor

$$K_{il} = D_l\, \eta_i - \Gamma_{il}^\beta\, \eta_\beta \qquad (20.55)$$

and, by using the form (20.54) of the unitary vector η_β,

$$K_{il} = -\Gamma_{li}^0 = K_{li}. \qquad (20.56)$$

Because the Christoffel symbol is symmetric with respect to the subscripts i and l, the tensor K_{il} is likewise a symmetric tensor.

The tensor K_{il} describes the curvature of the surface Σ ($x^0 = $ constant) seen from the four-dimensional space-time in which it is immersed.

$$K_{il} = K_{li} = -\Gamma_{il}^0 = -\frac{1}{2}\,\partial_0\,g_{il} = -\frac{1}{2}\,\dot{g}_{il} \quad . \tag{20.57}$$

By extrinsic curvature of the surface Σ we mean the scalar constructed from the fundamental second form:

$$\mathrm{Tr}\,K^2 - (\mathrm{Tr}\,K)^2 = K_{ij}\,K^{ij} - (K_i^i)^2 \quad . \tag{20.58}$$

5.1.2 Intrinsic Scalar Curvature

Let us evaluate the scalar $^{(4)}R_\mu^\mu$ by separating the spatial parts from the time parts, and by noting that, in comoving coordinates, the Christoffel symbols Γ_{00}^μ and $\Gamma_{0\mu}^0$ are zero. By intrinsic curvature we mean the curvature $^{(3)}R_i^i$ of the surface Σ and, using the expression for of the scalar curvature:

$$^{(4)}R_\mu^\mu = {}^{(4)}g^{\mu\nu}\,R_{\mu\nu} = {}^{(4)}g^{\mu\nu}\,[\partial_\nu\,\Gamma_{\mu\gamma}^\gamma - \partial_\gamma\,\Gamma_{\mu\nu}^\gamma + (\Gamma\Gamma)_{\mu\nu}],$$

and by introducing the tensors of the intrinsic and extrinsic curvature of the surface Σ, we obtain the so-called Gauss–Codazzi equation:

$$^{(4)}R_\mu^\mu = {}^{(3)}R_i^i + \mathrm{Tr}\,K^2 - (\mathrm{Tr}\,K)^2 + \partial_0\,\mathrm{Tr}\,K + g^{ij}\,\partial_0\,K_{ij} \quad .$$

$$\tag{20.59}$$

The space-time scalar curvature $^{(4)}R_\mu^\mu$ is therefore expressed by means of intrinsic curvature $^{(3)}R_i^i$ of the surface Σ, its extrinsic curvature $\mathrm{Tr}\,K^2 - (\mathrm{Tr}\,K)^2$ and the evolution of the tensor K by the terms $\partial_0\,K_i^i + g^{ij}\,\partial_0\,K_{ij}$.

5.2 Action of the Gravitational and Matter Fields

Consider a scalar matter field ϕ in the presence of a gravitational field and the Einstein–Hilbert action:

$$\tilde{S}(g_{\mu\nu},\phi) = -\frac{1}{2\chi}\left\{\int_{\partial M} 2\,\sqrt{h}\,K\,d^3x + \int_M (R - 2\Lambda)\sqrt{g}\,d^4x\right\}$$

$$+ \int L_{\mathrm{mat}}(g_{\mu\nu},\phi)\,\sqrt{g}\,d^4x , \tag{20.60}$$

in which the Einstein constant χ is expressed with the universal gravitation constant G or in natural units ($\hbar = c = 1$) with the Planck mass M_p:

$$\chi = \frac{8\pi G}{c^4} = \frac{8\pi}{M_p^2}.\tag{20.61}$$

The metric $^{(3)}g_{ij} = h_{ij}$ is that of a three-dimensional metric on a boundary ∂_M and $K = K_i^i = \text{Tr } K$ is the trace of the fundamental second form. The surface term is necessary to cancel out terms from the scalar curvature by the second derivatives of the metric tensor.

It can be shown that the Hamiltonian action $S_g = -(1/2\chi)\int \sqrt{g}R d^4x$ can be replaced with the Hilbert reduced action (Elbaz, 1992) $\bar{S}_g = -(1/2\chi)\int (\Gamma\Gamma)\sqrt{g}\, d^4x$. When we decompose the scalar curvature $^{(4)}R_\mu^\mu$ into its time and space parts, the terms in $\partial_0 K_{ij}$ and K_i^i will make no contribution to the action integral.

An especially important case is obtained when a compact three-dimensional subvariety divides the four-dimensional variety into two parts. This is what we considered during the folding by the surfaces Σ in (20.53). The space-time metric can then be locally described in the form

$$ds^2 = -(N^2 - N_i\, N^i)dt^2 + 2\, N_i\, dx^i\, dt + h_{ij}\, dx^i\, dx^j,\tag{20.62}$$

where $h_{ij} = {}^{(3)}g_{ij}$ is the three-dimensional metric on the surface Σ on which the points $x^i (i = 1, 2, 3)$ are located, and $h = \det|h_{ij}|$.

The Einstein–Hilbert action then becomes:

$$\tilde{S}(g_{\mu\nu,\phi}) = \frac{M_p^2}{16\pi}\int dt d^3x\, \sqrt{h}\, N$$

$$\times \left\{ K_{ij}\, K^{ij} - K^2 - {}^{(3)}R_i^i + 2\wedge - \frac{16\pi}{M_p^2} L_{\text{mat}}(\phi) \right\}.\tag{20.63}$$

Expressed in comoving coordinates, this amounts to introducing a reduced Lagrangian density for the gravitational field of the form

$$\tilde{S}_g = \frac{M_p^2}{16\pi}\int dt\, d^3\, x\, \sqrt{h} \left\{ \text{Tr } K^2 - (\text{Tr } K)^2 - {}^{(3)}R_i^i \right\}.\tag{20.64}$$

The term in brackets is the difference between the intrinsic and extrinsic curvatures of the surface Σ. It can also take the form

$$\tilde{S}_g = \frac{M_p^2}{16\pi}\int dt\, d^3x\, \sqrt{h} \left\{ (h^{ik}\, h^{jl} - h^{ij}\, h^{kl})K_{ij}\, K_{kl} - {}^{(3)}R_i^i \right\}$$

$$= \frac{M_p^2}{16\pi}\int dt\, d^3x\, \sqrt{h} \left\{ \frac{1}{4}\, (h^{ik}\, h^{jl} - h^{ij}\, h^{kl})\, \dot{h}_{ij}\, \dot{h}_{kl} - {}^{(3)}R_i^i \right\}.$$

$$\tag{20.65}$$

5.3 Superspaces

5.3.1 Supermetric Tensors

Let us use the space $\{x^i\}$ of the metric h_{ij} to construct a six-dimensional superspace with supermetric G^{ijkl}, the space corresponding to all three-dimensional spaces. To simplify the notation, we introduce a global index $i, j = a$, which gives

$$
\begin{aligned}
G^{ijkl} &= \frac{1}{2}\sqrt{h}\,\{h^{ik}\,h^{jl} + h^{il}\,h^{jk} - 2\,h^{ij}\,h^{kl}\}\,, \\
G^{ab} &= \sqrt{h}\,\{h^{ab} - h^a\,h^b\}\,.
\end{aligned}
\tag{20.66}
$$

The covariant form of the supermetric tensor is, for its part,

$$
G_{ab} = \frac{1}{\sqrt{h}}\,(h_{ab} - \frac{1}{2}\,h_a\,h_b)\,,
\tag{20.67}
$$

which leads to the closure relation:

$$
G_{ab}\,G^{ac} = G_b^c = \delta_b^c = \frac{1}{2}\,(\delta_k^m\,\delta_l^n + \delta_l^m\,\delta_k^n)\,.
\tag{20.68}
$$

It must be recalled here that the supermetric tensor is defined locally, that is, we should in fact write:

$$
G^{ab}(x, x') = \sqrt{h}\,\{h^{ab}(x, x') - h^a(x)\,h^b(x')\}\,\delta^3\,(x - x')\,.
\tag{20.69}
$$

Applied to a rank-two symmetrical tensor, the supermetric transforms into its codual according to relations that are easily directly verified:

$$
\begin{aligned}
G_{ab}\,T^b &= \hat{T}_a = \frac{1}{\sqrt{h}}\,\left(T_a - \frac{1}{2}\,h_a\,T\right)\,, \\
G^{ab}\,T_b &= \hat{T}^a = \sqrt{h}\,\left(T^a - h^a T\right)\,, \\
G_{ab}\,\hat{T}^b &= T_a \quad \text{and} \quad G^{ab}\,\hat{T}_b = T^a\,.
\end{aligned}
\tag{20.70}
$$

The trace of the tensor T is defined in the three-dimensional space or in the six-dimensional superspace:

$$
\begin{aligned}
T &= \operatorname{Tr} T = T_\mu^\mu = T_{\mu\nu}\,h^{\nu\mu} = T^{\mu\nu}\,h_{\mu\nu} \\
&= T_a\,h^a = T^a\,h_a\,.
\end{aligned}
\tag{20.71}
$$

Bearing in mind that in a three-dimensional space the trace of the metric tensor has the value

$$
\operatorname{Tr} h = h = h_\mu^\mu = 3\,,
\tag{20.72}
$$

we obtain the coduals

$$
\hat{h}_a = 2\,\sqrt{h}\,h_a \quad \text{and} \quad \hat{h}^a = -\frac{1}{2\sqrt{h}}\,h^a\,.
\tag{20.73}
$$

5.3.2 Conjugate Momentum

Let us introduce the conjugate momentum of the metric h_{ij} according to the usual definition and by using the global notation

$$\pi^{ij} = \frac{\partial \bar{\mathcal{L}}}{\partial \dot{h}_{ij}} \quad \text{or} \quad \pi^a = \frac{\partial \bar{\mathcal{L}}}{\partial \dot{h}_a}, \tag{20.74}$$

with form (20.65) of the Einstein–Hilbert action (we set $\bar{\mathcal{L}} = \frac{16\,\pi}{M_P^2}\,\bar{\mathcal{L}}g$):

$$\tilde{S}_g = \frac{M_P^2}{16\pi} \int dt\, d^3 x \, \sqrt{h} \, \{(h^{ik}\, h^{ij} - h^{ij}h^{kl})\, K_{ij}\, K_{kl} -^{(3)} R_i^i\}$$

$$= \int dt\, d^3 x \, \bar{\mathcal{L}}_g. \tag{20.75}$$

Now, let us write the Lagrangian density in a different form by introducing the supermetric tensor:

$$\bar{\mathcal{L}} = \frac{16\,\pi}{M_P^2}\,\bar{\mathcal{L}}_g = -\sqrt{h}^{\,(3)}R_i^i + \sqrt{h}\,\left\{\frac{1}{2}(h^{ik}\,h^{jl} + h^{il}\,h^{jk}) - h^{ij}\,h^{kl}\right\}\frac{1}{4}\,\dot{h}_{ij}\,\dot{h}_{kl}$$

$$= -\sqrt{h}\,^{(3)}R_i^i + \sqrt{h}\,\{h^{ab} - h^a\,h^b\}\,\frac{1}{4}\,\dot{h}_a\,\dot{h}_b$$

$$= -\sqrt{h}^{\,(3)}R_i^i + \frac{1}{4}\,G^{ab}\,\dot{h}_a\,\dot{h}_b = -\sqrt{h}^{\,(3)}R_i^i + \frac{1}{4}\,\dot{h}^{\,b}\,\dot{h}_b. \tag{20.76}$$

By applying relation (20.74), we easily obtain

$$\pi^a = \frac{1}{2}\,\dot{h}^{\,a} = -\hat{K}^a = -\sqrt{h}\,(K^a - h^a\,K). \tag{20.77}$$

The Einstein–Hilbert action can therefore be described in the following form:

$$\tilde{S}_g = \frac{M_P^2}{16\pi} \int_{t_1}^{t_2} (\pi^a\,\dot{h}_a - NX^0 - N_i X^i)\, d^3 x\, dt, \tag{20.78}$$

with, by definition,

$$X^0 \equiv \sqrt{h}\Big\{\text{Tr } K^2 - (\text{Tr } K)^2 +^{(3)} R_i^i\Big\},$$

$$X^i \equiv 2D_\ell\,(K^{il} - h^{il}\, K) = -2\,D_l\,\pi^{il}. \tag{20.79}$$

The quantities X^0 and X^i can be considered as Lagrange multipliers, and the classical field equation $\delta\tilde{S}_g/\delta N = 0$ is the Hamiltonian constraint of general relativity. By forcing constraints $X^0 = 0$ and $X^i = 0$ during the variation of $\tilde{S}_g$, we may consider the h_{ij}, N_i and N as independent variables. The constraint $X^i = 0$ corresponds to the conservation of the conjugate momentum and the constraint $X^0 = 0$ to the conservation of the total energy in the Universe considered as an isolated, closed system.

5.3.3 Intervals in the Superspace

Each point on the surface Σ is associated with the metric $h_{ij}(x) = h_a(x)$ in which the global index a assumes the values:

$$(ij) = (11)\ (22)\ (33)\ (12)\ (13)\ (23)$$
$$a = \ \ 1\ \ \ \ 2\ \ \ \ \ 3\ \ \ \ \ 4\ \ \ \ \ 5\ \ \ \ 6\,. \tag{20.80}$$

The space interval on $\sum$ is written

$$ds^2 = h_{ij}\ dx^i\ dx^j\,. \tag{20.81}$$

Because each metric tensor $h^{ij}(x) = h^a(x)$ becomes a point in the superspace, we introduce the interval corresponding to the superspace:

$$dL^2 = \int h_{ab}(x, x')\ dh^a(x)\ dh^b(x')\ d^3x\ d^3x'\,. \tag{20.82}$$

Because of the local definition (20.69) of the supermetric tensor,

$$h_{ab}(x, x') = h_{ab}(x)\ \delta^3(x - x')\,, \tag{20.83}$$

the interval dL^2, defined at each point on $\sum$, for all three-metrics $h_a(x)$ can also take the form

$$dL^2 = \int h_{ab}(x)\ dh^a(x)\ dh^b(x)\ d^3x\,. \tag{20.84}$$

It is convenient to generalize the Einstein summation convention by supplementing it with the integration over all space variables such as they appear in (20.82) and (20.83), giving

$$dL^2 = h_{ab}(x, x')\ dh^a(x)\ dh^b(x')\,. \tag{20.85}$$

We can use the three-metric of $\sum$ to define the supermetric tensor in the codual supermetric space in which the interval dS^2 will be written with the generalized Einstein convention

$$dS^2 = G^{ab}(x, x')\ dh_a(x)\ dh_b(x')\,. \tag{20.86}$$

5.3.4 Geodetics

The Lagrangian density (20.76) can also be written

$$\bar{\mathcal{L}} = \frac{16\pi}{M_p^2}\ \bar{\mathcal{L}}_g = -\sqrt{h}\ ^{(3)}R_i^i + \frac{1}{4}\ \dot{h}_a\ \overset{\cdot\,a}{h}$$
$$= -\sqrt{h}\ ^{(3)}R_i^i + G^{ab}\ K_a\ K_b$$
$$= -\sqrt{h}\ ^{(3)}R_i^i + \hat{K}^b\ K_b\,. \tag{20.87}$$

We use the intrinsic curvature of the surface Σ to introduce a potential energy by setting

$$V(x, t) = \sqrt{h} \; {}^{(3)}R_i^i \,, \tag{20.88}$$

thus making it possible to write the Lagrangian density in the classical form

$$\begin{aligned}
\bar{\mathcal{L}} &= \hat{K}^b \, K_b - V \\
&= \frac{1}{4} \dot{h}^{;b} \, \dot{h}_b - V = \frac{1}{4} \, G^{ab} \, \dot{h}_a \, \dot{h}_b - V \,.
\end{aligned} \tag{20.89}$$

Now let us write the latter form by using the interval dS^2 of the codual supermetric space defined in (20.86):

$$\bar{\mathcal{L}} = \frac{1}{4} \left(\frac{dS}{dt} \right)^2 - V \,. \tag{20.90}$$

We obtain with (20.87) and (20.76) the corresponding Einstein–Hilbert action

$$\tilde{S}_g = \int dt \, d^3x \, \bar{\mathcal{L}}_g = \frac{M_p^2}{16 \, \pi} \int dt \, d^3x \left\{ \frac{1}{4} \left(\frac{dS}{dt} \right)^2 - V \right\} \,. \tag{20.91}$$

We know that an extremum of the Hamiltonian action corresponds to a length extremum of the ordinary space, that is to a geodesic [see (20.34) and (20.35)].

The Einstein gravitation equations are obtained by seeking an extremum of the Einstein–Hilbert action

$$\tilde{S}_g = \frac{1}{2\chi} \int \bar{\mathcal{L}} \, d^3x \, dt \,. \tag{20.92}$$

They therefore correspond to a geodesic equation in the superspace of all three-metrics. They are called geodetics:

$$-\frac{\delta \, V}{\delta \, h^a} = \frac{d^2 \, h^a}{dt^2} + \Gamma_{bc}^a \frac{d \, h^b}{dt} \frac{d \, h^c}{dt} \,, \tag{20.93}$$

with the Christoffel super-symbols:

$$\Gamma_{bc}^a = \frac{1}{2} \, G^{ad} \left(\partial_g \, G_{cd} + \partial_c \, G_{bd} - \partial_d \, G_{bc} \right) \tag{20.94}$$

and the differentiation with respect to the metric

$$\partial_a = \frac{\partial}{\partial \, h^a} \,. \tag{20.95}$$

By choosing an adequate parametrization of the time $dt = f^2(t') \, dt$ such that the variation of the potential is zero:

$$\frac{\delta \, f^2 \, V}{\delta \, h^a} = 0 \,, \tag{20.96}$$

we obtain a geodetic equation without a second member:

$$\frac{d^2\,h^a}{dt'^2} + \Gamma^a_{bc}\,\frac{d\,h^b}{dt'}\,\frac{dh^c}{dt'} = 0\,. \tag{20.97}$$

We introduce the covariant differentiation of the superspace by setting

$$\mathcal{D}_b = \partial_b \pm \Gamma_b\,. \tag{20.98}$$

The Einstein gravitation equations will then be given by the geodetics of the superspaces:

$$\boxed{\mathcal{D}_b\,\hat{h}^a = 0}\,. \tag{20.99}$$

6. Fundamental Equation of Quantum Cosmology

6.1 Illustrative Example

As an application of the formula (20.53) of space-time folding, let us consider a homogeneous and isotropic Universe with curvature k, described by an FRW metric in the form

$$ds^2 = \sigma^2\left[-N^2\,dt^2 + e^{2\alpha(t)}\,d\Omega_3^2(k)\right]\,, \tag{20.100}$$

in which $\alpha(t)$ is a scale factor, $d\Omega_3^2(k)$ the metric of the space sections with curvature K and the coefficient $\sigma = 2/3\pi\,M_P^2$. The source of the matter field is the scalar field $\sqrt{2\pi}\,\sigma^2\,\phi(t)$ with potential $2\pi^2\sigma^2\,V(\phi)$. By using the form (20.65) of the Einstein–Hilbert action, we obtain the Lagrangian density

$$\bar{\mathcal{L}} = \frac{1}{2}Ne^{3\alpha}\left[-\frac{\dot{\alpha}^2}{N^2} + \frac{\dot{\phi}^2}{N^2} - V(\phi) + k\,e^{-2\alpha}\right]\,. \tag{20.101}$$

The variation with respect to the variables ϕ, N and α leads to the three Euler–Lagrange equations:

$$(\ddot{\phi} + 3\dot{\alpha}\,\dot{\phi})\,N - \dot{\phi}\,\dot{N} - \frac{1}{2}\,N^3V' = 0\,,$$

$$-(\ddot{\alpha} + 3\dot{\alpha}^2 + \frac{3}{2}\,\dot{\phi}^2)\,N + \dot{\alpha}\,\dot{N} + \left(\frac{3}{2}\,V - \frac{1}{2}k\,e^{-2\alpha}\right)\,N^3 = 0\,, \tag{20.102}$$

$$\dot{\alpha}^2 - \dot{\phi}^2 - N^2\,(V + k\,e^{2\alpha}) = 0\,.$$

Following re-arrangement and by choosing the gauge $N = 1$, we obtain the equations of motion and the constraint

$$\ddot{\phi} + 3\dot{\alpha}\,\dot{\phi} + \frac{1}{2}\,V'(\phi) = 0\,,$$

$$\ddot{\alpha} + \dot{\alpha}^2 + 2\,\dot{\phi}^2 - V(\phi) = 0\,, \qquad (20.103)$$

$$-\dot{\alpha}^2 + \dot{\phi}^2 + V(\phi) = k\,e^{-2\alpha}\,.$$

The general solution of this system of equations introduces three parameters: α, ϕ and V.

The potential $V(\phi)$ is inflationary-type if, for some values of ϕ, the potential $V(\phi)$ is large, whereas $|V'(\phi)/V(\phi)| \ll 1$.

The initial conditions leading to an inflationary scenario, in fact, require (Halliwel, 1990)

$$\dot{\phi} \simeq 0 \quad \text{and} \quad k = +1\,. \qquad (20.104)$$

Let us use definition (20.74) to determine the conjugate momenta:

$$\pi_\alpha = \frac{\partial\bar{\mathcal{L}}}{\partial\dot{\alpha}} = -e^{3\alpha}\,\frac{\dot{\alpha}}{N} \quad \text{and} \quad \dot{\alpha} = -N\,\pi_\alpha\,e^{-3\alpha}\,,$$

$$\pi_\phi = \frac{\partial\bar{\mathcal{L}}}{\partial\dot{\phi}} = e^{3\alpha}\,\frac{\dot{\phi}}{N} \quad \text{and} \quad \dot{\phi} = N\,\pi_\phi\,e^{-3\alpha}\,. \qquad (20.105)$$

We infer from this the canonical Hamiltonian

$$\begin{aligned} H_c &= \pi_\alpha\,\dot{\alpha} + \pi_\phi\,\dot{\phi} - \bar{\mathcal{L}} \\ &= \frac{1}{2}\,N\,e^{-3\alpha}\left[-\pi_\alpha^2 + \pi_\phi^2 + e^{6\alpha}V - e^{4\alpha}\right] \equiv N\,H\,. \quad (20.106) \end{aligned}$$

The Einstein–Hilbert action written with (20.101) in the Hamiltonian form by taking $\bar{\mathcal{L}}$ from the first equation in (20.106) then becomes

$$S = \int\,dt\left[\dot{\alpha}\,\pi_\alpha + \dot{\phi}\pi_\alpha - N\,H\right]. \qquad (20.107)$$

By choosing the curvature $k = +1$ and inserting (20.105) into expression (20.106) for H, we notice that N is a Lagrange multiplier imposing the constraint

$$H = 0\,, \qquad (20.108)$$

which in fact corresponds to the third equation of (20.103).

Now, let us quantize the system by introducing a wave function $\psi(\alpha, \phi, t)$ obeying a Schrödinger equation constructed with the canonical Hamiltonian:

$$i\,\partial_t\,\psi = H_c\,\psi\,. \qquad (10.109)$$

By using the constraint $H = 0$ and the correspondence principle for the variables α and ϕ,

$$\pi_\alpha = -i\,\partial_\alpha \quad \text{and} \quad \pi_\phi = -i\partial_\phi\,, \qquad (20.110)$$

we obtain a Schrödinger-type equation

$$H\,\psi = \frac{1}{2}e^{-3\alpha}\left[\frac{\partial^2}{\partial\,\alpha^2} - \frac{\partial^2}{\partial\,\phi^2} + e^{6\alpha}V - e^{4\alpha}\right]\psi = 0\,. \qquad (20.111)$$

The solution is time independent, which is an essential characteristic of WDW-type quantum cosmology wave functions.

We introduce a potential $U(\alpha)$ grouping the two terms:

$$U(\alpha) = e^{6\alpha}V(\phi) - e^{4\alpha}\,, \qquad (20.112)$$

and we examine the solutions that are weakly dependent upon ϕ, which is equivalent to seeking the WKB semi-classical solutions:

$$\text{for } U \ll 0 \quad \psi(\alpha,\phi) \simeq \exp\left[\pm\frac{1}{3V}(1 - e^{2\alpha}V)^{3/2}\right]\,,$$

$$\text{for } U \gg 0 \quad \psi(\alpha,\phi) \simeq \exp\left[\pm\frac{i}{3V}(e^{2\alpha}V - 1)^{3/2}\right]\,. \qquad (20.113)$$

The latter regime with a large scale factor is considered the classical zone. In addition, we will have to impose boundary conditions to each solution and link the two types of wave function, exponential and oscillating. This is exactly the procedure used to calculate the tunnel effect in quantum mechanics.

Let us introduce a function S of the variables in α and ϕ, and then seek solutions of the type

$$\psi = \exp(iS) \qquad (20.114)$$

Using this hypothesis in the WDW Eq. (20.111), we obtain

$$-\left(\frac{\partial S}{\partial\alpha}\right)^2 + \left(\frac{\partial S}{\partial\phi}\right)^2 + U(\alpha,\phi) = 0\,. \qquad (20.115)$$

This is a Hamilton–Jacobi equation. Rewritten with (20.106) and (20.112), the solution $H = 0$ becomes

$$-\pi_\alpha^2 + \pi_\phi^2 + U(\alpha,\phi) = 0\,, \qquad (20.116)$$

showing that the conjugate momenta are defined by the relations

$$\pi_\alpha = \frac{\partial\,S}{\partial\,\alpha} \quad \text{and} \quad \pi_\phi = \frac{\partial\,S}{\partial\,\phi}\,. \qquad (20.117)$$

By comparing (20.117) and (20.105), we notice that the preceding relation defines a set of possible trajectories in the plane (α, ϕ), solutions to the equations of motion and with a classical constraint (20.103). The solution $\psi = e^{iS}$ is strongly centered around a set of classical solutions of Eq. (20.103), making use of two out of the three free parameters. This shows that the boundary conditions of the WDW wave function imply special initial conditions for the semi-classical solutions of the wave function of the Universe.

For $e^{2\alpha}V \gg 1$, for example, we obtain with (20.113) the solution

$$\psi(\alpha,\ \phi) \simeq \exp\left[\pm\frac{i}{3V}e^{3\alpha}V^{3/2}\right],\tag{20.118}$$

which defines with (20.114) the function S:

$$S \simeq -\frac{1}{3}e^{3\alpha}V^{3/2}.\tag{20.119}$$

Using (20.117) and (20.112), it is possible to obtain $\dot{\alpha}$ as

$$\dot{\alpha} = -\frac{\partial S}{\partial \alpha}e^{-3\alpha} = V^{1/2},\tag{20.120}$$

and following integration, we obtain the scale factor

$$\alpha(t) = V^{1/2}(\phi)\ (t-t_0).\tag{20.121}$$

We further impose the initial condition $\dot{\phi} \simeq 0$, which thus defines

$$\exp(\alpha(t)) \simeq \exp(V^{1/2}(t-t_0))\ \ \text{and}\ \ \phi \simeq \phi_0 = \ \ \text{constant}.\tag{20.122}$$

The solutions are therefore parametrized from the two constants t_0 and ϕ_0, and the choice of these special initial conditions leads, following the form of the scale factor (20.121), to an inflationary-type solution.

6.2 Hamiltonian Formulation of General Relativity

The Lagrangian density $\bar{\mathcal{L}}$ described in (20.89) takes the form of a difference between a kinetic energy and a potential energy:

$$\bar{\mathcal{L}} = \hat{K}^b\ K_b - V = \pi^b\ \hat{\pi}_b - V$$
$$= G_{ab}\ \pi^a\ \pi^b - V.\tag{20.123}$$

We can infer from this a super-Hamiltonian denoted $\bar{H}$:

$$\bar{H} = \hat{\pi}^a\ \dot{h}_a - \bar{\mathcal{L}} = 2\ \hat{\pi}^a\ \pi_a - \bar{\mathcal{L}}$$
$$= \hat{\pi}^b\ \pi_b + V$$
$$= G_{ab}\ \pi^a\ \pi^b + V = G_{ab}\ \pi^a\ \pi^b + \sqrt{h}\ ^{(3)}R_i^i.\tag{20.124}$$

Constraining Eq. (20.79a) to $X^0 = 0$ therefore corresponds to $\bar{H} = 0$, that is, a zero total energy in the superspace.

Working from the general form (20.75) of the Einstein–Hilbert action:

$$\tilde{S} = \frac{M_p^2}{16\ \pi}\int dt\ d^3x\sqrt{h}\ N\ \{\text{Tr}\ K^2 - (\text{Tr}\ K)^2 - ^{(3)}R_i^i + 2\wedge - \frac{16\ \pi}{M_p^2}L_{\text{mat}}(\phi)\},\tag{20.125}$$

the Hamiltonian with neither Λ nor ϕ is obtained by variation of $\tilde{S}$ with respect to N:

$$\bar{H} \equiv \frac{\delta \tilde{S}}{\delta N}. \tag{20.126}$$

We infer from this that the super-Hamiltonian can be written in the integral form:

$$\bar{H} \equiv \int (N \, H_0 + N_i \, H^i) \, d^3x, \tag{20.127}$$

with the super-Hamiltonian H_0:

$$
\begin{aligned}
H_0 = \frac{16 \, \pi}{M_p^2} &\int G_{ab}(x, x') \pi^a(x) \, \pi^b(x') \, d^3x \\
&+ \int \sqrt{h} \, \delta^3(x - x') \left\{ \frac{M_p^2}{16 \, \pi} \left(-{}^{(3)} R_i^i + 2\Lambda \right) + T_{00}(\phi, \pi^\phi, h^a) \right\} d^3x'
\end{aligned}
\tag{20.128}
$$

and the super-Hamiltonian H^i defined by

$$H^i = 2 \, D_j \, \pi^{ij} - \sqrt{h} \, T^{0i}. \tag{20.129}$$

The functions $N(x)$ and $N_i(x)$ are considered as Lagrange multipliers such that the classical equations of the fields will be obtained by setting

$$H_0(x) = 0 \quad H^i(x) = 0 \quad \forall \, x, \tag{20.130}$$

or, in an equivalent manner, by using (20.127),

$$H = 0 \quad \forall \, N, N_i. \tag{20.131}$$

6.3 Canonical Quantization

We use the correspondence principle in the superspace by introducing the usual commutation relations of field theory:

$$
\begin{aligned}
[h_a(x), h_b(x')] &= 0, \\
[\pi^a(x), \pi^b(x')] &= 0, \\
[h_a(x), \pi^b(x')] &= i \, \delta_a^b \, \delta^3(x - x').
\end{aligned}
\tag{20.132}
$$

This amounts to using the correspondence principle for the conjugate momenta in terms of functional derivatives (and not partial derivatives with respect to the variables as is the case in Schrödinger's quantum mechanics):

$$
\begin{aligned}
\pi^a &= -i \, \delta_a = -i \frac{\delta}{\delta \, h_a}, \\
\pi^\phi &= -i \, \delta_\phi = -i \frac{\delta}{\delta \, \phi}.
\end{aligned}
\tag{20.133}
$$

The super-Hamiltonian $\bar{H} = \bar{E}$ and the correspondence principle also lead to the Wheeler–DeWitt equation for the wave function of the Universe:

$$\boxed{-G_{ab}\ \delta^2\psi/\delta h_a\ \delta\ h_b + V\ \psi = \bar{E}\ \psi} \ . \tag{20.134}$$

In a closed Universe, the total energy is zero, giving the Wheeler–DeWitt (WDW) equation in the absence of a matter field

$$-G_{ab}\ \frac{\delta^2\ \psi}{\delta\ h_a\ \delta\ h_b} + V\ \psi = 0. \tag{20.135}$$

The definition of the conjugate momentum (20.123) gives, by virtue of the conservation law,

$$D_l\ \frac{\delta\ \psi}{\delta\ h_{il}} = 0. \tag{20.136}$$

6.4 Example of a Wheeler–DeWitt Equation

Consider the Lagrangian density of a Friedmann–Robertson–Walker metric of the form

$$\bar{\mathcal{L}} = \alpha(R\ \dot{R}^2 - k\ R),$$

with the multiplier coefficient

$$\alpha = -\frac{3\ M_p^2}{8}\ \frac{r^2}{\sqrt{1 - kr}}.$$

This defines the Einstein–Hilbert action (20.75b) in the form

$$\tilde{S}_g = \int\ r^2\ dr\ \alpha(R\ \dot{R}^2 - kR)\ dt. \tag{20.137}$$

The conjugate momentum is, from its classical definition,

$$\pi = \frac{\partial\bar{\mathcal{L}}}{\partial\ \dot{R}} = 2\alpha\ R\ \dot{R}, \tag{20.138}$$

and the super-Hamiltonian takes the following form:

$$\begin{aligned}
\bar{H} = \pi\ \dot{R} - \bar{\mathcal{L}} &= 2\alpha\ R\ \dot{R}^2 - \alpha\ R\ \dot{R}^2 + \alpha\ kR \\
&= \alpha(R\ \dot{R}^2 + kR) \\
&= \alpha R\ \left(\frac{\pi}{2\alpha\ R}\right)^2 + \alpha\ kR \\
&= \frac{\pi^2}{4\alpha\ R} + \alpha\ kR.
\end{aligned} \tag{20.139}$$

The eigenvalue equation of the super-Hamiltonian $\bar{H}$ in this special case becomes

$$\left(\frac{\pi^2}{4\alpha\,R} + \alpha\,kR\right)\psi = \bar{E}\,\psi, \tag{20.140}$$

giving, following application of the correspondence principle (20.133),

$$\left(-\frac{1}{4\alpha\,R}\frac{\delta}{\delta\,h_a}\frac{\delta}{\delta\,h_b} + k\,\alpha\,R\right)\psi = \bar{E}\,\psi \tag{20.141}$$

and, in the R representation, we obtain the WDW wave equation

$$\left(\frac{1}{4\alpha}R^{-1/4}\frac{\partial}{\partial\,R}R^{-1/2}\frac{\partial}{\partial\,R}R^{-1/4} - \alpha\,k\,R\right)\psi = \bar{E}\,\psi. \tag{20.142}$$

6.5 Mini-Superspaces

By using the correspondence principle, we establish the general form of the WDW equation, the fundamental equation of quantum cosmology,

$$\left[-G_{ab}\frac{\delta^2}{\delta\,h_a\,\delta\,h_b} - \sqrt{h}\left(\,^{(3)}R(h) - 2\Lambda + \frac{16\pi}{M_p^2}T_{00}(\delta_\phi,\,\phi)\right)\right]\psi(h_a,\phi) = 0, \tag{20.143}$$

with the boundary condition

$$D_i\left(\frac{\delta\psi}{\delta\,h_{ij}}\right) = 0. \tag{20.144}$$

These two equations contain the dynamic information necessary to treat gravitation. However, the functional derivatives appearing make it practically impossible to use these equations in this form. We can reduce the number of degrees of freedom of h_{ij} to a finite number by using symmetrical cosmological models such as those of Friedmann–Robertson–Walker. In this case, the metric can be described by a finite and small number of functions of t (for example, the scale function $R(t)$ in a standard Friedmann–Robertson–Walker metric). In such a case, we say that the superspace has been reduced to a finite-dimensional mini-superspace.

6.6 Two-Dimensional Mini-Superspace

Consider the case in which the gravitational field is locally homogeneous and isotropic. The three-metric is then determined by the scale factor $R(t)$ and we can set

$$h_{ij} = h_a = \sigma^2\,R^2(t)\,\tilde{h}_a, \tag{20.145}$$

where the three-dimensional space has an intrinsic curvature

$$
\begin{aligned}
^{(3)}R_{ab} &= k \, \tilde{h}_{[ab]} \\
&= k \, \{\tilde{h}_{ik} \, \tilde{h}_{jl} - \tilde{h}_{il} \, \tilde{h}_{jk}\} \, .
\end{aligned}
\tag{20.146}
$$

We thus obtain $k = +1$ (a sphere of unit-radius), $k = 0$ (a torus or flat space), or $k = -1$ (a hyperbolic space). The coefficient σ is a normalization coefficient given by

$$
\sigma = \left(\frac{3M_p^2}{4\pi} \int \sqrt{\tilde{h}} \, d^3x\right)^{1/2} \, .
\tag{20.147}
$$

We introduce a constant matter scalar field ϕ on the homogeneity surfaces Σ by setting

$$
\phi = \left(\frac{3M_p^2}{4\pi}\right)^{1/2} \tilde{\phi} \, ,
\tag{20.148}
$$

and a self-interaction of this potential denoted $U(\phi)$:

$$
U(\phi) = \frac{3M_p^2}{4\pi} \tilde{U}(\tilde{\phi})
\tag{20.149}
$$

(a scalar field ϕ with mass M/σ will have a self-interaction $\tilde{U} = \frac{1}{2}m^2 \, \tilde{\phi}^2$, for example).

The mini-superspace thus defined is a two-dimensional space with coordinates $R(t)$ and $\tilde{\phi}(x)$, and its metric can be defined, for N constant on the surfaces Σ by setting

$$
ds^2 = \sigma \, N^{-1} \, (-RdR^2 + R^3 \, d \, \tilde{\phi}^2) \, .
\tag{20.150}
$$

Because the mini-superspace is flat for N independent of $\tilde{\phi}$, we choose the gauge $N = \sigma$, which defines an interval

$$
ds^2 = -R \, dR^2 + R^3 \, d\tilde{\phi}^2 \, .
\tag{20.151}
$$

In a closed universe, ψ is not explicitly time dependent and $\partial_t \psi = 0$, which leads to a time-independent WDW equation

$$
\left(R\frac{\partial}{\partial R}R\frac{\partial}{\partial R} - \frac{\partial^2}{\partial \tilde{\phi}^2} - k \, R^4 + 2 \, R^6 \, \tilde{U}\right) \psi(R, \tilde{\phi}) = 0 \, .
\tag{20.152}
$$

6.7 Wheeler–DeWitt Equation
in a Friedmann–Robertson–Walker Metric

Consider a scalar field $V(\phi)$ in interaction with the gravitational field. The Lagrangian density can be written in the form

$$\tilde{\mathcal{L}}(g_{\mu\nu}, \phi) = \sqrt{g}\left(-\frac{R_\mu^\mu}{2\chi} + \frac{1}{2}\partial_\mu\phi\,\partial^\mu\phi - V(\phi)\right). \tag{20.153}$$

In a Friedmann closed universe of comoving coordinates, the space-time interval is expressed with the relation

$$ds^2 = -N^2(t)\,dt^2 + R^2(t)\,d\Omega_3^2, \tag{20.154}$$

where $N(t)$ is an auxiliary function for defining the time scale and $d\Omega_3^2$ the length element in a three-dimensional space. We can infer from (20.153) and (20.154) an effective Lagrangian density of the form

$$\tilde{\mathcal{L}} = -\frac{3M_p^2\,\pi}{4}\left(\frac{\dot{R}^2\,R}{N} - NR\right) + 2\pi^2\,R^3\,N\left(\frac{\dot\phi^2}{2\,N^2} - V(\phi)\right). \tag{20.155}$$

The term $2\pi^2\,R^3$ corresponds to the integration over the space coordinates of $\sqrt{g}$ for a sphere of unit radius.

We begin by determining the conjugate momenta of the Lagrangian density $\tilde{\mathcal{L}}$ with definition (20.74):

$$\pi_\phi = \frac{\partial\tilde{\mathcal{L}}}{\partial\dot\phi} = \frac{2\pi^2\,R^3}{N}\dot\phi,$$

$$\pi_R = \frac{\partial\tilde{\mathcal{L}}}{\partial\dot{R}} = -\frac{3M_p^2\,\pi}{2N}\dot{R}\,R, \tag{20.156}$$

$$\pi_N = \frac{\partial\tilde{\mathcal{L}}}{\partial\dot{N}} = 0.$$

We infer from this the super-Hamiltonian in the three-dimensional minisuperspace:

$$\bar{H} = \pi_R\,\dot{R} + \pi_\phi\,\dot\phi + \pi_N\,\dot{N} - \tilde{\mathcal{L}}(R,\,\phi)$$

$$= -\frac{N}{R}\left(\frac{\pi_R^2}{3\pi\,M_p^2} + \frac{3\pi\,M_p^2}{4}\,R^2\right)$$

$$+ \frac{N}{R}\left(\frac{\pi_\phi^2}{4\pi^2\,R^2} + 2\pi^2\,R^4\,V(\phi)\right). \tag{20.157}$$

The value of the partial derivative of $\bar{H}$ with respect to N can be immediately determined, and following (20.124) and (20.128) is equal to zero:

$$0 = \frac{\partial \bar{H}}{\partial N} = \frac{\bar{H}}{N} = -\frac{1}{R}\left(\frac{\pi_R^2}{3\pi M_p^2} + \frac{3\pi M_p^2}{4}R^2\right) + \frac{1}{R}\left(\frac{\pi_\phi^2}{4\pi^2 R^2} + 2\pi^2 R^4 V\right).$$

(20.158)

This corresponds to the classical constraint mentioned in Sect. 6.2, giving $\bar{H} = 0$, that is a zero total energy in the mini-superspace for a closed Friedmann Universe.

The quantization of the foregoing equation yields the Universe's wave function in a form analogous to the Schrödinger wave function:

$$\boxed{i\partial\psi(R,\phi)/\partial t = \bar{H}\,\psi\,(R,\phi) = 0}.$$

(20.159)

By using the canonical conjugation rules, that is, the correspondence principle associated with the "variables" ϕ and R of the mini-superspace:

$$\pi_\phi \rightarrow -i\,\partial_\phi,$$
$$\pi_R \rightarrow -i\,\partial_R,$$

(20.160)

Eq. (20.158) gives the equivalent of the Schrödinger equation in the two-dimensional mini-superspace. This is the Wheeler DeWitt equation for determining the Universe's wave function in an FRW metric:

$$\left(-\frac{1}{3\pi M_p^2}\frac{\partial^2}{\partial R^2} + \frac{3\pi M_p^2}{4}R^2 + \frac{1}{4\pi^2 R^2}\frac{\partial^2}{\partial \phi^2} - 2\pi^2 R^4 V(\phi)\right)\psi(R,\phi) = 0.$$

(20.161)

Because of the commutation relations between R and π_R, there might be some form of ambiguity in writing the differential operator in the WDW equation. Some authors introduce a term $\frac{1}{R^p}\frac{\partial}{\partial R}R^p\frac{\partial}{\partial R}$ in place of $\frac{\partial^2}{\partial R^2}$ where p is a freely chosen parameter. Eq. (20.161) was written for $p = 0$, whereas Eq. (20.142) used the value $p = \frac{1}{4}$; other authors use the values $p = \pm 1$ and 3.

6.8 Semi-classical Solution

The WDW Eq. (20.161) giving the wave function of the Universe has different possible solutions and Hartle and Hawking (1983) have sought semi-classical solutions of the type

$$\psi(R,\varphi) \simeq N\,\exp(S_E(R,\varphi)),$$

(20.162)

where N is a normalization constant and $S_E(R, \varphi)$ the Euclidean action giving the equations of motion $R(\varphi(\tau), \tau)$ and $\varphi(\tau)$ for the boundary conditions:

$$R(\varphi(0), 0) = R(\varphi),$$
$$\varphi(0) = \varphi, \tag{20.163}$$

in a space with Euclidean metric.

For a slowly varying scalar field φ, the Euclidean solution $R(\varphi, \tau)$ of the Einstein equations is of the form (Linde, (1990)

$$
\begin{aligned}
R(\varphi, \tau) &= \frac{1}{H(\varphi)} \cos(H(\varphi)\tau) \\
&= R(\varphi) \cos(H(\varphi)\tau).
\end{aligned} \tag{20.164}
$$

The function $H(\varphi)$ and the Euclidean action $S_E(R, \varphi)$ are linked to the scalar potential $V(\varphi)$ by the relations

$$H(\varphi) = \sqrt{\frac{8\pi \, V(\varphi)}{3M_p^2}} \quad \text{and} \quad S_E(R, \varphi) = -\frac{3M_p^4}{16V(\varphi)}, \tag{20.165}$$

giving the semi-classical wave function of the Universe

$$
\begin{aligned}
\psi(R(\varphi), \varphi) &\simeq N \, \exp\left(\frac{3M_p^4}{16V(\varphi)}\right) = N \, \exp\left(\frac{\pi \, M_p^2}{2 \, H^2(\varphi)}\right) \\
&= N \, \exp\left(\frac{\pi \, M_p^2 \, R^2(\varphi)}{2}\right)
\end{aligned} \tag{20.166}
$$

The existence probability of a closed Universe in a state φ of the scalar field and of the coordinate $R(\varphi) = 1/H(\varphi)$ is therefore

$$P \sim \mid\psi\mid^2 \sim N^2 \, \exp\frac{3M_p^4}{8V(\varphi)} = N^2 \, \exp\left[\pi \, M_p^2 \, R^2(\varphi)\right]. \tag{20.167}$$

In the ground state $\varphi = \varphi_0$ and $0 < V(\varphi_0) \ll M_p^4$, the normalization function N^2 ensuring that the total probability is equal to 1 takes the value

$$N \simeq \exp\left(-\pi \, M_p^2 \, R_0^2\right) = \exp\left(-\frac{3 \, M_p^4}{8V(\varphi_0)}\right). \tag{20.168}$$

By inserting this result in (20.166), we obtain

$$P \simeq \exp\left\{\frac{3 \, M_p^4}{8}\left[\frac{1}{V(\varphi)} - \frac{1}{V(\varphi_0)}\right]\right\} \tag{20.169}$$

If the only role of φ is to produce a non-zero vacuum energy $V(\varphi)$ (cosmological term), then we can neglect the contributions made to the propagator by negative energy excitations by choosing

$$\psi(R, \varphi) \sim N \exp(S_E(R, \varphi)) \sim N \exp\left[-\frac{3\,M_p^4}{16V(\varphi)}\right]. \qquad (20.170)$$

This leads to the probability that the Universe exists in the state φ:

$$P \simeq |\psi|^2 = N^2 \exp\left[-\frac{3\,M_p^4}{8V(\varphi)}\right]. \qquad (20.171)$$

Attention is drawn to the difference in sign in the exponential between (20.166) and (20.170). The probability of detecting the Universe with a large potential $V(\varphi)$ is small.

7. Quantization in the Mini-Superspaces

J.J. Halliwel gave an excellent introduction to the theory of mini-superspaces within the framework of quantum cosmology at the Winter School of Theoretical Physics of Jerusalem in January 1990 (Halliwell, 1990). We shall be content here with recalling the major points, referring the interested reader to the proceedings of the school for a more complete description and a more detailed bibliography of quantum cosmology.

7.1 Hamiltonian Formulation

Consider the folding of a homogeneous and isotropic space described by a four-metric (3,1) with a homogeneous lapse $N = N(t)$:

$$ds^2 = -N^2(t)\,dt^2 + h_{ij}(x,t)\,dx^i\,dx^j. \qquad (20.172)$$

The Einstein–Hilbert action (20.64) is then written

$$S(h_{ij},\ N) = \frac{M_p^2}{16\pi} \int dt\,d^3x\,N\,\sqrt{h}\,(\mathrm{Tr}\ K^2 - (\mathrm{Tr}\ K)^2 - {}^{(3)}R_i^i + 2\Lambda). \qquad (20.173)$$

Following the introduction of the foregoing metric, it will take the form

$$S(q^\alpha(t), N(t)) = \int_0^1 dt\,N\left[\frac{1}{2N^2}f_{\alpha\beta}(q)\dot{q}^\alpha\dot{q}^\beta - U(q)\right] \equiv \int L\,dt. \qquad (20.174)$$

The $f_{\alpha\beta}(q)$ are covariant elements of the metric of the mini-superspace with a (+2) signature while the limits of integration are determined by an adequate change of variable and scale.

The Lagrangian obtained from (20.174), which is therefore written

$$L = \frac{1}{2N} f_{\alpha\beta} \frac{dq^\alpha}{dt} \frac{dq^\beta}{dt} - N\, U(q),\qquad (20.175)$$

resembles that of a particle in motion in a curved space in the presence of a potential. The variation with respect to the variables q^α hence leads to the equation of the geodesics in the space with metric $f_{\alpha\beta}$, and Christoffel symbols $\Gamma^\alpha_{\beta\gamma}$, that is,

$$\Gamma^\alpha_{\beta\gamma} = \frac{1}{2} f^{\alpha\lambda}(\partial_\beta f_{\gamma\lambda} + \partial_\gamma f_{\beta\lambda} - \partial_\lambda f_{\beta\gamma}).\qquad (20.176)$$

We obtain from (20.175) the equations of the geodesics in the mini-super-space:

$$\frac{\delta L}{\delta q^\alpha} = \frac{1}{N} \frac{d}{dt}\left(\frac{\dot q^\alpha}{N}\right) + \frac{1}{N^2}\Gamma^\alpha_{\beta\gamma}\, \dot q_\beta\, \dot q_\gamma + f^{\alpha\beta}\frac{\partial U}{\partial q^\beta} = 0.\qquad (20.177)$$

The variation with respect to N leads to the constraint

$$\frac{\delta L}{\delta N} = \frac{1}{2N^2} f_{\alpha\beta}(q)\dot q^\alpha\, \dot q^\beta + U(q) = 0.\qquad (20.178)$$

These equations ought to be equivalent to the Einstein equations

$$\mathcal{G}_{\mu\nu} - \Lambda\, g_{\mu\nu} = -\frac{8\pi}{M_p^2}T_{\mu\nu},\qquad (20.179)$$

the first for the (00) elements, and the second for the (ij) elements.
We next determine the conjugate canonical momenta with the definition

$$p_\alpha = \frac{\partial L}{\partial \dot q^\alpha} = f_{\alpha\beta}\frac{\dot q^\beta}{N}\qquad (20.180)$$

and the corresponding canonical Hamiltonian

$$H_c = p_\alpha\, \dot q^\alpha - L = N\left[\frac{1}{2}f^{\alpha\beta}\, p_\alpha\, p_\beta + U(q)\right] \equiv N\, H.\qquad (20.181)$$

The Hamiltonian form of the action,

$$S = \int_0^1 L\, dt = \int_0^1 (p_\alpha \dot q^\alpha - N\, H)\, dt,\qquad (20.182)$$

shows that the N lapse function is a Lagrange multiplier forcing the constraint (20.179), hence setting

$$H = \frac{1}{2}f^{\alpha\beta}\, p_\alpha p_\beta + U = 0.\qquad (20.183)$$

7.2 Canonical Quantization and the Wheeler–DeWitt Equation

The quantization is obtained with the time-independent WDW equation

$$H(q^\alpha, -i\partial_\alpha)\psi(q^\alpha) = 0,\qquad(20.184)$$

which takes the simple form

$$\left[-\frac{1}{2}\nabla^2 + U(q)\right]\psi(q) = 0.\qquad(20.185)$$

The curvature terms of the mini-superspace have been included in the potential $U(q)$.

Now, let us separate the coordinates q^α of the mini-superspace into a time-like coordinate q^0 with the remaining coordinates being space-like ones. The WDW equation in the mini-superspace then becomes

$$\left[\frac{\partial^2}{\partial\,q^{02}} - \frac{\partial^2}{\partial\,\mathbf{q}^2} + U(q^0,\mathbf{q})\right]\psi(q) = 0.\qquad(20.186)$$

7.3 Semi-classical Solution

The WDW Eq. (20.185) can be written, by introducing the Planck mass M_p, as

$$\left[-\frac{1}{2M_p^2}\nabla^2 + M_p^2\,U(q)\right]\psi(q) = 0.\qquad(20.187)$$

Let us determine the BKW solutions of the form

$$\psi(q) = C(q)\,\exp(-M_p^2 I(q)),\qquad(20.188)$$

in which $I(q)$ and $C(q)$ are complex functions. Inserting (20.187) into (20.186) and equating terms of the same power of M_p, we obtain two equations fixing the amplitude and phase functions, $C(q)$ and $I(q)$ respectively,

$$-\frac{1}{2}(\vec{\nabla}\,I)^2 + U(q) = 0,$$
$$2\vec{\nabla}\,I\cdot\vec{\nabla}C + C\,\nabla^2\,I = 0.\qquad(20.189)$$

Let us separate the real part $I_R(q)$ of the function $I(q)$ from the imaginary part $S(q)$ by setting:

$$I(q) = I_R(q) - iS(q).\qquad(20.190)$$

Equation (20.188) then leads to two coupled equations

$$-\frac{1}{2}(\vec{\nabla}I_R)^2 + \frac{1}{2}(\vec{\nabla}S)^2 + U(q) = 0,$$
$$\vec{\nabla}I_R\cdot\vec{\nabla}S = 0.\qquad(20.191)$$

When the imaginary part of $I(q)$ varies much more rapidly than the real part, that is, when

$$|\vec{\nabla}S(q)| \gg |\vec{\nabla}I_R(q)|,\qquad(20.192)$$

we can neglect the first term of the BKW Eq. (20.190) to obtain classical trajectories obeying a Hamilton–Jacobi equation

$$\frac{1}{2}(\vec{\nabla}S)^2 + U(q) = 0.\qquad(20.193)$$

The wave function $\psi(q)$ is of the oscillating type:

$$\psi(q) = C(q) \; \exp\left(-M_p^2 \; I_R(q)\right) \; \exp\left(iM_p^2 \; S(q)\right).\qquad(20.194)$$

By writing the Hamilton–Jacobi equation corresponding to (20.186) with the Planck mass,

$$\frac{1}{2M_p^4}p_\alpha^2 + U(q) = 0,\qquad(20.195)$$

we obtain, by comparing with (20.193), the conjugate momenta

$$p_\alpha = M_p^2 \frac{\partial S}{\partial q^\alpha},\qquad(20.196)$$

whereas relation (20.180) gives the value $p_\alpha = \dot{q}_\alpha$ for a gauge $N = 1$. There is therefore a strong correlation between the coordinates and the momenta. One can therefore show (Halliwell, 1990) that the wave function is strongly peaked around a subset n of the $(2n-1)$ parameters upon which the general solution depends. As we have already shown with the example of Sect. 6.1, the boundary conditions for the wave function of the Universe imply special initial conditions for the classical solutions of the WDW equations.

Condition (20.189) can also be written

$$\vec{\nabla} \cdot (|C|^2 \; \vec{\nabla}S) = 0.\qquad(20.197)$$

We can introduce the term $\exp(-2M_p^2 \; I_R)$ since, following (20.191) $\vec{\nabla}I_R \cdot \vec{\nabla}S = 0$ and thus we obtain the expression

$$\vec{\nabla} \cdot (\exp(-2M_p^2 \; I_R)|C|^2\vec{\nabla}S) = 0.\qquad(20.198)$$

This is the probability current conservation law

$$\vec{\nabla} \cdot \vec{J} = 0 \quad \text{with} \quad \vec{J} = \exp\left(-2M_P^2 \; I_R\right) |C|^2 \; \vec{\nabla}S,\qquad(20.199)$$

deriving from the usual definition $\vec{J} = \frac{i}{2}(\psi^*\vec{\nabla}\psi - \psi\vec{\nabla}\psi^*)$ applied to form (20.188) of the solution of the WDW equation for the Universis wave function.

The probability that the classical trajectories corresponding to the WKB wave functions are centered around a given configuration is described by $|\psi|^2$, that is, by the coefficient of $\vec{\nabla}S$ of the probability current J. The classical solutions consist in a convergence of classical trajectories of the mini-superspace with $\vec{\nabla}S$ as a tangent vector.

In conclusion then, in some areas of the mini-superspace and for adequate boundary conditions, the WDW equations have BKW solutions of the form (20.188), satisfying inequality (20.192). These solutions correspond to the classical space-time by the sheer fact that they are centered on the classical solutions given by the first integral (20.196). The problem lies in the choice of adequate inital conditions, or, as already noted in the introduction, adequate boundary conditions to define the present state the Universe.

Quantum cosmological models should interpret the probability of the Universe being created from an effective potential $V(R, \varphi)$ (determining the wave function of the Universe), which is zero at the initial instant, and the existence of a "classical" Universe.

Exercises

Chapter 1

1.1 The magnetic field $\vec{B} = -g\frac{\vec{r}}{r^3} = -g\frac{\vec{e}_r}{r^2}$ of a magnetic monopole with charge g gives a motion to a spinless massive particle with electric charge e. Find an expression for the kinematic momentum π and for the corresponding Poisson bracket $[\pi_i, \pi_j]$.

1.2 Write down the classical laws of motion of the particle of Ex. 1.1 with kinematical orbital momentum $\vec{L} = \vec{r} \wedge \vec{\pi}$ and the Lorentz force applied to it with the $\vec{\ell} = \vec{r} \wedge \vec{p}$ orbital momentum.

1.3 Show that the kinetic energy and the magnitude of the orbital momentum are constant in time while the orbital momentum $\vec{\ell}$ is not constant.

1.4 Show that if one introduces a classical intrinsic spin $\vec{s} = eg\vec{e}_r$ the total angular momentum $\vec{j} = \vec{\ell} + \vec{s}$ is constant and $\vec{\ell}$ and $\vec{s}$ are orthogonal.

1.5 Find an expression for the radial component $\vec{J} \cdot \vec{e}_r$ of the total kinetic momentum of the particle.

Answers and Hints

1.1 $\quad [\pi_i, \pi_j] = \dfrac{ge}{r^3}\, \varepsilon_{ijk}\, r_k$

1.2 $\quad \vec{\pi} = m\,\vec{r} \quad \dot{\vec{\pi}} = \dfrac{ge}{r^3}\,(\vec{r} \wedge \vec{\pi}) = \dfrac{ge}{r^3}\,\vec{L} \quad \vec{F} = \dfrac{ge}{mr^3}\,\vec{\ell}$

1.3 $\quad \dot{T} = \vec{v} \cdot \vec{F} = 0 \quad \dot{\vec{\ell}^2} = 0 \quad \dot{\vec{\ell}} = -ge\,\vec{e}_r$

1.4 $\quad \dot{\vec{J}} = 0 \quad \vec{\ell} \cdot \vec{s} = (\vec{r} \wedge \vec{p}) \cdot \vec{s} = 0$ since $\dot{\vec{r}} = \dot{r}\,\vec{e}_r + \dot{\vec{e}}_r$

1.5 $\quad \vec{J} \cdot \vec{\ell}_r = s$

Chapter 2

2.1 The wave function $\psi(\vec{x}, t)$ which solves the Schrödinger equation for a particle with mass m in a potential V is expressed with a real phase function $S(\vec{x}, t)$ and a real amplitude function $R(\vec{x}, t)$ in the form $\psi(\vec{x}, t) = R(\vec{x}, t) \exp\left[\frac{i}{\hbar} S(\vec{x}, t)\right]$. Express R and S in terms of ψ and ψ^*.

2.2 Express the wave function $\psi(\vec{x}, t)$ in terms of the probability density $\varrho(\vec{x}, t)$ and the phase function $S(\vec{x}, t)$.

2.3 Determine the coupled equations satisfied by $R(\vec{x}, t)$ and $S(\vec{x}, t)$ for a spinless massive particle in an external potential $V(\vec{x}, t)$.

2.4 Show that a scale tranformation leaves the quantum potential Q [see eq. (2.15)] invariant.

2.5 Write down the fundamental equation obeyed by a particle with mass m when a classical velocity $\vec{v} = \frac{1}{m} \vec{\nabla} S$ is introduced.

Answers and Hints

2.1 $R = (\psi^*\psi)^{1/2} \quad S = \dfrac{\hbar}{2i} \, \log(\psi/\psi^*)$

2.2 $\psi = \sqrt{\varrho} \, \exp\left(\dfrac{i}{\hbar} S\right)$

2.3 $\partial_t R + \vec{\nabla} \cdot (R^2 \vec{\nabla} S/m) = 0$ and $\partial_t S + 1/2m(\vec{\nabla} S)^2 + Q + V = 0$

2.4 $R \to aR \quad I \to a^2 I \quad Q \to Q$

2.5 $\partial_t \vec{v} + \vec{v} \cdot \vec{\nabla}\vec{v} = -1/m \, \vec{\nabla}(V + Q)$ since $\vec{\nabla}(\vec{\nabla} S)^2 = 2\vec{\nabla} S \cdot \vec{\nabla}(\vec{\nabla} S)$

Chapter 5

The Hilbert space spanned by the eigenvectors ψ^α of the Hamiltonian H is truncated into a smaller space and P is the projector onto this space, Q the projector onto the complementary space, while A_{21} denotes the projection QAP or $A_{12} = PAQ$.

5.1 Show that the $P\psi^\alpha$ quantum states in the truncated space are eigenvalues of the effective Hamiltonian $H_{\text{eff}} = H_{11}^0 + V_{11} + V_{12} \dfrac{1}{E_\alpha - H_{22}^0 - V_{22}} V_{21}$ if we set $H = H^0 + V$.

5.2 Consider the Hamiltonian H represented by the 2×2 matrix

$$H = \begin{pmatrix} 0 & 5 \\ 5 & -24 \end{pmatrix}$$

in energy units. Determine its eigenvalues and eigenfunctions.

5.3 The above Hamiltonian is separated in two parts as follows:

$$H = H^0 + V = \begin{pmatrix} -25 & 0 \\ 0 & 0 \end{pmatrix} + \begin{pmatrix} 25 & 5 \\ 5 & -24 \end{pmatrix} \text{ or } H = \begin{pmatrix} 0 & 0 \\ 0 & -24 \end{pmatrix} + \begin{pmatrix} 0 & 5 \\ 5 & 0 \end{pmatrix}.$$

Determine in each case the projector onto the truncated space and the effective Hamiltonian.

5.4 Show that $H_{\text{eff}} = E_\alpha$ defines the actual eigenvalues of the Hamiltonian.

Answers and Hints

5.1 $H\psi^\alpha = E_\alpha \psi^\alpha$ is multiplied by $\mathbb{1} = (P + Q)$ from the right-hand side and successively by P and Q from the left-hand side

5.2 $E_1 = -25$ $\psi^1 = 1/\sqrt{26} \begin{pmatrix} 1 \\ -5 \end{pmatrix}$

$\quad\ E_2 = \quad 1$ $\psi^2 = 1/\sqrt{26} \begin{pmatrix} 5 \\ 1 \end{pmatrix}$

5.3 $H_{\text{eff}} = \dfrac{25}{E + 24}$ in both cases

5.4 $H_{\text{eff}} = E$ gives the eigenvalues of H in both cases

Chapter 6

6.1 A potential $V(x)$ acts on a spinless particle with mass m. Find expressions for the quantum state $|\alpha t\rangle$ with the initial state $|\alpha t_0\rangle$ and the quantum state $|\tilde{\alpha} t\rangle$ when a potential V_0 is added to V. What is the wave function $\tilde{\psi}_\alpha(x, t)$ if the potential V_0 is time dependent?

6.2 A charged particle beam is split into two different paths perturbed by the potentials V_1 and V_2 and recombined in a detector D. We write down the wave functions $\psi_1(x, t)$ and $\psi_2(x, t)$ and the detection probability $|\psi(x, t)|^2$. What is the phase shift ΔS between the two beams?

6.3 The above phase shift if produced by an electromagnetic field characterized by a vector potential $A(x^\mu)$, produced by an electric current in a solenoid. Express the phase shift ΔS in terms of the magnetic flux through the solenoid.

6.4 The Aharonov–Bohm experiment consists in perturbing the paths of a charged electron beam in a two-slit experiment, by an electric current through a solenoid. What magnetic flux is necessary in order to cancel the shift of the interference pattern when the electric current is on?

Answers and Hints

6.1 $\quad |\alpha t\rangle = \exp\left(-\frac{i}{\hbar}(t - t_0)\left(\frac{p^2}{2m} + V\right)\right) |\alpha t_0\rangle$

$\quad |\tilde{\alpha} t\rangle = \exp\left(\frac{-i}{\hbar}(t - t_0) V_0\right) |\alpha t\rangle$

$\quad \tilde{\psi}_\alpha(x, t) = \exp\left(\frac{-i}{\hbar} \int_{t_0}^t V(t') dt'\right) \psi_\alpha(x, t)$

6.2 $\quad |\psi(x, t)|^2 = |\psi_1|^2 + |\psi_2|^2 + 2\,\mathrm{Re}(\psi_1 \psi_2^*) \exp\left(\frac{i}{\hbar}\Delta S\right)$

$\quad \frac{\Delta S}{\hbar} = \frac{1}{\hbar} \int_{t_i}^{t_f} (V_2(t') - V_1(t')) dt'$

6.3 $\quad \frac{\Delta S}{\hbar} = \frac{e}{\hbar} \oint A_\mu\, dx^\mu = -\frac{e}{\hbar c} \int \vec{H} \cdot \vec{ds} = -\frac{e}{\hbar c} \phi$

6.4 $\quad \phi = \frac{2\pi\hbar c}{e} n \qquad n$ integer

Chapter 7

7.1 Consider a one-dimensional wave function $\psi(x, 0)$ at initial time $t = 0$. What is the probability that this wave function peakes at a value $x = a$. What does this imply about $\dot{x}$?

7.2 Let the above wave function be a Gaussian function peaked at $x = a$ with a dispersion depending on time $\sigma^2(t)$, at instant $t = 0$:

$$\psi(x, 0) = (2\pi\,\sigma^2(0))^{-1/4} \exp\left[-\frac{(x - a)^2}{4\sigma^2(0)}\right].$$

What is the probability density $|\psi(x, 0)|^2$ in the limiting case with zero dispersion, $\sigma(0) = 0$?

7.3 Express the above wave function at instant t using expression (7.18) with the propagator of a free massive particle.

7.4   Determine the probability density at time t at the point x.

7.5   What conclusion can we draw for the evolution in time of the dispersion function $\sigma^2(t)$?

Answers and Hints

7.1 $|\psi(x,0)|^2 = (2\pi)^{-1}\, \delta(x-a)$ $\dot{x}$ is undetermined at $t=0$

7.2 $\sigma(0) \to 0 \Rightarrow |\psi(x,0)|^2 \to (2\pi)^{-1}\, \delta(x-a)$

7.3 $\psi(x,t) = \int G(x,t;x',0)\, \psi(x',0)dx'$

with $G(x,t;x',0) = \left(\dfrac{m}{2\pi i\hbar t}\right)^{1/2} \exp\left(\dfrac{im}{2\hbar}\dfrac{(x-x')^2}{t}\right)$

$$\psi(x,t) = \left(\frac{m}{2\pi i\hbar t}\right)^{1/2} (2\pi\, \sigma^2(0))^{-1/4} \left(\frac{4\pi\hbar\sigma^2(0)it}{\hbar it + 2m\sigma^2(0)}\right)^{1/2}$$
$$\times \exp\left(-\frac{(x-a)^2}{4\sigma^2(0) + 2\pi i\hbar t/m}\right)$$

7.4 $|\psi(x,t)|^2 = (2\pi\sigma^2(t))^{-1/2} \exp\left(-\dfrac{(x-a)^2}{2\sigma^2(t)}\right)$

7.5 $\sigma^2(t) = \sigma^2(0)\,[1 + \hbar^2 t^2/4m^2\, \sigma^4(0)]$. The Gaussian wave packet speads out in time

Chapter 8

8.1   What is the covariant derivative D_μ associated with the kinematic momentum of a spinless massive particle with electic charge e in an electromagnetic field for a Minkowski space-time with a metric signature $(+2)$ or (-2). Write down the corresponding Hamiltonian.

8.2   Determine the commutators of the operators D_i associated with the above covariant derivative $[D_i, D_j]$ and $[D_i, x_j]$ for a magnetic field produced by a magnetic monopole (cf. Ex. 1.1).

8.3   How may the total kinetic momentum $\vec{J}$ of such a particle be written? Verify the validity of the commutation rules $\vec{J} \wedge \vec{J}$.

Answers and Hints

8.1 (+2) signature $\pi_\mu = -i\hbar\, D_\mu$ and $D_\mu = \partial_\mu - ie/\hbar c\; A_\mu$
   (−2) signature $\pi_\mu = i\hbar\, D_\mu$ and $D_\mu = \partial_\mu + ie/\hbar c\; A_\mu$
$H = -\hbar^2/2m\; D^2 = -\hbar^2/2m\; D_\mu\, D^\mu$

8.2 $[D_i, D_j] = ieg/\hbar c\; \varepsilon_{ijk}\; r_k/r^3$ and $[D_i, x_j] = \delta_{ij}$

8.3 $\vec{J} = -i\hbar\, \vec{r} \wedge \vec{D} + s\vec{e}_r$ and $\vec{J} \wedge \vec{J} = i\hbar\, \vec{J}$

Chapter 9

9.1 By considering the quantum aspect of the motion of a charged particle in the magnetic field of a magnetic monopole (cf. Ex. 1.1) determine the commutation rules of the operators $\vec{\pi}$ and $\vec{L} = \vec{r} \wedge \vec{\pi}$ associated with the kinematic momentum.
Show that the intrinsic spin $\vec{s} = eg\vec{e}_r$ added to the orbital momentum $\vec{L}$ leads to a total angular momentum $\vec{J}$ satisfying the usual commutation rules $\vec{J} \wedge \vec{J} = i\hbar\, \vec{J}$.
Express the operator $\vec{\pi}$ with a transverse part π_t and a longitudinal part π_r and the operator π_t^2 in terms of J^2 and s^2. What then is the expression for the Hamiltonian H of the particle ?

9.2 Determine the commutator and anticommutator of the Pauli matrices $\vec{\sigma}$ and the product $\sigma_i \sigma_j$ of two matrices. Deduce from the above expression the products of operators $(\vec{\sigma} \cdot \vec{A})(\vec{\sigma} \cdot \vec{B})$, $(\vec{\sigma} \cdot \vec{p})^2$, $(\vec{\sigma}, \vec{\pi})^2$.

9.3 Determine the commutator of the scalar operators $[\vec{\sigma} \cdot \vec{A}, \vec{\sigma} \cdot \vec{B}]$

Answers and Hints

9.1 $\vec{\pi} \wedge \vec{\pi} = -\dfrac{i\hbar}{r^2}\, \vec{s}$ and $\vec{L} \wedge \vec{L} = i\hbar\, (\vec{L} - \vec{s})$
   $\vec{J} \wedge \vec{J} = i\hbar\, \vec{J}$
   $\pi_t^2 = (J^2 - s^2)/r^2$ and $H = \dfrac{1}{2m}\, (\pi_t^2 + \pi_r^2)$
   $H = -\dfrac{\hbar^2}{2m}\, \left(\dfrac{d^2}{dr^2} + \dfrac{2}{r}\dfrac{d}{dr}\right) + (J^2 - s^2)/2mr^2$

9.2 $[\sigma_i, \sigma_j] = 2i\, \varepsilon_{ijk}\, \sigma_k$ and $\{\sigma_i, \sigma_j\} = 2\delta_{ij}\, \mathbb{1}$
   $\sigma_i \sigma_j = \delta_{ij}\, \mathbb{1} + i\, \varepsilon_{ijk}\, \sigma_k$
   $(\vec{\sigma} \cdot \vec{A})(\vec{\sigma} \cdot \vec{B}) = \vec{A} \cdot \vec{B} + i\, \vec{\sigma} \cdot (\vec{A} \wedge \vec{B})$
   $(\vec{\sigma} \cdot \vec{p})^2 = p^2$ and $(\vec{\sigma} \cdot \vec{\pi})^2 = \pi^2 - \dfrac{e}{\hbar}\, \vec{\sigma} \cdot \vec{B}$

9.3 $[\vec{\sigma} \cdot \vec{A}, \vec{\sigma} \cdot \vec{B}] = 2(\vec{A} \wedge \vec{B}) \cdot \vec{\sigma}$

Chapter 10

10.1 Determine with the GSA rules the particular $9j$ coefficient in which one momentum (say j_1) takes a zero value.

10.2 Give the graphical representation and the corresponding analytical expression for the quantum states $|jm\rangle_A$ and $|jm\rangle_B$ corresponding to the A coupling schemes $(\vec{j}_1 + \vec{j}_2) + \vec{j}_3 = \vec{j}$ and to the B coupling scheme $(\vec{j}_1 + \vec{j}_3) + \vec{j}_2 = \vec{j}$.

10.3 Determine the scalar product of these states $_B\langle j'm'|jm\rangle_A$.

10.4 Express the $|jm\rangle_B$ state in terms of the $|jm\rangle\rangle_A$ state.

10.5 Use the pinching rule over three lines in the GSA to simplify the following expression and write it in terms of $3nj$ coefficients:

$$F = \sum_{\text{all } m} \begin{pmatrix} j_1 & m_2 & n_2 \\ m_1 & j_2 & \ell_2 \end{pmatrix} \begin{pmatrix} m_1 & \ell_1 & k_1 \\ j_1 & n_1 & \chi_1 \end{pmatrix} \begin{pmatrix} n_1 & k_2 & j_3 \\ \ell_1 & \chi_2 & m_3 \end{pmatrix}$$
$$\times \begin{pmatrix} m_3 & j_4 & \ell_2 \\ j_3 & m_4 & n_2 \end{pmatrix} \begin{pmatrix} m_4 & \chi_2 & \ell_3 \\ j_4 & k_2 & n_3 \end{pmatrix} \begin{pmatrix} n_3 & j_2 & \chi_2 \\ \ell_3 & m_2 & k_2 \end{pmatrix}.$$

10.6 Express the invariance property of the trace of a two-body operator in a coupled or in an uncoupled basis to get an expression for $\sum_J \langle (j_1 j_2)JM|T|(j_1 j_2)JM\rangle$ as a sum over a single magnetic momentum.

Answers and Hints

10.1 $\begin{Bmatrix} 0 & j_2 & j_3 \\ \ell_1 & \ell_2 & \ell_3 \\ k_1 & k_2 & k_3 \end{Bmatrix} = [j_2^{-1}\ell_1^{-1}]\,\delta_{\ell_1 k_1}\delta_{j_2 j_3}(-)^{k_1+k_2+j_2+\ell_3} \begin{Bmatrix} \ell_3 & k_3 & j_2 \\ k_2 & \ell_2 & k_1 \end{Bmatrix}$

10.2 $|jm\rangle_A = \sum \langle j_1 m_1 j_2 m_2|j_{12} m_{12}\rangle\langle j_{12} m_{12} j_3 m_3|jm\rangle|j_1 m_1 j_2 m_2 j_3 m_3\rangle$
$\qquad = \sum \langle j_1 m_1 j_2 m_2 j_3 m_3|(j_1 j_2 j_3)j_{12}; jm\rangle\,|j_1 m_1 j_2 m_2 j_3 m_3\rangle$
$\qquad = \sum [j_{12} j]\,(-)^{2j_2} \begin{pmatrix} m_1 & m_2 & m_3 \\ j_1 & j_2 & j_3 \end{pmatrix} \begin{pmatrix} jm \end{pmatrix}\,|j_1 m_1 j_2 m_2 j_3 m_3\rangle$

10.3 $_B\langle j'm'|jm\rangle_A = [j_{12} j_{13}]\,(-)^{j_2+j_3+j_{12}+j_{13}} \begin{Bmatrix} j_1 & j_{13} & j_3 \\ j & j_{12} & j_2 \end{Bmatrix} \delta_{jj'}\,\delta_{mm'}$

10.4 $|jm\rangle_B = \sum {}_A\langle jm|jm\rangle_B\,|jm\rangle_A$

10.5 $F = \begin{Bmatrix} j_1 & j_2 & \ell_2 \\ \ell_3 & \ell_1 & k_1 \end{Bmatrix} \begin{Bmatrix} \ell_1 & \ell_3 & \ell_2 \\ j_4 & j_3 & k_2 \end{Bmatrix} (-)^{2j_2+2k_2}$

10.6 $\sum_J \langle (j_1 j_2) JM|T|(j_1 j_2) JM \rangle$

$\qquad = \sum_{m_1} \langle j_1 m_1 j_2 M - m_1 |T| j_1 m_1 j_2 M - m_1 \rangle$

Chapter 11

11.1   Use the GSA rules to define the graphical representation of the group character $\chi_j(g) = \sum_m D^i_{mm}(g)$. Express the closure relation on rotation matrices $\sum_{mm'} D^j_{mm'}(g) D^{j\,*}_{mm'}(h)$ in terms of the character $\chi_j(gh^{-1})$.

11.2   Use the GSA summation rule over an angular momentum to express the sum over j of the above closure relation.

11.3   Use the particular value of the rotation matrix $D^0_{00}(g)$ to determine the integral $I_0 = \int D^j_{mm'}(g)\, dg$.

11.4   Use the contraction rule (11.66) over two rotation matrices to obtain the contraction rule (or its graphical representation) over n rotation matrices.

11.5   Use the two above results to define the integral In over n rotation matrices.

Answers and Hints

11.1 $\chi_j (gh^{-1})$

11.2 $\sum_j [j^2]\, \chi_j (gh^{-1}) = 8\pi^2\, \delta(g - h)$

11.3 $D^0_{00}(g) = \mathbb{1}$ leads to $I_0 = 8\pi^2\, \delta_{j0}\, \delta_{m0}\, \delta_{m'0}$

11.4 $D^{j_1}_{m_1 m_1'}(g) \ldots D^{j_n}_{m_n m_n'}(g)$

$\qquad = \sum \langle j_1 m_1 \ldots j_m m_n|(j_1 \ldots j_n)_A\, X_s; j_{n+1}\, m_{n+1}\rangle$

$\qquad\qquad \times D^{j_{n+1}}_{m_{n+1} m_{n+1}'}(g) \langle (j_1 \ldots j_n)_A\, X_s; j_{n+1} m_{n+1}'|j_1 m_1' \ldots j_n m_n'\rangle$

11.5 $In = \displaystyle\int D^{j_1}_{m_1 m_1'}(g) \ldots D^{j_n}_{m_n m_n'}(g)\, dg$

$\qquad = 8\pi^2 \displaystyle\sum [X_s^2] \begin{pmatrix} m_1 & \cdots & m_n \\ j_1 & \cdots & j_n \end{pmatrix}_{\!\!A}\!\! |X_s\rangle \begin{pmatrix} m_1' & \cdots & m_n' \\ j_1 & \cdots & j_n \end{pmatrix}_{\!\!A}\!\! |X_s\rangle$

Chapter 13

13.1   Use the particular value of $Y_{10}(\theta, \varphi)$ and the contraction rule (12.53) to determine the product $\cos\theta\, Y_{\ell 0}(\theta, \varphi) = F_L(\theta, \varphi)$. Determine its Reduced Matrix Element (RME).

13.2 Determine the RME $\langle(\ell'1/2)j'||Y_L||(\ell1/2)j\rangle$ with a marked $6j$ coefficient. Express it in terms of a particular $3jm$ coefficient.

13.3 Determine the RME of the tensor force S_{12} expressed in terms of $Y_{2q}(\Omega)$.

13.4 Determine the scalar and vector parts of the irreducible tensor operator

$$\pi_{kq}\,(\vec{A},\vec{B},\vec{C}) = ((A_{1m} \otimes B_{1n}) \otimes C_{1s})_{kq} \ .$$

13.5 Compute the RME $\langle(\ell1/2)j||\vec{\sigma}\cdot\vec{r}||(\ell'1/2)j'\rangle$ given in (18.167).

13.6 What is the tensor product of two spherical harmonics acting in the same space ?

Answers and Hints

13.1 $F_L(\theta,\varphi) = \cos\theta\, Y_{\ell0}(0\varphi) = [\ell L] \begin{pmatrix} 1 & \ell & L \\ 0 & 0 & 0 \end{pmatrix}^2 Y_{L0}(\theta\,\varphi)$

$\langle\ell'||F_L||\ell\rangle = \dfrac{(-)^\ell}{\sqrt{4\pi}}\,[\ell'\ell^2 L^2] \begin{pmatrix} 1 & \ell & L \\ 0 & 0 & 0 \end{pmatrix}^2 \begin{pmatrix} \ell' & L & \ell \\ 0 & 0 & 0 \end{pmatrix}$

13.2 $\langle(\ell'1/2)j||Y_L||(\ell1/2)j\rangle$

$\qquad = \dfrac{1}{\sqrt{4\pi}}\,[\ell'\ell jj'L]\,(-)^{L+1/2+j'} \begin{pmatrix} \ell & \ell' & L \\ 0 & 0 & 0 \end{pmatrix} \begin{Bmatrix} \ell & \ell' & L \\ j & j' & \frac{1}{2} \end{Bmatrix}$

$\qquad = \dfrac{1}{\sqrt{4\pi}}\,[\ell\ell'Ljj']\dfrac{1}{2}\left(1+(-)^{\ell+\ell'+L}\right)(-)^{L+\frac{1}{2}+j'} \times \begin{pmatrix} j & L & j' \\ \frac{1}{2} & 0 & -\frac{1}{2} \end{pmatrix}$

13.3 $\langle\alpha j||S_{12}||\alpha'j'\rangle$

$\qquad = \delta_{\alpha\alpha'}\,\delta_{jj'}\,\sqrt{\dfrac{10}{3}}\,(-)^{S-J}\,[LL'SS'] \begin{pmatrix} L & 2 & L' \\ 0 & 0 & 0 \end{pmatrix} \begin{Bmatrix} L & L' & 2 \\ S & S' & j \end{Bmatrix}$

$\qquad \times \begin{Bmatrix} S_1 & S_2 & S \\ S_1 & S_2 & S' \\ 1 & 1 & 2 \end{Bmatrix}$

13.4 $\pi_{00} = \sqrt{3}\,(\vec{A}\wedge\vec{B})\cdot\vec{C}$

$\qquad \pi_{1q} = (\vec{A}\cdot\vec{C})\,B_{1q}$ or equivalently $\vec{\pi} = (\vec{A}\cdot\vec{C})\,\vec{B}$

13.5 $\pi_{LM} = \dfrac{[\ell_1\ell_2]}{\sqrt{4\pi}} \begin{pmatrix} \ell_1 & \ell_2 & L \\ 0 & 0 & 0 \end{pmatrix} Y_{LM}\,(\Omega)$

Chapter 14

14.1 Determine the potential $V(t)$ from which the electric field

$$\vec{E} = \begin{cases} 0 & \text{for } t < 0 \\ E_0 \exp(-t/2)\, \vec{e}_z & \text{for } t \geq 0 \end{cases}$$

is derived. What is the corresponding potential energy?

14.2 Determine the transition probability from the $1S$ to the $2S$ state for the above potential.

14.3 Determine the transiton probability from the $1S$ to the $2P$ state when one sets $I_1 = \int_0^\infty R_{21}^*(r)\, r^3\, R_{10}(r)$.

Answers and Hints

14.1 $\vec{E} = -\vec{\nabla} V$ gives $V = -E_0 \exp\left(-\dfrac{t}{\tau}\right) z$

and $E_p = -eV = eE_0 \exp\left(-\dfrac{t}{\tau}\right) z$

14.2 $P_{1S \to 2S} = \dfrac{1}{\hbar^2} \left| \int_0^t dt'\, \exp\left(-\dfrac{it'}{\hbar}(E_{1S} - E_{2S})\right) \right| \langle 2S|E_p|1S\rangle|^2$

$\langle 2S|E_p|1S\rangle \, \alpha \, \langle 2S|z|1S\rangle = \int d^3r\, R_{20}^*\, Y_{00} \sqrt{\dfrac{4\pi}{3}}\, r\, Y_{10}\, R_{10}\, Y_{00} = 0$

14.3 $\langle 2P|E_p|1S\rangle = \dfrac{1}{\sqrt{3}}\, \delta_{m0}\, I_1$

$P_{1S \to 2P} = e^2\, E_0^2\, \dfrac{I_1^2}{3\hbar^2}\, \delta_{m0}\, \dfrac{\tau^2}{1 + \omega^2\, \tau^2}$ with $\hbar\omega = (E_{1S} - E_{2P})$

Chapter 15

15.1 Determine the matrix element of the screened electric potential $V(r) = (z_1 z_2 e^2/4\pi\varepsilon_0)\,(\exp(-\frac{r}{a})/r)$ in the Born approximation.

15.2 Express the squate length of the transferred K^2 momentum in an elastic scattering process for an incident energy E and a scattering angle θ.

15.3 Deduce for the above potential the elastic scattering amplitude $f(\theta)$ in the Born approximation.

15.4 Determine in the above case the differential cross section of an elastic scattering process.

15.5 How is it possible to get the usual Rutherford elastic scattering differential cross-section.

Answers and Hints

15.1 $V_{fi} = \langle f| V |i \rangle = \langle \vec{k}_f|V|\vec{k}_i \rangle$ with $\langle \vec{r}|\vec{k}\rangle = (2\pi)^{-3/2} \exp(i\vec{k}\cdot\vec{r})$
gives when setting
$$\vec{K} = \vec{k}_f - \vec{k}_i \text{ and } c = \frac{z_1 z_2 e^2}{4\pi\varepsilon_0} \frac{1}{(2\pi)^3} \qquad V_{fi} = 4\pi c \frac{1}{K^2 + (\frac{1}{a})^2}$$

15.2 $|\vec{k}_f| = |\vec{k}_i| = k \qquad E = \dfrac{\hbar^2 k^2}{2m} \qquad K^2 = \dfrac{2m}{\hbar^2} 4E \sin^2 \dfrac{\theta}{2}$

15.3 $f(\theta) = \dfrac{1}{4\pi}(2\pi)^3 \dfrac{2m}{\hbar^2} \quad \langle \vec{k}|V|\vec{k}\rangle = (2\pi)^3 \, c \, \dfrac{2m}{\hbar^2} \dfrac{1}{K^2 + (\frac{1}{a})^2}$

15.4 $\left(\dfrac{d\sigma}{d\Omega}\right)^2 = |f(\theta)|^2 = \left[\dfrac{z_1 z_2}{4\pi\varepsilon_0} c^2 \dfrac{1}{\frac{\hbar^2}{2ma^2} + 4E \sin^2 \frac{\theta}{2}}\right]^2$

15.5 If $a \rightarrow \infty$ the screening is removed and $(d\tau/d\Omega)$ becomes the cross section for Rutherford scattering.

Chapter 17

17.1 What is the Lagrangian density of two real free scalar fields ϕ_1 and ϕ_2? Deduce their Euler–Lagrange equation of motion.

17.2 Is the current $J_\mu = \phi_1 \, \partial_\mu \, \phi_2 - \phi_2 \, \partial_\mu \, \phi_1$ conserved ?

17.3 Are the vector current J_μ^ϱ defined by $J_\mu^\varrho = \partial^\varrho \, \phi_1 \, \partial_\mu \, \phi_2 - \phi_2 \, \partial_\mu \, \partial^\varrho \, \phi_1$ and the tensor current defined by $J_\mu^{\varrho\sigma} = \partial^\varrho \, \partial^\sigma \, \phi_1 \, \partial_\mu \, \phi_2 - \phi_2 \, \partial_\mu \, \partial^\varrho \, \partial^\sigma \, \phi_1$ conserved?

17.4 Find expressions for the vectorial and the tensorial charge corresponding to these currents.

Answers and Hints

17.1 $\mathcal{L} = \frac{1}{2} \, \partial^\mu \, \phi_1 \, \partial_\mu \, \phi_1 + \frac{1}{2} \, \partial^\mu \, \phi_2 \, \partial_\mu \, \phi_2$ and then $\partial^\mu \partial_\mu \, \phi_1 = 0 \; \partial^\mu \partial_\mu \, \phi_2 = 0$

17.2 $\partial^\mu \, J_\mu = 0$; the current is conserved

17.3 Since $\partial^\mu \partial_\mu \, \phi_i = 0$; $\partial^\mu \partial_\mu \partial^\alpha \, \phi_i = 0$ and $\partial_\mu \partial^\mu \partial^\alpha \partial^\beta \, \phi_i = 0$ and thus $\partial^\mu \, J_\mu^\varrho = 0$ and $\partial^\mu \, J_\mu^{\varrho\sigma} = 0$; these currents are conserved

17.4 $Q^\varrho = \int d^3x \, J_0^\varrho$ and $Q^{\varrho\sigma} = \int d^3x \, J_0^{\varrho\sigma}$

Chapter 18

18.1 Starting with the spinor ψ and the Dirac γ-matrices, we introduce the scalar $v = \bar{\psi}\psi$ and the pseudo-scalar $w = \bar{\psi}\,\gamma_5\,\psi$. Using of the Pauli–Krofink formula (18.64), prove that the currents $J^\mu = \bar{\psi}\,\gamma^\mu\,\psi$ and $I^\mu = \bar{\psi}\,\gamma^\mu\,\gamma_5\,\psi$ satisfy the relations

$$J^2 = J_\mu\, J^\mu = v^2 - w^2 \quad I^2 = I_\mu\, I^\mu = w^2 - v^2 \quad I\cdot J = I_\mu\, J^\mu = 0\,.$$

18.2 Determine the right-hand part r^μ and left-hand part ℓ^μ of the current J^μ and show that

$$r^\mu\, r_\mu = 0 \quad \ell^\mu\, \ell_\mu = 0 \quad r^\mu\, \ell_\mu = \frac{1}{2}\, J^2\,.$$

18.3 The metric $\gamma_{\mu\nu}$ of a Minkowski space-time (we set $\gamma = |\det\,\gamma_{\mu\nu}|$) associated with the current J^μ determines the effective metric $g_{\mu\nu}$ by the relations

$$\sqrt{g}\, g^{\mu\nu} = \sqrt{\gamma}\, (\gamma^{\mu\nu} + \alpha\, J^\mu\, J^\nu)\,,$$
$$\frac{1}{\sqrt{g}}\, g_{\mu\nu} = \frac{1}{\sqrt{\gamma}}\, (\gamma_{\mu\nu} + \beta\, J_\mu\, J_\nu)\,.$$

What constraints have to be imposed on α and β. Express such a constant in the particular case when $\alpha = \beta$.

Answers and Hints

18.1 Set $M = \mathbb{1}$ and $M = \gamma_5$, then multiply the right-hand side by $\bar{\psi}$ or by $\bar{\psi}\,\gamma_5$

18.2 $\eta^\mu = J^\mu + I^\mu = 2r^\mu \qquad m^\mu = J^\mu - I^\mu = 2\ell^\mu$
$\qquad m^\mu + n^\mu = 2J^\mu \qquad n^\mu\, h_\mu = 0 \ \ m^\mu m_\mu = 0 \qquad m^\mu\, \eta_\mu = 2J^2$

18.3 $g^{\mu\nu'}\, g_{\mu\nu} = \delta^{\nu'}_\nu$ gives $\beta = \dfrac{-\alpha}{1 + \alpha J^2}$ and $\alpha = -\dfrac{2}{J^2}$ for $\alpha = \beta$

Chapter 19

Consider a spinless particle with mass m in a one-dimensional potential $V_1(x)$.

19.1 Determine $V_1(x)$ in terms of the ground-state $\psi_0(x)$ and its second derivative $\psi_0''(x)$ for a ground-state energy E_0 equal to zero.

19.2 One introduces the operators $A^+ = -\frac{\hbar}{\sqrt{2m}}\,\partial_x + W_1(x)$ and $A = \frac{\hbar}{\sqrt{2m}}\,\partial_x + W_1(x)$. What must be the form of the potential $V_1(x)$ for the Hamiltonian $H_1 = -\frac{\hbar^2}{2m}\,\partial_x^2 + V_1(x)$ to be factorized as $H_1 = A^+A$.

19.3 Show that the superpotential $W(x)$ of the potential $V_1(x)$ is compatible with the value $W(x) = -\frac{\hbar}{\sqrt{2m}}\,\psi_0'/\psi_0$.

19.4 Determine the potential $V_2(x)$ such that $H_2 = -\frac{\hbar}{2m}\,\partial_x^2 + V_2(x) = A\,A^+$.

19.5 We denote by $\psi_n^{(1)}$ the eigenvectors of H_1 and by $\psi_n^{(2)}$ the eigenvectors of H_2. Show that $A\,\psi_n^{(1)}$ is eigenvector of H_2 and $A^+\,\psi_n^{(1)}$ an eigenvector of H_1.

19.6 One introduces the matrix

$$H = \begin{pmatrix} H_1 & 0 \\ 0 & H_2 \end{pmatrix}$$

and the operators

$$Q = \begin{pmatrix} 0 & 0 \\ A & 0 \end{pmatrix} \quad \text{and} \quad Q^+ = \begin{pmatrix} 0 & A^+ \\ 0 & 0 \end{pmatrix}.$$

Determine the commutation and anticommutation relations
$[H, Q]\ [H, Q^+]\ \{Q, Q^+\}\ \{Q, Q\}\ \{Q^+, Q^+\}$.

19.7 Compute the ground states of a particle in an infinite square well $V(a) = 0$ for $0 \le x \le L$ and $V(x) = \infty$ for $\infty < x < 0$ or $x > L$.

19.8 Determine the eigenstates of the Hamiltonian $H_1 = H - E_0$ and the superpotential $W(x)$ corresponding to the potential $V_2(x)$.

Answers and Hints

19.1 $V_1 = \dfrac{\hbar^2}{2m}\,\dfrac{\psi''}{\psi_0}$

19.2 $V_1 = W^2 - \dfrac{\hbar}{\sqrt{2m}}\,\partial_x W$

19.4 $V_2(x) = W^2 + \dfrac{\hbar}{\sqrt{2m}}\,\partial_x W$

19.6 $[H, Q] = [H, Q^+] = 0 \quad \{Q, Q^+\} = H \quad \{Q, Q\} = \{Q^+, Q^+\} = 0$

19.7 $\psi_0^{(1)} = \sqrt{\dfrac{2}{L}}\,\sin\dfrac{\pi}{L}x \quad \text{and} \quad E_0 = \dfrac{\hbar^2\,\pi^2}{2m\,\hbar^2}$

19.8 $E_n^{(1)} = \dfrac{n(n+2)}{2m\,L^2}\,\hbar^2\,\pi^2$

$W(x) = -\dfrac{\hbar}{\sqrt{2m}}\,\dfrac{\pi}{L}\,\cot\,\dfrac{\pi x}{L}$

$V_2(x) = \dfrac{\hbar^2\pi^2}{2mL^2}\left(2\,\mathrm{cosec}^2\,\dfrac{\pi x}{L} - 1\right)$

References

Abbot, L. F., Wise, M. B.: Am. J. Phys., 49 (1981) 37.

Aspect, A., Grangier, P.: Le Courier du CNRS (1984).

Badhuri, R. K., Cohler, l. E., Nogami, Y.: Il Nuovo Cimento, 65A (1981) 376.

Baranger, M.: Recent Progress in the Understanding of Finite Nuclei (Varenna Lectures, 1967).

Barsella, B., Fabri, E.: Suppl. al Nuovo Cimento, Vol. 11 (1970) 294.

Bell, J. S.: Physics 1 (1964) 195.

Bordarier, Y.: Contribution l'Emploi de Méthodes Graphiques en Spectroscopie Atomique, Thèse de Doctorat, Orsay, 1970.

Briggs, J. S.: Rev. Mod. Phys., Vol. 43 (2) (1971) 189.

Brink, D. R., Satchler, G. R.: Angular Momentum, Oxford Lib. of the Phys. Sc. (2nd Edition), (Oxford University Press, London, 1967).

Buttle, P. J. A. and Goldfarb, L. J. B.: Nucl. Phys., 78 (1966) 409.

Canning, G. P.: Phys. Rev. D, 8 (4) (1973) 1151.

Cartan, E.: La theorie des groupes et la geometrie, (Gauthier-Villard, Paris, 1952).

Chatterjee, R., Buckmaster, H. A., Shury, Y. H.: Phys. Stat. Sol., 13 (1972) 9.

Clauser, J. F., Horne, M. A., Shimony, A., Holt, R. A.: Phys. Rev. Lett., 23 (1969) 880.

Cohen-Tannoudji, C., Diu, B., Lalö, F.: Mécanique Quantique (Hermann, Paris, 1973).

Condon, E. U., Shortley, G. H.: The Theory of Atomic Spectra (Cambridge University Press, Cambridge, 1935).

Coope, J. A. R, Snider, R. F., Mac Court, F. R.: The Journal of Chemical Phys., 43 (7) (1965) 2269.

Coope, J. A. R., Snider, R. F.: J. of Math. Phys., 11 (3) (1970) 1003; J. of Math. Phys., 11 (5) (1970) 1591.

De Rafael, E.: Lectures on Quantum Electrodynamics, (Universidad de Barcelona, 1976).

De Witt, B. S.: Quant Theory of Gravity, Phys. Rev., 160 (1967).

Edmonds, A. L.: Angular Momentum in Quantum Mechanics, Princeton University Press, Princeton, CA, 1957).

Elbaz, E., Massot, j. N., Lafoucrière, J.: Nucl. Phys., 82 (1966) 189; Nucl. Phys., 83 (1966) 469; Nucl. Phys., 86 (1966) 625; J. de Phys., 27 (1966) 27; Rev. Mod. Phys., 39 (1967) 288.

Elbaz, E., Castel, B., Lafoucrière, J.: Nucl. Phys., B 2 (1967) 51; Nucl. Phys., B 2 (1967) 296.

Elbaz, E., Fayard, C., Lamot, G. H., Lafoucrière, J.: Nuovo Cimento, X 64A (1969) 13.

Elbaz, E., Castel, B.: Graphical Methods of Spin Algebra (Marcel Dekker, New York, 1972).

Elbaz, E., Meyer, J., Nahabetian, R.: Lett. Nuovo Cimento, 10 (1974) 417.

Elbaz, E., Nahabetian, R.: J. Phys. Math. Gen., A 10 (7) (1977) 1063.

Elbaz, E.: Traitement Graphique de l'Algèbre de Racah et de l'Analyse Vectorielle, Institut Physique Nucléaire, Lyon, 1979;

Traitement Graphique de l'Algèbre des Moments Angulaires (Masson, Paris, 1966);
Mécanique Quantique (Ellipse, Paris, 1985);
Algèbre de Racah et Analyse Vectorielle Graphique (Ellipse, Paris, 1985);
J. Math. Phys., 26 (4) (1985) 728;
De l'Electromagnétisme à l'Electrofaible, (Ellipse, Paris, 1989);
Cosmologie (Ellipse, Paris, 1992).

Elbaz, E., Meyer, J., Nahabetian, R.: Phys. Rev. D 23 (1981) 1170.

Elbaz, E., Meyer, J.: Lett. Nuovo Cimento, 3148 (1981) 291.

Everett, H.: Rev. Mod. Phys., 29 (1957) 45.

Fano, Racah, G.: Irreducible Tensorial Sets (Academic Press, New York, 1959).

Fayet, P.: Introduction to supersymmetric theories of particles and interactions. 4th Hellenic School on Elementary Particle Physics (Corfoue, 1993); 2nd Séminaire Rhodanien de Physique (Dolomieu, 1994).

Felden, M.: Le Modèle Géométrique de la Physique (Masson, 1992).

Gieres, F.: Séminaire d'analyses de l'Université Blaise Pascal (1993).

Gödel, K.: Monat. Math. Phys., 38 (1931) 173.

Goodison, J. M., Toms, D. T.: Phys. Rev. Lett., 71 (20) (1993) 3240.

Grosse, H.: Supersymmetric Quantum Mechanics Brasov Int. School (1989).

Gruber, C., Piron, C., Minh Tam, T., Weil, R.: Les Fondements de la Mécanique Quantique (Montana, 1983).

Guichon, P. A.,M.: Thèse de 3ème Cycle, Institut Physique Nucléaire, Université Lyon 1, 1975.

Halliwel, J. J.: Quantum Cosmology and Baby Universes (World Scientific, 1990).

Harper, E. P., Kim, Y. E., Tubis, A.: Phys. Rev. C, 2 (3) (1970) 877.

Hartle, J. B.: General Relativity and Gravitation, (C. R. Cordoba, Argentine 1993).

Hartle, J. B., Hawking, S. W.: Phys. Rev., D 28 (1983) 2560.

Hawking, S. W.: Nucl. Phys. B, 239 (1984) 257;
A brief History of Time (Bantam, 1995).

Holland, P. R.: The Quantum Theory of Motion (Cambridge Univ. Press, Cambridge, 1993).

Jayant, V., Narlikar, N., Padmanathan, T.: Gravity Gauge Theories and Quantum Cosmology (Reidel, 1986).

Kibler, M., Elbaz, E.: International J. of Quantum Chemistry, XVI (1979) 1161.

Kibler, M.: "Introduction to Quantum Algebras" in Symmetry and Structural Properties of Condensed Matter, Ed. Florek Lipinki Lulek (World Scientific, Singapore, 1993).

Le Bellac, M.: Des Phénomènes Critiques aux Champs de Jauge (Inter Editions, 1988).

Levy-Leblond, J. M., Balibar, F.: Quantique (Rudiments) (Intereditions CNRS Paris, 1984), Ciel et Espace 229 (1988) 40.

Leader, E., Predazzi, E.: An introduction to Gauge Theory and the New Physics (Cambridge University Press, Cambride, 1982)

Linde, A.: Particle Physics and Inflationary Cosmology Harwood (Ac. Pub., 1990).

Levinson, I. B.: Trudy Fiziko-Teknicheskogo Instituta, Akademii Nauk Litovskoi, SSR 2. 17, 1957.

Mandelbrot, B.: Les objets fractals (Flammarion, Paris, 1975).

Messiah, A.: Mécanique Quantique (Dunod, Paris, 1963).

Nelson, E.: Phys. Rev., 150 (44) (1966) 1079.

Nottale, L.: Fractal Space-Time and Microphysics: Towards a Theory of Scale Relativity (World Scientific, Signapore, 1993).

Normand, J. M.: Rotations in Quantum Mechanics (North-Holland, Amsterdam, 1980).

Phillipidis, C., Dewdney, C., Hiley, B. J.: Il Nuovo Cimento, 52B (1979) 15.

Prugovecki, E.: Quantum Mechanics in Hilbert Space (Academic Press, New York, 1971).

Rabitz, H. A., Gordon, R.: J. of Math. Phys., 11 (12) (1970) 3339.

Racah, G.: Theory of complex spectra
  I Phys. Rev., 61 (1941) 186;
  II Phys. Rev., 62 (1942) 438;
  III Phys. Rev., 63 (1943) 367;
  IV Phys. Rev., 76 (1949) 1352.

Rose, M., E.: Multipole Fields, (Wiley, New York, 1955).

Rotenberg, Bivens, R., Metropolis, M., Wooten, J. K.: The 3-j and 6-j Symbols (The Technology Press, MIT Cambridge, MA, 1959).

Stedman, G.E.: J. Phys. A., 9 (1976) 1999.

Stone, A. J.: J. Phys. A., 9 (4) (1976) 485.

Sylvestre-Brac, B.: 9ème Session d'Etudes Biennales de Physique Nucléaire, Aussois, 1987.

Tamura, T., Low, K. S.: Phys. Rev. Lett., 31 (1973) 1356.

Tamura, T.: Phys. Rep., C, 14 (2) (1974) 60.

Tonomura, A., Endo, J., Matsuda, T., Kawasaki, T., Ezawa, H.: AM. J. of Phys., 57 (1989) 127.

Wigner, E. P.: Group Theory (Academic Press, New York, 1959).

Yutsis, A. P., Levinson, I. B., and Vanagas, V.V.: The Theory of Angular Momentum, Israël, Prog. for Scient. Transl. (1960).

Zee, A.: Unity of Forces in the Universe (World Scientific, Singapore, 1982).

Zurek, W. H.: Phys. Rev., D 24 (1981) 1516,
  Phys. Rev., D 26 (1982) 1862,
  Quantum Theory and Measurement, (Princeton University Press, Princeton, CA, 1981).

Author Index

Subject Index

Springer
and the
environment

At Springer we firmly believe that an international science publisher has a special obligation to the environment, and our corporate policies consistently reflect this conviction.

We also expect our business partners – paper mills, printers, packaging manufacturers, etc. – to commit themselves to using materials and production processes that do not harm the environment. The paper in this book is made from low- or no-chlorine pulp and is acid free, in conformance with international standards for paper permanency.

 Springer